Deep Oil Spills

Steven A. Murawski • Cameron H. Ainsworth
Sherryl Gilbert • David J. Hollander
Claire B. Paris • Michael Schlüter
Dana L. Wetzel

Editors

Deep Oil Spills

Facts, Fate, and Effects

Editors
Steven A. Murawski
College of Marine Science
University of South Florida
St. Petersburg, FL, USA

Sherryl Gilbert
College of Marine Science
University of South Florida
Saint Petersburg, FL, USA

Claire B. Paris
Rosenstiel School of Marine
& Atmospheric Science
University of Miami
Miami, FL, USA

Dana L. Wetzel
Mote Marine Laboratory
Sarasota, FL, USA

Cameron H. Ainsworth
College of Marine Science
University of South Florida
St. Petersburg, FL, USA

David J. Hollander
College of Marine Science
University of South Florida
St. Petersburg, FL, USA

Michael Schlüter
Hamburg University of Technology
Hamburg, Germany

ISBN 978-3-030-11604-0   ISBN 978-3-030-11605-7   (eBook)
https://doi.org/10.1007/978-3-030-11605-7

This Springer imprint is published by the registered company Springer Nature Switzerland AG
The registered company address is: Gewerbestrasse 11, 6330 Cham, Switzerland

Foreword and Dedication

Eric Brown, a British aircraft engineer, described structural engineering as *"the art of molding materials we do not really understand into shapes we cannot really analyze, so as to withstand forces we cannot really assess, in such a way that the public does not really suspect"* (quote from Broad 2010). There is much to learn from engineering failures regarding the fragility of such structures and systems (Love et al. 2011), no more so than those of oil rig blowouts such as *Deepwater Horizon* (DWH) and Ixtoc 1, the two largest accidental blowouts in world history.

While the forensics of engineering and systems failures in the *Deepwater Horizon* case are well documented (National Commission on the BP *Deepwater Horizon* Oil Spill and Offshore Drilling 2011; Boebert and Blossom 2016), perhaps less well understood were the failures of regulators, legislative oversight, and science to anticipate, plan for, and understand the risks involved in such a catastrophic failure. Quantifiable risk is the product of both the probability of events happening and the consequences of such an event should it happen. In the case of deepwater blowouts, the event has an exceedingly low probability of occurrence but a very large potential consequence. A system subject to a critical single point of failure combined with the inability to contain the ensuing blowout for 87 days points to systematic breakdown in regulatory as well as industrial oversight systems for risk reduction. Doubtlessly, the oil and gas industries have learned from these spectacular engineering failures and put in place what they believe to be appropriate risk reduction measures. Similarly, additional government regulation, inspection, and oversight have been forthcoming.

The DWH event also clearly pointed out the dearth of scientific information necessary to make informed decisions once the blowout occurred, including deciding on appropriate response measures and calculating the impacts of that event in the milieu that is the Gulf of Mexico. Previous research was insufficient to confidently evaluate the risks and trade-offs of, for example, using chemical dispersants injected into the stream of oil and gas emanating from the blown-out well. Likewise, the lack of systematic contaminant baselines for nearly all biota and habitats in the Gulf of Mexico made assessing the damage from that disaster more difficult than it needed to be if such baselines had been available. The lack of specific information points to a larger failure of science and science administration to adequately assess the risks involved in deepwater oil and gas production and to organize a research program of sufficient rigor and scope to have answers – or at least a plausibly narrow set of outcomes – that would guide such a response. The science necessary for informed decision-making regarding oil spills is well documented in the "wish lists" of government agencies (ICCOPR 2015), but the industry, various federal and state administrations, and government agencies were unable to muster the political will and resources to close these gaps.

Much changed in the funding and direction of oil spill-related research following DWH. Through a $500 million grant from BP (British Petroleum, for whom the Macondo well was being drilled), the Gulf of Mexico Research Initiative was established. This ambitious 10-year program has produced an enormous body of research spanning the physical, geological, chemical, engineering, biological, and human health sciences. Additionally, significant funding spent by the Natural Resources Damage Assessment (NRDA), the Gulf Restoration Council, and the National Academy of Sciences Gulf Research Program (GRP) is also contributing to the wealth of new science informing oil spill risk reduction, preparedness, and assessment. The scope of the research programs supported by these funds has been both broad and deep, with many of the fundamental uncertainties of what, how, and why of oil spill science being addressed. As with any science portfolio, the GoMRI-funded research spanned the theoretical to applied science continuum and that of high risk-high reward to incremental.

This volume synthesizes a considerable portion of GoMRI-sponsored research and that commissioned by government, industry, and other entities. Many of the chapter authors are members of the Center for Integrated Modeling and Analysis of Gulf Ecosystems (C-IMAGE) and a number of other GoMRI-funded science centers (ADDOMEx, ECOGIG, RECOVER). Additional authorship includes researchers working for the federal government, in academia, and in private industry. The goal of this book is to synthesize what has been learned from these research investments and to identify additional or as yet unanswered research questions going forward. Given the considerable wealth of new information generated during the 9 plus years after DWH, it is contingent on researchers and regulators to put this information into practical application. The challenge will be for the industry and regulators to assimilate and use this information to devise more risk-averse oil and gas exploration and exploitation strategies and to implement more agile, targeted, and effective response strategies in the event of future accidents. One of the particular barriers for more complete integration of new science into the current industrial regulatory frameworks supporting oil and gas production will be that much of the new, relevant research has been generated by academics not traditionally affiliated with industry or government. The aversion to research "not invented here" by those outside the historical institutional relationships supporting the industry is thus a concern. However, while academic scientists may not understand the history or details of industrial applications, science conducted by independent researchers has the positive attribute of being unencumbered by dogma. New, more expansive, and more productive working relationships among the tripartite science community (industry-government-academia) need to be nurtured.

This volume is dedicated to our mentors, C-IMAGE colleagues, and friends Drs. John W. ("Wes") Tunnell, Jr., John E. Reynolds, and Benjamin ("Ben") Flower.

John W. ("Wes") Tunnell, Jr. Wes had many roles during his illustrious career in marine research, focusing on Gulf of Mexico studies. Working from his home institution at Texas A&M-Corpus Christi, Wes conducted wide-ranging and important studies of the natural history of the Gulf. As a young researcher, Wes was literally on the spot of the Ixtoc 1 oil well blowout in 1979–1980 along the Campeche coast of Mexico. His studies located oil deposition centers along the Campeche, Veracruz, and Tabasco coasts, northward to south Texas. He and his students and collaborators revisited these locations over the next 30 years. In 2016, the C-IMAGE-II consortium undertook the "Tunnell Trek" visiting these deposition locations to apply new methods for understanding oil weathering over a nearly 40-year interval.

Wes was a central figure in conceiving the "OneGulf" concept to encourage multinational research among scientists from Cuba, Mexico, and the United States. Without his enthusiasm, patience, and sense of purpose, the international collaborations sponsored by GoMRI – many documented in this volume – would not have occurred. We are forever grateful to our friend and colleague for his humor, generosity, dedication, and sage advice.

John E. Reynolds Dr. John Elliott Reynolds, III, was an icon in the world of marine mammal science and conservation. The volume and value of his science, and his rare ability to understand, inspire, and lead those around him, will ensure his lasting legacy. At the very young age of 36, John was appointed by President George H. W. Bush as Chair of the US Marine Mammal Commission and was retained under Presidents Clinton, G. W. Bush, and Obama. He had a keen interest in helping to recover the health and integrity of the Gulf of Mexico ecosystem after the *Deepwater Horizon* spill, using sound conservation decisions that came both from the head and the heart. John was the epitome of a "gentleman scholar" with his humor and gentle nature integrated with his incredible knowledge and experience. The conservation world, and indeed the world in general, is a lesser place without him.

Benjamin P. Flower We also dedicate this volume to our friend, mentor, and colleague Benjamin ("Ben") P. Flower. Ben's paleoceanographic research focused on the role of ocean circulation in past global climate change on decadal through orbital timescales. He was a pioneer in recognizing the value of pairing foraminifera to determine the past oxygen isotopic composition of seawater. Despite his propensity for seasickness, Ben participated in eight oceanographic research cruises, including expeditions in the Gulf of Mexico following the DWH oil spill. He was a key player in early work to assess the impact of the *Deepwater Horizon* oil spill on the sediments and deepwater communities of the West Florida Shelf and Slope. He coined the terms "flocculent blizzard" and "dirty bathtub ring" referring to two mechanisms for oil residue sedimentation. He also initiated a high-resolution sediment sampling approach, which proved to be essential for detecting the *Deepwater Horizon* in the sediments of the northern Gulf of Mexico. Although his scientific accomplishments were substantial, Ben was also a loving and involved father and an accomplished athlete. He played tennis competitively at Brown, was an avid soccer and ultimate Frisbee player, and was a member of the National Champion Santa Barbara Condors ultimate team. In ultimate, players are responsible for playing fairly, refereeing themselves, and upholding the "spirit of the game." Ben was a special person: kind, caring, hardworking, honest, and dedicated to his family, friends, and colleagues. In Ben's personal and professional life, he truly embodied the "spirit of the game."

Wes, John, and Ben were ardent scientists with a passion for the natural world. Their contributions to GoMRI and C-IMAGE and to science and society were significant and long lasting. They will be missed.

St. Petersburg, FL, USA	Steven A. Murawski
St. Petersburg, FL, USA	Cameron H. Ainsworth
St. Petersburg, FL, USA	Sherryl Gilbert
St. Petersburg, FL, USA	David J. Hollander
Miami, FL, USA	Claire B. Paris
Hamburg, Germany	Michael Schlüter
Sarasota, FL, USA	Dana L. Wetzel

Acknowledgement and Attribution Much of the research presented in this volume was made possible by grants from the Gulf of Mexico Research Initiative (GoMRI) as noted in the individual chapters. Data resulting from original research supported by GoMRI are publicly available through the Gulf of Mexico Research Initiative Information & Data Cooperative (GRIIDC) at https://data.gulfresearchinitiative.org with specific locations (DOI) provided in individual chapters. Some chapters present preliminary findings, therefore finalized datasets will be made available through GRIIDC when available for submission.

References

Boebert E, Blossom JM (2016) *Deepwater Horizon*: a systems analysis of the Macondo disaster. Harvard University Press, Cambridge, MA, 290 pp

Broad WJ (2010) Taking lessons from what went wrong. New York Times, July 19, 2010

Interagency Coordinating Committee on Oil Pollution Research (ICCOPR) (2015) Oil pollution research and technology plan: fiscal years 2015–2021, 270 pp. https://www.bsee.gov/sites/bsee_prod.opengov.ibmcloud.com/files/bsee-interim-document/statistics/2015-iccopr-research-and-technology-plan.pdf

Love PED, Lopez R, Goh YM, Tam CM (2011) What goes up, shouldn't come down: learning from construction and engineering failures. Procedia Eng 14:844–850

National Commission on the BP *Deepwater Horizon* Oil Spill and Offshore Drilling (2011) Deep water: the Gulf oil disaster and the future of offshore drilling. Report to the president, 398 pp. https://www.gpo.gov/fdsys/pkg/GPO-OILCOMMISSION/pdf/GPO-OILCOMMISSION.pdf

Contents

Part I
Introduction

Curtis Whitwam
Fever
Watercolor on Aquaboard
20" × 16"

Chapter 1
Introduction to the Volume

Steven A. Murawski, Cameron H. Ainsworth, Sherryl Gilbert, David J. Hollander, Claire B. Paris, Michael Schlüter, and Dana L. Wetzel

Abstract Over half of the US supply of marine-derived crude oil now comes from wells deeper than 1500 meters (one statute mile) water depth – classified by industry and government regulators as "ultra-deep" production. A number of factors make ultra-deep exploration and production much more challenging than shallow-water plays, including strong ocean currents, extremely high pressures and low temperatures at the sea bottom, varied sub-bottom rock and sediment strata, and high oil and gas reservoir pressures/temperatures. All of these factors, combined with the extremely high production costs of ultra-deep wells, create enormous challenges to explore, develop, and produce from ultra-deep oil and gas extraction facilities safely and with minimal environmental damage. In the wake of the *Deepwater Horizon* and other well blowouts, a considerable body of scientific research on the fate of spilled oil and the resulting environmental effects of deep blowouts has emerged. This and a companion volume, published by Springer, *Scenarios and Responses to Future Deep Oil Spills: Fighting the Next War*, are intended to contribute to the ongoing and important task of synthesizing what we know now and identifying critical "known-unknowns" for future investigation. How can society minimize the risks and make informed choices about trade-offs in the advent of another ultra-deep blowout? Also, what research questions, experiments, and approaches remain to be undertaken which will aid in reducing risk of similar incidents and their

S. A. Murawski (✉) · C. H. Ainsworth · S. Gilbert · D. J. Hollander
University of South Florida, College of Marine Science, St. Petersburg, FL, USA
e-mail: smurawski@usf.edu; ainsworth@usf.edu; sherryl@usf.edu; davidh@usf.edu

C. B. Paris
University of Miami, Department of Ocean Sciences, Rosenstiel School of Marine & Atmospheric Science, Miami, FL, USA
e-mail: cparis@rsmas.miami.edu

M. Schlüter
Hamburg University of Technology, Institute of Multiphase Flows, Hamburg, Germany
e-mail: michael.schlueter@tu-harburg.de

D. L. Wetzel
Mote Marine Laboratory, Sarasota, FL, USA
e-mail: dana@mote.org

ensuing impacts should ultra-deep blowouts reoccur? It is to these questions that this volume intended to contribute.

Keywords Ultra-deep oil and gas · Ixtoc 1 · *Deepwater Horizon* · Oil spill response

1.1 Background

The *Deepwater Horizon* (DWH) oil spill in 2010 (Lubchenco et al. 2012) challenged the essence of what industry, government, scientists, and the public perceived at the time about marine oil spills, and revealed the tremendous technological challenges – and risks – being undertaken to maintain hydrocarbon supplies globally. Currently, offshore oil from the Gulf of Mexico accounts for >90% of US marine production and about 20% of total US oil production (terrestrial and marine). Over half of the US supply of marine-derived crude oil now comes from wells >1500 meters (one statute mile) water depth – classified by industry and government regulators as "ultra-deep" production. The technologies to exploit ultra-deep and highly productive formations within them have developed rapidly since 2000 (Murawski et al. 2020b) when no ultra-deep wells existed anywhere in the world. Deep water drilling no longer involves derricks standing on the sea bottom, but rather ships tethered to anchoring systems with extended drilling and production pipe strings from the sea surface to blowout preventers (BOPs) resting on the seafloor. A number of factors make ultra-deep exploration and production much more challenging than shallow-water plays, including strong ocean currents, extremely high pressures and low temperatures at the sea bottom, varied sub-bottom rock and sediment strata, and high oil and gas reservoir pressures/temperatures. All of these factors, combined with the extremely high production costs of ultra-deep wells, create enormous challenges to explore, develop, and produce from ultra-deep oil and gas extraction facilities safely and with minimal environmental damage.

1.2 Introduction to the Volume

In the United States, offshore oil and gas exploration, development, and production are primarily regulated by the federal government, under conditions specified by the Outer Continental Shelf Lands Act (OCSLA), originally signed into law in 1953. Additional regulations and applicable statutes related to marine oil and gas production include the Clean Water Act (1972), the Marine Mammal Protection Act (1972), and the Endangered Species Act of 1973, among others. The latter two statutes are primarily applicable to the exploration phase (regulating the use of seismic testing for sub-bottom profiling) and, with respect to oil spills, damages that may be incurred to species and their habitats subject to law's jurisdictions. The response to marine oil spills is managed by the US Coast Guard acting with other related federal and applicable state "trustee" agencies, and the responsible parties, using authorities granted to them under the Oil Pollution Act of 1990 (OPA-90). None of these regulatory regimes were crafted, nor did their congressional or administration framers anticipate, the rapid development of ultra-deep exploration and production, and the unique issues associated with them. In the wake of the 1989 *Exxon Valdez* tanker spill in Prince William Sound, Alaska (Peterson et al. 2003), new regulations were forthcoming the following year – OPA-90. In contrast, in the 9-plus years following DWH, no new federal legislation specifically addressing the unique issues associated with ultra-deep drilling oversight, production, and spill response has been forthcoming. Administratively, the then Minerals Management Service (MMS) of the Department of the Interior was subsequently split into two agencies following DWH – the Bureau of Ocean Energy Management (BOEM) and the Bureau of Safety and Environmental Enforcement (BSEE). The stated goal of the reorganization was to separate "… conflicting missions of promoting resource development, enforcing safety regulations, and maximizing revenues from offshore operations."[1] Under these new agencies, a brief moratorium on additional new drilling permits for the Gulf of Mexico was enacted (and has since been lifted) and additional requirements for BOP design and inspection promulgated (Krupnick and Echarte 2018). The essential question that remains, however, is: Has enough been done both by the industry on its own volition, and through regulatory processes, to lower the risk of another catastrophic ultra-deep well failure, and if such a spill occurs again, have the lessons learned from previous spills been incorporated into spill response to minimize impacts on humans and the environment both from the spill itself and the mitigation measures employed?

Globally, the landscape for fossil-derived energy sources is changing rapidly. Notwithstanding the effects of burning fossil fuels on the global climate system, maintaining and expanding the use of hydrocarbon-based fuels for transportation, home heating, and industrial purposes has resulted in novel applications of science and technologies to produce from "tight" formations by the use of hydraulic fracturing of shale rock ("fracking") to free natural gas. Injection of fluids and gasses has allowed once abandoned oil and gas sources to yet again yield economically recoverable quantities. The quest for hydrocarbons to supply the ever-growing human population of the earth (>7.5 billion) has increased the urgency to explore new frontier areas where oil and gas might exist. Thus, marine oil and gas operations now extend to water depths >3000 m (2 miles) and will likely to continue into yet deeper waters. Recent industry projections are that exploration and ultimately production from deep frontier operations will occur off six continents (Murawski et al. 2020b), with the most likely finds in the "golden triangle" between West Africa, Brazil, and the Gulf of Mexico. Other areas will also be explored, including the Arctic, which presents yet other unique challenges.

The number of marine oil and gas-related tanker accidents has declined steadily for several decades following several catastrophic and highly publicized groundings and collisions (Ramseur 2010). The decline in tanker accidents resulted from more stringent vessel construction standards, development of more precise global positioning, navigation and tracking aids, and a concomitant rise in the use of pipelines and offshore unloading facilities minimizing tanker traffic close to shores. Resultantly, the risks of

---

[1] https://www.boem.gov/Reorganization.

serious tanker accidents have declined substantially both in terms of the number of accidents and the number of barrels of spilled oil. With the advent of ultra-deep exploration and production, however, the risks of another serious blowout remain essentially incalculable because of the myriad of factors that influence the integrity of deep subsurface technologies, operations, ocean and formation conditions, and the degree of training of operators to deal with unique and rapidly changing situations. The industry, and society in general, can ill-afford another ultra-deep blowout of the magnitude of DWH (Lubchenco et al. 2012). That single accident resulted in approximately \$60 billion in cleanup costs, fines and penalties, and compensation payments (Bomey 2016). The DWH accident likewise shook people's faith in the ability of engineering solutions to solve environmental problems quickly with minimal damage to people and wildlife. That the equipment to cap and contain such a catastrophic blowout had never been developed or tested prior to the accident is a monumental failure to anticipate and prepare for a truly "worst case" scenario (National Commission on the BP *Deepwater Horizon* Oil Spill and Offshore Drilling 2011; Boebert and Blossom 2016). Subsequent to the DWH accident, billions of dollars have been invested in technology and scientific research to better understand the conditions and environmental consequences of DWH and, by inference, ultra-deep regions of the world that are likely targets for frontier oil and gas development (Global Industry Response Group 2011).

In addition to the DWH spill, there are a number of deepwater blowouts and other spills that can provide relevant lessons learned bearing upon strategic planning for spill prevention and response strategies. The Ixtoc 1 spill off Campeche, Mexico, occurred in 1979–1980 and ran unchecked for 9 months before being shut down (Soto et al. 2014). While only at 54 m water depth, it was, prior to DWH, the largest unintentional marine oil spill in history (about 2/3 the volume of DWH). It is relevant to these discussions because of the extended spatial footprint of the spill, the use of massive quantities of Corexit® dispersant, and because large quantities of Ixtoc 1 oil came to rest in ultra-deep waters (Chap. 13). In 2000 a unique set of controlled release experiments – called DeepSpill – occurred off the Norwegian coast in 844 m of water (Johansen et al. 2003). DeepSpill is elucidating because of the intensive monitoring of fuel oil and natural gas from this series of releases, the results of which were subsequently used to calibrate models of spill behavior. In 2009 the Montara spill in the Timor Sea was uncontrolled for 10 weeks before a relief well successfully controlled oil releases. Similarly, in November of 2011, Chevron had a deep well control failure off Brazil resulting in releases through the underlying rock formations. This volume makes use of research following accidents and investigative reviews in their aftermath. Thus, there are a series of field-scale experiments and monitoring studies, laboratory-based experiments, and monitoring of accidental releases of oil and gas, from which significant research has been generated and summarized in this volume.

We consider the physics, chemistry, and ecological characteristics surrounding deep frontier oil and gas operations, with special emphasis on information obtained from the multiple spill sources of information listed above. From all of these spills, the appropriateness of mitigation techniques and lessons learned from them bear on the seminal question of whether the inherent risks are balanced by the rewards of ultra-deep oil and gas production. After the DWH well was capped, we initiated a scoping effort to characterize factors bearing on deep spill dynamics and how they would impact various sub-ecosystems potentially impacted by a deep spill (Fig. 1.1). The resulting schematic (Fig. 1.1) considers three important domains affecting the outcomes from any set of deep spill circumstances: (1) the oil spill scenario – that is, the specifics of the surrounding oceanographic setting, type, and characteristics of oil and gas being released, the geometry of the subsurface casualty (e.g., BOP failure, casing rupture, formation failure), etc. These factors and responses to them all set the stage determining (2) the fate of oil and gas released into the environment. Much has been learned about oil and gas fate from deep spills from experiments, field observations, and models. The DeepSpill experiment (Johansen et al. 2003), for example, demonstrated that natural gas would dissolve nearly completely prior to surfacing – a result demonstrated in field measurements and models of DWH. This is in contrast to the Ixtoc 1 experience (shallow water) where gas was transported and ignited at the sea surface (Soto et al. 2014).

When the schematic (Fig. 1.1) was constructed in 2011, we had only a rudimentary understanding of the potential impacts of a number of the mechanisms influencing the fate of oil and gas from an ultra-deep blowout. In particular, the relative contribution of the use of subsurface dispersant injection (SSDI) vs. the formation of small oil droplets from the sudden, rapid degassing of gas-saturated, highly pressurized oil has proved to be one of the most fundamental and significant issues from a response planning perspective. It is also an issue of continuing vigorous scientific debate and investigation. This volume considers the issue of SSDI and the state of the science surrounding it, in detail. Similarly, the finding that significant quantities

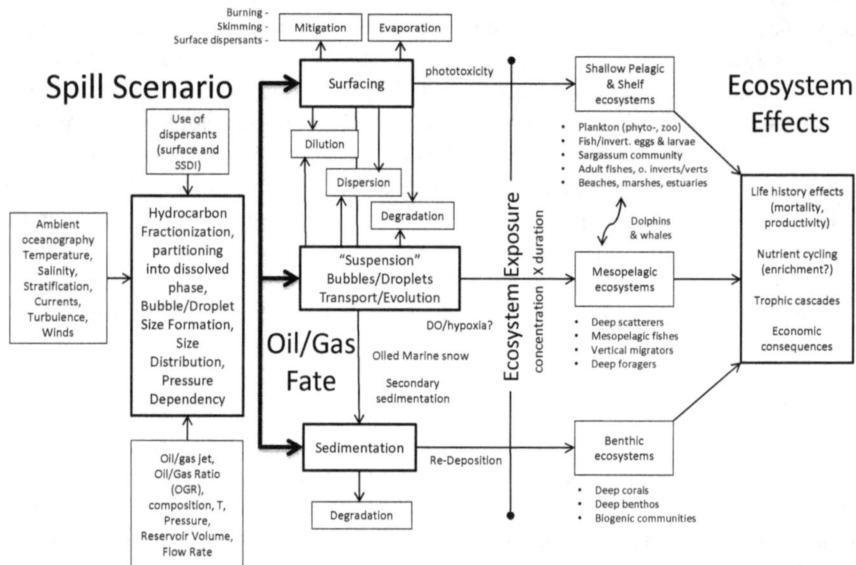

Fig. 1.1 Schematic of the potential mechanisms affecting the circumstances, fate, and effects of a hypothetical deep-water oil and gas spill. (Graphic courtesy of the Center for Integrated Analyses of Gulf Ecosystems – C-IMAGE)

of weathered oil would be transported to the deep sea bottom via a complex set of mechanisms is an important and, prior to DWH, a poorly understood phenomenon.

Ecosystem-level effects (3) from an ultra-deep spill integrate the full range of benthic, water column, sea surface, and coastal ecosystems (Fig. 1.1). This is due to the complex transport mechanisms affecting the fate of oil and gas. Because oil in various states of weathering (fresh crude, dissolved, weathered oil components, emulsified oil, etc.) will be transported both vertically and horizontally in a large ultra-deep spill, the potential environmental and human impacts are much more diverse and complex than in typical surface oil spills. Because response measures can influence the fate and thus the exposure vectors of spilled oil, spill response managers are thus faced with the challenge of balancing trade-offs among ecosystem components, recognizing that there are no benign options in oil spill mitigation and cleanup. For example, the use of SSDI in large quantities is likely to toxify deep benthic and mesopelagic realms where highly diverse, but relatively unproductive communities exist (Fisher et al. 2016; Romero et al. 2018). However, allowing oil to surface in large droplet sizes may increase the oil volume affecting surface-dwelling and coastal animals and plants (French-McCay et al. 2018). The ultimate choice of response measures for a particular spill must be made with transparency and forethought and not simply left to ad hoc, situational decision-making, particularly since the nature and implications of such choices are becoming more clear with additional research.

The concept of this two-volume series (e.g., this book, *Deep Oil Spills, Facts, Fate and Effects*, and the companion volume – *Scenarios and Responses to Future Deep Oil Spills: Fighting the Next War*) is to synthesize some of the salient research examining key issues that have emerged since the DWH casualty and from other deep oil spills and experiments. Much of the scientific research summarized in these volumes was sponsored by the Gulf of Mexico Research Initiative (GoMRI), funded by a grant of $500 million from BP. Additionally, under the DWH Response and Natural Resource Damage Assessment (NRDA) programs following DWH, a large number of studies examined the impacts on wildlife and lost human uses of the environments affected by the spill. Much of that research has now been widely disseminated in reports and scientific publications (e.g., *Deepwater Horizon* Natural Resource Damage Assessment Trustees 2016). The GoMRI Research Board, chaired by Eminent Biologist Dr. Rita Colwell, overseen by Chief Scientist Dr. Charles "Chuck" Wilson, and comprised of experts in many scientific domains, has consistently encouraged synthesis of research findings as a way to improve the rigor and relevance of the research applied to real-world issues. These books are meant to contribute to the ongoing and important task of synthesizing what we know now and for identifying critical "known-unknowns" for future investigation. How can society minimize the risks and make informed choices about trade-offs, such as the use of subsurface injection of dispersants at the wellhead in the advent of another ultra-deep blowout? Finally, what research questions, experiments, and approaches remain to be undertaken which will aid in reducing risk if similar incidents and their ensuing impacts should ultra-deep blowouts reoccur? It is to these questions that this volume intended to contribute.

This book is organized into eight thematic sections, generally following the fate and effect scheme outlined in Fig. 1.1. Each section is introduced with a brief chapter outlining the significance of issues addressed by the chapters within that section. Part 8 provides an overall summary of major science and management-related themes emerging from the book as well as a reprise of significant "lessons learned."

References

Boebert E, Blossom JM (2016) *Deepwater Horizon*: a systems analysis of the Macondo disaster. Harvard University Press, Cambridge, MA, 290 pp

Bomey N (2016) BP's *Deepwater Horizon* costs total $62B. USA TODAY, July 14. https://www.usatoday.com/story/money/2016/07/14/bp-deepwater-horizon-costs/87087056/

Deepwater Horizon Natural Resource Damage Assessment Trustees (2016) *Deepwater Horizon* oil spill: final programmatic damage assessment and restoration plan and final programmatic environmental impact statement. Retrieved from: http://www.gulfspillrestoration.noaa.gov/restoration-planning/gulf-plan

Fisher CR, Montagna PA, Sutton TT (2016) How did the *Deepwater Horizon* oil spill impact deep-sea ecosystems? Oceanography 29:182–195

French-McCay D, Crowley D, Rowe JJ, Bock M, Robinson H, Wenning R, Hayward Walker A, Joeckel J, Nedwed TJ, Parkerton TF (2018) Comparative risk assessment of spill response options for a deepwater oil well blowout: part 1. Oil spill modeling. Mar Pollut Bull 133:1001–1015

Global Industry Response Group (2011) Capping and containment. Global industry response group recommendations. International Association of Oil and Gas Producers. Report 464, 44 pp.

Johansen Ø, Rye H, Cooper C (2003) DeepSpill—field study of a simulated oil and gas blowout in deep water. Spill Sci Technol Bull 8:433–443

Krupnick AJ, Echarte I (2018) The 2016 blowout preventer systems and well control rule: should it stay or should it go? Resources for the Future (RFF) report series: the costs and benefits of eliminating or modifying US oil and gas regulations. http://www.rff.org/files/document/file/RFF-Rpt-Oil%26GasRegs-BSEE%20BOP%20rule.pdf

Lubchenco J, McNutt MK, Dreyfus G, Murawski SA, Kennedy DM, Anastas PT, Chu S, Hunter T (2012) Science in support of the *Deepwater Horizon* response. Proc Natl Acad Sci 109:20212–20221

Murawski SA, Ainsworth C, Gilbert S, Hollander D, Paris CB, Schlüter M, Wetzel D (eds) (2020a) Scenarios and responses to future deep oil spills – fighting the next war. Springer, Cham

Murawski SA, Hollander D, Gilbert S, Gracia A (2020b) Deep-water oil and gas production in the Gulf of Mexico, and related global trends. In: Murawski SA, Ainsworth C, Gilbert S, Hollander D, Paris CB, Schlüter M, Wetzel D (eds) Scenarios and responses to future deep oil spills – fighting the next war. Springer, Cham

National Commission on the BP *Deepwater Horizon* Oil Spill and Offshore Drilling (2011) Deep water: the Gulf oil disaster and the future of offshore drilling. Report to the President, 398 pp. https://www.gpo.gov/fdsys/pkg/GPO-OILCOMMISSION/pdf/GPO-OILCOMMISSION.pdf

Peterson CH, Rice SD, Short JW, Esler D, Bodkin JL, Ballachey BE, Irons DB (2003) Long-term ecosystem response to the *Exxon Valdez* oil spill. Science 302:2082–2086

Ramseur JL (2010) Oil spills in U.S. coastal waters: background, governance, and issues for congress. U.S. Congressional Research Service 7-5700, RL33705, 34 pp.

Romero IC, Sutton T, Carr B, Quintana-Rizzo E, Ross SW, Hollander DJ, Torres JJ (2018) Decadal assessment of polycyclic aromatic hydrocarbons in mesopelagic fishes from the Gulf of Mexico reveals exposure to oil- derived sources. Environ Sci Technol 52(19):10985–10996. https://doi.org/10.1021/acs.est.8b02243

Soto LA, Botello AV, Licea-Durán S, Lizárraga-Partida M, Yáñez-Arancibia A (2014) The environmental legacy of the Ixtoc 1 oil spill in Campeche Sound, Southwestern Gulf of Mexico. Front Mar Sci 1:1–9

Part II
Physics and Chemistry of Deep Oil Well Blowouts

Tessa Wilson
Perfect Plumes
India ink, Gold ink, Motor oil on Canvas
18" × 26"

Chapter 2
The Importance of Understanding Fundamental Physics and Chemistry of Deep Oil Blowouts

William Lehr and Scott A. Socolofsky

Abstract The science of deep oil spills is complex, spanning several engineering and scientific specialties, different environmental conditions, and a large range of dimensional scales. It is also an applied and pragmatic field where research is expected to produce a better assessment and improved response of any spill incident. This summary chapter reviews at an introductory level the important physical and chemical factors relevant to deep oil blowouts, beginning with the nature of the crude oil itself. While knowledge of the necessary properties of such a mixture of hydrocarbons is often limited, this chapter describes the information typically known to spill responders, expressed in units common to the industry. Material characteristics of the leaking reservoir are defined along with some special features of the subsurface release and plume that are not present in surface spills. Finally, the early far-field behavior of the submerged oil is described.

Keywords Bulk oil properties · API · Oil viscosity · Oil reservoir · Porosity · Gas hydrates · Oil plumes · Oil jets

2.1 Introduction

A common expression, "challenging as rocket science," is occasionally used to describe a complex topic. Analysis of oil releases at great water depth is not rocket science. It is in fact much more imposing than rocket science, spanning numerous fields of science and technology and great extremes of environmental conditions. Consider, for example, the extreme ranges of temperature and pressure to which the spilled oil is subjected. For the *Deepwater Horizon* (DWH) oil spill, pressure ranged

W. Lehr
National Oceanic and Atmospheric Administration, Office of Response & Restoration, Seattle, WA, USA
e-mail: Bill.Lehr@noaa.gov

S. A. Socolofsky (✉)
Texas A&M University, Zachry Department of Civil Engineering, College Station, TX, USA
e-mail: socolofs@tamu.edu

S. A. Murawski et al. (eds.), *Deep Oil Spills*,
https://doi.org/10.1007/978-3-030-11605-7_2

from 1 bar at the water surface to 817 bars in the reservoir; water temperature varied from a few degrees °C at the wellhead to117 °C in the reservoir (Blunt 2013).

The range of spatial and temporal scales affecting the behavior of oil spills is also large. Transport of surface oil may involve oceanographic features that span thousands of kilometers while "weathering" processes such as oil-water dissolution operate at the molecular scale. Evaporative loss rate for oil that reaches the water surface is greatly reduced after the first day, while biodegradation of beached oil or oil dispersed in the water column may take months or even years.

The spilled oil itself presents an insurmountable problem for complete description from first principles. Rather than being a single substance, crude oil is a mixture of thousands of hydrocarbons with traces of non-hydrocarbon molecules that are known to the spill responder in only the most cursory manner. Moreover, the deep water spill release is a multiphase process of gas and liquid hydrocarbons along with entrained water that changes rapidly over short spatial and temporal ranges.

Fortunately for those researchers that have chosen oil spill science as their field of study, it is also a pragmatic subject. Unlike rocket science, high precision is often not required. Answers need only be as accurate as necessary to make educated cleanup or damage assessment decisions. Approximations are allowed, in fact required, and a great amount of spill literature involves discussion of the better approximation for a particular spill phenomenon.

Several chapters in this book examine specific recent improvements in our understanding of spill mechanisms. Much of this new research is a result of the Gulf of Mexico Research Initiative (GoMRI) launched as a consequence of the DWH spill. A herculean task for those responsible for funding such investigations is estimating for any individual study its importance and applicability to the pragmatic goals of the entire field of oil spill science. Assessing a particular tree's value requires knowing its connection to the entire forest. Therefore, this introductory review chapter will discuss at least part of the forest, as it relates to our understanding of the fundamental physics and chemistry of deep oil well blowouts. A word of warning to the reader: the chapter must neglect some important topics and cover others at a level that may seem too basic to the specialist. However, the specialist needs to recognize that outside of their specialty, they are themselves a novice.

2.2 The Oil

Deep oil spills involve crude oil, as opposed to petroleum products generated by the world's refineries and transported in the marine environment by pipeline or, more commonly, surface shipping. Deep oil spills may involve a single reservoir or a network of reservoirs connected to a common platform. In either case, the oil that is extracted will change its characteristics over time, sometimes resulting in slightly different behaviors if spilled.

Industry characterizes crude oils by properties that are important for commercial purposes, but these properties do not provide a complete description for environmental impact. For example, while the information available for DWH was more extensive than a typical spill incident, certain weathering characteristics remained unknown. Crude oil has a typical elemental composition of 83–87% carbon, 10–14% hydrogen, 0.1–2% nitrogen, 0.05–6% sulfur, 0.05–1.5% oxygen, and less than 1% other trace elements (Speight 2007). Industry traditionally separates oil by structure into one of four somewhat arbitrary categories, labeled SARA after the first letter in the category name: saturates, aromatics, resins, and asphaltenes.

The first two groups, saturates and aromatics, represent the largest mass fraction of most crude oils. The saturate group is nonpolar oil molecules without double bonds that include linear, branched, and cyclic saturated hydrocarbons. The group name refers to the fact that the carbon atoms are "saturated" with the maximum number of hydrogen atoms. Smaller saturate molecules (less than seven carbons) are relatively volatile and are mostly lost to evaporation in surface oil spills. However, they have extremely low solubility in seawater, even when compared to other hydrocarbons of similar molecular weight. For example, while hexane and benzene (smallest aromatic) have similar molar masses and limited solubility, benzene is nevertheless the much more soluble by more than an order of magnitude. This distinction between the two hydrocarbon categories has consequences for deep oil spills where dissolution replaces evaporation as an important weathering process and toxicity is a concern.

The aromatic group hydrocarbons have at least one benzene ring and often play the lead role with regard to toxic impacts from the oil while being generally less biodegradable than saturates of the same carbon number. Fingas and Fieldhouse (2012), based on laboratory results, claim that the ratio of aromatics to saturates plays a role in the formation of stable water-in-oil emulsions. However, the quantity and ratio of the two remaining groups, resins and asphaltenes, are even more important for emulsion stability. Resins are large hydrocarbon molecules with one to three sulfur, oxygen, or nitrogen atoms per molecule. Resins can dissolve in oil, an important factor in the initiation of emulsification where they prevent escape of water droplets until the larger asphaltene molecules can migrate to the oil-water interface. Asphaltenes are not uniquely defined in the literature although a common definition might be very large hydrocarbon molecules that have one to three sulfur, oxygen, or nitrogen atoms per molecule but do not dissolve in oil. The ambiguity in asphaltene classification between laboratories complicates the task of devising computer models of spill weathering, particularly emulsification onset. Added to this complication is the inherent limitation of the SARA classification scheme itself, as it does not record important oil characteristics such as the detailed structure and degree of polarization in the larger hydrocarbons, particularly the asphaltenes. This is an active area of research (Groenzin and Mullins 2007) with good prospects for improved oil characterization databases in the future.

Certain bulk oil properties are not dependent on the individual hydrocarbons in the oil but still have a major impact on the fate and behavior of the spill. One obvious example is density. Industry reports density in API degrees:

$$\text{API} = \left\{ \frac{141.5}{\text{sg}} \right\} - 131.5 \tag{2.1}$$

where sg = specific gravity at stock tank conditions (1 bar, 16 C). The number constants are selected so that pure water has an API of 10. Oils with API less than 10 would be non-buoyant in freshwater but might be slightly buoyant in seawater. While spilled oil density usually increases with weathering, it is rare that the resulting density change will cause a buoyant oil to sink. Instead, other processes may interact to cause submergence. During DWH, aggregation with marine snow (Passow and Ziervogel 2016) may have played a major role in causing oil to settle on the bottom.

Industry refers to oil that does not contain dissolved gases as "dead oil." Such oils show relatively little change to density as pressure increases. However, deep well-released oils such as that from DWH are "live oils" and contain significant amounts of dissolved gases, mainly methane, ethane, propane, and butane. Some dissolved gases will escape during the pressure drop from reservoir to water surface. The oil formation volume factor calculates the change in oil volume from reservoir conditions to the resulting volume of liquid and gas if it were directly brought to stock tank conditions. For DWH, the oil formation volume factor was about two and a third (Hsieh 2010). This does not represent the actual observed change between reservoir volume and surface spill volume since dissolved gas release, other weathering processes, and incorporation of surrounding seawater are not included.

Like most fluids, the density of the oil increases as the temperature decreases. The increase parameter is a nonlinear function of temperature and density (ASTM 2007) but can be approximated as linear over conditions outside the reservoir.

Another important bulk property is viscosity. Unfortunately, the term relates to two different properties with different dimensional units. Kinematic viscosity has dimensions of area/time with its SI unit being the stoke. Dynamic viscosity, sometimes called absolute viscosity, is kinematic density multiplied by the oil density. Its SI unit is the poise. The oil industry traditionally used neither but instead would use the time for an oil sample to flow through a certain type of measuring viscometer (e.g., Saybolt universal second). Fortunately, this is less common today, and most large oil property libraries store viscosity in one hundredth of poise or centipoise.

Oil viscosity is highly sensitive to temperature change. Past practice in older surface spill models was to utilize the Eyring's equation (1935) to calculate (kinematic) viscosity

$$v_0 = v_{ref} \exp\left[k_v \left(\frac{1}{T_0} - \frac{1}{T_{ref}} \right) \right]$$
(2.2)

at the environmental temperature T_0 (K) by extrapolating from some measured laboratory viscosity v_{ref} and temperature T_{ref}. Choice of k_v varied depending on the model, but a typical value, expressed in Kelvin, would be 5000 K (Bobra and Callaghan 1990). Industry itself needed more accurate estimates over greater environmental extremes, so more complex methods have been developed (e.g., Orbey and Sandler 1993; Abu-Eishah 1999; Hemmati-Sarapardeh et al. 2013) but are not necessarily imported into current spill modeling.

While mass loss through evaporation (for surface oil) or dissolution will also increase viscosity, the most significant cause of viscosity increase for susceptible oils is water-in-oil emulsification where the emulsified oil viscosity can increase by more than an order of magnitude. The Moody equation (MacKay et al. 1982) remains the most common method used to calculate an estimated increased emulsion viscosity value:

$$v_{emul} = v_0 \exp\left[\frac{k_1 f_w}{1 - k_2 f_w}\right] \tag{2.3}$$

where k_1, k_2 are empirically determined constants and f_w is the water fraction of the emulsion, which can be as much as 90%. Some newer models have replaced the Moody equation by the approach of Pal and Rhodes (1989).

A final bulk property introduced in this chapter is important in part because of the use of chemical surfactants as a cleanup device. Oil-water interfacial tension, the tension that holds the surface of each liquid phase together, has been measured by industry for a century (Johnson 1924). However, the actual range of measured interfacial tension values for most crudes is small, typically between 20 and 40 dynes/cm. Application of chemical surfactants at the interface drastically reduces interfacial tension by one or more orders of magnitude. If the oil is simultaneously subjected to turbulent energy, numerous small oil droplets will be produced that can then disperse over a wider aqueous domain and increase certain weathering processes such as dissolution and biodegradation. Chemical surfactants, both on the surface and subsurface at the oil release point, were widely used in DWH.

2.3 The Reservoir

To be economically viable, oil from deep well reservoirs should be under high pressure. As previously mentioned, the DWH reservoir fluid was at 817 bars. At this pressure with a corresponding high temperature, the liquid-gas mixture is a critical fluid, meaning that the gas and liquid are indistinguishable from each other. In spite of a common public misconception, the fluid does not exist as a uniform subterranean pool but is instead interspersed in pore spaces of rock structures such as those in sandstone (e.g., DWH) or dolomite.

Two key characteristics that define a reservoir potential are the porosity of the reservoir rock (fraction of open space) and permeability, a measure of the reservoir fluid capability to pass through the rock pores: the connectivity of the rock pores. Porosity can be surprisingly high for a productive reservoir. DWH was reported to have better than 20% average porosity (Bommer 2010), and porosities of 30% are not uncommon (Ehrenberg et al. 2009). Porosity varies both spatially and temporally. Different rock layers may demonstrate different porosities, and not all rock pores may be filled with fluid. As the reservoir fluid is extracted, the rock is compressed down, reducing the size of the pore spaces. The relative change in volume (V) per unit change in pressure (p) is called compressibility, c:

$$c = -\frac{1}{V} \cdot \frac{\partial V}{\partial p} \tag{2.4}$$

Compressibility varies for the rock, the reservoir fluid, and any connate water. As with many measured quantities, the oil industry sometimes uses nonstandard units to measure compressibility, in this case the microsip. A pressure change of 70 bars will result in approximately a 1% change in volume for a material with 10 microsips of compressibility. Typical sandstone has a compressibility of about three microsips, similar to that of water (Newman 1973). The reservoir fluid itself can show higher compressibility. Blunt (2013) estimated DWH values of about 14 microsips. Total compressibility, the sum of oil, rock, and water, was a matter of dispute for DWH since it could be connected to an estimate of the oil released. Oil released from the broken well was postulated to be a product of the amount of available oil in the reservoir, the total compressibility, and the pressure difference between the reservoir and ocean floor.

Also in dispute for DWH was the connectivity of the rock pores or permeability of the reservoir. Pores that are not connected cannot share contained fluid. Fluid flow in the reservoir follows some form of Darcy's law, discovered by Henry Darcy in 1856. According to the simplest form, flow in the reservoir can be estimated as

$$\dot{V} = A \cdot \frac{\mu_r}{\eta_{\text{oil}}} \cdot \nabla p \tag{2.5}$$

where A = flow cross section, μ_r = permeability, η_{oil} = oil dynamic viscosity, and ∇p = pressure gradient. Once again, units employed in the oil industry do not match standard SI choices. Permeability is recorded in millidarcy (mD). A Darcy is approximately 10^{-12} square meters. A typical permeability number might be a few hundred mD. However, permeability determination in reservoirs can be quite challenging and is often subject to large uncertainty.

2.4 Subsurface Release

As introduced above, a subsurface oil spill may occur from an individual well tapping a single reservoir or may originate at a platform that connects multiple reservoirs. Leaks may form at a single wellhead or anywhere along a subsea pipeline or riser, thus involving either live crude oil or separated hydrocarbon and gas. The most likely types of accidents are an offshore blowout or subsea pipeline leak. A blowout is defined as any time the operator loses control of the flow rate in a well or pipeline.

Oil spilled from a subsea reservoir could originate from distributed or point sources. A distributed source could be formed if a well bore lost integrity below the seafloor such that oil percolated up through the sediments near the seabed, a matter of concern to DWH responders that fortunately did not happen. In such a case, the release to the water column would cover a larger area and may not include a rigorous jet of oil or rising buoyant plume. In the case of a point release, the gas and liquid petroleum exit a small orifice, initially breaking up into gas bubbles and oil droplets that form a buoyant plume that rises together with entrained seawater through the ocean water column. During DWH, point releases occurred at the end of the broken riser pipe, from holes that formed in the kink above the blowout preventer (BOP) and finally directly from the open pipe above the BOP after the riser was removed in preparation for attaching a containment system.

During the early days of DWH, British Petroleum (BP) attempted to control the oil plume by confining it in a cofferdam, a large dome placed over the broken riser. The attempt failed because the cofferdam rapidly filled with gas hydrates. These hydrates increased the buoyancy of the structure to such an extent that it became unstable and had to be removed. Natural gas hydrates, discovered by Sir Humphrey Davy in 1810, are formed when water molecules are linked through hydrogen bonding and create cavities that enclose hydrocarbon gas molecules. However, no chemical bonding takes place between the host water molecules and the enclosed hydrocarbon. The resulting solid has a snow-like appearance with density less than ice. The amount of naturally occurring hydrates, 3000 trillion cubic meters according to Chong et al. (2016), dwarfs known traditional gas reserves. Hydrates first became of interest to the hydrocarbon industry in 1934, when they were observed blocking pipelines. Blockage in offshore production has become more severe with the increase of the water depth (Sinquin et al. 2004). Hydrate formation is favored by low temperature and high pressure such as found at the sea floor.

If hydrocarbons escape from the reservoir through a point release, the transport of gas and liquid petroleum occurs in four distinct stages, each governed by slightly different physics and each occurring at later times such that the gas and liquid phases may have different composition and physical/chemical characteristics. These stages are (1) the initial jet breakup, where the petroleum fluids are the most fresh; (2) the ensuing buoyant plume, where much of the dissolution of the soluble compounds occurs; (3) a region where the density stratification of the ocean arrests the upward buoyant plume, forming a horizontal intrusion layer rich in dissolved petroleum compounds; and (4) the final transport phase, where gas bubbles and oil droplets rise independently, moved by their own buoyancy and by the currents and turbulence of the ocean.

The jet break-up region is strongly turbulent, dominated by momentum coming from the pipeline fluids, and occurs immediately downstream of the spill orifice. Seawater rapidly mixes into the gas and liquid petroleum stream, breaking the immiscible oil up into individual gas bubbles and oil droplets. The velocity of the oil and gas in this region is similar to that in the leaking pipeline at the orifice, and the strong turbulence is responsible for creating the initial size distribution of gas bubbles and oil droplets.

Shortly after release, the strong buoyancy of the light gas and liquid petroleum fluids dominates the dynamics, and a vertically rising buoyant plume forms. The strength of this plume is characterized by the initial buoyancy flux coming from the release, given

$$B = g'Q_0 \qquad (2.6)$$

where g' is the reduced gravitational acceleration of the oil through the water and Q_0 is the initial gas/oil volume flow rate from the release point. Because of its net positive buoyancy flux, the plume lifts the gas and liquid droplets faster through the water than individual bubbles or droplets would rise on their own, similar to the behavior of buoyant smoke from a smokestack in the atmosphere. As it ascends, seawater is mixed into the plume, while any horizontal ocean currents may also cause it to bend. Generally, ocean water is density stratified, with colder, saltier water near the bottom and warmer water of lower salinity being higher in the water column. Oceanographers and meteorologists define water or air stability related to this stratification by the value of the Brunt-Vaisala frequency:

$$N = \sqrt{g \frac{\partial \ln(\rho_w)}{\partial \zeta}} \tag{2.7}$$

where, for oceans, ρ_w is the potential (adiabatically brought to 1 bar reference pressure) seawater density and ζ is water depth. The seawater entrained near the source becomes relatively heavier than its surroundings as it is lifted by the plume into progressively lighter ambient seawater. Plume buoyancy and upward flow are reduced. Eventually, the seawater cannot be lifted any higher, and the buoyant plume stops rising, descending to a level of neutral buoyancy. The location of this transition can be estimated using the plume characteristic velocity:

$$u_c = \sqrt[4]{BN} \tag{2.8}$$

which can be compared to the horizontal current velocity and the individual droplet buoyant velocity to determine at what depth the plume will be arrested, usually a few hundred meters above the release point (Socolofsky and Adams 2002).

Since DWH, numerical models have been developed that can predict the complicated jet and plume dynamics together with the chemical kinetics to simulate the behavior of the oil and gas between the release and the intrusion layer. These generally fall into two categories, simple response models based on an average cross section of the plume and more complex three-dimensional models employing computational fluid dynamics.

2.5 Early Far-Field Fate

Once the buoyant plume stops rising, it spreads horizontally, forming an intrusion layer that continues to move downstream with the ocean currents. The entrained seawater becomes trapped in the intrusion layer, and this water carries with it petroleum compounds that dissolved out of the rising gas bubbles and liquid droplets during the buoyant plume ascent. Small oil droplets that have negligible rise velocity of their own may also reside in the intrusion layer for a significant distance downstream of the intrusion formation (see, e.g., Chan et al. 2014).

Outside of the coherent plume, the released oil becomes a collection of oil droplets of varying size. Determination of the droplet size distribution is necessary to ascertain this oil fate and behavior. According to Stokes' law, buoyant droplets will quickly accelerate to a terminal velocity where frictional drag balances gravitation.

Assuming spherical shape, valid for smaller dead oil droplets, the droplet terminal velocity can be calculated as

$$v_T = \frac{gd^2 \Delta\rho}{18\eta} \qquad (2.9)$$

where $\Delta\rho$ is the density difference between the oil and water, d is the particle diameter, and η is the of water dynamic viscosity. One consequence is that a spherical droplet with twice the diameter would rise at four times the speed. Larger, non-spherical droplet rise can be estimated by similar empirical equations (Clift et al. 1978; Zheng and Yapa 2000). Whether the droplet reaches the surface or weathers subsurface because of slow rise velocity depends critically on its initial size.

Calculating droplet size distribution is a challenging task. Hinze (1955) claimed droplet breakup occurred due to two exclusive processes, laminar breakup and turbulent breakup. Laminar breakup occurs due to shear in the water phase. This tends to twist and elongate the droplet until it breaks, usually into more or less equal size droplets. Laminar breakup has been well studied in the literature, and the governing equations in certain circumstances have closed solutions (Taylor 1934). Turbulent breakup conversely occurs due to pressure variation within the droplet overcoming a binding force. If the oil has a low viscosity, surface tension would be the dominant force holding the droplet together, and the droplet breaks when the Weber number exceeds a critical value. The droplet tends to burst and to generate daughter droplets of non-equal sizes. If the oil has a high viscosity, shear forces within the droplet tend to be the forces that resists breakup.

Currently, the research community is divided on the best technique for estimating droplet size distribution (Nissanka and Yapa 2018). One school of thought (Zhao et al. 2014; Bandara and Yapa 2011) believes the best way is to begin with first principles and dynamically determine size distribution by including transient populations in the calculation. These populations may further break up due to turbulence but are resisted by interfacial tension and viscosity. They may also coalesce as some droplets combine during the process. The results often show an initial bimodal droplet distribution that evolves over time to a lognormal distribution. An alternative school of thought begins with the lognormal or similar distribution, fitting the necessary model parameters using modified versions of Reynolds, Weber, and/or other engineering numbers (Johansen et al. 2013).

Both of the new approaches consider the impact of reduced surface tension on reducing mean droplet size, something not considered by classical methods (Delvigne and Sweeney 1988). During DWH, surfactants were injected directly into the outgoing plume in large amounts, and at least some evidence indicates that it resulted in reduced surface oil expression due to a reduction in mean droplet size (Zhao et al. 2014, 2015). However, turbulence was quite high in the exiting oil jet, and some researchers claim that large energy dissipation at the release point was a major influence on creating smaller droplets that would have occurred even without massive surfactant use (Paris et al. 2012).

Subsurface oil droplets show a different weathering pattern than surface oil. In the short term, dissolution substitutes for evaporation as a removal mechanism that changes the characteristics of the remaining oil (Stevens et al. 2015). Since dissolution is directly dependent on the oil-water interfacial area, droplet size distribution becomes critical to estimating this process. For live oil, dissolved gas, particularly methane, can alter density and droplet diameter (Malone et al. 2018) and, along with it, rise velocity. From a practical response point of view, it matters little if non-surfacing oil is dispersed, dissolved, and biodegraded or interacts with mineral or organic particles since it is no longer subject to normal response treatment. Generally, only floating or beached oil can be further removed by common cleanup techniques (Lehr et al. 2010). However, the biological impacts from the spilled oil will of course depend on its particular final fate. As this chapter only considers fundamental physics and chemistry processes, the reader will need to look elsewhere in this text for discussion on that matter.

Hopefully, the reader now has a clearer picture of the fascinating complexity found in the physical and chemical science that describes deep water oil releases. If the authors have met their task, the reader should better appreciate those chapters that cover specialized topics only briefly mentioned and be able to put those topics into proper context when considering the entire subject matter.

Disclaimer The information in this chapter reflects only the professional views of the authors and do not necessarily reflect the official positions or policies of the US Government or Texas A&M University.

References

Abu-Eishah S (1999) A new correlation for prediction of the kinematic viscosity of crude oil fractions as a function of temperature, API gravity, and 50% boiling point. Int J Thermophys 20:1425–1434

ASTM (2007) Manual of petroleum measurement standards, Chapter 11, Section 1, Addendum 1, ASTM D 1250-04

Bandara U, Yapa P (2011) Bubble sizes, breakup, and coalescence in deepwater gas/oil plumes. J Hydraul Eng 137(7):729–738

Blunt M (2013) Modelling Macondo, Report prepared on behalf of BP Exploration & Products Inc. 209 pp

Bommer P (2010) Appendix 2: reservoir fluid studies, in deepwater horizon release estimate of rate by PIV. Plume Team, Flow Rate Technical Group, Washington, DC

Bobra M, Callaghan S (1990) A catalog of crude oil and oil product properties. Environment Canada EE-125, Ottawa

Chan G, Chow A, Adams E (2014) Effects of droplet size on intrusion of sub-surface oil spills. Environ Fluid Mech 15(5):959–973

Chong Z, Yang S, Babu P, Linga P, Li X (2016) Review of natural gas hydrates as an energy resource: prospects and challenges. Appl Energy 162:1633–1652

Clift R, Grace J, Weber M (1978) Bubbles, drops, and particles. Academic Press, New York

Delvigne G, Sweeney C (1988) Natural dispersion of oil. Oil Chem Pollut 4:281–310

Ehrenberg S, Nadeau P, Steen O (2009) Petroleum reservoir porosity versus depth: influence of geological age. AAPG Bull 93:1281–1296

Eyring H (1935) The activated complex in chemical reactions. J Chem Phys 3(2):107–115

Fingas M, Fieldhouse B (2012) Studies on water-in-oil products from crude oils and petroleum products. Mar Pollut Bull 64:272–283

Groenzin H, Mullins O (2007) Asphaltene molecular size and weight by time-resolved fluorescence depolarization. In: Mullins OC, Sheu EY, Hammami A, Marshall AG (eds) Asphaltenes, heavy oils, and petroleomics. Springer, New York

Hemmati-Sarapardeh A, Khishvand M, Naseri A, Mohammadi A (2013) Toward reservoir oil viscosity correlation. Chem Eng Sci 90:53–68

Hinze J (1955) Fundamentals of the hydrodynamic mechanism of splitting in dispersion processes. Am Inst Chem Eng J 1:289–295

Hsieh P (2010) Computer simulation of reservoir depletion and oil flow from the Macondo well following the Deepwater Horizon blowout: U.S. Geological Survey Open-File Report 2010–1266, 18 p

Johansen O, Brandvik P, Farooq U (2013) Droplet breakup in subsea oil releases-part2: predictions of droplet size distributions with and without injection of chemical dispersants. Mar Pollut Bull 73(1):327–335

Johnson E (1924) The interfacial tension between petroleum products and water. Ind Eng Chem 16(2):132–135

Lehr B, Bristol S, Possolo A (2010) Oil budget calculator—deepwater horizon. Federal Interagency Solutions Group, Oil Budget Calculator Science and Engineering Team. Published online at http://www.crrc.unh.edu/publications/OilBudgetCalcReport_Nov2010.pdf

MacKay D, Shiu WY, Hossain D, Stiver W, McCurdy D, Paterson S (1982) Development and calibration of an oil spill behavior model. Report CG-D-27-83, USCG R&D Center, Groton Conn

Malone K, Pesch S, Schlüter M, Krause D (2018) Oil droplet size distribution in deep-sea blowouts: influence of pressure and dissolved gases. Environ Sci Technol 52(11):6326–6333

Newman G (1973) Pore volume compressibility of consolidated, friable, and unconsolidated reservoir rocks under hydrostatic loading. J Pet Technol 25:129–134

Nissanka I, Yapa P (2018) Calculation of oil droplet size distribution in ocean oil spills: a review. Mar Pollut Bull 135:723–734

Orbey H, Sandler S (1993) The prediction of the viscosity of liquid hydrocarbons and their mixtures as a function of temperature and pressure. Can J Chem Eng 71:437–446

Pal R, Rhodes E (1989) Viscosity/concentration relationships for emulsions. J Rheol 33:1021–1045

Paris C, Le Henaff M, Aman Z, Subramaniam A, Helgers J, Wang D, Kourafalou V, Srinivasan A (2012) Evolution of the Macondo well blowout: simulating the effects of the circulation and synthetic dispersants on the subsea oil transport. Environ Sci Technol 46(24):13293–13302

Passow U, Ziervogel K (2016) Marine snow sedimented oil released during the deepwater horizon spill. Oceanography 29:118–125

Sinquin A, Palermo T, Peysson Y (2004) Rheological and flow properties of gas hydrate suspensions. Oil Gas Sci Technol – Rev IFP 59:41–57

Speight J (2007) The chemistry and technology of petroleum, 4th edn. CRC Press, Boca Raton

Socolofsky S, Adams E (2002) Multi-phase plumes in uniform and stratified crossflow. J Hydraul Res 40(6):661–672

Stevens C, Thibodeaux L, Overton E, Valsaraj K, Nandakamar K, Rao A, Walker N (2015) Sea surface oil slick light component vaporization and heavy residue sinking: binary mixture theory and experimental proof of concept. Environ Eng Sci 32(8):694–702

Taylor G (1934) The formation of emulsions in definable fields of flows. Proc R Soc Lond A 146:501–523

Zhao L, Boufadel M, Socolofsky S, Adams E, King T, Lee K (2014) Evolution of droplets in subsea oil and gas blowouts: development and validation of the numerical model VDROP-J. Mar Pollut Bull 83(1):58–69

Zhao L, Boufadel MC, Adams EE, Socolofsky SA, King T, Lee K, Nedwed T (2015) Simulation of scenarios of oil droplet formation from the deepwater horizon blowout. Mar Pollut Bull 101(1):304–319

Zheng L, Yapa PD (2000) Buoyant velocity of spherical and nonspherical bubbles/droplets. J Hydraul Eng 126(11):852–854

Chapter 3
Physical and Chemical Properties of Oil and Gas Under Reservoir and Deep-Sea Conditions

Thomas B. P. Oldenburg, Philip Jaeger, Jonas Gros, Scott A. Socolofsky, Simeon Pesch, Jagoš R. Radović, and Aprami Jaggi

Abstract Petroleum is one of the most complex naturally occurring organic mixtures. The physical and chemical properties of petroleum in a reservoir depend on its molecular composition and the reservoir conditions (temperature, pressure). The composition of petroleum varies greatly, ranging from the simplest gas (methane), condensates, conventional crude oil to heavy oil and oil sands bitumen with complex molecules having molecular weights in excess of 1000 daltons (Da). The distribution of petroleum constituents in a reservoir largely depends on source facies (original organic material buried), age (evolution of organisms), depositional environment (dysoxic versus anoxic), maturity of the source rock (kerogen) at time of expulsion, primary/secondary migration, and in-reservoir alteration such as biodegradation, gas washing, water washing, segregation, and/or mixing from different oil charges. These geochemical aspects define the physical characteristics of a petroleum in the reservoir, including its density and viscosity. When the petroleum is released from the reservoir through an oil exploration accident like in the case of the *Deepwater Horizon* event, several processes are affecting the physical and chemical

T. B. P. Oldenburg (✉) · J. R. Radović · A. Jaggi
University of Calgary, PRG, Department of Geoscience, Calgary, Canada
e-mail: toldenbu@ucalgary.ca; Jagos.Radovic@ucalgary.ca; aprami.jaggi@ucalgary.ca

P. Jaeger
Eurotechnica GmbH, Bargteheide, Germany
e-mail: jaeger@eurotechnica.de

J. Gros (✉)
GEOMAR Helmholtz-Zentrum für Ozeanforschung Kiel, Kiel, Germany
e-mail: jogros@geomar.de

S. A. Socolofsky
Texas A&M University, Zachry Department of Civil Engineering, College Station, TX, USA
e-mail: ssocolofsky@civil.tamu.edu

S. Pesch
Hamburg University of Technology, Institute of Multiphase Flows, Hamburg, Germany
e-mail: simeon.pesch@tuhh.de

© Springer Nature Switzerland AG 2020
S. A. Murawski et al. (eds.), *Deep Oil Spills*,
https://doi.org/10.1007/978-3-030-11605-7_3

properties of the petroleum from the well head into the deep sea. A better understanding of these properties is crucial for the development of near-field oil spill models, oil droplet and gas bubble calculations, and partitioning behavior of oil components in the water. Section 3.1 introduces general aspects of the origin of petroleum, the impact of geochemical processes on the composition of a petroleum, and some molecular compositional and physicochemical background information of the Macondo well oil. Section 3.2 gives an overview over experimental determination of all relevant physicochemical properties of petroleum, especially of petroleum under reservoir conditions. Based on the phase equilibrium modeling using equations of state (EOS), a number of these properties can be predicted which is presented in Sect. 3.3 along with a comparison to experimental data obtained with methods described in Sect. 3.2.

Keywords Physicochemical oil properties · Molecular oil composition · Deep sea · Petroleum reservoir · Phase equilibria modeling

3.1 Molecular Composition and Physical Properties of Petroleum in Reservoirs

Petroleum is one of the most complex naturally occurring organic mixtures containing in some cases over 100,000 different chemical molecules (including isomers). It originates from sedimentary organic matter which itself is derived from once-living organisms. The molecular composition of petroleum varies greatly, ranging from the simplest gas (methane), condensates, conventional crude oil to heavy oil and oil sands bitumen with complex molecules such as asphaltenes with molecular weights in excess of 1000 daltons (Da). The composition of oil is commonly divided into four fractions based on liquid chromatography behavior of the oil constituents. Two of these fractions comprise the saturated and aromatic hydrocarbons, whereas the other two non-hydrocarbon fractions are named resins and asphaltenes. The latter two fractions contain one or more heteroatoms in the form of functional groups per molecule. The majority of heteroatoms (defined as atoms which are not carbon or hydrogen) in oil constituents are typically nitrogen, sulfur, and oxygen. In addition, metals such as nickel and vanadium are common in porphyrinic structures. The proportions of these hydrocarbon and non-hydrocarbon fractions vary greatly from non-hydrocarbon fractions containing less than 1% in some condensates to over 50% in highly viscous extra-heavy oils and bitumen from the oil sands. In conventional crude oils, the proportion of the non-hydrocarbon fractions is typically below 20%. The composition and the interaction of the oil constituents determine the physical and chemical properties of petroleum such as density, viscosity, water solubility, evaporation, and emulsion-forming potential among others. For example, some oil sands bitumens have densities greater than that of water and viscosities of over 1000,000 centipoise (cP) at ambient conditions, whereas the viscosities of some condensates are equal or below that of water (1 cP) with densities as low as 750 kg m^{-3}.

3.1.1 Source Dependence, Generation, Accumulation, and Alteration of Petroleum

The molecular distribution of petroleum in a reservoir depends on several factors. This chapter cannot cover all the details of these important factors, and the reader is referred to one of the great summary books of petroleum/organic geochemistry and the references therein (e.g., Tissot and Welte 1984; Peters et al. 2005). Briefly, the origin/source of the organic material (OM) has strong impact on the composition of expelled oil from source rocks, the OM-containing rock after burial and thermal alteration. Sources include terrestrial, lacustrine, and marine organic matter and mixtures thereof, derived from the domains of life which are prokaryotes (eubacteria and archaea) and eukaryotes in different aquatic ecosystems. The extent of preservation of OM depends on the depositional environment during sedimentation. High sedimentation rates, fine-grained mud, and an anoxic aqueous environment during OM sedimentation are beneficial for the preservation of OM. These conditions minimize the occurrence of aerobic microorganisms, the fast degraders of organic material, and their contact time with the OM prior to its settling. In addition, the redox conditions have an impact on the chemical alteration of OM during sedimentation. After sedimentation, the OM is further altered through chemical, physical, and biological processes during burial up to temperatures of 60 °C (diagenesis) yielding a complex, high molecular weight polymer named kerogen. When buried to greater depths (60–150 °C; oil window; catagenesis), heat/thermal maturity becomes the main OM alteration process transforming the kerogen into petroleum. At the early oil window stages, large kerogen structures are cracked into smaller ones leading to the first expelled oil which contains typically larger molecules with high abundances of heteroatoms in the oil mixture. The average molecular weight of newly expelled oil from the kerogen decreases with increasing thermal maturity at later stages in the oil window accompanied with an increase in the proportion of the hydrocarbon fractions. At the end of the oil window, mainly gas (C1 to C5 hydrocarbons) is expelled from the kerogen. The oil expelled from the kerogen at the different stages in the oil window migrates through a carrier bed (porous and permeable layer) to a trap (the reservoir) primarily governed by buoyancy forces. The accumulated petroleum in the reservoir might be single charged or more likely a mixture of oil from the same source rock but from various maturity stages of the kerogen or even a mixture of expelled oil from more than one source rock. In most cases of petroleum systems, the complexity of the petroleum composition is further increased by in-reservoir alteration such as biodegradation, gas washing, water washing, segregation, and mixing of altered petroleum with fresh charged expelled oil. The most common and in many cases most impactful alteration process affecting the petroleum composition in a reservoir is biodegradation. This slow anaerobic microbial process at the interface between oil and water occurs over a multimillion year timescale by degrading preferred small hydrocarbons and leaving a residual heavy oil (Head et al. 2003; Bennett et al. 2013). In addition, many of the larger molecules are transformed by partial oxidation processes leading to undesirable high acidic oil (Oldenburg et al. 2017). These compositional changes have strong impact on the physical and chemical properties of the oil and influence the transport, interfacial, and corrosion properties. More details on in-reservoir oil biodegradation can be found in Chap. 9.

3.1.2 Macondo Well Oil Molecular Characteristics

The Macondo well (MW) petroleum expelled into the Gulf of Mexico after the explosion on the *Deepwater Horizon* rig in 2010 was a mixture of oil and gas. Reddy et al. (2012) sampled directly above the Macondo well during the blowout and determined a gas-to-oil ratio of 1600 standard cubic feet per petroleum barrel. Valentine et al. (2010) and Reddy et al. (2012) both reported the gas being composed of mainly methane (87.5% and 82.5%, respectively) and smaller amounts of ethane (abundance just above 8%) and propane (around 5%). Reddy et al. (2012) assessed the Macondo oil as a light oil (API gravity 40° with a density of 820 g L^{-1}), whereas Daling et al. (2014) reported that the oil collected through the riser insertion tube tool (RITT) on the Discoverer Enterprise on May 22, 2010, had a density of 833 kg m^{-3}, a pour point of −27 °C, and interfacial tension of 20 mN m^{-1}. The initial viscosity of the Macondo oil was evaluated to be 3.9 cP at 32 °C (Daling et al. 2014). A maturity assessment of the MW oil based on the relative distribution of compound groups of the pyrrolic N1 heteroatom class, namely, DBE 9 (alkylated carbazoles), DBE 12, and DBE 15 (alkylated benzo- and dibenzocarboazoles, respectively) as established by Oldenburg et al. (2014) indicates a maturity level of 0.9% vitrinite reflectance equivalent (%Re) for the MW oil (Fig. 3.1).

Liquid chromatography (TLC-SARA) measurements revealed that the non-biodegraded MW oil is dominated by saturated hydrocarbons (74%), followed by aromatic hydrocarbons (16%) with the non-hydrocarbon (polar) fraction containing 10% (Reddy et al. 2012). The authors reported that the GC-MS amenable MW oil composition (C5 to C38 saturated and aromatic hydrocarbons) is predominated by branched alkanes (26%), followed by cycloalkanes (16%) and *n*-alkanes (15%). Aromatic species such as alkylbenzenes and indenes (9%) and polycyclic aromatic hydrocarbons (4%) are less abundant (Reddy et al. 2012). The GC-MS amenable content of polar oil constituents is 10% (e.g., dibenzothiophenes) with sulfur and nitrogen elemental abundances of the MW oil assessed as 0.4% each. As only a very small percentage of the non-hydrocarbon (polar) fraction is GC-amenable due to the high boiling point of those heteroatom-containing oil species, Fourier-transform ion cyclotron resonance mass spectrometry (FTICR-MS) is the analytical instrument of choice to study this oil fraction. McKenna et al. (2013) characterized more than 30,000 acidic, basic, and nonpolar unique neutral elemental compositions for the MW crude oil. Figure 3.2 shows the compound class distribution of the NIST SRM 2779 Macondo oil measured in electrospray-positive (ESI-P) and electrospray-negative (ESI-N) ion mode as well as in atmospheric pressure photoionization in

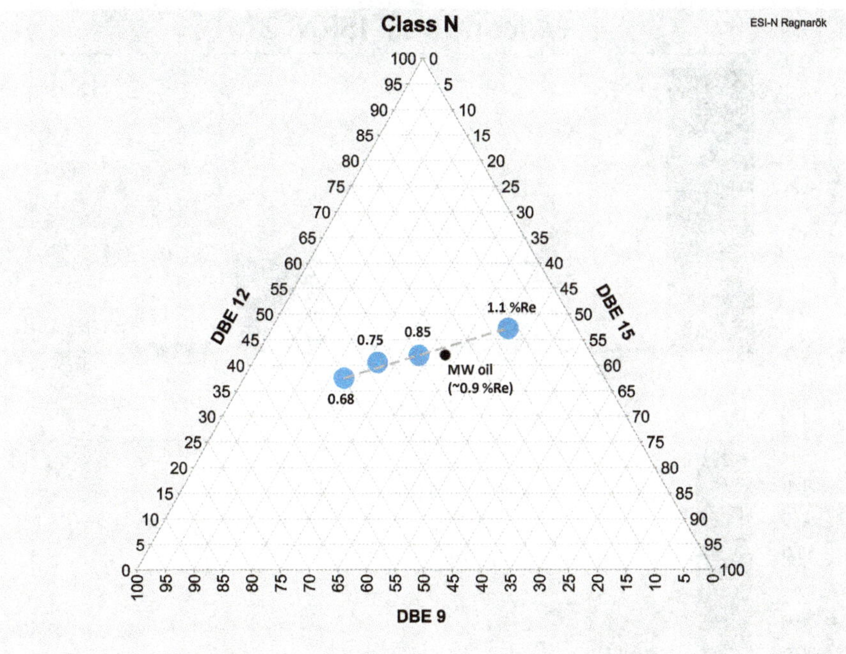

Fig. 3.1 Maturity assessment of the Macondo well oil (MW oil) based on the relative distribution of DBE 9 (alkylated carbazoles), DBE 12, and DBE 15 (alkylated benzo- and dibenzocarboazoles, respectively) of nitrogen-containing compounds measured in electrospray ionization negative ion mode (ESI-N) as established by Oldenburg et al. (2014) indicates a maturity level of 0.9% vitrinite reflectance equivalent (%Re)

positive ion mode (APPI-P) at the University of Calgary with a Bruker 12 T SolariX FTICR-MS.

Whereas nitrogen-containing pyrrolic species such as alkylcarbazoles and their benzannulated homologues strongly dominate the acidic ESI-N-amenable fraction of the MW oil, other nitrogen-containing species with pyridinic structures (e.g.,(benzo)quinolines) are highly abundant in the fraction of the oil constituents which is ionized through protonation (ESI-P). The nonpolar fraction (APPI-P) reveals a strong predominance of hydrocarbons (70%) followed by nitrogen- and sulfur-containing species (~12% each). Despite the broad range of molecules with various functional chemical groups, the overall carbon number range in all three ionization modes are similar (ESI-N, C12 to C76; ESI-P, C16 to C85; APPI-P, C12 to C83) as illustrated in Fig. 3.3. An analogous observation was made when comparing the consolidated double bond equivalent (DBE) distribution of all species measured in the three ionization modes (Fig. 3.3).

Whereas the overall DBE distribution and averaged molecular mass (weighted, AMM) of the acidic fraction (ESI-N) and nonpolar fraction (APPI-P) are very similar (ESI-N, DBE 1 to 29, AMM 517; APPI-P, DBE 1 to 31, AMM 524), the protonated fraction (ESI-P) is composed of the broadest DBE range and highest AMM

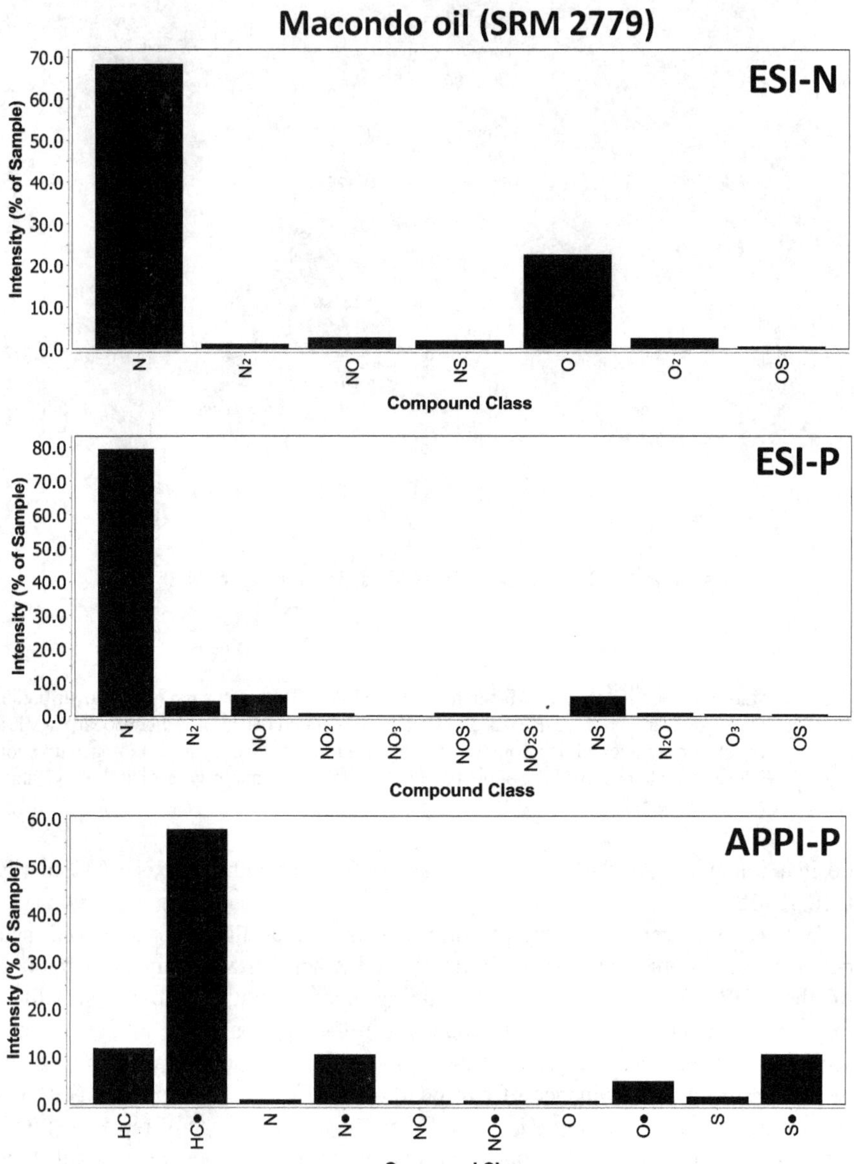

Fig. 3.2 Compound class distributions of the Macondo oil (NIST SRM 2779) measured with a Bruker 12 T SolariX FTICR-MS in three ionization modes: electrospray ionization in negative (ESI-N) and positive (ESI-P) ion mode and atmospheric pressure photoionization in positive (APPI-P) ion mode

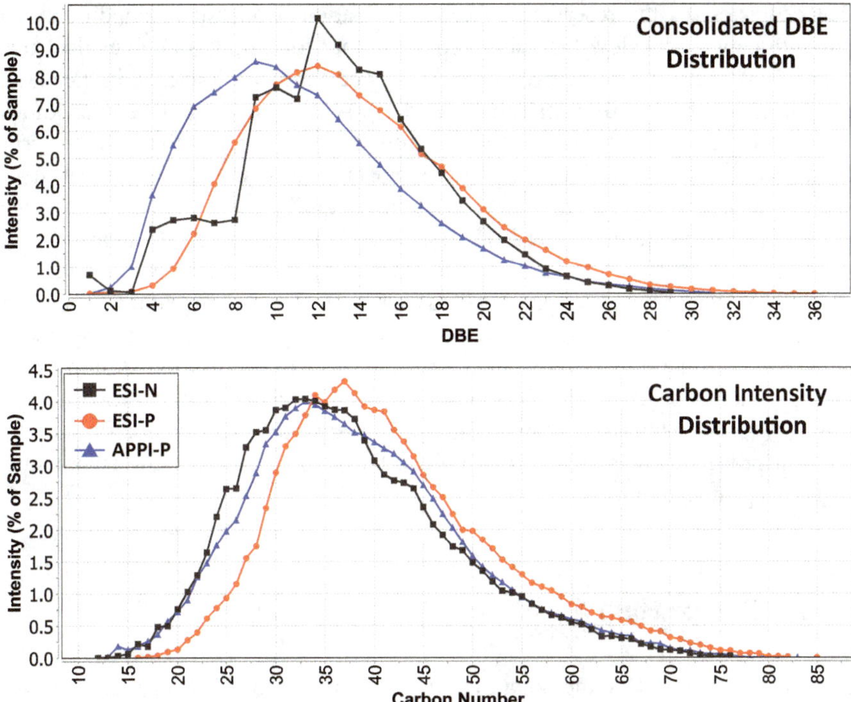

Fig. 3.3 Consolidated double bond equivalent (DBE) and carbon intensity distribution of the Macondo oil (NIST SRM 2779) measured with a Bruker 12 T SolariX FTICR-MS in three ionization modes: electrospray ionization in negative (ESI-N) and positive (ESI-P) ion mode and atmospheric pressure photoionization in positive (APPI-P) ion mode

(DBE 1 to 36, AMM 580). The molecular composition and the interactions between all the oil constituents are responsible for the physicochemical properties of an oil under given pressure and temperature conditions. Important physicochemical oil properties will be discussed in the following sections.

3.2 Physical Properties of Oil Under Deep Ocean Conditions

In order to reflect the real situation in the oil reservoir, the production well, and possible events like oil spills or emergency shutdowns, the properties of the so-called live fluids shall be taken into account. Under reservoir, deep-sea, or production conditions, commonly a certain amount of gas is present in the coexisting petroleum liquid and aqueous phases. The bubble point of a live oil resembles the saturation pressure below which a gas phase theoretically separates from the hydrocarbon liquid. Live oils may be sampled directly from the field or may be recombined in the laboratory from dead oil and the respective gas mixture knowing the bubble point that characterizes the reservoir under consideration. Under field conditions, the pressure may drop below the bubble point without forming gas bubbles due to an energy barrier that has to be overcome during nucleation as described in the classical nucleation theory (Blander and Katz 1975). A large pressure difference known as capillary or Laplace pressure impedes growth of nuclei below the so-called critical nucleation radius. Different approaches exist to predict the related critical supersaturation, i.e., the pressure difference to the equilibrium conditions, at which nucleation sets on (Kalikmanov 2013; Bauget and Lenormand 2002). In most cases, the interfacial tension is used as a crucial property for describing the nucleation phenomena, also because of practical reasons. Transferred to the situation of deepsea spills, gas-saturated droplets undergo a decreasing pressure on their way toward the sea surface and may at some point start to degas which is discussed in this book in Chap. 5 and in Pesch et al. (2018) and subsequently change further properties like viscosity and density. In continuation, methods are explained for determining the relevant properties of real fluid mixtures at the predominating elevated pressures under deep-sea conditions.

3.2.1 Bubble Point

Determination of the bubble point of a live oil is part of the PVT analysis. The volume inside the sampling cylinder, mostly containing a piston to separate the oil from a hydraulic fluid, is increased step-by-step while continuously recording the pressure. At the bubble point, pressure reduction slows down due to formation of a compressible second phase (Danesh 1998).

3.2.2 Gas Saturation

The capacity of the petroleum liquid to dissolve gas at given conditions may be determined by different ways:

1. Completely decompress the live oil from the bubble point down to atmospheric pressure while capturing the amount of gas previously being dissolved (Jaggi et al. 2017; Chap. 8).
2. Gravimetrically, using a microbalance for detecting the gas uptake of the liquid phase at given pressure and temperature until reaching equilibrium (Knauer et al. 2017).

In Sect. 3.3.2 we report solubility data determined by both methods along with modeled data. There is a deviation among both experimental approaches: at lower pressures the second method seems to give more reliable data, whereas at higher pressures, the gravitational method (2) clearly underestimates the solubility due to the extraction of lighter components and subsequent loss of mass (see also Sect. 3.3.2).

3.2.3 Density and Swelling

The densities of petroleum liquids are usually a nearly linear function of pressure. The most common method to measure liquid densities at elevated pressures is the oscillating tube (Eggers 2012). A bended tube is filled with the liquid under investigation and excited. The response in terms of the periodic time is related to the density. The system requires thorough calibration in order to deliver reliable results. It is even more challenging to determine the density of liquids that contain dissolved gases. The influence of pressure in this case depends on the type of gas. In Sect. 3.3.3 we report the density of Louisiana sweet crude (LSC) oil comparing the pure liquid to the methane-containing liquid, including modeled data using the Peng-Robinson EOS.

3.2.4 Viscosity

There are a large number of methods available for determining the viscosity at elevated pressures (Eggers 2012). Mainly two points have to be taken into account: Newtonian flow and gas saturation. The main methods used are:

1. Vibrating crystal.
2. Capillary tube.
3. Rotational rheometer.
4. Rolling ball/oscillating piston.

In Fig. 3.4 the viscosity as determined by a capillary tube viscometer is depicted as a function of pressure. For determining the viscosity of gas-saturated oils, a piston accumulator similar to the one used to determine the bubble point is used to prepare the mixture.

3.2.5 Diffusivity

Applying the setup described above, the diffusion of gas into the liquid phase can be followed as a function of time until the liquid is saturated with gas (Knauer et al. 2017). The microbalance described in Sect. 3.2.2 is used to determine this sorption kinetics. The analytical solution to the nonstationary Fick's diffusion (Crank 1975) is adjusted to the experimental data in order to deduce the diffusion coefficient. Because of the higher viscosity, the diffusion coefficient of methane in crude oil takes lower values than in an aqueous phase. Diffusivities at a pressure of 5 MPa are found in the order of $1*10^{-9}$ m^2 s^{-1} for LSC compared to around $5*10^{-9}$ m^2 s^{-1} in water, which is slightly higher than values reported in literature for atmospheric pressure (Hayduk and Laudie 1974).

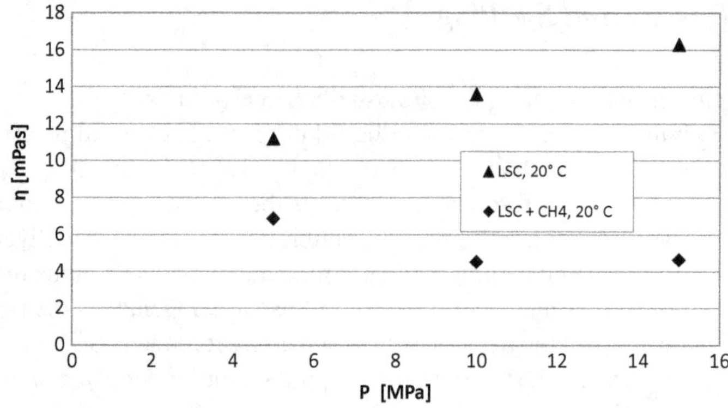

Fig. 3.4 Viscosity of LSC with and without CH$_4$, as a function of (gauge) pressure, 20 °C. Similar to the density, there is a reducing effect of dissolved gases on the viscosity. The dissolved gas acts like a lubricant among the large petroleum molecules

3.2.6 Interfacial Tension

The pendant drop method has proved to be the most feasible way to measure the interfacial tension at elevated pressures. The respective experiments are carried out in so-called high-pressure view cells (Jaeger et al. 2010). The axisymmetric shapes of pendant drops of a liquid within the surrounding compressed phase are recorded and evaluated based on the Laplace pressure difference across curved phase boundaries (Bashford and Adams 1883). Under deep-sea conditions, it is especially of interest to determine the interfacial tension between a gas-saturated (live) oil and seawater. In view of the complexity of this type of system, well-defined procedures need to be followed in order to minimize artifacts caused by chemical reactions and excessive formation of surfactants. Figure 3.5 shows the interfacial tension of live crude oil drops in seawater as a function of the drop age for different pressures.

In all cases the interfacial tension drops to a quasi-equilibrium value. Due to interface-active compounds in the crude oil, the interfacial tension decreases within the first 20 minutes. The methane content in this liquid-liquid system has a substantial effect on density and composition (Sect. 3.4) and is therefore assumed to play a decisive role in this particular behavior of increasing the interfacial tension as a function of pressure.

3.3 Modeling Phase Equilibria (Gas-Oil-Water) in Oil Reservoirs and in the Deep Sea and Oil Constituent Partitioning

Phase equilibria play a major role in understanding the state and behavior of petroleum fluids. Petroleum fluids within oil reservoirs have had millions of years to reach equilibrium. During a deep-sea release of petroleum fluids, new pressure and

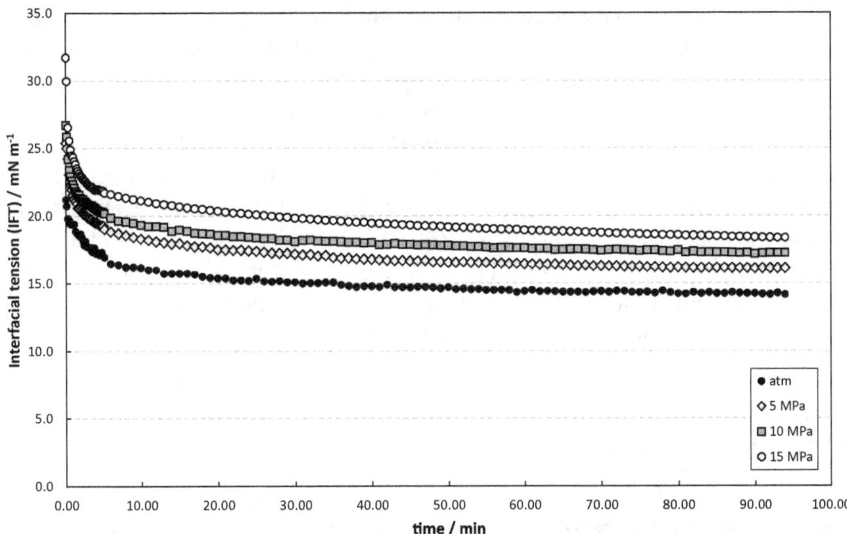

Fig. 3.5 IFT of recomposed LSC live oil in seawater as a function of drop age and (gauge) pressure, 20 °C

temperature conditions lead to disruption of the pre-existing thermodynamic equilibrium. Under turbulent conditions, equilibration of the petroleum compounds in two phases of natural gas and petroleum liquid is thought to proceed within a short time frame. The kinetics of aqueous dissolution of petroleum compounds is partly driven by equilibrium partitioning properties between petroleum phases and seawater (McGinnis et al. 2006; Schwarzenbach et al. 2003) and proceeds on a variety of time scales depending on the flow conditions and specific area of the interface, as well as the equilibrium aqueous solubility of each specific compound (Aeppli et al. 2013; Bartha et al. 2015; Gros et al. 2017, 2014; Wardlaw et al. 2008). For example, >99% of the released methane was observed to remain subsurface during the 2010 *Deepwater Horizon* disaster, whereas low-solubility compounds were not affected (<1%) by aqueous dissolution during ascent to the sea surface (Ryerson et al. 2012). Knowledge of equilibrium partitioning properties under relevant conditions of temperature, pressure, salinity, and composition is crucial for explaining these field observations (Gros et al. 2016, 2017).

Gas-liquid equilibrium partitioning of petroleum fluids is a common task performed by petroleum engineers, necessary for the evaluation, design, and operation of extraction wells, and the transport and separation of reservoir fluids (Danesh 1998; McCain 1990; Schou Pedersen et al. 2006). However, industry needs usually don't include an understanding of the equilibrium between petroleum fluids and seawater under the high-pressure and low-temperature conditions prevailing in the deep sea. The gas-liquid equilibrium partitioning of petroleum fluids is frequently predicted using equations of state (such as Peng-Robinson or Soave-Redlich-Kwong), which are solved following ad hoc procedures (Michelsen and Mollerup 2007). For industry needs, petroleum fluids are frequently modeled using a set of

10–20 components representing individual compounds (e.g., CH_4 or CO_2) or group of poorly identified compounds (e.g. C_{10}–C_{14}, all compounds between normal decane and normal tetradecane). Chemical properties of these petroleum fractions (including critical properties, acentric factors, and molecular weight) need to be collected from literature or estimated (Riazi 2005). Predictions of equations of state are usually validated with laboratory data obtained with reservoir fluid samples, and equation of state "tuning" is also a frequently used procedure to improve prediction ability when enough data is available (Zick 2013). This tuning consists of a careful modification of pseudo-component properties or binary interaction parameters performed to minimize the discrepancy with available laboratory data. Predictions of selected properties are discussed in turn below.

3.3.1 Bubble Point

The bubble point can be predicted from EOSs using published procedures (e.g. (Michelsen and Mollerup 2007)). This calculation is relatively straightforward at low-pressure conditions. However, a careful procedure at higher pressures is necessary to avoid convergence of the algorithm to a false solution.

3.3.2 Gas Saturation

Gas saturation of a petroleum mixture can be calculated with an equation of state. The amount of "gas" (e.g., CH_4) is gradually increased until the equation of state predicts two phases rather than one.

At high-volume ratios of gas to petroleum liquid, the gas phase will deplete the petroleum liquid from its lighter components, affecting the equilibrium content of light C_1–C_5 components within the petroleum liquid. Equation of state calculations for such systems highlight the importance of taking this effect into consideration when designing laboratory measurement procedure. Figure 3.6 shows solubility data determined by the two methods described in Sect. 3.2.2 along with modeled data.

3.3.3 Density and Swelling

Density of both gas and petroleum liquid phases are readily predicted from cubic equations of states, though they are known to lead to poor predictions of the densities of liquid petroleum phases (Ahmed 2010). However, the so-called "volume translation" procedures have been developed to improve prediction abilities of equations of states (Lin and Duan 2005; Péneloux et al. 1982). Very good prediction abilities for binary mixtures can be achieved when using volume translation (e.g., Fig. 3.7).

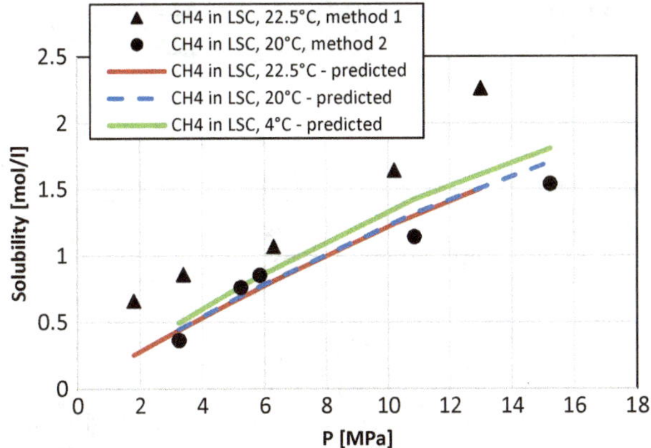

Fig. 3.6 Measured solubility of methane in Louisiana sweet crude oil at 0–16 MPa gauge pressures. Filled symbols are measurements conducted in the laboratory according to method 1 and method 2 described in Sect. 3.2.2, whereas the solid lines are predicted values using the Peng-Robinson equation of state with volume translation, after tuning

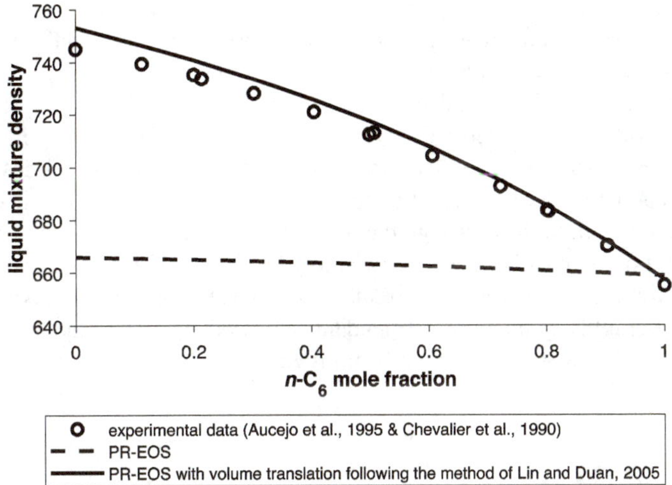

Fig. 3.7 Densities of mixtures of normal hexane (n-C_6) and normal dodecane (n-C_{12}) at 1 atmosphere (101325 Pa) and 25 °C, as measured in the laboratory (open circles) and predicted with the Peng-Robinson equation of state (PR-EOS) with (solid line) and without (dashed line) volume translation

Figure 3.8 shows the measured density of Louisiana sweet crude (LSC) oil comparing the dead oil liquid to the liquid saturated with methane and includes modeled data using the Peng-Robinson EOS and scaled to the data given in Fig. 3.6 (Gros et al. 2016).

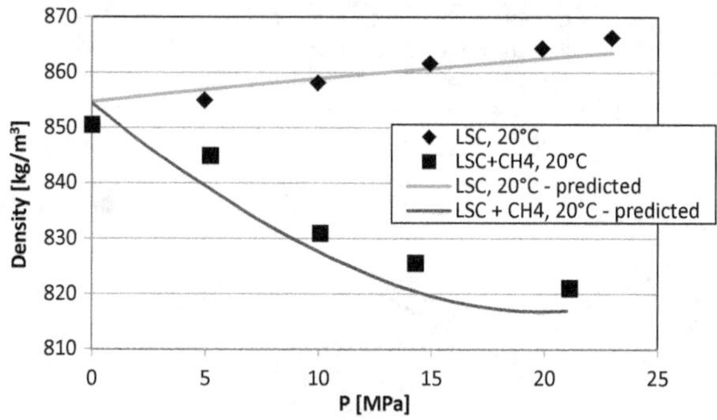

Fig. 3.8 Density of Louisiana sweet crude oil dead oil (LSC) and Louisiana sweet crude oil saturated with methane (LSC + CH4) at 0–23 MPa gauge pressures. Filled symbols are measurements conducted in the laboratory, whereas the solid lines are predicted values using the Peng-Robinson equation of state with volume translation, after tuning

3.3.4 Viscosity

Equations of state don't readily predict phase viscosities, and additional models have to be used to derive this property. Several methods have been presented, and prediction errors of 100% are deemed not unusual (Riazi 2005). Such uncertainties on the viscosities of petroleum fluids are usually acceptable for oil spill modeling, as the processes simulated are only weakly dependent on viscosity. Viscosity is also an important parameter for evaluation of reservoir, as it plays a key role in determining the ease for oil to flow through the permeable reservoir rock. Viscosity of oil is dependent on composition and varies widely between different oils and increases dramatically upon evaporation (Sebastião and Guedes Soares 1995). Additionally, viscosity is highly temperature dependent, and (water-in-oil) emulsification can lead to orders-of-magnitude increases of oil viscosity (Lehr et al. 2002).

3.3.5 Diffusivity

Several methods exist to estimate the diffusivity of petroleum compounds in petroleum gas and liquid phases (Riazi 2005) and in water (Schwarzenbach et al. 2003). For example, the diffusivity of both gaseous and liquid petroleum compounds in water can be estimated by use of the Hayduk-Laudie formula (Hayduk and Laudie 1974):

$$D_i = \frac{13.26 \cdot 10^{-9}}{\left(\eta_w\left(T\right) \cdot 10^3\right)^{1.14} \cdot \left(\overline{V}_{bp,i} \cdot 10^6\right)^{0.589}} \tag{3.1}$$

Where D_i is the diffusion coefficient of compound i in water, in $m^2\ s^{-1}$; η_w is the viscosity of water at temperature T (Sharqawy et al. 2010), in Pa s; and $\overline{V}_{bp,i}$ is the molar volume of the solute at its normal boiling point, in $m^3 \, mol^{-1}$. (note: this is the correct version of the Hayduk-Laudie formula as expressed in base SI units; the reader is warned that the abstract of the original article by Hayduk and Laudie (1974) contains a typo in one of the coefficients.) The effects of both pressure and salinity on diffusivities in water are yet unclear as no definite picture emerges from the scarce data available in the scientific literature. Yet, these data indicate that these effects are likely limited for conditions of environmental significance.

3.3.6 Interfacial Tension

Interfacial tension between petroleum phases and seawater can be extrapolated from one measured value; however it is notably difficult to predict it in the absence of any measured value. A method has, for example, been presented by Firoozabadi and Ramey (1988), but its prediction ability is very limited unless a measured value can be used for scaling of predictions (Danesh 1998). Such measurements have been made in the aftermath of the *Deepwater Horizon* disaster (Schlüter and Pesch unpublished data; Abdelrahim and Rao 2014; Venkataraman et al. 2013). However, this information is usually lacking before an accidental release occurs, as highlighted by the paucity of such data reported in oil libraries such as the NOAA ADIOS oil library.

3.4 Summary

In summary, this chapter describes geochemical aspects confining the molecular composition and physicochemical properties of a petroleum in the reservoir with a specific focus on the Macondo well oil. Important physicochemical properties of petroleum under reservoir conditions and when released into the seawater are discussed and compared between experimental and simulated data. Several of these physical and chemical petroleum properties affected by processes involved in *Deepwater Horizon* oil spill will be further discussed throughout this book.

Acknowledgment This research was made possible by a from the Gulf of Mexico Research Initiative/C-IMAGE. Data are publicly available through the Gulf of Mexico Research Initiative Information and Data Cooperative (GRIIDC) at https://data.gulfresearchinitiative.org/ (DOI: 10.7266/n7-xpgb-g817, DOI: 10.7266/N7DF6PQK, DOI: 10.7266/N7HX19QW, DOI: 10.7266/N7J38R2F) and at http://gulfresearchinitiative.org/hydrocarbon-intercalibration-experiment/

References

Abdelrahim MA, Rao DN (2014) Measurement of interfacial tension in hydrocarbon/water/dispersant systems at deepwater conditions. Oil Spill Remediat. Wiley Online Books. https://doi.org/10.1002/9781118825662.ch14

Aeppli C, Reddy CM, Nelson RK, Kellermann MY, Valentine DL (2013) Recurrent oil sheens at the *Deepwater Horizon* disaster site fingerprinted with synthetic hydrocarbon drilling fluids. Environ Sci Technol 47:8211–8219. https://doi.org/10.1021/es4024139

Ahmed T (2010) Reservoir Engineering Handbook, 4th edn. Elsevier, Burlington

Aucejo A, Burguet MC, Munoz R, Marques JL (1995) Densities, viscosities, and refractive indices of some n-alkane binary liquid systems at 298.15K. J Chem Eng Data 40:141–147

Bartha A, De Nicolais N, Sharma V, Roy SK, Srivastava R, Pomerantz AE, Sanclemente M, Perez W, Nelson RK, Reddy CM, Gros J, Arey JS, Lelijveld J, Dubey S, Tortella D, Hantschel T, Peters KE, Mullins OC (2015) Combined petroleum system modeling and comprehensive two-dimensional gas chromatography to improve understanding of the crude oil chemistry in the llanos basin, Colombia. Energy Fuel 29:4755–4767. https://doi.org/10.1021/acs.energyfuels.5b00529

Bauget F, Lenormand R (2002) Mechanisms of bubble formation by pressure decline in porous media SPE 77457

Bashford FB, Adams JC (1883) An attempt to test the theories on capillary action. University Press, Cambridge

Blander M, Katz JL (1975) Bubble nucleation in liquids. AICHE J 21(5):833–848

Bennett B, Adams JJ, Gray ND, Sherry A, Oldenburg TBP, Huang H, Larter SR, Head IM (2013) The controls on the composition of biodegraded oils in the deep subsurface – part 3. The impact of microorganism distribution on petroleum geochemical gradients in biodegraded petroleum reservoirs. J Org Geochem 56:94–105

Chevalier JLE, Petrino PJ, Gaston-Bonhomme YH (1990) Viscosity and density of some aliphatic, cyclic, and aromatic hydrocarbons binary liquid mixtures. J Chem Eng Data 35:206–212

Crank J (1975) The mathematics of diffusion, 2nd edn. Oxford University Press, New York, pp 44–89

Daling PS, Leirvik F, Almas IK, Brandvik PJ, Hansen AL, Reed M (2014) Surface weathering and dispersibility of Macondo crude oil. Mar Pollut Bull 87:300–310. https://doi.org/10.1016/j.marpolbul.2014.07.005

Danesh A (1998) PVT and phase behaviour of petroleum reservoir fluids. Elsevier. ed, Developments in Petroleum Science, Amsterdam

Eggers R (ed) (2012) High pressure applications in enhanced crude oil recovery, high pressure processes. Wiley. Chapter 3.2

Firoozabadi A, Ramey HJ (1988) Surface tension of water-hydrocarbon systems at reservoir conditions. J Can Pet Technol 27:41–48. https://doi.org/10.2118/88-03-03

Gros J, Nabi D, Würz B, Wick LY, Brussaard CPD, Huisman J, van der Meer JR, Reddy CM, Arey JS (2014) First day of an oil spill on the open sea: early mass transfers of hydrocarbons to air and water. Environ Sci Technol 48:9400–9411. https://doi.org/10.1021/es502437e

Gros J, Reddy CM, Nelson RK, Socolofsky SA, Arey JS (2016) Simulating gas−liquid−water partitioning and fluid properties of petroleum under pressure: implications for deep-sea blowouts. Environ Sci Technol 50:7397–7408. https://doi.org/10.1021/acs.est.5b04617

Gros J, Socolofsky SA, Dissanayake AL, Jun I, Zhao L, Boufadel MC, Reddy CM, Arey JS (2017) Petroleum dynamics in the sea and influence of subsea dispersant injection during *Deepwater Horizon*. Proc Natl Acad Sci U S A 114:201612518. https://doi.org/10.1073/pnas.1612518114

Hayduk W, Laudie H (1974) Prediction of diffusion coefficients for nonelectrolytes in dilute aqueous solutions. AICHE J 20:611–615. https://doi.org/10.1002/aic.690200329

Head IM, Jones DM, Larter SR (2003) Biological activity in the deep subsurface and the origin of heavy oil. Nature 426:344–352

Jaeger P, Alotaibi M, Nasr-El-Din H (2010) Influence of compressed carbon dioxide on the capillarity of the gas-crude oil-reservoir water system. J Eng Data 55:5246–5251

Jaggi A, Snowdon RW, Stopford A, Radovic JR, Oldenburg TBP, Larter SR (2017) Experimental simulation of crude oil-water partitioning behavior of BTEX compounds during a deep submarine oil spill. Org Geochem 108:1–8

Kalikmanov VI (2013) Nucleation theory. Lecture notes in physics. Springer, Dordrecht

Knauer S, Schenk M, Köddermann T, Reith D, Jaeger P (2017) Interfacial tension and related properties of ionic liquids in CH4 and CO2 at elevated pressures: experimental data and molecular dynamics simulation. JCED J Eng Data 62(8):2234–2243

Lehr W, Jones R, Evans M, Simecek-Beatty D, Overstreet R (2002) Revisions of the ADIOS oil spill model. Environ Model Softw 17:189–197. https://doi.org/10.1016/S1364-8152(01)00064-0

Lin H, Duan Y-Y (2005) Empirical correction to the Peng–Robinson equation of state for the saturated region. Fluid Phase Equilib 233:194–203. https://doi.org/10.1016/j.fluid.2005.05.008

McCain WD (1990) The properties of petroleum fluids. PennWell Books

McGinnis DF, Greinert J, Artemov Y, Beaubien SE, Wüest A (2006) Fate of rising methane bubbles in stratified waters: how much methane reaches the atmosphere? J Geophys Res Oceans 111:C09007. https://doi.org/10.1029/2005JC003183

McKenna AM, Nelson RK, Reddy CM, Savory JT, Kaiser NK, Fitzsimmons JE, Marshall AG, Rodgers RP (2013) Expansion of the analytical window for oil spill characterization by ultrahigh resolution mass spectrometry: beyond gas chromatography. Environ Sci Technol 47:7530–7539

Michelsen ML, Mollerup JM (2007) Thermodynamic models: fundamentals & computational aspects, 2nd edn. Tie-Line Publications, Holte

Oldenburg TBP, Brown M, Bennett B, Larter SR (2014) The impact of thermal maturity level on the composition of crude oils, assessed using ultra-high resolution mass spectrometry. Org Geochem 75:151–168. https://doi.org/10.1016/j.orggeochem.2014.07.002

Oldenburg TBP, Jones M, Huang H, Bennett B, Shafiee NS, Head I, Larter SR (2017) The controls on the composition of biodegraded oils in the deep subsurface- part 4. Degradation and production of high molecular weight aromatic and polar species during in-reservoir biodegradation. Org Geochem 114:57–80

Péneloux A, Rauzy E, Fréze R (1982) A consistent correction for Redlich-Kwong-soave volumes. Fluid Phase Equilib 8:7–23. https://doi.org/10.1016/0378-3812(82)80002-2

Pesch S, Jaeger P, Jaggi A, Malone K, Hoffmann M, Krause D, Oldenburg TBP, Schlüter M (2018) Rise velocity of live-oil droplets in deep-sea oil spills. Environ Eng Sci 35:289–299. https://doi.org/10.1089/ees.2017.0319

Peters KE, Walters CC, Moldowan JM (2005) The biomarker guide, volume 1. Cambridge University Press, Cambridge, UK

Reddy CM, Arey JS, Seewald JS, Sylva SP, Lemkau KL, Nelson RK, Carmichael CA, McIntyre CP, Fenwick J, Ventura GT, Van Mooy BAS, Camilli R (2012) Composition and fate of gas and oil released to the water column during the *Deepwater Horizon* oil spill. Proc Natl Acad Sci U S A 109:20229–20234

Riazi MR (2005) Characterization and properties of petroleum fractions. ASTM International, West Conshohocken

Ryerson TB, Camilli R, Kessler JD, Kujawinski EB, Reddy CM, Valentine DL, Atlas E, Blake DR, de Gouw J, Meinardi S, Parrish DD, Peischl J, Seewald JS, Warneke C (2012) Chemical data quantify *Deepwater Horizon* hydrocarbon flow rate and environmental distribution. Proc Natl Acad Sci U S A 109:20246–20253. https://doi.org/10.1073/pnas.1110564109

Schou Pedersen K, Kristensen PL, Azeem Shaikh J (2006) Phase behavior of petroleum reservoir fluids. CRC Press, Boca Raton

Schwarzenbach RP, Gschwend PM, Imboden DM (2003) Environmental organic chemistry, 2nd edn. John Wiley & Sons, Inc, Hoboken

Sebastião P, Guedes Soares C (1995) Modeling the fate of oil spills at sea. Spill Sci Technol Bull 2:121–131. https://doi.org/10.1016/S1353-2561(96)00009-6

Sharqawy MH, Lienhard JH, Zubair SM (2010) Thermophysical properties of seawater: a review of existing correlations and data. Desalination Water Treat 16:354–380. https://doi.org/10.5004/dwt.2010.1079

Tissot BP, Welte DH (1984) Petroleum formation and occurrence. Springer, Heidelberg

Valentine DL, Kessler JD, Redmond MC, Mendes SD, Heintz MB, Farwell C, Hu L, Sinnaman FS, Yvon-Lewis S, Du M, Chan EW, Tigreros FG, Villanueva CJ (2010) Propane respiration jump-starts microbial response to a deep oil spill. Science 330:208–211

Venkataraman P, Tang J, Frenkel E, McPherson GL, He J, Raghavan SR, Kolesnichenko V, Bose A, John VT (2013) Attachment of a Hydrophobically modified biopolymer at the oil–water Interface in the treatment of oil spills. ACS Appl Mater Interfaces 5:3572–3580. https://doi.org/10.1021/am303000v

Wardlaw GD, Arey JS, Reddy CM, Nelson RK, Ventura GT, Valentine DL (2008) Disentangling oil weathering at a marine seep using GCxGC: broad metabolic specificity accompanies subsurface petroleum biodegradation. Environ Sci Technol 42:7166–7173. https://doi.org/10.1021/es8013908

Zick AA (2013) Equation-of-state fluid characterization and analysis of the Macondo reservoir fluids. (Expert report prepared on behalf of the United States No. TREX-011490R)

Chapter 4
Jet Formation at the Spill Site and Resulting Droplet Size Distributions

Karen Malone, Zachary M. Aman, Simeon Pesch, Michael Schlüter, and Dieter Krause

Abstract The size distribution of oil droplets and gas bubbles forming at the exit geometry of a deep-sea blowout is one of the key parameters to understand its propagation and fate in the ocean, whether with regard to rising time to the surface, drift by ocean currents, dissolution or biodegradation. While a large 8 mm droplet might rise to the sea surface within minutes or hours, microdroplets <100 µm may take weeks or months to surface, if at all. On the other hand, a microdroplet or bubble dissolutes faster due to its larger surface to volume ratio and is also more available for biodegrading bacteria. To be able to properly model these effects, it is necessary to understand the drop formation processes near the discharge point and to predict the evolving droplet size distribution (DSD) for the specific conditions.

In this chapter, the general breakup mechanisms and flow regimes of an oil-in-water jet are discussed in Sect. 4.1. Section 4.2 focuses on the different approaches to determine the DSD in the laboratory and field settings and critically reviews the existing datasets. State-of-the-art models for the prediction of the DSD of a subsea oil discharge are presented alongside a new approach based on the turbulent kinetic energy (TKE) in Sect. 4.3, while Sect. 4.4 takes a closer look at the specific effects of the deep sea on the DSD. Based on this, Sect. 4.5 discusses the advantages and limitations of subsea dispersant injection. Section 4.6 provides a summary of the chapter and gives an outlook to unresolved questions.

K. Malone (✉) · D. Krause
Hamburg University of Technology, Institute of Product Development and Mechanical Engineering Design, Hamburg, Germany
e-mail: karen.malone@tuhh.de; krause@tuhh.de

Z. M. Aman
University of Western Australia, Department of Chemical Engineering, Perth, WA, Australia
e-mail: zachary.aman@uwa.edu.au

S. Pesch · M. Schlüter
Hamburg University of Technology, Institute of Multiphase Flows, Hamburg, Germany
e-mail: simeon.pesch@tuhh.de; michael.schluter@tuhh.de

© Springer Nature Switzerland AG 2020
S. A. Murawski et al. (eds.), *Deep Oil Spills*,
https://doi.org/10.1007/978-3-030-11605-7_4

Keywords Jet formation · Droplet size distribution · Live oil · Median drop size · Flow regime · Turbulent kinetic energy · Turbulence · Droplet breakup · Near-field · In situ measurements · Dissolved gas · Outgassing · Phase change · Dispersants

Nomenclature

Latin

A	Empirical coefficient in the modified Weber number scaling
B	Empirical coefficient in the modified Weber number scaling
CDF	Cumulative distribution function
D	Nozzle/discharge diameter
d_{32}	Sauter diameter
d_{n50}	Median diameter of number distribution
d_p	Drop/particle diameter
d_{v50}	Median diameter of volume distribution
DOR	Dispersant-to-oil ratio
DSD	Drop size distribution
erf(x)	Gauss error function
exp(x)	Exponential function
IFT	Interfacial tension
k_i	Scaling factor
M	Oil mass inside the nozzle
Oh	Ohnesorge number
p	Pressure
Δp	Pressure drop at the nozzle
Q	Volume flow rate
Re	Reynolds number
u_1	Exit velocity of dispersed liquid phase
Vi	Viscosity number
We	Weber number
We*	Modified Weber number

Greek

α	Spreading factor of the Rosin-Rammler distribution function
ε	Turbulent energy dissipation rate
ε_u	Turbulent energy dissipation rate caused by the exit velocity
ε_{pd}	Turbulent energy dissipation rate caused by pressure drop at the nozzle
η_l	Dynamic viscosity of dispersed liquid phase
ρ_l	Density of dispersed liquid phase
ρ_c	Density of continuous phase
σ	Spreading factor of the log-normal distribution function
σ_l	Interfacial tension (IFT) between dispersed liquid phase and continuous phase

4.1 Introduction

When crude oil enters into seawater from a pipeline or a well, drop formation will occur in different ways based on the characteristics of the discharge. In general, oil drop formation is governed by the relation of stabilizing forces (interfacial tension, viscosity) versus destabilizing forces such as turbulence within the oil and the water phase, friction and cavitation (Lefebvre and McDonell 2017). There are several studies distinguishing a number of different flow regimes, ranging from the formation of single drops at the exit geometry to full atomization of a jet (Masutani and Adams 2001; Boxall et al. 2012; Ohnesorge 1936; Lefebvre and McDonell 2017).

Originally, Ohnesorge (1936) identified four different regimes (see Fig. 4.1):

  0. Formation of individual drops at the nozzle without jet formation.
  I. Breakup of a cylindrical jet caused by Rayleigh instability.
 II. Breakup caused by sinuous waves along the jet.
III. Atomization of the jet due to turbulent breakup.

The different regimes are distinguished using the relation between the dimensionless Reynolds number (Re)

$$Re = \frac{D \cdot \rho_1 \cdot u_1}{\eta_1} \tag{4.1}$$

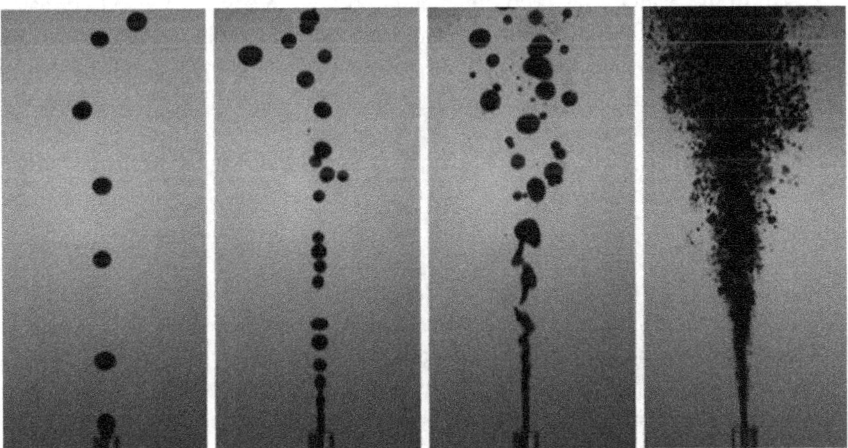

Fig. 4.1 Different breakup/flow regimes of Louisiana sweet crude oil discharged into artificial seawater at 150 bar, 20 °C from a 1.5 mm nozzle. Volume flow increases from left to right. Far left: regime (0), individual drop formation. Mid left: regime (I), Rayleigh instability. Mid right: regime (II), sinuous wave breakup. Far right: regime (III), atomization, turbulent breakup

and Ohnesorge number (Oh)

$$Oh = \frac{\eta_1}{D \cdot \rho_1 \cdot \sigma_1} \qquad (4.2)$$

with D as the discharge diameter, ρ_1 is the oil density, σ_1 is the oil-water interfacial tension (IFT), η_1 is the dynamic viscosity of the oil and u_1 is the oil exit velocity at the nozzle. In each regime, the breakup of the dispersed oil into drops is governed by different mechanisms. While the drop diameter is mainly determined by the exit geometry, buoyancy and interfacial tension between oil and water in regime (0), the atomization regime is mainly governed by turbulence and the shear forces acting between oil and water. Since the original work of Ohnesorge, several authors have sought to specify the borders of these and additional sub-regimes (e.g. Tang 2004; Hsiang and Faeth 1992; Masutani and Adams 2001; Adams and Socolofsky 2004; Lefebvre and McDonell 2017), especially with regard to the transition from Rayleigh instability to full atomization. In a major oil well blowout, it can, however, be assumed that the discharged liquid will be well beyond the atomization boarder (Masutani and Adams 2001). In the following therefore only the drop formation processes in a fully atomized jet will be considered. As it represents the upper limit of the different approaches for classification of the flow regimes, Tang's expression (Tang 2004)

$$Oh \geq 24.9548 \cdot Re^{-1.0027} \qquad (4.3)$$

is chosen to mark the transition to full atomization and turbulent breakup.

The breakup of a turbulent jet consists of two phases, the primary and secondary breakup. In the primary breakup, ligaments form at the edge of the cylindrical jet due to shear between the continuous phase and the jet, which eventually break up into drops. The secondary breakup describes the further disintegration of these initial drops into smaller drops due to the ambient flow field. The primary breakup stage has a major influence on the final droplet size distribution (DSD) and is mainly controlled by the discharge conditions at the nozzle (Lefebvre and McDonell 2017). The hydrodynamics taking place at this stage of the jet formation are very complex and not yet fully understood (Zuzio et al. 2013).

As a result of the turbulent nature of an atomized jet, the formation and size of individual drops cannot be predicted. The entirety of all drops in an atomized jet can, however, approximately be described using a characteristic average diameter and a size distribution function. Depending on the area of application, different characteristic diameters are used. In oil spill modelling, the median diameters of the number (d_{n50}) or volume distribution (d_{v50}) are most commonly used, describing the 50th percentile of all drops by number and total oil volume, respectively. With regard to mass transfer between the dispersed and the continuous phase, the Sauter mean diameter (d_{32}) has been established. It describes a drop with the same volume to surface ratio as the overall oil phase and can be calculated using:

$$d_{32} = \frac{\sum_{i=1}^{n} d_i^3}{\sum_{i=1}^{n} d_i^2} \tag{4.4}$$

The DSD of a dispersed oil phase is widely discussed in the literature. Experimental studies mostly found either a Rosin-Rammler distribution (e.g. Johansen et al. 2013) according to

$$\mathrm{CDF}(d) = 1 - \exp\left(-k_i \left(\frac{d}{d_i}\right)^{\alpha}\right) \tag{4.5}$$

or a log-normal distribution function (e.g. Malone et al. 2018)

$$\mathrm{CDF}(d) = \frac{1}{2} + \frac{1}{2} \cdot \mathrm{erf}\left(\frac{\ln d - \ln d_i}{\sqrt{2} \cdot \sigma}\right) \tag{4.6}$$

with d_i as median diameter, α and σ spreading coefficients and $k_i = -\ln(0.5) = 0.69$. Both functions describe a unimodal size distribution, whereas the Rosin-Rammler distribution is slightly biased towards larger drop diameters in direct comparison of the two.

4.2   Determination of Drop Size Distributions in Laboratory and Field Settings

Today, the knowledge on drop formation processes in a subsea oil discharge is based mainly on small-scale experiments in the lab. A single full-scale experiment ("DeepSpill") has been performed off the Norwegian coast in 2000, and some measurements were taken during the *Deepwater Horizon* (DWH) oil spill in 2010 a large distance from the wellhead.

For a better understanding of the available datasets and the possibilities to assess the DSD experimentally in the lab during a future spill, the different available experimental setups are discussed. Several possibilities for in situ measurements during a spill are presented, and the existing datasets are critically reviewed.

4.2.1   Pilot-Scale Jet Experiments

Several studies have been performed to determine drop size distributions from downscaled oil jets entering into seawater at surface conditions, especially since the DWH oil spill. Masutani and Adams (2001) were one of the first to perform in-depth surveys of the DSD of a vertical oil discharge for several exit velocities and discharge diameters ranging from 1 to 5 mm. A pulse-free gear pump was used to inject the oil from a pressure-free reservoir into a water depth of approximately 1 m. Droplet sizes were determined using a Phase Doppler Particle Analyzer (PDPA). Brandvik et al. (2013) performed a series of experiments using the "Tower Basin" facility at SINTEF, Norway, to measure oil droplet sizes with and without subsea dispersant injection (SSDI). The water tank used was 3 m wide and 6 m deep; oil was injected vertically upward by pressurizing the oil reservoir with nitrogen and releasing the pressurized oil through a controlled valve. Size measurements were taken with both a LISST-100X laser diffractometer and a macro camera 2 m above the nozzle. On the same scale, Belore (2014) investigated the effect of varying dispersant-to-oil ratio (DOR) and gas void fraction on the DSD at Ohmsett Wave Tank. In this study, two LISST-100X laser diffractometers were used to measure the DSD. Zhao et al. (2016) performed experiments at a larger scale also at Ohmsett Wave Tank in 2016 with a discharge of 6.3 L/s from a 1 inch horizontal nozzle; DSD was again determined using two LISST-100X.

These and other similar studies provide key information on the DSD of an oil-only discharge in shallow waters. But because of the limitations of the facilities, the effects of a deep-sea environment such as in the DWH spill could not be investigated. A deep-sea oil discharge will alter from one in shallow waters not only with regard to the water temperature and hydrostatic pressure but also to the high amount of short-chained, gaseous hydrocarbons dissolved in the oil and a pressure difference between the oil reservoir and the surrounding seawater. Recently, two research groups investigated these effects by generating jets under artificial deep-sea conditions. Brandvik et al. (2017) used a holocam (Davies et al. 2017) to determine the DSD of "live oil", i.e. oil saturated with natural gas, with and without an additional gas void fraction and/or chemical dispersants. The experiments took place in a 2.3-m-wide pressure tank at the Southwest Research Institute (SwRI) at elevated hydrostatic pressures between 59 and 172 bar; jets were generated by overpressure in the oil reservoir. Malone et al. (2018) at Hamburg University of Technology performed several studies at 151 bar hydrostatic pressure to quantify the effects of gas dissolution, outgassing and sudden pressure loss at the nozzle. Oil DSD was determined from an endoscopic imaging system at the centreline of the jet. Oil flow was generated either isobarically or with a defined pressure difference using an equal-volume cylinder (Seemann et al. 2014), enabling a wide range of spill scenarios.

4.2.2 Stirrer Cells

A different approach to determine the DSD of an accidental oil discharge is based on the turbulent nature of such a discharge. This approach uses a stirrer cell to simulate not the oil jet itself, but rather its turbulent flow field by stirring an oil-in-water emulsion at a certain speed. This approach avoids upscaling over several orders of magnitude but allows only the investigation of an equilibrated system.

Among others, Boxall et al. (2012) measured the DSD of a water-in-oil emulsion with a combination of the focused beam reflectance method (FBRM) and an endo-scopic imaging system for a wide range of oil blends and stirrer speeds. Aman et al. (2015) used the same approach to investigate the DSD under high hydrostatic pres-sure up to 110 atm in a methane-saturated system using high-speed video image analysis.

4.2.3   DeepSpill: Field Experiment in the Deep Sea

In June 2000, SINTEF performed a field experiment at the Helland-Hansen site off the coast of Norway in a water depth of 844 m. Four discharges of dyed seawater, crude oil and marine diesel fuel in combination with LNG (liquefied natural gas) or nitrogen were conducted in order to test numerical spill propagation models and equipment for spill monitoring and surveillance (Johansen et al. 2000). The liquids were discharged at a rate of 60 m³/h for 60 min each from a 120 mm nozzle; gas flow was between 0.6 and 0.7 Sm³/s. Droplet sizes were evaluated manually from ROV video data for one discharge of diesel fuel in combination with LNG at four different heights over the discharge point. Images of the drops were taken by a colour video camera (resolution 460 TV lines) with a ruler mounted in the fore-ground to provide scale. The reported volume median diameter varies from four distinct values between 3 and 7 mm for the four measurement points (d_{v50} increasing with increasing distance from the discharge point) (Johansen et al. 2000; Socolofsky et al. 2015) to a single value of 5.5 mm (Brandvik et al. 2017).

4.2.4   Equipment for Field Measurements

In situ measurement of the DSD of a deep-sea oil spill remains a challenge. Apart from the high-pressure environment, the equipment needs to detect a particle size range of three orders of magnitude, from <10 μm to several millimetres. In addition, the oil fraction in the jet near the discharge point can be very high, resulting in poor light transmittance.

Although originally designed for the quantification of solid particles suspended in the water column, the commercially available LISST laser diffractometers by *Sequoia Scientific Inc.* have been widely used in the past years to measure oil drop-let sizes as well (see Sect. 4.2.1). While the system is easy to deploy on ROVs for in situ measurements and also available in a deep-sea configuration (depth rating 3000 m), its measurement range is limited to a maximum drop diameter of 500 μm, a maximum particle concentration of 750 mg/L (less for large particles) and an optical transmission rate >30%. The technology is therefore well suited for measurements in the far-field plume or intrusion layer but less for determination of the initial DSD next to the discharge site due to the larger drop diameters and high oil concentration that are to be expected there.

As an alternative, video imaging systems can be used for droplet size measurements. A promising approach by Davies et al. (2017) uses a holographic camera system to size particles in sample volume with 3–50 mm path length. The large path length and adjustable magnification enable sizing in the range of 28 μm to several millimetres. Based on backlighted images of small sample volumes, oil concentration is again a limiting factor in this approach as images should ideally contain no overlapping particles in the view plane to enable automated image analysis and sizing. In addition, a narrow path length and small sample volume poses the risk of under-sampling large particles (Davies et al. 2017).

At the cost of generating images that might not be automatically evaluable, endoscopic systems with incident lighting can handle very high oil concentrations over a large diameter range in the order of magnitude from 10^1 to 10^3 μm (Maaß et al. 2011; Malone et al. 2018).

4.2.5 Critical Review of Datasets

The confident translation of laboratory or pilot-scale experiments to the field remains an outstanding research objective. Multiple authors have proposed correlations with the claim of accurately representing the field scale, yet in all cases the empirical or semiempirical correlations employed require *extrapolation*. Fundamentally, the scientific method imposes an upper limit on the confidence of any extrapolation, particularly in the case it includes any empirical contribution. While several scaling quantities may be considered to unite laboratory and pilot-scale data, including the Reynolds and Weber numbers, it should be recognized that these quantities are fundamentally *approximations* that balance contributions in *simple* turbulent systems. To be clear, the Weber and Reynolds numbers *are not defined for a turbulent plume* and *are not defined for a flow system containing dispersed phases (oil, gas and potentially hydrate in this case)*. As a consequence, the attempts to unify the datasets reported in Table 4.1 within a single correlation have failed to date. One attempt has shown limited success – as discussed in Sect. 4.3.3 – where turbulence is evaluated at a more *fundamental* level than the Weber and Reynolds numbers provide.

4.3 Modelling Approaches

Based on the experimental and field data presented in the previous section and Table 4.1, different models were developed to predict the initial droplet size distribution of an accidental subsea oil discharge. These models can be divided into two

Table 4.1 Experimental measurements of oil droplet size distributions

Study	DeepSpill (Johansen et al. 2000)	Masutani and Adams (2001)	Brandvik et al. (2013)	Belore (2014)	Aman et al. (2015)	Zhao et al. (2016)	Brandvik et al. (2017)	Malone et al. (2018)
No. of datasets (no. of dispersant cases thereof)	1	71	15 (8)	56 (56)	8	1	24 (12)	8
Varied parameter(s)	–	Oil type, Q, D	Q, D, DOR	Oil type, Q, D, GOR, DOR	Turning speed	–	Gas void fraction, DOR, p, T	Gas saturation, oil type, Q
Oil type(s)	Marine diesel fuel	4 crudes, silicone oil	Oseberg blend	Endicott crude, dorado crude	Louisiana sweet crude	JP 5 (jet fuel oil)	Oseberg blend	Louisiana sweet crude, n-decane
Hydrostatic pressure	85 atm	1 atm	1.6 atm	1 atm	110 atm	1 atm	59–172 bar	151 atm
Gas-saturated oil?	No	No	No	No	Yes	No	Yes	Yes, partly
Gas void fraction	Yes	No	Partly	Yes	No	No	Yes, partly	No
Nozzle diameter	120 mm	1 mm; 2 mm; 5 mm	0.5 to 3 mm	1.5 mm; 4.5 mm	n/a	25.4 mm	3 mm	1.5 mm
Oil volume flow	60 m^3/h	0.05 to 3 L/min	0.2 to 5 L/min	0.285 to 1.5 L/min	n/a	378 L/min	1.5 L/min	1.1–2.1 L/min
Dispersant	No	No	Yes, partly	Yes, all	No	No	Yes, partly	No
Reynolds number (only no dispersant cases)	Approx. $3.9*10^4$	23 to 795	$1.4–8.9*10^3$	n/a	$1.1–5.4*10^3$	$5.9*10^4$	$8.4–9.7*10^3$	$0.16–3.0*10^4$
Weber number (only no dispersant cases)	Approx. $2.5*10^4$	40 to 1860	$0.7–7.6*10^4$	n/a	50–300	$1.6*10^5$	$1.3–1.5*10^3$	$0.22–2.2*10^4$

principal groups: models that use characteristic flow parameters for up- or down-scaling of a median diameter and models that calculate a complete droplet size distribution through mechanistic modelling of the flow. Both groups differ widely with regard to the computational effort but also with regard to the level of detail of the result.

4.3.1 Scaling-Based Models Using Dimensionless Numbers

Probably the most widely used scaling approach in the oil spill community is the modified Weber number scaling by Johansen et al. (2013). Based on the model of Wang and Calabrese (1986) for stirred-tank reactors, they proposed the implicit scaling law for the volume median diameter d_{v50}

$$\frac{d_{v50}}{D} = A \cdot \mathrm{We}^{*-3/5} \tag{4.7}$$

with the modified Weber number We^*

$$\mathrm{We}^* = \frac{\mathrm{We}}{\left[1 + B \cdot \mathrm{Vi} \cdot \left(\frac{d_{v50}}{D} \right)^{\frac{1}{3}} \right]} \tag{4.8}$$

where $\mathrm{We} = \dfrac{D \cdot \rho_1 \cdot u_1^2}{\sigma_1}$ is the Weber number, $\mathrm{Vi} = \dfrac{\mathrm{We}}{Re}$ the viscosity number and A and B are empirical coefficients. Those coefficients were calculated using the dataset of Brandvik et al. (2013) to be $A = 15$ and $B = 0.8$; a later work based on a larger dataset updates these to $A = 24.8$ and $B = 0.08$ (Socolofsky et al. 2015). The data used to calibrate the model span a range of approximately $10^3 \leq \mathrm{We} \leq 10^4$ and jets with and without additional dispersant injection. The authors claim the model to be applicable to multiphase discharges of oil and gas as well by adjusting the exit velocity u_1 to account for the gas phase and its buoyancy.

A second model, called the unified droplet size model, by Li et al. (2017) scales the volume median diameter d_{v50} with a combination of Weber and Ohnesorge number in an explicit equation:

$$\frac{d_{v50}}{D} = r \cdot \left(1 + 10 \cdot \mathrm{Oh} \right)^p \cdot \mathrm{We}^q \tag{4.9}$$

where $r = 14.05$, $p = 0.460$ and $q = -0.518$ are empirically derived coefficients for a liquid-liquid jet. The model is proposed for determining both the d_{v50} of a jet and of wave entrainment at the surface. The coefficients p and q were determined based on 28 wave entrainment datasets, whereas r was calculated from the d_{v50} of the Norwegian DeepSpill experiment. The model was tested against both laboratory data and field measurements by a Holocam during the DWH spill (Li et al. 2015).

Both models provide a median volume diameter at very little computational cost. As input parameters, both require the same information about the physical properties of the oil, oil exit velocity and exit diameter. Especially with regard to the physical properties of "live" oil under deep-sea conditions (high pressure, low temperature), there are often no measured data, and the properties can only be calculated using empirical correlations (Lake and Fanchi 2006) or numerical models (Gros et al. 2016). In addition, they have only been validated for "dead oil" in a limited range of We and Oh and require extrapolation over several orders of magnitude beyond these limits to provide an estimation for a major spill like DWH.

The scaling laws described above only provide a steady-state volume median diameter and no actual size distribution. However, as the size distribution may differ widely for different blowout scenarios (Malone et al. 2018), information on the spreading factor of the underlying distribution factor is of considerable importance for a realistic near- and far-field modelling (see Vaz et al. 2020; Perlin et al. 2020).

4.3.2 Mechanistic Modelling

A different approach by Zhao et al. (2014, 2017) uses a hydrodynamic model of the jet to predict the complete drop size distribution. The Lagrangian model, called VDROP-J, calculates the DSD of a small portion of the jet based on a population balance of drop breakup and coalescence in a given time step. A single large drop size is taken as initial input and tracked downstream while the jet widens and takes in water, thereby reducing the oil fraction in the considered portion of the jet. Breakage rate is determined stepwise by the probability of a drop to collide with a turbulent eddy with sufficient energy to cause breakup of this drop; coalescence rate is defined by the probability of two drops colliding and coalescing due to turbulence.

As an outcome, this model provides the full DSD of a jet at different positions downstream of the orifice, albeit at significant computational cost. Because the DSD is directly calculated from the discharge conditions without the need for upscaling by several orders of magnitude, it is less sensitive to miscorrelation of experimental data than scaling-based models. However, a crucial point in the calculation of the DSD is turbulent dissipation rate in the jet, which can be seriously affected by reactions and interactions of the multiphase flow of oil, gas, water and possibly hydrates that are discharged in a major subsea spill like DWH, and which are not yet fully understood.

4.3.3 Novel Applications of Energy Dissipation Metrics to Understand Droplet Sizes from Experimental Data

To date, the prediction of live oil droplet size distributions has been treated with two approaches: (i) the use of a modified Weber number scaling, proposed by Johansen et al. (2013), and (ii) the use of Reynolds number, proposed by Aman et al. (2015). In *both* cases, the studies define their respective dimensionless quantities based on the physical conditions at the contact point of the blowout preventer (BOP) stack and the seawater; that is, the internal *oil pipe diameter* was employed as the singular property to define the length scale of the problem. This understanding clarifies why both methods have failed to unify the available data: oil droplets are not created at the exit point of the pipe, but rather in the near-field plume. As such, the research community must understand and characterize turbulence *at the point of droplet creation* using a more fundamental basis of turbulence that can incorporate the *known* contributions.

One such approach may be derived by evaluating both the turbulent kinetic energy (TKE) and turbulence dissipation rate (TDR), which, respectively, describe the energetic content of the eddies in the flow and the rate at which eddies transfer TKE down the so-called energy cascade. As droplets are generated through the *transfer* of TKE, the use of TDR-based scaling provides an attractive opportunity to compare the available datasets. To this end, models must consider the three relevant contributions to TDR in the context of a subsea blowout: (i) the momentum of the gas/oil jet itself, defined between the exit of the pipe and the top of the near-field plume; (ii) the additional momentum generated *inside* the BOP stack, where fluids experienced an 84-bar orifice pressure drop imposed by the annular preventer within a few diameters of the exit point; and (iii) additional turbulence introduced by the evolution of free gas from live oil droplets. To date, computational fluid dynamics (CFD) approaches have been unable to rationalize all three contributions together. However, the first two contributions may be estimated from experimental laboratory studies, to estimate the range of probable TDR in the field case.

Zhao et al. (2014) summarized experimental TDR estimates for *single-phase jets* (the first contribution above), deriving a correlation as a function of jet exit diameter, velocity and distance into the plume. The study demonstrated that, within 10 diameters of the exit, the TDR remains constant and decreases monotonically thereafter; as such, the maximum TDR corresponding to the jet momentum may be conservatively estimated. This TDR method can be applied to data collected on the apparatus at SINTEF, SWRI and TUHH (as described above). Through the use of CFD to map stirred cells, Booth et al. (2018) have further demonstrated the ability to correlate TDR for autoclave systems employed at the University of Western Australia. For the range of volumetric flowrates reported in the DWH blowout and for the relevant range of oil and gas properties, the TDR correlation from Zhao et al. (2014) suggests a TDR range of between 10^{-3} and 10^{-5} m^2 s^{-3}, which only represents the first contribution identified above. For reference, most autoclave experiments are performed between 10^{-2} and 10^2 m² s⁻³, while most data from pilot-scale jets corresponds to TDRs between 10^3 and 10^5 m² s⁻³.

The DWH jet was also influenced by a partially closed annular preventer, which closely resembles a classical orifice and imposed an 84-bar pressure drop within a few diameters of the exit point. This contribution of this orifice drop to the TDR has largely been neglected to date. Kundu et al. (2016) provide a well-established relation from fluid mechanics to estimate TDR contribution from a restricted pressure drop:

$$\varepsilon_{pd} = \frac{\Delta p \cdot Q}{\rho_c \cdot V}$$

(4.10)

where Δp is the permanent pressure drop applied by the restriction, Q is the volumetric flowrate through the restriction, ρ_c is the continuous phase density and V is the volume of energy dissipation (e.g. the volume of the BOP stack). With an 84-bar pressure drop, an average fluid density of 700 kg/m³ and a 2 m height of the BOP stack with a 0.5 m internal diameter, this orifice contribution to TDR is estimated at 2700 m² s⁻³.

The third contribution above – from live oil degassing – has not been studied in sufficient detail to quantitatively inform contributions to TDR. However, the first two contributions identified above demonstrate the elegant rationale as to why Weber- and Reynolds-based scaling arguments have failed to unify the available data: such approaches *severely underrepresent the turbulence of the Macondo plume, because they only account for the first of two important contributions.* That is, the TDR contribution from the orifice pressure drop aligns well with the magnitude of TDRs captured at high mixing speed in autoclaves or in most pilot-scale jet experiments. Importantly, neither Weber- nor Reynolds-based methods are *able* to capture the additional TDR contributions from the orifice pressure drop, resulting in the discrepancy of predicted droplet sizes heretofore.

4.4   Effects of Deep-Sea Blowout Characteristics

In case of a large-scale, deep-sea oil discharge from an uncontrolled well, several factors must be considered in addition to the modelling approaches presented in Sect. 4.3. These are mainly:

- Gaseous components (C_1-C_5, N_2, CO_2, H_2S) dissolved in the crude oil ("live" instead of "dead" oil).
- A multiphase flow (oil, gas, water) inside the borehole and at the wellhead.
- High hydrostatic pressure and low temperature of the surrounding seawater.
- Rapid pressure and temperature changes at the wellhead, inducing phase changes in the oil.

For the DWH spill, the fluid exiting the wellhead consisted of approx. 40–50 vol% gaseous components (mainly methane) and 50–60 vol% oil, which was by itself saturated with dissolved gases (Reddy et al. 2012; Boufadel et al. 2018). The fluid experienced a pressure drop of approximately 86 bar (Lehr et al. 2010) within the BOP and cooled from approximately 105 to 4.3 °C within a few meters distance from the discharge point (Reddy et al. 2012; Gros et al. 2016).

4.4.1 Influence of Dissolved Gases on the Droplet Size Distribution

As the solubility of many gases in oil rises linearly with pressure (Jaggi et al. 2017), the amount of short-chained hydrocarbons such as C_1 to C_5 dissolved in crude oil can be hundredfold at reservoir or deep-sea conditions compared to sea surface conditions (Lehr and Socolofsky 2020). For the DWH spill, Gros et al. (2016) calculated the amount of methane dissolved to be 141 times higher at the wellhead than at the sea surface. A crude oil with such an amount of dissolved and volatile components, such as it exists inside the reservoir itself, is commonly called a "live oil", in comparison to a "dead oil" at ambient conditions with no or just very little dissolved gases (Ahmed 2010). While "dead" oil is a nearly incompressible liquid and is comparatively unaffected by an elevated hydrostatic pressure, the properties of "live oil" will change significantly with increasing pressure due to the increased amount of dissolved gas. For more details on the effect of deep-sea conditions on physical and chemical properties of oil and gas, please see Oldenburg et al. (2020).

Two recent studies performed by SINTEF and Hamburg University of Technology investigated the effect of dissolved gas on the DSD of a crude oil under elevated hydrostatic pressure (Brandvik et al. 2017; Malone et al. 2018). As Brandvik et al. (2017) also reported the formation of gas bubbles in their experiments, the "live oil" must have undergone a phase change during the discharge, which means that the results cannot be attributed to the "live oil" properties only. They do, however, report an underestimation of volume median diameter by the modified We-scaling model by approximately 10%.

At Hamburg University of Technology, a direct comparison between two "dead" and "live" oils (Louisiana sweet crude oil and n-decane) was performed using the same setup and experimental conditions (Malone et al. 2018). Oil jets were generated quasi-isobaric by using an equal-volume cylinder in order to exclude any side effects of a pressure change on the oil (Seemann et al. 2014, Malone et al. 2018). Median diameters of "live oil" were increased by 74% to 97% compared to "dead oil" under otherwise unchanged conditions. The experimental data was compared with the models by Johansen and Li (Johansen et al. 2013, Li et al. 2017), which account for changes of the physical properties density, viscosity and IFT by use of the dimensionless numbers We and Re (see Sect. 4.3.1). Both models predicted a very similar d_{v50} for "live" and "dead" oil in contrast to the experimental results.

A possible explanation for this unexpected size increase could be the increased compressibility of the "live oil" (Satter and Iqbal 2016) which would reduce the TKE of the jet by elastic deformation of individual drops rather than breakup.

In addition, the size distribution of the "live oil" was significantly broader compared to the dead oil. While both "live" and "dead" oil discharges followed a log-normal distribution function, the spreading factor σ varied widely for the different oils and increased from "dead" to "live" oil by over 30%.

4.4.2 Influence of Rapid Pressure Loss at the Wellhead and Phase Changes of the Oil

According to Lehr et al. (2010), the pressure of the oil at the DWH spill site changed rapidly from 241 bar measured inside the BOP to 154 bar ambient hydrostatic pressure after exiting the wellhead. This change took place over a height of only 16.4 m. As a result from this rapid pressure loss and the previous, slow decompression along the borehole, a multiphase flow of "live oil" and gas is discharged from the wellhead. Due to the pressure-dependent solubility of gaseous components in the crude oil (Chap. 3 and Sect. 4.4.1), such a massive pressure drop can lead to an oversaturation of the oil and therefore outgassing of C_1 to C_5 in addition to the expansion of the already existing gas phase. This outgassing and gas expansion might affect the drop formation in different ways. First, the expansion of a separate gas phase will add to the overall TKE of the multiphase plume by its expansion energy. Secondly, outgassing from the oversaturated oil will lead to the formation of a gas phase within the oil drops, which will significantly alter and destabilize the drop. The gas microbubbles might either leave the oil drop at its surface, thereby removing parts of the oil from the "mother drop" or expand within the drop, thereby forming a two-phase particle with different breakup and rising characteristics (see also Pesch et al. 2020 on the rise velocity of live oil droplets). Both effects are depicted in Fig. 4.2.

To assess the effect of such a pressure drop on the DSD, oil jets of both "live" and "dead" oil were generated at Hamburg University of Technology with a defined pressure difference between oil reservoir pressure and hydrostatic pressure of the seawater. For this purpose, a throttle valve was added in front of the nozzle of the experimental setup described in Seemann et al. (2014) and Malone et al. (2018) to generate the required pressure drop. The oil reservoir was pressurized to 161 bar, while the seawater basin was kept at 151 bar pressure like in the earlier experiments. The DSD was determined from manually evaluated images of an endoscopic camera system (Malone et al. 2018).

With this pressure drop of $\Delta p = 10$ bar, the "dead oil" droplet sizes were distributed log-normally as expected, whereas the "live oil" was distributed bimodally. This bimodal distribution could very well be described by superposing two log-normal distributions. It is assumed that this bimodality is caused by the processes hypothesized above and depicted in Fig. 4.2. Under this assumption, the larger

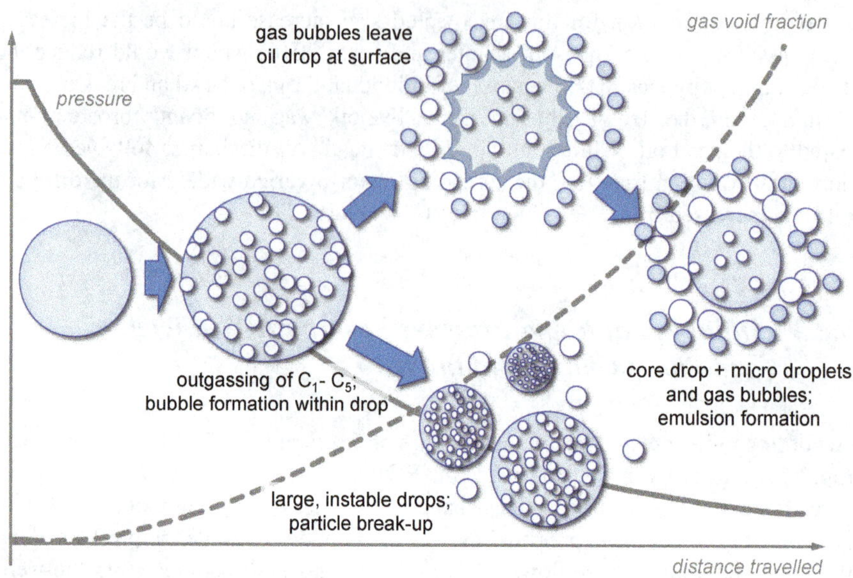

Fig. 4.2 Possible effects of oversaturation and outgassing on the drop formation and drop size distribution after a rapid pressure drop (not to scale)

mode is formed by those drops where no or very few gas bubbles nucleated. The smaller mode consists of drops that were split up when the gas bubbles nucleated, expanded and exited the oil drop. That gas bubbles did in fact form and expand prior to the measurement point is evident from the footage of the surveillance camera (Fig. 4.3), though those bubbles were not captured by the measurement system in sufficient numbers to allow for quantification. This interpretation of the bimodality is supported when plotting the median volume diameter of both modes over the modified Weber number (Fig. 4.4). While the d_{v50} of the larger mode lies in approximately the same range as the quasi-isobaric live oil from Malone et al. (2018) and is significantly enhanced compared to the "dead" oil and the model prediction by Johansen et al. (2013), the d_{v50} of the smaller mode is significantly smaller than either the "live" or the "dead oil". In terms of the TKE, the pressure drop from the oil reservoir and subsequent outgassing of methane from the oil provide an additional energy source in the jet leading to further breakup of the oil droplet and consequently a smaller median drop diameter.

Plotting the volume median diameters from both the quasi-isobaric and the "pressure drop" experiments at Hamburg University of Technology over the maximum TKE according to Zhao et al. (2014)

$$\varepsilon_u = 0.003 \cdot u_1^3 / D \tag{4.11}$$

reveals the good correlation of the different data sets to the assumptions on the effects of gas dissolution and pressure drop/outgassing (Fig. 4.5).

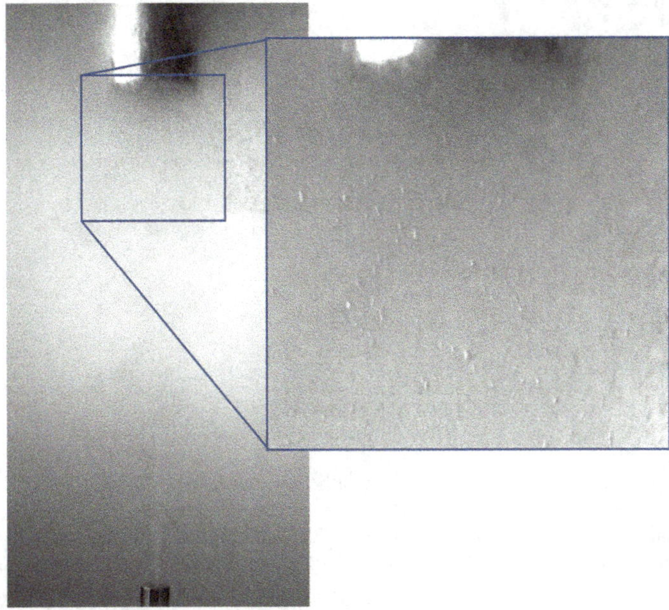

Fig. 4.3 Methane-saturated n-decane discharged into artificial seawater at 150 bar ambient pressure, 20 °C. Methane bubbles form from the oversaturated n-decane after a pressure drop of 10 bar from the oil reservoir

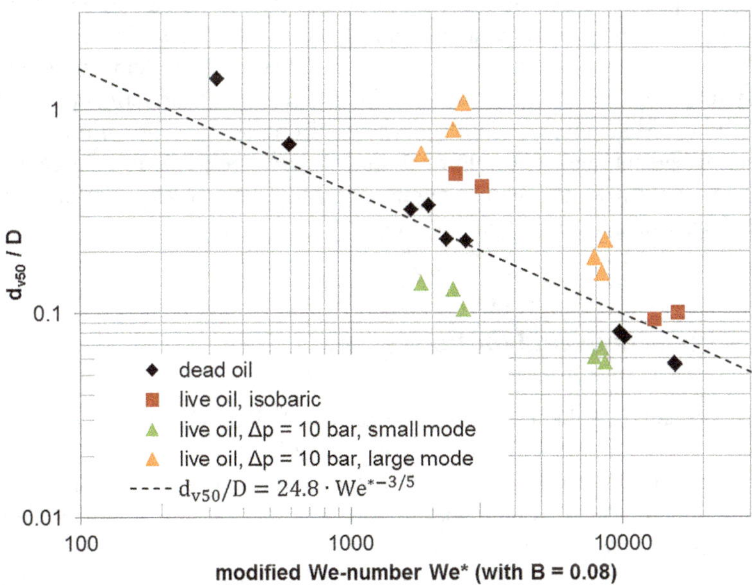

Fig. 4.4 Comparison of experimental result from jet experiments at Hamburg University of Technology to We*-scaling model proposed by Johansen et al. (2013)

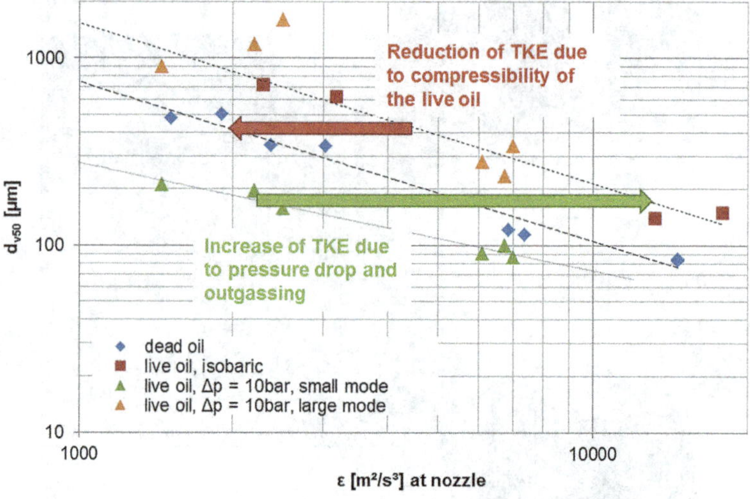

Fig. 4.5 Effects of deep-sea oil characteristics on the volume median diameter and the maximum turbulent kinetic energy. Dashed lines are provided to guide the eye

4.5 Capabilities and Limits of Subsea Dispersant Injection

The above-referenced autoclave and pilot-scale jet studies have also considered the effect of dispersant injection, typically at dispersant-to-oil ratios of between 1:100 and 1:20. Interestingly, the results demonstrate minimal dependence between the reported average droplet size and the maximum TDR in any system, including wave tank studies from the EPA (Fig. 4.6). Within the context of the estimated TDR contributions for Macondo, the data show that oil droplet sizes are similar with and without dispersant application. Further studies across multiple mixing geometries, DORs and gas/oil properties are required to fully contextualize the critical TDR range under which dispersant application may benefit droplet size.

4.6 Conclusions and Outlooks

The determination of the initial size distribution of oil drops and gas bubbles is still a major challenge in modelling of deep-sea oil spills.

The state-of-the-art knowledge on drop formation is mainly based on small-scale lab experiments of liquid-liquid jets at ambient conditions. For a better understanding of future oil spills, investigations at deep-sea conditions as well as in situ measurements with capable equipment are critical. Despite numerous attempts, a reliable translation of laboratory or pilot-scale experiments to the field conditions including a turbulent, multiphase plume remains an outstanding research objective.

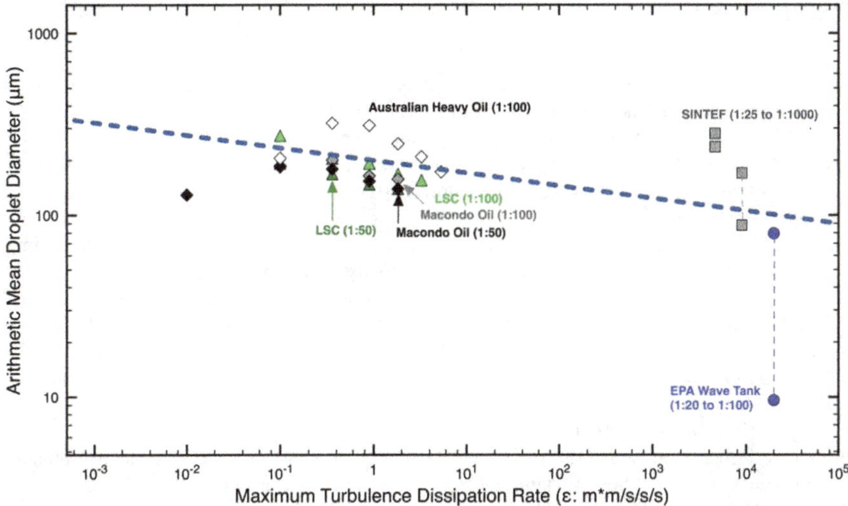

Fig. 4.6 Average reported droplet size as a function of maximum TDR for sapphire autoclave measurements (left-hand grouping), wave tank studies (blue circles) and pilot-scale blowout jets (grey squares) containing dispersant; DORs relative to each study are shown in parentheses. The dashed line is provided to guide the eye

A new approach to predict the median drop diameter based on the turbulent energy dissipation rate was presented in this chapter. To be able to generalize this approach, further studies across multiple mixing geometries, DORs and gas/oil properties are required.

Recent studies at artificial deep-sea conditions at Hamburg University of Technology showed a significant influence of dissolved gases and outgassing on the drop size distribution. In a first attempt, these influences could be modelled with good accuracy using the turbulent energy dissipation rate. Especially the outgassing of short-chained hydrocarbons from the oil might lead to a significant decrease in the median drop diameter, as oil drops are broken up by expanding gas bubbles.

To be better prepared for a possible future spill in the deep-sea, it is necessary to investigate the high-pressure, multiphase plume near the exit at a more detailed level and at a larger scale than heretofore. Only by a thorough understanding of the drop formation processes and turbulent conditions in this multiphase plume is it possible to find a knowledge-based mitigation strategy, which might or might not include subsea dispersant injection.

Acknowledgments This research was made possible by a from the Gulf of Mexico Research Initiative/C-IMAGE. Data are publicly available through the Gulf of Mexico Research Initiative Information and Data Cooperative (GRIIDC) at https://data.gulfresearchinitiative.org/ (DOIs: 10.7266/n7-jjqd-pa77, 10.7266/n7-eha7-tv03, 10.7266/N7V69H19, 10.7266/N77D2SM2).

References

Adams EE, Socolofsky SA (2004) Review of deep oil spill modeling activity supported by the DeepSpill JIP and offshore operators committee: Final Report. 26 pp

Ahmed TH (2010) Reservoir engineering handbook, 4th edn. Gulf Professional, Oxford

Aman ZM, Paris CB, May EF, Johns ML, Lindo-Atichati D (2015) High-pressure visual experimental studies of oil-in-water dispersion droplet size. Chem Eng Sci 127:392–400. https://doi.org/10.1016/j.ces.2015.01.058

Belore R (2014) Subsea chemical dispersant research. In: Proceedings of the 37th AMOP technical seminar on environmental contamination and response, Canmore, Alberta

Booth CP, Leggoe JW, Aman ZM (2018) The use of computational fluid dynamics to predict the turbulent dissipation rate and droplet size in a stirred autoclave. Chem Eng Sci. In Press

Boufadel MC, Gao F, Zhao L, Özgökmen T, Miller R, King T, Robinson B, Lee K, Leifer I (2018) Was the Deepwater Horizon well discharge churn flow?: implications on the estimation of the oil discharge and droplet size distribution. Geophys Res Lett 45:2396–2403. https://doi.org/10.1002/2017GL076606

Boxall JA, Koh CA, Sloan ED, Sum AK, Wu DT (2012) Droplet size scaling of water-in-oil emulsions under turbulent flow. Langmuir 28:104–110. https://doi.org/10.1021/la202293t

Brandvik PJ, Johansen Ø, Leirvik F, Farooq U, Daling PS (2013) Droplet breakup in subsurface oil releases – part 1: Experimental study of droplet breakup and effectiveness of dispersant injection. Mar Pollut Bull 73:319–326. https://doi.org/10.1016/j.marpolbul.2013.05.020

Brandvik PJ, Davies EJ, Storey C, Leirvik F, Krause DF (2017) Subsurface oil releases – verification of dispersant effectiveness under high pressure using combined releases of live oil and natural gas, SINTEF report no: A27469. Trondheim Norway 2016. ISBN: 978-821405857-4

Davies EJ, Brandvik PJ, Leirvik F, Nepstad R (2017) The use of wide-band transmittance imaging to size and classify suspended particulate matter in seawater. Mar Pollut Bull 115:105–114. https://doi.org/10.1016/j.marpolbul.2016.11.063

Gros J, Reddy CM, Nelson RK, Socolofsky SA, Arey JS (2016) Simulating gas–liquid–water partitioning and fluid properties of petroleum under pressure: implications for deep-sea blowouts. Environ Sci Technol 50:7397–7408. https://doi.org/10.1021/acs.est.5b04617

Hsiang L-P, Faeth GM (1992) Near-limit drop deformation and secondary breakup. Int J Multiphase Flow 18:635–652

Jaggi A, Snowdon RW, Stopford A, Radović JR, Oldenburg TB, Larter SR (2017) Experimental simulation of crude oil-water partitioning behavior of BTEX compounds during a deep submarine oil spill. Org Geochem 108:1–8. https://doi.org/10.1016/j.orggeochem.2017.03.006

Johansen Ø, Rye H, Melbye AG, Jensen HV, Serigstad B, Knutsen T (2000) Deep spill JIP - experimental discharges of gas and oil at Helland Hansen – June 2000, Technical Report

Johansen Ø, Brandvik PJ, Farooq U (2013) Droplet breakup in subsea oil releases--part 2: Predictions of droplet size distributions with and without injection of chemical dispersants. Mar Pollut Bull 73:327–335. https://doi.org/10.1016/j.marpolbul.2013.04.012

Kundu PK, Cohen IM, Dowling DR (2016) Fluid mechanics, Sixth edition. Academic Press, Oxford

Lake LW, Fanchi JR (2006) Petroleum engineering handbook. Society of Petroleum Engineers, Richardson

Lefebvre AH, McDonell VG (2017) Atomization and sprays, Second edition. CRC Press, Boca Raton

Lehr W, Socolofsky SA (2020) The importance of understanding fundamental physics and chemistry of deep oil blowouts (Chap. 2). In: Murawski SA, Ainsworth C, Gilbert S, Hollander D, Paris CB, Schlüter M, Wetzel D (eds) Deep oil spills: facts, fate, effects. Springer, Cham

Lehr B, Aliseda A, Bommer P, Espina P, Flores O, Lasheras JC, Leifer I, Possolo A, Riley J, Savas O, Shaffer F, Wereley S, Yapa PD (2010) Deepwater Horizon Release: estimate of rate by PIV, July 21, 2010, Accessed on October 29, 2018. https://www.doi.gov/sites/doi.gov/files/migrated/

deepwaterhorizon/upload/Deepwater_Horizon_Plume_Team_Final_Report_7-21-2010_comp-corrected2.pdf

Li Z, Bird A, Payne JR, Vinhateiro N, Kim Y, Davis C, Loomis N (2015) Technical reports for Deepwater Horizon water column injury assessment: oil particle data from the Deepwater Horizon oil spill. https://www.fws.gov/doiddata/dwh-ar-documents/946/DWH-AR0024715.pdf. Accessed 28 Sept 2018

Li Z, Spaulding M, French McCay D, Crowley D, Payne JR (2017) Development of a unified oil droplet size distribution model with application to surface breaking waves and subsea blowout releases considering dispersant effects. Mar Pollut Bull 114:247–257. https://doi.org/10.1016/j.marpolbul.2016.09.008

Maaß S, Wollny S, Voigt A, Kraume M (2011) Experimental comparison of measurement techniques for drop size distributions in liquid/liquid dispersions. Exp Fluids 50:259–269. https://doi.org/10.1007/s00348-010-0918-9

Malone K, Pesch S, Schlüter M, Krause D (2018) Oil droplet size distributions in deep-sea blowouts: influence of pressure and dissolved gases. Environ Sci Technol 52:6326–6333. https://doi.org/10.1021/acs.est.8b00587

Masutani S, Adams EE (2001) Experimental study of multiphase plumes with application to deep ocean oil spills: final report to U.S. Dept. of the Interior

Ohnesorge WV (1936) Die Bildung von Tropfen an Düsen und die Auflösung flüssiger Strahlen. Z Angew Math Mech 16:355–358. https://doi.org/10.1002/zamm.19360160611

Oldenburg TBP, Jaeger P, Gros J, Socolofsky SA, Pesch S, Radović J, Jaggi A (2020) Physical and chemical properties of oil and gas under reservoir and deep-sea conditions (Chap. 3). In: Murawski SA, Ainsworth C, Gilbert S, Hollander D, Paris CB, Schlüter M, Wetzel D (eds) Deep oil spills: facts, fate, effects. Springer, Cham

Perlin N, Paris CB, Berenshtein I, Vaz AC, Faillettaz R, Aman ZM, Schwing PT, Romero IC, Schlüter M, Liese A, Noirungsee N, Hackbusch S (2020) Far-field modeling of a deep-sea blowout: sensitivity studies of initial conditions, bio-degradation, sedimentation and subsurface dispersant injection on surface slicks and oil plume concentrations (Chap. 11). In: Murawski SA, Ainsworth C, Gilbert S, Hollander D, Paris CB, Schlüter M, Wetzel D (eds) Deep oil spills: facts, fate, effects. Springer, Cham

Pesch S, Schlüter M, Aman ZM, Malone K, Krause D, Paris CB (2020) Behavior of rising droplets and bubbles – impact on the physics of deep-sea blowouts and oil fate (Chap. 5). In: Murawski SA, Ainsworth C, Gilbert S, Hollander D, Paris CB, Schlüter M, Wetzel D (eds) Deep oil spills: facts, fate, effects. Springer, Cham

Reddy CM, Arey JS, Seewald JS, Sylva SP, Lemkau KL, Nelson RK, Carmichael CA, McIntyre CP, Fenwick J, Ventura GT, van Mooy BAS, Camilli R (2012) Composition and fate of gas and oil released to the water column during the Deepwater Horizon oil spill. Proc Natl Acad Sci 109:20229–20,234. https://doi.org/10.1073/pnas.1101242108

Satter A, Iqbal GM (2016) Reservoir engineering: the fundamentals, simulation, and management of conventional and unconventional recoveries. Elsevier/Gulf Professional Publishing, Amsterdam

Seemann R, Malone K, Laqua K, Schmidt J, Meyer A, Krause D, Schlüter M (2014) A new high-pressure laboratory setup for the investigation of deep-sea oil spill scenarios under in-situ conditions. In: Proceedings of the seventh International Symposium on Environmental Hydraulics, pp 340–343

Socolofsky SA, Adams EE, Boufadel MC, Aman ZM, Johansen Ø, Konkel WJ, Lindo D, Madsen MN, North EW, Paris CB, Rasmussen D, Reed M, Rønningen P, Sim LH, Uhrenholdt T, Anderson KG, Cooper C, Nedwed TJ (2015) Intercomparison of oil spill prediction models for accidental blowout scenarios with and without subsea chemical dispersant injection. Mar Pollut Bull 96:110–126. https://doi.org/10.1016/j.marpolbul.2015.05.039

Tang L (2004) Cylindrical liquid-liquid jet instability. Ph. D. Thesis

Vaz AC, Paris CB, Dissanayake AL, Socolofsky SA, Gros J, Boufadel MC (2020) Dynamic coupling of near-field and far-field models (Chap. 9). In: Murawski SA, Ainsworth C, Gilbert S, Hollander D, Paris CB, Schlüter M, Wetzel D (eds) Deep oil spills: facts, fate, effects. Springer, Cham

Wang CY, Calabrese RV (1986) Drop breakup in turbulent stirred-tank contactors. Part II: relative influence of viscosity and interfacial tension. Am Inst Chem Eng J 32:667–676. https://doi.org/10.1002/aic.690320417

Zhao L, Boufadel MC, Socolofsky SA, Adams E, King T, Lee K (2014) Evolution of droplets in subsea oil and gas blowouts: development and validation of the numerical model VDROP-J. Mar Pollut Bull 83:58–69. https://doi.org/10.1016/j.marpolbul.2014.04.020

Zhao L, Shaffer F, Robinson B, King T, D'Ambrose C, Pan Z, Gao F, Miller RS, Conmy RN, Boufadel MC (2016) Underwater oil jet: hydrodynamics and droplet size distribution. Chem Eng J 299:292–303. https://doi.org/10.1016/j.cej.2016.04.061

Zuzio D, Estivalezes J-L, Villedieu P, Blanchard G (2013) Numerical simulation of primary and secondary atomization. Comptes Rendus Mécanique 341:15–25. https://doi.org/10.1016/j.crme.2012.10.003

Chapter 5
Behavior of Rising Droplets and Bubbles: Impact on the Physics of Deep-Sea Blowouts and Oil Fate

Simeon Pesch, Michael Schlüter, Zachary M. Aman, Karen Malone, Dieter Krause, and Claire B. Paris

Abstract The rise behavior of oil droplets and natural gas bubbles is of major importance for understanding the physical behavior of dispersed oil in the aftermath of a deep-sea blowout. It is also indispensable for the development and adjustment of oil fate modeling tools that estimate the subsea distribution of oil masses. So far, to estimate the oil distribution throughout the water column and the expected surfacing times, a couple of well-known correlations for the calculation of the rise velocities of single particles with fluidic interfaces in stagnant media are available from process engineering applications. Besides the physical properties of live oil under environmental conditions, the size of the gas bubbles and oil droplets are the most crucial parameters that determine rise velocities.

Keywords Rise velocity · Initial shape deformation · Live oil droplets · Degassing · High pressure

S. Pesch (✉) · M. Schlüter
Hamburg University of Technology, Institute of Multiphase Flows, Hamburg, Germany
e-mail: simeon.pesch@tuhh.de; michael.schlueter@tuhh.de

Z. M. Aman
University of Western Australia, Department of Chemical Engineering, Perth, WA, Australia
e-mail: zachary.aman@uwa.edu.au

K. Malone · D. Krause
Hamburg University of Technology, Institute of Product Development and Mechanical Engineering Design, Hamburg, Germany
e-mail: karen.malone@tuhh.de; krause@tuhh.de

C. B. Paris
University of Miami, Department of Ocean Sciences, Rosenstiel School of Marine & Atmospheric Science, Miami, FL, USA
e-mail: cparis@rsmas.miami.edu

5.1 Introduction

The rise behavior of oil droplets and natural gas bubbles is of major importance for understanding the physical behavior of dispersed oil in the aftermath of a deep-sea blowout. It is also indispensable for the development and adjustment of oil fate modeling tools that estimate the subsea distribution of oil masses. So far, to estimate the oil distribution throughout the water column and the expected surfacing times, a couple of well-known correlations for the calculation of the rise velocities of single particles with fluidic interfaces in stagnant media are available from process engineering applications. Besides the physical properties of live oil under environmental conditions, the size of the gas bubbles and oil droplets are the most crucial parameters that determine rise velocities.

However, the distribution and fate of oil, thoroughly discussed in the third section of this book (Part III: Transport and Degradation of Oil and Gas from Deep Spills), are much more complicated than just calculating the rise velocities of single particles of a given size and composition. Swarm effects within the multiphase plume, ocean currents that transport the hydrocarbon mixture horizontally, gas hydrate formation due to the high-pressure low-temperature conditions in the deep sea, and the partitioning of certain oil components affect the oil distribution and render the modeling of the oil fate extremely challenging. The buoyancy-driven ascent of the bubbles and droplets forms the basis for any oil spill modeling. In this chapter, we investigate and discuss the rise behavior of fluid particles in realistic oil-gas-seawater mixtures under deep-sea conditions.

5.2 Correlations for the Rise Velocity of Single Fluid Particles

The most important parameters that influence the rise velocity u_p of buoyant particles (i.e., gas bubbles and oil droplets) are the density of both the ambient seawater (continuous aqueous phase) and the oil and/or gas (dispersed phase) as well as the median particle diameter d_p of the particle size distribution in oil and gas plumes (described in Chap. 4). The difference between the density of the continuous aqueous phase ρ_c and the density of the dispersed oil or gas phase ρ_d defines the driving gradient for the buoyant ascent of the fluid particles. Other important physical properties influencing the rise velocity are the dynamic viscosity of the continuous phase μ_c and the interfacial tension between the continuous and the dispersed phase σ, determining the interaction between the phases and the size-dependent particle's shape. These thermodynamic properties depend on water depth (i.e., pressure and temperature) and on the exact composition and mutual solubility of the oil and gas that also vary with the ambient pressure and temperature. An extensive overview on experimental and modeling results of physical properties and gas-in-oil solubility under simulated reservoir and deep-sea conditions can be found in Chap. 3.

For the purpose of better comparability and general validity, most of the correlations are established by means of dimensionless numbers. A very important one that is widely used in fluid mechanics is the Reynolds number Re, which relates inertial forces to viscous forces. The particle Reynolds number that is relevant for the movement of a rising droplet or bubble is defined as:

$$Re = \frac{u_p \cdot d_p \cdot \rho_c}{\mu_c}.$$

(5.1)

The Eötvös number Eo (also referred to as Bond number) relates gravitational forces to interfacial tension forces and is therefore important for buoyancy-driven processes. It is defined as:

$$Eo = \frac{g \cdot \Delta\rho \cdot d_p^2}{\sigma},$$

(5.2)

with $\Delta\rho = \rho_c - \rho_d$ and g the gravitational acceleration. Together with Eo, the Morton number Mo characterizes the shape of fluid particles. It is defined as a dimensionless combination of physical properties and the gravitational acceleration as:

$$Mo = \frac{g \cdot \mu_c^4 \cdot \Delta\rho}{\rho_c^2 \cdot \sigma^3}.$$

(5.3)

Some correlations also use the Archimedes number Ar, which can be interpreted as the ratio of buoyancy force to friction force for the distinction of different validity regimes of the correlations and the calculation of the corresponding rise velocities:

$$Ar = \frac{\rho_c \cdot \Delta\rho \cdot g \cdot d_p^3}{\mu_c^2}.$$

(5.4)

It is important to mention that the particle diameter d_p (or the median diameter of the particle size distribution) has to be calculated as volume-equivalent diameter for non-spherical fluid particles. An extensive collection of correlations for the rise behavior of fluid particles can be found in Clift et al. (1978) for various particle sizes and shapes, flow regimes, and physical property ranges, including clean and contaminated systems. A prevalent correlation that is suitable for both gas bubbles and oil droplets of ellipsoidal shape with "contaminated" phase boundaries (i.e., with suspended matter, hydrates, chemical, or bio-surfactants) rising in a liquid continuous phase is depicted in the subsequent paragraph.

The correlation between d_p and the buoyant velocity is typically described by an integrated approach of regimes defined by the size and shape of the fluid particles (gas bubbles and liquid droplets, Zheng and Yapa 2000). This approach has been validated against a vast amount of literature data and is used in several oil and gas spill models, e.g., by the groups of Yapa (Zheng et al. 2003; Zheng and Yapa 2000),

Boufadel (Zhao et al. 2016), Schlüter (Pesch et al. 2018), and Paris (Paris et al. 2012; Lindo-Atichati et al. 2016; Perlin et al. 2020).

In this physics-based empirical correlation, a dimensionless group H is introduced (Grace et al. 1976; Zheng and Yapa 2000):

$$H = \frac{4}{3} \cdot Eo \cdot Mo^{-0.149} \cdot \left(\frac{\mu_c}{\mu_w}\right)^{-0.14}, \tag{5.5}$$

where $\mu_w = 0.0009\ Pas$ is the dynamic viscosity of pure water. Depending on the magnitude of H, two regimes for a second dimensionless quantity J are introduced:

$$J = 0.94 \cdot H^{0.757}\ \left(for\ 2 < H \leq 59.3\right), \tag{5.6}$$

$$J = 3.42 \cdot H^{0.441}\ \left(for\ H > 59.3\right). \tag{5.7}$$

The stationary rise velocity u_p of an ellipsoidal fluid particle of intermediate size (typically 1 mm < d_p < 18 mm) with a contaminated interface and negligible wall effects is then calculated according to:

$$u_p = \frac{\mu_c}{\rho_c \cdot d_p} \cdot Mo^{-0.149} \cdot \left(J - 0.857\right). \tag{5.8}$$

These equations are valid for $Mo < 10^{-3}$, $Eo < 40$ and $Re > 0.2$ (Grace et al. 1976).

5.3 Gas Bubble Behavior: Theoretical and Experimental Insights

In contaminated systems of gas bubbles with surface-active substances (e.g., suspended matter in seawater, hydrates, bio- or chemical surfactants) dispersed in a continuous liquid phase, the contaminants accumulate at the boundary between the two phases and lead to an immobile interface. That is, the interface impedes internal circulation and causes higher drag coefficients in the range of those of solid particles. In this case, the bubbles rise much slower than in clean systems, and the rise velocities can be calculated using the abovementioned correlation of Grace et al. (1976).

By contrast, in pure gas-liquid systems of gas bubbles dispersed in a continuous liquid phase without any surface-active contaminants, the interface is supposed to be mobile. In other words, the free interface between the two phases enables internal circulations inside the bubble that lead to a reduced drag coefficient and therefore a higher rise velocity. Different regimes corresponding to different bubble diameters have been proposed by Mersmann (1977) and Räbiger and Schlüter (2002). The two most relevant regimes for the characterization of gas bubbles in subsea oil spills are those describing the ascent of spherical bubbles and of somewhat larger ellipsoidal bubbles with internal circulations, respectively. The former equation was proposed by Brauer (1971) and is valid in the range of:

$$7.2 < Ar < 125 \cdot Mo^{-\frac{1}{4}}. \tag{5.9}$$

The rise velocity in this regime is calculated according to the following equation:

$$u_\mathrm{p} = 0.136 \cdot \left(\frac{\mu_\mathrm{c} \cdot g \cdot \Delta\rho}{\rho_\mathrm{c}^2} \right)^{\frac{1}{3}} \cdot Ar^{0.4266}. \tag{5.10}$$

The correlation for pure bubbles of ellipsoidal shape was introduced by Peebles and Garber (1953) and is valid in the range of:

$$125 \cdot Mo^{-\frac{1}{4}} < Ar \leq 22.6 \cdot Mo^{-\frac{1}{2}}. \tag{5.11}$$

The rise velocity in this regime is calculated according to the following equation:

$$u_\mathrm{p} = 1.91 \cdot \left(\frac{\mu_\mathrm{c} \cdot g \cdot \Delta\rho}{\rho_\mathrm{c}^2} \right)^{\frac{1}{3}} \cdot \left(Mo \cdot Ar \right)^{-\frac{1}{6}}. \tag{5.12}$$

The rise velocity is decreasing with increasing diameter in this regime because of the increased drag force of the deformed bubbles. Due to the presence of seawater and suspended matter and the possibility of hydrate formation under deep-sea conditions, it seems natural to apply the correlation of Grace et al. (1976) comprising the assumption of contaminated interfaces for the calculation of the rise velocity of gas bubbles during subsea oil spills (Maini and Bishnoi 1981; Rehder et al. 2002, 2009).

Laqua et al. (2016) investigated the rise velocity of methane bubbles under simulated deep-sea conditions in a high-pressure laboratory facility (Gust et al. 2012). It consists of a high-pressure steel vessel with a volume of 99 L that can withstand pressures up to 55 MPa. The vessel contains an inner module made of acrylic glass with an inner diameter of 190 mm and a height of 600 mm that is connected to the outer vessel via a flexible membrane in order to enable pressure equalization (Seemann et al. 2014). Temperature and pressure are monitored to ensure reproducible experimental conditions. Methane is used as gas phase, since it is the most abundant component of natural gas. Artificial seawater with a salt concentration of 3.5% is used as continuous phase. In some experiments deionized water is used instead in order to enable comparison concerning the contamination of the phase boundary. The bubbles are generated via 1/16″ capillaries with inner diameters d_N of 0.25 and 0.5 mm. A high-speed camera captures the rising bubbles against the light. From the gray value distribution of the images, the bubble displacement over time is detected automatically, and the terminal rise velocities of the bubbles are

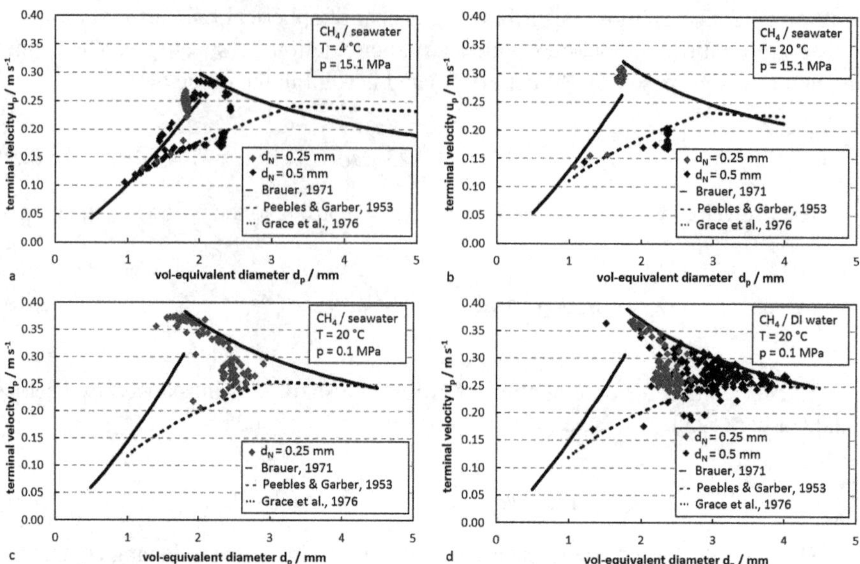

Fig. 5.1 Experimental results and theoretical correlations for the rise velocity of methane bubbles as function of the volume-equivalent diameter: (**a**) 4 °C and 15.1 MPa, artificial seawater; (**b**) 20 °C and 15.1 MPa, artificial seawater; (**c**) 20 °C and 0.1 MPa, artificial seawater; (**d**) 20 °C and 0.1 MPa, deionized water (Laqua et al. 2016)

calculated using a custom-made Java script. Experiments have been carried out for absolute pressures of 0.1 and 15.1 MPa, temperatures of 4 and 20 °C, and artificial seawater and deionized water.

In Fig. 5.1 the terminal rise velocities of methane bubbles of different sizes are plotted over their equivalent diameter alongside the curves of the three correlations for clean and contaminated interfaces for different temperature/pressure conditions and for both kinds of water. Surprisingly, not all bubbles follow the correlation of Grace et al. (1976) assuming a contaminated interface. Regarding Fig. 5.1a one could speculate that some bubbles are hydrate-coated and therefore follow the curve of Grace et al. (1976), while others are uncoated and follow the curves of Brauer (1971) or Peebles and Garber (1953) for this reason. But the same bifurcation can be observed in Fig. 5.1b, where hydrate formation can be excluded due to the high temperature of 20 °C. The composition of the seawater can be ruled out as a possible reason for the case of contaminated interfaces, too, since the scatter of rise velocities, found in both Fig. 5.1c and d, provides a direct comparison between the two types of water, while the other experimental conditions remain identical. In fact, for all tested experimental conditions, the rise velocities scatter between the correlations for clean and contaminated phase boundaries as approximate upper and lower boundary, respectively. The correlations by Brauer (1971) or Peebles and Garber (1953) can thus be used as upper boundaries and by Grace et al. (1976) as lower boundaries in order to define the range of expected bubble rise velocities.

The immobilization of the bubble surface is obviously not the determining factor, and other physical processes might influence the rise velocity.

A reason for different rise velocities of equally sized bubbles has been given by Tomiyama et al. (2002). They explained that the initial conditions during the bubble formation force the bubbles on either a zigzag or a helical rise path. Bubbles rising on the helical path are much faster than those on the zigzag path because the latter are accelerating and decelerating repeatedly, while those on the helical path are moving with a more or less constant velocity. Initial shape deformations lead to different sphericities of the bubbles. Nearly spherical bubbles show an acceleration and deceleration behavior and are therefore quite slow. More ellipsoidal bubbles (of the same volume-equivalent diameter) rise on a helical trajectory and are consequently much faster. Due to the random intensity of bubble deformation, the measured rise velocities are distributed over a quite large range (Fig. 5.1). The results of Laqua et al. (2016) agree very well with the correlation between sphericity and terminal rise velocity given by Tomiyama et al. (2002). It can therefore be concluded that the initial shape deformation and the resulting rise trajectories of the bubbles are the governing factors for the terminal rise velocities (Laqua et al. 2016). During an actual blowout, the trajectories of the bubbles are influenced by the high level of turbulence and collisions with other bubbles and droplets. However, either a helical or a zigzag trajectory will establish once the bubbles leave the high-turbulence high-particle-density regime. Again, the well-known correlations for clean and contaminated systems seem to provide a good estimate for the upper and lower boundaries of the bubble rise velocities, respectively.

5.4   Oil Droplet Behavior: Theoretical and Experimental Insights

For the investigation of the rise velocities of crude oil droplets, a high-pressure experimental facility with a volume of 35 mL and optical access via a windowed slot is applied at Hamburg University of Technology (TUHH) in Hamburg, Germany. The droplets are generated using a 1/16″ capillary with an inner diameter d_N of 0.18 mm leading to spherical to slightly ellipsoidal droplets with volume-equivalent diameters between 1.8 and 2.5 mm. During their ascent the droplets are captured with a high-speed camera against the light. Deionized water and artificial seawater are used as continuous phase, while Louisiana Sweet Crude (LSC) oil serves as dispersed phase. The temperature of 20 °C and pressure levels between 0.1 and 15.1 MPa are monitored to ensure reproducible experimental conditions. From the gray value distribution of the images, the droplet displacement over time is detected automatically using a self-written Java script, and the terminal rise velocities of the bubbles are calculated. All results are in good accordance with the empirical correlation of Grace et al. (1976), which can therefore be applied for the calculation of crude oil droplet rise velocities in a quiescent water phase. The correlation slightly overestimates the experimental values by up to 5.6% depending on the respective experimental conditions. It is worth noting that the terminal rise velocity of a droplet is higher at ambient pressure than that of an equally sized droplet at high pressure. This can be explained by the higher compressibility of crude oil as compared to water leading to a smaller density difference at elevated pressure. Also, the rise velocity is higher in seawater than in deionized water because of the higher density difference in the former case. For instance, a 2 mm LSC droplet at a pressure of 12.1 MPa has a terminal rise velocity of about 0.069 m/s in seawater and of about 0.065 m/s in deionized water. A 1 mm LSC oil droplet would have a rise time of approximately 13.44 hours through 1.5 km of water column corresponding the depth of the *Deepwater Horizon* (DWH) blowout at a mean rise velocity of 0.031 m/s (Pesch et al. 2018).

A modified setup (see Fig. 5.2) is applied in order to capture the oil droplets over any desired time period, setting pressure gradients to simulate a droplet's ascent from the deep sea to shallower depths and finally to the sea surface. Inside the high-pressure cell, an hourglass-shaped glass tube is positioned between the observation windows. A gear pump is applied to generate a laminar downward circular flow of artificial seawater inside this tube in order to fix a single oil droplet in a vertical position beneath the tapering of the tube. A cooling circulator is used to adjust the desired temperature. The droplet's size and shape are captured against the light using a high-speed camera and are evaluated automatically with the aid of the image

Fig. 5.2 Experimental high-pressure facility for the investigation of the oil droplet rise behavior in countercurrent flow with oil presaturation at Hamburg University of Technology. The shown setup corresponds to the flow diagram in Fig. 5.4 without the part inside the dashed box

processing software ImageJ. Pressure and temperature are monitored to ensure reproducible experimental conditions. Again, LSC oil droplets are generated using a capillary at the bottom of the cell. While the evolution of the droplet volume and the respective equivalent diameter is measurable, the rise velocity is not directly accessible in this setup. When pressure is released from an initial value of 15.1 MPa to ambient pressure (0.1 MPa) with a decompression rate of 1 MPa/min at a constant temperature of 20 °C, the single droplets are shrinking by 3–5% of their initial volume-equivalent diameter. This shrinking is accounted for by mass transfer or dissolution of fairly water-soluble components into the surrounding seawater (Pesch et al. 2018).

5.5 How Reservoir and Deep-Sea Conditions Change Everything: Rise Behavior of Live Oil Droplets

During deep submarine blowouts like in the case of the DWH accident, the oil exits a reservoir with a very high pressure into the deep sea at a still fairly high pressure level. From there, pressure is steadily decreasing along the rise path of ascending oil droplets. As detailed in Chap. 3, tremendous amounts of the present natural gas dissolve in the crude oil under reservoir conditions. As pressure decreases during the blowout, the amount of gas that can stay dissolved in the oil decreases, too. The initially gas-saturated, so-called live oil becomes supersaturated and starts to degas. The majority of the gas components is nonpolar and therefore tends to stay inside the oil, forming a second, gaseous phase: The pressure decline and supersaturation of the oil cause formation and growth of tiny bubbles inside the liquid oil droplets, giving rise to the term "internal degassing." A simplified modeling approach by Pesch et al. (2018) uses methane as model component, like has been done for the experimental determination of live oil properties in Chap. 3. The internal degassing leads to three complementary effects. First, the droplet volume and therefore its volume-equivalent diameter d_p increase as the internal gas holdup ϵ increases. Second, the mean density ρ_p of the composite bubble/droplet decreases with an increasing amount of free gas. And third, the methane density is decreasing tremendously with decreasing pressure. This triad of changes in the droplet's properties causes an increment of buoyancy and hence an acceleration of the droplet rise.

For the calculation of the droplet rise velocity u_p, measured physical properties for all involved fluids as detailed in Chap. 3 and the correlation of Grace et al. (1976) are applied. For instance, the rise of a gas-saturated LSC oil droplet with an initial diameter d_p of 1 mm, rising from a depth of 1.5 km (rise height z, 0 m; 15.1 MPa) to the sea surface (rise height z, 1500 m; 0.1 MPa) takes approximately 6.08 hours at a mean rise velocity u_p of 0.069 m/s. Notably, this is less than half of the rise time of a 1 mm dead LSC oil droplet. The calculated curves of the gas void fraction, the droplet diameter d_p and the mean droplet density ρ_p, depicted in a dimensionless form by dividing them by the respective initial values, are displayed in Fig. 5.3a. The resulting rise velocity u_p of a live oil droplet with an initial diameter $d_{p,0}$ of

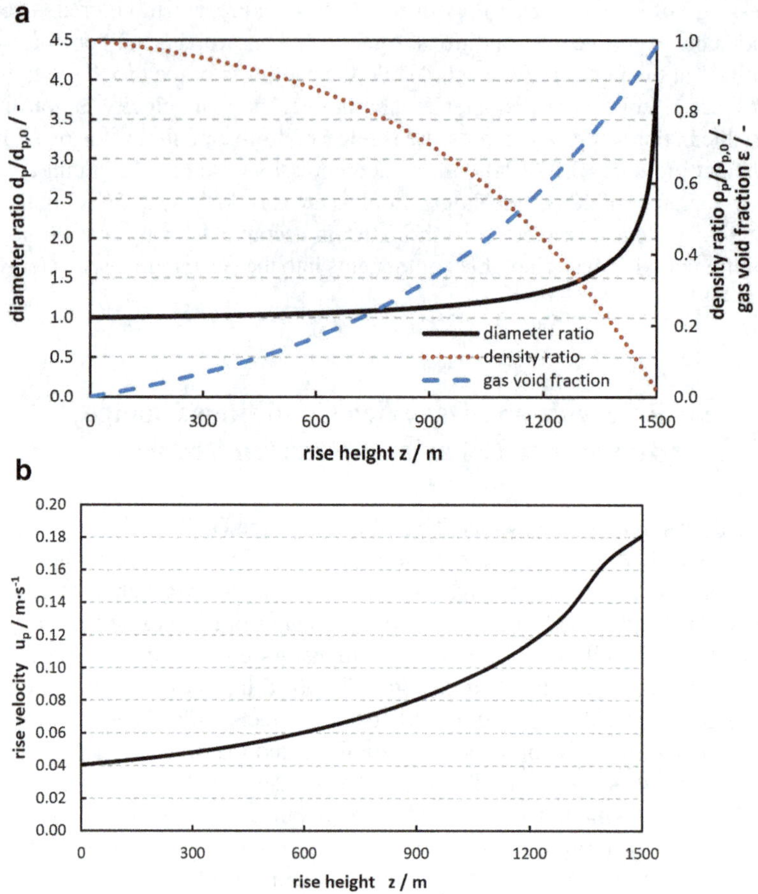

Fig. 5.3 Modeling results for the rise behavior of a methane-saturated LSC oil droplet through 1500 m of water column: (**a**) increasing droplet diameter d_p, depicted in a dimensionless form by dividing it by its initial value $d_{p,0}$; decreasing mean droplet density ρ_p, depicted in a dimensionless form by dividing it by its initial value $\rho_{p,0}$; and increasing gas void fraction ϵ dependent on the rise height z; (**b**) rise velocity evolution of a droplet with an initial diameter of 1 mm dependent on the rise height z (Pesch et al. 2018)

1 mm is shown in Fig. 5.3b. As pressure decreases along the rise path, the rise velocity is steadily increasing.

To account for the aforementioned effects experimentally, a setup with adjustable pressure is required that enables the investigation of rising gas-saturated crude oil droplets over an extended period of time. For this purpose, the high-pressure cell with the hourglass-shaped glass tube that has been used for the investigation of dead oil droplets before is extended in such a way as to allow for saturation of the crude oil with a gas at a defined saturation pressure. LSC oil is brought into contact with

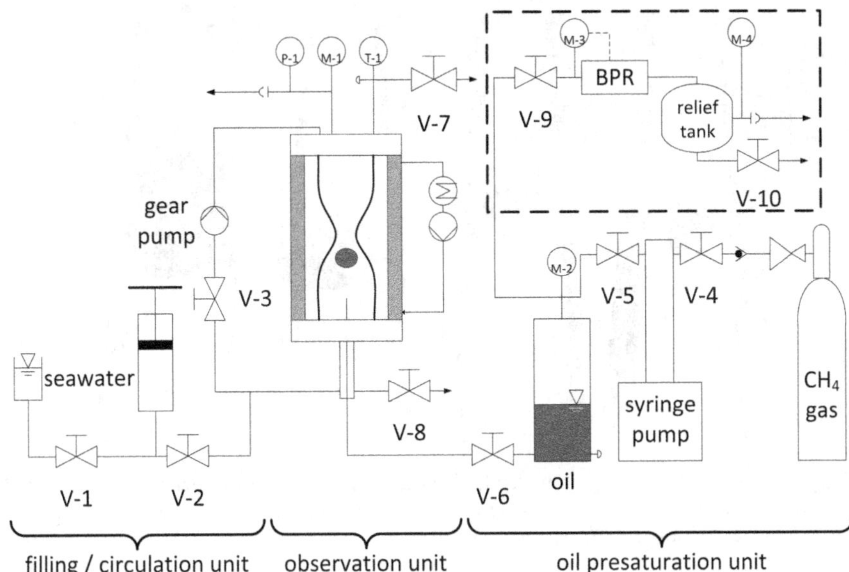

Fig. 5.4 Flow diagram of the high-pressure setup for the droplet rise observation experiments, depicting the three main units of the experimental setup: filling/circulation unit, observation unit, oil presaturation unit. The part inside the dashed box is added for the experiments with initial depressurization only (Pesch et al. 2018)

methane using a precision high-pressure syringe pump. The pressure is maintained at 15.1 MPa over a couple of days, and the amount of dissolved gas is monitored. Once the saturation is complete, a single oil droplet is transferred into the counter-current flow cell via the capillary. The subsequent execution of the experiment and the data acquisition and evaluation are identical to that of the dead oil droplet treatment. Different decompression rates and temperatures can be adjusted. The described setup is depicted in Figs. 5.2 and 5.4 (without the part inside the dashed box). While the droplets are slightly shrinking in the beginning, they are vigorously expanding at the end of the experiments, as can be seen in Fig. 5.5. The volume-equivalent droplet diameters are increasing by factors as high as 1.9, confirming the predicted behavior qualitatively. In comparison to the curve of the diameter ratio $d_p/d_{p,\,0}$ that is displayed in Fig. 5.3a, the growth of the droplets is delayed in the described experiments: The initial shrinking of the droplets in combination with the delay in degassing can be explained by poor (homogeneous) nucleation of methane bubbles within the liquid oil. As explained in Chap. 3, a certain degree of supersaturation has to be overcome to enable gas bubble nucleation. Obviously, this critical supersaturation or, more specifically, critical pressure difference is only reached toward the end of the experiments. The quite high degree of supersaturation before the onset of degassing enhances mass transfer of methane molecules across the phase boundary into the surrounding seawater, which in turn lowers the corresponding degree of supersaturation and therefore increases the observed time delay.

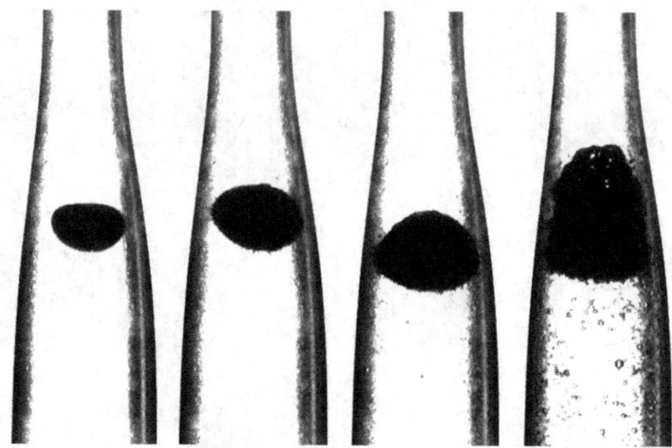

Fig. 5.5 Growth of a methane-saturated LSC oil droplet during depressurization from 15.1 MPa (left) to 0.1 MPa (right), held in a countercurrent flow of artificial seawater; decompression rate, 1 MPa/min; temperature, 20 °C (Pesch et al. 2018)

In contrast, during an actual deep-sea blowout, the crude oil is supposed to be already highly supersaturated and contain bubble nuclei at the blowout site. In the case of the DWH blowout, the reservoir pressure at which the oil was saturated with natural gas was at least about 70 MPa (Hickman et al. 2012). As the oil flows upward, the pressure decreases leastwise 55 MPa from the bubble point curve into the two-phase region – this is certainly enough supersaturation for the creation of bubble nuclei and actual gas bubbles, which is besides proved by the huge amounts of free gas during the blowout (Satter 2016). Furthermore, over the last section of the broken riser stub, the so-called blowout preventer, a sudden pressure drop of approximately 8.6 MPa occurred, and the flow conditions were highly turbulent, facilitating bubble nucleation (Aliseda et al. 2010). The calculated curves correspond to an actual blowout with the presence of bubble nuclei due to high supersaturation and shear forces.

The extension of the experimental plant at Hamburg University of Technology, indicated by the dashed box in Fig. 5.4, accounts for the high pressure levels inside oil reservoirs and the pressure release before the oil exits the broken riser into the water column. A back pressure regulator enables a pressure release from a saturation pressure of 25.1 MPa to the respective initial pressure of the experiments that are conducted and evaluated like before. With this extended setup, no initial shrinking is observed, but the droplets are growing increasingly from the beginning of the experiments during the continuous pressure release. Apparently, the nucleation barrier is already overcome when the droplets are generated. The experimental results represent the predicted curve quite well with much higher diameter ratios $d_p/d_{p,\,0}$ than before that now lie within the range predicted by means of the presented procedure.

5.6 Swarm Effects, Mass Transfer, and Gas Hydrates

In oil and gas plumes, the rise velocity of single droplets and bubbles through stagnant water is not the only parameter that governs the distribution and fate of the hydrocarbons. Layers of different salinity and density and the interactions of ocean currents with the seafloor topography, upwelling and subduction currents, and turbulent eddy diffusion have to be considered to gain reliable plume modeling results (see Sect. 5.3). But also more fundamental physical phenomena that alter the rise velocity locally like swarm effects, mass transfer of gas components, partitioning of hydrocarbons, and the formation of gas hydrates should be taken into account. Some of those are discussed in separate chapters of this book. In the following, a brief overview of the aforementioned effects is given.

The rise or sink velocity of a particle swarm in stagnant media is usually smaller than that of single particles due to displacement and a hereby induced opposing flow of the continuous phase. In principle, this process also holds for fluid particles like droplets and bubbles (Richardson and Zaki 1954; Brauer 1971). However, in dense droplet and bubble swarms, clustering of the fluid particles, i.e., the formation of a group of particles that is rising collectively, causes the reverse effect. These clusters can cause an upwelling flow of the surrounding water that allows the droplets and bubbles to rise significantly faster. This effect has been observed for rising methane bubble swarms with and without oil coating (MacDonald et al. 2002). The rise velocity of a dense particle swarm is therefore supposed to be enhanced as compared to single rising particles, especially when it contains large amounts of gas bubbles. In deeper water and in close vicinity to the blowout site, the rise velocity is hence supposed to be higher than one would calculate from the correlations for single particles presented above. In this region, breakup and coalescence of bubbles and droplets can occur, too. In shallower regions and farther away from the spill site, most of the pure gas bubbles are dissolved, and the distance between the droplets is much larger. Therefore, swarm effects are negligible here and the aforementioned correlations apply.

Due to the decreasing pressure along their rise path, gas bubbles tend to expand, since their compressibility and specific volume are increasing. On the other hand, mass transfer of gas molecules across the phase boundary and subsequent dissolution into the seawater counteract this effect. For pure bubbles rising in pure water, dissolution dominates over expansion, so the bubbles are shrinking (Rehder et al. 2009; Zhao et al. 2016). However, mass transfer is governed by the respective driving gradient and thus by the concentration of the gas component in the surrounding water. Particularly close to the blowout site and inside the intrusion layers caused by an oil spill, where hydrocarbons are accumulated, high background concentrations of methane and other gas components diminish the dissolution. Moreover, for gas bubbles that are oil-coated, the mass transfer resistance of the interface is significantly higher, leading to much longer bubble lifetimes and larger rise distances – also resulting in faster rising of the oil entrained, as known from natural oil and gas seeps (MacDonald et al. 2002).

The dissolution of components of liquid (live) oil is governed by the pressure-dependent oil-to-water partitioning coefficient that is thoroughly discussed in Chap. 8. When gas bubbles are present inside methane-(super)saturated oil droplets as explained above, a certain amount of the dissolved methane will leave the droplet by dissolution into the surrounding water; but the large share of methane will rather enter the bubbles due to energy reasons. In turn, the oil acts as a mass transfer barrier for the gas molecules much like at the surface of oil-coated bubbles. Overall, the common practice regarding oil and gas plumes as a combination of liquid-only oil droplets and pure gas bubbles is questionable. There is no explicit reason why the gaseous and liquid hydrocarbon phases should evolve and rise separately. Actually, an oil droplet that contains free gas or an amount of oil that is rising at the surface of a coated gas bubble rises much faster than an oil droplet containing the same amount of oil (but no gas), because of the different particle size and average particle density.

Deepwater blowout conditions further deliver a nearly ideal environment for the formation of clathrate hydrates, which are icelike solids and are formed when molecular cages of water enclathrate low-molecular-weight species; guest molecules can include methane, ethane, propane, isobutene, and carbon dioxide. The internal diameter of the molecular water cages provides a geometric constraint on the species that can be successfully enclathrated, where the severity of the high-pressure and low-temperature conditions depends on the availability of guest species. Sloan and Koh (2007) discuss the array of hydrate structures – based in combinations of cages with distinct internal diameters – that are found in both nature and laboratory applications; structures I and II (respectively, "sI" and "sII") are the most common, where sII hydrate is typically generated from the light hydrocarbon composition found in conventional oil and gas reservoirs.

While hydrate equilibrium is typically identified by a boundary in temperature-pressure space, this representation assumes both water and hydrate-forming guests are available in excess. Deepwater blowouts invoke a unique constraint on equilibrium, as the continuous seawater is depleted in methane. As such, hydrate cannot immediately form upon a rising natural gas bubble until the *local* seawater has reached equilibrium with the light hydrocarbon components. Subsea currents can reach the order of 0.1 m/s, so the successful formation of hydrates fundamentally requires seawater to be entrained within the plume at sufficient residence times to approach equilibrium. In regions where the subsea currents are effectively mitigated – either due to mechanical placement of oil and gas equipment or through boundary effects near the ocean floor – hydrate formation may also be achievable. It should be noted that, in the context of a deepwater blowout, most models predict gas bubble diameters significantly larger than 100 microns; visual experiments from the Colorado School of Mines (Taylor et al. 2007; Brown and Koh 2016; Davies et al. 2010) have demonstrated that the formation of hydrate at the boundary of a hydrate guest phase (e.g., methane) and water will result in an initial "shell" of approximately 50 microns in thickness. In a bubble swarm, this hydrate shell will enclose residual high-pressure gas inside, where further hydrate growth requires diffusion of either water and/or gas across the shell.

Hydrate investigation in the laboratory can prove difficult, due to the severe pressure conditions often required to mimic the natural blowout. Chen et al. (2014) constructed a novel high-pressure water tunnel (HPWT) at the Colorado School of Mines Center for Hydrate Research, where counterflowing conditioned seawater was used to suspend high-pressure gas bubbles in a view window. Chen et al. (2014) noted that Multiflash (Infochem 2012) and CSMGem (Ballard and Sloan 2004) – common phase equilibrium tools in hydrate research – correctly identified the methane saturation required to achieve hydrate formation from 70 to 139 bar. At 173 bar, hydrate formation was observed with as little as one fifth of the methane content predicted by phase equilibrium calculators; the authors speculate that, if the chemical potential driving force *toward* hydrate formation outstrips the dissolution rate of light hydrocarbons in the seawater, hydrate formation may be achievable. Chen et al. (2014) further noted that, below the equilibrium saturation, the hydrate shell was not observed to grow to completion, but rather may be characterized by dendritic needles or plates on the gas-water interface.

Within the above context, computational fluid dynamics approaches may provide a tractable method by which to estimate the rate of hydrate formation considering both (i) the chemical potential driving force for hydrate formation across the range of bubble sizes generated and (ii) the rate balance in the near-field plume between fresh seawater entering and partially saturated seawater exiting. In the limit that a partial or complete hydrate shell is *able* to form on the gas bubble surface, the consequential rate of gas diffusion to the seawater will decrease, as demonstrated by Davies et al. (2010). Although simplistic approaches may treat this effect as constant, the complexity of the above-described process should inform caution with modeling efforts. That is, the thickness, porosity, and permeability of the hydrate "shell" are distributed properties, due to both anisotropy in the extent of gas saturation within the plume and the well-known stochasticity (May et al. 2018) of hydrate crystal nucleation. As such, the estimation of near-field behavior – including the rise velocity of hydrate-encrusted bubbles – should ultimately yield a distributed output to appropriately represent the consequences of hydrate formation.

5.7 Conclusion

Several correlations for the calculation of bubble or droplet rise velocities for both "clean" and "contaminated" interfaces are available in literature. Surprisingly, even though the environment during an oil spill is characterized by saltwater, multicomponent oil droplets/bubbles, suspended matter, and possibly gas hydrates, the experimental results for rising methane bubbles do not necessarily follow the correlations for contaminated interfaces. Instead, the results scatter between the correlations for clean and contaminated interfaces as upper and lower boundary, respectively. This can be explained by initial shape deformations and the resulting rise trajectories of the bubbles.

The experimental results of the stationary dead oil rise velocities are in good agreement with the correlations for contaminated interfaces. When live oil droplets are considered, the actual pressure conditions inside the reservoir and the deep sea play an important role. Due to a reduced gas-in-oil solubility when pressure decreases, degassing can occur inside the oil droplets, leading to an elevated buoyancy and consequently an increasingly fast ascent. Rise times can therefore be substantially smaller for live oil droplets than predicted with the available correlations using the initial droplet diameters. Swarm effects, ocean currents, mass transfer, and gas hydrate formation can have additional effects on the oil rise velocities and render the oil fate modeling extremely challenging.

In conclusion, it must be noted that the correlations for the rise velocity, although applicable for the calculation of individual particle rise velocities, should not be used uncritically. Also, the inference of median particle sizes from observed rise times is deceptive. While droplet size distributions, as detailed in Chap. 4, can cover a broad spectrum of particle sizes with corresponding distributions of rise velocities, the combined release and rise of oil and gas lead to swarm effects and two-phase particles, both leading to a bubble-mediated, accelerated ascent of the oil.

References

Aliseda A, Bommer P, Espina P, Flores O, Lasheras JC, Lehr B, Leifer I, Possolo A, Riley J, Savas O, Shaffer F, Wereley S, Yapa P (2010) Deepwater horizon release estimate of rate by PIV: report to the Flowrate Technical Group

Ballard AL, Sloan ED (2004) The next generation of hydrate prediction. IV: a comparison of available hydrate prediction programs. Fluid Phase Equilib 216(2):257–270

Brauer H (1971) Grundlagen der Einphasen- und Mehrphasenströmungen. Grundlagen der Chemischen Technik. Sauerländer, Aarau und Frankfurt am Main

Brown EP, Koh CA (2016) Micromechanical measurements of the effect of surfactants on cyclopentane hydrate shell properties. Phys Chem Chem Phys 18(1):594–600

Chen L, Levine JS, Gilmer MW, Sloan ED, Koh CA, Sum AK (2014) Methane hydrate formation and dissociation on suspended gas bubbles in water. J Chem Eng Data 59(4):1045–1051

Clift R, Grace JR, Weber ME (1978) Bubbles, drops, and particles. Academic Press, New York

Davies SR, Sloan ED, Sum AK, Koh CA (2010) In situ studies of the mass transfer mechanism across a methane hydrate film using high-resolution confocal Raman Spectroscopy. J Phys Chem C 114(2):1173–1180

Grace JR, Wairegi T, Nguyen TH (1976) Shapes and velocities of single drops and bubbles moving freely through immiscible liquids. Trans Inst Chem Eng 54:167–173

Gust G, Meyer A, Koepke D, Eggers R (2012) Mass transfer processes of a hydrate-covered deep CO_2 lake. Int J Greenhouse Gas Control 9:312–321. https://doi.org/10.1016/j.ijggc.2012.04.010

Hickman SH, Hsieh PA, Mooney WD, Enomoto CB, Nelson PH, Mayer LA, Weber TC, Moran K, Flemings PB, McNutt MK (2012) Scientific basis for safely shutting in the Macondo well after the April 20, 2010 Deepwater horizon blowout. Proc Natl Acad Sci 109:20268–20273. https://doi.org/10.1073/pnas.1115847109

Infochem (2012) Multiflash for windows 4.1. Infochem Computer Services Ltd, London

Laqua K, Malone K, Hoffmann M, Krause D, Schlüter M (2016) Methane bubble rise velocities under deep-sea conditions—influence of initial shape deformation. Colloids Surf A Physicochem Eng Asp 505:106–117. https://doi.org/10.1016/j.colsurfa.2016.01.041

Lindo-Atichati D, Paris CB, Le Hénaff M, Schedler M, Valladares Juárez AG, Müller R (2016) Simulating the effects of droplet size, high-pressure biodegradation, and variable flow rate on the subsea evolution of deep plumes from theMacondo blowout. Deep-Sea Research II 129(2016):301–310

MacDonald IR, Leifer I, Sassen R, Stine P, Mitchell R, Guinasso N (2002) Transfer of hydrocarbons from natural seeps to the water column and atmosphere. Geofluids 2:95–107. https://doi.org/10.1046/j.1468-8123.2002.00023.x

Maini BB, Bishnoi PR (1981) Experimental investigation of hydrate formation behaviour of a natural gas bubble in a simulated Deep Sea environment. Chem Eng Sci 36:183–189

May EF, Lim VW, Metaxas PJ, Du J, Stanwix PL, Rowland D, Johns ML, Haandrikman G, Crosby D, Aman ZM (2018) Gas hydrate formation probability distributions: the effect of shear and comparisons with nucleation theory. Langmuir 34(10):3186–3196. https://doi.org/10.1021/acs.langmuir.7b03901

Mersmann A (1977) Auslegung und Maßstabsvergrößerung von Blasen- und Tropfensäulen. Chemie Ingenieur Technik 49:679–691

Paris CB, Le Henaff M, Aman ZM, Subramaniam A, Helgers J, Wang D-P, Kourafalou VH, Srinivasan A (2012) Evolution of the Macondo well blowout: simulating the effects of the circulation and synthetic dispersants on the subsea oil transport. Environ Sci Technol 46:13293–13302. https://doi.org/10.1021/es303197h

Peebles FN, Garber HJ (1953) Studies on the motion of gas bubbles in liquids. Chem Eng Prog 49:88–97

Perlin N, Paris CB, Berenshtein I, Vaz AC, Faillettaz R, Aman ZM, Schwing PT, Romero IC, Schlüter M, Liese A, Noirungsee N, Hackbusch S (2020) Far-field modeling of a deep-sea blowout: sensitivity studies of initial conditions, bio-degradation, sedimentation and subsurface dispersant injection on surface slicks and oil plume concentrations (Chap. 11). In: Murawski SA, Ainsworth C, Gilbert S, Hollander D, Paris CB, Schlüter M, Wetzel D (eds) Deep oil spills: facts, fate, effects. Springer, Cham

Pesch S, Jaeger P, Jaggi A, Malone K, Hoffmann M, Krause D, Oldenburg TB, Schlüter M (2018) Rise velocity of live-oil droplets in deep-sea oil spills. Environ Eng Sci 35:289–299. https://doi.org/10.1089/ees.2017.0319

Räbiger N, Schlüter M (2002) Bildung und Bewegung von Tropfen und Blasen. In: VDI-Wärmeatlas: Berechnungsblätter für den Wärmeübergang, Neunte, überarbeitete und erweiterte Auflage. Springer Berlin Heidelberg, Berlin, Heidelberg, s.l., Lda 1-Lda 15

Rehder G, Brewer PG, Peltzer ET, Friederich G (2002) Enhanced lifetime of methane bubble streams within the deep ocean. Geophys Res Lett 29:21-1-21-4. https://doi.org/10.1029/2001GL013966

Rehder G, Leifer I, Brewer PG, Friederich G, Peltzer ET (2009) Controls on methane bubble dissolution inside and outside the hydrate stability field from open ocean field experiments and numerical modeling. Mar Chem 114:19–30. https://doi.org/10.1016/j.marchem.2009.03.004

Richardson JF, Zaki WN (1954) The sedimentation of a suspension of uniform spheres under conditions of viscous flow. Chem Eng Sci 3(2):65–73. https://doi.org/10.1016/0009-2509(54)85015-9

Satter A (2016) Reservoir engineering: the fundamentals, simulation, and management of conventional and unconventional recoveries. Gulf Professional Publishing, Waltham

Seemann R, Malone K, Laqua K, Schmidt J, Meyer A, Krause D, Schlüter M (2014) A new high-pressure laboratory setup for the investigation of deep-sea oil spill scenarios under in-situ conditions. In: Proceedings of the 7th international symposium und environmental hydraulics, pp 340–343

Sloan ED, Koh CA (2007) Clathrate hydrates of natural gases. CRC Press, Boca Raton

Taylor CJ, Miller KT, Koh CA, Sloan ED (2007) Macroscopic investigation of hydrate film growth at the hydrocarbon/water interface. Chem Eng Sci 62(23):6524–6533

Tomiyama A, Celata GP, Hosokawa S, Yoshida S (2002) Terminal velocity of single bubbles in surface tension force dominant regime. Int J Multiphase Flow 28:1497–1519

Zhao L, Boufadel MC, Lee K, King T, Loney N, Geng X (2016) Evolution of bubble size distribution from gas blowout in shallow water. J Geophys Res Oceans 121:1573–1599. https://doi.org/10.1002/2015JC011403

Zheng L, Yapa PD (2000) Buoyant velocity of spherical and nonspherical bubbles/droplets. J Hydraul Eng 126:852–854

Zheng L, Yapa PD, Chen F (2003) A model for simulating Deepwater oil and gas blowouts – part I: theory and model formulation. J Hydraul Res 41:339–351. https://doi.org/10.1080/00221680309499980

Part III
Transport and Degradation of Oil and Gas
from Deep Spills

Teri Navajo
A Speck in the Sea
Acrylic and ink on Canvas
24" × 24"

Chapter 6
The Importance of Understanding Transport and Degradation of Oil and Gasses from Deep-Sea Blowouts

Karen J. Murray, Paul D. Boehm, and Roger C. Prince

Abstract Assessing the environmental impact of spilled oil and gas requires an understanding of both the petroleum's movement and the physical and chemical changes it undergoes after the release. Deep subsea releases differ from surface releases primarily because of the extended interaction of the oil and gas with the water column prior to atmospheric exposure. Furthermore, the remote locations and uncertain volumes of deep subsea oil and gas releases result in logistical challenges to data collection both during and after a release. However, there have been multiple subsea releases with large-scale environmental investigations, and the extensive investigation undertaken during and after the Deepwater Horizon (DWH) spill resulted in an especially comprehensive empirical data set. The specific chemical and physical properties of the oil and gas mixtures in subsea releases differ between spills, and the DWH investigation and others have highlighted the importance of understanding biodegradation rates and transport to offshore sediments. The field and experimental data collected during and after a spill can improve oil fate prediction when incorporated into comprehensive oil spill models.

Keywords Oil spills · Biodegradation · Deepwater Horizon · NRDA

6.1 Introduction

Deep subsea oil and gas release scenarios are quite different from typical surface releases, such as those from tankers, shallow platforms, or pipeline leaks of liquid oil. The volume of a subsurface release is often uncertain, as oil is released from a large reservoir usually over an extended period, and the reconstruction of those releases is prone to large uncertainties because of uncertainties in various data sets

K. J. Murray (✉) · P. D. Boehm
Exponent, Inc., Maynard, MA, USA
e-mail: kmurray@exponent.com; pboehm@exponent.com

R. C. Prince
Stonybrook Apiary, Pittstown, NJ, USA

© Springer Nature Switzerland AG 2020
S. A. Murawski et al. (eds.), *Deep Oil Spills*,
https://doi.org/10.1007/978-3-030-11605-7_6

86

that are and have been used in a variety of methods to estimate the volumes released. In addition, the usual remoteness of a subsea release—both in terms of depth of releases and remote geographical areas where many deep exploration prospects are developed—presents many logistical and technical challenges to data collection. These data collection challenges need to be addressed both during spill response and mitigation and during environmental injury assessment. To understand the overall impact of released oil and gas, it is important to overcome these challenges to understand how the oil moves and the physical and chemical changes it undergoes once it is released (Table 6.1).

The initial priority during and immediately after an oil spill is the response, which aims to minimize damage and protect resources. Concurrent with the emergency response is the urgent need to collect "ephemeral" data that will be needed to assess environmental injuries (Boehm et al. 2013). In the following days and weeks, the effort shifts to understanding the environmental impact and potential injury from the oil, and key to that is understanding the fate and transport of the oil. Besides providing information about the geographical extent and temporal trends of the chemical concentrations of oil and oil compounds for use in environmental damage assessments, the sampling efforts are aimed at characterizing the changes in the surface and subsurface distribution and composition of the oil over time. Sampling is usually combined with fate and transport modeling to provide a three-dimensional understanding (Spaulding et al. 2017).

The *Deepwater Horizon* (DWH) spill resulted in an unprecedented scope of investigations of the fate and transport of the released oil. However, studies of the fate, transport, and degradation (also known as "natural attenuation") of oil and gas have been important topics in oil spill research and environmental assessment for many years, and there are many examples of environmental investigations of oil spills from the past 50 years where such extensive environmental studies were also undertaken. Although there is variability in the focus and intensity of the studies, due to both trends over time and the relative magnitudes of the releases, more recently, comprehensive data sets in combination with modeling have provided a picture of the impact of each spill to the environment.

For example, in 1978, the tanker *Amoco Cadiz* ran aground near the shore of France and released approximately 230,000 tonnes of light crude oil (API gravity 32.8). Studies from this release showed that the weather resulted in the formation of a stable emulsion and tarballs on the ocean surface, and evaporation resulted in the loss of a large fraction of the oil (Gundlach et al. 1983; Mille et al. 1998). The remaining oil that reached shorelines was highly weathered, and periodic follow-up studies in 7 years after the release showed further degradation of the remaining petroleum and led to the description of stages of weathering (Page et al. 1989). In March 1989, the *Exxon Valdez* tanker ran aground in Prince William Sound, Alaska, spilling 35,470 tonnes of Alaska North Slope Crude (API gravity 32.3) over a 6-hour period. A storm 2 days after the release resulted in the mixing of the slick and the formation of mousse and tarballs, which were distributed over a geographical area more than 500 miles in length. The resulting environmental investigation was one of the largest in US history and led to the development of standardized procedures for

Table 6.1 Comparison of weathering processes in surface vs. subsurface release

Process	Subsurface blowout		Surface release	
	Timeframe	Notes	Timeframe	Notes
Dissolution	Begins immediately	Smaller droplet sizes formed at depth can increase dissolution and result in dissolved components at depth, as opposed to surface waters	Begins immediately	Governed by droplet size, which may be larger in a surface spill—less dissolution likely than for a spill at depth
Dispersion	Begins immediately	High pressure at release point can result in smaller droplet sizes which are not positively buoyant. This may result in "subsurface plumes" as seen in the Deepwater Horizon	Begins immediately	Waves and wind can initially control
Surface spreading	Begins when oil surfaces and forms a slick—could be minutes to hours, or may not happen	The conditions of the release will control the transit time to the surface, which may be some horizontal distance from the release point due to currents	Begins immediately	The direction of surface spreading from the release point will be primarily controlled by surface winds
Evaporation	Begins when oil surfaces—could be minutes to hours, or may not happen	Because in subsurface spills there is a lag before the oil is exposed to the atmosphere, some compounds which would otherwise evaporate can be dissolved in the water column	Begins immediately, lasts up to a few days	Loss depends on specific product released (could be 10–75% of total mass)
Photodegradation	Begins if/when oil reaches photic zone		Can begin immediately	
Mousse (emulsions)	Can form at surface or at depth. May incorporate gas		Can happen shortly after release	Physical mixing at the sea surface produces a stable emulsion resistant to dispersion

(continued)

Table 6.1 (continued)

Process	Subsurface blowout		Surface release	
	Timeframe	Notes	Timeframe	Notes
Biodegradation	Begins within the first hours	Can occur within the water column, sediment, or at the surface. Rates are dependent on site conditions	Begins within the first hours	Can occur within the water column, sediment, or at the surface
Sedimentation	Time scale dependent on mechanism	Oil may sink after loss of lighter compounds or due to interactions with organic or inorganic particles Deepwater blowouts may be distant from shoreline, but interactions with particulate plumes near the continental shelf or at shorelines may occur	Time scale dependent on mechanism	Oil may sink after loss of lighter compounds or due to interactions with organic or inorganic particles

characterizing shoreline oiling, SCAT (Owens and Sergy 2003), and the first well-documented use of bioremediation to cleanse oiled shorelines (Bragg et al. 1994). While the water column aspects of the spill were relatively short-lived, the focus of longer-term fate and degradation studies turned to the shorelines where oil weathering and biodegradation (Boehm et al. 2008) and bioavailability (Neff et al. 2006) of buried oil residues dominated investigations for almost two decades after the spill.

In 2010, a large and complex set of investigations drove multiple field sampling efforts immediately following the DWH wellhead blowout in the Gulf of Mexico (GoM) and continued for many months after the well was capped. The DWH well blowout at Mississippi Canyon Lease Block 252 (MC252) occurred on April 20, 2010, at approximately 1500 m water depth and approximately 103 km from Southwest Pass on the Louisiana coast. Over an 87-day period, until the well was capped on July 15, 2010, large volumes of natural gas and oil (API gravity 37 (SL Ross Environmental Research 2010)) were released into the waters of the GoM. To lessen the ecological impact from surface oiling and prevent the fouling of shorelines, Corexit 9500, a chemical dispersant, was injected into the oil stream at the release point at depth and applied aerially to surface slicks (Houma ICP Aerial Dispersant Group 2010; U.S. Senate 2010). The chemical dispersion at depth, combined with the physical dispersion resulting from the force of the gas and oil escaping from the well, resulted in the formation of oil droplets in a range of sizes. The larger droplets of oil rose to the surface, while finer droplets remained suspended within the water column where they were neutrally buoyant (Camilli et al. 2010). In the water column, soluble components of the oil dissolved and were degraded by microbial populations (Atlas and Hazen 2011; Hazen et al. 2010). At the ocean surface, the oil was also subjected to evaporation and photodegradation of susceptible components. The overall magnitude and duration of the DWH incident prompted investigations to characterize the environmental distributions, the degradation, and the fate of the released petroleum. More than 100 separate cruises were involved in collecting more than 10,000 water samples in the offshore waters of the GoM as part of the collaboration between the US government and BP during the response to the spill and as part of Natural Resource Damage Assessment (NRDA) phases of the DWH investigation.

6.2   How Deep Subsea Spills Differ from Surface Releases

The fate of the DWH oil differed from the *Amoco Cadiz*, the *Exxon Valdez*, and most other well-studied spills because of the depth of the release and the transit of the oil through the water column before surfacing. Nevertheless, several subsea spills occurred before the DWH oil spill, and in each case, studies provided insight into the specific oil properties and environmental conditions that controlled the ultimate fate of the oil. On January 28, 1969, a leak occurred during the drilling of well A-21 on Platform A, approximately 6 miles off the coast of Santa Barbara, California, and oil and gas were released from approximately 190 ft. of water depth (Swift et al. 1969). Oil reached the surface of the ocean within 14 minutes, forming a slick, and despite occurring in relatively deep water, the oil from the release reached nearby shorelines a week later. The well was capped on February 7, but oil continued to seep from the sediments at increased rates in the general area for months. Because of the extended and variable nature of the release, the exact spill volume is not well understood but has been estimated to be between 4,000 and 9,600 tonnes (NOAA 2018b). Evaluations of the impact of the oil were complicated by the presence of naturally occurring hydrocarbon seeps in the area, which have been estimated to contribute more than 12 tonnes of oil per day to the local ecosystem (Hornafius et al. 1999). Additionally, local weather conditions at the time of the spill resulted in higher than average coastal runoff, which may have contributed to increased oil sedimentation through the presence of denser-than-water particulate matter (Straughan and Kolpack 1971).

The 1979 Ixtoc 1 well blowout in the Gulf of Campeche, Mexico, released more than 3 million barrels (360,000 tonnes) of a light crude oil (API gravity 34.5) at high pressure at a depth of approximately 50 m. The released material formed a water/gas/oil emulsion that reached the ocean surface where the lighter components were lost via burning and evaporation. The resulting oil/water mixture interacted with particles, and ultimately, a fraction became heavier than the surrounding water and was removed from the water column through sedimentation (Boehm and Fiest 1980). There was debate over the amount of oil which reached the sediments, which was estimated by difference (i.e., the unaccounted for oil) at up to 25% of the initial oil (Jernelöv and Lindén 1981). However, direct investigation using sediments and sediment trap results suggests that much less of the oil reached the seafloor via sedimentation (ca. 1–3%), and analysis of the alkane composition of material in the sediment traps found that sinking material was actually scouring previously dissolved lighter alkanes from the water column, as opposed to the heavier fraction found in the surfaced oil, emphasizing the importance of examining the specific composition of petroleum components to understand oil transport (Boehm and Fiest 1979).

However, not all spills from offshore platforms involve subsea releases. The Montara H1 well in the Timor Sea off the northwest coast of Australia blew out in 2009, releasing approximately 4,000 tonnes of oil over 74 days (Spies et al. 2017) and producing a surface slick that was reported to travel approximately 35 km from the location of the release (Burns and Jones 2016). However, unlike the previously discussed deepwater releases, the release of this oil was from the platform above the surface of the water, and hot oil and associated gas were exposed to the air before reaching the water surface. Despite the released oil being classified as a light crude (API gravity 34.6), the oil had a high wax content resulting in a "waxy" surface accumulation of oil. This spill exhibited the typical challenges of platform blow-outs—a very remote location, undetermined spill rate and volume, and the logistical challenges needed to stop the flow of oil, which required subsea drilling. Chemical dispersants also played an important role in reducing the amount of surface oiling and contributed to the lack of any observed impacts on Ashmore and Cartier Reefs (Storrie 2011). In the Montara spill, as with the Ixtoc blowout, only small amounts of oil were measured to have been transported to the deep sea sediment (Burns and Jones 2016).

The 1977 blowout of the Bravo well in the Ekofisk field of the North Sea also released oil about 20 m above the sea surface (NOAA 2018a). This release was of approximately 24,000 tonnes of oil (API gravity 35.7) (Audunson 1980). The release occurred from the platform about 20 m above the sea surface, and despite mild temperatures, extensive evaporation occurred as the oil was exposed to the atmosphere before entering the water, resulting in an estimated 30–40% loss and a measurable increase in density.

In contrast to the shallower spills, the high rate of release of oil and gas in deep waters in the DWH oil spill and its extended release period drove a large investigative focus on the fate, transport, partitioning, and degradation of oil in the water column and transport to the sea bottom (e.g., Boehm et al. 2016; Camilli et al. 2010; Hazen et al. 2010; Stout et al. 2016; Valentine et al. 2010; Wade et al. 2016; Reddy et al. 2012).

6.3 Properties of Oil Related to Fate and Transport

Oil and gasses released during a well blowout are a complex mixture of thousands of hydrocarbon and non-hydrocarbon compounds with distinct properties. Though oils released during blowouts have many chemical components and characteristics in common, the specific chemical mixture that makes up the oil and gas and the physical properties of the mixture of these components released during a subsurface spill differ based on the oil reservoir from which the release occurs, the specific properties of the oil released, and the release scenario (i.e., water depth and pressure, rate of release, temperature, etc.). These differences in characteristics affect the partitioning of the oil into dissolved and droplet phases, droplet sizes, the related movement, and the physical behavior of the different oil phases. These in turn influence the processes that affect the fate of the chemical components contained therein. Oils targeted during subsea drilling ideally have a high API gravity to recoup investment promptly as the refining of crude oil to gasoline requires fewer steps for high API gravity oils compared to low API gravity oils (American Petroleum Institute 2011). However, there is large variation in API gravity between wells. For example, oil released during the DWH spill was relatively light, with an API gravity of 37, compared to the 1969 Santa Barbara release, which came from a reservoir with an API gravity of approximately 26, resulting in a density closer to water (Fingas and Fieldhouse 2004). These differences can affect the buoyancy of the oil, as the lighter oils need to be relatively more weathered before they reach and exceed the density of water and begin to sink.

The individual components of the oil can be grouped into four categories based on chemical properties: saturates, aromatics, resins, and asphaltenes. Of these, the saturates and aromatics are of greatest concern in terms of human and ecological toxicity, while the asphaltenes play a major role in the emulsification of oil as the lighter oil components are lost to weathering (Fingas and Fieldhouse 2004). Chemicals in the saturated and aromatic fractions are more soluble and mobile in the environment to varying degrees and are more amenable to degradation. Resins and asphaltenes are relatively insoluble and recalcitrant.

At the point of the release in the subsea, petroleum will segregate into different phases based on chemical properties as well as temperature and pressure. Components can remain in the liquid oil (droplets), form a gas phase, form hydrates, or dissolve into the aqueous phase. These phases are affected differently by transport processes and can result in the differential transport of oil components. The physical processes governing this transport are discussed in detail in Lehr and Socolofsky (2020).

6.4 Fate of Oil and Gas: Understanding Where Oil Goes

The high pressures at the point of a deep-sea release result in the turbulent expulsion of oil and gas mixtures, as fast-rising gas bubbles interact with the oil to increase dispersion and decrease droplet size. In cases of minimal dispersion due to turbulence, oil will appear at the surface close to the blowout, undergoing minimal alteration (Gros et al. 2014). If the oil is dispersed, either via a high-energy release or through the application of chemical dispersants near the release point, the oil will form small microdroplets. These droplets alter the fate of the oil, changing both the transport of the droplet itself and the fate of the specific chemical components. Despite the oil being lighter than water, microdroplets are effectively neutrally buoyant, as the rise of the droplet through the water column is opposed by the resistance against its relatively large surface area. The result of these forces is that the droplets rise more slowly; are advected laterally; undergo dissolution, dilution, and biodegradation (i.e., partitioning and weathering); and may never reach the ocean surface. For droplets and aggregations of droplets that do eventually surface, the resulting surface slick will be more diffuse as the droplets are transported laterally during their rise. Droplets that never reach the surface can form a subsurface anomaly or diffuse "plume" of oil, which travels within the water column at a density boundary.

In contrast to shallower or surface releases, gasses released at depth are likely to dissolve into the water column and not be available to evaporate from a surface slick. For example, Reddy et al. (2012) summarize multiple studies which confirm that during the DWH oil spill, nearly all methane present in the released material was dissolved and remained present in the water column at approximately 1100 m depth and use that information to examine the retention of ethane and propane, which were also nearly completely dissolved at depth (Reddy et al. 2012). The longer contact time with the water column and higher surface area also result in the increased dissolution of volatile and soluble components of the liquid oil, resulting in higher dissolved concentrations and less release to the atmosphere. Larger droplets of oil rose rapidly to the surface, while smaller neutrally buoyant oil microdroplets (approximately 10–60 microns in diameter) remained in the 1000–1200 m depth range advecting with the water currents (Camilli et al. 2010). The enhanced formation of these small droplets resulting from dispersant application promoted the dissolution of the most soluble components of the oil (Reddy et al. 2012), and microbial action further degraded the petroleum hydrocarbons in the water column (Atlas and Hazen 2011; Hazen et al. 2010).

Spills that have been studied extensively resulting in large data collections, such as the DWH, *Ixtoc*, and *Exxon Valdez* spills, have shown that chemicals associated with oil spills, such as polycyclic aromatic hydrocarbons (PAHs), reach maximum water concentrations early after release and decrease rapidly with time and distance from the release point (Boehm and Fiest 1982; Boehm et al. 2007; Boehm et al. 2016). Within the water column, aggregated chemical parameters such as the "total concentration of PAHs" or sums of other oil components neither completely nor adequately characterize the important chemical exposure to organisms. As components dissolve into the aqueous fraction, the resulting aqueous profile is controlled by the relative solubilities in addition to the composition of the original oil. This phenomenon results in different profiles based on the relative amounts of dissolved and particulate oil, even from a single source. For example, filtration of two water samples collected during the DWH oil spill contained approximately 10 parts per billion (ppb) total PAH (TPAH) had very different distributions in the aqueous and particulate (filtered) phases (Fig. 6.1).

As discussed above, the vast majority of crude oil compounds have a specific density less than the surrounding water, meaning they will naturally float on the surface. However, weathering processes lead to the removal or alteration of the lighter/smaller compounds leaving the relatively heavier, more resistant, compounds behind. Although these compounds may still be less dense than the

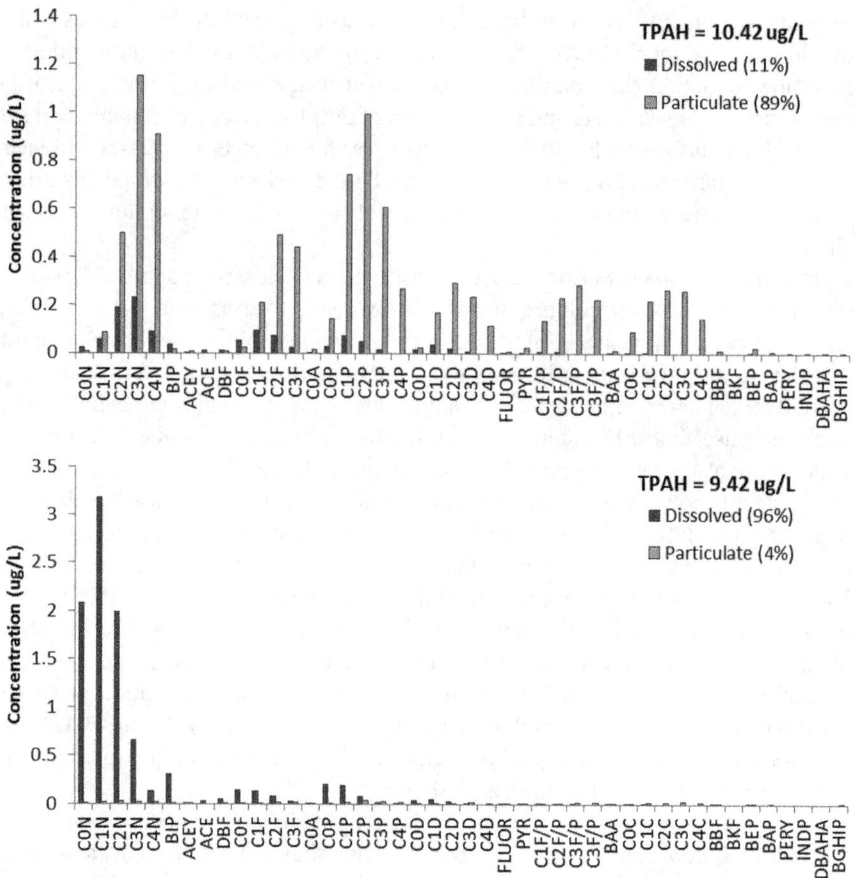

Fig. 6.1 Comparison of PAH profiles in two samples with similar TPAH concentrations but differing dissolved/particulate fractionation. (Reprinted with permission from IOSC (Murray and Boehm 2017))

surrounding water, they may be actively or passively absorbed onto particulate matter, either inorganic or biological.

Organic-rich particulate matter that forms and sinks within the water column is described as "marine snow" and is an important pathway for carbon cycling in the ocean. Marine snow aggregates can comprise materials such as microbial cells and their extracellular polymers, detritus, fecal material, and inorganic particles. During oil spills, oil can associate with marine snow aggregates, which may exceed 1 cm in diameter (Passow 2016). If the aggregate material exceeds the density of the surrounding water, it may settle out of the water column and oil may be transported to the sediments (Loh et al. 2014). For oil associated with marine snow or other settling materials to reach the seafloor, the sedimentation rate needs to be fast enough that natural degradation processes are unable to metabolize or remove the oil compounds during transit. The settling velocity depends on the density of the particles and their hydrodynamic diameter along with the ambient water temperature. Typical settling velocities are between 10 and 150 m day^{-1} in marine environments (McDonnell and Buesseler 2010), leading to close coupling between the surface and deep waters (Asper et al. 1992). The magnitude of this process has been debated, and no clear formula is available to estimate the likely flux from a given spill, although it is likely to be small compared to the other attenuation processes.

Several field measurement techniques have been employed to quantify the flux, density, and settling velocity of marine snow. Sediment traps are the most direct way to measure flux, as they are deployed for a known time and collect settling particles in the trap for analysis. However, sediment traps can be affected by degradation, mobile animals, and hydrodynamic forces (Buesseler et al. 2007). Marine snow flux can also be measured using a photographic apparatus in which a camera captures images at regular intervals and the images are processed to determine the number of unique particles in each image. More than 800 studies of marine snow flux measurements using sediment traps, filtration systems, or other techniques were published between 1979 and 2015 (Turner 2002, 2015). Laboratory studies have attempted to quantify the production and flux of marine snow in the presence of oil. For example, Passow and her colleagues (Passow 2016) used filtered seawater and natural bacterial assemblages in a bottle experiment to generate marine snow in the presence of weathered crude. However, other similar experiments were not able to produce marine snow, although oil was degraded (Ziervogel et al. 2014), highlighting the importance of the specific conditions. Additionally, in laboratory experiments, the long incubation times (weeks) and relatively high concentrations may not be relevant to field conditions, where mixing may prevent marine snow formation.

Measuring the amount of oil that has reached the sediments in an offshore oil spill provides information on the overall mass balance of the spill, as well as the potential injury to natural resources, but can be logistically challenging. For example, several different approaches were used to estimate the amount of oil reaching the sediments during and after the DWH spill, resulting in different estimates of the impact of oil on the deep sediments. In all cases, the analyses relied on obtaining sediment data from deep-sea locations, a process which is prone to uncertainty due to the patchiness of the oil transported to the seafloor and the very large number of samples required to understand this distribution over such a large geographical area. Direct approaches to estimating the oil sedimentation used comprehensive fingerprinting of sediment oil residues to identify locations containing oil and to confirm that the oil present was consistent with oil released from the Macondo wellhead. The oil fingerprinting was necessary to separate the spilled oil from other petroleum products present in these areas, such as seep oil or spilled oil from other sources. Independent analyses using this approach and a common, publicly available data set found very similar footprints of oil on the seafloor (Stout et al. 2016; Murray et al. 2017). Other approaches used proxies for the oil to estimate the seafloor impact. These methods can be less analytically intensive and can provide an estimate of the impact of oil, but differ from the fingerprinting methods in their specificity. The use of proxy measurements relies on assumptions that may not be applicable under all conditions. For example, hopane, a relatively recalcitrant petroleum biomarker, was used to delineate the seafloor impact of the DWH spill (Valentine et al. 2014). This approach was based on the assumptions that hopane does not degrade and that the only hopane present in the GoM sediments came from the DWH spill. However, the GoM has many active hydrocarbon seeps which contribute hopane to the water column and sediments (MacDonald et al. 2015), and using a single hopane compound to trace oil does not distinguish between spill and seep sources. If background hopane never degrades, 100% of the hopane ever released from the seeps would be present in the GoM, confounding interpretations. However, hopane has been suggested to degrade in this environment (Bagby et al. 2017), potentially complicating interpretations. As we will discuss below, the material containing the hopane on the seafloor is certainly not chemically similar to fresh crude oil—it is the highly degraded residue of that oil, and studies show that degradation began before deposition (Hazen et al. 2010; Valentine et al. 2010). Assuming that the hopane in the sediments represents fresh oil substantially overestimates the amount of oil that reached the seafloor. Another approach uses carbon isotopes to estimate oil flux to the seafloor based on the radiocarbon content of sediment samples as in the approach used by Chanton et al. (Chanton et al. 2012; Chanton et al. 2014). In these studies, the authors used the radioactive tracer ^{14}C to differentiate between relatively modern inputs that contain ^{14}C and fossil inputs such as crude oil that no longer contain any ^{14}C. However, this method does not discriminate between different fossil sources and cannot be reliably used to calculate oil mass balance and overall sedimentation. The presence of oil-derived carbon does not necessarily indicate the presence of any oil component in the sediment and may be the result of fully metabolized oil that has been incorporated into biomass that has sunk. This distinction is very important because one represents a potential injury (sediments containing toxic components of oil) and one does not (degraded oil-derived carbon in biomass).

6.5 Fate of Oil and Gas: Understanding the Degradation of Oil Components

One of the most significant findings of the early sampling from the DWH spill was the discovery of deep-sea anomalies or diffuse "plumes" of oil and dissolved gas (Camilli et al. 2010; Spier et al. 2013) and the rate of biodegradation of the dispersed and dissolved oil and gas within those plumes (Hazen et al. 2010; Kessler et al. 2011). The existence of "plumes" was not completely unexpected since the ocean, like the atmosphere, is often stratified, but the rate of biodegradation—oil half-lives of 1.2–6.1 days—was substantially faster than many expected. While later analysis suggests these rates may have been overestimated (Bagby et al. 2017), laboratory experiments have confirmed that such rates for very dilute oil are not unreasonable (Prince et al. 2017) and that the much slower rates measured after oil has beached, for example, following the *Exxon Valdez* spill (Bragg et al. 1994), are due to the limited surface area of more concentrated oil. The key determinants are the oil concentration (Prince et al. 2017), which was exceptionally dilute in the subsurface—almost all samples were lower than 250 ppb total petroleum hydrocarbon (TPH), 1 ppb TPAH (the sum of all the two- to six-ring aromatic hydrocarbons including their alkylated forms detectable by gas chromatography-mass spectrometry) (Wade et al. 2016; Boehm et al. 2016). Surface area and surface area (S)/volume(V) played an important role in the biodegradation process in that small oil droplets afford a large surface area (high S/V) for biodegradation to take place, while, by contrast, the surface oil slick's S/V was much smaller. For dilute dispersed oil with high S/V, seawater provides the necessary biologically available nutrients that oil does not provide—nitrogen, phosphorus, iron, etc.—as well as the requisite oxygen for aerobic biodegradation. It was no surprise that biodegradation was only marginally slower at great depths and low temperatures than nearer the ocean surface (Prince et al. 2016; Nguyen et al. 2018; Marietou et al. 2018). Biodegradation of oil at the surface was probably negligible until the S/V increased after the oil reached the shoreline (Atlas et al. 2015).

Determining the extent of dissolution of aromatic hydrocarbons from oil droplets, and distinguishing this from biodegradation in field samples, remains an area of active research and investigation. It is well known that the small aromatics benzene, toluene, ethylbenzene, and the xylenes dissolve out of oils and gasolines (Poulsen et al. 1992), and the oil/water partition coefficient increases significantly at high pressure in the presence of methane, so-called live oil (Jaggi et al. 2017). This presumably slows dissolution at depth, although of course the equilibrium partition coefficient does not reveal much about the kinetics of dissolution or about the effect of the essentially infinite dilution available as hydrocarbons leave an oil droplet for the aqueous phase. Modeling (Gros et al. 2016) suggests that high pressures dramatically favor the aqueous dissolution of $C_1 - C_4$ hydrocarbons and also influence the buoyancies of bubbles and droplets. These authors believe solubility was a far more important process than dispersion (Gros et al. 2017), although clearly both depend on the generation of small droplets, which can be facilitated by dispersant injection at source. What is clear is that the vast majority of hydrocarbons in dispersed crude oils are biodegraded within a few weeks when oil concentrations are a few parts per million (Prince et al. 2017), irrespective of their initial presence in solution or in droplets. This is the great benefit of using chemical dispersants (Prince 2015)—not only do dispersants protect birds and shorelines from oiling by surface slicks, they also deliver oil to the microbes that provide the only process that will remove the spilled oil from the biosphere. It is now clear that dispersants have no significant influence on hydrocarbon biodegradation rates once the oil is dispersed (Brakstad et al. 2018b) and that they themselves are degraded at similar rates to the hydrocarbons (Brakstad et al. 2018c; McFarlin et al. 2018).

It is well known that the GoM has thousands of oil seeps (MacDonald et al. 2015), so it is no surprise that a large "standing crop" of oil-degrading microbes was available to utilize the dispersed oil. A succession of genera proliferated as the spill progressed (Redmond and Valentine 2011; Dubinsky et al. 2013; Hu et al. 2017;

Yang et al. 2016), and their appetites altered the oil in a predictable fashion—*n*-alkanes and small aromatics were degraded most rapidly, followed by branched alkanes and unsubstituted PAH, followed by alkylated PAH (Prince et al. 2017; Brakstad et al. 2015; Prince et al. 2013; Wang et al. 2016). The biodegradation of the asphaltenes is less clear, not least because there are no good tools for characterizing their individual molecular structures. It may be that biodegradation is restricted to alkyl side chains (Hernández-López et al. 2015).

Among the last molecules to be degraded are the hopanes, and individual hopanes can serve as conserved internal markers of how much oil was originally present (Prince et al. 1994). The family of these molecules also provides a fingerprint that allows Macondo oil to be distinguished from other indigenous oils in the GoM (Stout et al. 2016; Murray et al. 2017). It is thus possible to clearly identify biodegradation (and other specific chemical processes, such as evaporation and photooxidation (Prince et al. 2003) in an oil sample—information that would not be available if the analysis was restricted to simple quantitation of analytes.

Biodegradation has profound effects on the density of the remaining oil. Most hydrocarbons are lighter than water and tend to float (Fig. 6.2), but some are denser than water and thus tend to sink—it is the average densities of the hydrocarbons and rather denser asphaltenes (Barrera et al. 2013) that determines the overall density of a crude oil. The n-alkanes are the least dense components (average density of nC_{15}–nC_{40} = 0.79), while unsubstituted aromatics and the asphaltenes are the densest (e.g., chrysene = 1.27, asphaltenes 1.2). Alkyl aromatics are less dense than their parents, at least as predicted from Fig. 6.2. For example, alkylation beyond four

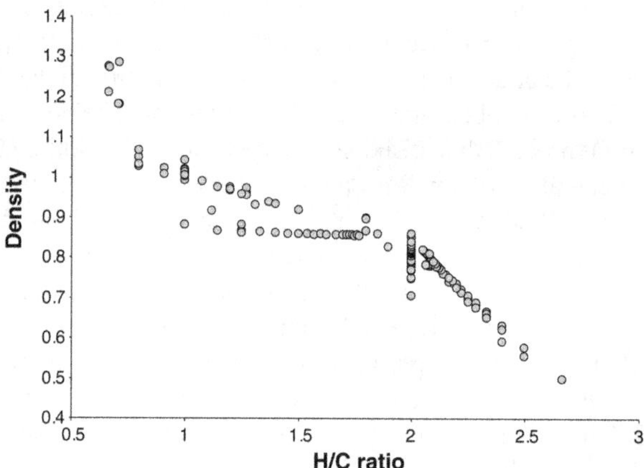

Fig. 6.2 Densities of petroleum hydrocarbons as a function of the H/C ratio. Density is complicated because it depends on molar volume, which in turn is dependent on structure (Wakeham et al. 2002). The densest hydrocarbon known is cuneane (C_8H_8) with a density of 1.58, but cubane and barrelene have densities of 1.29 and 1.03, while cyclooctatetraene has a density of 0.93 and styrene has a density of 0.91—all are C_8H_8 molecules, but since none are present in detectable amounts in crude oils, they are not included here. Benzene, C_6H_6, is the lightest molecule with an H/C ratio of 1—its density is 0.88. The lightest hydrocarbon on this graph is propane (H/C = 2.67) with a density of 0.50; the heaviest is chrysene (H/C of 0.67) with a density of 1.27

pendant carbons on phenanthrene likely lowers the density below 1, while six substituents likely do the same for chrysene. Although GC/MS only usually detects alkyl aromatics with up to four alkyl carbon substituents, the typical phenanthrene, for example, is said to have a total of 18 alkyl carbons; presumably with such diversity that they are not individually resolved by gas chromatography (Cho et al. 2017).

The initial Macondo oil had a density of 0.8, but as the alkanes were degraded, the average density increased. This was partially offset by the preferential degradation of parent PAH before alkylated forms, but the overall effect of biodegradation was to preferentially remove the least dense hydrocarbons, leaving a denser residue that would be neutrally buoyant or possibly sink. This is not a simple phenomenon. While the buoyant density of large drops of oil led them to rise to the surface and form a slick, the similar intrinsic buoyancy of smaller droplets was clearly overwhelmed by viscosity and turbulence to entrain them in the deep sea. Similarly, even though intrinsically denser after the biodegradation of the alkanes and alkyl aromatics, such tiny droplets may well have taken time to sink unless they became entrained with dense material such as suspended sediment particles (Sørensen et al. 2014) or marine snow (Brakstad et al. 2018a; Gutierrez et al. 2018; Passow 2016). As expected, oil residues in the sediment of the GoM within 5–8 km of the wellsite are very biodegraded (Stout and Payne 2016; Stout et al. 2016; Bagby et al. 2017). Quite likely some of the later biodegradation occurred after the partially degraded droplets were incorporated in the sediment, perhaps even anaerobically (Kimes et al. 2014), but the amount of oil residue was quite limited: if the residual oil is 80% biodegraded as suggested in Stout et al. (2016), their subsequent quantitation of the amount of oil residue in the sediment extrapolates to 4.4 g/m^2 (Stout et al. 2017).

Oil that promptly arrived at the surface formed large slicks, and biodegradation was likely minimal while photooxidation began (Aeppli et al. 2012). The latter was an important process for undispersed oil (Ward et al. 2018), but it is important to remember that a major goal of the response was to encourage dispersion and hence biodegradation—and most of the oil was dispersed as evidenced by its prompt disappearance from the surface after oil flow from the damaged well was stopped (Houma ICP Aerial Dispersant Group 2010). It is not yet clear what effect photooxidation has on the biodegradability of the oil; the polymerization of the aromatics (Prince et al. 2003) likely slows the process, but the unexpected generation of photooxidized saturates (Aeppli et al. 2012) may make them more bioavailable and speed biodegradation. The chemical signature related to the photodegradation of specific oil components is also helpful in tracing oil fate; for example, it can be identified in areas such as sediments, helping to define the pathway that the oil took before sedimentation.

An important ecological goal is to understand the food chain that develops after the hydrocarbons are initially degraded. The bloom of oil-degrading microbes is reasonably well characterized (Redmond and Valentine 2011; Dubinsky et al. 2013; Yang et al. 2016; Hu et al. 2017), but we do not know what fraction of the hydrocarbon was mineralized to water and CO_2 and what fraction was assimilated to microbial biomass. Under nutrient-limited conditions, methane can be converted to polyhydroxybutyrate with a yield of 45% on a weight basis (Helm et al. 2008), so a working assumption might be that primary metabolism converts 50% of the oil to biomass and that this conversion ratio is repeated as the carbon proceeds up the food web. Isotopic evidence indicates some of this petrocarbon was still circulating in the microbial food web for several years and also suggested that some methane metabolism stimulated nitrogen fixation at depth (Fernández-Carrera et al. 2016). In the long run, it is quite possible that the DWH spill will lead to more fish (Levy and Lee 1988).

6.6 Tracing the Fate of Oil in the Deep Sea: The Mass Balance

Ultimately, oil released in an offshore oil spill can form a slick, be transported to shorelines, volatilize to the atmosphere, dissolve or disperse within the water column, or be transported to the sediment (Fig. 6.3). Within each of these compartments, components of the oil can be degraded either by microbial action or photochemistry and/or may cause injury to the local ecosystem. To understand injury, in addition to knowing how much oil reached the sediment, it is important to understand the form of the oil which has reached the sediment and also to distinguish oil reaching the bottom from oil-derived carbon reaching the bottom. While, historically, the amount of oil reaching ocean sediments has been calculated using a

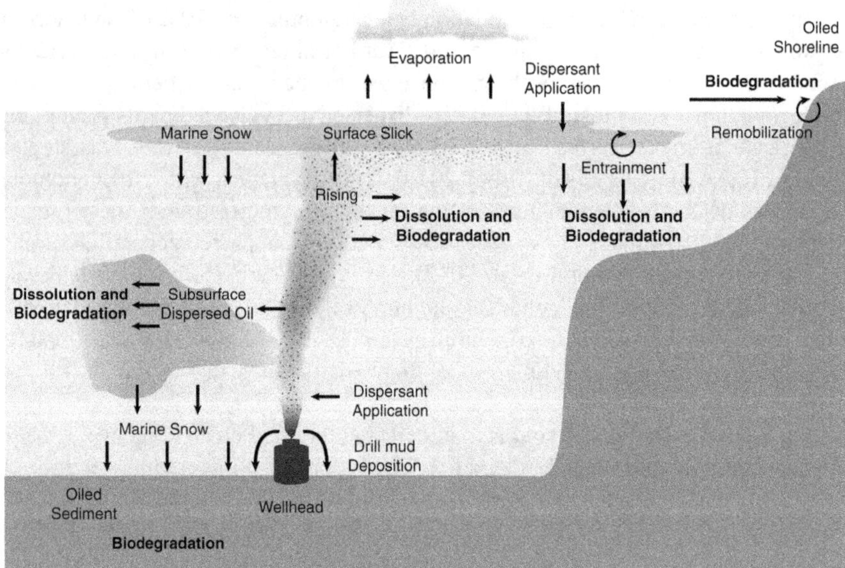

Fig. 6.3 Conceptual model of the fate and transport processes affecting oil in a subsea spill

mass balance approach (i.e., by difference), the large uncertainty in some measurements means that using the remainder to estimate sediment impacts is not useful for injury analysis, for example, in the Ixtoc overestimation by Jernelöv and Lindén (Jernelöv and Lindén 1981). Laboratory measurements can be used to test the mechanisms of marine snow formation but should be interpreted with caution, especially in cases where field concentrations of biological material or oil are exceeded. In the field, sediment traps can be used to determine the amount and nature of sinking oiled marine snow and can provide an estimate of the flux of oil to the sediment. While logistically challenging, it is also beneficial to directly measure petroleum in sediments to determine whether injury has occurred, and it is important that methods clearly differentiate between petroleum and petrocarbon.

In addition to discrete sample collection and laboratory analysis, new methods have emerged to follow oil in the ocean, including remote sensing techniques and oil spill models. Complex computational models are increasingly important in understanding the fate of spilled oil, as they can fill in gaps in time and space for which data are unavailable or unobtainable. However, these tools are limited based on the quality of the input data and ultimately rely on incorporating the field and experimental measurements discussed above. Models must recognize the speed of biodegradation of dispersed oil—oceanographic models that treat oil as if it were a persistent dye will massively mislead the public, even if traces of very biodegraded oil do indeed reach as far as the models suggest (Maltrud et al. 2010). While it is impossible to predict the exact biodegradation rate of a spilled oil in a given place and time, good experimental data, using realistic temperatures, organisms, and nutrient levels, are available to incorporate biodegradation processes into models, resulting in improved prediction of oil fates.

References

Aeppli C, Carmichael CA, Nelson RK, Lemkau KL, Graham WM, Redmond MC, Valentine DL, Reddy CM (2012) Oil weathering after the Deepwater Horizon disaster led to the formation of oxygenated residues. Environ Sci Technol 46(16):8799–8807

American Petroleum Institute (2011) Robust summary of information on substance: crude oil, CAS No. 8002-05-9

Asper VL, Deuser W, Knauer G, Lohrenz SE (1992) Rapid coupling of sinking particle fluxes between surface and deep ocean waters. Nature 357(6380):670

Atlas RM, Hazen TC (2011) Oil biodegradation and bioremediation: a tale of the two worst spills in US history. Environ Sci Technol 45:6709–6715

Atlas RM, Stoeckel DM, Faith SA, Minard-Smith A, Thorn JR, Benotti MJ (2015) Oil biodegradation and oil-degrading microbial populations in marsh sediments impacted by oil from the Deepwater horizon well blowout. Environ Sci Technol 49(14):8356–8366

Audunson T (1980) The fate and weathering of surface oil from the Bravo blowout. Mar Environ Res 3(1):35–61

Bagby SC, Reddy CM, Aeppli C, Fisher GB, Valentine DL (2017) Persistence and biodegradation of oil at the ocean floor following Deepwater Horizon. Proc Natl Acad Sci 114(1):E9–E18

Barrera D, Ortiz D, Yarranton H (2013) Molecular weight and density distributions of asphaltenes from crude oils. Energy Fuel 27(5):2474–2487

Boehm PD, Fiest DL (1979) Surface water column transport and weathering of petroleum hydrocarbons during the IXTOC-I blowout in the Bay of Campeche and their relation to surface oil and microlayer compositions. In: Proceedings of a symposium on preliminary results from the September, 1979. pp 267–338

Boehm PD, Fiest DL (1980) Aspects of the transport of petroleum hydrocarbons to the offshore benthos during the IXTOC-I blowout in the Bay of Campeche. In: Proceedings of Conference on Researcher/Pierce IXTOC-I Cruises. National Oceanographic and Atmospheric Administration, AOML, Miami

Boehm PD, Fiest DL (1982) Subsurface distributions of petroleum from an offshore well blowout. The Ixtoc I blowout, Bay of Campeche. Environ Sci Technol 16(2):67–74. https://doi.org/10.1021/es00096a003

Boehm PD, Neff JM, Page DS (2007) Assessment of polycyclic aromatic hydrocarbon exposure in the waters of Prince William Sound after the Exxon Valdez oil spill: 1989–2005. Mar Pollut Bull 54(3):339–356. https://doi.org/10.1016/j.marpolbul.2006.11.025

Boehm PD, Page DS, Brown JS, Neff JM, Bragg JR, Atlas RM (2008) Distribution and weathering of crude oil residues on shorelines 18 years after the Exxon Valdez spill. Environ Sci Technol 42(24):9210–9216

Boehm PD, Gundlach ER, Page DS (2013) The phases of an oil spill and scientific studies of spill effects. Cambridge University Press, New York

Boehm PD, Murray KJ, Cook LL (2016) Distribution and attenuation of polycyclic aromatic hydrocarbons in Gulf of Mexico seawater from the Deepwater Horizon oil accident. Environ Sci Technol 50(2):584–592

Bragg JR, Prince RC, Harner EJ, Atlas RM (1994) Effectiveness of bioremediation for the Exxon Valdez oil spill. Nature 368(6470):413

Brakstad OG, Nordtug T, Throne-Holst M (2015) Biodegradation of dispersed Macondo oil in seawater at low temperature and different oil droplet sizes. Mar Pollut Bull 93(1–2):144–152

Brakstad OG, Lewis A, Beegle-Krause C (2018a) A critical review of marine snow in the context of oil spills and oil spill dispersant treatment with focus on the Deepwater Horizon oil spill. Mar Pollut Bull 135:346–356

Brakstad OG, Ribicic D, Winkler A, Netzer R (2018b) Biodegradation of dispersed oil in seawater is not inhibited by a commercial oil spill dispersant. Mar Pollut Bull 129(2):555–561

Brakstad OG, Størseth TR, Brunsvik A, Bonaunet K, Faksness L-G (2018c) Biodegradation of oil spill dispersant surfactants in cold seawater. Chemosphere 204:290–293

Buesseler KO, Antia AN, Chen M, Fowler SW, Gardner WD, Gustafsson O, Harada K, Michaels AF, Rutgers van der Loeff M, Sarin M (2007) An assessment of the use of sediment traps for estimating upper ocean particle fluxes. J Mar Res 65(3):345–416

Burns KA, Jones R (2016) Assessment of sediment hydrocarbon contamination from the 2009 Montara oil blow out in the Timor Sea. Environ Pollut 211:214–225

Camilli R, Reddy CM, Yoerger DR, Van Mooy BA, Jakuba MV, Kinsey JC, McIntyre CP, Sylva SP, Maloney JV (2010) Tracking hydrocarbon plume transport and biodegradation at Deepwater Horizon. Science. https://doi.org/10.1126/science.1195223

Cho Y, Birdwell JE, Hur M, Lee J, Kim B, Kim S (2017) Extension of the analytical window for characterizing aromatic compounds in oils using a comprehensive suite of high-resolution mass spectrometry techniques and double bond equivalence versus carbon number plot. Energy Fuel 31(8):7874–7883

Chanton J, Cherrier J, Wilson R, Sarkodee-Adoo J, Bosman S, Mickle A, Graham W (2012) Radiocarbon evidence that carbon from the Deepwater Horizon spill entered the planktonic food web of the Gulf of Mexico. Environ Res Lett 7(4):045303

Chanton J, Zhao T, Rosenheim BE, Joye S, Bosman S, Brunner C, Yeager KM, Diercks AR, Hollander D (2014) Using natural abundance radiocarbon to trace the flux of petrocarbon to the seafloor following the Deepwater Horizon oil spill. Environ Sci Technol 49(2):847–854

Dubinsky EA, Conrad ME, Chakraborty R, Bill M, Borglin SE, Hollibaugh JT, Mason OU, M. Piceno Y, Reid FC, Stringfellow WT (2013) Succession of hydrocarbon-degrading bacteria in the aftermath of the Deepwater Horizon oil spill in the Gulf of Mexico. Environ Sci Technol 47(19):10860–10867

Fernández-Carrera A, Rogers K, Weber S, Chanton J, Montoya J (2016) Deep Water Horizon oil and methane carbon entered the food web in the Gulf of Mexico. Limnol Oceanogr 61(S1):S387

Fingas M, Fieldhouse B (2004) Formation of water-in-oil emulsions and application to oil spill modelling. J Hazard Mater 107(1–2):37–50

Gros J, Nabi D, Würz B, Wick LY, Brussaard CP, Huisman J, van der Meer JR, Reddy CM, Arey JS (2014) First day of an oil spill on the open sea: Early mass transfers of hydrocarbons to air and water. Environ Sci Technol 48(16):9400–9411

Gros J, Reddy CM, Nelson RK, Socolofsky SA, Arey JS (2016) Simulating gas–liquid– water partitioning and fluid properties of petroleum under pressure: implications for deep-sea blowouts. Environ Sci Technol 50(14):7397–7408

Gros J, Socolofsky SA, Dissanayake AL, Jun I, Zhao L, Boufadel MC, Reddy CM, Arey JS (2017) Petroleum dynamics in the sea and influence of subsea dispersant injection during Deepwater Horizon. Proc Natl Acad Sci 114(38):10065–10070

Gundlach ER, Boehm PD, Marchand M, Atlas RM, Ward DM, Wolfe DA (1983) The fate of Amoco Cadiz oil. Science 221(4606):122–129

Gutierrez T, Teske A, Ziervogel K, Passow U, Quigg A (2018) Microbial exopolymers: sources, chemico-physiological properties, and ecosystem effects in the marine environment. Front Microbiol 9:1822

Hazen TC, Dubinsky EA, DeSantis TZ, Andersen GL, Piceno YM, Singh N, Jansson JK, Probst A, Borglin SE, Fortney JL (2010) Deep-sea oil plume enriches indigenous oil-degrading bacteria. Science 330(6001):204–208

Helm J, Wendlandt KD, Jechorek M, Stottmeister U (2008) Potassium deficiency results in accumulation of ultra-high molecular weight poly-β-hydroxybutyrate in a methane-utilizing mixed culture. J Appl Microbiol 105(4):1054–1061

Hernández-López E, Ayala M, Vazquez-Duhalt R (2015) Microbial and enzymatic biotransformations of asphaltenes. Pet Sci Technol 33(9):1017–1029

Hornafius JS, Quigley D, Luyendyk BP (1999) The world's most spectacular marine hydrocarbon seeps (Coal Oil Point, Santa Barbara Channel, California): quantification of emissions. J Geophys Res Oceans 104(C9):20703–20711

Houma ICP Aerial Dispersant Group (2010) After action report Deepwater horizon MC252 aerial dispersant response

Hu P, Dubinsky EA, Probst AJ, Wang J, Sieber CM, Tom LM, Gardinali PR, Banfield JF, Atlas RM, Andersen GL (2017) Simulation of Deepwater Horizon oil plume reveals substrate specialization within a complex community of hydrocarbon degraders. Proc Natl Acad Sci 114(28):7432–7437

Jaggi A, Snowdon RW, Stopford A, Radović JR, Oldenburg TB, Larter SR (2017) Experimental simulation of crude oil-water partitioning behavior of BTEX compounds during a deep submarine oil spill. Org Geochem 108:1–8

Jernelöv A, Lindén O (1981) Ixtoc I: a case study of the world's largest oil spill. Ambio 10(6):299–306

Kessler JD, Valentine DL, Redmond MC, Du M, Chan EW, Mendes SD, Quiroz EW, Villanueva CJ, Shusta SS, Werra LM, Yvon-Lewis SA, Weber TC (2011) A persistent oxygen anomaly reveals the fate of spilled methane in the deep Gulf of Mexico. Science 331(6015):312–315. https://doi.org/10.1126/science.1199697

Kimes NE, Callaghan AV, Suflita JM, Morris PJ (2014) Microbial transformation of the Deepwater Horizon oil spill—past, present, and future perspectives. Front Microbiol 5:603

Lehr W, Socolofsky SA (2020) The importance of understanding fundamental physics and chemistry of deep oil blowouts (Chap. 2). In: Murawski SA, Ainsworth C, Gilbert S, Hollander D, Paris CB, Schlüter M, Wetzel D (eds) Deep oil spills: facts, fate, effects. Springer, Cham

Levy E, Lee K (1988) Potential contribution of natural hydrocarbon seepage to benthic productivity and the fisheries of Atlantic Canada. Can J Fish Aquat Sci 45(2):349–352

Loh A, Shim WJ, Ha SY, Yim UH (2014) Oil-suspended particulate matter aggregates: Formation mechanism and fate in the marine environment. Ocean Sci J 49(4):329–341

MacDonald IR, Garcia-Pineda O, Beet A, Asl SD, Feng L, Graettinger G, French-McCay D, Holmes J, Hu C, Huffer F (2015) Natural and unnatural oil slicks in the Gulf of Mexico. J Geophys Res Oceans 120(12):8364–8380

Maltrud M, Peacock S, Visbeck M (2010) On the possible long-term fate of oil released in the Deepwater Horizon incident, estimated using ensembles of dye release simulations. Environ Res Lett 5(3):035301

Marietou A, Chastain R, Scoma A, Hazen TC, Bartlett DH (2018) The effect of hydrostatic pressure on enrichments of hydrocarbon degrading microbes from the Gulf of Mexico following the Deepwater Horizon oil spill. Front Microbiol 9:808

McDonnell AM, Buesseler KO (2010) Variability in the average sinking velocity of marine particles. Limnol Oceanogr 55(5):2085–2096

McFarlin KM, Perkins MJ, Field JA, Leigh MB (2018) Biodegradation of crude oil and Corexit 9500 in arctic seawater. Front Microbiol 9:1788

Mille G, Munoz D, Jacquot F, Rivet L, Bertrand J-C (1998) The Amoco Cadiz oil spill: evolution of petroleum hydrocarbons in the Ile Grande salt marshes (Brittany) after a 13-year period. Estuar Coast Shelf Sci 47(5):547–559

Murray KJ, Boehm PD (2017) Towards an understanding of the evolution (Fate and Transport) of the 2010 Deepwater Horizon oil spill. In: International oil spill conference proceedings, 2017. International oil spill conference, pp 536–555

Murray KJ, Brown JS, Cook LL, Boehm PD (2017) Fingerprinting of weathered oil residues from the Deepwater Horizon oil spill: the importance of multiple lines of investigation. In: International oil spill conference proceedings, 2017. International oil spill conference, pp 3051–3070

Neff JM, Bence AE, Parker KR, Page DS, Brown JS, Boehm PD (2006) Bioavailability of polycyclic aromatic hydrocarbons from buried shoreline oil residues thirteen years after the Exxon Valdez oil spill: a multispecies assessment. Environ Toxicol Chem 25(4):947–961

Nguyen UT, Lincoln SA, Juárez AGV, Schedler M, Macalady JL, Müller R, Freeman KH (2018) The influence of pressure on crude oil biodegradation in shallow and deep Gulf of Mexico sediments. PLoS One 13(7):e0199784

NOAA (2018a) Ekofisk Bravo oil field. https://incidentnews.noaa.gov/incident/6237. Accessed 19 Aug 2018

NOAA (2018b) Santa Barbara well blowout. https://incidentnews.noaa.gov/incident/6206#!513759. Accessed 19 Aug 2018

Owens EH, Sergy GA (2003) The development of the SCAT process for the assessment of oiled shorelines. Mar Pollut Bull 47(9–12):415–422

Page DS, Foster JC, Fickett PM, Gilfillan ES (1989) Long-term weathering of Amoco Cadiz oil in soft intertidal sediments. In: International oil spill conference, 1989. American Petroleum Institute, pp 401–405

Passow U (2016) Formation of rapidly-sinking, oil-associated marine snow. Deep Sea Res Part II Top Stud Oceanogr 129:232–240. https://doi.org/10.1016/j.dsr2.2014.10.001

Poulsen M, Lemon L, Barker JF (1992) Dissolution of monoaromatic hydrocarbons into groundwater from gasoline-oxygenate mixtures. Environ Sci Technol 26(12):2483–2489

Prince RC (2015) Oil spill dispersants: boon or bane? Environ Sci Technol 49(11):6376–6384

Prince RC, Elmendorf DL, Lute JR, Hsu CS, Haith CE, Senius JD, Dechert GJ, Douglas GS, Butler EL (1994) 17. alpha.(H)-21. beta.(H)-hopane as a conserved internal marker for estimating the biodegradation of crude oil. Environ Sci Technol 28(1):142–145

Prince RC, Garrett RM, Bare RE, Grossman MJ, Townsend T, Suflita JM, Lee K, Owens EH, Sergy GA, Braddock JF (2003) The roles of photooxidation and biodegradation in long-term weathering of crude and heavy fuel oils. Spill Sci Technol Bull 8(2):145–156

Prince RC, McFarlin KM, Butler JD, Febbo EJ, Wang FC, Nedwed TJ (2013) The primary biodegradation of dispersed crude oil in the sea. Chemosphere 90(2):521–526

Prince RC, Nash GW, Hill SJ (2016) The biodegradation of crude oil in the deep ocean. Mar Pollut Bull 111(1–2):354–357

Prince RC, Butler JD, Redman AD (2017) The rate of crude oil biodegradation in the sea. Environ Sci Technol 51(3):1278–1284

Reddy CM, Arey JS, Seewald JS, Sylva SP, Lemkau KL, Nelson RK, Carmichael CA, McIntyre CP, Fenwick J, Ventura GT (2012) Composition and fate of gas and oil released to the water column during the Deepwater Horizon oil spill. Proc Natl Acad Sci 109(50):20229–20234

Redmond MC, Valentine DL (2011) Natural gas and temperature structured a microbial community response to the Deepwater Horizon oil spill. Proceedings of the National Academy of Sciences: 201108756

SL Ross Environmental Research (2010) Spill related properties of MC 252 crude oil, sample ENT-052210-178. Appendix 8 to oil budget calculator: Deepwater Horizon. Technical documentation. A report to the national incident command

Sørensen L, Melbye AG, Booth AM (2014) Oil droplet interaction with suspended sediment in the seawater column: influence of physical parameters and chemical dispersants. Mar Pollut Bull 78(1–2):146–152

Spaulding M, Li Z, Mendelsohn D, Crowley D, French-McCay D, Bird A (2017) Application of an integrated blowout model system, OILMAP DEEP, to the Deepwater Horizon (DWH) spill. Mar Pollut Bull 120(1–2):37–50

Spier C, Stringfellow WT, Hazen TC, Conrad M (2013) Distribution of hydrocarbons released during the 2010 MC252 oil spill in deep offshore waters. Environ Pollut 173:224–230

Spies RB, Mukhtasor M, Burns KA (2017) The montara oil spill: a 2009 well blowout in the Timor Sea. Arch Environ Contam Toxicol 73(1):55–62

Storrie J (2011) Montara wellhead platform oil spill–a remote area response. In: International oil spill conference proceedings (IOSC), 2011. American Petroleum Institute, p abs159

Stout SA, Payne JR (2016) Macondo oil in deep-sea sediments: part 1–sub-sea weathering of oil deposited on the seafloor. Mar Pollut Bull 111(1–2):365–380

Stout SA, Payne JR, Ricker RW, Baker G, Lewis C (2016) Macondo oil in deep-sea sediments: Part 2—distribution and distinction from background and natural oil seeps. Mar Pollut Bull 111(1–2):381–401

Stout SA, Rouhani S, Liu B, Oehrig J, Ricker RW, Baker G, Lewis C (2017) Assessing the footprint and volume of oil deposited in deep-sea sediments following the Deepwater Horizon oil spill. Mar Pollut Bull 114(1):327–342

Straughan D, Kolpack RL (1971) Biological and oceanographical survey of the Santa Barbara Channel oil spill, 1969–1970, vol 1. Allan Hancock Foundation, University of Southern California

Swift W, Touhill C, Haney W, Nakatani R, Peterson P (1969) Review of Santa Barbara Channel Oil Pollution Incident. BATTELLE-NORTHWEST RICHLAND WASH PACIFIC NORTHWEST LAB

Turner JT (2002) Zooplankton fecal pellets, marine snow and sinking phytoplankton blooms. Aquat Microb Ecol 27(1):57–102

Turner JT (2015) Zooplankton fecal pellets, marine snow, phytodetritus and the ocean's biological pump. Prog Oceanogr 130:205–248

U.S. Senate (2010) Review of the use of dispersants in response to the Deepwater Horizon oil spill. Subcommittee of the Committee on Appropriations, One Hundred Eleventh Congress, Second Session edn. U.S. Government Printing Office, Washington, DC

Valentine DL, Kessler JD, Redmond MC, Mendes SD, Heintz MB, Farwell C, Hu L, Kinnaman FS, Yvon-Lewis S, Du M (2010) Propane respiration jump-starts microbial response to a deep oil spill. Science 330(6001):208–211

Valentine DL, Fisher GB, Bagby SC, Nelson RK, Reddy CM, Sylva SP, Woo MA (2014) Fallout plume of submerged oil from Deepwater Horizon. Proc Natl Acad Sci 111(45):15906–15911

Wade TL, Sericano JL, Sweet ST, Knap AH, Guinasso NL Jr (2016) Spatial and temporal distribution of water column total polycyclic aromatic hydrocarbons (PAH) and total petroleum hydrocarbons (TPH) from the Deepwater Horizon (Macondo) incident. Mar Pollut Bull 103(1–2):286–293

Wakeham WA, Cholakov GS, Stateva RP (2002) Liquid density and critical properties of hydrocarbons estimated from molecular structure. J Chem Eng Data 47:559–570

Wang J, Sandoval K, Ding Y, Stoeckel D, Minard-Smith A, Andersen G, Dubinsky EA, Atlas R, Gardinali P (2016) Biodegradation of dispersed Macondo crude oil by indigenous Gulf of Mexico microbial communities. Sci Total Environ 557:453–468

Ward CP, Sharpless CM, Valentine DL, French-McCay DP, Aeppli C, White HK, Rodgers RP, Gosselin KM, Nelson RK, Reddy CM (2018) Partial photochemical oxidation was a dominant fate of Deepwater Horizon surface oil. Environ Sci Technol 52(4):1797–1805

Yang T, Nigro LM, Gutierrez T, Joye SB, Highsmith R, Teske A (2016) Pulsed blooms and persistent oil-degrading bacterial populations in the water column during and after the Deepwater Horizon blowout. Deep-Sea Res II Top Stud Oceanogr 129:282–291

Ziervogel K, D'souza N, Sweet J, Yan B, Passow U (2014) Natural oil slicks fuel surface water microbial activities in the northern Gulf of Mexico. Front Microbiol 5:188

Chapter 7
Biodegradation of Petroleum Hydrocarbons in the Deep Sea

Joel E. Kostka, Samantha B. Joye, Will Overholt, Paul Bubenheim, Steffen Hackbusch, Stephen R. Larter, Andreas Liese, Sara A. Lincoln, Angeliki Marietou, Rudolf Müller, Nuttapol Noirungsee, Thomas B. P. Oldenburg, Jagoš R. Radović, and Juan Viamonte

Abstract The *Deepwater Horizon* (DWH) discharge is unique in that it represents the first large spill that occurred in the deep sea, and unparalleled volumes of chemical dispersant were applied during emergency response efforts. Thus, the DWH incident raised new challenges with regard to predictions of petroleum hydrocarbon (PHC) biodegradation and the fate of discharged hydrocarbons in the deep sea, which is permanently cold (~4 °C) and exposed to high hydrostatic pressure (1 MPa per 100 m). Although extensive information is available on the rates and controls of

J. E. Kostka (✉)
Georgia Institute of Technology, School of Biology and Earth & Atmospheric Sciences, Atlanta, GA, USA
e-mail: joel.kostka@biology.gatech.edu

S. B. Joye
University of Georgia, Department of Marine Sciences, Athens, GA, USA
e-mail: mjoye@uga.edu

W. Overholt
Friedrich Schiller University Jena, Institute of Biodiversity, Jena, Germany

P. Bubenheim · S. Hackbusch · A. Liese · R. Müller · N. Noirungsee
Hamburg University of Technology, Institute of Technical Biocatalysis, Hamburg, Germany
e-mail: paul.bubenheim@tuhh.de; steffen.hackbusch@tuhh.de; liese@tuhh.de; ru.mueller@tu-harburg.de; nuttapol.noirungsee@tuhh.de

S. R. Larter · T. B. P. Oldenburg · J. R. Radović · J. Viamonte
University of Calgary, PRG, Department of Geoscience, Calgary, AB, Canada
e-mail: slarter@ucalgary.ca; toldenbu@ucalgary.ca; jagos.radovic@ucalgary.ca; juan.viamonte@tuhh.de

S. A. Lincoln
The Pennsylvania State University, Department of Geosciences, University Park, PA, USA
e-mail: slincoln@psu.edu

A. Marietou
Aarhus University, Department of Bioscience-Microbiology, Aarhus, Denmark
e-mail: a.marietou@bios.au.dk

S. A. Murawski et al. (eds.), *Deep Oil Spills*,
https://doi.org/10.1007/978-3-030-11605-7_7

PHC biodegradation in marine environments, relatively few studies have been conducted under conditions resembling the deep sea. In particular, hydrostatic pressure is a key environmental parameter that has been largely overlooked in biodegradation studies, due to methodological challenges and the difficulty to obtain samples. Considering the rapid expansion of oil and gas drilling into deeper waters, there is an urgent need to improve understanding of the influence of low temperature and high pressure on biodegradation in order to better constrain the fate of hydrocarbons in the deep sea. This chapter addresses the current understanding of deep sea PHC biodegradation, highlighting discoveries made during the scientific response to the DWH disaster.

Keywords Biodegradation · Deep sea · Pressure · Temperature · Petroleum hydrocarbon · Hydrocarbon-degrading bacteria

7.1 Introduction

Similar to the breakdown of natural organic matter, biodegradation mediated by indigenous microbial communities is the ultimate fate of the majority of petroleum hydrocarbons (oil and gas) that enter the marine environment (Leahy and Colwell 1990; Hazen and Prince 2015; Joye et al. 2016). The fate and transport of discharged oil are in turn determined by a complex interplay between hydrocarbon chemistry, the microbial food web, and ambient oceanographic processes including dispersion, dilution, dissolution, advection by ocean currents, particle flocculation and aggregation, sedimentation, and evaporation, along with biodegradation (Fig. 7.1). A suite of environmental parameters limits the capacity and efficiency of microbial communities to degrade petroleum hydrocarbons in marine ecosystems. All of the same environmental pressures that impact microbial food webs are in play during oil discharge including temperature, pressure, oxygen and nutrient availability, and physical or chemical form of the oil (Head et al. 2006).

Defined as the region below 200 m of water depth, the deep sea is cold, dark, and exposed to incredibly high hydrostatic pressures. Although it comprises the largest habitat for life on the planet and covers the majority of Earth's surface area, the lack of light precludes photosynthetic production, and biological communities are dependent upon sedimented organic matter originating from the surface ocean. Further, temperatures at the seafloor approach 3–4 °C, and hydrostatic pressure increases with depth (1.0 MPa per 100 m) (Jørgensen and Boetius 2007). Thus, in general, the deep ocean can be considered as a biological desert as well as an extreme environment for life.

The *Deepwater Horizon* (DWH) disaster represents the largest accidental marine oil spill in history. One of the unique characteristics of the DWH spill is the depth at which it occurred, approximately one mile below the sea surface. Spanning 3 months, approximately 4.1–4.9 million barrels of crude oil were discharged into

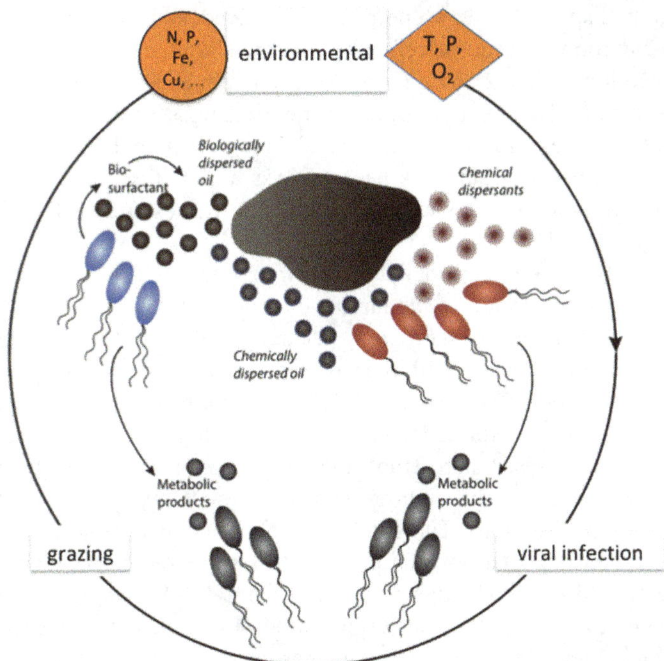

Fig. 7.1 Ecology of oil, dispersed oil, and dispersant biodegradation. Hydrocarbon-oxidizing microbes with the capability to produce biosurfactants to facilitate oil degradation are shown in blue. Environmental parameters, shown in orange, that regulate biodegradation include temperature, pressure, and nutrient and electron acceptor availability. It remains a question as to whether the activity of these microorganisms is stimulated or inhibited by chemical dispersants. Different types of hydrocarbon degraders, shown in red, have the ability to degrade chemically dispersed oil as well as dispersants (e.g., *Colwellia* sp. RC25). Secondary metabolite consumers of compounds produced during oil biodegradation, for which dispersant impacts are largely unknown, are shown in gray. Parts of this network (nutrient availability, viruses, and grazers) likely influence all the above types of microorganisms. (Illustration based on Head et al. (2006) and modified from Joye et al. (2016))

the deep ocean along with gaseous hydrocarbons which represented ~30–50% of the mass of liquid hydrocarbons, primarily as methane (McNutt et al. 2012; Joye 2015). Approximately half of all discharged hydrocarbons were entrained in the deep ocean in several intrusion layers, termed "deep plumes," from 900 to 1200 m water depth due to kinetic fractionation (Valentine et al. 2012). These deep subsurface plumes contained all the soluble gaseous hydrocarbons, monoaromatic hydrocarbons (BTEX), as well as larger molecular weight insoluble compounds that were sequestered in small neutrally buoyant droplets (Valentine et al. 2012). Furthermore, emergency responders injected 2.1 million gallons of chemical dispersant into the rising hydrocarbon plume at the wellhead. Of the oil discharged from MC252, 17% was recovered by drill or skimming ships, 29% was naturally or chemically dispersed, 23% evaporated or dissolved, and the remaining 23% was left in the environment sequestered in both deep benthic and shallow coastal sediments (McNutt et al. 2012). Within the water column and seafloor sediments, dissolved and dispersed hydrocarbons induced rapid changes in microbial communities (Kleindienst et al. 2015a; Joye et al. 2016). While the microbial response to DWH oil in planktonic ecosystems has been extensively studied, less information is available on the microbial response in sediments (Kimes et al. 2014; King et al. 2015; Joye et al. 2016).

A large body of literature, including laboratory and field studies, is devoted to hydrocarbon biodegradation in the marine environment (Hazen and Prince 2015). However, despite tremendous progress inspired by the DWH spill, the fate and transport of oil are underexplored in deep ocean ecosystems in comparison to their shallow-water counterparts. Historically, the majority of hydrocarbon biodegradation studies have been performed in the laboratory, with pure cultures or enrichment cultures under conditions that resemble the surface ocean, and relatively few studies have been conducted under high-pressure and low-temperature conditions that mimic deepwater conditions. This fundamental gap in the understanding of microbial hydrocarbon degradation is in contrast to the petroleum industry's trend of increasing oil and gas production from ultradeep (>1500 m) wells and the risk of another deep sea oil well blowout. Application of chemical dispersants and their influence on biodegradation or weathering has yet to be interrogated across the full range of oceanographic conditions observed in areas of oil exploration/production. In addition, more information is available on the environmental controls of hydrocarbon degradation under conditions relevant to marine water columns in comparison to the seafloor or sediments. This lack of knowledge of the impacts of oceanographic controls, especially in the deep sea, acts as a critical obstacle to the effective parameterization of oil plume models.

7.2 Biodegradation in the Water Column

7.2.1 Rate of Liquid and Gaseous Hydrocarbon Biodegradation in the Water Column

Hydrocarbon degradation in the water column is limited by a complex and interacting suite of factors (Fig. 7.1). Hydrocarbon solubility exerts a strong regulatory force on hydrocarbon degradation rates because hydrocarbons are not soluble in the aqueous phase (Joye et al. 2018). In the presence of ample hydrocarbons, genomic potential, electron acceptor (i.e., oxygen) availability, and nutrient and metal availability regulate microbial abundance and, hence, activity (Joye et al. 2016, 2018). During the DWH spill, oil and low molecular weight alkanes, like methane, were concentrated in the deepwater plume (Joye 2015), and additional oil was present in surface oil slicks. No direct rates of oil degradation in the environment are available (Joye et al. 2014; Joye 2015), which makes it extremely difficult to assess the magnitude and efficiency of oil biodegradation. Methane oxidation rates were measured rather extensively, and available data suggests that microbial methane oxidation rates were sluggish and ineffective (Valentine et al. 2010; Crespo-Medina et al. 2014).

Estimates of hydrocarbon oxidation rates derived from deepwater oxygen anomalies support the notion that hydrocarbon degradation rates were inefficient; e.g., measured hydrocarbon loads in the deepwater plumes could have driven more extensive oxygen depletion, had there been efficient oxidation (Joye et al. 2011). Given the paucity of field data on hydrocarbon degradation rates (i.e., direct measurements), it is extremely difficult to constrain the fate of oil and gas discharged during the DWH spill (Joye 2015). Best estimates of water column hydrocarbon oxidation suggest that between 43% and 61% of the discharged hydrocarbons were oxidized (Joye 2015). In the case of a future offshore marine oil spill, it is imperative that cutting-edge approaches be employed to constrain hydrocarbon oxidation rates across the water column (e.g., Sibert et al. 2017).

7.2.2 Microbial Community Changes During the Spill, Pre-spill, and Post-spill

The "baseline" microbial community in the Gulf water column contains a significant fraction of hydrocarbon-degrading microorganisms, mostly affiliated with the *Gammaproteobacteria* (Kleindienst et al. 2015c). This community is perpetuated by widespread, diffuse natural seepage of hydrocarbons into the water column (Joye et al. 2014). Prior to the oil spill, *Gammaproteobacteria* accounted for up to ~23% of SSU rRNA gene sequence reads, and about 7% of those sequences affiliated specifically with hydrocarbon-degrading microorganisms (Yang et al. 2016; Kleindienst et al. 2015a). The metabolic requirements of oil-degrading microbes vary, but early in the incident within the deepwater plume, *Gammaproteobacteria* became highly enriched, increasing in abundance to account for 43–98% of sequence reads. The vast majority of the organisms that bloomed were hydrocarbon degraders.

The most abundant *Gammaproteobacteria* early in the event were *Oceanospirillales*, *Colwellia*, and *Cycloclasticus* (Hazen et al. 2010; Mason et al. 2012). Methylotrophs and methanotrophs were less abundant, but *Methylophaga* reached levels up to 7% of all sequence reads by September 2010. The *Bacteroidetes* also increased in abundance late in the incident. The *Methylophaga* and *Bacteroidetes* are likely secondary consumers of intermediate metabolites instead of primary degraders of hydrocarbons. The ability of gammaproteobacterial oil degraders to respond rapidly to oil infusion underscores their critical environmental role. Even though their population biomass is constrained under "background" conditions, sufficient metabolic diversity and functional efficiency exist to allow a rapid response to increased substrate availability.

The majority of studies used patterns of SSU rRNA gene sequence abundance to infer patterns of metabolism (Kessler et al. 2011; Dubinsky et al. 2013). While some microbial metabolic clades vary in ways that could suggest changes in substrate abundance (Dubinsky et al. 2013), a metagenomic approach allows for specific microbial populations to be linked to specific metabolic pathways of hydrocarbon degradation (Rodriguez-R et al. 2015). As noted previously, the rates of hydrocarbon degradation in the water column are debated. Camilli et al. (2010) suggest very slow rates of hydrocarbon metabolism, while Hazen et al. (2010) suggest rapid turnover. Since degradation rates of major hydrocarbon classes were not determined directly under in situ (temperature and pressure) conditions and were instead inferred or estimated using indirect approaches, assessments of structure-function relationships are problematic.

7.2.3   The Influence of Dispersants on Microbial Community and Biodegradation

The impact of chemical dispersants on rates of hydrocarbon degradation is hotly debated, and the literature contains conflicting reports (Kleindienst et al. 2015a, b; Ferguson et al. 2017; Joye et al. 2016; Rahsepar et al. 2016). While dispersants increase the amount of oil partitioned into the aqueous phase, it is unclear whether dispersants increase oil biodegradation rates, which is a requirement for their efficacy. A consistent body of evidence documenting dispersant-mediated stimulation of oil degradation under realistic environmental conditions is lacking. Thus, a great deal of this variability likely stems from the different experimental approaches that were employed, such as addition of nutrients to samples that would otherwise be nutrient limited. Given the inconsistency in available reports, it is imperative to develop new approaches that allow investigations to be carried out under conditions that closely match those in the environment. The oil used in the experiment also plays an important role and may explain discrepancies between published reports. For example, few studies matched DWH field deepwater plume dissolved dispersant (DOSS) and oil concentrations (as TPH) (e.g., Kleindienst et al. 2015a), but when conditions mimicked in situ conditions, no stimulation of oil biodegradation by dispersant addition was observed.

7.3   Biodegradation in Sediments

A unique aspect of the DWH spill is that it injected large amounts of oil into the deep sea, 1.5 km below the sea surface, with approximately 14% of the oil deposited onto the sediments (Chanton et al. 2012; Valentine et al. 2014; Bagby et al. 2017;

Romero et al. 2017). Similar to planktonic environments, the biodegradation of hydrocarbons in deep sea sediments is controlled by the abovementioned suite of environmental parameters including temperature, pressure, oxygen and nutrient availability, and physical or chemical form of the oil (Fig. 7.1) (Head et al. 2006). Oxygen is consumed rapidly by microbial activity in sediments, and it is resupplied slowly by diffusion or intermittently by bioturbation in the deep sea. Thus, sediments differ in that petroleum hydrocarbons (PHCs) become buried in an environment that is limited in the supply of oxygen and other electron acceptors. In addition, rates of biodegradation were shown to inversely correlate with oil concentration, and since sediments serve as a repository for oil deposition, oil concentrations are elevated, and slower rates are to be expected (Prince et al. 2017).

Understanding of the metabolic pathways of hydrocarbon degradation under anoxic conditions remains in its infancy (Meckenstock et al. 2016), and this represents a key knowledge gap for predicting the long-term fate of recalcitrant oil compounds in fine-grained sediments that cover much of the seafloor (Shin et al. 2019; Shin 2018). Activation of hydrocarbon molecules during degradation is energetically demanding for microorganisms in the absence of oxygen, and thus anaerobic hydrocarbon degradation is much slower in comparison to aerobic metabolism. Since aerobic hydrocarbon-degrading bacteria preferentially utilize short-chain n-alkanes or simple polycyclic aromatic hydrocarbons (PAHs) more readily, PHCs buried in anoxic sediments tend to be enriched in more recalcitrant compounds, further exacerbating degradation processes (Shin et al. 2019; Shin 2018). Once buried in the anoxic zone, microorganisms metabolize hydrocarbons through respiration pathways using nitrate, iron, or sulfate as their terminal electron acceptor or by fermentation. Since sulfate is the most abundant electron acceptor present in marine sediments, anaerobic hydrocarbon degradation coupled to sulfate reduction is presumably quantitatively more important than degradation coupled to other electron acceptors.

The schematic in Fig. 7.2 illustrates how long-term oil degradation is controlled by oxygen depletion in surficial sediments. The rate-limiting steps are likely mass transport (principally diffusion), both of petroleum species within the oil-contaminated zone and also nutrients and electron acceptors within the more highly water-saturated burnout zones above and below oil-saturated sediments. Thicker oil-saturated zones will degrade much more slowly than thin ones, with degradation timescales likely scaling as the square of oil-saturated zone thickness and scaling inversely with diffusion coefficients for species transfer within the oil zone itself, which will decrease with increasing net levels of biodegradation. Very high oil saturations in near-surface sediments are likely enhanced by a so-called Marine Oil Snow Sedimentation and Flocculent Accumulation (MOSSFA), aggregation of oil with mineral particles, extracellular polymers, and biomass, which caused sinking of a significant amount of oil following major oil spills (Brooks et al. 2015; Daly et al. 2016; Quigg et al. 2020).

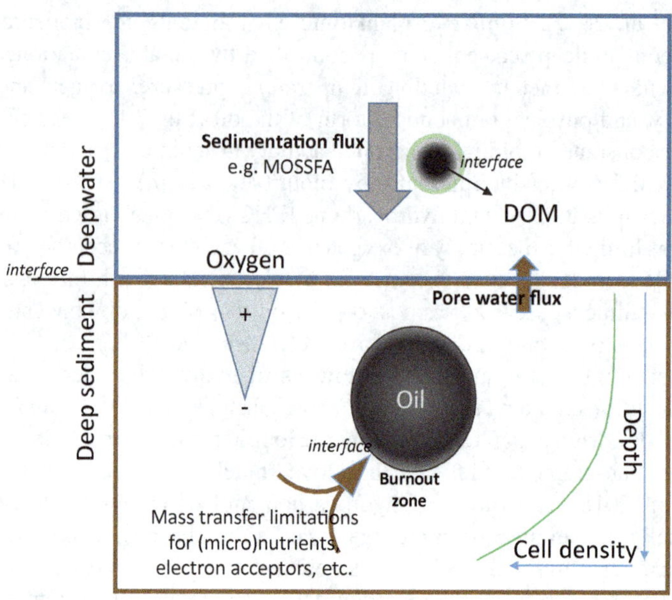

Fig. 7.2 Conceptual model for oil biodegradation in the deep sea. Depending on the amount of deposited oil, due to physical mass transfer limitations of many key variables such as oxygen, electron acceptors, and others, occurring at interfaces, there is a potential for long-term preservation of oil in the sediment, even when hydrocarbon-degrading microorganisms are present – so-called burnout effect

7.3.1 Biodegradation of Petroleum in Marine Sediments (DWH and Other Case Studies)

Oil is transported to the deep seafloor via two main mechanisms, through subsurface intrusions impinging on bottom sediments (Brooks et al. 2015; Romero et al. 2017) and through the sedimentation of oiled marine snow across large areas of the northern Gulf of Mexico (Passow et al. 2012; Brooks et al. 2015; Romero et al. 2017; Quigg et al. 2020). Estimates for the amount of DWH oil sequestered in deep ocean sediments range from 2% to 14% (Chanton et al. 2012; Valentine et al. 2014; Bagby et al. 2017; Romero et al. 2017) and vary in footprint size from 3200 km^2 to 32,000 km^2 (Valentine et al. 2014; Romero et al. 2017). As for the spatial distribution of DWH-sourced hydrocarbons in the sediments, results suggest that the deposition was focused in the deep bathymetric depo-centers and negligibly on the continental shelf (Bagby et al. 2017; Romero et al. 2017).

The chemistry of deposited oil in sediments indicated that biodegradation rates were dependent of the degree of oiling, and consistent with previous findings, the smallest molecular weight compounds exhibited the fastest decay rates (Fig. 7.2) (Bagby et al. 2017). Slower degradation/longer preservation of oil is most likely due to mass transfer limitations of electron acceptors. Similar trends both of structure-dependent biodegradation susceptibility and oil concentration-driven degradation rates have also been observed in biodegraded oil reservoirs, in which the biodegradation most rapidly occurs at the oil-water transition zone which is most abundant with auxiliary chemical species needed for petroleum biodegradation (Head et al. 2014). Moving away from the active zone of oil contamination, biodegradation rate likely decreases, due to mass transfer limitations and concentration gradients – a "burnout effect" of a kind. In particular, biodegradation would be inhibited when a large amount of petroleum is deposited, relative to the available reactants involved in various biodegradation pathways. In such cases, one could assume that deposited oil will be preserved over longer timescales, Fig. 7.1. An additional control is interfacial/phase phenomena, as suspended oil components are subjected to more pronounced (faster) degradation, contrary to oil deposited to sediment, which was degraded at slower rates (Fig. 7.2).

7.3.2 Microbial Community Response in Deep Sea Sediments

A substantial challenge in determining the impacts of oil deposition into deep ocean sediments is the nearly complete lack of studies on benthic microbial communities performed prior to the DWH discharge (Overholt 2018). Due to interest in petroleum exploration, previous studies in the Gulf focused almost entirely on deep-subsurface sediments with the shallowest samples collected at ~4 m below the seafloor or on sediments impacted by natural hydrocarbon seeps (Biddle et al. 2011; Nunoura et al. 2009; Orcutt et al. 2010). Due to this limitation in the knowledge of surficial microbial communities, the majority of studies on deep ocean sediments associated with the DWH event used samples that were collected outside of impacted areas as controls (Kimes et al. 2014; Mason et al. 2014; Yang et al. 2016; Liu et al. 2017). Sediments were primarily impacted as deepwater oil intrusion layers impinged on the seafloor and by a large sedimentation event of marine oil-derived snow (Brooks et al. 2015; Passow et al. 2012; Valentine et al. 2014; Romero et al. 2017). While the datasets on benthic microbes are limited to relatively few samples (Overholt 2018), results indicate that microbial communities responded quickly to oil perturbation, and shifts in community composition as well as total metabolic potential were observed (Kimes et al. 2014; Mason et al. 2014; Yang et al. 2016).

In areas immediately surrounding the DWH wellhead that were heavily contaminated to above EPA limits, microbial communities shifted to a different state dominated by known hydrocarbon degraders of the *Colwellia* group as well as uncultured *Gammaproteobacteria* that comprised up to 18% of the total community (Mason et al. 2014; Overholt 2018). Total metabolic potential was enriched in both aliphatic and PAH degradation pathways as well in anaerobic respiratory metabolism (Mason et al. 2014). Yang et al. (2016) used a clone library approach to follow the succession in deep benthic communities, including a pulse in sulfate-reducing bacterial populations followed by a longer persistence of the PAH-degrading genus *Cycloclasticus* (Yang et al. 2016). Although reports are contradictory and based on few samples (<10), sedimentary microbial communities may have returned to baseline conditions within 1 year following the spill (Yang et al. 2016; Liu et al. 2017). The inferred rates of recovery suggested by these studies exceed estimates of decay rates in contaminating hydrocarbons. This is potentially due to a dilution effect from sedimented natural organic matter, whereby few contaminated oil particles settling to the seafloor result in low to moderately contaminated sediments (Bagby et al. 2017).

7.4 Effect of High Pressure on Microbially Mediated Hydrocarbon Degradation

Organisms with optimal growth rates above 0.1 MPa (atmospheric pressure) are defined as piezophiles (from the Greek "piezo," meaning "to press") (Yayanos 1995) and span a wide range of bacterial and archaeal lineages (Horikoshi 1998). While our understanding of the mechanisms by which piezophiles adapt and thrive at high pressures is limited, available evidence points to adjustments in gene replication, protein expression, membrane chemistry, production of osmolytes (Simonato et al. 2006), and antioxidant defense response (Xie et al. 2018). The most well-characterized response of piezophiles to pressure is their ability to increase the proportion of unsaturated fatty acids in their membrane lipids at increasing pressure in order to preserve membrane fluidity (Simonato et al. 2006).

Almost 70 years ago, Zobell and Johnson (1949) reported that sulfate-reducing bacteria isolated from several thousand meters deep terrestrial oil wells were metabolically more active at elevated pressures of 40 MPa or 60 MPa (Zobell and Johnson 1949). Subsequently, hydrocarbon-degrading microorganisms have been isolated from numerous deep sea environments, thus suggesting that hydrocarbon degradation can proceed at increasing pressure (Tapilatu et al. 2010a, b). As offshore deep sea drilling operations are expected to increase in the future, understanding the effect of pressure on microbial hydrocarbon degradation is becoming increasingly important.

7.4.1 *Ex Situ Incubations of Enriched Seawater and Sediments*

The earliest reports on the effects of pressure on microbial hydrocarbon degradation were published in the late 1970s (Schwarz et al. 1974). Schwarz et al. (1974) provided the first evidence that the indigenous microbial community was able to degrade hydrocarbons at high pressure using respiration to CO_2 as a proxy

(Schwarz et al. 1974). Sediment samples collected off the coast of Florida from a depth of 4940 m were enriched in a mineral salts medium containing *n*-hexadecane, and biodegradation was monitored at either atmospheric or elevated pressure (51 MPa = 500 atm). Growth and degradation rates were comparable between the two pressure treatments. However, the experimental conditions did not adequately resemble the deep sea as initial enrichments were performed at room temperature. Subsequently, Schwarz et al. conducted a similar enrichment culture experiment at 4 °C (Schwarz et al. 1975). Despite the lower cell density and the significantly lower degradation rates (tenfold decrease) observed at high pressure, the observed total amount of degraded n-hexadecane was similar between the two tested pressures (Schwarz et al. 1975). Lowering the temperature amplified any effects of pressure on the microbial metabolism and subsequently lowered the rates of hydrocarbon degradation, indicating that low temperature and high pressure act synergistically to slow the rates of hydrocarbon degradation. In these early studies, preincubation at atmospheric pressure may have inhibited the growth and metabolism of microbial populations that are adapted to high pressure.

Few studies of PHC biodegradation were conducted at pressures relevant to the deep sea from the 1970s until the DWH spill. Since that time, accumulating evidence points to a threshold in pressure effects, whereby hydrocarbon degradation is only moderately affected at 10–15 MPa, and more dramatic effects are observed at extremely high pressures approaching 50 MPa. Prince et al. (2016) reported that hydrocarbon degradation rates diminished by approximately one-third at 15 MPa in comparison to atmospheric pressure in incubations of seawater collected from a depth of 8 m off the coast of Newfoundland (Canada). The Prince et al. (2016) study is notable in that weathered crude oil (3 ppm) and dispersant (dispersant/oil ratio of 1:15) were amended at close to in situ concentrations expected during response efforts. The majority of n-alkanes were degraded within the first 7 days of the incubation, while the aromatic fraction was only partially degraded by day 35 at 15 MPa (>100 day half-life for aromatic fraction).

Results from seawater enrichments were corroborated by investigations of deep sea sediments. Nguyen et al. (2018) incubated sediments collected from a range of water depths (62–1520 m) in the Gulf of Mexico amended with crude oil at in situ pressure (0.1–15 MPa) and temperature (4, 10, and 20 °C). The authors confirmed an inhibitory effect of pressure on hydrocarbon degradation at a pressure of 15 MPa. Total n-alkanes were degraded under all tested conditions following 18 days of incubation; however the extent of biodegradation was inversely proportional to pressure with greater degradation observed at shallower sites. For every 1 MPa of pressure increase, a 4% decrease in the rate of alkane degradation was estimated (Nguyen et al. 2018). Polycyclic aromatic hydrocarbons were also degraded but to a smaller extent (3–60%) compared to the observed alkane degradation (90%).

Marietou et al. (2018) investigated the impacts of pressure on microbial communities along with hydrocarbon degradation in enrichments at in situ temperature (4 °C) and a range of pressures (0.1–30 MPa) in seawater samples collected at 1070 m water depth from the Gulf of Mexico. Using the abundance of hydrocarbon degradation genes as a proxy for hydrocarbon degradation potential, it was concluded that while alkane degradation was not affected, a significant decrease was observed in aromatic hydrocarbon degradation potential at increasing pressure. While growth declined with increasing pressure, no differences in microbial community composition were observed across pressure treatments. Community analysis of the incubations showed that known psychrotolerant hydrocarbon-degrading bacteria affiliated with *Oleispira antarctica* dominated the communities at all pressures (Marietou et al. 2018). Similarly, Fasca et al. (2018) observed minor effects of pressure (22 MPa) on the microbial community of microcosms prepared with seawater collected off the coast of Brazil from an oil extraction platform.

Pressure is thought to impact hydrocarbon biodegradation through effects on solubility, especially for gaseous hydrocarbons. For example, a number of studies have demonstrated the effect of methane partial pressure, i.e., concentration of dissolved methane, on rates of anaerobic methane oxidation (Kallmeyer and Boetius 2004; Deusner et al. 2009; Zhang et al. 2010; Bowles et al. 2011; Timmers et al. 2015). In sediments from a number of cold seep sites including the Mississippi Canyon in the Gulf of Mexico, Bowles et al. (2011) observed 6–10 times higher rates of anaerobic methane oxidation at 10 MPa compared to atmospheric pressure. In agreement with Bowles et al., Timmers et al. (2015) reported that high methane partial pressure stimulated initial anaerobic methane oxidation activity of sediment from Eckernförde Bay, Germany. Total pressure and methane concentration appeared to epistatically influence the rate of anaerobic methane oxidation (Bowles et al. 2011). Therefore, total pressure and methane concentration should be considered independently in the investigation of pressure effects on methanotrophy.

State-of-the-art technology, which can independently control the total pressure and methane concentration, can be used to separate out the effects of solubility (Deusner et al. 2009; Bowles et al. 2011; Schedler et al. 2014; Valladares Juárez et al. 2015; Case et al. 2017). High-pressure reactors were constructed at the Hamburg University of Technology in collaboration with industry partners (Technikservice A. Meyer, Eurotechnica, and PreSens, Germany) (Fig. 7.3). Pressurization is achieved with inert gas, enabling independent control of total pressure of the system and partial pressure of another gas of interest, for example, oxygen for aerobic biodegradation and gaseous hydrocarbons such as methane for methanotrophy studies. The system permits a higher biodegradation extent than conventional high-pressure reactors without a headspace, allowing detailed investigation of hydrocarbon degradation while avoiding toxicity associated with increased oxygen partial pressure. The reactors are equipped with an oxygen sensor for continuous monitoring as a proxy for biodegradation. Additionally, a new optical system has been developed based on an optical flow cell for near infrared (NIR) spectroscopic analysis of gas mixtures at high pressure (Norton et al. 2014). The optical system was directly mounted onto a high-pressure reactor. Thus, gaseous hydrocarbon biodegradation can be monitored at high pressure online in real time via removal of the gas from the headspace while bypassing the subsampling of high-pressure gas mixtures for off-line analysis (Fig. 7.3).

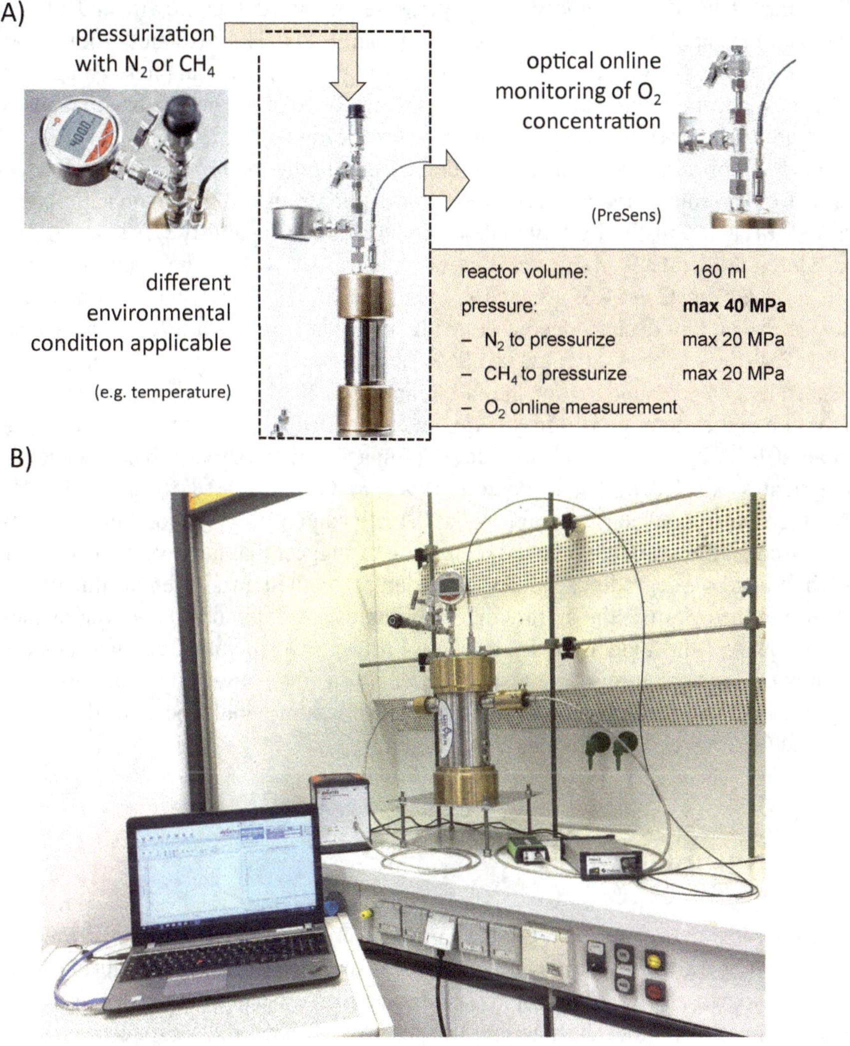

Fig. 7.3 High hydrostatic pressure characteristic of the deep sea can be simulated within these reactors allowing the examination of biological processes under deep sea conditions. (**a**) The reactors made from stainless steel and bronze can withstand maximum pressure of 400 bar. Reactors are equipped with oxygen sensors to continuously monitor oxygen and/or (**b**) NIR optical system to simultaneously monitor biodegradation of hydrocarbon gases

7.4.2 Pure Culture Studies

Results from pure cultures of hydrocarbon-degrading bacteria are equivocal with respect to the influence of pressure on growth and hydrocarbon degradation potential. Elevated partial pressure of methane and oxygen, but not total pressure, is inhibitory to *Methylococcus capsulatus* (Bath), isolated from sediment and water, and *Methylomicrobium album* BG8, isolated from freshwater (Soni et al. 1998).

Schedler et al. (2014) reported that pressure impacted the growth and alkane-degradation capability of *Rhodococcus qingshengii* TUHH-12 isolated from surface seawater of the Norwegian Arctic. Growth during the exponential phase and the rate of *n*-hexadecane degradation were approximately twofold higher at atmospheric pressure (0.1 MPa) in comparison to pressures mimicking the deep sea (15 MPa). No growth of aromatic hydrocarbon-degrading *Sphingobium yanoikuyae* B1, isolated from polluted stream, was observed when the strain was grown on naphthalene at 13.9 MPa. Interestingly, naphthalene biodegradation capability was still observed at a pressure of 13.9 MPa, albeit slower than at 0.1 MPa. The growth yield of *Alcanivorax dieselolei* KS_293 and *A. jadensis* KS_339, both isolated from surface seawater, on n-dodecane decreased with increasing hydrostatic pressure (5 and 10 MPa) (Scoma et al. 2016a), and 95% of expressed genes were downregulated, especially genes related to translation (Scoma et al. 2016a). The authors posited that pressure is the major constraint of *Alcanivorax* distribution in the deep sea (Scoma et al. 2016b). In contrast, pressure did not impact growth rate or *n*-hexadecane bio-degradation of *Marinobacter hydrocarbonoclasticus*, isolated from 3475 m bsl Mediterranean seawater, when grown at 0.1 MPa and 35 MPa (Grossi et al. 2010).

Based on the available data, it is tempting to suggest that the investigated strains of hydrocarbon-degrading bacteria are adapted to their in situ environment, and therefore the origin of the culture determines pressure tolerance. However, as mentioned above, the field is in its infancy. Further investigations are warranted to improve understanding of pressure effects on the physiology of deep sea hydrocarbon-degrading microorganisms, and pressure should be considered as a crucial factor in deep sea biodegradation studies.

7.5 Conclusions

The *Deepwater Horizon* (DWH) discharge is unique in that it occurred in the deep sea, and unparalleled volumes of chemical dispersant were applied during emergency response efforts. A lack of knowledge on biodegradation in the deep sea hampered efforts to predict the fate and transport of released oil. Despite tremendous progress, many knowledge gaps remain with regard to biodegradation under the low-temperature and high-pressure conditions found in the deep ocean. Since biodegradation rates were not determined directly under in situ conditions, the role of biodegradation as well as dispersant application on the fate and transport of petroleum hydrocarbons remains under debate. Hydrostatic pressure was shown to substantially impact biodegradation rates in both the water column and sediments of the deep sea. However, as mentioned above, the field is in its infancy. Further investigations are warranted to improve understanding of pressure effects on the physiology of deep sea hydrocarbon-degrading microorganisms, and pressure should be considered as a crucial factor in deep sea biodegradation studies. Direct rate measurements and determination of oceanographic controls such as pressure are sorely needed to improve the effective parameterization of oil plume models. Ongoing technological advances show potential to address these knowledge gaps.

References

Bagby SC, Reddy CM, Aeppli C, Fisher GB, Valentine DL (2017) Persistence and biodegradation of oil at the ocean floor following *Deepwater Horizon*. Proc Natl Acad Sci 114(1):E9

Biddle JF, White JR, Teske AP, House CH (2011) Metagenomics of the subsurface Brazos-Trinity Basin (IODP site 1320): comparison with other sediment and pyrosequenced metagenomes. ISME J 5:1038–1047

Bowles MW, Samarkin VA, Joye SB (2011) Improved measurement of microbial activity in deep-sea sediments at in situ pressure and methane concentration. Limnol Oceanogr Methods 9(10):499–506. https://doi.org/10.4319/lom.2011.9.499

Brooks GR, Larson RA, Schwing PT, Romero I, Moore C, Reichart GJ, Jilbert T, Chanton JP, Hastings DW, Overholt WA, Marks KP, Kostka JE, Holmes CW, Hollander D (2015) Sedimentation pulse in the NE Gulf of Mexico following the 2010 DWH blowout. PLoS One 10:1–24

Camilli R, Reddy CM, Yoerger DR, Van Mooy BAS, Jakuba MV, Kinsey JC, McIntyre CP, Sylva SP, Maloney JV (2010) Tracking hydrocarbon plume transport and biodegradation at Deepwater Horizon. Science 330(6001):201–204. https://doi.org/10.1126/science.1195223

Case DH, Ijiri A, Morono Y, Tavormina P, Orphan VJ, Inagaki F (2017) Aerobic and anaerobic methanotrophic communities associated with methane hydrates exposed on the seafloor: a high-pressure sampling and stable isotope-incubation experiment. Front Microbiol 8:2569

Chanton JP, Cherrier J, Wilson RM, Sarkodee-Adoo J, Bosman S, Mickle A, Graham WM (2012) Radiocarbon evidence that carbon from the Deepwater Horizon spill entered the planktonic food web of the Gulf of Mexico. Environ Res Lett 7:045303

Crespo-Medina M, Meile CD, Hunter KS, Diercks AR, Asper VL, Orphan VJ, Joye SB (2014) The rise and fall of methanotrophy following a Deepwater oil-well blowout. Nat Geosci 7(6):423–427. https://doi.org/10.1038/ngeo2156

Daly KL, Passow U, Chanton J, Hollander D (2016) Assessing the impacts of oil-associated marine snow formation and sedimentation during and after the Deepwater Horizon oil spill. Anthropocene 13:18–33. https://doi.org/10.1016/j.ancene.2016.01.006

Deusner C, Meyer V, Ferdelman TG (2009) High-pressure systems for gas-phase free continuous incubation of enriched marine microbial communities performing anaerobic oxidation of methane. Biotechnol Bioeng:105. https://doi.org/10.1002/bit.22553

Dubinsky EA, Conrad ME, Chakraborty R, Bill M, Borglin SE, Hollibaugh JT, Mason OU, M. Piceno Y, Reid FC, Stringfellow WT, Tom LM (2013) Succession of hydrocarbon-degrading bacteria in the aftermath of the Deepwater Horizon oil spill in the Gulf of Mexico. Environ Sci Technol 47(19):10860–10867

Fasca H, de Castilho LV, de Castilho JFM, Pasqualino IP, Alvarez VM, de Azevedo Jurelevicius D, Seldin L (2018) Response of marine bacteria to oil contamination and to high pressure and low temperature deep sea conditions. MicrobiologyOpen 7(2):e00550

Ferguson RMW, Gontikaki E, Anderson JA, Witte U (2017) The variable influence of dispersant on degradation of oil hydrocarbons in subarctic deep-sea sediments at low temperatures (0–5 °C). Sci Rep 7(1):2253. https://doi.org/10.1038/s41598-017-02475-9

Grossi V, Yakimov MM, Ali BA, Tapilatu Y, Cuny P, Goutx M, LaCono V, Giuliano L, Tamburini C (2010) Hydrostatic pressure affects membrane and storage lipid compositions of the piezotolerant hydrocarbon-degrading Marinobacter hydrocarbonoclasticus strain #5. Environ Microbiol 12(7):2020–2033. https://doi.org/10.1111/j.1462-2920.2010.02213.x

Hazen TC, Prince RC (2015) Marine oil biodegradation. Environ Sci Technol 50:2121–2129

Hazen TC, Dubinsky EA, DeSantis TZ, Andersen GL, Piceno YM, Singh N, Jansson JK, Probst A, Borglin SE, Fortney JL, Stringfellow WT, Bill M, Conrad ME, Tom LM, Chavarria KL, Alusi TR, Lamendella R, Joyner DC, Spier C, Baelum J, Auer M, Zemla ML, Chakraborty R, Sonnenthal EL, D'haeseleer P, Holman HY, Osman S, Lu Z, Van Nostrand JD, Deng Y, Zhou J, Mason OU (2010) Deep-sea oil plume enriches indigenous oil-degrading Bacteria. Science 330(6001):204–208. https://doi.org/10.1126/science.1195979

Head IM, Jones DM, Röling WF (2006) Marine microorganisms make a meal of oil. Nat Rev Microbiol 4(3):17

Head IM, Gray ND, Larter SR (2014) Life in the slow lane; biogeochemistry of biodegraded petroleum containing reservoirs and implications for energy recovery and carbon management. Front Microbiol 5:566. https://doi.org/10.3389/fmicb.2014.00566

Horikoshi K (1998) Barophiles: deep-sea microorganisms adapted to an extreme environment. Curr Opin Microbiol 1(3):291–295

Jørgensen BB, Boetius A (2007) Feast and famine — microbial life in the deep-sea bed. Nat Rev Microbiol 5:770–781

Joye SB (2015) Deepwater Horizon, 5 years on. Science 349(6248):592–593. https://doi.org/10.1126/science.aab4133

Joye SB, MacDonald IR, Leifer I, Asper V (2011) Magnitude and oxidation potential of hydrocarbon gases released from the BP oil well blowout. Nat Geosci 4(3):160–164. http://www.nature.com/ngeo/journal/v4/n3/abs/ngeo1067.html#supplementary-information

Joye SB, Teske AP, Kostka JE (2014) Microbial dynamics following the Macondo oil well blowout across Gulf of Mexico environments. Bioscience 64(9):766–777. https://doi.org/10.1093/biosci/biu121

Joye SB, Kleindienst S, Gilbert JA, Handley KM, Weisenhorn P, Overholt WA, Kostka JE (2016) Responses of microbial communities to hydrocarbon exposures. Oceanography 29(1):136–149

Joye S, Kleindienst S, Peña-Montenegro TD (2018) SnapShot: microbial hydrocarbon bioremediation. Cell 172(6):1336–1336.e1331. https://doi.org/10.1016/j.cell.2018.02.059

Kallmeyer J, Boetius A (2004) Effects of temperature and pressure on sulfate reduction and anaerobic oxidation of methane in hydrothermal sediments of Guaymas Basin. Appl Environ Microbiol 70(2):1231–1233. https://doi.org/10.1128/AEM.70.2.1231-1233.2004

Kessler JD, Valentine DL, Redmond MC, Du M, Chan EW, Mendes SD, Quiroz EW, Villanueva CJ, Shusta SS, Werra LM, Yvon-Lewis SA, Weber TC (2011) A persistent oxygen anomaly reveals the fate of spilled methane in the deep Gulf of Mexico. Science 331(6015):312–315. https://doi.org/10.1126/science.1199697

Kimes NE, Callaghan AV, Suflita JM, Morris PJ (2014) Microbial transformation of the Deepwater Horizon oil spill – past, present, and future perspectives. Front Microbiol 5:603

King GM, Kostka JE, Hazen TC, Sobecky PA (2015) Microbial responses to the Deepwater Horizon oil spill: from coastal wetlands to the deep sea. Annu Rev Mar Sci 7:377–401. https://doi.org/10.1146/annurev-marine-010814-015543

Kleindienst S, Seidel M, Ziervogel K, Grim S, Loftis K, Harrison S, Malkin SY, Perkins MJ, Field J, Sogin ML, Dittmar T, Passow U, Medeiros PM, Joye SB (2015a) Chemical dispersants can suppress the activity of natural oil-degrading microorganisms. Proc Natl Acad Sci 112(48). https://doi.org/10.1073/pnas.1507380112

Kleindienst S, Paul JH, Joye SB (2015b) Using dispersants after oil spills: impacts on the composition and activity of microbial communities. Nat Rev Microbiol 13(6):388–396. https://doi.org/10.1038/nrmicro3452

Kleindienst S, Grim S, Sogin M, Bracco A, Crespo-Medina M, Joye SB (2015c) Diverse, rare microbial taxa responded to the Deepwater Horizon deep-sea hydrocarbon plume. ISME J. https://doi.org/10.1038/ismej.2015.121

Leahy JG, Colwell RR (1990) Microbial degradation of hydrocarbons in the environment. Microbiol Rev 54(3):305–315

Liu J, Bacosa HP, Liu Z (2017) Potential environmental factors affecting oil-degrading bacterial populations in deep and surface waters of the northern Gulf of Mexico. Front Microbiol 7:2131

Marietou A, Chastain R, Beulig F, Scoma A, Hazen TC, Bartlett DH (2018) The effect of hydrostatic pressure on enrichments of hydrocarbon degrading microbes from the Gulf of Mexico following the Deepwater Horizon oil spill. Front Microbiol 9:808

Mason OU, Hazen TC, Borglin S, Chain PSG, Dubinsky EA, Fortney JL, Han J, Holman H-YN, Hultman J, Lamendella R, Mackelprang R, Malfatti S, Tom LM, Tringe SG, Woyke T, Zhou J, Rubin EM, Jansson JK (2012) Metagenome, metatranscriptome and single-cell sequencing reveal microbial response to Deepwater Horizon oil spill. ISME J 6(9):1715–1727. https://doi.org/10.1038/ismej.2012.59

Mason OU, Scott NM, Gonzalez A, Robbins-Pianka A, Bælum J, Kimbrel J, Bouskill NJ, Prestat E, Borglin S, Joyner DC, Fortney JL, Jurelevicius D, Stringfellow WT, Alvarez-Cohen L, Hazen TC, Knight R, Gilbert JA, Jansson JK (2014) Metagenomics reveals sediment microbial community response to Deepwater Horizon oil spill. ISME J 8(7):1464–1475

McNutt MK, Camilli R, Crone TJ, Guthrie GD, Hsieh PA, Ryerson TB, Savas O, Shaffer F (2012) Review of flow rate estimates of the Deepwater Horizon oil spill. Proc Natl Acad Sci U S A 109:20260–20267

Meckenstock RU, Boll M, Mouttaki H, Koelschbach JS, Cunha Tarouco P, Weyrauch P, Dong X, Himmelberg AM (2016) Anaerobic degradation of benzene and polycyclic aromatic hydrocarbons. J Mol Microbiol Biotechnol 26:92–118. https://doi.org/10.1159/000441358

Nguyen UT, Lincoln SA, Juárez AGV, Schedler M, Macalady JL, Müller R, Freeman KH (2018) The influence of pressure on crude oil biodegradation in shallow and deep Gulf of Mexico sediments. PloS one 13(7):e0199784

Norton CG, Suedmeyer J, Oderkerk B, Fieback TM (2014) High pressure and temperature optical flow cell for Near-Infra-Red spectroscopic analysis of gas mixtures. Rev Sci Instrum 85(5):053101. https://doi.org/10.1063/1.4873195

Nunoura T, Soffientino B, Blazejak A, Kakuta J, Oida H, Schippers A, Takai K (2009) Subseafloor microbial communities associated with rapid turbidite deposition in the Gulf of Mexico continental slope (IODP Expedition 308). FEMS Microbiol Ecol 69:410–424

Orcutt BN, Joye SB, Kleindienst S, Knittel K, Ramette A, Reitz A, Samarkin V, Treude T, Boetius A (2010) Impact of natural oil and higher hydrocarbons on microbial diversity, distribution, and activity in Gulf of Mexico cold-seep sediments. Deep-Sea Res II 57:2008–2021

Overholt W (2018) The response of marine benthic microbial populations to the Deepwater Horizon oil spill, Ph.D. Dissertation, Georgia Institute of Technology, pp 248

Passow U, Ziervogel K, Asper V, Diercks A (2012) Marine snow formation in the aftermath of the Deepwater Horizon oil spill in the Gulf of Mexico. Environ Res Lett 7(3):035301

Prince RC, Nash GW, Hill SJ (2016) The biodegradation of crude oil in the deep ocean. Mar Pollut Bull 111(1):354–357

Prince RC, Butler JD, Redman AD (2017) The rate of crude oil biodegradation in the sea. Environ Sci Technol 51(3):1278–1284

Quigg A, Passow U, Daly KL, Burd A, Hollander DJ, Schwing PT, Lee K (2020) Marine Oil Snow Sedimentation and Flocculent Accumulation (MOSSFA) events: learning from the past to predict the future (Chap. 12). In: Murawski SA, Ainsworth C, Gilbert S, Hollander D, Paris CB, Schlüter M, Wetzel D (eds) Deep oil spills: facts, fate, effects. Springer, Cham

Rahsepar S, Smit MPJ, Murk AJ, Rijnaarts HHM, Langenhoff AAM (2016) Chemical dispersants: oil biodegradation friend or foe? Mar Pollut Bull 108(1):113–119. https://doi.org/10.1016/j.marpolbul.2016.04.044

Rodriguez-r LM, Overholt WA, Hagan C, Huettel M, Kostka JE, Konstantinidis KT (2015) Microbial community successional patterns in beach sands impacted by the Deepwater Horizon oil spill. ISME J

Romero IC, Toro-Farmer G, Diercks A-R, Schwing P, Muller-Karger F, Murawski S, Hollander DJ (2017) Large-scale deposition of weathered oil in the Gulf of Mexico following a deep-water oil spill. Environ Pollut 228:179–189. https://doi.org/10.1016/j.envpol.2017.05.019

Schedler M, Hiessl R, Valladares Juarez A, Gust G, Muller R (2014) Effect of high pressure on hydrocarbon-degrading bacteria. AMB Express 4(1):77

Schwarz JR, Walker JD, Colwell RR (1974) Deep-sea bacteria: growth and utilization of hydrocarbons at ambient and in situ pressure. Appl Microbiol 28(6):982–986

Schwarz JR, Walker JD, Colwell RR (1975) Deep-sea bacteria: growth and utilization of n-hexadecane at in situ temperature and pressure. Can J Microbiol 21(5):682–687

Scoma A, Barbato M, Hernandez-Sanabria E, Mapelli F, Daffonchio D, Borin S, Boon N (2016a) Microbial oil-degradation under mild hydrostatic pressure (10 MPa): which pathways are impacted in piezosensitive hydrocarbonoclastic bacteria? Sci Rep 6:23526

Scoma A, Barbato M, Borin S, Daffonchio D, Boon N (2016b) An impaired metabolic response to hydrostatic pressure explains Alcanivorax borkumensis recorded distribution in the deep marine water column. Sci Rep 6:31316. https://doi.org/10.1038/srep31316

Shin B (2018) Hydrocarbon degradation under contrasting redox conditions in shallow coastal sediments of the northern Gulf of Mexico, Ph.D. Dissertation, Georgia Institute of Technology, pp 160

Shin B, Kim M, Zengler K, Chin KJ, Overholt WA, Gieg LM, Konstantinidis KT, Kostka JE (2019) Anaerobic degradation of hexadecane and phenanthrene coupled to sulfate reduction by enriched consortia from northern Gulf of Mexico seafloor sediment. Sci Rep 9(1):1239

Sibert R, Harrison S, Joye SB (2017) Protocols for radiotracer estimation of primary hydrocarbon oxidation in oxygenated seawater. In: McGenity TJ, Timmis KN, Nogales B (eds) Hydrocarbon and lipid microbiology protocols: field studies. Springer, Berlin, Heidelberg, pp 263–276

Simonato F, Campanaro S, Lauro FM, Vezzi A, D'Angelo M, Vitulo N, Valle G, Bartlett DH (2006) Piezophilic adaptation: a genomic point of view. J Biotechnol 126(1):11–25

Soni BK, Conrad J, Kelley R, Srivastava V (1998) Effect of temperature and pressure on growth and methane utilization by several methanotrophic cultures. Appl Biochem Biotechnol 70–72(1):729–738

Tapilatu Y, Acquaviva M, Guigue C, Miralles G, Bertrand JC, Cuny P (2010a) Isolation of alkane-degrading bacteria from deep-sea Mediterranean sediments. Lett Appl Microbiol 50(2):234–236

Tapilatu YH, Grossi V, Acquaviva M, Militon C, Bertrand J-C, Cuny P (2010b) Isolation of hydrocarbon-degrading extremely halophilic archaea from an uncontaminated hypersaline pond (Camargue, France). Extremophiles 14(2):225–231. https://doi.org/10.1007/s00792-010-0301-z

Timmers PHA, Gieteling J, Widjaja-Greefkes HCA, Plugge CM, Stams AJM, Lens PNL, Meulepas RJW (2015) Growth of anaerobic methane-oxidizing archaea and sulfate-reducing bacteria in a high-pressure membrane capsule bioreactor. Appl Environ Microbiol 81(4):1286–1296

Valentine DL, Kessler JD, Redmond MC, Mendes SD, Heintz MB, Farwell C, Hu L, Kinnaman FS, Yvon-Lewis S, Du M, Chan EW, Garcia Tigreros F, Villanueva CJ (2010) Propane respiration jump-starts microbial response to a deep oil spill. Science 330(6001):208–211. https://doi.org/10.1126/science.1196830

Valentine DL, Mezić I, Maćešić S, Črnjarić-Žic N, Ivić S, Hogan PJ, Fonoberov VA, Loire S (2012) Dynamic autoinoculation and the microbial ecology of a deep water hydrocarbon irruption. Proc Natl Acad Sci U S A 109:20286–20291

Valentine DL, Fisher GB, Bagby SC, Nelson RK, Reddy CM, Sylva SP, Woo MA (2014) Fallout plume of submerged oil from Deepwater Horizon. Proc Natl Acad Sci 111(45):15906–15911. https://doi.org/10.1073/pnas.1414873111

Valladares Juárez AG, Kadimesetty HS, Achatz DE, Schedler M, Muller R (2015) Online monitoring of crude oil biodegradation at elevated pressures. IEEE J Sel Top Appl Earth Obs Remote Sens 8(2):872–878. https://doi.org/10.1109/JSTARS.2014.2347896

Xie Z, Jian H, Jin Z, Xiao X (2018) Enhancing the adaptability of the deep-sea bacterium Shewanella piezotolerans WP3 to high pressure and low temperature by experimental evolution under H2O2 stress. Appl Environ Microbiol 84(5):e02342–e02317

Yang T, Nigro LM, Gutierrez T, D'Ambrosio L, Joye SB, Highsmith R, Teske A (2016) Pulsed blooms and persistent oil-degrading bacterial populations in the water column during and after the Deepwater Horizon blowout. Deep-Sea Res II Top Stud Oceanogr 129:282–291. https://doi.org/10.1016/j.dsr2.2014.01.014

Yayanos AA (1995) Microbiology to 10,500 meters in the Deep Sea. Annu Rev Microbiol 49(1):777–805

Zhang Y, Henrie JP, Bursens J, Boon N (2010) Stimulation of in vitro anaerobic oxidation of methane rate in a continuous high-pressure bioreactor. Bioresour Technol 101(9):3132–3138. https://doi.org/10.1016/j.biortech.2009.11.103

ZoBell CE, Johnson FH (1949) The influence of hydrostatic pressure on the growth and viability of terrestrial and marine bacteria. J Bacteriol 57(2):179

Chapter 8
Partitioning of Organics Between Oil and Water Phases with and Without the Application of Dispersants

Aprami Jaggi, Ryan W. Snowdon, Jagoš R. Radović, Andrew Stopford, Thomas B. P. Oldenburg, and Steve R. Larter

Abstract Immediately following an oil spill, more hydrophilic and toxic oil compounds such as benzene, toluene, ethylbenzene, and xylene (BTEX) partition from the oil into the water phase. The partitioning behavior of individual organic compounds between petroleum and water phases is influenced by their molecular properties and by the pressure, the temperature, and the composition of bulk oil and surrounding water. The traditional shake flask technique for determining oil-water partition ratios (equilibrium $[X]_{oil}/[X]_{water}$) cannot accurately assess the extremes of high pressure and low water temperatures found in deep submarine oil spill conditions. To address that challenge, an oil-water partitioning device was constructed to experimentally simulate the partition behavior of BTEX compounds under submarine oil spill conditions, using simulated live oil (methane charged) with saline waters, over a range of pressure (2–15 MPa) and temperature (4–20 °C). Within the investigated ranges, the partition ratios of BTEX compounds increase proportionally with an increase in methane charging pressure (oil saturation pressure) and the degree of alkylation within the BTEX compound group. The increase in experimental temperature, however, resulted in a decrease in the partition ratios of BTEX compounds. The change of the partition ratio values, due to changes in system pressure and increasing methane concentration, is much more significant than the changes that are due to varying temperature over the range studied.

The customized system was also operated with chemical dispersant, which is often applied as a spill response option to enhance the natural dispersion of oil following spillage, to understand its effect on the partitioning of oil at an oil/dispersant ratio of 1000:1. The addition of dispersants was found to increase the extent of BTEX compound partitioning from the oil into water. The increase observed

A. Jaggi (✉) · R. W. Snowdon · J. R. Radović · A. Stopford · T. B. P. Oldenburg · S. R. Larter
University of Calgary, PRG, Department of Geoscience, Calgary, AB, Canada
e-mail: aprami.jaggi@ucalgary.ca; rwsnowdo@ucalgary.ca; jagos.radovic@ucalgary.ca; toldenbu@ucalgary.ca; slarter@ucalgary.ca

© Springer Nature Switzerland AG 2020
S. A. Murawski et al. (eds.), *Deep Oil Spills*,
https://doi.org/10.1007/978-3-030-11605-7_8

was higher at near surface conditions, while being within the experimental error limits at the higher pressure conditions. These data may be used in near-field and far-field distribution modeling of the environmental fate of highly toxic BTEX compounds derived from submarine oil spills and their impact on the ecosystem. The parameters will also aid in the prediction of oil migration and dispersion away from the spill thus helping to improve response strategies.

Keywords Submarine oil spill · High pressure · Partitioning · BTEX · Experimental simulation · Live oil

8.1 Introduction

An oil spill is defined as the accidental or intentional release of petroleum into the environment. Over one million metric tonnes of petroleum enter the marine environment annually from marine transport, industrial sources, natural oil seeps, and oil spills (GESAMP 2007). The environmental impacts from the release of these large volumes of oil range from oiled shorelines and marine wildlife to intoxication of organisms that ingest oil or are exposed to the organic compounds that partition from oil to water phase. This chapter focuses on estimating the extent of partitioning of organics between the oil and water phases following a submarine oil spill.

Chemical/physical partitioning occurs when a compound distributes itself to an equilibrium state within a multiphase system, as a function of its interaction with each individual phase. The extent of partitioning may be quantified in a two-phase system using the partition ratio (Rice et al. 1993), which is the equilibrium ratio of the compound concentration in the two phases (Sangster 1989; Bennett et al. 2003). In terms of oil-water systems, the magnitude of a partition ratio is influenced by the thermodynamics of dissolution, the physicochemical properties of the oil (Bennett et al. 2007), and the bulk properties of the water system (e.g., salinity, temperature; Bennett and Larter 1997). This study measures the partitioning of benzene, toluene, ethylbenzene, and xylene (BTEX) as the organic compounds to measure the partitioning behavior between the oil and water phases. BTEX compounds are abundant in most crude oils, and their environmental partitioning is particularly important due to their toxic and carcinogenic properties (Mehlman 2006). In the past, the environmental fate of these compounds was mainly studied in surface oil spills, during which the main portion of BTEX is rapidly transported to the atmosphere via evaporation (National Research Council (NRC) 2003). In the aftermath of the *Deepwater Horizon* (DWH) oil spill, which occurred at the Macondo well in offshore Louisiana, Gulf of Mexico, at ~1500 m depth (Camilli et al. 2010), the BTEX compounds were tracked in a subsurface plume at a depth of 1100 m over 35 km away from the wellhead, shifting the focus to the subsurface partitioning behavior of BTEX compounds from oil, under low-temperature and high-pressure (i.e., deep water) conditions.

The application of chemical dispersants to oil-contaminated areas has long been used as a primary response to mitigate the impacts of the spill. The estimate of chemical dispersants used has ranged from 5500 gallons of dispersant used during the 1989 Exxon Valdez spill to 2.1 million gallons of dispersant used during the DWH spill (Kujawinski et al. 2011). Chemical dispersants aid in breaking down larger oil slicks into smaller sized oil droplets by reducing the surface tension between oil and water particles, thereby enhancing the natural dispersion. With surfactants as the key component of their chemistry, dispersants emulsify oil and help break down larger clumps of high molecular weight viscous oil, reducing the risk of surface slicks to birds and mammals and preventing the stranding of oil on sensitive shorelines. While the physical effect of breaking oil into smaller droplets is beneficial, there are conflicting views concerning the potential risks for human and environmental health posed by the use of dispersants (Bostrom et al. 2015; Kleindienst et al. 2015; Prince et al. 2016a). Considering the high volume of dispersant previously used as a critical response to oceanic oil spills and the novel deep-sea application during DWH, it becomes imperative to understand the impact and effectiveness of dispersant addition on the partitioning of the organics along the water column gradients.

The traditional methods for determining oil-water partition ratios use the shake flask technique (Xie et al. 1984; Shiu et al. 1990; Taylor et al. 1997), which is performed using "dead" oil (i.e., oil at ambient pressure and temperature [PT] conditions, devoid of dissolved solution gas). This approach cannot account for the changes in composition of the liquid hydrocarbon phase under the large variations in pressure and the low temperatures found in deep submarine oil spill conditions. In addition, such procedures commonly lead to the loss of volatile BTEX components due to evaporation. To experimentally simulate the partition behavior for submarine oil spill conditions, a customized instrument to measure the partition ratios of organic solutes between methane-charged "live" crude oil and water under high-pressure and low-temperature conditions typical of a deep submarine oil spill was built (Figs. 8.1 and 8.2). The data obtained from this study help to quantify the partitioning behavior of BTEX compounds under varying environmental conditions and could be used to predict their partitioning trends in any future submarine oil spill, as well as for pollutant plume migration and toxicity modeling.

8.2 Partition Device

The in-house-built instrument was adapted from the concepts of Bennett et al. (2003), where partition ratios for alkylphenols in an oil-water system were investigated under reservoir conditions. This approach was modified to design and build the *Partition Device*, capable of simulating deep-sea conditions of high pressure and low temperature by adding a temperature-controlled separator region and pressurized sampling loops. The partition device comprises five main

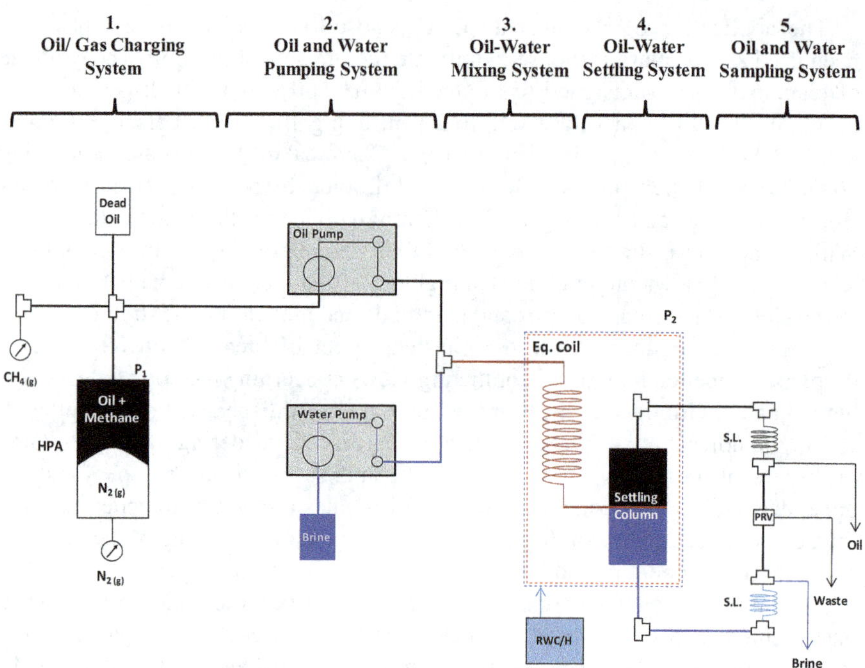

Fig. 8.1 Block diagram of the partition device demonstrating the five main segments of the system – oil-gas charging, oil and water pumping, oil-water mixing, oil-water settling, and an oil and water sampling system (*HPA* hydropneumatic accumulator, *Eq. Coil* equilibration coil, *RCW/H* recirculating water chiller/heater, *PRV* pressure release valve, P_1 methane charging pressure, P_2 methane operating pressure, *SL* sampling loops) (Jaggi et al. 2017). (Reproduced with permission from Organic Geochemistry)

sections: oil-gas charging, oil and water pumping, oil-water mixing, oil-water settling, and an oil and water sampling system (Figs. 8.1 and 8.2). The construction is described in detail in Jaggi et al. (2017).

The experiments were conducted using the device at various pressures (2, 4, 7, 12, and 15 MPa) and temperatures (4, 12, and 20 °C) to measure the partitioning trends of BTEX compounds corresponding to water depths of about 1500 m vertically through the water column to near surface depths (200 m). The oil was charged with methane in the hydropneumatic accumulator, following which the oil and the water were pumped at equal and low flow rates (0.1 mL/min) into the equilibration coil. The flow rate was kept low to promote a slow flow and shearing equilibrium of water and methane-charged oil segments in the coil. After passing through the coil, the mixed oil and water effluent entered the settling column and separated due to the density difference. A series of BTEX concentration vs. time profiles were plotted to confirm presence of steady-state conditions in the instrument during this travel time. The separated oil and water phases were collected in separate glass vials from the sampling loops connected to the top and bottom of the settling column. The separated oil and water phases were maintained under the respective system pressure conditions as they entered the sampling loops. It was only at the time of sampling of

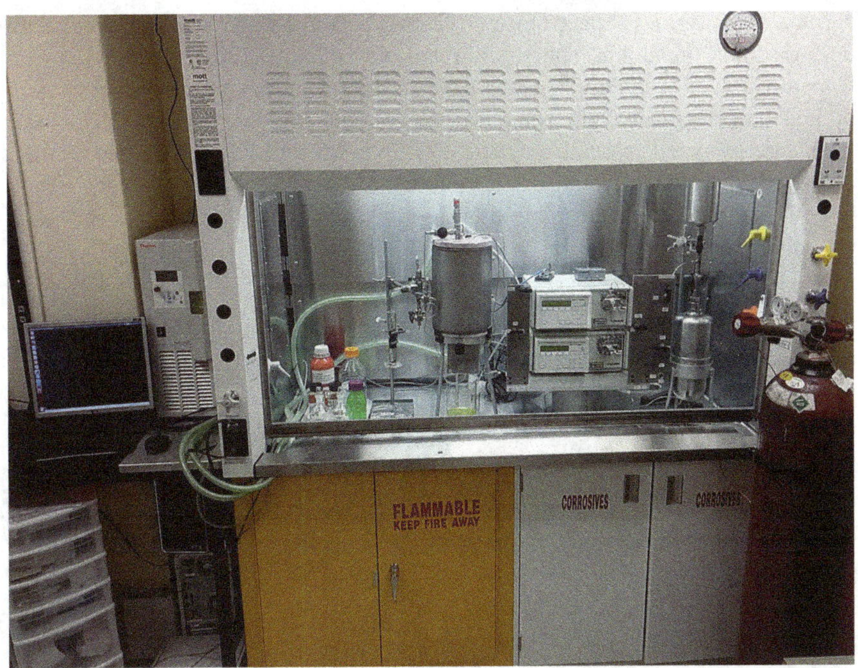

Fig. 8.2 Picture of the partition device showcasing the device, along with the computer controller for the thermometers and pumps, nitrogen and methane cylinders, and the recirculating water chiller

the separated discrete phases that the oil and water samples were brought down to ambient atmospheric pressure and temperature conditions.

All experiments were repeated with dispersant added to the system, to evaluate its effect on the partitioning of organics in water. The oil/dispersant ratio for the experiments was determined by conducting several benchtop experiments with varying dispersant concentrations to determine the time of breaking of the formed emulsion. Among the oil/dispersant ratios of 10:1, 100:1, 500:1, and 1000:1 evaluated, the emulsion breaking time ranged from 24 hours for 10:1 to 15 minutes for 1000:1. The ratio of 1000:1 was selected to be used for the experiments being conducted on the partitioning device to have a reasonable total experimental time for each run. The dispersant was mixed with the oil and added to the hydropneumatic accumulator prior to charging with methane. The rest of the experimental operations were conducted as per normal treatments.

The water samples from the partition device were filtered immediately following collection using Whatman 0.45 μm GD/X glass microfiber filters to reduce the interaction between any emulsified oil and the free water as the pressure was released. The filtration ensured that only the water-soluble organics are considered while removing the emulsified oil-in-water droplets by filtration (Liu and Kujawinski 2015; Girling 2013). The filtered water samples were then spiked with ethylbenzene-D10 as an internal standard and extracted with an aliquot of dichloromethane. The whole dichloromethane extracts were then directly analyzed by gas chromatography-mass spectrometry (GC-MS) to minimize any evaporative loss during sample handling. The equilibrated oil samples from the partition device were also spiked with ethylbenzene-d10, with 50 µL of this oil subsequently being diluted with 150 µL of dichloromethane and directly analyzed by GC-MS. All samples were analyzed in triplicate. The BTEX compounds were analyzed using an Agilent 6890 N gas chromatograph equipped with a quadrupole mass spectrometer (HP5973) using selected-ion monitoring mode (m/z 78, 91, and 98). The instrumental method is described in Jaggi et al. (2017).

8.3 Results

8.3.1 Partition Ratio Calculations

The partition ratio values are assigned to the respective analyte based upon their relative concentration in each phase using the formula:

$$P_{ow} = \frac{[X]_o}{[X]_w} \tag{8.1}$$

where $[X]$ denotes the equilibrium concentration of the analyte in oil (o) and water fraction (w) after partitioning at any given conditions. The partition ratio of a compound should not be confused with its saturation solubility in water when a pure compound phase is in contact with water. The BTEX compounds in the water and oil phase were quantified by calculating the area under the peak from the GC-MS and comparing to calculated response factors. The change in volume of oil after depressurization to the ambient conditions during sampling was measured and found to be insignificant. The equilibrium concentrations of BTEX compounds (mg/mL) in the water and oil phase were measured in triplicate for each pressure and temperature combination. The concentrations were then averaged to give the respective partition ratios using Eq. 8.1 at varying pressure and temperature conditions with equal volumes of oil and water (Table 8.1). The errors were assessed based on the standard deviation of the measured values.

The partition ratios measured from the partition device are listed in Jaggi et al. (2017) and are also publicly available through the Gulf of Mexico Research Initiative Information and Data Cooperative (GRIIDC) at https://data.gulfresearchinitiative. org/data/R4.x267.179:0004.

8.3.2 Partition Ratios Measured with the Application of Dispersant

Table 8.1 Partition ratios of benzene, toluene, ethylbenzene, *m*- and *p*-xylene, and *o*-xylene following equilibration (1:1, vol/vol) with live oil (saturated with methane gas), water, and dispersant (at oil/dispersant of 1000:1), at varying pressures (3–11 MPa) and temperatures (4–20 °C), measured in triplicates. Data are publicly available through the Gulf of Mexico Research Initiative Information and Data Cooperative (GRIIDC) at https://data.gulfresearchinitiative.org/data/R4.x267.179:0004

System conditions			Partition ratios				
Methane charging pressure (MPa)	Operating pressure (MPa)	Temperature (°C)	Benzene	Toluene	Ethylbenzene	*m*- and *p*-Xylene	*o*-Xylene
2.4	3.0	4.0	19.02	65.84	409.20	437.70	324.50
		12	17.00	68.51	370.19	392.69	296.65
		20	15.44	65.15	346.69	376.85	269.43
4.8	6.0	4.0	25.20	104.37	541.02	614.95	409.27
		12	22.63	97.38	490.49	556.65	368.14
		20	19.81	90.28	432.05	486.16	314.97
7.2	9.0	4.0	31.85	132.28	685.00	766.35	479.34
		12	27.92	123.11	604.89	692.26	416.87
		20	25.68	118.33	533.02	604.78	366.04
9.1	11	4.0	41.66	167.39	846.65	926.30	582.37
		12	36.87	152.82	770.66	822.00	534.08
		20	32.81	138.71	675.04	717.90	465.45

8.4 Discussion

8.4.1 Effects of Pressure, Temperature, and Alkylation on Partitioning of Organics

During a submarine oil spill as the oil moves buoyantly through the water column, the gases dissolved in the oil (methane, ethane, propane, etc.) exsolve, creating a three-phase system of oil, gas, and water containing distributed oil and gas components, whose concentration relationships are defined by a variety of partition ratios: $[X]_{oil}/[X]_{water}$, $[X]_{oil}/[X]_{gas}$, and $[X]_{water}/[X]_{gas}$. This chapter considers the partitioning behavior of the biphasic equilibria between gas-charged oil and the water phase under submarine oil spill conditions. The partition ratios (P_{ow}) of BTEX compounds were found to increase up to three times, with increase in pressure from 2 to 15 MPa (Jaggi et al. 2017) both with and without dispersant application. However, only minor variations were observed in the P_{ow} values (within the error bars) for the experiments conducted with dead oil, as the pressure was increased from 0.5 to 2.5 MPa at 20 °C (Jaggi et al. 2017).

Using the pressure/volume/temperature simulation software PVT-SIM, the methane content in the Macondo oil at a saturation pressure of 15 MPa was determined to be 22 mol% methane which swells the volume of the dead oil. The difference in the volume between live oil and dead oil was calculated to be up to 33% at 20 °C. Since the differences between the calculated partition ratios of BTEX between live and dead oil at 15 MPa were measured to be greater than 33%, the swelling cannot account for the entirety of this behavior. The observed approximate tripling of partition ratios at 15 MPa with live oil relative to the partition ratio at atmospheric pressure suggests that beyond swelling, there are other factors at play. As expected from engineering considerations, with an increase in gas charging pressure in the hydropneumatic accumulator, the amount of dissolved solution gas (in this case methane) in the oil increases. The pressure-dependent addition of methane to oil would principally reduce the oil polarity, increasing the polarity contrast between oil and water. This increased contrast would then reduce the tendency of investigated organics to concentrate in the water phase, causing an increase in their partition ratios with higher system pressure. As indicated however, this effect would primarily be driven by pressure-dependent oil compositional change, as much more gas dissolves in the oil than in the water with increasing system pressure. The increase of observed P_{ow} values with increasing pressure is not linear (Fig. 8.3), but to determine the true extent of this trend higher pressure measurements would be required.

The results summarized in Table 8.1 show a decrease in P_{ow} values with increase in temperature from 4 °C to 20 °C (Table 8.1, Fig. 8.3). With the increase in temperature, the solvation of BTEX compounds in the water phase relative to the oil phase increases, causing a decrease in the P_{ow} values. In addition, an observed increase in P_{ow} values, with the decrease in system temperature from 20 °C to 4 °C, is less marked compared to that with the increase in pressure from 2 to 15 MPa. This behavior is in line with the previous findings by Leo et al. (1971), who reported that the effect of temperature on P_{ow} values is not significant if the two solvents are not

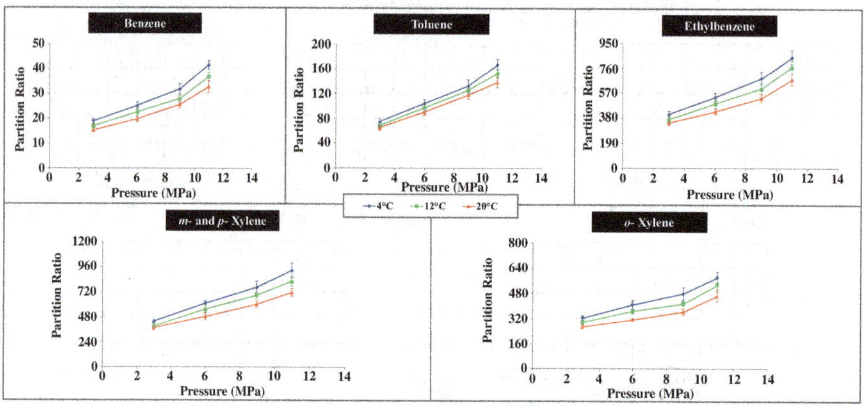

Fig. 8.3 Partition ratios of benzene, toluene, ethylbenzene, *m*- and *p*-xylene, and *o*-xylene following equilibration (1:1, vol/vol) with live oil (saturated with methane gas), water, and dispersant (at oil/dispersant of 1000:1), at varying pressures (3–11 MPa) and temperatures (4–20 °C), measured in triplicate. The error bars show a single standard deviation on either side of the mean

miscible with each other. The change in dissolved concentration of gas (methane) in oil with pressure was therefore determined to be the primary control on the partition behavior of the studied oil constituents.

In addition to the pressure and temperature, the extent of aromatic hydrocarbon alkylation also influences the partition ratios. The partition ratios of BTEX compounds were found to increase by approximately four times with the addition of each methyl group to the aromatic ring, from benzene to toluene and toluene to xylene (Table 8.1). The observed increase in organic solvent/water partition ratios with the addition of alkyl groups to the benzene core, in homologous series, is a well-recognized trend (Leo et al. 1971). With the addition of each methyl group, the strength of nonpolar intermolecular forces (London dispersion) increases, offsetting the polar interactions of the aromatic benzene ring with water (Leo et al. 1971). This leads to decreased interactions with water, lowering the concentration of the molecules in water and increasing the partition ratio values (Table 8.1).

8.4.2 Equilibrium Partition Ratio Along the Water Column

Figure 8.4 shows the equilibrium partition ratio variation of BTEX compounds in gas-saturated oils, corresponding to the pressure and temperature profile (Table 8.1) along a 1500 m depth water column in the Gulf of Mexico, with and without dispersant Corexit added (at oil/dispersant of 1000:1, Table 8.1). During the plume migration in an actual oil release, upward through the water column, the decreasing pressure contributes to the progressive loss of methane from the oil, which in turn enhances the dissolution of BTEX compounds in the water phase. Methane exsolution may also mechanically aid partition between the phases. The cumulative effect of decreasing pressure with increasing temperature, as the plume migrates to shallower waters, has a strong impact on the P_{ow} behavior of the BTEX compounds, especially for the more highly alkylated homologues (Fig. 8.4). At these conditions, as the oil constituents move upward toward the surface (decreasing pressure, increasing temperature), the P_{ow} values of BTEX compounds keep decreasing, i.e., there is an increase in their partitioning to the water phase.

With the addition of dispersant in the partitioning experiments, at an oil/dispersant concentration of 1000:1, the partition ratio values were observed to shift to lower values (Fig. 8.4) relative to the partition ratios measured without any dispersant in the system. This shift was within the error bars at higher pressure conditions (9 and 11 MPa) while being slightly higher at lower pressures (3 MPa). This effect indicates that the extent of the BTEX compounds partitioning into the water column increases with the addition of dispersant to the system, even if it is at low concentrations (1000:1). The addition of dispersants to an oil-water system promotes the formation of smaller oil droplets, contributing to an increased surface area of contact between oil and water and in turn resulting in an increased rate of dissolution of organic species. Since the partition ratios were measured at equilibrium conditions both with and without dispersant, the smaller droplets are not

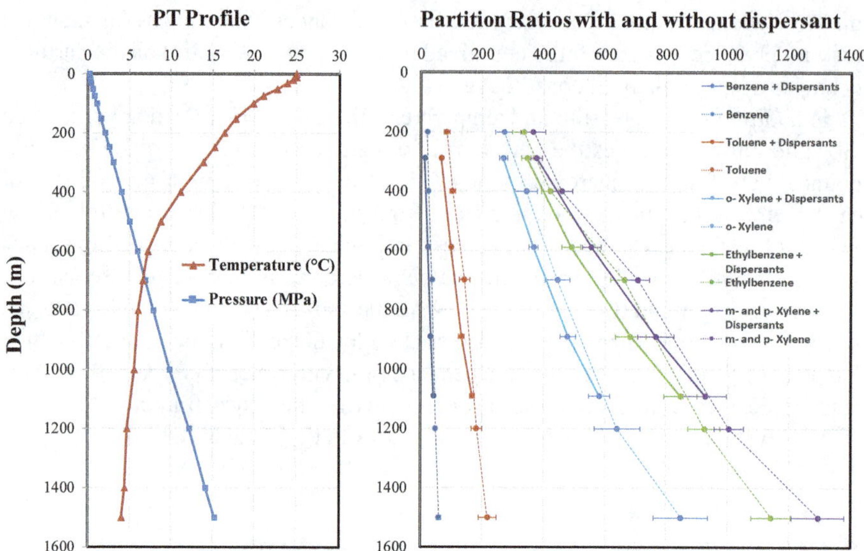

Fig. 8.4 Partition ratio trends with depth (pressure) in the with and without dispersant (at oil/dispersant ratio of 1000:1). Partition ratios (P_{ow}, the equilibrium ratio of the analyte concentration in oil (o) to the analyte concentration in water (w)) for BTEX compounds have been plotted at typical pressures along a hypothetical marine water column, with methane-charged oil and water (1:1, vol/vol) with dispersant (at oil/dispersant of 1000:1) at varying pressure (2–15 MPa) and temperature (4–20 °C) conditions as they correspond to the water column conditions at the DWH blowout site in the Gulf of Mexico. Oil at the local bubble point is assumed. (PT Profile*: The pressure and temperature profile for the Gulf of Mexico water column was built from the National Oceanic and Atmospheric Administration database for the Gulf of Mexico (Byron and Deepwater Horizon Oil Spill Water Column Technical Working Group 2015))

expected to contribute to the increased partitioning. In addition, the filtration of the water sample collected from the partition device ensured the absence of a water accommodated fraction from the smaller oil droplets, and the analysis represents only the dissolved BTEX fraction. Therefore, the results might indicate that the addition of dispersant actually changes the equilibrium conditions of the system, resulting in an increased partitioning of organics from the oil into the water phase.

As the plume migrates upward buoyantly, the exsolving methane would set up a triphasic equilibrium of the oil, water, and gas components partitioning between each other: $[X]_{oil}/[X]_{water}$, $[X]_{oil}/[X]_{gas}$, and $[X]_{water}/[X]_{gas}$. With the experimental design used in this study, only two phases are experimentally considered: methane-charged oil and water. In order to understand and address the influence of this variation on the partitioning values from the experimental setup to a "real" water column, PVT-SIM (pressure/volume/temperature simulation software, Mort A, personal communication) software was used to simulate the partitioning of BTEX into the gaseous phase with changing pressure and temperature conditions. The extent of BTEX partitioning from the oil to the vapor phase was simulated to range between 1% and

8 wt%, along the considered hypothetical water column. It is important to note here that the partitioning behavior of organics at the time of the spill is very complex, with factors such as plume dynamics and oil droplet size distribution influencing the physics and kinetics of the processes. This study focuses primarily on the effects of variations in the dissolved gas composition of oil and temperature on the dissolution of organics into the water.

8.4.3 Use of Dispersants as a Spill Response Method

The ecological impact of addition of dispersants on human and environmental health, following a spill has long been a point of contention (Kleindienst et al. 2016; Prince et al. 2016b; Rahsepar et al. 2016). Dispersants are often used as a response method to promote natural dispersion of oil, by breaking up larger oil volumes into smaller droplets and preventing the formation of oil slicks. While the dispersants themselves contribute little to the toxicity, it is the oil that gets dispersed as a result of their application that largely governs the toxicity (Bobra et al. 1989; Ramachandran et al. 2004).

The dispersants, however, can remain in the environment for long periods of time. Kujawinski et al. (2011) detected the DOSS (dioctyl sodium sulfosuccinate) signature, a key component in the dispersant (Corexit) in the oil plume that was formed at 1000–1200 m water depth near the Macondo wellhead, implying that applied dispersant stayed in the water without appreciably degrading for 64 days after the dispersant application ceased during the DWH oil spill. In certain environments, the addition of dispersants has been found to increase the absorption of poly-aromatic hydrocarbons (PAHs) in aquatic species (Bobra et al. 1989; Ramachandran et al. 2004). The dispersants also influence the microbial activity in the marine environment and have been known to both increase (Prince et al. 2016b) and suppress the activity of specific oil-degrading microorganisms (Kleindienst et al. 2015).

The application of dispersants following an oil spill needs to be carefully evaluated in each environmental setting by weighing both merits and demerits and the efficacy of alternative response methods.

8.5 Conclusions

The partitioning of organics between the oil and water phase is affected by changing pressure and temperature trajectories during deepwater submarine oil spill scenarios. As gas-charged oil from a deep, high-pressure reservoir is injected into cold, relatively lower-pressure water, dramatic changes in oil solution gas content, oil polarity, and density will occur as gas exsolves from the oil and continues to exsolve as the oil plume moves to shallower water. The traditional methods of assessing oil-water partition measurements fail to reflect changes in gas concentrations in live oils, which were addressed by using an in-house constructed device, capable of measuring the partitioning behavior of BTEX compounds between gas-charged live oil and cold high-pressure water.

The partition ratios of BTEX compounds between oil and water decrease substantially as the plume migrates to lower pressure values near the sea surface (from 15 to 2 MPa) due to the decreasing concentration of dissolved methane gas in oil at lower pressures. The partition ratios of BTEX compounds increase proportionally with an increase in alkylation and with decrease in temperature over the range of 4–20 °C, but the decrease in pressure dominates the net behavior as oils migrate to shallower depths. These observed effects of pressure and temperature act together to increase the equilibrium concentration of BTEX compounds in the water phase as the plume migrates toward the surface. However, as the plume loses BTEX compounds during its upward migration, the actual behavior becomes much more complex. The effects of pressure and temperature are similar when comparing systems with and without dispersants. The addition of dispersants, however, adds an additional effect of increasing the extent of organics partitioning into the water, especially at equilibrium under lower-pressure conditions.

The results of this study will help improve both near-field and far-field distribution modeling of the environmental fate of toxic crude oil components, by predicting component migration pathways from potential oil spills, as well as plume behavior assessments from deep-water blowouts. Overall, these models will help to understand and predict the impact of oil spills on marine ecosystems and to improve oil spill response strategies.

For additional insights, it will be useful to use different oil types to assess the impact of changing oil chemistry on the partitioning behavior of xenobiotics through the hypothetical water column. This study was conducted to measure the partition ratios at PT conditions for ~1500 m of water depth, as per the Macondo setting. However, with planned deeper future drilling operations in the Gulf of Mexico at double that depth (~3000 m), there is also a need to expand the data set to higher pressures, either experimentally or via experimentally calibrated models. The information gathered from these deeper well models, when combined with the findings of this report, could be used to ensure that safer and more environmentally friendly, mitigation strategies are undertaken, should an incident similar to the Macondo blowout occur.

Acknowledgments This research was made possible in part by a grant from The Gulf of Mexico Research Initiative/C-IMAGE II and in part by the Canada Foundation for Innovation (CFI), the Natural Sciences and Engineering Research Council of Canada (NSERC) and Canada Research Chairs (CRC), PRG, and the University of Calgary.

References

Bennett B, Aplin AC, Larter SR (2003) Measurement of partition coefficients of phenol and cresols in gas-charged crude oil/water systems. Org Geochem 34:1581–1590

Bennett B, Larter SR (1997) Partition behaviour of alkylphenols in crude oil/brine systems under subsurface conditions. Geochim Cosmochim Acta 61:4393–4402

Bennett B, Noke KJ, Bowler BF, Larter SR (2007) The accurate determination of C0-C3 alkylphenol concentrations in crude oils. Int J Environ Anal Chem 87:307–320

Bobra AM, Shiu WY, Mackay D, Goodman RH (1989) Acute toxicity of dispersed fresh and weathered crude oil and dispersants to *Daphnia Magna*. Chemosphere 19:1199–1222

Bostrom A, Walker AH, Scott T, Pavia R, Leschine TM, Starbird K (2015) Oil spill response risk judgments, decisions, and mental models: findings from surveying U.S. stakeholders and coastal residents. Hum Ecol Risk Assess 21:581–604. https://doi.org/10.1080/10807039.2014.947865

Byron S and Deepwater Horizon Oil Spill Water Column Technical Working Group (2015) NRDA-processed CTD data from the OCEAN VERITAS in the Gulf of Mexico, Cruise 4 Leg 1, collected from 2010-06-14 to 2010-06-16, associated with the Deepwater Horizon Oil Spill event (NCEI Accession 0128170)

Camilli R, Reddy CM, Yoerger DR, VanMooy BAS, Jakuba MV, Kinsey JC, McIntyre CP, Sylva SP, Maloney JV (2010) Tracking hydrocarbon plume transport and biodegradation at Deepwater Horizon. Science 330:201–204

Girling AE (2013) WAF evaluation for toxicity analysis. J Chem Inf Model 53:1689–1699

Jaggi A, Snowdon RW, Stopford A, Radovic JR, Oldenburg TBP, Larter SR (2017) Experimental simulation of crude oil-water partitioning behavior of BTEX compounds during a deep submarine oil spill. Org Geochem 108:1–8. https://doi.org/10.1016/j.orggeochem.2017.03.006

Joint Group of Experts on the Scientific Aspects of Marine Environmental Protection (GESAMP) (2007) Estimates of oil entering the marine environment from sea-based activities. Int Maritime Organ

Kleindienst S, Seidel M, Ziervogel K, Grim S, Loftis K, Harrison S, Malkin SY, Perkins MJ, Field J, Sogin ML, Dittmar T, Passow U, Medeiros PM, Joye SB (2015) Chemical dispersants can suppress the activity of natural oil-degrading microorganisms. Proc Natl Acad Sci 112:14900–14905. https://doi.org/10.1073/pnas.1507380112

Kleindienst S, Seidel M, Ziervogel K, Grim S, Loftis K, Harrison S, Malkin SY, Perkins MJ, Field J, Sogin ML, Dittmar T, Passow U, Medeiros PM, Joye SB (2016) Reply to Prince et al.: ability of chemical dispersants to reduce oil spill impacts remains unclear. Proc Natl Acad Sci 113:201600498. https://doi.org/10.1073/pnas.1600498113

Kujawinski EB, Kido Soule MC, Valentine DL, Boysen AK, Longnecker K, Redmond MC (2011) Fate of dispersants associated with the Deepwater Horizon oil spill. Environ Sci Technol 45:1298–1306

Leo A, Hansch C, Elkins D (1971) Partition coefficients and their uses. Chem Rev 71:525

Liu Y, Kujawinski EB (2015) Chemical composition and potential environmental impacts of water-soluble polar crude oil components inferred from esi FT-ICR MS. PLoS One 10:1–18

Mehlman MA (2006) Dangerous and cancer-causing properties of products and chemicals in the oil refining and petrochemical industries. Part XXX: causal relationship between chronic myelogenous leukemia and benzene-containing solvents. Ann N Y Acad Sci 1076:110–119

National Research Council (NRC) (2003) Oil in the sea III: inputs, fates, and effects. National Academies Press, Washington DC

Prince RC, Coolbaugh TS, Parkerton TF (2016a) Oil dispersants do facilitate biodegradation of spilled oil. Proc Natl Acad Sci 113:E1421–E1421. https://doi.org/10.1073/pnas.1525333113

Prince RC, Nash GW, Hill SJ (2016b) The biodegradation of crude oil in the deep ocean. Mar Pollut Bull 111:6–9. https://doi.org/10.1016/j.marpolbul.2016.06.087

Rahsepar S, Smit MPJ, Murk AJ, Rijnaarts HHM, Langenhoff AAM (2016) Chemical dispersants: oil biodegradation friend or foe? Mar Pollut Bull 108:113–119. https://doi.org/10.1016/j.marpolbul.2016.04.044

Ramachandran SD, Hodson PV, Khan CW, Lee K (2004) Oil dispersant increases PAH uptake by fish exposed to crude oil. Ecotoxicol Environ Saf 59:300–308. https://doi.org/10.1016/j.ecoenv.2003.08.018

Rice NM, Irving HMNH, Leonard MA (1993) Nomenclature for liquid-liquid distribution (solvent extraction) (IUPAC Recommendations 1993). Pure Appl Chem 65:2373–2396

Sangster J (1989) Octonol water partition coefficients of simple organic compounds. J Phys Chem Ref Data 18:1111–1229

Shiu WY, Bobra M, Bobra AM, Maijanen A, Suntio L, Mackay D (1990) The water solubility of crude oils and petroleum products. Oil Chem Pollut 7:57–84

Taylor P, Larter S, Jones M, Dale J, Horstad I (1997) The effect of oil-water-rock partitioning on the occurrence of alkylphenols in petroleum systems. Geochim Cosmochim Acta 61:1899–1910

Xie TM, Hulthe B, Folestad S (1984) Determination of partition coefficients of chlorinated phenols, guaiacols and catechols by shake-flask GC and HPLC. Chemosphere 13:445–459

Chapter 9
Dynamic Coupling of Near-Field and Far-Field Models

Ana C. Vaz, Claire B. Paris, Anusha L. Dissanayake,
Scott A. Socolofsky, Jonas Gros, and Michel C. Boufadel

Abstract Deepwater spills pose a unique challenge for reliable predictions of oil transport and fate, since live oil spewing under very high hydrostatic pressure has characteristics remarkably distinct from oil spilling in shallow water. It is thus important to describe in detail the complex thermodynamic processes occurring in the near-field, meters above the wellhead, and the hydrodynamic processes in the far-field, up to kilometers away. However, these processes are typically modeled separately since they occur at different scales. Here we directly couple two oil prediction applications developed during the *Deepwater Horizon* blowout operating at different scales: the near-field Texas A&M Oilspill Calculator (TAMOC) and the far-field oil application of the Connectivity Modeling System (oil-CMS). To achieve this coupling, new oil-CMS modules were developed to read TAMOC output, which consists of the description of distinct oil droplet "types," each of specific size and pseudo-component mixture that enters at a given mass flow rate, time, and position into the far field. These variables are transformed for use in the individual-based framework of CMS, where each droplet type fits into a droplet size distribution (DSD). Here we used 19 pseudo-components representing a large range of hydrocarbon compounds and their respective thermodynamic properties. Simulation

A. C. Vaz (✉) · C. B. Paris
University of Miami, Department of Ocean Sciences, RSMAS, Miami, FL, USA
e-mail: avaz@rsmas.miami.edu; cparis@rsmas.miami.edu

A. L. Dissanayake
RPS Group North America, Ocean Sciences, South Kingstown, RI, USA
e-mail: Anusha.Dissanayake@rpsgroup.com

S. A. Socolofsky
Texas A&M University, Department of Ocean Sciences, College Station, TX, USA
e-mail: socolofs@tamu.edu

J. Gros
GEOMAR Helmholtz Centre for Ocean Research Kiel, Kiel, Germany
e-mail: jogros@geomar.de

M. C. Boufadel
The New Jersey Institute of Technology, John A. Reif, Jr. Department of Civil and Environmental Engineering, Newark, NJ, USA

© Springer Nature Switzerland AG 2020
S. A. Murawski et al. (eds.), *Deep Oil Spills*,
https://doi.org/10.1007/978-3-030-11605-7_9

results show that the dispersion pathway of the different droplet types varies significantly. Indeed, some droplet types remain suspended in the subsea over months, while others accumulate in the surface layers. In addition, the decay rate of oil pseudo-components significantly alters the dispersion, denoting the importance of more biodegradation and dissolution studies of chemically and naturally dispersed live oil at high pressure. This new modeling tool shows the potential for improved accuracy in predictions of oil partition in the water column and of advancing impact assessment and response during a deepwater spill.

Keywords Far-field model · Near-field model · Coupling · Deepwater well blowout · Coupled near-field and far-field models · Oil transport prediction · Model parameterization

9.1 Introduction

During deepwater oil spills, it is fundamentally important to account for both the complex dynamic processes occurring in the near-field, a couple hundred of meters above the wellhead, and the continuing transport and biological processes occurring in the far field, up to kilometers away from the source. The near-field refers to the dynamic, buoyant mixture of oil, gas, entrained seawater, and potentially gas hydrate ascending through the water column after release from a broken wellhead, leaking pipe, or distributed seafloor release. In this chapter we focus on localized sources, and as this near-field plume ascends in the water column, the lighter fractions of the gas bubbles and live oil droplets dissolve. Eventually, the ocean density stratification arrests the buoyant plume, forming an intrusion layer, and petroleum fluids having a weathered composition enter the far-field domain, where ocean currents and particle motion dominates. Tracking of the oil droplets beyond the near-field is referred to as far-field modeling. These domains are typically modeled separately since they occur at different spatiotemporal scales. However, to accurately simulate the fate and transport of oil and gas emitted from deep-sea blowouts, it is necessary to enable a smooth transition between near- and far-field models.

When gas and oil are released from an accidental deepwater oil well blowout or similar highly localized source, models must track the dispersal and dynamics of the live oil, simulating the generation of the turbulent, buoyant jet caused by the mixture of ambient seawater and gas and oil existing the source, and the subsequent breakup of the jet into small gas bubbles and oil droplets (Bandara and Yapa 2011; Johansen et al. 2013; Zhao et al. 2014, 2015, 2017; Nissanka and Yapa 2016; Li et al. 2017; Wang et al. 2018). After the initial jet breakup, a buoyant plume of gas bubbles, oil droplets, and entrained seawater develops (Zheng et al. 2003; Socolofsky and Dissanayake 2016; Dissanayake et al. 2018). The spatial scale of such plumes is small, in the order of meters; thus, specialized models are adapted to simulate processes at these small scales. Beginning at the oil spill source, near-field models track the evolution of spilled oil and gas throughout the region dominated by local, non-hydrostatic buoyant forcing. Models to predict the initial conditions to near-field models must also capture changes occurring before the petroleum was released in the seawater, such as phase changes occurring during ascent from the petroleum reservoir to the seafloor (Zick 2013; Gros et al. 2016), the complex multiphase flows within crippled oil well pipelines (Boufadel et al. 2018), and the breakup of gas and liquid petroleum into bubbles and droplets in the initial jet at the spill orifice. Near-field models may also account for the live oil chemical composition and the wide range of thermodynamic and chemical processes it undergoes. At high pressure and temperature, the petroleum fluids are non-ideal, and as pressure and temperature change throughout their transport, the phase equilibrium between gas and liquid petroleum is constantly evolving (Gros et al. 2016, 2017; Lehr and Socolofsky 2020). For a deepwater release, the liquid-phase petroleum may contain a large amount of dissolved gases, giving a lighter, quite soluble live oil. The gas phase also has significant fractions of heavier petroleum molecules dispersed within it, so that after dissolution of the volatile constituents, gas-phase bubbles may condense to liquid drops (Gros et al. 2017). Gradually, the buoyant near-field plume is arrested by the density stratification in the ocean and is bent over in the downstream direction of the currents, eventually forming a lateral intrusion layer of dissolved petroleum and fine oil droplets mixed with seawater (Socolofsky et al. 2011; Gros et al. 2017; Dissanayake et al. 2018).

Once the oil reaches the far field, the remaining petroleum fluids rise out of the intrusion layer. Depending on the spill, these may include gas bubble, condensed gas droplets, live oil droplets, and weathered oil droplets, each advected by its own buoyant rise velocity and the ambient currents and dispersed by turbulent diffusion of the ocean. These droplets and bubbles can then be represented by individual elements in Lagrangian models and are subjected to transport and dispersion by local hydrodynamic processes. Physical processes, varying on spatial scales from centimeters to tens of kilometers, and the continuing, complex biogeochemical processes influence the final distribution and fate of the oil. In far-field numerical models of oil dispersal, these physical processes are represented by the advection, driven by the deterministic velocity from hydrodynamic models, by the parametrization of sub-grid-scale turbulence, using a random displacement (Paris et al. 2013), and by the droplet terminal velocity, given by the droplet sizes and buoyancy (Zheng et al. 2003). During their transport, the characteristics (diameter, density, composition) of each droplet evolve continually due to biogeochemical processes, such as dissolution, biodegradation, particle flocculation and aggregation, sedimentation, and evaporation (see Boehm et al. 2020; Bubenheim et al. 2020; Jaggi et al. 2020; Le Hénaff et al. 2012; North et al. 2015; Paris et al. 2012; Yapa et al. 2010). Ambient conditions will influence the rate at which these processes take place, and these interactions need to be accounted for in the model.

Furthermore, the response actions selected for containment or mitigation of a deepwater oil spill may also alter the chemical and thermodynamics characteristics of the oil itself, both in the near- and far-field domains. Burning of surface oil, a common measure for containment of superficial oil, can result in the creation of heavy, pyrogenic hydrocarbons in cases of incomplete combustion. Surface application of chemical dispersants, a common practice used to break up fresh surface oil slicks, may also enhance the formation of oil-mineral aggregates, resulting in transport of oil to the seafloor by marine snow. Subsea dispersant injection (SSDI) also alters not only the droplet size distribution at the orifice but also the kinetics of aqueous dissolution of oil compounds, biodegradation rates, and aggregation and flocculation processes. In the case of SSDI, jet breakup and near-far-field models must account for these processes to accurately predict the oil distribution and fate. For surface burning and surface dispersant application, far-field models must account for the evolving oil composition and transport characteristics. Hence, oil spill models must account for diverse intervention methods to accurately predict oil fate from accidental deepwater oil spills.

Here, we present an overview of plume and oil droplet dynamics in both the near- and far-field domains, describing mechanisms underlying the biogeochemical and physical processes influencing the oil fate. We also discuss what the main processes are that a coupled near-field/far-field model must present in order to successfully improve the hindcast and forecast of oil transport. Finally, we present a proof-of-concept model, which couples two numerical tools developed since the *Deepwater Horizon* (DWH) blowout, each operating at different scales and coupled together: the near-field Texas A&M Oilspill Calculator (TAMOC) and the far-field oil Connectivity Modeling System (oil-CMS). During deepwater spills, it is fundamental to forecast the formation and propagation of lateral intrusion layers and the impact of response scenarios; thus, we focus our results on the vertical distribution of hydrocarbons and on quantifying the effects of biodegradation and SSDI.

9.2　Models Description and Coupling

9.2.1　Near-Field Modeling

In this chapter, we confine near-field modeling to those models designed to predict the initial formation of gas bubbles and oil droplets in the jet breakup region immediately following the release at the orifice and those that predict the ensuing buoyant dynamics of the near-field gas, oil, and entrained seawater plume. Two approaches are used for jet breakup, and these include empirical equations that predict a characteristic bubble or droplet size at the end of the dynamic breakup region (e.g., Johansen et al. 2013; Li et al. 2017; Wang et al. 2018) and those that solve for the competing physical and chemical processes regulating breakup and coalescence, allowing for a dynamic solution of the entire bubble and droplet size distribution throughout the jet break-up region (Bandara and Yapa 2011; Zhao et al. 2014, 2015, 2017). Likewise, for the buoyant plume, two types of models have been developed to predict the multiphase oil spill plume dynamics at field scale: integral models, which solve the cross-sectionally averaged flow along the centerline trajectory of the plume (Johansen et al. 2003; Chen and Yapa 2003; Yapa et al. 2001), and three-dimensional computational fluid dynamics (CFD) models, based on large eddy simulation (LES, Fraga et al. 2016). Since CFD modeling approaches are computationally costly to implement, comprehensive models designed to guide response during an oil spill still use the integral modeling approach (Johansen et al. 2003; Zheng et al. 2003; Gros et al. 2017; Dissanayake et al. 2018).

The initial conditions to each of near-field models we consider include the oil and gas composition, physicochemical properties, flow rate, orifice characteristics, water column conditions, and depth. Thermodynamic calculations partition the released petroleum fluid into gas and liquid phases and predict their properties throughout the near-field region. These models consider the main fate processes affecting the gas bubbles and oil droplets. Dissolution is the primary fate process in the near-field, and gas dissolution should be modeled to accurately predict the intrusion location (Socolofsky et al. 2015). Dissolution from liquid-phase petroleum can also be significant for the fate of light hydrocarbons, especially for deepwater releases of live oil (Lehr and Socolofsky 2020; Oldenburg et al. 2020; Malone et al. 2020, Pesch et al. 2020; Gros et al. 2016, 2017; Dissanayake et al. 2018). Other important processes to consider are heat transfer and the density equation of state of gas and liquid petroleum as the oil composition changes and the pressure and temperature evolve. For breakup, equations are needed to predict interfacial tension and viscosity. Each of these processes is significant during the rapid cooling near the release and in the early stages of dissolution. Once the oil and gas droplets reach the intrusion layer, however, much of the rapid dissolution has completed, and biodegradation begins to compete with dissolution for the dominant fate processes.

When the petroleum fluids exit the orifice and enter the ocean, break-up models simulate the formation of bubbles and droplets. To predict the evolving sizes and complete size distribution throughout the turbulent jet, physics-based models are needed (Zhao et al. 2014, 2015, 2017). These physical models compare the resistive forces from interfacial tension and viscosity for each bubble and droplet in a flow to the destructive forces from turbulence that work to break the gas and liquid into bubbles and droplets. These models respond to the rapidly varying turbulent dissipation rate in a plume and have been developed to handle gas, oil, and the effects of injected subsea dispersant, both in the near-field, turbulence-dominated regime and in the tip-streaming-dominated breakup, occurring in the buoyant plume, and potentially in the intrusion layer and far-field domain.

Following the momentum-dominated jet break-up region near the release, a plume develops that is controlled by the collective positive buoyancy of the oil and gas, the negative buoyancy of the entrained, stratified ocean water, and the drag resulting from the lateral currents. Integral models simulate these effects by tracking the trajectory of a control volume, integrated laterally across the plume and tracked along the plume centerline. Eulerian (Jirka 2004) and Lagrangian (Lee and Cheung 1990; Lee and Chu 2003) integral models exist, with all blowout models being of the Lagrangian integral plume model type. In these models, entrainment occurs by shear entrainment at the plume edge (Turner 1968) and by engulfment of water entering the plume from the crossflow (Lee and Chu 2003). Because oil and gas particles advect within the control volume with their own rise velocity, they may leave the plume on the upstream side. This effect has been observed in laboratory experiments (Socolofsky and Adams 2002) and is considered by mainstream blow-out models (Johansen et al. 2003; Chen and Yapa 2003; Dissanayake et al. 2018). Eventually, the negative buoyancy of the entrained seawater will be greater than the buoyant force from the remaining gas bubbles and oil droplets, the plume will stop rising, and it will fall to form a lateral intrusion at a level of neutral buoyancy in the density-stratified ocean. This region where the intrusion layer forms is generally considered the end of the near-field and the beginning of the far-field oil and gas transport (Socolofsky and Dissanayake 2016).

At the end of the near-field, the simulation results are used to provide initial conditions to the far-field model. These initial conditions include the oil droplet and gas bubble sizes, their composition, density, locations, and mass flow rates at the end of the near-field domain. The near-field model also tracks the dissolved-phase petroleum within the integral plume model control volume, and the mass flow rates of dissolved hydrocarbons at the intrusion layer may also be passed to a far-field Lagrangian transport model.

9.2.2　Far-Field Lagrangian Modeling

Once in the far field, the gas bubbles and oil droplets are apportioned to individual Lagrangian particles. These models track the individual particles (i.e., gas bubbles and oil droplets, each representing a given mass flow rate) geographical coordinates, depths, and compositions, while simultaneously simulating the physical and biogeochemical processes influencing their distribution and fate. The explicit composition and characteristics of each oil droplet and gas bubble needs to be considered. The final petroleum distribution is primarily driven by the advection and dispersion by the oceanic hydrodynamics, including currents, fronts, turbulence, and internal waves. However, other important processes influence their distribution. For instance, droplets reaching the far field originate from many different types (i.e., median size, hydrocarbon compositions), resulting in different phases, densities, rates of ensuing fate processes, notably dissolution, and biodegradation. The physicochemical characteristics of each droplet will determine their terminal velocity and vertical distribution (Pesch et al. 2020). Larger droplets ascend the water column faster, while smaller ones can remain at depth, driven mostly by ocean stratification, upwelling, and subduction currents.

Within the far field, the advection of individual droplets is represented by the integration of the deterministic velocity field from hydrodynamic models, most commonly using a spatiotemporal 4th order Runga-Kutta scheme (Paris et al. 2013). The sub-grid-scale turbulent diffusion is resolved by adding a random displacement scaled by a diffusivity coefficient derived from observations (Okubo 1987). The wind-induced drift needs to be incorporated in the velocity field and explicitly considered to accurately predict the surface slicks (Reed et al. 1999; Le Hénaff et al. 2012; Perlin et al. 2020). The droplet terminal velocity is obtained from a relationship involving their buoyancy as a function of density, size, and shape (i.e., Reynolds number) and the ambient fluid viscosity and density (Clift et al. 1978; Zheng et al. 2003; Chen and Yapa 2003; Yapa et al. 2010). The oil-CMS uses the shape classification from Clift et al. (1978) modified by Zheng et al. (2003), working with particles in the ellipsoidal and spherical regimes (Paris et al. 2012, Perlin et al. 2020). Consequently, the droplet volume median diameter (d_{50}) provided by the near-field to the far-field model is a critical parameter for reliable predictions of the distribution and fate of petroleum compounds.

To couple near- and far-field models, it is also necessary to convert the near-field model output into the far-field Lagrangian input while conserving mass balance. Commonly, near-field models output the flow rate given as a mass flow rate (kg/s) for each droplet size (Gros et al. 2017; Dissanayake et al. 2018). This mass flow rate is then converted into a number of individual particles (oil droplets and gas bubbles) simulated within the far-field framework (Eq. 9.1).

$$Number\_particles_{droplet\_type_i} = \frac{\left(Mass\_flow\_rate_{droplet\_type_i}\right)dt_{CMS}}{\left(Mass_{droplet\_type_i}\right)dt_{near\text{-}field}} \quad (9.1)$$

Similar to the near-field, gas and oil droplets undergo continuous changes while transported in the far field. Dissolution continues to take place, and biodegradation of the dissolved components become a dominant fate process. Biodegradation rates are dependent on the local microbial fauna, the composition of the petroleum fluids, and the environmental conditions since local temperature, oxygen, nutrient availability, and pressure change the rate of oil degradation (see Bubenheim et al. 2020; Joye et al. 2016). Once droplets reach the ocean surface, evaporation affecting low molecular components will take place (Stout et al. 2016; Drozd et al. 2015; Romero et al. 2015). Depending on the pressure drop between the reservoir and the wellhead, droplet buoyancy may be affected by degassing (Malone et al. 2020), which would result in droplet of larger diameters and of lower densities as their rise in the water column (Pesch et al. 2020). Droplets can aggregate with organic and inorganic matter and form marine oil snow (MOS), which presents a higher density than droplets and subsequently settles on the seafloor (Dissanayake et al. 2018; Daly et al. 2020). These processes are accounted for in far-field models, either as temporally-spatially explicit processes or by the use of constant parameters.

9.3 Coupled Near-Field and Far-Field Model

The work of the C-IMAGE consortium resulted in a coupled near-field and far-field modeling system based on VDROP-J and TAMOC for the near-field model and oil-CMS for the far-field model. This dynamic integrated modeling system simulates droplets with realistic compositions, accounting for important thermodynamic and biochemical processes in both the near and far fields. The dynamic coupling of TAMOC and oil-CMS uses observed in situ temperature, salinity, pressure, and velocity profiles in the near-field computations to capture the formation of the deep intrusion that formed near 1100 m depth. As the live oil rises in the water column, TAMOC tracks the evolving composition of droplets and passes their mass flow rates and properties off to the far-field oil-CMS model as initial conditions for the far-field computations of individual droplets transport and fate, taken from an assumed droplet size distribution (DSD). TAMOC simulations are described in Gros et al. (2016, 2017). Here the original 279 pseudo-components were reduced to 19, including 5 soluble pseudo-components that showed significant extent of aqueous dissolution during the DWH accident (Reddy et al. 2012; Ryerson et al. 2012), to reduce the computational cost of simulations. Similarly to the Gros et al. (2017) paper, the model includes two different scenarios, one with SSDI and one without.

TAMOC output describing distinct droplet types is converted into the individual-based Lagrangian framework of oil-CMS following Eq. 9.1. Because TAMOC considers a finite number of particle sizes and oil-CMS initializes particles from a continuous size distribution, the discrete droplet diameters from TAMOC are fit to a CMS probabilistic droplet size distribution (DSD) while ensuring that mass is conserved. The pseudo-component mixture is represented in oil-CMS by multiple fractions (pseudo-components) within individual droplet (Lindo-Atichati et al. 2016), and processes affecting oil fate (biodegradation, sedimentation, landfall) are calculated for each fraction. Droplet movement is a result of their displacement due to the deterministic velocity field, turbulent mixing (Paris et al. 2013), and buoyancy (Zheng et al. 2003). For these DWH simulations, environmental variables, such as horizontal and vertical velocity, temperature, and salinity, are from the Gulf of Mexico HYCOM hindcast (0.04 degree horizontal resolution, 20 vertical layers). Values for the ranges of the biodegradation rates for each pseudo-component are compiled from the list of half-lives of chemical compounds from Prosser et al. (2016). For pseudo-components with a small number of data points ($N < 6$), the minimum and maximum half-lives were used for the faster and slower biodegradation rates, respectively, while for pseudo-components with a larger number of data points ($n > 6$), the 25th and 75th percentiles were used.

Different droplet sizes released in the far field at the same time form contrasting horizontal dispersion and vertical layers (Fig. 9.1), indicating that the initial density and diameter of droplets are a major driver of their vertical distributions and, ultimately, oil fate. Smaller and denser droplets (Fig. 9.1a), such as the microdroplets added in the near-field model to account for tip-streaming by dispersant injection, remain at depth, following the deep intrusion layer which extends over 300 m of the water column near 1100 m depth. In contrast, larger, less dense droplets ascend into the water column rapidly (Fig. 9.2b–d). We further observe that oil droplets sequestered in the deep intrusion layer are only formed for the chemically dispersed case (CD, i.e., including SSDI), which is expected given the large initial DSD from TAMOC used in our simulations for the naturally dispersed case (ND, i.e., without SSDI) (Figs. 9.2 and 9.3).

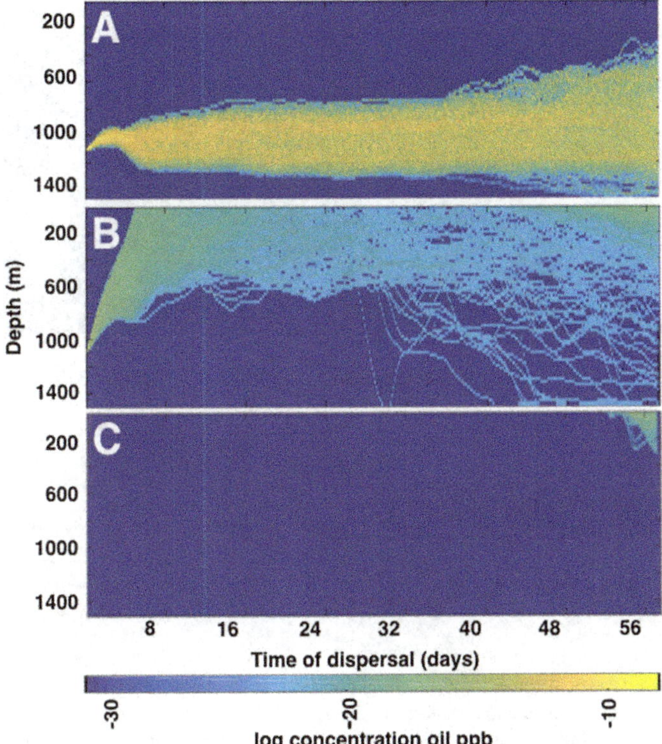

Fig. 9.1 Daily vertical concentrations (log ppb) of oil from our coupled near-far-field model (TAMOC, Gros et al. 2017, and the oil-CMS, Paris et al. 2012). The hydrocarbon plume is released from the Macondo fallen riser on 5/2/10 00 hr., chemically dispersed with subsea dispersant injection, subjected to slower biodegradation rates, and dispersed for 60 days. Droplets are grouped into classes based on their characteristics (density, diameter, chemical composition, mass flow rate). Daily vertical concentrations averaged over a 3D regularly spaced grid spanning from 25°N to 30°N and from 93°W to 84°W are shown for specific droplet categories: (**a**) microdroplets added in the far field to account for tip-streaming from subsea dispersant injection (diameter < 50 µm, density 875.5 kg/m^3); (**b**) droplets with diameter 70 µm-0.12 mm, density 875.5 kg/m^3; and (**c**) droplets with diameter 0.9 mm -1.02 mm, density 867.9 kg/m^3

When considering different biodegradation rates, a faster biodegradation rate reduces oil droplet concentrations in the water column at the end of the 60-day simulations compared to simulations with slower rates. For the simulations in Fig. 9.3, oil concentrations (in ppb) are reduced by ~4% for both the ND and the CD cases in the faster-rate simulations compared to the slow-rate ones. However, since the ND plume is constrained to the surface, changes in biodegradation rates affect primarily surface oil, while for the CD cases, biodegradation can change oil distributions throughout the water column (Figs. 9.2e, f and 9.3). In our simulations, the small oil droplets in the deep intrusion remains mostly bellow 1000 m for the faster biodegradation scenario, while they spread from 500 m to 1000 m for the slower biodegradation case. The differences observed highlight the need for predictive

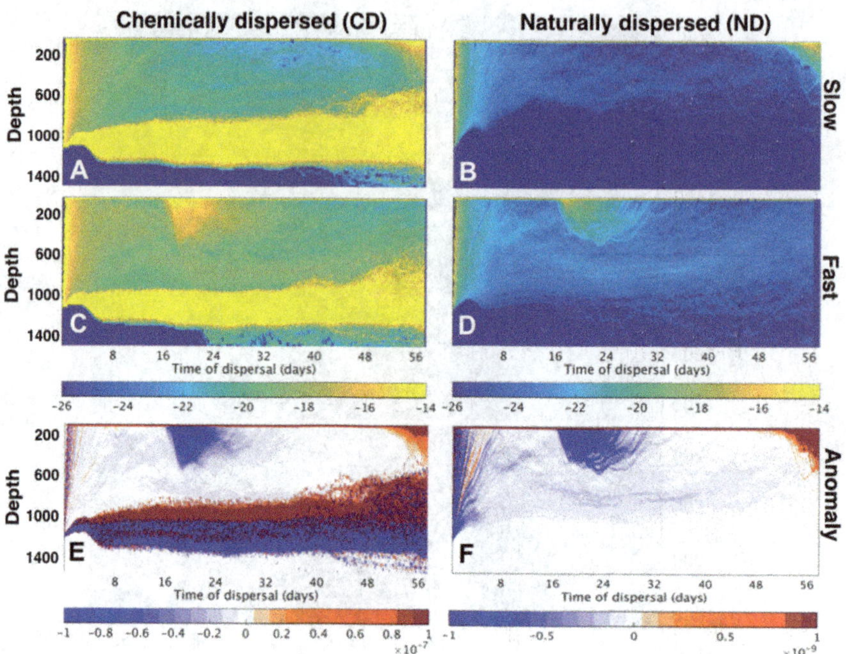

Fig. 9.2 Daily vertical concentrations (log ppb) of oil from our coupled near-far-field model (TAMOC, Gros et al. 2017, and the oil-CMS, Paris et al. 2012). The hydrocarbon plume is released from the Macondo fallen riser on 5/2/10 00 hr. Daily vertical concentrations averaged over a 3D regularly spaced grid spanning from 25°N to 30°N and from 93°W to 84°W are shown for the ensemble of all droplet categories, where the right panel (**a** and **c**) represents the plume chemically dispersed with subsea dispersant injection and the left panel (**b** and **d**) the naturally dispersed plume. The upper panels (**a** and **b**) are plumes subject to a slower biodegradation rate, mid panels (**c** and **d**) are subject to faster biodegradation rates, and bottom panels (**e** and **f**) reflect the difference between the biodegradation scenarios (log ppb)

models to consider a more realistic range of droplet characteristics. It also supports the inclusion of individual droplet composition, as achieved by oil-CMS in the form of fractions for pseudo-components, so that simulated oil biogeochemical processes (i.e., biodegradation, evaporation, and dissolution) are parameterized within appropriate rates for different chemical components.

Therefore, the results of our coupled near- and far-field model highlight the importance of an accurate estimation of the initial DSD: if a higher initial DSD is used, predictive and hindcast models predict a larger concentration of hydrocarbon at surface, while they predict lower quantities of oil droplets retained within deep waters, as it is clear in the difference between the ND and CD intrusion layer concentrations (Fig. 9.3). Current approaches to calculate initial DSDs are being improved in order to capture droplet atomization and breakup of live oil subject to a high hydrostatic pressure drop at the wellhead, combined with a high turbulent

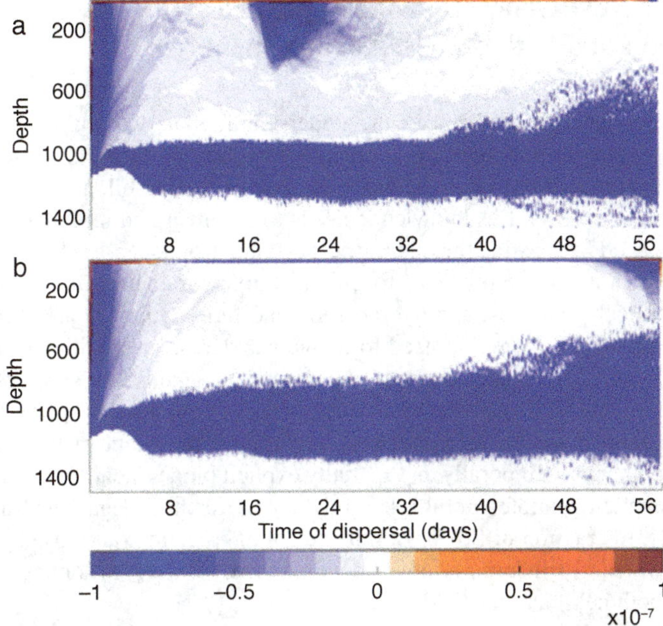

Fig. 9.3 Difference (log ppb) between daily vertical concentrations of oil from our coupled near-far-field model (TAMOC, Gros et al. 2017, and the oil-CMS, Paris et al. 2012) between a plume chemically dispersed with subsea dispersion injection (SSDI) and naturally dispersed. The hydrocarbon plume was released from the Macondo fallen riser on 5/2/10 00 hr. The upper panel (**a**) represents the difference for the plumes subjected to a slower biodegradation rate and the bottom panel (**b**) the plume subject to faster biodegradation rates

dissipation rate (Malone et al. 2020; Pesch et al. 2020), degassing, and variable partitioning with SSDI.

Our coupled model also exemplifies the importance of adequately representing biodegradation rates in predictive models and capture vertical concentrations of oil (Paris et al. 2018). Deep ocean conditions alter the growth rate of oil-degrading bacteria, their metabolic functions, hydrocarbon utilization, and thus degradation rates of oil (Schedler et al. 2014; Scoma et al. 2016; Bubenheim et al. 2020; Perlin et al. 2020). SSDI will further alter biodegradation rates, but how the interaction of dispersant, low temperature, and high pressure affects oil-degrading bacteria is still a topic of investigation. Laboratory studies suggest that the rate of dispersants applied at the DWH are enough to inhibit biodegradation for up to 10 days on surface conditions (Rahsepar et al. 2016), suggesting that its effects at depth might be significant. These processes need to be incorporated in oil models to evaluate the tradeoffs of novel cleanup strategies such as SSDI versus traditional containment response measures.

9.4 The Next Generation of Coupled Near-Field and Far-Field Models: Advancements

Post-DWH science has advanced the understanding of two-phase oil dynamics released at variable pressure and temperature conditions in the deep sea and has considered a breadth of biogeochemical processes affecting oil, gas, and dispersant mixture fate. However, this knowledge needs to be integrated into models used for hindcasting and forecasting the oil transport and fate from deep-sea blowouts. While best numerical modeling practices for initialization, parameterization, inclusion of biogeochemical processes, and hydrocarbon and fluid dynamics are still evolving (Perlin et al. 2020), there is a need to define standards for sensitive analyses and identify datasets for verification of these models against in situ observations (Berenshtein et al. 2020). We identify three areas of continued progress: (1) model parameterizations and initial conditions, including refinement of the initial DSD and development of temporally and spatially explicit biodegradation rates; (2) simulation of explicit spatiotemporal live oil dynamics, such as degassing, dissolution, and mostly, interaction of SSDI with flocculent aggregation; and (3) physical processes, particularly incorporation of detailed smaller-scale dynamics and turbulence, on both near- and far-field models.

The successful integration of TAMOC with the oil-CMS applied to the DWH accident serves as building blocks of the next generation of 4D cutting-edge models of oil transport and fate under high pressure and low temperature. Such modeling advancements can enhance model accuracy and maximize its applications, particularly for risk assessment, response planning, and estimating trade-offs of environmental impacts and risks to responders.

Acknowledgments This research was made possible by a from the Gulf of Mexico Research Initiative/C-IMAGE.

References

Bandara UC, Yapa PD (2011) Bubble sizes, breakup, and coalescence in Deepwater gas/oil plumes. J Hydraul Eng 137(7). https://doi.org/10.1061/(ASCE)HY.1943-7900.0000380

Berenshtein I, Perlin N, Ainsworth C, Ortega-Ortiz J, Vaz AC, Paris CB (2020) Comparison of the spatial extent, impacts to shorelines and ecosystem, and 4-dimensional characteristics of simulated oil spills (Chap. 20). In: Murawski SA, Ainsworth C, Gilbert S, Hollander D, Paris CB, Schlüter M, Wetzel D (eds) Scenarios and responses to future deep oil spills: fighting the next war. Springer, Cham

Boehm P, Prince R, Murray K (2020) The importance of understanding transport and degradation of oil and gasses from deep-sea blowouts (Chap. 6). In: Murawski SA, Ainsworth C, Gilbert S, Hollander D, Paris CB, Schlüter M, Wetzel D (eds) Deep oil spills: facts, fate, effects. Springer, Cham

Boufadel MC, Gao F, Zhao L, Özgökmen T, Miller R, King T, Leifer I (2018) Was the Deepwater Horizon well discharge churn flow? Implications on the estimation of the oil discharge and droplet size distribution. Geophys Res Lett 45(5):2396–2403. https://doi.org/10.1002/2017GL076606

Bubenheim P, Hackbusch S, Joye S, Kostka J, Larter SR, Liese A, Lincoln S, Marietou A, Müller R, Noirungsee N, Oldenburg TBP, Radović J, Viamonte J (2020) Biodegradation of hydrocarbons in deep water and sediments (Chap. 7). In: Murawski SA, Ainsworth C, Gilbert S, Hollander D, Paris CB, Schlüter M, Wetzel D (eds) Deep oil spills: facts, fate, effects. Springer, Cham

Chen F, Yapa PD (2003) A model for simulating deep water oil and gas blowouts – Part II: comparison of numerical simulations with "Deepspill" field experiments. J Hydraul Res 41(4): 353–365. https://doi.org/10.1080/00221680309499981

Clift R, Grace J, Weber ME (1978) Bubbles, drops, and particles. Dover Publications Inc., Mineola

Daly K, Vaz AC, Paris CB (2020) Physical processes influencing the sedimentation and lateral transport of MOSSFA in the northeast Gulf of Mexico (Chap. 18). In: Murawski SA, Ainsworth C, Gilbert S, Hollander D, Paris CB, Schlüter M, Wetzel D (eds) Scenarios and responses to future deep oil spills: fighting the next war. Springer, Cham

Dissanayake AL, Gros J, Socolofsky SA (2018) Integral models for bubble, droplet, and multiphase plume dynamics in stratification and crossflow. Environ Fluid Mech:1–36. https://doi.org/10.1007/s10652-018-9591-y

Drozd GT, Worton DR, Aeppli C, Reddy CM, Zhang H, Variano E, Goldstein AH (2015) Modeling comprehensive chemical composition of weathered oil following a marine spill to predict ozone and potential secondary aerosol formation and constrain transport pathways. J Geophys Res Oceans 120:7300–7315. https://doi.org/10.1002/2015JC011093

Fraga B, Stoesser T, CCK L, Socolofsky SA (2016) A LES-based Eulerian-Lagrangian approach to predict the dynamics of bubble plumes. Ocean Model 97:27–36. https://doi.org/10.1016/j.ocemod.2015.11.005

Gros J, Reddy CM, Nelson RK, Socolofsky SA, Arey JS (2016) Simulating gas-liquid-water partitioning and fluid properties of petroleum under pressure: implications for deep-sea blowouts. Environ Sci Technol 50(14):7397–7408. https://doi.org/10.1021/acs.est.5b04617

Gros J, Socolofsky SA, Dissanayake AL, Jun I, Zhao L, Boufadel MC, Reddy CM, Arey JS (2017) Petroleum dynamics in the sea and influence of subsea dispersant injection during Deepwater Horizon. Proc Natl Acad Sci 114(38):10065–10070. https://doi.org/10.1073/pnas.1612518114

Jaggi A, Snowdon RW, Radović J, Stopford A, Oldenburg TBP, Larter SR (2020) Partitioning of organics between oil and water phases with and without the application of dispersants (Chap. 8). In: Murawski SA, Ainsworth C, Gilbert S, Hollander D, Paris CB, Schlüter M, Wetzel D (eds) Deep oil spills: facts, fate, effects. Springer, Cham

Jirka GH (2004) Integral model for turbulent buoyant jets in unbounded stratified flows Part 2: plane jet dynamics resulting from multiport diffuser jets. Environ Fluid Mech 6(1):43–100. https://doi.org/10.1007/s10652-005-4656-0

Johansen Ø, Rye H, Cooper C (2003) DeepSpill-Field study of a simulated oil and gas blowout in deep water. Spill Sci Technol Bull 8(5–6):433–443. https://doi.org/10.1016/S1353-2561(02)00123-8

Johansen Ø, Brandvik PJ, Farooq U (2013) Droplet breakup in subsea oil releases – Part 2: predictions of droplet size distributions with and without injection of chemical dispersants. Mar Pollut Bull 73(1):327–335. https://doi.org/10.1016/j.marpolbul.2013.04.012

Joye SB, Bracco A, Özgökmen TM, Chanton JP, Grosell M, MacDonald IR, Cordes EE, Montoya JP, Passow U (2016) The Gulf of Mexico ecosystem, six years after the Macondo oil well blowout. Deep-Sea Res II 129:4–19. https://doi.org/10.1016/j.dsr2.2016.04.018

Le Hénaff M, Kourafalou VH, Paris CB, Helgers J, Aman ZM, Hogan PJ, Srinivasan A (2012) Surface evolution of the Deepwater Horizon oil spill patch: combined effects of circulation and wind-induced drift. Environ Sci Technol 46(13):7267–7273. https://doi.org/10.1021/es301570w

Lee JHW, Cheung V (1990) Generalized Lagrangian model for buoyant jets in current. J Environ Eng 116:1085–1106

Lee JHW, Chu VH (2003) Turbulent jets and plumes: a Lagrangian approach. Kluwer Academic Publishers Group, Dordrecht

Lehr W, Socolofsky S (2020) The importance of understanding fundamental physics and chemistry of deep oil blowouts (Chap. 2). In: Murawski SA, Ainsworth C, Gilbert S, Hollander D, Paris CB, Schlüter M, Wetzel D (eds) Deep oil spills: facts, fate, effects. Springer, Cham

Li Z, Spaulding ML, French-McCay D (2017) An algorithm for modeling entrainment and naturally and chemically dispersed oil droplet size distribution under surface breaking wave conditions. Mar Pollut Bull 119(1):145–152. https://doi.org/10.1016/j.marpolbul.2017.03.048

Lindo-Atichati D, Paris CB, Le Hénaff M, Schedler M, Valladares Juárez AG, Müller R (2016) Simulating the effects of droplet size, high-pressure biodegradation, and variable flow rate on the subsea evolution of deep plumes from the Macondo blowout. Deep-Sea Res II Top Stud Oceanogr 129:301–310. https://doi.org/10.1016/j.dsr2.2014.01.011

Malone K, Krause D, Pesch S, Schlüter M, Aman Z, Boufadel M (2020) Jet formation at the spill site and resulting droplet size distributions (Chap. 4). In: Murawski SA, Ainsworth C, Gilbert S, Hollander D, Paris CB, Schlüter M, Wetzel D (eds) Deep oil spills: facts, fate, effects. Springer, Cham

Nissanka ID, Yapa PD (2016) Calculation of oil droplet size distribution in an underwater oil well blowout. J Hydraul Res 54(3):307–320. https://doi.org/10.1080/00221686.2016.1144656

North EW, Adams EE, Thessen AE, Schlag Z, He R, Socolofsky S, Masutani SM, Peckham SD (2015) The influence of droplet size and biodegradation on the transport of subsurface oil droplets during the Deepwater Horizon spill: a model sensitivity study. Environ Res Lett 10:024016

Okubo A (1987) Fantastic voyage into the deep: marine biofluid mechanics. In: Teramoto E, Yumaguti M (eds) Mathematical topics in population biology, morphogenesis and neurosciences. Lecture notes in biomathematics, vol 71. Springer, Berlin, Heidelberg

Oldenburg TBP, Jaeger P, Gros J, Socolofsky S, Radović J, Jaggi A, Larter SR (2020) Physical and chemical properties of oil and gas under reservoir and deep sea conditions (Chap. 3). In: Murawski SA, Ainsworth C, Gilbert S, Hollander D, Paris CB, Schlüter M, Wetzel D (eds) Deep oil spills: facts, fate, effects. Springer, Cham

Paris CB, Le Hénaff M, Aman ZM, Subramaniam A, Helgers J, Wang DP, Srinivasan A (2012) Evolution of the Macondo well blowout: simulating the effects of the circulation and synthetic dispersants on the subsea oil transport. Environ Sci Technol 46(24):13293–13302. https://doi.org/10.1021/es303197h

Paris CB, Helgers J, van Sebille E, Srinivasan A (2013) Connectivity modeling system: a probabilistic modeling tool for the multi-scale tracking of biotic and abiotic variability in the ocean. Environ Model Softw 42:47–54. https://doi.org/10.1016/j.envsoft.2012.12.006

Paris CB, Berenshtein I, Trillo ML, Faillettaz R, Olascoaga MJ, Aman ZM, Schlüter M, Joye SB (2018) BP Gulf Science Data Reveals Ineffectual Sub-Sea Dispersant Injection for the Macondo Blowout, Frontiers in Marine Science, section Marine Pollution, accepted October 4, 2018, Manuscript ID: 385346

Perlin N, Berenshtein I, Vaz AC, Faillettaz R, Schwing PT, Romero IC, Schlüter M, Liese A, Viamonte J, Noirungsee N, Gros J, Paris CB (2020) Far-field modeling of a deep-sea blowout: sensitivity studies of initial conditions, biodegradation, sedimentation, and subsurface dispersant injection on surface slicks and oil plume concentrations (Chap. 11). In: Murawski SA, Ainsworth C, Gilbert S, Hollander D, Paris CB, Schlüter M, Wetzel D (eds) Deep oil spills: facts, fate, effects. Springer, Cham

Pesch S, Schlüter M, Aman Z, Malone K, Krause D, Paris CB (2020) Behavior of rising droplets and bubbles – impact on the physics of deep-sea blowouts and oil fate (Chap. 5). In: Murawski SA, Ainsworth C, Gilbert S, Hollander D, Paris CB, Schlüter M, Wetzel D (eds) Deep oil spills: facts, fate, effects. Springer, Cham

Prosser CM, Redman AD, Prince RC, Paumen ML, Letinski DJ, Butler JD (2016) Evaluating persistence of petroleum hydrocarbons in aerobic aqueous media. Chemosphere 155:542–549. https://doi.org/10.1016/j.chemosphere.2016.04.089

Rahsepar S, MPJ S, Murk AJ, Rijnaarts HHM, Langenhoff AAM (2016) Chemical dispersants: oil biodegradation friend or foe? Mar Pollut Bull 108(1–2):113–119. https://doi.org/10.1016/j.marpolbul.2016.04.044

Reddy CM, Arey JS, Seewald JS, Sylva SP, Lemkau KL, Nelson RK, Carmichael CA, McIntyre CP, Fenwick J, Ventura GT, Van Mooy AS, Camilli R (2012) Composition and fate of gas and oil released to the water column during the Deepwater Horizon oil spill. Proc Natl Acad Sci 109(50):20229–20234. https://doi.org/10.1073/pnas.1101242108

Reed M, Johansen O, Brandvik PJ, Daling P, Lewis A, Fiocco R, Mackay D, Prentki R (1999) Oil spill modeling towards the close of the 20th century: overview of the state of the art. Spill Sci Technol Bull 5:3–16. https://doi.org/10.1016/S1353-2561(98)00029-2

Romero IC, Schwing PT, Brooks GR, Larson RA, Hastings DW, Ellis G, Goddard EA, Hollander DJ (2015) Hydrocarbons in deep-sea sediments following the 2010 Deepwater Horizon blowout in the northeast Gulf of Mexico. PLoS ONE 10(5):e0128371. https://doi.org/10.1371/journal.pone.0128371

Ryerson TB, Camilli R, Kessler JD, Kujawinski EB, Reddy CM, Valentine DL, Atlas E, Blake DR, de Gout J, Meinardi S, Parrish DD, Peischl J, Seewald JS, Warneke C (2012) Chemical data quantify Deepwater Horizon hydrocarbon flow rate and environmental distribution. Proc Natl Acad Sci 109(50):20246–20253. https://doi.org/10.1073/pnas.1110564109

Schedler M, Hiessl R, Valladares Juárez AG, Gust G, Müller R (2014) Effect of high pressure on hydrocarbon-degrading bacteria. AMB Express 4(1):1–7. https://doi.org/10.1186/s13568-014-0077-0

Scoma A, Barbato M, Hernandez-Sanabria E, Mapelli F, Daffonchio D, Borin S, Boon N (2016) Microbial oil-degradation under mild hydrostatic pressure (10 MPa): which pathways are impacted in piezosensitive hydrocarbonoclastic bacteria? Sci Rep 6:1–14. https://doi.org/10.1038/srep23526

Socolofsky SA, Adams EE (2002) Multi-phase plumes in uniform and stratified crossflow. J Hydraul Res 40(6):661–672

Socolofsky SA, Dissanayake AL (2016) Bent plume model source code for the Texas A&M Oilspill Calculator (TAMOC). Center for the Integrated Modeling and Analysis of Gulf Ecosystems II (C- IMAGE II), Gulf of Mexico Research Initiative Information and Data Cooperative (GRIIDC). https://data.gulfresearchinitiative.org/data/R4.x267.000:000333

Socolofsky SA, Adams EE, Sherwood CR (2011) Formation dynamics of subsurface hydrocarbon intrusions following the deepwater horizon blowout. Geophys Res Lett 38(9):L09602

Socolofsky SA, Adams EE, Boufadel MC, Aman ZM, Johansen Ø, Konkel WJ, Lindo D, Madsen MN, North EW, Paris CB, Rasmussen D, Reed M, Rønningen P, Sim LH, Uhrenholdt T, Anderson KG, Cooper C, Nedwed TJ (2015) Intercomparison of oil spill prediction models for accidental blowout scenarios with and without subsea chemical dispersant injection. Mar Pollut Bull 96(1–2):110–126. https://doi.org/10.1016/j.marpolbul.2015.05.039

Stout SA, Payne JR, Emsbo-Mattingly SD, Baker G (2016) Weathering of field-collected floating and stranded Macondo oils during and shortly after the Deepwater Horizon oil spill. Mar Pollut Bull 105(1):7–22. https://doi.org/10.1016/j.marpolbul.2016.02.044

Turner JS (1968) The influence of molecular diffusivity on turbulent entrainment across a density interface. J Fluid Mech 33:639–656

Wang B, Socolofsky SA, CCK L, Adams EE, Boufadel MC (2018) Behavior and dynamics of bubble breakup in gas pipeline leaks and accidental subsea oil well blowouts. Mar Pollut Bull 131:72–86. https://doi.org/10.1016/j.marpolbul.2018.03.053

Yapa PD, Zheng L, Chen F (2001) A model for Deepwater oil/gas blowouts. Mar Pollut Bull 43(7–12):234–241. https://doi.org/10.1016/S0025-326X(01)00086-8

Yapa PD, Dasanayaka LK, Bandara UC, Nakata K (2010) A model to simulate the transport and fate of gas and hydrates released in deepwater. J Hydraul Res 48(5):559–572. https://doi.org/10.1080/00221686.2010.507010

Zhao L, Boufadel MC, Socolofsky SA, Adams EE, King T, Lee K (2014) Evolution of droplets in subsea oil and gas blowouts: development and validation of the numerical model VDROP-J. Mar Pollut Bull 83(1):58–69. https://doi.org/10.1016/j.marpolbul.2014.04.020

Zhao L, Boufadel MC, Adams EE, Socolofsky SA, King T, Lee K, Nedwed T (2015) Simulation of scenarios of oil droplet formation from the Deepwater Horizon blowout. Mar Pollut Bull 101(1):304–319. https://doi.org/10.1016/j.marpolbul.2015.10.068

Zhao L, Boufadel MC, King T, Robinson B, Gao F, Socolofsky SA, Lee K (2017) Droplet and bubble formation of combined oil and gas releases in subsea blowouts. Mar Pollut Bull 120 (1–2):203–216. https://doi.org/10.1016/j.marpolbul.2017.05.010

Zheng L, Yapa PD, Chen F (2003) A model for simulating deepwater oil and gas blowouts - part I: theory and model formulation. J Hydraul Res 41(4):339–351. https://doi.org/10.1080/00221680309499980

Zick AA (2013) Equation-of-State Fluid Characterization and Analysis of the Macondo Reservoir Fluids; Expert report prepared on behalf of the United States; TREX-011490R; (evidence in the United States of America v. BP Exploration & Production, Inc., et al. case). (30)

Chapter 10
Effects of Oil Properties and Slick Thickness on Dispersant Field Effectiveness and Oil Fate

Marieke Zeinstra-Helfrich and Albertinka J. Murk

Abstract One of the available oil spill response options is to enhance the natural dispersion process by the addition of dispersants known as chemical dispersion. An informed decision for such response requires insight about the added effects of treatment with dispersants on the oil slick fate. To provide such insight, a mathematical model for oil slick elongation as a result of dispersion was developed including the effects of oil viscosity, dispersed oil droplet sizes, and oil layer thickness. This chapter briefly revisits this oil slick elongation model to explain the consequences of different key parameters on dispersion, vertical droplet size distribution, formation of a comet-like tail and oil slick (dis)appearance, as well as the implications of the results for future decision-making. The model outcomes indicate that wind speed is a very dominant factor in dispersion and subsequent slick behavior. More surprising, the influence of oil type on the elongation process is only limited. The increased density of the high-viscosity oil types allows larger droplets to be stably suspended. High-viscosity oil, however, was found to benefit less from a decrease in interfacial tension than a low-viscosity oil. Weighing estimated risks for adverse effects in the water column with a reduced surface oil presence allows for future dispersant decisions based on a thorough Spill Impact Mitigation Analysis (SIMA).

Keywords Dispersion · Dispersant · Spill modelling · Wind speed · Oil type

M. Zeinstra-Helfrich (✉)
NHL Stenden University of Applied Sciences, Leeuwarden, The Netherlands
e-mail: m.zeinstra@nhl.nl

A. J. Murk
Wageningen University, Marine Animal Ecology Group, Wageningen, The Netherlands
e-mail: Tinka.murk@wur.nl

© Springer Nature Switzerland AG 2020
S. A. Murawski et al. (eds.), *Deep Oil Spills*,
https://doi.org/10.1007/978-3-030-11605-7_10

10.1 Introduction

The decision to apply chemical dispersants on surface oil is a trade-off between surface effects (impact of floating and stranded oil) and subsurface effects (direct impact of suspended oil and potential for sedimentation and sinking). Making an informed decision regarding such response requires insight into the induced changes in fate and adverse effects of the oil.

Natural dispersion calculations in the currently available oil fate and transport models are based on the empirical results of Delvigne and Sweeney, although it is generally agreed these can be improved (Delvigne and Sweeney 1988; Reed et al. 1999; National Research Council of the National Academies 2005). Furthermore, these calculations do not permit prediction of chemical dispersion. The calculation of chemical dispersion in the current oil fate and transport models requires knowledge of the estimated effectiveness of dispersant application rather than providing output thereof (National Research Council of the National Academies 2005). Estimating dispersant effectiveness in advance of commencing such oil spill response relies heavily on expert judgment.

A small group of researchers under the C-IMAGE project (as part of the Gulf of Mexico Research Initiative) set out to create insight into the processes governing natural and chemical dispersion of spilled surface oil and to provide a strategy to assess and predict the added value of chemical dispersion for specific spill conditions and oil qualities. The tiered experimental design allowed for structured analysis of available laboratory results and the identification of unknown parameters (Zeinstra-Helfrich et al. 2015b). Additionally, the process of entrainment, or initial submergence of the oil, can be examined more thoroughly in a laboratory environment to quantitatively capture the influence of layer thickness and oil properties on vertical oil droplet size distribution (Zeinstra-Helfrich et al. 2015a, 2016). Ultimately, a model was proposed that calculates the evolution of the surface oil slick as a result of the dispersion process (Zeinstra-Helfrich et al. 2017).

This chapter examines the oil slick elongation model (Sect. 10.2) in order to explain the influence of different key parameters on dispersion and oil slick (dis) appearance (Sect. 10.3) and the implications of the results for future decision-making (Sect. 10.4).

10.2 How Natural or Chemical Dispersion Affects Oil Slick Fate

Dispersion is not a singular event but rather a combination of several processes (Fig. 10.1). Breaking waves temporarily submerge oil through entrainment. The entrained oil is broken up into droplets, and their size distribution in the water column is dependent on mixing energy, oil properties, and slick thickness. As most oil is still lighter than water, the droplets rise back to the surface at a rate dependent on

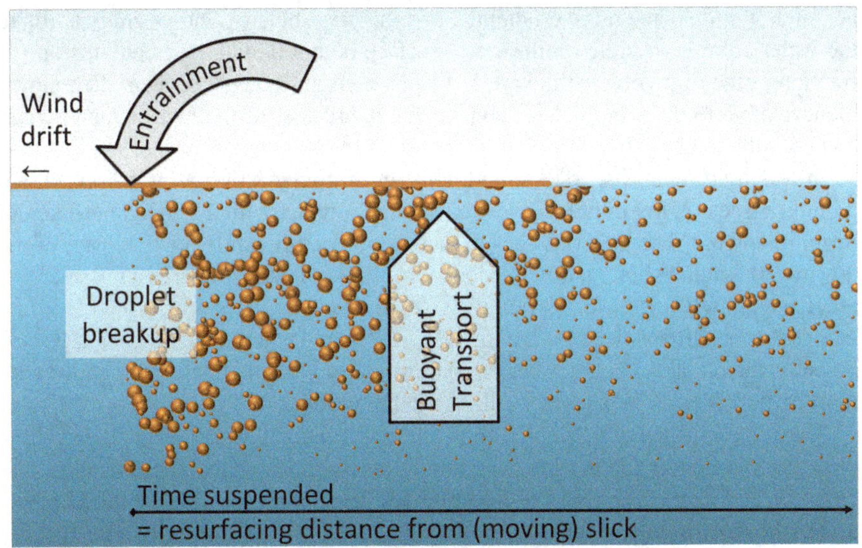

Fig. 10.1 Visualization of the mixing processes. Under a breaking wave, oil is entrained, broken into droplets of various sizes, and distributed over a certain depth. While the oil droplets are suspended, the slick is pushed along by the wind. Droplets resurface upwind from their original location in the moving slick, of which the distance depends on the time in suspension. As a droplet's rise velocity is determined by its diameter, the residence time in the water column is determined by its mixing depth and droplet diameter. (Reprinted from Zeinstra-Helfrich et al. (2015b), Copyright (2015), with permission from Elsevier)

their size and the oil's density. Some of these droplets reach the water surface relatively quickly, while others are remixed by subsequent breaking wave events. As the floating oil moves downwind at a faster rate than the suspended droplets, some of the oil resurfaces upwind of the slick, elongating and diluting the slick.

The oil dispersion model (Zeinstra-Helfrich et al. 2017) uses a number of relatively simple input parameters (wind speed, oil viscosity, oil density, oil-water interfacial tension, and initial slick thickness and length) to calculate the entrainment and subsequent resurfacing of oil over a predefined period of time, yielding the slick length, displacement, and mass distribution between slick and water column and across the slick length.

The modelling outcomes reveal that the mechanism by which the oil slick disappears from the water surface ultimately determines the nature of the oil slick over time. In ideal dispersion conditions, identified by very high wind speed paired with low-viscosity oil and dispersants, the disappearing of the oil slick is a vertical process. The abundance of breaking waves provides ample "entrainment events," where a substantial amount of oil is temporarily transferred to the water column. High mixing energy combined with oil properties that enable easy breakup results in a large portion of the entrained oil, now in small droplets, to remain suspended in the water column for a long period of time before resurfacing. As a result, a substantial fraction of the oil is quickly moved to the water column. The resulting thinner slick will disperse more efficiently, and the mass balance increasingly shifts to the water column. Visible on the water surface is an oil slick that contains only a fraction of the oil, with the thickest part being central in the slick. This slick movement relative to the wind is less than expected, as most of the mass is underwater and not affected by wind.

In less than ideal dispersion conditions, characterized by low wind speed and high oil viscosity, the initial slick remains largely intact with a long comet-like tail forming upwind. The tail continues to increase in length, while the thickness of the downwind "main" slick gradually decreases. While thickness decreases, dispersion efficiency slowly increases as relatively more small droplets are formed, and the mass balance shifts slowly toward the water column. If the thickness decreases sufficiently, the oil slick behavior begins to transition toward that described above with most of the oil mass in the water column and not affected by the wind.

These findings are consistent with in situ observations, where long slicks with a thick portion downwind are found in conditions with low wind speeds and/or viscous oils and more favorable conditions create smaller slicks with the thickest portion in the center (Reed et al. 1994).

10.3 Influence of Individual Key Parameters on Dispersion and Oil Slick Elongation

With model application, we can investigate the theoretical influence of different parameters on the dispersion process and the development of the oil slick over time with its eventual disappearance from the water surface.

The volume fraction of oil that is broken up into droplets that are sufficiently small to be stably suspended can be calculated from oil properties, wind speed, and initial oil layer thickness (Zeinstra-Helfrich et al. 2017). This variable, named dispersibility factor, represents the combined result of the different dispersion processes and provides a good indicator of dispersion success. In this chapter, the term dispersibility refers to the favorability of conditions for creating small stable droplets without considering oil slick behavior over time.

When discussing oil slick behavior, the most relevant aspects are thickness profile, slick length, and the lifetime of the oil slick. This surface expression of a spill can be summarized with the "time-integrated length of oil slick exceeding the effects threshold of 25 µm" (Zeinstra-Helfrich et al. 2017). With equal oil slick dimensions, this parameter correlates well with dispersibility factor: A highly dispersible slick will disappear quickly and have a low surface expression value. In case of low dispersibility, a slick is present on the surface for a long period of time and increases in size before disappearing, resulting in a high value for surface expression.

In the following paragraphs, the influence of the different key parameters on the oil dispersibility as well as oil slick behavior is discussed in more detail.

10.3.1 Main Oil Properties

The viscosity of the oil is considered one of the most important parameters in the dispersion process. As viscosity is a counteracting force to droplet breakup (Walstra 2005), high-viscosity oil results in larger oil droplets. The influence of viscosity as a factor in oil droplet breakup is commonly reported in small- and large-scale dispersion tests (Zeinstra-Helfrich et al. 2015b) as oil droplet sizes increase with oil viscosity.

Following Stokes' law, the oil density is known to increase oil droplet rise speed (Robbins et al. 1995) and bias dispersibility tests that incorporate a settling step as a consequence (SL Ross Environmental Research LTD and MAR Incorporated 2011). However, in current oil spill modelling, often a fixed droplet size is assumed and considered to be stably suspended, regardless of environmental conditions or oil density (Reed et al. 1999).

In laboratory experiments, oil type did not significantly influence the volume of oil entrained, with the exception of high-viscosity oil (above 5 Pa.s), of which 60–80% less was entrained than expected based on oil layer thickness (Zeinstra-Helfrich et al. 2016). Images of the entrainment process during plunge impact provided evidence demonstrating that the high-viscosity oil layer may prevent droplets from being sheared off.

A similar entrainment limitation in a plunging jet test was observed by other researchers (Reed et al. 2009). Reed et al. (2009) reported a twice as high maximum viscosity but also used a two times greater plunge height. An increasing maximum viscosity with plunge/wave height is logical as the impact from larger wave heights is expected to be more successful in separating oil from the floating layer. More elaborate experiments could provide experimental data to further calculate the entrainment thresholds for larger wave heights.

Although oil properties clearly influence the process of droplet breakup and thus droplet size, dispersion model calculations reveal that oil type hardly affects the overall dispersibility and the oil slick behavior outputs of the model (Fig. 10.2), partly because weather conditions are a more dominant factor in the model outcome and partly because of the correlation between oil viscosity and density resulting in opposite effects. Although high-viscosity oil results in a larger mean droplet size, these high-density droplets rise slower to the water surface than equally sized low-viscosity (and low-density) oil droplets. The mean droplet rise velocity of the full droplet size distribution for different oil types therefore only shows little variation (Fig. 10.3, left panel). In the unrealistic scenario with equal densities for the chosen three oil types (Fig. 10.3, right panel), mean rise velocity would be much more affected by oil type. Evidently, the density-viscosity correlation strongly reduces the influence of viscosity on the mean droplet rise velocity. This means that, when modelling dispersion, both viscosity and density should be separate inputs, as a disproportion between these qualities can seriously affect the dispersibility.

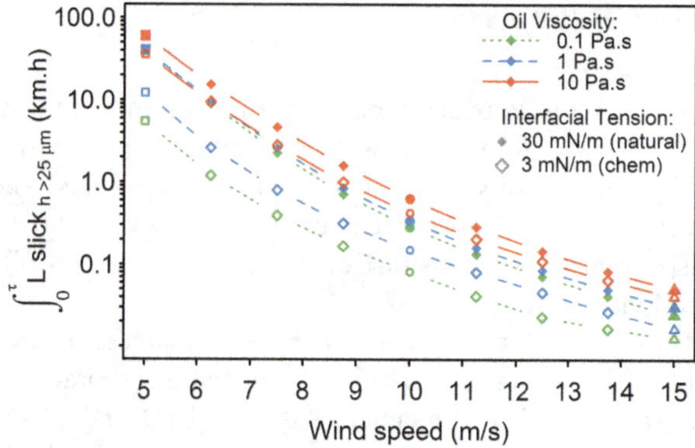

Fig. 10.2 "Time-integrated length (km.h) of the slick part with a thickness >25 μm", as a function of oil viscosity, interfacial tension, and wind speed (Note the logarithmic Y axis.). Model calculations using method and inputs as described in Zeinstra-Helfrich et al. (2017) for a larger number of wind speed settings. Initial oil slick length was 250 m and thickness 1 mm

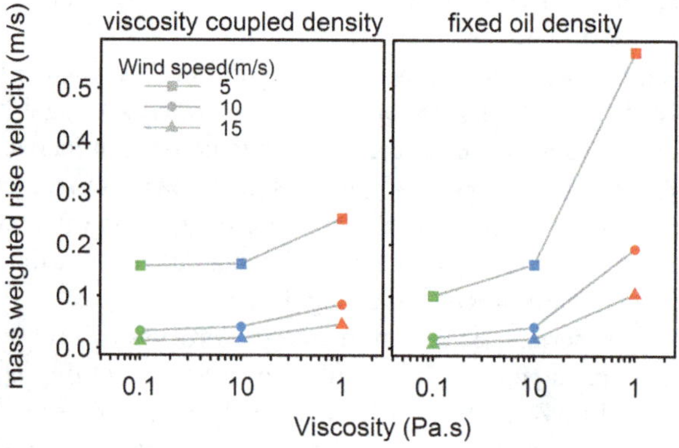

Fig. 10.3 Calculated mean droplet rise velocity for droplet size distributions formed from of different oil types (colors) and under different wind speeds (symbols). In the model inputs, the oil density either matched viscosity (left) or was kept equal for the three oil types (right). Droplet size distributions are calculated using a Weber and Reynolds relation (Zeinstra-Helfrich et al. 2017), assuming an oil layer thickness of 0.5 mm and no dispersant

Summarizing, viscosity above a wave height-dependent threshold impedes oil entrainment and will limit the dispersion and elongation of the slick. Below this viscosity threshold, oil droplet size is positively related with viscosity, yet the resulting effect on droplet suspension time is limited due to the compensating influence of higher density on droplet rising velocity.

10.3.2   Oil Layer Thickness

In operational guidelines for chemical dispersion of oil, oil layer thickness is not considered very relevant for oil fate modelling. It mainly is seen as a parameter relevant for dispersant dosages: spraying dispersants on a slick that's too thin would cause the dispersant to fall through and be lost to the water column (Tamis et al. 2012). Spraying the thicker areas of the slick is advised; however, for too high thicknesses, multiple spray passes are advised to reach the effective dosage (EMSA 2009).

Using a plunging jet setup (Zeinstra-Helfrich et al. 2015a, 2016), the importance of the oil layer thickness in both entrainment and droplet breakup was revealed: The volume of oil entrained increases proportionally with layer thickness. The availability of oil per unit surface area (oil layer thickness) clearly determines the volume entrained per "mixing event." With increasing oil layer thickness, the mean oil droplet size increases, but due to the enhanced entrainment volume, the absolute amount of oil in small droplets also increases. Still, thin layers have a higher dispersibility than thick layers due to the larger *relative* portion of small droplets produced by an impact. This influence of layer thickness on dispersibility also is crucial for the modelled behavior of the slick over time. With the same oil mass, a longer, thin slick is dispersed faster than a short thicker slick (Zeinstra-Helfrich et al. 2017). Such rapid removal of the thin slick areas, while thicker parts remain, has also been observed in situ (Lewis et al. 1998).

These observations confirm that aiming any response at the thick slick portions (National Research Council of the National Academies 2005) is most effective as the thin slick part will naturally disperse more easily. Additionally, for mechanical recovery, removal rates are higher in thick oil.

Based on these outcomes, one can hypothesize that mechanical dispersion on thicker portions of the slick to enhance spreading could be effective as the resulting thinner slick will subsequently disperse more easily.

10.3.3   Initial Slick Size

The model study (Zeinstra-Helfrich et al. 2017) only briefly examined the influence of initial slick length on the dispersion process. It is, however, possible to make some prognoses based on the different elongation mechanisms.

For favorable conditions (dispersibility factor >0.4), the initial slick length will not have a large influence on the dispersion process: In these situations, the oil mass will be rapidly entrained into the water phase. For a larger slick oil will simply move to the water phase over the larger area.

For less favorable conditions, initial slick length is expected to influence the outcome. The absolute elongation is minimally affected by initial slick length: A longer initial slick means that resurfacing oil from the downwind edge "feeds" the main slick instead of the tail. The oil resurfacing back into the main slick slows down the decrease in thickness that is necessary to transition into the more efficient vertical dispersion regime. In this case, dispersant application could assist by enhancing dispersibility and transitioning into the vertical regime more quickly.

10.3.4   Wind Speed

Wind speed is considered an important variable in the dispersion process, as it indirectly provides the energy for the dispersion to occur.

In the dispersion model studies (Zeinstra-Helfrich et al. 2017), indeed, wind speed is a very dominant factor in the outcome. This input parameter plays a role in several process parameters. Firstly, the amount of entrainment and energy levels indeed depend on wind speed:

- The area fraction agitated increases $A_{mix} \sim U_{wind}^{2.26}$, resulting in a proportional increase of volume of oil entrained.
- The plunge height increases $H_{plunge} \sim U_{wind}^{2}$, causing smaller oil droplets to be formed.

As the mixing depth increases, $z_i \sim U_{wind}^{2}$, and the time between breaking waves decreases, $T_{bw} \sim U_{wind}^{-2.26}$, as a consequence, much larger droplets can remain suspended until the next breaking wave hits (Fig. 10.4). Thus, in contrast to commonly assumed in current models (Reed et al. 1999), the droplet size that is stably suspended is wind speed dependent: in more energetic conditions, larger droplets can successfully remain suspended.

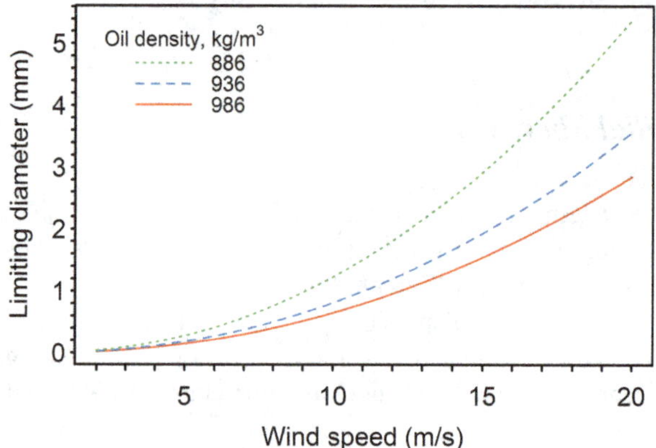

Fig. 10.4 Limiting diameter, largest droplet diameter that can remain suspended until the next breaking wave hits, as a function of wind speed. Calculated based on the equations given in Zeinstra-Helfrich et al. (2017)

In addition, at a higher wind speed, the slick moves faster causing more of the suspended droplets to resurface in the tail instead of in the main slick area, thus increasing the relatively thin slick area with more effective natural dispersion.

10.3.5 Dispersants

It is self-evident that addition of dispersants enhances the dispersion process. The main mode of dispersant action is to reduce the oil-water interfacial tension, thereby allowing smaller droplets to be formed.

In the slick elongation model, dispersant application is simulated by a decrease of the input value for oil-water interfacial tension. This parameter affects the droplet breakup process by decreasing the resulting droplet sizes but has no impact on the other steps in the dispersion and elongation process.

The droplet sizes of the entrained oil are calculated with a Weber and Reynolds number relation fitted to the experimental results (Zeinstra-Helfrich et al. 2017). Using this calculation, the absolute decrease of mass median droplet size with decreasing interfacial tension hardly depends on oil viscosity (Fig. 10.5). As the relative dispersant-induced decrease in droplet size is much smaller for high-viscosity oils, the increase of dispersibility is lower for these oil types.

Apart from affecting droplet size, dispersants can influence entrainment in specific situations: In the absence of breaking waves, oil with dispersant dosages of 1:20 or more was entrained with unintentional and less energetic vertical input (SL Ross Environmental Research LTD et al. 2006). This is because such dispersant dosages can cause the oil-water interfacial tension to drop down to values close to 10^{-6} N/m (Khelifa and So 2009), at which minimal energy is needed to commence droplet formation (Walstra 1993). This potential extra entrainment is currently not accounted for in the oil slick elongation model, as it is unlikely that it will significantly influence oil slick behavior in at-sea conditions. For such highly dispersible oil, even at 5 m/s winds, the weak and infrequent breaking wave impact already

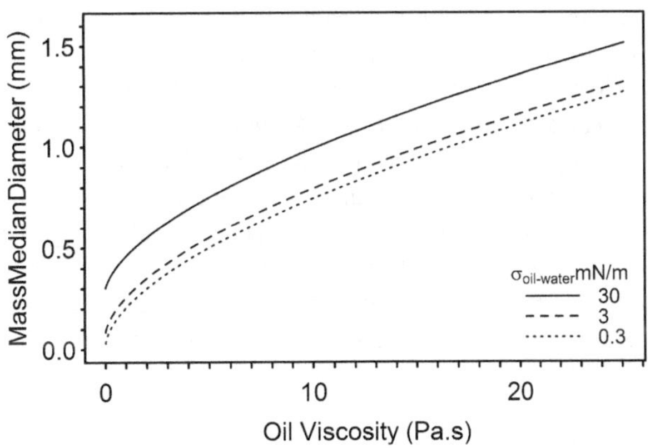

Fig. 10.5 Mass median oil droplet diameter as a function of oil viscosity, calculated using the Weber and Reynolds number relation (Zeinstra-Helfrich et al. 2017). Input data used: wind speed of 10 m/s and oil layer thickness of 0.4 mm

causes such a stable dispersion (when $\mu_{oil} = 0.1\,Pas$ and $\sigma_{oil} = 10{-}6\ N/m$, dispersibility factor: 0.63) that additional entrainment only minimally impacts mass balance. Although breaking waves decline even further at lower wind speeds, so does the presence of other sea surface features that could cause additional entrainment, for instance, Langmuir circulation (Moum and Smyth 1994).

The oil slick elongation model requires input of the oil-water interfacial tension with or without dispersant application. The oil-water interfacial tension after dispersant application depends on success of dispersant application (logistics, targeting, incorporation into the slick, dispersant effect on oil properties). Currently no exact data on the effective dispersant dosage in the field and resulting oil-water-interfacial tension are readily available, partly as it depends on so many factors (falling of the dispersant droplet, oil slick skin formation and the composition of the oil and the dispersant) and partly as interfacial tension measurements in the lab are highly influenced by the test settings and are not in agreement with each other. Therefore, expert judgment is still required to parametrize this effect.

More information on the chemical effectiveness of dispersants could be obtained via small-scale laboratory tests such as the baffled flask dispersibility test. These easy and repeatable tests provide a measure of the susceptibility to breakup of the dispersant-oil combination; therefore the test results provide an indication of the change in dispersibility after addition of dispersants. The effect of dispersants can be determined from the change in results between treated and untreated oil.

10.4   Decision-Making About Application of Chemical Dispersion

For selecting an oil spill response strategy, either in preparation or in active spill situation, the implications of available response options are considered. An explicit and structured approach to analyzing the trade-offs between strategies was formerly referred to as NEBA (net environmental benefit analysis) and recently renamed SIMA (spill impact mitigation assessment) (IPIECA 2017). Because the implications of dispersant application span multiple environmental compartments, SIMA can be a particularly useful tool for this response technique.

Assessment of the extent to which application of dispersants aids mitigation of the impacts considers three factors: (1) the logistical feasibility, (2) the effectiveness, and (3) the effects or environmental benefit. The logistical feasibility of chemical dispersion is a practical consideration based on dispersant and equipment stocks and the travel time to reach the slick location and will not be discussed in this chapter. The following paragraphs summarize how to assess the effectiveness of chemical dispersion as well as the potential environmental benefit.

10.4.1   *Effectiveness*

When applying dispersants on a surface oil slick with the goal of mitigating the effects of the floating oil, the term "effectiveness" should express that aspect. Assuming a minimal oil layer thickness above which impacts occur (in this case 25 μm), potential for oil slick impact can be inferred from the size of the oil slick above this threshold and its lifetime. The parameter "time-integrated length of slick exceeding the effects threshold of 25 μm" was devised for this purpose. This parameter allows for comparison of potential impact in different conditions, as well as for direct calculation of dispersant success (Fig. 10.6). For wind speeds exceeding 10 m/s, the slick area with a thickness of 25 μm or more is very small and will disappear very quickly (Fig. 10.2; L slick$_{h > 25\ \mu m}$ < 1 km.h). These conditions create a small symmetrical surface slick, while most mass is in the water column. Dispersants will speed up the displacement of oil to the water column, yet the absolute effect is only slight as the surface area of such slicks is already relatively small.

At lower wind speeds, both the absolute and relative impacts of dispersant application are much larger (Fig. 10.6). Although the theoretical dispersant effectiveness seems highest for the lowest wind speed shown here (5 m/s), one should be aware of the possibly limited improvement of the situation. In the worst-case scenario (heavy oil, 5 m/s wind), our simulated dispersant addition to the natural dispersion drops the "time-integrated slick length with H$_{oil}$ > 25μm" from 60.2 to 35.9 km.h,

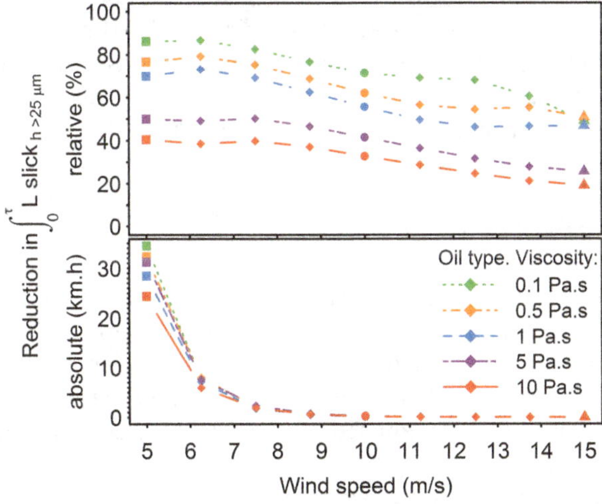

Fig. 10.6 Modelled decrease in "time-integrated slick length with H$_{oil}$ > 25 μm" as a result of a factor 10 reduction in oil-water interfacial tension, simulating dispersant addition. Shown as a percentage of the untreated slick (top) or as absolute decrease (in km.h) (bottom). Model calculations as described in Zeinstra-Helfrich et al. (2017) for a larger number of wind speed and oil-type settings. Initial oil slick length was 250 m and thickness 1 mm

meaning that the treated oil slick will still have substantial size and lifetime. The response decision should include whether the result is worth the effort and costs for the specific situation.

As expected, a certain reduction in interfacial tension is most effective on low-viscosity oil, as the relative change in droplet size is higher (Sect. 10.3.5). Together with the observation that high viscosity hinders dispersant incorporation into the oil slick (Canevari 1984), this means that chemical dispersion is relatively and absolutely less effective on higher-viscosity than on low-viscosity oils.

Sub-optimal natural dispersion causes the oil slick to elongate substantially in the wind direction, where the thick slick portion is followed by a long tail in the upwind direction. Dispersants can be beneficial in these situations, accelerating the spreading of the thick slick part and the subsequent transition into the phase where size is slowly reduced as more mass is moved to the water column. The chemical as well as operational effectiveness of dispersants does decrease with increasing viscosity.

The added benefit of dispersants is limited in conditions that are optimal for natural dispersion, resulting in only slight slick elongation and with a slick thickness that is symmetrical in the wind direction.

10.4.2   Effects

As explained, the environmental benefit of dispersant application lies mainly in removing oil from the water surface: Depending on the effectiveness, dispersants can have a positive influence on mitigation of surface effects. However, the enhanced volumes of suspended/dissolved oil can enhance adverse effects in other compartments of the environment (O'Sullivan and Jacques 2001).

10.4.2.1   Water Column

Although the dispersants themselves are designed to be low in toxicity, the resulting enhanced exposure to hydrocarbons can potentially cause toxic effects to water column biota. Where physical surface oil impacts are mass independent for thicknesses above the effects threshold (Jongbloed et al. 2002), potential water column effects are more transient and are determined by exposure time and maximum concentration (Lee et al. 2015).

The increased exposure to hydrocarbons by enhanced dispersion is only brief, as concentrations are expected to diminish quickly due to dilution (Lee et al. 2015). This dilution is less evident in case of continuous dispersion of large or prolonged spills such as the *Deepwater Horizon* incident or in shallow or enclosed waters (BfR Wissenschaft 2016). Assessment of potential toxic effects encompasses uncertainties such as species-dependent sensitivity. Moreover, the precise exposure of the organisms over time depends on the specific spill conditions, oil qualities, as well as the behavior of the organisms (Tamis et al. 2012). Different laboratory tests are available simulating various exposure regimes (Redman and Parkerton 2015), yet these may not necessarily match exposure regimes at sea.

In addition to chemical dispersion aiming to "dilute" the pollutant quicker, it is also considered to enhance biodegradation of the suspended oil. The role of biodegradation in natural and chemically dispersed oil is unclear. Theoretically dispersed oil is more bioavailable, but bacteria may also experience more acute toxic effects (Rahsepar et al. 2016). Conditions are heterogenic in space and time and difficult to replicate in a laboratory. As a result, conflicting information is available indicating either biodegradation is enhanced, decreased, or indifferent of chemical dispersion (Lee et al. 2013; Kleindienst et al. 2016; Rahsepar et al. 2016).

10.4.2.2   Benthic

Dispersed oil can also reconcentrate on the seafloor, either as a result of suspended sediments adhering to oil droplets and sinking (Khelifa et al. 2008) or via the observed so-called MOSSFA mechanism (Marine Oil Snow Sedimentation and Flocculent Accumulation). This effect is enhanced in situations with phytoplankton blooms combined with particulate matter in the water column (Vonk et al. 2015; Daly et al. 2016; van Eenennaam et al. 2016). This MOSSFA could cause long-lasting effects as the oil will be concentrated on/in the sediment followed by reduced biodegradation under oxygen-limited conditions (Langenhoff et al. 2020) and may cause prolonged local exposure to hydrocarbons (Lee et al. 2015). This applies even more so for "marine oil snow," which was found to cause more adverse effects in benthic invertebrates than oil-sediment aggregates (van Eenennaam et al. 2018; Schwing et al. 2020).

10.5   Concluding Remarks

The notion that dispersants should not be applied if their effectiveness is unsure stands to reason. Apart from the limited cost-effectiveness in such a case, the resulting change in oil properties can hamper effectiveness of mechanical removal methods (ITOPF 2012).

However, for a given spill situation, if we can prove that addition of dispersants can successfully mitigate potential for effects on the water surface and coastlines, as well as demonstrate that risks for adverse effects in other compartments are acceptable, dispersants provide a useful response strategy that can work in conditions where other methods are unsuccessful.

Acknowledgments This research was made possible by a from the Gulf of Mexico Research Initiative/C-IMAGE. Data are publicly available through the Gulf of Mexico Research Initiative Information and Data Cooperative (GRIIDC) at https://data.gulfresearchinitiative.org/ (doi:10.7266/n7-8hmm-kx37).

References

BfR Wissenschaft (2016) The use of dispersants to combat oil spills in Germany at sea. Federal Institute for Risk Assessment, Berlin, Germany

Canevari GP (1984) A review of the relationship between the characteristics of spilled oil and dispersant effectiveness. In: Allen TE (eds) Oil spill chemical dispersants: research, experience, and recommendations. American Society for Testing and Materials Philadelphia, pp 87–93

Daly KL, Passow U, Chanton J, Hollander D (2016) Assessing the impacts of oil-associated marine snow formation and sedimentation during and after the Deepwater Horizon oil spill. Anthropocene 13:18–33. https://doi.org/10.1016/j.ancene.2016.01.006

Delvigne GAL, Sweeney CE (1988) Natural dispersion of oil. Oil Chem Pollut 4:281–310

van Eenennaam JS, Wei Y, Grolle KCF, Foekema EM, Murk AJ (2016) Oil spill dispersants induce formation of marine snow by phytoplankton-associated bacteria. Mar Pollut Bull 104:294–302. https://doi.org/10.1016/j.marpolbul.2016.01.005

van Eenennaam JS, Rahsepar S, Radović JR, Oldenburg TBP, Wonink J, Langenhoff AAM, Murk AJ, Foekema EM (2018) Marine snow increases the adverse effects of oil on benthic invertebrates. Mar Pollut Bull 126:339–348. https://doi.org/10.1016/j.marpolbul.2017.11.028

EMSA (2009) Manual on the applicability of oil spill dispersants

IPIECA (2017) Guidelines on implementing spill impact mitigation assessment (SIMA)

ITOPF (2012) Use of dispersants to treat oil spills. In: Technical Information Papers

Jongbloed RH, Tamis JE, Holthaus KIE, van der Veen DPC, van der Velde I, Blankendaal VG, Goedhart PC, Jak RG, Koops W (2002) Chemicals in combating oil spills. A literature review in perspective of the Dutch situation. Report No. R 2002/640. Den Helder, The Netherlands: TNO

Khelifa A, Fingas M, Brown C (2008) Effects of Dispersants on Oil-SPM Aggregation and Fate in US Coastal Waters, Final Report Grant Number: NA04NOS4190063

Khelifa A, So LLC (2009) Effects of Chemical Dispersants on Oil Brine Interfacial Tension and Droplet Formation. In: AMOP Technical Seminar on Environmental Contamination and Response. Ottawa (ON), pp 383–396

Kleindienst S, Seidel M, Ziervogel K, Grim S, Loftis K, Harrison S, Malkin SY, Perkins MJ, Field J, Sogin ML, Dittmar T, Passow U, Medeiros P, Joye SB (2016) Reply to Prince et al.: ability of chemical dispersants to reduce oil spill impacts remains unclear. Proc Natl Acad Sci 113(11):E1422–E1423. https://doi.org/10.1073/pnas.1600498113

Langenhoff AAM, Rahsepar S, van Eenennaam JS, Radović JR, Oldenburg TBP, Foekema E, Murk ATJ (2020) Effect of marine snow on microbial oil degradation (Chap. 18). In: Murawski SA, Ainsworth C, Gilbert S, Hollander D, Paris CB, Schlüter M, Wetzel D (eds) Deep oil spills: facts, fate, effects. Springer, Cham

Lee K, Nedwed T, Prince RC, Palandro D (2013) Lab tests on the biodegradation of chemically dispersed oil should consider the rapid dilution that occurs at sea. Mar Pollut Bull 73:314–318. https://doi.org/10.1016/j.marpolbul.2013.06.005

Lee K, Boufadel M, Chen B, Foght J, Hodson P, Swanson S, Venosa A (2015) Expert panel report on the behaviour and environmental impacts of crude oil released into aqueous environments. Royal Society of Canada, Ottawa. isbn:978-1-928140-02-3

Lewis A, Crosbie A, Davies L, Lunel T (1998) Dispersion of emulsified oil at sea, AEA technology report. AEAT-3475, June 1998

Moum JNN, Smyth WDD (1994) Upper Ocean Mixing Processes. Encyclopedia of Ocean Sciences 3093–3100. https://doi.org/10.1006/rwos.2001.0156

National Research Council of the National Academies (2005) Understanding oil spill dispersants: efficacy and effects. National Academies Press, Washington, DC

O'Sullivan AJ, Jacques TG (2001) Impact reference system: effects of oil in the marine environment: impact of hydrocarbons on fauna and flora. European Commission, Brussels, Belgium, orid. ed. 1991

Rahsepar S, Smit MPJ, Murk AJ, Rijnaarts HHM, Langenhoff AAM (2016) Chemical dispersants: oil biodegradation friend or foe? Mar Pollut Bull 108:113–119. https://doi.org/10.1016/j.marpolbul.2016.04.044

Redman AD, Parkerton TF (2015) Guidance for improving comparability and relevance of oil toxicity tests. Mar Pollut Bull 98(1–2):156–170. https://doi.org/10.1016/j.marpolbul.2015.06.053

Reed M, Turner C, Odulo A (1994) The role of wind and emulsification in modelling oil spill and surface drifter trajectories. Spill Sci Technol Bull 1:143–157. https://doi.org/10.1016/1353-2561(94)90022-1

Reed M, Johansen Ø, Brandvik PJ, Daling P, Lewis A, Fiocco R, Mackay D, Prentki R (1999) Oil spill modeling towards the close of the 20th century: overview of the state of the art. Spill Sci Technol Bull 5:3–16. https://doi.org/10.1016/S1353-2561(98)00029-2

Reed M, Leirvik F, Johansen Ø, Brørs B (2009) Numerical algorithm to compute the effects of breaking waves on surface oil spilled at sea. Trondheim, Norway. October, 2009, pp 1–127. https://crrc.unh.edu/sites/crrc.unh.edu/files/final_report_sintef_natural_dispersion_october-2009.pdf. Accessed on: 29 Oct 2018

Robbins ML, Varadaraj R, Bock J, Pace SJ (1995) Effect of stokes' law settling on measuring oil dispersion effectiveness. Int Oil Spill Conf Proc 1995:191–196. https://doi.org/10.7901/2169-3358-1995-1-191

Schwing PT, Hollander DJ, Brooks GR, Larson RA, Hastings DW, Chanton JP, Lincoln SA, Radović JR, Langenhoff A (2020) The sedimentary record of MOSSFA events in the Gulf of Mexico: a comparison of the Deepwater Horizon (2010) and Ixtoc 1 (1979) oil spills (Chap. 13). In: Murawski SA, Ainsworth C, Gilbert S, Hollander D, Paris CB, Schlüter M, Wetzel D (eds) Deep oil spills: facts, fate, effects. Springer, Cham

SL Ross Environmental Research LTD, A. Lewis Oil Spill Consultancy, MAR Incorporated (2006) Chemical dispersibility of U.S. outer continental shelf crude oils in non-breaking waves, Ottawa, Canada. September 2006, pp 1–48. https://www.bsee.gov/sites/bsee.gov/files/osrr-oil-spill-response-research//546aa.pdf. Accessed 4 Mar 2019

SL Ross Environmental Research LTD, MAR Incorporated (2011) Comparison of large-scale (Ohmsett) and small-scale dispersant effectiveness test results, Ontario, Canada. June 2011, pp 1–55. https://www.bsee.gov/sites/bsee.gov/files/osrr-oil-spill-response-research//638ab.pdf. Accessed on: 29 Oct 2018

Tamis JE, Jongbloed RH, Karman CC, Koops W, Murk AJ (2012) Rational application of chemicals in response to oil spills may reduce environmental damage. Integr Environ Assess Manag 8:231–241. https://doi.org/10.1002/ieam.273

Vonk SM, Hollander DJ, Murk AJ (2015) Was the extreme and wide-spread marine oil-snow sedimentation and flocculent accumulation (MOSSFA) event during the Deepwater Horizon blowout unique? Mar Pollut Bull 100(1):5–12. https://doi.org/10.1016/j.marpolbul.2015.08.023

Walstra P (1993) Principles of Emulsion formation. Chemical Engineering Science 48:333–349. https://doi.org/10.1016/0009-2509(93)80021

Walstra P (2005) 8 emulsions. In: Lyklema J (ed) Fundamentals of interface and colloid science. Elsevier, pp 1–94

Zeinstra-Helfrich M, Koops W, Dijkstra K, Murk AJ (2015a) Quantification of the effect of oil layer thickness on entrainment of surface oil. Mar Pollut Bull 96:401–409. https://doi.org/10.1016/j.marpolbul.2015.04.015

Zeinstra-Helfrich M, Koops W, Murk AJ (2015b) The NET effect of dispersants — a critical review of testing and modelling of surface oil dispersion. Mar Pollut Bull 100:102–111. https://doi.org/10.1016/j.marpolbul.2015.09.022

Zeinstra-Helfrich M, Koops W, Murk AJ (2016) How oil properties and layer thickness determine the entrainment of spilled surface oil. Mar Pollut Bull 110:184–193. https://doi.org/10.1016/j.marpolbul.2016.06.063

Zeinstra-Helfrich M, Koops W, Murk AJ (2017) Predicting the consequence of natural and chemical dispersion for oil slick size over time. J Geophys Res Oceans 122:7312–7324. https://doi.org/10.1002/2017JC012789

Chapter 11
Far-Field Modeling of a Deep-Sea Blowout: Sensitivity Studies of Initial Conditions, Biodegradation, Sedimentation, and Subsurface Dispersant Injection on Surface Slicks and Oil Plume Concentrations

Natalie Perlin, Claire B. Paris, Igal Berenshtein, Ana C. Vaz, Robin Faillettaz, Zachary M. Aman, Patrick T. Schwing, Isabel C. Romero, Michael Schlüter, Andreas Liese, Nuttapol Noirungsee, and Steffen Hackbusch

Abstract Modeling of large-scale oil transport and fate resulting from deep-sea oil spills is highly complex due to a number of bio-chemo-geophysical interactions, which are often empirically based. Predicting mass-conserved total petroleum hydrocarbon concentrations is thus still a challenge for most oil spill models. In addition, dynamic quantification and visualization of spilled oil concentrations are necessary both for first response and basin-wide impact studies. This chapter presents a new implementation of the Connectivity Modeling System (CMS) oil application that tracks individual multi-fraction oil droplets and estimates oil

N. Perlin (✉) · C. B. Paris · I. Berenshtein · A. C. Vaz · R. Faillettaz
University of Miami, Rosenstiel School of Marine and Atmospheric Science,
Miami, FL, USA
e-mail: nperlin@miami.edu; cparis@rsmas.miami.edu; iberenshtein@rsmas.miami.edu;
avaz@rsmas.miami.edu; robin.faillettaz@rsmas.miami.edu

Z. M. Aman
The University of Western Australia, Chemical Engineering, School of Engineering,
Perth, WA, Australia
e-mail: zachary.aman@uwa.edu.au

P. T. Schwing · I. C. Romero
University of South Florida, College of Marine Science, Petersburg, FL, USA
e-mail: pschwing@mail.usf.edu; isabelromero@mail.usf.edu

M. Schlüter
Hamburg University of Technology, Institute of Multiphase Flows, Hamburg, Germany
e-mail: michael.schlueter@tuhh.de

A. Liese · N. Noirungsee · S. Hackbusch
Hamburg University of Technology, Institute of Technical Biocatalysis, Hamburg, Germany
e-mail: andreas.liese@tuhh.de; nuttapol.noirungsee@tuhh.de; steffen.hackbusch@tuhh.de

© Springer Nature Switzerland AG 2020
S. A. Murawski et al. (eds.), *Deep Oil Spills*,
https://doi.org/10.1007/978-3-030-11605-7_11

concentrations and oil mass in a 3D space grid. We used the *Deepwater Horizon* *(DWH)* blowout as a case study and performed a sensitivity analysis of several modeling key factors, such as biodegradation, sedimentation, and alternative initial conditions, including droplet size distribution (DSD) corresponding to an untreated and treated live oil from subsurface dispersant injection (SSDI) predicted experimentally under high pressure and by the VDROP-J jet-droplet formation model. This quantitative analysis enabled the reconstruction of a time evolving three-dimensional (3D) oil plume in the ocean interior, the rising and spreading of oil on the ocean surface, and the effect of SSDI in shifting the oil to deeper waters while conserving the mass balance. Our modeling framework and analyses are thus important technical advances for understanding and mitigating deep-sea blowouts.

Keywords Deep-sea blowout · Subsea oil spill · Oil transport modeling · Oil fate modeling · Biogeophysical oil modeling · Connectivity Modeling System · Subsurface dispersant injection (SSDI) effects

11.1 Far-Field Modeling of Oil Spills

Modeling of a deep-sea oil spill can be divided into few stages that represent the dominant processes of oil and gas during a deep-sea blowout. The initial stage deals with the buoyant jet of oil and gas mixture, where the turbulence is dominant and where the coalescence of bubbles and droplets is the key process in the plume affecting the droplet and bubble size distribution (Johansen et al. 2003; Bandara and Yapa 2011; Aman et al. 2015; Malone et al. 2020). This jet plume model is sometimes followed by near-field modeling, which deals with oil and gas mixture separation, hydrate formation, loss of buoyancy, and formation of intrusion layers (Vaz et al. 2020). The latter occurs in stratified water column conditions, dictating the trap height of oil and gasses above the oil well and establishing a relative equilibrium in a characteristic droplet size distribution (DSD). The information about the attained DSD and the trap height can be further used for three-dimensional (3D) far-field modeling of oil transport and fate in the ocean at scales of hundreds and thousands of kilometers.

The far-field modeling typically employs Lagrangian-based methods to advance oil droplets in the horizontal and vertical directions based on environmental conditions and droplet buoyancy, using 3D hydrodynamic models supplying ocean state input data and wave and surface wind data to support surface transport modeling. The oil fate is modeled through individual-based algorithms of weathering processes such as biodegradation, dissolution, adsorption, sedimentation, degassing, surface evaporation, and photooxidation (Spaulding et al. 2017; Li et al. 2017). These algorithms are often empirically derived, and their accuracy is limited to the laboratory or field observations of available case studies. The validation of hindcast model results is often limited to quantitative comparison with observational data such as the surface oil slick extent detected by remote sensing

(Le Hénaff et al. 2012) or of oxygen anomalies indicating the presence of deep intrusions (Paris et al. 2012). Numerical sensitivity experiments of meaningful parameter ranges are a key tool for estimating the uncertainty associated with the outcome of each variable and allow the gauging of the relative importance of each variable in terms of outcome, such as large-scale surface expression (Le Hénaff et al. 2012), sedimentation (North et al. 2015), and the formation of the deep plume (Paris et al. 2012).

11.2 Laboratory Experiments and Observational Data for Numerical Modeling Support

Knowledge of the equilibrium droplet size distribution (DSD) is likely one of the most important characteristics of the jet plume to understand and to model the far-field oil trajectory and dispersal (Zhao et al. 2017). At present, several concepts and existing models in the literature discuss the representative DSDs for the DWH blow-out conditions (Socolofsky et al. 2011; Boxall et al. 2012; Paris et al. 2012, Adams et al. 2013; Aman and Paris 2013; Aman et al. 2015; Zhao et al. 2014; Li et al. 2017; Malone et al. 2018), in particular, the droplet mean diameter (d_{50}) and diameter range, including the maximum droplet diameter. Most of the concepts seem to agree that the DSDs expect to follow log-normal or Rosin-Rammler distribution shapes. Some of the field studies of oil droplet formation under various hydrodynamic conditions and oil types do support the possibility of larger droplets (order of one to several millimeters, e.g., Gros et al. 2017), although the diving experiments during the DWH incident were more conservative in the droplet size estimates and reported maximum droplet diameters near or slightly above 400 m (Davis and Loomis 2014; Li et al. 2017 and Figs. 11.4 and 11.5 therein). The high-pressure and cold-temperature experiments reported in Aman et al. (2015) are specifically designed to replicate the deep-sea DWH blowout conditions and provide support for the DSD choices adapted for the modeling studies presented in this chapter.

11.2.1 Droplet Formation in Deep-Sea Conditions

Fundamentally, the dispersion of one fluid – oil or gas in this context – in a secondary fluid phase invokes a balance between turbulent shear stresses, which act to rupture the interface and generate smaller dispersed particles and interfacial restorative forces, as illustrated in fundamental studies by Kolmogorov (1949). Hinze (1955) characterized this balance in terms of the turbulence dissipation rate (TDR), which is characterized by the *rate* at which turbulent energy is disseminated through the cascade of eddies. Hinze (1955) also demonstrated that under *simple* mixing conditions, the TDR contribution may be approximated through the use of Weber number, where the constraints and required assumptions of this approach are detailed in Malone et al. (2020) and Pesch et al. (2020). Both explicit TDR- and empirical Weber-based approaches have been deployed to assess the likely range of oil droplet sizes generated at the Macondo blowout, which naturally result in a wide distribution of behavior. However, the fundamental balance between turbulence and interfacial tension provides key insight as to droplet formation: smaller droplets can be generated by (i) increasing turbulence and/or (ii) decreasing interfacial tension. The latter effect provides a basis for considering subsea dispersant injection (SSDI) during deepwater blowouts but only in the limit of fully understanding and characterizing the inherent turbulence of the deepwater plume.

Deepwater conditions provide a unique reality for droplet formation, as the *natural momentum* of the blowout will be insufficient for droplets to reach the surface. That is, droplet buoyancy – as controlled by droplet diameter and fluid properties – is the primary surfacing mechanism in deepwater blowouts. To this end, the simulation community has adopted a bifurcated approach to characterize surfacing behavior: (i) fundamental or semi-empirical methods are adopted to relate turbulence to the oil DSD, whereafter (ii) far-field simulations assess the vertical and lateral migration of droplets in the water column. Malone et al. (2020) provides further insight as to the mechanics of droplet breakup and dependence on TDR.

11.2.2 Biodegradation of Hydrocarbons in the Water Column

Removal of oil compounds from the water column due to bacterial activity (biodegradation) occurs at different rates, depending on the strain of bacteria affecting certain groups of hydrocarbons, and may be subject to variations in fate and partitioning under deep-sea conditions (Bubenheim et al. 2020; Jaggi et al. 2020).

High-pressure reactors for biological experiments were constructed at the Hamburg University of Technology in cooperation with the companies Technik Service A. Meyer and Eurotechnica. High hydrostatic pressure characteristic of the deep sea can be simulated within these reactors allowing the examination of various biological processes under deep-sea conditions. The reactors made from stainless steel and bronze can withstand a maximum pressure of 40 MPa. Reactors are equipped with sensor and near-infrared (NIR) spectrometer to continuously monitor oxygen as a proxy for aerobic biodegradation and gaseous hydrocarbons, respectively (Valladares Juárez et al. 2015). To overcome the limitation of oxygen availability for aerobic biodegradation, the reactor setup has a headspace, which acts as a reservoir of oxygen that can replenish oxygen used in aerobic biodegradation in the water phase. Nitrogen gas is used as a pressurizing medium to avoid the toxicity of increased oxygen partial pressure occurring when the gas phase in the headspace of the reactor is compressed via mechanical pressurization with a spindle press. Experiments clearly indicate that pressure influences different taxa of microorganisms unequally. For example, the growth rate of alkane-degrading bacteria *Rhodococcus qingshengii* TUHH-12 is slightly affected by pressure. The growth rate at 15 MPa decreased to 0.16 h^{-1}, in comparison to 0.36 h^{-1} at atmospheric pressure (0.1 MPa) (Schedler et al. 2014). *Rhodococcus* PC20, isolated from the deep-sea sediment of the Gulf of Mexico (GoM), has comparable growth at 15 MPa and atmospheric pressure. However, in the presence of dispersant COREXIT 9500A, the growth further diminishes, suggesting an enhanced toxicity of the dispersant at elevated pressure (Hackbusch et al. submitted). On the other hand, growth of aromatic hydrocarbon-degrading *Sphingobium yanoikuyae B1* is strongly impacted by pressure as low as 8.8 MPa (Schedler et al. 2014). Pressure also exhibits an inhibitory effect on hydrocarbon-degrading indigenous microbial community under nutrient-replete conditions, with a 4% decrease in *n*-alkane degradation for every 1 MPa increase (Nguyen et al. 2018). Application of high-pressure hydrocarbon degradation rates into a model increases accuracy of predictions with time, suggesting that high-pressure biodegradation is one of the mechanisms for the persistence of deepwater plume (Lindo-Atichati et al. 2014).

11.2.3 Sediment Analysis

A large-scale spatial model of the deposition of petroleum hydrocarbons following the DWH spill was generated to provide a better guidance for future accidental releases of oil contaminants from deep waters (Romero et al. 2017). In that study, sediment cores ($n = 2613$) were collected and chemically analyzed by different institutions (University of South Florida, US Government, and BP P.L.C.) from coastal to deep-sea areas in the GoM (total area coverage: 194000 km^2). A wide range of compounds were included in the spatial analysis to include the different hydrocarbon mixtures produced during the DWH spill due to partitioning of the oil within the water column (subsurface intrusion layers, oil slick, oil-mineral aggregates). In addition, due to multiple hydrocarbon sources in the GoM, hydrocarbon concentrations were compared within sediment core layers to distinguish background in each site to post-spill levels.

Results indicate that ~9.1 ± 4.1 × 10^4 metric tons of DWH oil was deposited in ~110,000 km^2, containing 21% on average (up to 47%) of the total amount of oil discharged and not recovered from the DWH spill, similar to the unaccounted amount of oil reported previously (McNutt et al. 2012a). Relative to the oil mass that remained in the environment, a large deposit of oil occurred in coastal (up to 43% in ~34,000 km^2) and deep-sea (up to 4% in ~33,000 km^2) areas, while negligible deposition was observed on the continental shelf (up to 0.5% in ~43,000 km^2, behaving as a transition zone in the northern GoM). Spatial trends in the amount and composition of deposited hydrocarbons in sediments demonstrated that weathering occurred mostly before deposition of DWH oil via various natural processes (evaporation, dissolution, degradation, photooxidation) and anthropogenic (burning of surface slick), depending on the area of deposition.

Schwing et al. (2017) employed a commonly used radioisotope tracer (excess ^{210}Pb or ^{210}Pb$_{xs}$) in sediment cores to characterize the spatial extent in the deep-sea area of the large sedimentation event occurred after the DWH spill (Brooks et al. 2015; Larson et al. 2020). The overall seafloor spatial extent of the large sedimentation event, as calculated by increased $^{210}Pb_{xs}$ flux after 2010, ranged from 12,805 to 35,425 km^2. The benefit of using $^{210}Pb_{xs}$ flux as a complimentary tool to petroleum tracers for validation of modeling results is the constraint of the overall spatial extent of increased flocculent mass flux from the water column, beyond the oil fraction.

Specifically for the deep sea, both studies demonstrated a large deposition event following the DWH spill of petroleum hydrocarbons (Romero et al. 2017) and sediments (Schwing et al. 2017) over ~34,000 km^2, mostly around the wellhead. This unexpected and prolonged deposition of oil and sediments to the seafloor has been referred as MOSSFA (marine oil snow sedimentation and flocculent accumulation, Daly et al. 2016) and was initially demonstrated in the DeSoto Canyon (Romero et al. 2015; Brooks et al. 2015). These studies demonstrate the need for multidisciplinary approaches for better understating of complex processes such as submerged oil spills.

11.3 Numerical Simulation Description

11.3.1 Modeling and Experimental Setup

Our study implements the existing oil application (Paris et al. 2012) of the Connectivity Modeling System (CMS; Paris et al. 2013), to improve the expected transport and fate of the live oil (or gas-saturated oil) spilled during the DWH accident and for providing a better tool for future deep-sea blowouts. The CMS oil application performs Lagrangian particle tracking of oil droplets released at the trap height above the wellhead. Particle transport calculations take into account 3D ocean currents, temperature, salinity, multi-fractional droplet buoyancy, droplet evolution due to biodegradation, and surface oil evaporation. The 4th order Runge-Kutta integration scheme forms the basis for particle advection in the model. Computations of the vertical terminal velocity of a droplet are based on its density and size, its Reynolds number, as well as on other the ambient conditions such as water temperature, salinity, density, and kinematic viscosity (Zheng et al. 2003).

The model output is saved every 2 hours and includes oil droplets' effective density, size, depth, and geographic location. Here the CMS model grid adapted horizontal grid spacing of 0.04 degrees and 20 layers in the vertical. The biodegradation dynamics of the present study is based on high-pressure experiments and assumes a different decay rate for the each of the oil droplet fractions. In the hydrocarbon multi-fraction approach implemented in the oil module, all the pseudo-components are now in the same oil droplet. This allows to account for dissolution processes where the droplet shrinks with the partitioning of the oil compounds in the water column (Jaggi et al. 2017). Post-processing algorithms further translate model output into oil concentrations (see Sect. 11.3.3). The DSD profiles for the untreated oil and for the oil treated with Corexit 9500®, the chemical dispersant deployed at the

Macondo wellhead during DWH, are based on prior research (Li et al. 2008 and Fig. 8 therein; Aman et al. 2011; Aman and Paris 2013). The DSD is binned to set out a log-normal distribution adapted from Paris et al. (2012) by extending the oil diameter range from 300 μm to 500 μm, resulting in d_{50} of 103 μm and 85 μm for the untreated and chemically treated live oil, respectively. Addition of the chemical dispersant (i.e., treated oil) reduces the interfacial tension between oil and water (Rewick et al. 1984) and shifts the mean droplet size (d_{50}) toward smaller values.

Here, because oil concentrations are estimated by post-processing, multiple scenarios of droplet size distribution (DSD) could be carried out using the same model output, i.e., the trajectories output of the deep-sea blowout control run representing the DWH spill that starts on April 20, 2010 (*DB_control*). Fluid flow includes a release of 3000 oil droplets at the trap height every 2 hours for 87 days until July 15, 2010 – the day of successful containment of the spill using a capping stack. The oil droplets are tracked for total of 167 days from the initial blowout date. The release location is 28.736°N, 88.365°W at 1222 m, i.e., the estimated height of oil and gas separation 300 m above the wellhead (Socolofsky et al. 2011). Initial droplet diameters are determined at random by the CMS model in the range from 1 μm to 500 μm. The post-processing algorithm then utilized the change-of-variable technique for the probability density functions to simulate various DSDs for the scenarios with (chemically treated) and without (untreated) SSDI. Each droplet deployed in the CMS model contains three pseudo-components (fractions) accounting for the differential petroleum hydrocarbons volatility as follows: 10% of light molecular weight with the density of 800 kg/m^3, 75% of medium molecular weight with 840 kg/m^3, and 15% of heavy molecular weight with 950 kg/m^3 density. The biodegradation rates for the oil fractions are based on laboratory experiments for the first-order decay at high pressure and are set to 30 h for the light fraction and 40 h for the intermediate oil fraction (Schedler et al. 2014; Lindo-Atichati et al. 2014). Half-life decay rate for the heavy fraction is set to 180 h based on the observational study of the degradation rates in the GoM (Hazen et al. 2010, Supplementary material Table S7). Additional decay of the oil occurred at the surface due to evaporation, which half-life was set to 250 h in the control case (see discussion in De Gouw et al. 2011 on evaporation rates for different hydrocarbon types).

Ocean hydrodynamic forcing for the present study uses daily output from the HYbrid Coordinate Ocean Model (HYCOM; Chassignet et al. 2003, Halliwell 2004) for the GoM region on a 0.04 degree horizontal grid and provides diagnostic model output at 40 vertical levels spanning from the surface down to 5500 m. HYCOM model employs data assimilation using the Navy Coupled Ocean Data Assimilation (NCODA; Cummings 2005), which assimilates available satellite altimeter observations, satellite and in situ sea surface temperature (SST) observations, as well as available in situ temperature and salinity profiles from moored buoys, XBTs, and Argo floats. HYCOM output variables used for CMS simulations included horizontal and vertical velocity components, temperature, and salinity.

The experiments included parameterization of the effects of the surface wind drift, the importance of which was emphasized by Le Hénaff et al. (2012). Wind stress components from the 0.5 degree Navy Operational Global Atmospheric

Prediction System (NOGAPS, Hogan and Rosemond 1991) are interpolated into HYCOM GoM 0.04 degree grid, and 3% of their values were added to the horizontal components of the upper ocean level velocity, taking into account wind stress rotation. The corrected ocean velocity fields are then used as a hydrodynamic forcing in the oil-CMS control experiment.

11.3.2 Suite of Numerical Case Studies

Additional model experiments of the deep blowout (*DB*) are conducted for the sensitivity studies (summarized in Table 11.1). These experiments aim to study the effects of (1) added chemical dispersant at the Macondo wellhead or SSDI (*DB_treated_oil*, *DB_FALL_treated*), (2) modified subsurface oil droplet landing conditions (*DB_noSubSfcLanding*), (3) added temperature-dependent biodegradation (*DB_tempBiodegrad*), (4) the oil transport under different environmental conditions (*DB_FALL*, *DB_FALL_treated*), and (5) different DSDs estimated for the DWH scenario, for the untreated and treated oil (*DB_VDROPJ_untreated*, *DB_VDROPJ_treated*).

The simulation *DB_FALL* assumes a blowout at the same location as the control run but occurring later in the season starting September 1, 2010, under a different ocean state and development of the Loop Current in the GoM. Oil droplets are released during the 87 days using the same release scheme as in the control run, and the droplets are tracked for the next 90 days.

The scenarios for the treated oil, *DB_treated* and *DB_FALL_treated*, use the same CMS output trajectories as their corresponding *untreated* cases but differ in their post-processing algorithm with the DSD peak shifted toward smaller values. The scenario *DB_noSubSfcLanding* modified subsurface landing by only allowing for the particles to land at shallow depths (i.e., 20 m, near the coastline). *DB_tempBiodegrad* scenario turns on temperature-dependent biodegradation scheme, in addition to the existing decay rates for each fraction.

Due to an ongoing debate in the literature about DSD for the DWH blowout, we included two additional simulations that use DSD predicted by a jet-droplet forma-

Table 11.1 Summary of CMS oil model numerical simulation suite

Model simulation label	Differences from the control experiment
DB_control (untreated)	167-day run, untreated oil, DSD range 1–500 μm
DB_treated	SSDI-treated oil, DSD range 1–500 μm
DB_noSubSfcLanding	No subsurface particle landing
DB_tempBiodegrad	Temperature-dependent biodegradation added
DB_FALL	Fall blowout scenario: 1 Sep. 2010, 90-day run, untreated oil
DB_FALL_treated	Fall blowout scenario: 1 Sep. 2010, 90-day, SSDI-treated oil
DB_VDROPJ_untreated	VDROP-J DSD for untreated oil, range 1–8000 μm
DB_VDROPJ_treated	VDROP-J for SSDI-treated oil, DSD range 1–2400 μm

tion model (VDROP-J), *DB_VDROPJ_untreated* and *DB_VDROPJ_treated*. VDROP-J predicts initial stabilized size distribution for the droplets and gas bubbles (Zhao et al. 2014). We used the DSD for the liquid droplets as reported in Gros et al. (2017), with the range of droplets of 1–8000 m for the untreated oil and 1–2400 m for the SSDI-treated oil. It has been discussed, however, in Sect. 11.2 that the DSD from the diving observations during the DWH incident indicated a maximum droplet size of approximately 400 μm. Nevertheless, we included these options of larger droplets predicted by the jet-droplet formation model VDROP-J to allow model validation and comparison of corresponding far-field modeling results against those done with more conservative DSD estimates.

11.3.3 Model Output and Post-processing Variables

The CMS model output consists of trajectory files that contain the following variables at a specified output frequency (2 hours) for each released droplet: droplet horizontal (latitude, longitude) and vertical (depth) location, its average density and average droplet diameter, and whether the droplet is still suspended in a water column. Averaging is done across multiple fractions within each droplet. The post-processing algorithm outlined below translates these output variables into oil concentrations or oil mass in each grid box of a 3D output grid. To compute the sedimented oil mass, only particles that have been landed (no longer transported by the flow) are accounted for. These oil mass or oil concentrations could be used for the quantitative analysis and model verification.

An estimated 4.9 million of stock tank barrels or about 730,000 tons of gas and oil mixture were spilled into the GoM during the 87-day period of relief efforts (McNutt et al. 2012b; Griffiths 2012). A total of 3.132 million droplets are released by the CMS model during the 87 days of the simulated DWH incident, which need to represent the total amount of crude oil that spilled into the water from the oil well. To conserve the oil mass balance of the total spilled oil, the droplet mass is scaled up to obtain a representative amount of oil for each droplet.

The CMS model generates a uniform DSD at the droplet release time. Droplet oil mass, m, is computed as $m = (\pi/6)\rho d^3$, where $\pi=3.14$, ρ is the droplet average density, and d is the droplet diameter. If the probability density function (PDF) of a continuous random variable d, $P(d)$, is known on a range of diameters d_{min} to d_{max}, then the PDF of the oil mass, $P(m)$, could be determined by employing a change-of-variable technique for a random variable (Pishro-Nik 2014). Knowing the oil flow rate and the number of released droplets, we could further use the $P(m)$ to determine a fraction of oil that each droplet needs to represent to achieve the mass balance; under a uniform DSD assumption, this amounts to 233 kg for each released droplet regardless of its size.

These values differ for the DSD corresponding to the untreated oil and for the treated oil (Sect. 11.3.1), and we assume the droplets in the same bin represent similar mass of oil per droplet. Furthermore, while the size and mass of released droplets changes with time, each droplet is still representative of the same amount of oil assigned at the release time. We then compute the scaling factor, sf, for each droplet from its initial mass m_{t0} and the mass of oil it represents, m_R, as follows: $sf = m_R/m_{t0}$. This scaling factor is unique for each droplet and remains invariant during the run. Oil mass of a droplet at each time, m_t, is estimated from the droplet diameter and effective density at the corresponding times; it is then multiplied by the scaling factor to obtain the effective oil mass: $m_t^{Ef} = sf \cdot m_t$. This effective oil mass is further summed for all the droplets found in each post-processing domain 3D grid box and at a given time.

The 3D post-processing domain has 0.02-degree × 20-m grid boxes in the horizontal and vertical directions, respectively, spanning from the surface down. For concentration estimates in the surface layer, the vertical boundaries are taken between 0 and 1 m. After the oil mass is computed for a given droplet and given output time, the corresponding post-processing domain grid box is determined, and the effective oil mass of all the droplets found in that grid box at a given time are summed up. After looping over all the droplets and all the times, we yield the cumulative effective oil mass in the time-space field. Concentrations are obtained by normalizing the total oil mass to the mass of water in the corresponding grid box, and the daily averages are further determined from the 2-hourly output products. Bathymetry is taken into the account to compute volumes of the grid boxes; the volume of oil is assumed to be much smaller than the volume of water in a 3D grid box.

11.4 Modeling Results and Analyses

11.4.1 Surface Oil Expression

The extent of oil slicks on the ocean surface was closely monitored during the spill using satellite observations and can thus be used for qualitative model verification. Surface oil concentrations from the control run in the top surface layer 0–1 m for May 13, 2010, show similar features as the Roffer's Ocean Fishing Forecasting System (ROFFS) analysis (Fig. 11.1a). However, there is not much oil reported by the observations along the coastal areas, which are present in model results (Fig 11.1b). Reported oiled coastal areas (Nixon et al. 2016) agree with Romero et al. (2017) and the CMS output. Surface oil concentrations for May 13, 2010, indicate wider spread of the oil slick around the blowout location and a well-defined extent of the oil presence south of it. The oil slick is further split around 27°N and 88.5°W into eastward and westward branches; the eastward branch follows the strong Loop Current, while westward propagation results from inclusion of wind drift effects (see also Le Hénaff et al. 2012, Fig. 11.2 with similar surface oil behavior).

The *DB_VDROPJ_treated* simulation yields greater surface concentrations than the *DB_control* scenario in most of the areas, except of the oil slick branch extending eastward along the 28°N. *DB_VDROPJ_untreated* (not shown) resulted in even higher concentrations due to larger droplets surfacing quickly.

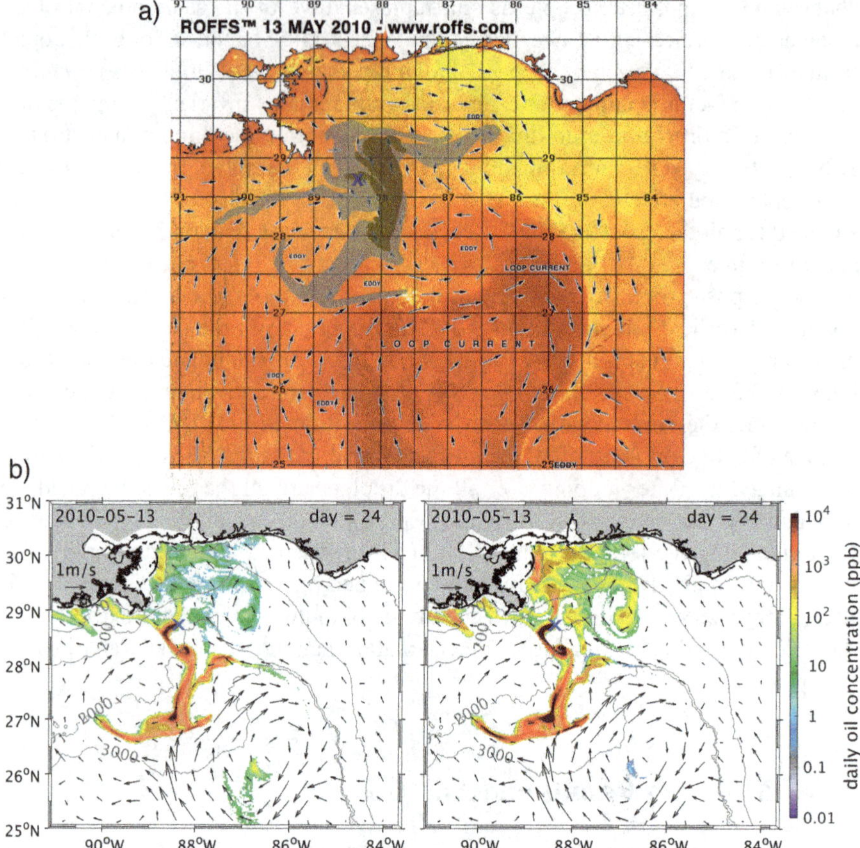

Fig. 11.1 (**a**) ROFFS composite image showing oil presence at the surface on May 13, 2010. ROFFS image is adapted from the similar image used by Le Hénaff et al. (2012), retrieved from http://roffs.com/deepwaterhorizon.html; the colors refer to sea surface temperatures, while the gray- and olive green-shaded areas indicate surface oil presence inferred from different observations (detailed explanations can be found at http://www.roffs.com/wp-content/uploads/2012/12/ROFFSExplanation.pdf). Vectors represent the daily average ocean surface currents. (**b**) Daily average oil concentration estimates for the same date from the *DB_control* (left panel) and *DB_VDROPJ_treated* (right panel) simulations. Panels for the ROFFS and CMS modeling results cover similar areas for each corresponding day. Blue "x" mark indicates a blowout location; vectors are the daily average ocean surface currents with the surface wind adjustment (see Sect. 11.3.1); day after a blowout is shown in the top right-hand side of the panels

A closer comparison of the ROFFS image and the modeled data reveals a discrepancy in surface currents along the northern Gulf Coast east of Mississippi River delta: eastward flow in the ROFFS image but largely west-northwestward effective currents in the model data in the same area. A likely source of the differences in surface currents is Mississippi River outflow discharge (Kourafalou and Androulidakis 2013) that was not taken into the account by the HYCOM model and which could lead to inaccuracies in oil slick location estimates in the coastal regions.

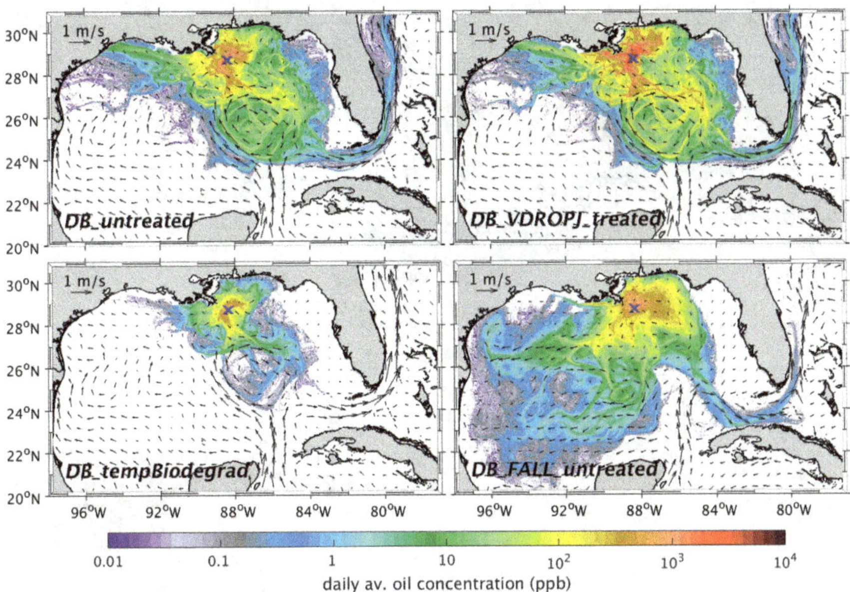

Fig. 11.2 Surface expression of the oil viewed in 90-day average surface oil concentrations (0–1 m) for the four simulations as indicated in the lower left corner of each panel. Vectors are average surface currents for the same 90-day period, with the 3% surface wind correction (see Sect. 11.3.1), and the blue "x" mark indicates a blowout location

Surface expression of the oil (Fig. 11.2) for the 90-day period shows high similarity between the *DB_control* and *DB_VDROPJ_treated* scenarios in oil spatial extent, except some higher values in the latter case. In the *DB_tempBiodegrad* case, not only are the oil concentrations smaller, but the special extent of the oil is notably diminished: less oil propagating westward and impinging onto the coastline in the north-northwest part of the GoM. In contrast, the surface oil expression in *DB_FALL_treated* case shows the oil spreading much farther westward into the GoM interior but staying off the West Florida Shelf and having less oil escaping into the Atlantic basin. Surface oil transport appears to be primarily dependent on the state of GoM currents, the development of the Loop Current, and the local coastal and shelf circulation and eddy formation. Variations in the DSDs produce first-order numerical differences in surface oil concentrations, but not qualitative changes of the oil slick.

11.4.2   Oil Distribution in the Water Column and SSDI Effect

The effects of the dispersants on the oil distribution in the water column are estimated from the horizontally cumulative oil mass for the different DSDs (Fig. 11.3). In the *DB_control* case, the vertical structure of horizontally cumulative oil mass

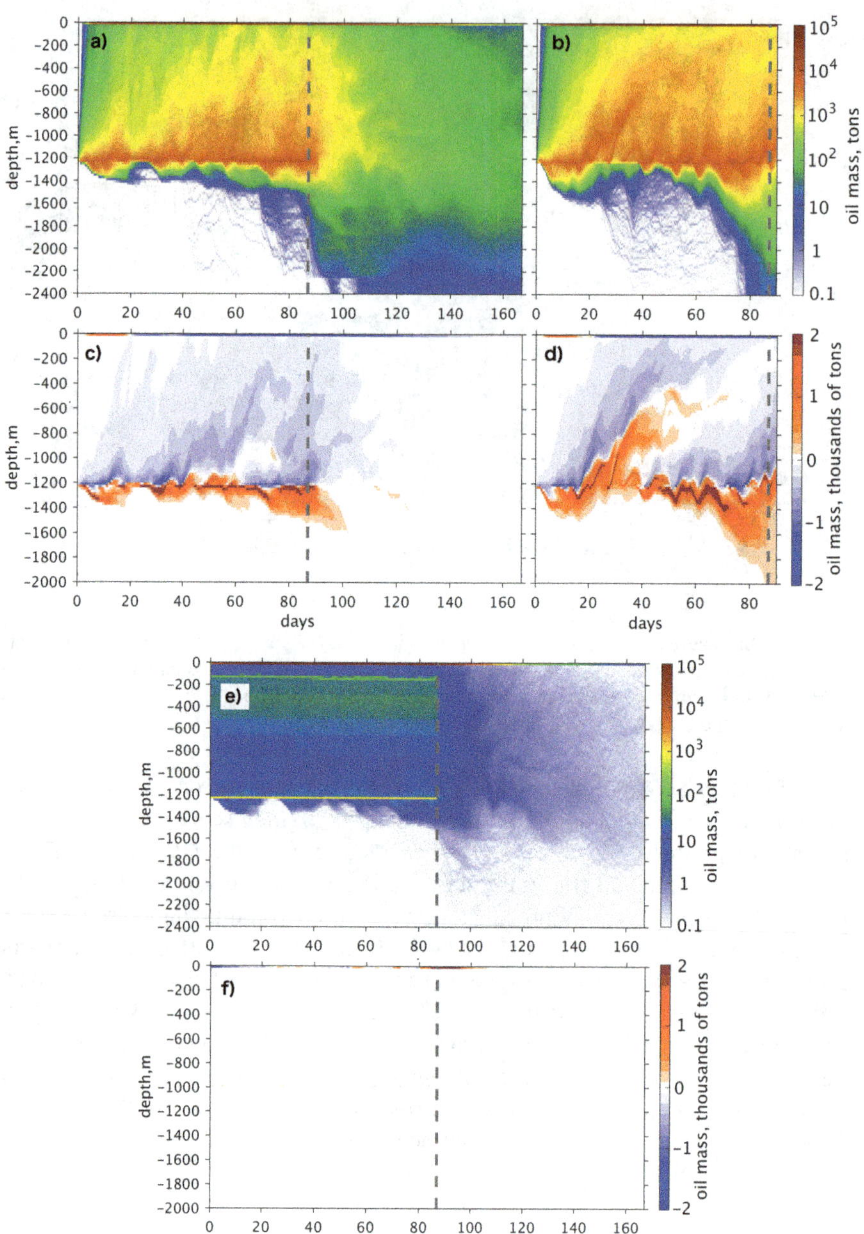

Fig. 11.3 (**a–d**) Vertical distribution of area-cumulative oil mass for the untreated oil in the (**a**) *DB_control* and (**b**) *DB_FALL* experiments. The differences between the oil treated with chemical dispersants and the untreated oil, for the (**c**) (*DB_treated – DB_control*) and (**d**) (*DB_FALL_treated – DB_FALL*). (**e–f**) Vertical distribution of area-cumulative oil mass for the untreated oil in the (**e**) *DB_VDROPJ_untreated* and (**f**) area-cumulative differential oil mass for (*DB_VDROPJ_treated – DB_VDROPJ_untreated*)

for the untreated oil (Fig. 11.3a) indicates that within few days after the onset of the simulation, the highest oil content is found in the subsea and upper 0–20 m layer, which corroborates observations from the BP Gulf Science Data (Wade et al. 2016; Paris et al. 2018). Bulk of weathered oil in excess of a thousand tons (shades of red) near the release depth gradually spreads up in the water column to around 400 m; the lower boundary of the bulk of oil mass shows subduction as well. The bulk of the subsurface oil decreases rapidly after the oil release stops on day 87, yet hundreds of tons of oil remain in the water column, including at depths below the release height until through the end of 167-day simulation. Oil content in the surface layer gradually diminishes as well, yet remaining at higher values than in the ocean interior.

The SSDI-treated oil in *DB_treated* simulation leads to higher oil content below the trap height but also initially in the 0–20 surface layer until about day 18 (Fig. 11.3c). At later times, the treated oil yields consistently lower oil mass in the surface layer, which is then in agreement with the expected behavior of the DSD; but this difference is by only a few thousand tons as compared to the untreated oil content. Below the surface layer, treated oil yields lower (higher) amounts in the water column interior above (below) the trap height most of the times due to down-welling currents, with some vertical fluctuations of the negative (positive) differences around that depth due to upwelling currents.

For the *DB_FALL* scenario, the major distinction in differential mass calculations from the *DB_control* case is seen in the intermediate depths, where a region of positive-difference oil mass propagating upward from the release depth is observed from approximately day 15 until after day 50 following the blowout (Fig. 11.3d), after which the vertical extent and the upward "tongue" of positive differences gradually diminish. This is also found in *DB_control*, yet at a notice-able lesser extent, on days 60–80 between about 900 and 1200 m. This distinctive behavior of vertical oil content is explained by the different ocean circulation patterns in the *DB_FALL* case, where the oil is transported upslope or along the conti-nental shelf slope. The positive differences likely result from smaller droplets being involved in bathymetry-controlled upslope or upwelling movement that would oth-erwise stay at greater depths. Similar behavior of oil residues deposited on the continental shelf slope was reported in Romero et al. (2017); another example in Romero et al. (2015) shows chemically the impingement of the intrusion on the continental shelf slope.

The vertical oil distribution and differential amounts in *DB_VDROPJ_untreated* and *DB_VDROPJ_treated* simulations (Fig. 11.3e–f) look noticeably different from the other cases. The primary reason is significantly larger droplet sizes up to 8 mm in diameter that cause faster surfacing of the oil. Regardless of chemical dispersion, the surface layer of the entire GoM and the trap height where oil droplets are released at a single location 300 m above Macondo contains the highest oil mass during the active blowout phase. The distinct time uniformity until the spill is con-tained on day 87 (Fig. 11.3e) is the combined result of large droplets surfacing too fast to be affected by the lateral transport, the unconventional distribution of the droplet size implemented from Gros et al. (2017). The vertical layering reflects the 2-hourly output frequency of the model and the droplet release frequency being

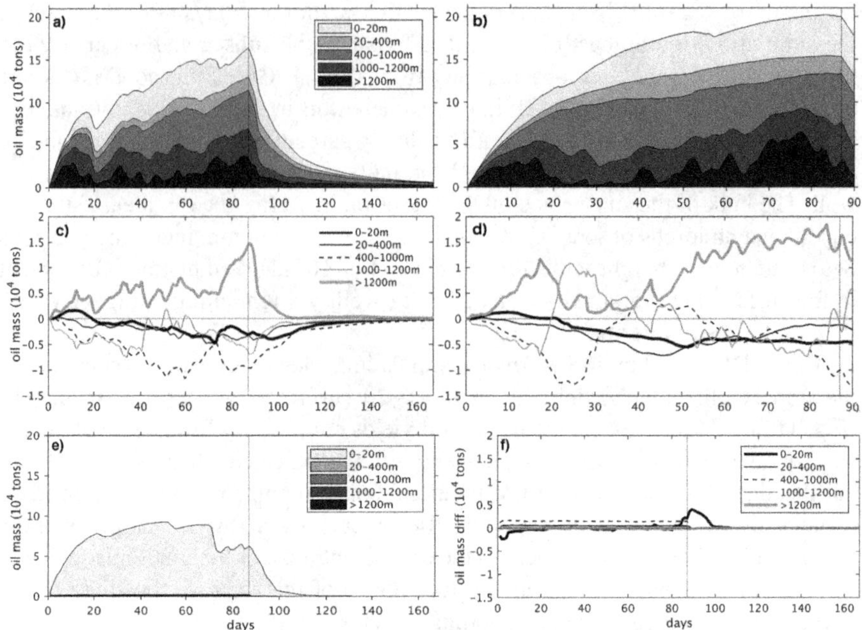

Fig. 11.4 (**a–d**) Area-cumulative oil mass in five vertical layers for the untreated oil in the (**a**) *DB_control* and (**b**) *DB_FALL* experiments, for every 2-h output time intervals, over the duration of the run (days in the x-axis). Oil mass in layers are stacked, so the sum of all the layers shows the total oil mass in a water column. (**c–d**) Treated versus untreated oil differences in mass in the five layers for (**c**) *DB_treated_oil – DB_control* and (**d**) *DB_FALL_treated – DB_FALL*. The "zero" difference line is marked as a dotted horizontal line for the reference. Dotted vertical line marks the day 87, the last oil release day in the model, corresponding to July 15, 2010, when the severed Macondo oil wellhead was capped. (**e–f**) Similar to (**a**) and (**b**), except for *DB_VDROPJ_untreated* and *DB_VDROPJ_treated_oil – DB_VDROPJ_untreated* differences, correspondingly. The temporal resolution on the x-axis is daily average

identical; however, this model configuration is the same for all scenarios. Clearly, *DB_VDROPJ_untreated* scenario is unrealistic with no secondary intrusions or oil in the water column (Diercks et al. 2010; Wade et al. 2016). Differential oil content demonstrates non-negligible values in the top 20-m layer only as well as being less consistent with time progression. The lack of distinct plume-like vertical structure in the interior also calls into question the validity of the DSDs produced by VDROP-J model for the DWH scenario.

The dispersant effect becomes more evident when the oil budget is computed through time for the several vertical layers in *DB_control* and *DB_FALL* scenarios (Fig. 11.4a–d). While the bulk of the oil remains submerged, the largest positive difference (more oil resulting in corresponding SSDI case) is found in the layer >1200 m. Very little subsurface oil or differential oil amounts result in the *DB_VDROPJ* cases (Fig. 11.4e–f), with most of the oil residing in 0–20 m layer, which is again unrealistic when compared to water column BP Gulf Science Data (Wade et al. 2016, Paris et al. 2018).

Despite differences in ocean circulation and ocean state condition in *DB_FALL* and *DB_control* scenarios, oil content and differential oil values in the surface and most of interior layers are qualitatively similar. The cumulative effect of the SSDI treatment thus appears in reducing the oil marginally in the surface layer but shifting significantly the bulk of oil to greater depths.

11.4.3 *Modeled Oil Residue Sedimentation*

The sinking of oil residues, such as weathered oil capable of sinking, may eventually be deposited on the ocean floor of the GoM or encounter a coast or a bathymetric slope. The oil application of the CMS model tracks and labels such "landed" droplets that effectively indicate oil residue sediments. The oil mass sedimented during the entire simulation period for each of the case studies is shown in Fig. 11.5.

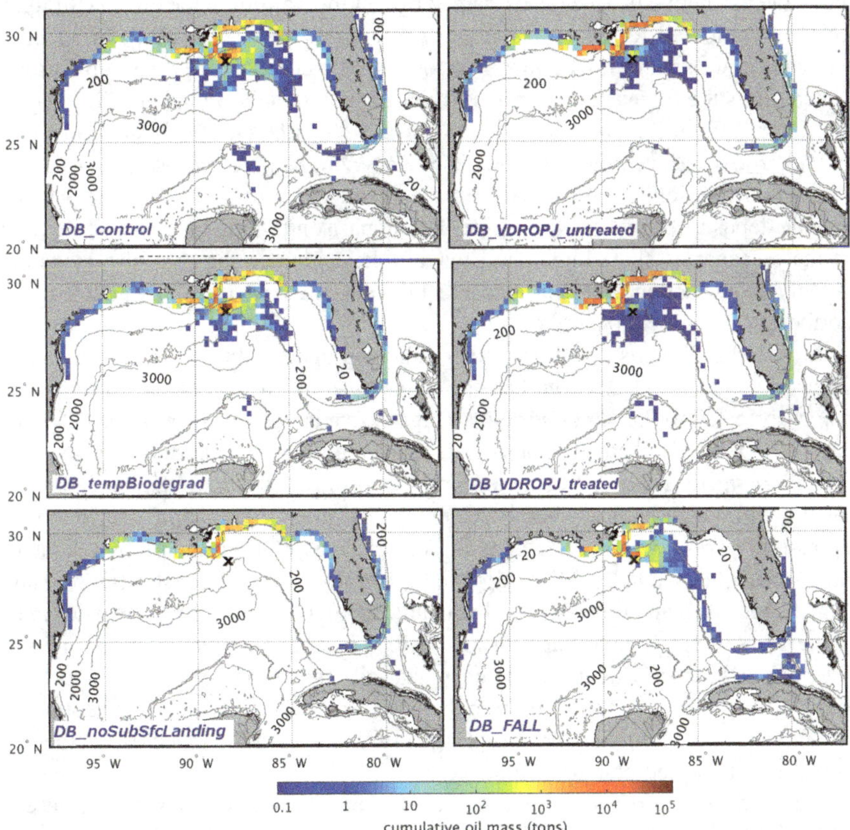

Fig. 11.5 Sedimented oil mass on a 0.5° horizontal grid, in the end of the model simulations as indicated in each lower left corner of each panel. The length of all simulations is 167 days, except for *DB_FALL* (90 days)

Our results indicate that for the *DB_control* scenario, highest oil residue deposits are found at depths around the deep plume, especially north, northeast, and east of the DWH site, including the DeSoto Canyon. Modeled oil residue is consistently deposited around the GoM shoreline from western shore, along Florida Peninsula and Florida Keys, and some deeper GoM interior regions along the bathymetric slopes. Note the distant regions with elevated oil residue deposits along the shoreline near 94–95°W and the south to southeastern Florida coast.

The *DB_noSubSfcLanding* scenario shows a picture similar to that of *DB_control*, except deeper regions, and highlights the coastal areas with oil residue deposits. In general, sedimented oil residue in *DB_tempBiodegrad* case illustrates surprising similarity with the control case in the coastal areas and some reduction in deposits only around the DWH site and in the GoM interior along the bathymetric slopes. Ocean circulation changes that dictate the outcome in *DB_FALL* case particularly limit the oil deposits along eastern GoM shore and Florida Keys (no deposits along the western Florida coast), no deposits in the north-northwest corner of the GoM shoreline or south-southwest of the DWH site. Note the sediment formation along the northern Cuba, easily explained by southward shift in the Florida Straights position compared to the conditions in *DB_control* case (Fig. 11.2). More oil is transported westward to the GoM interior with the eddies from the Loop Current, which limits the oil available for deposition around the DWH blowout site.

Minor differences were found between the *DB_VDROPJ_untreated* and *DB_VDROPJ_treated* scenarios, in which the *DB_VDROPJ_treated* case produces only some spatial variability and some increase in spatial coverage of the areas with oil residue deposits around the DWH site. In comparison to the *DB_control* case, both of these scenarios show highest oil residue deposition in the coastal regions and notably reduced deposition around the blowout site and DeSoto Canyon to the northeast of that location.

Modeled sediments could be further validated against the observational studies discussed in Sect. 11.2.3. The base-scale analysis of oil residues in surface sediments corrected for background values (Fig. 11.6; adapted from Fig. 11.4 in Romero et al. 2017) indicated contamination was observed up to a distance of 180 km from the DWH rig in the deep-sea area (depths >200 m) and up to 517 km in coastal (including inshore habitats) and continental shelf areas. The spatial distribution of oil residues show impacted regions in coastal areas of Louisiana (e.g., Barataria Bay, Chandeleur Islands), Mississippi (e.g., Horn and Petit Bois Islands), Alabama (e.g., Cat Island), and Florida (e.g., Panhandle area), and for offshore deep-sea areas in the Mississippi Canyon, up- and downslope of the DWH well, and, to a lesser extent, the DeSoto Canyon. In the deep sea, oil residue deposition was greater at 1300–1600 m depth up to 30 km from the DWH rig and at 1000–1300 m depth from 30 km to 175 km from the DWH rig (Fig. 11.6) following the trajectory of the seafloor and submerged plumes, respectively.

Schwing et al. 2017 found that relative to pre-DWH conditions, an increase in $^{210}Pb_{xs}$ flux occurred in two distinct deep-sea areas of the northern GoM: (1) on an east-northeast to west-southwest axis, stretching 230 km southwest and 140 km northeast of the DWH wellhead, and (2) on a 70 km northeast to southwest axis near the DeSoto Canyon (Fig. 11.7).

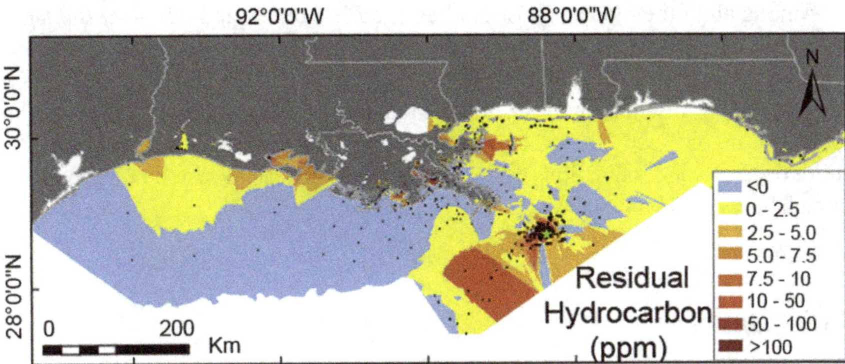

Fig. 11.6 The spatial extent of oil residues deposited on the seafloor as a result of the DWH spill (Reprinted from Romero et al. 2017, with permission from Elsevier). Residual hydrocarbons refer to the amount of oil residues deposited on the surface of the sediments after the spill minus background concentrations (pre-spill values). Hydrocarbon concentrations include the sum of aliphatics (n-alkanes C10–40, isoprenoids, branched alkanes), polycyclic aromatic (2–6 rings, including alkylated homologues), hopanoid (C27–35), sterane (C27–29), and triaromatic steroid (C26–28) compounds. Gray lines indicate bathymetry (m). Green star: location of the DWH site

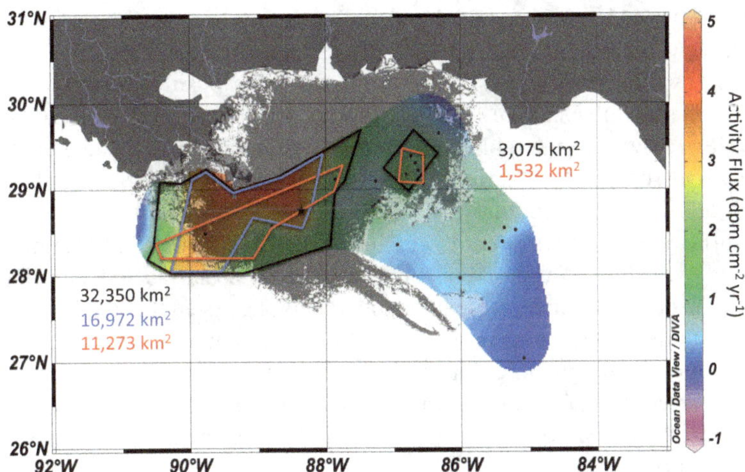

Fig. 11.7 A DIVA gridded contour map of the differential (difference of post-DWH and pre-DWH values) $^{210}Pb_{xs}$ flux. Polygons were constructed to determine the spatial extent of differential $^{210}Pb_{xs}$ flux. The polygons surround differential $^{210}Pb_{xs}$ flux greater than 3 dpm cm^{-2} yr.$^{-1}$ (blue), 1.5 dpm cm^{-2} yr.$^{-1}$ (black), and a minimum span between coring sites (red) in the western extent of the study area and greater than 1 dpm cm^{-2} yr.$^{-1}$ (black) and minimum span between coring sites (red) in the eastern extent of the study area. The gray overlay is the cumulative total extent of surface petroleum coverage (ERMA). The star represents the location of the DWH wellhead. This figure adapted with permission from Schwing et al. 2017; Copyright (2017) American Chemical Society

Among all of the modeled case studies, the *DB_control* and *DB_tempBiodegrad* scenarios (Fig. 11.5) seem to agree best with the observational studies in several aspects: first, in modeled high oil residue deposits around the DWH site and reflecting the SW-NE orientation of the area with the elevated deposition levels including DeSoto Canyon; second, elevated deposits along the northern GoM coastline; and third, distinctively, indication of higher values of oil residue patch in the coastline stretch between 94 and 95°W.

11.5 Summary

This chapter presented modeling case studies of deep-sea oil spills based on the *Deepwater Horizon* 2010 incident. The oil transport and fate application of the CMS trajectory model results were analyzed using well-founded post-processing algorithms and allowed for a quantitative estimate of oil concentrations and mass in both 3D space and time. CMS oil application model configuration parameters are backed by earlier laboratory and field studies. In addition to the deep-blowout control, *DB_control*, scenario, several additional case studies were conducted to investigate effects of modified initial conditions, different DSDs, and several other model parameters affecting oil weathering and transport.

One of the major findings of the modeling results was that the differences in interior GoM circulation, such as Loop Current development during different seasons, mesoscale ocean eddies, and Florida Straights position, appear to have a primary effect on the surface oil spread. Depending on circulation, an oil droplet could either travel considerable distances away from the oil spill site or be trapped within small-scale eddies. Variations in the DSD primarily yield first-order effects in surface of oil concentrations, but did not have qualitative effects on surface oil slick formation. Furthermore, the cumulative effect of the SSDI treatment thus appeared to slightly reduce oil content at and below the surface but contributed to shifting the deep oil plume to greater depths. This suggested that the use of dispersants is unlikely to help remove oil from the environment more rapidly. Furthermore, good agreement in spatial details of the modeled oil residue sediments in the control DWH case with the observational studies further confirms the robustness of the CMS oil application and the analysis methods to simulate deep-sea blowout scenario.

Acknowledgments This research was made possible by a grant from The Gulf of Mexico Research Initiative/C-IMAGE II to Steve Murawski. Data are publicly available through the Gulf of Mexico Research Initiative Information & Data Cooperative (GRIIDC) at https://data.gulfresearchinitiative.org (doi: https://doi.org/10.7266/N7KD1WDB).

The authors are thankful to Matthieu Le Hénaff for the discussion on the wind drift effects and to Paul Bubenheim and Juan Viamonte for their insight into biodegradation laboratory experiments and studies.

References

Adams EE, Socolofsky SA, Boufadel M (2013) Comment on "evolution of the Macondo well blowout: simulating the effects of the circulation and synthetic dispersants on the subsea oil transport". Environ Sci Technol 47:11905

Aman ZM, Paris CB (2013) Response to comment on "Evolution of the Macondo well blowout: simulating the effects of the circulation and synthetic dispersants on the subsea oil transport". Environ Sci Technol 47:11906–11907

Aman ZM, Brown EP, Sloan ED, Sum AK, Koh CA (2011) Interfacial mechanisms governing cyclopentane clathrate hydrate adhesion/cohesion. Phys Chem Chem Phys 13:19796–19806

Aman ZM, Paris CB, May EF, Johns ML, Lindo-Atichati D (2015) High-pressure visual experimental studies of oil-in-water dispersion droplet size. Chem Eng Sci 127:392–400. https://doi.org/10.1016/j.ces.2015.01.058

Bandara UC, Yapa PD (2011) Bubble sizes, breakup, and coalescence in Deepwater gas/oil plumes. J Hydraul Eng 137(7):729–738

Boxall JA, Koh CA, Sloan ED, Sum AK, Wu DT (2012) Droplet size scaling of water-in-oil emulsions under turbulent flow. Langmuir 28:104–110

Brooks GR, Larson RA, Schwing PT, Romero I, Moore C, Reichart GJ, Jilbert T, Chanton JP, Hastings DW, Overholt WA, Marks KP, Kostka JE, Holmes CW, Hollander D (2015) Sedimentation pulse in the NE Gulf of Mexico following the 2010 DWH blowout. PLoS One 10:1–24. https://doi.org/10.1371/journal.pone.0132341

Bubenheim P, Hackbusch S, Joye S, Kostka J, Larter SR, Liese A, Lincoln S, Marietou A, Müller R, Noirungsee N, Oldenburg TBP, Radović J, Viamonte J (2020) Biodegradation of hydrocarbons in deep water and sediments (Chap. 7). In: Murawski SA, Ainsworth C, Gilbert S, Hollander D, Paris CB, Schlüter M, Wetzel D (eds) Deep oil spills: facts, fate, effects. Springer, Cham

Chassignet EP, Smith LT, Halliwell GR, Bleck R (2003) North Atlantic simulations with the hybrid coordinate ocean model (HYCOM): impact of vertical coordinate choice, reference pressure and thermobaricity. J Phys Oceanogr 33:2504–2526

Cummings JA (2005) Operational multivariate ocean data assimilation. Q J R Meteorol Soc 131:3583–3604. https://doi.org/10.1256/qj.05.105

Daly KL, Passow U, Chanton J, Hollander D (2016) Assessing the impacts of oil associated marine snow formation and sedimentation during and after the Deepwater Horizon oil spill. Anthropocene 13:18–33

Davis CS, Loomis NC (2014) Deepwater Horizon Oil Spill (DWHOS) water column technical working group image data processing plan: Holocam, description of data processing methods used to determine oil droplet size distributions from in situ holographic imaging during June 2010 on cruise M/V Jack Fitz 3. Woods Hole Oceanographic Institution; MIT/WHOI Joint Program in Oceanography. 15 Pages + Appendices

De Gouw JA, Middlebrook AM, Warneke C, Ahmadov R, Atlas EL, Bahreini R, Blake DR, Brock CA, Brioude J, Fahey DW, Fehsenfeld FC, Holloway JS, Le Hénaff M, Lueb RA, McKeen SA, Meagher JF, Murphy DM, Paris C, Parrish DD, Perring AE, Pollack IB, Ravishankara AR, Robinson AL, Ryerson TB, Schwarz JP, Spackman JR, Srinivasan A, Watts LA (2011) Organic aerosol formation downwind from the Deepwater Horizon oil spill. Science 331(6022):1295–1299. https://doi.org/10.1126/science.1200320

Diercks AR, Highsmith RC, Asper VL, Joung DJ, Zhou Z, Guo L, Shiller AM, Joye SB, Teske AP, Guinasso N, Wade TL, Lohrenz SE (2010) Characterization of subsurface polycyclic aromatic hydrocarbons at the Deepwater Horizon site. Geophys Res Lett 37:L20602. https://doi.org/10.1029/2010GL045046

ERMA (Environmental Response Management Application) (2019) ERMA Deepwater Gulf Response. https://erma.noaa.gov/gulfofmexico/erma.html. Accessed 11 Mar 2019

Griffiths ST (2012) Oil release from Macondo well MC252 following the Deepwater Horizon oil accident. Environ Sci Technol 46:5616–5622

Gros J, Socolofsky SA, Dissanayake AL, Jun I, Zhao L, Boufadel MC, Reddy CM, Arey JS (2017) Petroleum dynamics during Deepwater Horizon. Proc Natl Acad Sci 114(38):10065–10070. https://doi.org/10.1073/pnas.1612518114

Halliwell GR (2004) Evaluation of vertical coordinate and vertical mixing algorithms in the HYbrid-Coordinate Ocean Model (HYCOM). Ocean Model 7:285–322

Hazen TC, Dubinsky EA, DeSantis TZ, Andersen GL, Piceno YM, Singh N, Jansson JK, Probst A, Borglin SE, Fortney JL, Stringfellow WT et al (2010) Deep-Sea oil plume enriches indigenous oil-degrading bacteria. Science 330:204–208

Hinze JO (1955) Fundamental of the hydrodynamic mechanism of splitting in dispersion processes. AICHE J 1(3):289–295

Hogan TF, Rosemond TE (1991) The description of the navy operational global atmospheric prediction system. Mon Weather Rev 119:1786–1815

Jaggi A, Snowdon RW, Stopford A, Radović JR, Oldenburg TBP, Larter R (2017) Experimental simulation of crude oil-water partitioning behavior of BTEX compounds during a deep submarine oil spill. Org Geochem 108. https://doi.org/10.1016/j.orggeochem.2017.03.006

Jaggi A, Snowdon RW, Radović J, Stopford A, Oldenburg TBP, Larter SR (2020) Partitioning of organics between oil and water phases with and without the application of dispersants (Chap. 8). In: Murawski SA, Ainsworth C, Gilbert S, Hollander D, Paris CB, Schlüter M, Wetzel D (eds) Deep oil spills: facts, fate, effects. Springer, Cham

Johansen Ø, Rye H, Cooper C (2003) DeepSpill – field study of a simulated oil and gas blowout in deep water. Spill Sci Technol Bull 8:433–443. https://doi.org/10.1016/s1353-2561(02)00123-8

Kolmogorov A (1949) On the disintegration of drops in turbulent flow. Dokl Akad Nauk SSSR 66:825–828

Kourafalou VH, Androulidakis YS (2013) Influence of Mississippi River induced circulation on the Deepwater Horizon oil spill transport. J Geophys Res Oceans 118:3823–3842. https://doi.org/10.1002/jgrc.20272

Larson RA, Brooks GR, Schwing PT, Diercks AR, Holmes CW, Chanton JP, Diaz-Asencio M, Hollander DJ (2020) Characterization of the sedimentation associated with the Deepwater Horizon blowout: depositional pulse, initial response, and stabilization (Chap. 14). In: Murawski SA, Ainsworth C, Gilbert S, Hollander D, Paris CB, Schlüter M, Wetzel D (eds) Deep oil spills: facts, fate, effects. Springer, Cham

Le Hénaff M, Kourafalou VH, Paris CB, Helgers J, Aman ZM, Hogan PJ, Srinivasan A (2012) Surface evolution of the Deepwater Horizon oil spill patch: combined effects of circulation and wind-induced drift. Environ Sci Technol 46:7267–7273. https://doi.org/10.1021/es301570w

Li Z, Lee T, King T, Boufadel MC, Venosa AD (2008) Oil droplet size distribution as a function of energy dissipation rate in an experimental wave tank. Int Oil Spill Conf Proc 2008(1):621–626

Li Z, Spaulding M, French McCay D, Crowley D, Payne JR (2017) Development of a unified oil droplet size distribution model with application to surface breaking waves and subsea blowout releases considering dispersant effects. Mar Pollut Bull 114:247–257

Lindo-Atichati D, Paris CB, Le Hénaff M, Schedler M, Valladares Juárez AG, Müller R (2014) Simulating the effects of droplet size, high-pressure biodegradation, and variable flow rate on the subsea evolution of deep plumes from the Macondo blowout. Deep-Sea Res II Top Stud Oceanogr. https://doi.org/10.1016/j.dsr2.2014.01.011

Malone K, Pesch S, Schlüter M, Krause D (2018) Oil droplet size distributions in deep-sea blowouts: influence of pressure and dissolved gases. Environ Sci Technol 52:6326–6333

Malone K, Aman Z, Pesch S, Schlüter M, Krause D (2020) Jet formation at the spill site and resulting droplet size distributions (Chap. 4). In: Murawski SA, Ainsworth C, Gilbert S, Hollander D, Paris CB, Schlüter M, Wetzel D (eds) Deep oil spills – facts, fate and effects. Springer, Cham

McNutt MK, Chu S, Lubchenco J, Hunter T, Dreyfus G, Murawski SA, Kennedy DM (2012a) Applications of science and engineering to quantify and control the Deepwater Horizon oil spill. Proc Natl Acad Sci U S A 109:20222e20228. https://doi.org/10.1073/pnas.1214389109

McNutt MK, Camilli R, Crone TJ, Guthrie GD, Hsieh PA, Ryerson TB (2012b) Review of flow rate estimates of the Deepwater Horizon oil spill. Proc Natl Acad Sci U S A 109:20260–20267

Nguyen UT, Lincoln SA, Valladares Juárez AG, Schedler M, Macalady JL, Müller R, Freeman KH (2018) The influence of pressure on crude oil biodegradation in shallow and deep Gulf of Mexico sediments. PLoS One 13(7):e0199784. https://doi.org/10.1371/journal.pone.0199784

Nixon Z, Zengel S, Baker M, Steinhoff M, Fricano G, Rouhani S, Michel J (2016) Shoreline oiling from the Deepwater Horizon oil spill. Mar Pollut Bull 107:170–178. https://doi.org/10.1016/j.marpolbul.2016.04.003

North EW, Adams EE, Thessen AE, Schlag Z, He R, Socolofsky SA, Masutani SM, Peckham SD (2015) The influence of droplet size and biodegradation on the transport of subsurface oil droplets during the Deepwater Horizon spill: a model sensitivity study. Environ Res Lett 10:024016. https://doi.org/10.1088/1748-9326/10/2/024016

Paris CB, Le Hénaff M, Aman ZM, Subramaniam A, Helgers J, Wang D-P, Kourafalou VH, Srinivasan A (2012) Evolution of the Macondo well blowout: simulating the effects of the circulation and synthetic dispersants on the subsea oil transport. Environ Sci Technol 46:13293–13302. https://doi.org/10.1021/es303197h

Paris CB, Helgers J, van Sebille E, Srinivasan A (2013) Connectivity Modeling System: a probabilistic modeling tool for the multi-scale tracking of biotic and abiotic variability in the ocean. Environ Model Softw 42:47–54

Paris CB, Berenshtein I, Trillo ML, Faillettaz R, Olascoaga MJ, Aman ZM, Schlueter M, Joye SB (2018) BP Gulf Science Data reveals ineffectual subsea dispersant injection for the Macondo blowout. Front Mar Sci. https://doi.org/10.3389/fmars.2018.00389

Pesch S, Schlüter M, Aman ZM, Malone KM, Krause D, Paris CBP (2020) Behavior of rising droplets and bubbles – impact on the physics of deep-sea blowouts and oil fate (Chap. 5). In: Murawski SA, Ainsworth C, Gilbert S, Hollander D, Paris CB, Schlüter M, Wetzel D (eds) Deep oil spills – facts, fate and effects. Springer, Cham

Pishro-Nik H (2014) Introduction to probability, statistics and random processes. Pubs. Kappa Research, LLC, Blue Bell. 744 pp.

Rewick RD, Sabo KA, Smith JH (1984) The drop-weight interfacial tension method for predicting dispersant performance. In: Allen TE (ed) Oil spill chemical dispersants: research, experience, and recommendations, STP 840. American Society for Testing and Materials, Philadelphia, pp 94–107

Romero IC, Schwing PT, Brooks GR, Larson RA, Hastings DW, Ellis G, Goddard EA, Hollander DJ (2015) Hydrocarbons in deep-sea sediments following the 2010 Deepwater Horizon blowout in the Northeast Gulf of Mexico. PLoS One 10(5):e0128371. https://doi.org/10.1371/journal.pone.0128371

Romero IC, Toro-Farmer G, Diercks A-R, Schwing P, Muller-Karger F, Murawski S, Hollander DJ (2017) Large-scale deposition of weathered oil in the Gulf of Mexico following a deep-water oil spill. Environ Pollut 228:179–189

Schedler M, Hiessl R, Valladares Juárez AG, Gust G, Müller R (2014) Effect of high pressure on hydrocarbon-degrading bacteria. AMB Express 4:77. https://doi.org/10.1186/s13568-014-0077-0

Schwing PT, Brooks GR, Larson RA Holmes CW, O'Malley BJ, Hollander DJ (2017) Constraining the spatial extent of the Marine Oil Snow Sedimentation and Accumulation (MOSSFA) following the DWH event using an excess ^{210}Pbxs flux approach. Environ Sci Technol 51:5962–5968. https://doi.org/10.1021/acs.est.7b00450

Socolofsky SA, Adams EE, Sherwood CR (2011) Formation dynamics of subsurface hydrocarbon intrusions following the Deepwater Horizon blowout. Geophys Res Lett 38:L09602. https://doi.org/10.1029/2011GL047174

Spaulding M, Li Z, Mendelsohn D, Crowley D, French-McCay D, Bird A (2017) Application of an integrated blowout model system, OILMAP DEEP, to the Deepwater Horizon (DWH) spill. Mar Pollut Bull 120:37–50. https://doi.org/10.1016/j.marpolbul.2017.04.043

Valladares Juárez AG, Kadimesetty HS, Achatz DE, Schedler M, Müller R (2015) Online monitoring of crude oil biodegradation at elevated pressures. IEEE J Sel Top Appl Earth Obs Remote Sens 8:872–878

Vaz AC, Paris CBP, Dissanayake AL, Socolofsky SA, Gros J, Boufadel MC (2020) Dynamic coupling of near-field and far-field models (Chap. 9). In: Murawski SA, Ainsworth C, Gilbert S, Hollander D, Paris CB, Schlüter M, Wetzel D (eds) Deep oil spills – facts, fate and effects. Springer, Cham

Wade T, Sericano JL, Sweet ST, Knap AH, Guinasso NL Jr (2016) Spatial and temporal distribution of water column total polycyclic aromatic hydrocarbons (PAH) and total petroleum hydrocarbons (TPH) from the Deepwater Horizon (Macondo) incident. Mar Pollut Bull 103:286–293

Zhao L, Boufadel MC, Socolofsky SA, Adams E, King T, Lee K (2014) Evolution of droplets in subsea oil and gas blowouts: development and validation of the numerical model VDROP-J. Mar Pollut Bull 83:58–69

Zhao L, Boufadel MC, King T, Robinson B, Gao F, Socolofsky SA, Lee K (2017) Droplet and bubble formation of combined oil and gas releases in subsea blowouts. Mar Pollut Bull 120:203–216

Zheng L, Yapa PD, Chen FH (2003) A model for simulating deepwater oil and gas blowouts – Part I: theory and model formulation. J Hydraul Res 41:339–351

Part IV

Oil Spill Records in Deep Sea Sediments

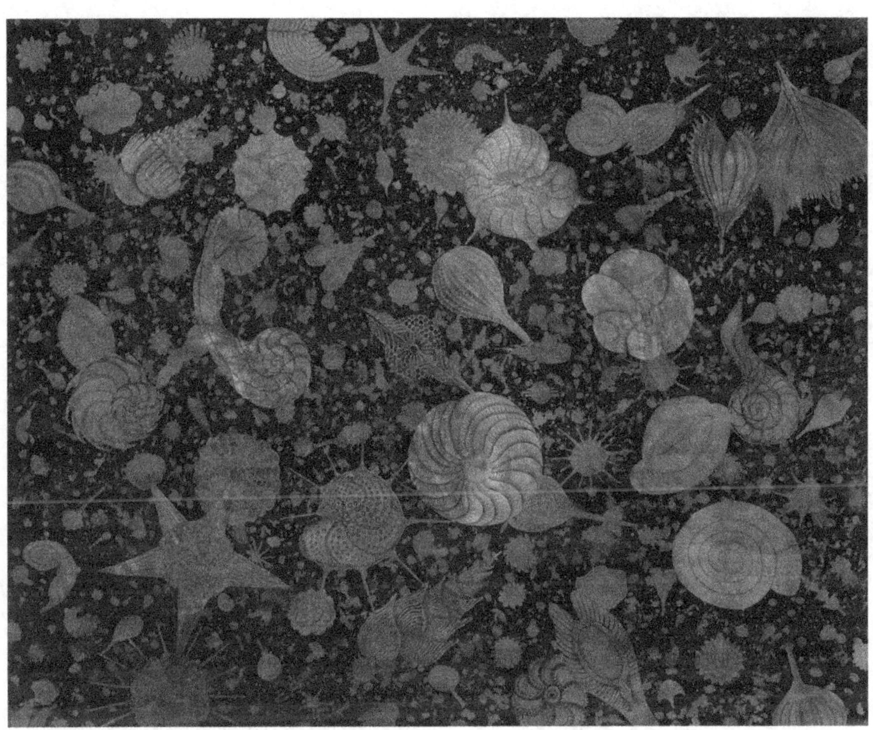

Curtis Whitwam
Foraminifera-Marine Snow
Watercolor on Aquaboard
20" × 16"

Chapter 12
Marine Oil Snow Sedimentation and Flocculent Accumulation (MOSSFA) Events: Learning from the Past to Predict the Future

Antonietta Quigg, Uta Passow, Kendra L. Daly, Adrian Burd, David J. Hollander, Patrick T. Schwing, and Kenneth Lee

"I see always the steady, unremitting, downward drift of materials from above, flake upon flake, layer upon layer — a drift that has continued for hundreds of millions of years, that will go on as long as there are seas and continents. For the sediments are the materials of the most stupendous snowfall the earth has ever seen…." (from Rachel Carson, The Sea Around Us, 1951)

"Marine snow is a nearly ubiquitous phenomenon in oceanic waters. The individual particles of marine snow are often fragile, difficult to sample intact, and provide distinctive microenvironments that support unique biological, chemical, and physical processes." (from Mary Silver 2015)

"Marine snow formation, incorporation of oil, and subsequent gravitational settling to the seafloor (i.e., MOSSFA: Marine Oil Snow Sedimentation and Flocculent Accumulation) was a significant pathway for the distribution and fate of oil, accounting for as much as 14% of the total oil released" as a result of the Deepwater Horizon oil spill in the Gulf of Mexico in 2010. (from Daly et al. 2016)

A. Quigg (✉)
Texas A&M University at Galveston, Department of Marine Biology, Galveston, TX, USA
e-mail: quigga@tamug.edu

U. Passow
Memorial University, Ocean Sciences Centre, Logy Bay, NL, Canada
e-mail: uta.passow@mun.ca

K. L. Daly · D. J. Hollander · P. T. Schwing
University of South Florida, College of Marine Science, St. Petersburg, FL, USA
e-mail: kdaly@mail.usf.edu; davidh@usf.edu; pschwing@mail.usf.edu

A. Burd
University of Georgia, Department of Marine Sciences, Athens, GA, USA
e-mail: adrianb@uga.edu

K. Lee
Fisheries and Oceans Canada, Ecosystem Science Directorate, Ottawa, ON, Canada
e-mail: Ken.Lee@dfo-mpo.gc.ca

© Springer Nature Switzerland AG 2020
S. A. Murawski et al. (eds.), *Deep Oil Spills*,
https://doi.org/10.1007/978-3-030-11605-7_12

196

Abstract Despite interest as early as in the 1880s, it was not until 1953 that Tokimi Tsujita (Seikai Fisheries Research Laboratory, Japan) was able to carefully collect and describe the matrix of microorganisms embedded in suspended organic matter (Tsujita, J Oceanogr Soc Jpn 8:1–14, 1953) that today we call marine snow. Subsequent studies reported that marine snow consisted of phytoplankton, small zooplankton, fecal material, and other particles (Nishizawa et al., Bull Fac Fish, Hokkaido Univ. 5:36–40, 1954). Across the ocean, Riley (Limnol Oceanogr 8:372–381, 1963) called this material "organic aggregates" which in addition to the organic material included nonliving material that was a "substrate for bacterial growth." More than a decade later, Silver et al. (Science 201:371–373, 1978) quantified the abundance of marine snow, and its contribution to the total community in situ, and showed that marine snow particles were "metabolic hotspots," with concentrations of microorganisms 3–4 orders of magnitude greater than those in the surrounding seawater. Alldredge and Cohen (Science 235:689–691, 1987) emphasized the importance of marine snow as unique chemical and physical microhabitats. The importance of transparent exopolymer particles (TEP), which form the matrix that embeds the individual component particles of marine snow, were described and quantified in the early 1990s (Alldredge et al., Deep-Sea Res I 40: 1131–1140, 1993; Passow and Alldredge, Mar Ecol Prog Ser 113:185–198, 1994; Passow et al., Deep-Sea Res Oceanogr Abstr 41:335–357, 1994).

The long-held belief that marine snow was both a specialized habitat and potential food source for those living in the deep ocean was also demonstrated at that time (Silver and Gowing, Prog Oceanogr 26:75–113, 1991). More recently it was confirmed that marine snow does indeed contribute significantly to the metabolism of the deep sea and provides hotspots of microbial diversity and activity at depth (e.g., Burd et al., Deep-Sea Res II 57:1557–1571, 2010; Bochdansky et al., Sci Rep 6:22633, 2016). Moreover, marine snow is now considered a transport vehicle for its biota and associated particulate matter (Volk and Hoffert, The carbon cycle and atmospheric CO: natural variations archean to present. American Geophysical Union, Washington, D.C., pp. 99–110, 1985; Alldredge and Gotschalk, Limnol Oceanogr 33:339–351, 1988). Rapidly sinking marine snow is important in the marine carbon cycle as it is responsible for vertical (re)distribution and remineralization of carbon. The transport of carbon from the surface to the deep sea is known as the "biological carbon pump" (De La Rocha and Passow, Deep Sea Res II 54:639–658, 2007; De La Rocha and Passow, Treatise on Geochemistry. Vol. 8, Elsevier, Oxford, 2014). This pump, which leads to the uptake and sequestration of atmospheric CO_2 (e.g., Volk and Hoffert, The carbon cycle and atmospheric CO: natural variations archean to present. American Geophysical Union, Washington, D.C., pp. 99–110, 1985; Finkel et al., J Plankton Res 32:119–137, 2010; Zetsche and Ploug, Mar Chem 175:1–4, 2015), also plays an important role in the biogeochemical cycling of elements (e.g., Quigg et al., Nature 425:291–294, 2003; Quigg et al., Proc R Soc: Biol Sci 278:526–534, 2011). How climate change will change these processes is the subject of intense interest but beyond the scope of this chapter.

Keywords Marine snow · Marine oil snow · MOSSFA · Deepwater Horizon · OMA · OSA · TEP · EPS

12.1   Defining of Marine Snow: An Operational Approach

Given the heterogeneous nature of marine snow, "operationally" definitions based upon their characteristics, size(s), and methods of quantification are used by the community (see Quigg et al. 2016). Particulate organic matter (POM) exists along a size continuum, from nano- and microgels (Chin et al. 1998; Verdugo et al. 2004), to TEP and other forms of extracellular polymeric substances (EPS), to large aggregates (>500 µm) called marine snow (Engel 2000; Passow 2002; Alldredge and Silver 1988; Grossart and Simon 1993). TEP are a class of free-floating extracellular polymeric particles ubiquitous in aquatic environments (e.g., Discart et al. 2015; Bar-Zeev et al. 2015). These acidic polysaccharide-rich particles are observable after staining with the dye Alcian Blue. TEP have gel-like properties and are known for their high stickiness (Engel 2000; Xavier et al. 2017), which make them a main driver for aggregation and sedimentation (Logan et al. 1995). Since their first description (Alldredge et al. 1993), their ubiquity and importance for carbon flux and the sedimentation of suspended mineral particles has been extensively documented (Kranck 1973; Bar-Zeev et al. 2015; Silver 2015). In particular, their role as the matrix within marine snow aggregates, such as those associated with declining diatom blooms, has been investigated in detail (e.g., Passow et al. 1994; Passow and Alldredge 1995; Passow 2002; De La Rocha and Passow 2007), but many questions remain regarding their composition, formation, and functional role. An extensive review of EPS was recently published by Quigg et al. (2016). Marine snow particles are typically >500 µm (Alldredge and Silver 1988) and either produced de novo by mucus-producing marine zooplankton, e.g., the feeding structures cast off by appendicularians (larvaceans) or pteropods (Alldredge 2005), or as the result of biologically enhanced physical aggregation of different components (e.g., phytoplankton, feces, detritus, mineral grains) glued together by EPS, a sticky form of POM produced by bacteria and phytoplankton (Alldredge and Silver 1988; Verdugo et al. 2004; Quigg et al. 2016).

12.2   Oil-Particle Interactions

Crude oil, as well as the oil residues remaining or forming due to biodegradation, evaporation, photooxidation, and other weathering processes, interacts with marine particles and dissolved substances forming different types of particles (Figs. 12.1 and 12.2). Interest on the ecological significance of oil-particle interactions following oil spills can be largely attributed to the studies following the Exxon Valdez oil spill which described "clay-oil flocculation" as a natural process responsible for the removal of oil stranded on low-energy shoreline environments (Bragg and Yang 1995; Bragg and Owens 1995). As this process was found to occur with a range of fine mineral particles besides those classified as clay, the term oil-mineral aggregate (OMA) was subsequently used to describe microaggregates formed between fine-grained sediment and oil (Lee et al. 1998; Stoffyn-Egli and Lee 2003). Since then

Oil-Particle Interactions

OMAs, OSAs, OPAs, oil-SPM:
- Small, e.g. 20 - 100 µm
- Neutrally buoyant or sinking, non-fractal
- 3 types: droplet, solid, flake
- Oil residue dominates (V/V), coat of inorganic particles, traces of fresh organic matter

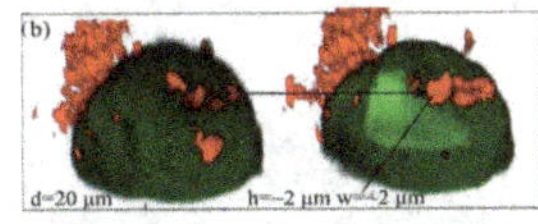

Micrograph: oil=green, mineral grains = red; Zhao et al. 2017

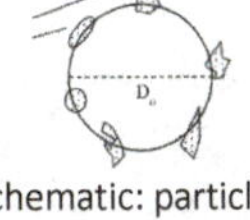

Schematic: particles coat oil droplet; Zhao et al. 2016

MOS, ORMS

- formed via aggregation:
- Oil droplets trapped in MOS matrix
- Larger 500 µm, TEP matrix
- Sinking at 100's m d^{-1}
- Porous & Fractal
- Oil < 50% (V/V) of organic and inorganic particles

- formed via biol. production:
- Larger 500 µm, EPS matrix
- Sinking at 100's m d^{-1}

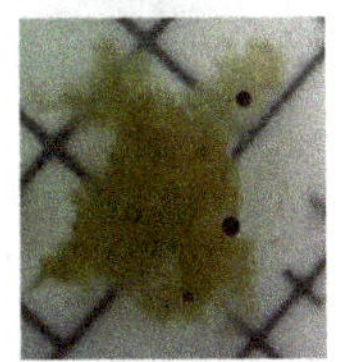

Cyanobacteria snow, (1.5 cm) oil=black, cells=green; Wirth et al. 2018

Small diatom snow (500 µm); Ziervogel, unpubl.

Microbial micro-aggregations/ mucus snow:
- Mucus production as response to oil exposure
- Formation of biofilm-like structures for efficient hydrocarbon degradation
- Microbial consortia – interspecies interactions
- Small (50 µm) or large (>500 µm)

Mucus snow (2 cm); Passow 2014

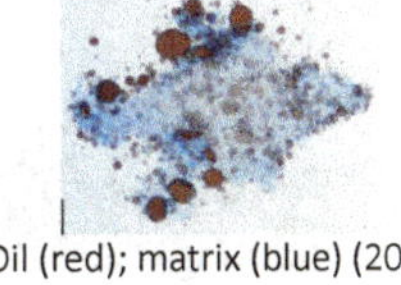

Oil (red); matrix (blue) (20 µm); Gutierrez et al. 2013b

Fig. 12.1 Summary of known oil-particle interactions

the terms oil-sediment aggregates (OSA) (Bandara et al. 2011; Cai et al. 2017) and oil-suspended particulate matter aggregates (OSMAs) (Khelifa et al. 2008a, b; Gong et al. 2014; Loh et al. 2014) have also been used to describe the association of oil droplets in or near the coastal zone, where inorganic mineral particles are predominant and ubiquitous. The term oil-particle aggregates (OPAs) is now commonly used to describe oil-particle interactions involving both inorganic and organic matter (Fitzpatrick et al. 2015; Zhao et al. 2016, 2017) and is defined by Gustitus and Clement (2017) to describe the aggregates that are on a scale of 1 mm or less. OMA/OPAs may have neutral buoyancy and float at or below the water surface or may sink and become entrained in sediments where they may also become buried and resuspended by turbulence (Lee et al. 2008; Gong et al. 2014; Fitzpatrick et al. 2015; Gustitus and Clement 2017). In contrast to these microscopic OPAs, the interaction of oil with coarse sediments along sandy marine shorelines may also result in the formation of macroscopic sediment-oil agglomerates (SOA, cm sized) or even massive sediment-oil mats (SOM, meter), which although similar in composition to their microscopic counterparts have different fates and effects on ecosystems (Gustitus and Clement 2017). A summary of known oil-particle interactions is provided in Fig. 12.1.

Studies on the mechanism of oil-particle aggregate formation have been largely focused on physicochemical forces at the surface of the oil droplets and mineral particles (Menon et al. 1987; Levine et al. 1989; Aveyard et al. 2003; Sterling et al.

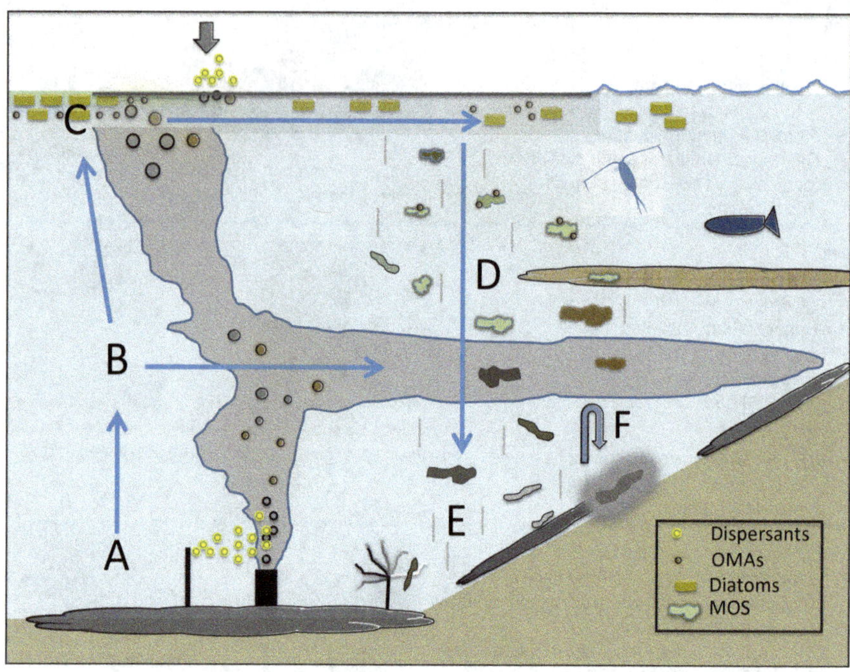

Fig. 12.2 Conceptual diagram of MOS-related processes from the source of oil discharge to the fate of hydrocarbons in sediments. The release of oil at the wellhead and application of dispersants (a) and rising oil droplets and gas bubbles and the formation of a deep oil plume (b). Surface processes influencing the formation of MOS include wind impacts, a diatom bloom, and application of surface dispersants, oil transformation due to UV light and evaporation, as well as the role of aerosols and oil burning in creating new material sources and processes impacting sinking MOS particles in surface waters (c) and particles sinking through a benthic nepheloid layer and deep oil plumes (d). Benthic sedimentation of MOS and flocculation onto corals (e) and resuspension of oiled sediments due to turbulence (f). See the text for a more detailed explanation (Daly et al. (2016)). (Reprinted from Daly et al. 2016 with permission from Elsevier)

2004). Stoffyn-Egli and Lee (2003) compared the performances of kaolinite, quartz, and montmorillonite of the same median particle size for the formation of OMAs. They found the higher cation exchange capacity of kaolinite leads to higher surface hydrophobicity and higher tendency of forming small OMAs. Wang et al. (2011) examined the interactions between oil and three types of solids, i.e., kaolin, modified kaolin, and diatomite. They found solids with higher hydrophobicity have more oil-mineral attraction. Stoffyn-Egli and Lee (2003) identified three distinct types of OMAs, namely, droplets, solid, and flake aggregates. Droplet OMAs are fine droplets surrounded by solid particles (scale of a few μm). Solid OMAs are a mixture of oil and solid bodies of irregular shapes (scale of tens of μm). Flake OMAs are the least common and have only been observed in laboratory conditions when montmorillonite and fumed silica were used as the mineral fines. Recently, Zhao et al. (2016) noted that in addition to their adsorption onto the surface of oil droplets, particles could actually penetrate into oil droplets, due to hydrodynamic forces. While small oil droplets coated by sediment or mineral particles become more stable, it was noted that penetration by mineral particles could result in the formation of larger clusters (Zhao et al. 2016).

The concentration and size of solids involved in the interaction is an important factor affecting the effectiveness of OMA formation. Payne et al. (2003) reported a positive correlation between OMA formation and the concentration of suspended sediments. Zhang et al. (2010) found smaller particle sizes of solids with larger surface to volume ratios favored the formation of OMAs. The formation of OMAs has been found to be more effective in marine environments than in freshwater environments (Fitzpatrick et al. 2015). While Gustitus and Clement (2017) found that formation of OMAs was depressed when oil was weathered, others reported that weathered oils tended to form OMAs more readily (Bragg and Yang 1995; Wood et al. 1998).

The effectiveness of oil dispersion is positively correlated to the formation of OMA/OPAs as smaller oil droplets require fewer suspended particles to form stable aggregates (Gong et al. 2014; Gustitus and Clement 2017). Oil with higher viscosity is more resistant to dispersing and form OMAs (Le Floch et al. 2002). Lee et al. (1998) found the viscosity of oil needed to be less than 9500 mPa-s to form OMAs. The addition of oil dispersants results in smaller oil droplets (Li et al. 2008; Zhao et al. 2014, 2017) that favor the formation of OMA/OPAs. Hydrodynamic conditions that influence energy dissipation influence the dispersion of oil slicks (Li et al. 2008) and thus OMA formation (Omotoso et al. 2002; Wincele et al. 2004; Ma et al. 2008; Sun et al. 2010; Sun et al. 2014). Ma et al. (2008) demonstrated that once OMAs are formed, they tend to be stable against continued turbulence. However, Zhao et al. (2017) found that under turbulence, the OMAs can be further broken down and recoalesce to form larger OMAs composed of multiple small droplets.

In terms of oil persistence, when suspended in the water column, the additional reactive surface area afforded by the minerals or organic matter favors enhanced rates of microbial oil biodegradation (Lee et al. 1997; Weise et al. 1999; Zhao et al. 2017). Thus, a number of experiments have been conducted to assess and demonstrate the feasibility of actively enhancing the formation of OMA as an alternative response measure. Sediment relocation (surf washing) involves the transport of oiled material from one section of a beach to another (i.e., the surf zone) where higher mixing energy and subsurface mineral fines associated with the excavated sediment accelerate the removal of oil from the sediments by the formation of OMA which would disperse and enhance the degradation of the residual oil. This in situ technique was demonstrated within the Arctic (Owens and Lee 2003; Lee et al. 2003a) and validated during actual oil spill response operations following the Sea Empress oil spill in the UK (Lunel et al. 1996). The scope of this oil translocation concept was further developed and tested as a potential strategy to remediate oil spills in ice-infested waters (Lee et al. 2011). While temperature affects the formation of OMAs mostly by influencing the viscosity and adhesion properties of oil, the formation of OMAs has been considered and demonstrated as a response to oil spills in low-temperature conditions (-1 to -4 °C) in laboratorial batch scale tests (Lee et al. 2012a; Wang et al. 2013), pilot-scale flume tank studies (Jézéquel et al. 2018), and field trials in ice-infested waters under Arctic conditions (Lee et al. 2011).

Toxicity studies and numerical models were used to demonstrate habitat recovery due to dispersion and biodegradation of the residual oil in these oil translocation-based studies (Lee et al. 2003b; Niu and Lee 2013).

The *Deepwater Horizon* (DwH) accident was the first oil spill in deep waters where natural sediment and/or mineral particle concentrations are too low to cause the formation of ubiquitous amounts of OPAs. Thus sedimentation of oil to depth was not expected. However, significant amounts of oil were observed to accumulate at the seafloor (e.g., Valentine et al. 2014; Chanton et al. 2015; Schwing et al. 2015; Passow and Hetland 2016), resulting in the realization that marine snow can lead to the vertical transport of large amounts of oil.

Significant sedimentation of hydrocarbons in association with phytoplankton had already been observed about 30 years earlier in mesocosm studies (Lee and Anderson 1977; Lee et al. 1978, 1985), but sorption and active uptake of oil compounds were considered the main causes (Lee et al. 1978), although whole oil droplets trapped in phytoplankton aggregates were also described (Lee et al. 1985). These earlier observations were not pursued, likely because such phytoplankton-oil associations were thought to be relatively rare and quantitatively unimportant, in the nearshore environments where oil exploration and production typically took place. The potential of oil incorporation into marine snow became important, once drilling in deep waters started, as sediment particles are too rare in these environments to form OPAs. However, natural marine snow formation may be significant in such environments and as was observed during the DwH oil spill. Marine oil snow (MOS) differs from OMAs, OSAs, or oil-SPM, in that marine snow naturally forms in the absence of oil, is per definition large (>0.5 mm), contains high amounts of organic particles, and is fractal. The inclusion of oil compounds in marine snow results in the formation of MOS.

12.3 Marine "Oil" Snow

Arguably, one of the most important discoveries about the fate of the DwH oil spilled in the Gulf of Mexico in 2010 is the role of the interactions of marine snow and oil and the consequences for the short- and long-term fate of oil chemicals and dispersants and their deposition. MOS associated with the DwH spill consisted largely of organic particles, including bacteria, phytoplankton, fecal material, feeding structures and detritus (Passow and Ziervogel 2016; Daly et al. 2016), and oil-related compounds from the Macondo crude oil (OSAT 2010), all embedded in a mucus-rich matrix formed from EPS/TEP (Fig. 12.2). Depending on the specific oil compound, hydrocarbons may be incorporated into marine snow via sorption to particles and as droplets caught in the marine snow matrix (Wirth et al. 2018). However, incorporation of oil compounds into marine snow also depends on weathering processes – including evaporation, photooxidation, and biodegradation (Brakstad and Faksness 2000) – as well as on dispersion and dissolution into the water. Incorporation of oil into marine snow is a function of both available oil compounds and marine snow, as predicted from coagulation theory. Overall, the fate of the oil, including its incorporation into marine snow, then depends on these factors, plus the physical and chemical properties of the oil, the oil release conditions, and the environment (Daling et al. 2014).

Application of dispersants on oil spills, extensively applied after the DwH, results in tiny oil droplets being generated within the water column as they break away from the surface slick following application from aircraft or vessels at sea or from the riser following subsurface injection (Fig. 12.2). Thus more oil may be trapped by marine snow, when dispersants are applied (see, e.g., Passow et al. 2017). However, the dispersant Corexit can also disperse organic matter, especially TEP, with the consequence that marine snow concentrations are reduced in the presence of Corexit. The net effect of Corexit on the oil transport via marine snow depends on the relative strength of these opposing processes (Passow et al. 2017). Additionally, dispersants frequently change the dissolution of specific oil compounds, thus impacting the chemical composition of oil residues. Toxicity with dispersant increases due to the enhanced bioavailability of the oil itself (Wirth et al. 2018). The formation of marine oil snow is of central importance for the fate of oil, because as marine oil snow sinks, it transports oil to depth and the seafloor.

12.4 MOS: Microhabitat and Entry Point to the Food Web

Marine snow and MOS represent nutrient-rich substrates: the microbial community composition on marine snow found in surface waters has been shown to differ greatly from that in the bathypelagic and further, as there are distinct differences between particle-associated and freely suspended microbes in both realms (e.g., Salazar et al. 2015). Differences in polysaccharide-hydrolyzing enzyme activities between MOS and the surrounding seawater suggest different bacterial communities between both in the presence of oil as well (Arnosti et al. 2016).

Marine snow harbors a highly variable eukaryotic microbial community, which is affected by depth, water mass, as well as the number of prokaryotes (Pernice et al. 2015). In general, prokaryotes numerically exceeded eukaryotes on marine snow (Bochdansky et al. 2017), and picoplankton, bacteria, and viruses are ubiquitous inhabitants of marine snow (Volkman and Tanoue 2002) and likely MOS. Eukaryotic microbes belonging to the fungi and the labyrinthulomycetes were found to dominate overall biomass of marine snow collected from ~ 1000 to 3900 m (Bochdansky et al. 2017). Of the labyrinthulomycetes, thraustochytrids are often associated with dead phytoplankton debris at the end of a bloom. Being tolerant to cold temperature and high hydrostatic pressure, these saprotrophic organisms have the potential to significantly contribute to the degradation of organic matter in the deep sea. Even less is known about the role of fungi on MOS, but preliminary evidence indicates they are more common in the presence of oil alone than oil plus Corexit (pers. obs).

Responses of heterotrophic prokaryotes to oil are complex and involve a multitude of interspecies interactions and successional stages as different oil components and degradation products are being metabolized (McGenity et al. 2012). Studies on the microbial response to the DwH spill elucidated a dramatic shift in microbial community composition (Gutierrez et al. 2013a; Joye et al. 2014) and the development of a multifunctional microbial assemblage containing primary oil-degrading and exopolysaccharide-producing prokaryotes (Arnosti et al. 2016). Such exopolysaccharides released in the presence of oil, e.g., by *Halomonas* spp., exhibit amphiphilic properties, which allows these macromolecules to interface with hydrophobic substrates, such as hydrocarbons (Gutierrez et al. 2013b).

The presence of oil droplets leads to the formation of micron-scale aggregates (20–100 μm), where microbial cells surround droplets of oil (Doyle et al. 2018). Such microaggregates promote the development of diverse, interacting bacterial communities (Doyle et al. 2018). The proximity of the cells to each other and reduced diffusion within the aggregate would allow direct exchange of substrates, efficient use of exoenzymes, and quorum sensing. The formation of such structures, which function as floating biofilms, allows the degradation of complex hydrocarbon mixtures, which requires the nonredundant capabilities of a diverse oil-degrading community (Dombrowski et al. 2016). The formation of such microbial microagglomerations is a function of oil concentration (Doyle et al. 2018). As degradation of oil products proceeds, the substrates change leading to a succession of bacterial communities. The microbial community of the deep plume (1000 m) that formed during DwH responded to the addition of specific substrates resembling bacterial EPS and oil degradation products, suggesting that similar bacterial transformations of oil degradation by-products also contributed to microbial activity inside the deepwater plume (Ziervogel et al. 2014). Experiments show the formation of microscopic bacterial oil aggregates at depth, similar to those observed in surface waters (Baelum et al. 2012).

The relationship between such microscopic bacteria, oil droplet agglomerations, and large (cm-sized) mucus-rich MOS that formed at the surface during the DwH and also consisted of exudates produced from oil carbon and bacteria (Passow et al. 2012; Passow and Ziervogel 2016) is currently unresolved. Two pathways from small bacteria-oil aggregates to large bacterial marine snow may be envisioned. First, possibly the specific conditions during DwH resulted directly in the formation of such exceptionally large bacterial MOS. Second, the bacterial microagglomerations are incorporated into marine snow-sized aggregates that form from detritus, feces, or algae. The importance of the synthesis of bacterial EPS and a diverse, shifting bacterial community, for the formation of MOS, was also confirmed by experiments conducted with water from the Faroe-Shetland Channel (Suja et al. 2017), emphasizing that MOS formation is a more general phenomenon.

Likely different factors drive the formation of the different types of MOS. Thus when predicting the likelihood of the formation of MOS, the type of MOS needs to be considered. But all MOS appears to sink eventually (after aging in some cases), and once MOS reaches the seafloor, it severely impacts the benthic microbial community (Yang et al. 2016a, b), as well as larger organisms (Baguley et al. 2015, Reuscher et al. 2017, Washburn et al. 2017, Schwing et al. 2018). Resuspension of MOS from the seafloor reactivates microbial degradation and activities (Ziervogel et al. 2016) and redistributes MOS horizontally (Diercks et al. 2018).

It is uncertain how the presence of oil impacted zooplankton's role in the formation and alteration of marine snow. Zooplankton contribute to the formation of marine snow through the release of fecal pellets, crustacean molts, dead bodies, and feeding structures, such as larvaceans' houses, and may break up or feed on marine snow as it sinks through the water column (Alldredge and Silver 1988). Zooplankton grazing may fragment marine snow, breaking it into smaller particles with lower sinking velocities (Dilling and Alldredge 2000). Some zooplankton taxa (e.g., dinoflagellates, gelatinous doliolids, copepods) ingest oil and egest oil compounds within fecal pellets, which sink rapidly to the seafloor (Lee et al. 2012b; Almeda et al. 2014a, c). In addition, oil may adhere to zooplankton and be passively absorbed or ingested, thus contributing to bioaccumulation of PAHs to upper trophic-level animals (Mitra et al. 2012). Carbon isotopic depletion in suspended particulate matter and zooplankton indicated that oil carbon was also incorporated into the lower trophic food web through biodegradation by bacteria (Graham et al. 2010; Chanton et al. 2012). Oil and dispersants may have lethal and sublethal effects on some zooplankton (Lee et al. 1985; Almeda et al. 2014b; Buskey et al. 2016), while other zooplankton, such as *Noctiluca* spp. (heterotrophic dinoflagellates), may increase in abundance after oil spills (Févre 1979) and ingest oil, depending on the oil concentration (Almeda et al. 2014a). Indeed, relatively high abundances of *Noctiluca* spp. were observed during the DwH spill in the water column and on MOS particles (Remsen et al. 2015).

12.5 MOS: Sedimentation and Flocculent Accumulation

Gravitational settling removes marine snow and MOS from the upper ocean and transports them to depth. Sinking velocities are generally described as a function of aggregate size, excess density of component particles, and aggregate porosity. The relatively large size of marine snow (>500 µm) thus makes marine snow and MOS important vehicles for vertical flux (De La Rocha and Passow 2007; Ploug et al. 2008). Measurements of laboratory-made MOS and marine snow indicate that sinking velocities of MOS do not appear lower than of similarly sized marine snow. Possibly the oil droplets allow for tighter packaging, so that an increased number of dense particles compensate for the low density of the oil in MOS (Passow et al. 2019).

During and for months after the DwH spill, sedimentation of MOS was likely spatially and temporally very variable, as different types of marine snow formed and incorporated oil compounds during their formation or sinking (Passow and Hetland 2016). Flux attenuation due to grazing, bacterial degradation, and physical fragmentation was likely high, and only a small fraction of the MOS (~10%) sinking out of the euphotic zone would have reached the seafloor at >1000 m. Once at the seafloor, horizontal redistribution, especially downhill, resuspension, and the associated enhanced degradation of oil compounds and snow would have changed the composition and drastically reduced the amount of oil compounds found weeks to months after the spill. Estimates of sedimented oil based on sediment cores are thus underestimates of the oil that sank, except if the core was collected in a low point on the seafloor (more below). The delivery of oil chemicals to the benthos and surface sediments is important to the fate of oil in marine ecosystems and its effects on benthic organisms or those that feed on benthic organisms.

MOSSFA (Marine Oil Snow Sedimentation and Flocculent Accumulation) is a term coined in October 2013 (see Daly et al. 2016) to evaluate the processes influencing the formation and fate of oil-associated marine snow as summarized in Fig. 12.2. Participants on oil spill response cruises in May and June of 2010 observed elevated marine snow particles both at the sea surface and throughout the water column (Passow et al. 2012; Passow 2016; Daly et al. 2016). Sedimentation of MOS to the seafloor was later documented by sediment traps (Yan et al. 2016) and sediment cores (Montagna et al. 2013; Valentine et al. 2014; Brooks et al. 2015; Romero et al. 2015). The mass deposition of MOS occurred over a 4–5-month period during and after the oil spill and far exceeded pre-spill sediment accumulation rates (Brooks et al. 2015). Deep-sea sediment traps revealed MOS "flocs" (Stout and German, 2015; Yan et al. 2016) up to 8 km from the well (Stout and Payne 2016a, b). Valentine et al. (2014) used hopanes as a biomarker tracer to estimate that 1.8–14% of the oil was transported to the seafloor, while Chanton et al. (2015) estimated the amount to be between 0.5% and 9% using radiocarbon distributions. A reevaluation of sedimentary geochemical data revealed that $21 \pm 10\%$ of the total amount of oil discharged, and not recovered from the DwH spill, sedimented to the seafloor (Romero et al. 2017). Thus, the DwH MOSSFA event was a significant pathway for the distribution and fate of spilled oil. Information on the processes impacting MOSSFA are needed.

Owing to intense scientific interest in the impacts of MOS, the Gulf of Mexico Research Initiative funded the first MOSSFA Workshop in October 2013, with the goal to evaluate what was known on three topics: (1) factors affecting the formation and sinking of MOS in the water column; (2) the deposition, accumulation, and biogeochemical fate of MOS on the seafloor; and (3) the ecologic impacts of MOS on pelagic and benthic species and communities. The results of the MOSSFA Workshop are reported in Daly et al. (2016), as well as a summary of published results at that time. There are many factors that are thought to affect the formation and alteration of MOS aggregates in the water column (Fig. 12.2). One major surface process relevant during the DwH was the elevated and extended Mississippi River discharge, which enhanced phytoplankton production and suspended particle concentrations, zooplankton grazing, and enhanced microbial mucus formation. Freshwater diversions along the lower Mississippi River were opened for several months in order to reduce the impact of the oil spill on estuaries and wetlands (Bianchi et al. 2011). As a result, a low salinity lens of river water created a shallow mixed layer, which extended over a large area to the east of the DwH wellhead (O'Connor et al. 2016). Due to the river water and possibly the presence of oil, unusually large phytoplankton blooms occurred in the region. Specifically, the elevated phytoplankton concentrations were detected by satellite as a >1 mg m^{-3} chlorophyll a anomaly to the east of the wellhead, covering more than 11,000 km^2

during August 2010 (Hu et al. 2011). A sediment trap revealed that high particle flux rates relative to other years were due to the sinking of a large *Skeletonema* sp. bloom (Yan et al. 2016), a cosmopolitan diatom that thrives under brackish conditions and is tolerant of the presence of oil. In addition to the MOSSFA data, consideration should also be given to the transport of OMA given the high concentration of suspended sediments in the coastal waters of the Gulf of Mexico due to sediment transport from the Mississippi River and lower than expected levels of oil reaching shore relative to the volume spilled. In fact, after the DwH spill, sinking MOS was rich in lithogenic material (Yan et al. 2016), which likely played a significant role in the formation and sinking (ballasting) of MOS (Brooks et al. 2015).

Another unique feature of the DwH oil spill was the presence of persistent subsurface oil plumes that occurred primarily between 1000 and 1400 m depth (Camilli et al. 2010; Diercks et al. 2010) (see Fig. 12.2). Marine snow, which formed at the surface and sedimented to the seafloor, likely interacted with rising oil droplets and/or the oil plumes at depth (Valentine et al. 2014). Surface bacteria and phytoplankton-affiliated gene sequences were found in sediments at and below the subsurface plume (Mason et al. 2014), validating the notion that surface-formed MOS sank to the seafloor. MOS also may have formed in the deepwater plumes, as many bacteria were active in those regions (Hazen et al. 2010); experiments using bacteria from the deep plume support this finding (Baelum et al. 2012; Kleindienst et al. 2015).

Cumulative evidence suggests that the large DwH MOSSFA episode occurred due to a nexus of events: (1) the DwH site was located in one of the most productive regions of the Gulf of Mexico governed by Mississippi River outflow, (2) the spill occurred during spring and summer when bacteria and phytoplankton densities were at a maximum and surface productivity rates were relatively high, and (3) there was greater microbial mucus formation, especially in the presence of weathered oil.

12.6   MOSSFA: Unique to the Deepwater Horizon Oil Spill?

Preservation of the DwH MOSSFA event in sediments and its long-term impact on the postdepositional chemical environment and the benthic ecosystem have been well documented (Brooks et al. 2015; Chanton et al. 2015; Romero et al. 2015, 2017; Schwing et al. 2015, 2017, 2018; Hastings et al. 2015). Vonk et al. (2015) performed a meta-analysis of 52 large, historical oil spills in an effort to investigate whether MOSSFA occurred and to identify the main drivers of oil sedimentation. Similar MOSSFA events were reported for the Tsesis, Ixtoc 1, and other oil spills (Johansson et al. 1980; Boehm and Fiest 1980; Jernelöv and Lindén 1981; Patton et al. 1981; Teal and Howarth 1984), and possibly even in the Santa Barbara spill in 1969 (Doyle et al. 2018), but these were not reported because sedimentation of oil was not expected and the seafloor was not monitored. The formation of MOSSFA in these cases is expected due to the presence of particulate matter, the proximity to river outflow (i.e., clay minerals and phytoplankton biomass), and the presence of EPS-producing biota (phytoplankton and oil-degrading bacteria; Vonk et al. 2015).

In a mesocosm study involving the subsurface injection of Corexit 9527 dispersed Prudhoe Bay crude oil, Lee et al. (1985) noted the association of oil droplets with biogenic material (phytoplankton and detritus) associated with a phytoplankton bloom, stimulatory effects on small zooflagellates, and increased mortality on ciliates and appendicularians. Results further showed that Corexit and dispersed Corexit stimulated bacterial productivity rates by serving as substrates and/or by inducing the release of organic compounds from the indigenous microbial population. A mass balance for ^{14}C-hexadecane added to the test oil revealed that within 22 days, 3% was recovered in the suspended particulate fraction and 36% was respired as CO_2, 1% in the dissolved organic pool, and 10% as sedimentary material.

Spatial gradients of MOSSFA are expected with distance from the predominant river systems (nutrients, clay minerals, salinity) and the center of oil release (Daly et al. 2016). Gradients in input parameters including flocculent thickness (Fig. 12.3a), clay particles/minerals, algal and photoautotrophic bacterial communities and associated EPS/TEP, and oiled sediments (e.g., PAHs) control postdepositional response. Natural spatial gradients exist in deltaic systems; however, MOSSFA inputs lead to intensified postdepositional response in redoxcline, particulate organic carbon flux, lithogenic flux, and benthic community impact and response. The formation of MOSSFA is therefore not unique to the Deepwater Horizon oil spill and is likely to occur in the event of any oil spill that occurs in an area that satisfies the aforementioned criteria.

12.7 MOS/MOSSFA: Modeling

Modeling of MOS formation during MOSSFA events and the processes affecting these pathways can be used to predict the magnitude and duration of the downward flux of oil. The DWH experience strongly indicates that MOSSFA events need to be factored into the fate-and-effect models for future oil spills and targeted for field measurements during response assessments and research. This is particularly important because surface oil budgets cannot be properly closed without an accurate, quantitative estimate of the amount of oil leaving the surface waters as MOS or OPAs.

Whereas MOS formed via bacterial exudate production or zooplankton activity is driven by these biological processes, a large portion of MOS forms via the biologically mediated coagulation of smaller marine particles, like diatoms. Such aggregates form through collision and adhesion of the component particles. Early models concentrated on the important process of aggregation of oil droplets and mineral or sediment particles, e.g., the formation of OMAs (Sterling et al. 2005; Sun and Zheng 2009; Sun et al. 2013). Some of these models used conventional aggregation theory (Burd and Jackson 2009) to predict the formation of oil-mineral aggregates (OMAs). The numerical model A-DROP (Zhao et al. 2016, 2017) models the aggregation of oil droplets onto the surface of mineral particles and predicts

A. Core Photographs

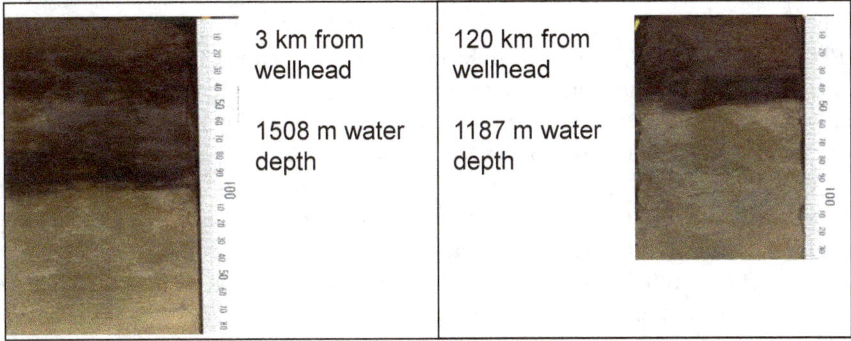

B. Gradients

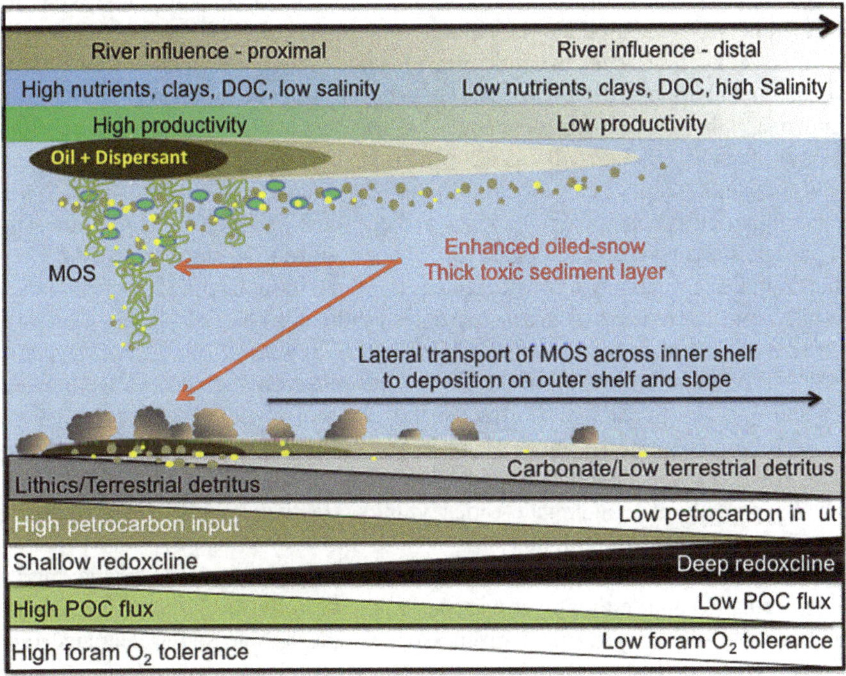

Fig. 12.3 Linking MOSSFA events from the water column to the sediments: assessing spatial and temporal gradients and patterns. Diagram illustrates (**a**) thickness of MOSSFA flocculent material with distance from the wellhead and (**b**) the environmental gradients of material properties and fluxes associated with a point source of oil released in regions influenced by river outflow compared to offshore regions not influenced by riverine processes. Gradient shifts include the concentration and composition of suspended particles (clays to carbonate), the magnitude of particulate organic carbon (POC) and petrochemical fluxes to the seafloor, the depth of the sediment redoxcline, and the tolerance of benthic organisms, such as foraminifera, to different oxygen levels in sediments. Oil-mineral aggregations (OMAs) may sediment separately or in association with marine oil snow (MOS). These environmental gradients overlap and interact with gradients generated by oil spills, e.g., oil and dispersant distributions, causing a complex temporal-spatial distribution of interactive effects. (Reprinted from Daly et al. 2016 with permission from Elsevier)

the amount of oil trapped in OMAs, allowing estimates of the consequences of the presence of mineral fines (Lee et al. 2008). However, during the DwH event, MOS was observed that consisted of oil incorporated into phytoplankton aggregates (Passow et al. 2012; Passow 2016). Modeling the formation of such large and heterogeneous MOS, which incorporates oil, but does not require oil to form, calls for information on all particle concentrations and sizes (mineral particles and biological particles such as phytoplankton and fecal pellets).

Models of marine snow formation and sinking using aggregation theory have been previously developed to study the export of carbon from the surface to the deep ocean (Jackson 1990; Burd and Jackson 2009; Jokulsdottir and Archer 2016). Typically, three collisional processes are considered: Brownian motion, which is the dominant collision mechanism for particles less than 1 µm in size; fluid shear; and differential sedimentation where a faster settling particle collides with a slower settling one. For particles of size between approximately 1 and 20 µm, fluid shear and differential sedimentation can be equally important, but for large aggregates, differential sedimentation is the dominant collisional process (Burd and Jackson 2009; Lambert and Variano 2016). The theoretical descriptions of these processes are well known and have been experimentally verified in a wide range of disciplines (Elimelech et al. 1995; Friedlander 2000; Pruppacher and Klett 2010).

Aggregation theory has been used to model the aggregation of oil droplets with mineral particles such as clay and silica and derive collision efficiencies (a measure of particle stickiness) and aggregate fractal dimensions (Sterling et al. 2004). These studies indicate relatively high efficiencies of ~0.4–0.6, indicating a high stickiness of oil. Typical stickiness of marine particles (without oil or TEP) is less than 0.01, whereas stickiness in the presence of TEP varies between 0.1 and 0.8 (Mari et al. 2017). The fractal dimension of the OMAs was relatively high and varied between 2.6 and 3.0. In comparison, the fractal dimension of phytoplankton aggregates is lower, usually between 1.3 and 1.9 (Logan and Alldredge 1989; Laurenceau-Cornec et al. 2015). The fractal dimension is an important parameter because it determines how the porosity of an aggregate varies with its size, thereby affecting particle collision rates and sinking velocities.

A first attempt at modeling MOSSFA used aggregation theory to predict marine snow formation and settling rates and then calculated oil-scavenging rates using the abundance of oil droplets of a given size and the marine snow clearance rates (Francis et al. 2017). This model was used to estimate oil removal rates from surface waters under different turbulence conditions and phytoplankton concentrations and showed that a strong phytoplankton bloom was able to remove all the oil from the surface waters in 0.5–4 days, depending on the turbulence strength. The model predicted the flux of oil and marine carbon in August/September 2010 as measured via a sediment trap, reasonably well, and indicated that the trap caught only the tail end of the bloom, missing the first part of the MOSSFA event.

A significant problem with modeling MOS and MOSSFA events is that MOS are formed from dynamic populations of a variety of different particle types, and it is unclear how the properties of an aggregate depend on those of the particles that comprise it. In particular, it is unclear how to estimate the fractal dimension and stickiness of the resultant MOS aggregate. Traditional marine coagulation models overcome this problem by effectively ignoring it and considering all particles to have the same stickiness and fractal dimension (Burd and Jackson 2009). Jackson (1998) used a scaling argument to calculate the fractal dimension of a composite particle, and a similar technique was used by Sterling et al. (2005). However, this scaling technique might not be applicable when considering aggregates formed from oil droplets and solid organic particles.

A different approach to modeling aggregation of heterogeneous particles uses a stochastic Lagrangian approach (Jokulsdottir and Archer 2016; Dissanayake et al. 2018). This employs computational and stochastic techniques to represent individual particles and the collisions between them and is able to calculate the time evolution of the aggregate size distribution throughout the water column. The properties of each individual aggregate can then be calculated from the properties of the constituent particles, all except the fractal dimension which is held constant. For example, Dissanayake et al. (2018) used different formulations for calculating aggregate stickiness based on the volume fractions of oil, TEP, sediment particles, and organic matter within each aggregate. The model was initialized with observed particle size distributions in the surface waters and reproduced observed size distributions to approximately 150 m at five locations near the DwH site reasonably well. The model made predictions for the amount of oil settling to the sea floor that were within the bounds of estimates made using observed tracer distributions. The model results were sensitive to the value chosen for the fractal dimension and to the method used to calculate the aggregate stickiness.

The theory of particle aggregation is well established and can provide valuable insights into MOS formation and MOSSFA events. However, for it to become a predictive tool will require a better understanding of key parameters such as aggregate fractal dimension and stickiness in the presence of oil and dispersants. If this can be done, there is considerable potential for aggregation theory to become an invaluable tool for first responders and oil spill research.

Acknowledgments This research was made possible by a grant from the Gulf of Mexico Research Initiative to Quigg (ADDOMEx), Passow (ADDOMEx, ECOGIG, FOMOSA), Daly (C-IMAGE, FOMOSA), Burd (FOMOSA), and Schwing/Hollander (C-IMAGE). Research support was also provided by the University of South Florida Division of Sponsored Research and Florida Institute of Oceanography to Daly and by the Multi-Partner Research Initiative, via the Department of Fisheries and Oceans, Canada, to Passow.

References

Alldredge AL (2005) The contribution of discarded appendicularian houses to the flux of particulate organic carbon from oceanic surface waters. In: Gorsky G, Youngbluth MJ, Deibel D (eds) Response of marine ecosystems to global change: ecological impact of appendicularians. Éditions Scientifiques, Paris, 435 pp. ISBN:2-8470-302-9-8

Alldredge AL, Passow U, Logan BE (1993) The abundance and significance of a class of large, transparent organic particles in the ocean. Deep-Sea Res I 40:1131–1140

Alldredge AL, Silver MW (1988) Characteristics, dynamics and significance of marine snow. Prog Oceanogr 20:41–82

Alldredge AL, Gotschalk C (1988) In situ settling behavior of marine snow. Limnol Oceanogr 33:339–351

Alldredge AL, Cohen Y (1987) Can microscale chemical patches persist in the sea?: microelectrode study of marine snow, fecal pellets. Science 235:689–691

Almeda R, Connelly T, Buskey EJ (2014a) Novel insight into the role of heterotrophic dinoflagellates in the fate of crude oil in the sea. Nat Sci Rep 4:7560. https://doi.org/10.1038/srep07560

Almeda R, Hyatt C, Buskey EJ (2014b) Toxicity of dispersant Corexit 9500A and crude oil to marine microzooplankton. Ecotoxicol Environ Saf 106:76–85

Almeda R, Baca S, Hyatt C, Buskey EJ (2014c) Ingestion and sublethal effects of physically and chemically dispersed crude oil on marine planktonic copepods. Ecotoxicology 23:988–1003

Arnosti C, Ziervogel K, Yang T, Teske A (2016) Oil-derived marine aggregates – hot spots of polysaccharide degradation by specialized bacterial communities. Deep-Sea Res II Top Stud Oceanogr 129:179–186

Aveyard R, Binks BP, Clint JH (2003) Emulsions stabilized solely by colloidal particles. Adv Colloid Interf Sci 100–102:503–546

Baelum J, Borglin S, Chakraborty R, Fortney J, Lamendella R, Mason O, Auer M, Zemla BM, Conrad M, Malfatti S, Tringe S, Holman H, Hazen T, Jansson J (2012) Deep-sea bacteria enriched by oil and dispersant from the Deepwater Horizon spill. Environ Microbiol 14:2405–2416

Bandara UC, Yapa PD, Xie H (2011) Fate and transport of oil in sediment laden marine waters. J Hydro Environ Res 5:145–156

Baguley J, Montagna P, Cooksey C, Hyland JL, Bang HW, Morrison C, Kamikawa A, Bennetts P, Saiyo G, Parsons E, Herdener M, Ricci M (2015) Community response of deep-sea soft-sediment metazoan meiofauna to the Deepwater Horizon blowout and oil spill. Mar Ecol Prog Ser 528:127–140. https://doi.org/10.3354/meps11290

Bar-Zeev E, Passow U, Romero-Vargas Castrillón S, Elimelech M (2015) Transparent exopolymer particles: from aquatic environments and engineered systems to membrane biofouling. Environ Sci Technol 49:691–707

Bianchi TS, Cook RL, Perdue EM, Kolic PE, Green N, Zhang Y, Smith RW, Kolker AS, Ameen A, King G, Ojwang LM, Schneider CL, Normand AE, Hetland R (2011) Impacts of diverted freshwater on dissolved organic matter and microbial communities in Barataria Bay, Louisiana, U. S. A. Mar Environ Res 72:248–257

Bochdansky AB, Clouse MA, Herndl GJ (2017) Eukaryotic microbes, principally fungi and labyrinthulomycetes, dominate biomass on bathypelagic marine snow. ISME J 11:362–373. https://doi.org/10.1038/ismej.2016.113

Bochdansky AB, Clouse MA, Herndl GJ (2016) Dragon kings of the deep sea: marine particles deviate markedly from the common number-size spectrum. Sci Rep 6:22633

Boehm PD, Fiest DL (1980) Aspects of the transport of petroleum hydrocarbons to the offshore benthos during the Ixtoc-I blowout in the Bay of Campeche. In: Proceedings of the Symposium on the Preliminary Results from the September, 1979 Pierce/Research IXTOC-1Cruises. Key Biscayne, Florida, June 9-10, 1980. Publications Office, NOAA/RD/MP3, Office of Marine Pollution Assessment, NOAA, US Dept. of Commerce, 325 Broadway, Boulder, CO 80303, USA

Bragg JR, Owens EH (1995) Shoreline cleansing by interactions between oil and fine mineral particles. International Oil Spill Conference Proc: February–March 1995, 1995, pp 219–227

Bragg JR, Yang SH (1995) Clay-oil flocculation and its role in natural cleansing in Prince William sound following the Exxon Valdez oil spill. ASTM STP 1219:178–214

Brakstad OG, Faksness L-G (2000) Biodegradation of water-accommodated fractions and dispersed oil in the seawater column. In: Proceedings of the International Conference on Health, Safety and Environment in Oil and Gas Exploration and Production, Stavanger, 26–28

Brooks GR, Larson RA, Schwing PT, Romero I, Moore C, Reichart G-J, Jilbert T, Chanton JP, Hastings DW, Overholt WA, Marks KP, Kostka JE, Holmes CW, Hollander D (2015)

Sedimentation pulse in the NE Gulf of Mexico following the 2010 DWH Blowou. PLoS One 10:e0132341. https://doi.org/10.1371/journal.pone.0132341

Burd AB, Hansell DA, Steinberg DK, Anderson TR, Arístegui J, Baltar F, Beaupré SR, Buesseler KO, DeHairs F, Jackson GA, Kadko DC, Koppelmann R, Lampitt RS, Nagata T, Reinthaler T, Robinson C, Robison BH, Tamburini C, Tanaka T (2010) Assessing the apparent imbalance between geochemical and biochemical indicators of meso- and bathypelagic biological activity: what the @$#! is wrong with present calculations of carbon budgets? Deep-Sea Res II 57:1557–1571

Burd AB, Jackson GA (2009) Particle aggregation. Annu Rev Mar Sci 1:65–90

Buskey EJ, White HK, Esbaugh AJ (2016) Impact of oil spills on marine life in the Gulf of Mexico: effects on plankton, nekton, and deep-sea benthos. Oceanography 29:174–181

Cai Z, Fu J, Liu W, Fu K, O'Reilly SE, Zhao D (2017) Effects of oil dispersants on settling of marine sediment particles and particle-facilitated distribution and transport of oil components. Mar Pollut Bull 114:408–418

Camilli R, Reddy CM, Yoerger DR, Van Mooy BAS, Jakuba MV, Kinsey JC, McIntyre CP, Sylva SP, Maloney JV (2010) Tracking hydrocarbon plume transport and biodegradation at Deepwater Horizon. Science 330:201–204

Carson RL (1951) The sea around us. Chapter 6, "The long snowfall". Oxford University Press, New York

Chanton J, Zhao T, Rosenheim BE, Joye S, Bosman S, Brunner C, Yeager KM, Diercks AR, Hollander D (2015) Using natural abundance radiocarbon to trace the flux of petrocarbon to the seafloor following the Deepwater Horizon oil spill. Environ Sci Technol 49:847–854

Chanton JP, Cherrier J, Wilson RM, Sarkodee-Adoo J, Bosman S, Mickle A, Graham WM (2012) Radiocarbon evidence that carbon from the Deepwater Horizon spill entered the planktonic food web of the Gulf of Mexico. Environ Res Lett 7:045303. https://doi.org/10.1088/1748-9326/7/4/045303

Chin WC, Orellana MV, Verdugo P (1998) Spontaneous assembly of marine dissolved organic matter into polymer gels. Nature 391:568–572

Daling PS, Leirvik F, Almås IK, Brandvik PJ, Hansen BH, Lewis A, Reed M (2014) Surface weathering and dispersibility of MC252 crude oil. Mar Pollut Bull 87:300–310

Daly KL, Passow U, Chanton J, Hollander D (2016) Assessing the impacts of oil-associated marine snow formation and sedimentation during and after the Deepwater Horizon oil spill. Anthropocene 13:18–33. https://doi.org/10.1016/j.ancene.2016.01.006

De La Rocha CL, Passow U (2014) The biological pump. In: Turekian KK, Holland HD (eds) Treatise on Geochemistry, vol 8. Elsevier, Oxford

De La Rocha C, Passow U (2007) Factors influencing the sinking of POC and the efficiency of the biological carbon pump. Deep Sea Res II 54:639–658

Diercks A-R, Dike C, Passow U, Ziervogel K, DiMarco SF, Asper VL (2018) Scales of seafloor sediment resuspension in the northern Gulf of Mexico. Elementa Sci Anthrop 6:32. https://doi.org/10.1525/elementa.285

Diercks AR, Highsmith RC, Asper VL, Joung D, Zhou Z, Guo L, Shiller AM, Joye SB, Teske AP, Guinasso N, Wade TL, Lohrenz SE (2010) Characterization of subsurface polycyclic aromatic hydrocarbons at the Deepwater Horizon site. Geophys Res Lett 37:L20602. https://doi.org/10.1029/2010GL045046

Dilling L, Alldredge AL (2000) Fragmentation of marine snow by swimming macrozooplankton: a new process impacting carbon cycling in the sea. Deep-Sea Res I 47:1227–1245

Discart V, Bilad M, Vankelecom IF (2015) Critical evaluation of the determination methods for transparent exopolymer particles, agents of membrane fouling. Crit Rev Environ Sci Technol 45:167–192

Dissanayake AL, Burd AB, Daly KL, Francis S, Passow U (2018) Numerical modeling of the interaction of oil, marine snow, and riverine sediments in the ocean. J Geophys Res Oceans 123:5388. https://doi.org/10.1029/2018JC013790

Dombrowski N, Donaho JA, Gutierrez T, Seitz KW, Teske AP, Baker BJ (2016) Reconstructing metabolic pathways of hydrocarbon-degrading bacteria from the Deepwater Horizon oil spill. Nat Microbiol 1:16057

Doyle SM, Whitaker EA, De Pascuale V, Wade TL, Knap AH, Santschi PH, Quigg A, Sylvan JB (2018) Rapid formation of microbe-oil aggregates and changes in community composition in coastal surface water following exposure to oil and Corexit. Front Microbiol 9:689. https://doi.org/10.3389/fmicb.2018.00689

Elimelech M, Gregory J, Jia X, Williams RA (1995) Particle deposition and aggregation: measurement, modelling and simulation. Butterworth-Heinemann, Woburn, MA

Engel A (2000) The role of transparent exopolymer particles (TEP) in the increase in apparent particle stickiness (α) during the decline of a diatom bloom. J Plankton Res 22:485–497

Févre JL (1979) On the hypothesis of a relationship between dinoflagellate blooms and the 'Amoco Cadiz' oil spill. J Mar Biol Assoc U K 59:525–528

Finkel ZV, Beardall J, Flynn KJ, Quigg A, Rees TAK, Raven JA (2010) Phytoplankton in a changing world: cell size and elemental stoichiometry. J Plankton Res 32:119–137

Fitzpatrick FA, Boufadel MC, Johnson R, Lee KW, Graan TP, Bejarano AC, Zhu Z, Waterman D, Capone DM, Hayter E, Hamilton SK (2015) Oil-particle interactions and submergence from crude oil spills in marine and freshwater environments: review of the science and future research needs (No. 2015–1076). US Geological Survey

Francis S, Burd AB, Daly KL, Passow U (2017) An aggregation model to estimate oil removal rate by sinking marine snow: a decision support tool. Gulf of Mexico oil spill and ecosystem science conference, New Orleans

Friedlander SK (2000) Smoke, dust, and haze: fundamentals of aerosol dynamics, 2nd edn. Oxford University Press, Oxford

Gong Y, Zhao X, Cai Z, O'Reilly SE, Hao X, Zhao D (2014) A review of oil, dispersed oil and sediment interactions in the aquatic environment: influence on the fate, transport and remediation of oil spills. Mar Pollut Bull 79:16–33

Graham WM, Condon RH, Carmichael RH, D'Ambra I, Patterson HK, Linn LJ, Hernandez FJ Jr (2010) Oil carbon entered the coastal planktonic food web during the Deepwater Horizon oil spill. Environ Res Lett 5:045301. https://doi.org/10.1088/1748-9326/5/4/045301

Grossart HP, Simon M (1993) Limnetic macroscopic organic aggregates (lake snow): occurrence, characteristics, and microbial dynamics in lake constance. Limnol Oceanogr 38:532–546

Gustitus SA, Clement TP (2017) Formation, fate, and impacts of microscopic and macroscopic oil-sediment residues in nearshore marine environments: a critical review. Rev Geophys 55(4):1130–1157. https://doi.org/10.1002/2017RG000572

Gutierrez T, Singleton DR, Berry D, Yang T, Aitken MD, Teske A (2013a) Hydrocarbon-degrading bacteria enriched by the Deepwater Horizon oil spill identified by cultivation and DNA-SIP. ISME J 7:2091

Gutierrez T, Berry D, Yang T, Mishamandani S, McKay L, Teske A, Aitken M (2013b) Role of bacterial exopolysaccharides (EPS) in the fate of the oil released during the Deepwater Horizon oil spill. PLoS One 8:e67717

Hastings DW, Schwing PT, Brooks GR, Larson RA, Morford JL, Roeder T, Quinn KA, Bartlett T, Romero IC, Hollander DJ (2015) Changes in sedimentary redox conditions following the BP DwH blowout event. Deep-Sea Res II 129:167–178. https://doi.org/10.1016/j.dsr2.2014.12.009

Hazen TC, Dubinsky EA, DeSantis TZ, Andersen GL, Piceno YM, Singh N, Jansson JK, Probst A, Borglin SE, Fortney JL, Stringfellow WT, Bill M, Conrad MS, Tom LM, Chavarria KL, Alusi TR, Lamendella R, Joyner DC, Spier C, Baelum J, Auer M, Zemla ML, Chakraborty R, Sonnenthal EL, D'Haeseleer P, Holman H-YN, Osman S, Lu Z, Van Nostrand JD, Deng Y, Zhou J, Mason OU (2010) Deep-sea oil plume enriches indigenous oil-degrading bacteria. Science 330:204–208

Hu C, Weisberg RH, Liu Y, Zheng L, Daly KL, English DC, Zhao J, Vargo GA (2011) Did the northeastern Gulf of Mexico become greener after the Deepwater Horizon oil spill? Geophys Res Lett 38:L09601. https://doi.org/10.1029/2011GL047184

Jackson GA (1990) A model of the formation of marine algal flocs by physical coagulation processes. Deep-Sea Res 37:1197–1211

Jackson GA (1998) Using fractal scaling and two-dimensional particle size spectra to calculate coagulation rates for heterogeneous systems. J Colloid Interface Sci 202:20–29

Jernelöv A, Lindén O (1981) Ixtoc I: a case study of the world's largest oil spill. Ambio 10:299–306

Jézéquel R, Receveur J, Nedwed T, Le Floch S (2018) Evaluation of the ability of calcite, bentonite and barite to enhance oil dispersion under arctic conditions. Mar Pollut Bull 127:626–636

Johansson S, Larsson U, Boehm P (1980) The Tsesis oil spill impact on the pelagic ecosystem. Mar Pollut Bull 11:284–293

Jokulsdottir T, Archer D (2016) A stochastic, Lagrangian model of sinking biogenic aggregates in the ocean (SLAMS 1.0): model formulation, validation and sensitivity. Geosci Model Dev 9:1455–1476

Joye SB, Teske AP, Kostka JE (2014) Microbial dynamics following the Macondo oil well blowout across Gulf of Mexico environments. Bioscience 64:766–777

Khelifa A, Fingas M, Brown C (2008a) Effects of dispersants on Oil-SPM aggregation and fate in US coastal waters. Report submitted to the Coastal Response Research Center, University of New Hampshire, July 2008, Project Number: 06-090, 57pp

Khelifa A, Fieldhouse B, Wang Z, Yang C, Landriault M, Brown CE, Fingas M (2008b) Effects of chemical dispersant on oil sedimentation due to oil-SPM flocculation: experiments with the NIST-1941b standard reference material. Proc IOSC 2008:627–631

Kleindienst S, Paul JH, Joye SB (2015) Using dispersants after oil spills: impacts on the composition and activity of microbial communities. Nat Rev Microbiol 13:388–396. https://doi.org/10.1038/nrmicro3452

Kranck K (1973) Flocculation of suspended sediment in the sea. Nature 246:348–350

Lambert RA, Variano EA (2016) Collision of oil droplets with marine aggregates: effect of droplet size. J Geophys Res Oceans 121:3250–3260. https://doi.org/10.1002/2015JC011562

Laurenceau-Cornec EC, Trull TW, Davies DM, De La Rocha CL, Blain S (2015) Phytoplankton morphology controls on marine snow sinking velocity. Mar Ecol Prog Ser 520:35–56

Lee K, Zheng Y, Merlin FX, Li Z, Niu H, King T, Doane R (2012a) Combining mineral fines with chemical dispersants to disperse oil in low temperature and low mixing environments, including the Arctic Rep. US Department of the Interior, Bureau of Safety and Environmental Enforcement (BSEE)

Lee RF, Köster M, Paffenhöfer GA (2012b) Ingestion and defecation of dispersed oil droplets by pelagic tunicates. J Plankton Res 34:1058–1063

Lee K, Li Z, Robinson B, Kepkay PE, Blouin M, Doyon B (2011) Oil spill countermeasures in the Arctic. Proceedings of the International Conference on Oil Spill Risk Management: preparedness, Response and Contingency Planning in the Shipping and Offshore Industries. 7–9 March, Malmo Borshus, Sweden, Neil Bellefontaine and Olaf Linden (eds.), WMU Publications, pp 93–108

Lee K, Li Z, King T, Kepkay P, Boufadel M, Venosa A, Mullin J (2008) Effects of chemical dispersants and mineral fines on partitioning of petroleum hydrocarbons in natural seawater. Proceedings of the 2008 International Oil Spill Conference, Savannah, Georgia, USA, May 4–8, 2008, pp 633–638. https://doi.org/10.7901/2169-3358-2008-1-633

Lee K, Stoffyn-Egli P, Tremblay GH, Owens EH, Sergy GA, Guénette CC, Prince RC (2003a) Oil-mineral aggregate formation on oiled beaches: natural attenuation and sediment relocation. Spill Sci Technol Bull 8:285–296

Lee K, Wohlgeschaffen G, Tremblay GH, Johnson BT, Sergy GA, Prince RC, Guénette CC, Owens EH (2003b) Toxicity evaluation with the Microtox test to assess the impact of in-situ oiled shoreline treatment options: natural attenuation and sediment relocation. Spill Sci Technol Bull 8:273–284

Lee K, Stoffyn-Egli P, Wood P, Lunel T (1998) Formation and structure of oil-mineral fine aggregates in coastal environments. Proceedings 21st Arctic and Marine Oilspill Program (AMOP) Technical Seminar. June 10–12, 1998, Edmonton, Alberta, pp 911–921

Lee K, Lunel T, Wood P, Swannell R, Stoffyn-Egli P (1997) Shoreline cleanup by acceleration of clay-oil flocculation processes. In International oil spill conference (1997, No. 1, pp 235–240). American Petroleum Institute, Washington, DC

Lee K, Wong CS, Cretney WJ, Whitney FA, Parsons TR, Lalli C, Wu J (1985) Microbial response to crude oil and Corexit 9527: SEAFLUXES enclosure study. Microb Ecol 11:337–351

Lee RF, Gardner WS, Anderson JW, Blaylock JW, Barwell-Clarke J (1978) Fate of polycyclic aromatic hydrocarbons in controlled ecosystem enclosures. Environ Sci Technol 12:832–838

Lee RF, Anderson JW (1977) Fate and effect of naphthalenes: controlled ecosystem pollution experiment. Bull Mar Sci 27:127–134

Le Floch S, Guyomarch J, Merlin FX, Stoffyn-Egli P, Dixon J, Lee K (2002) The influence of salinity on oil–mineral aggregate formation. Spill Sci Technol Bull 8:65–71

Levine S, Bowen BD, Partridge SJ (1989) Stabilization of emulsions by fine particles II. capillary and van der Waals forces between particles. Colloids Surf 38:345–364

Li Z, Lee K, King KT, Boufadel M, Venosa AD (2008) Assessment of chemical dispersant effectiveness in a wave tank under regular non-breaking and breaking wave conditions. Mar Pollut Bull 56:903–912

Logan BE, Passow U, Alldredge AL, Grossart H-P, Simon M (1995) Rapid formation and sedimentation of large aggregates is predictable from coagulation rates (half-lives) of transparent exopolymer particles (TEP). Deep-Sea Res II 42:203–214

Logan BE, Alldredge AL (1989) Potential for increased nutrient uptake by flocculating diatoms. Mar Biol 101:443–450

Loh A, Shim WJ, Ha SY, Yim UH (2014) Oil-suspended particulate matter aggregates: formation mechanism and fate in the marine environment. OSJ 49:329–341

Lunel T, Lee K, Swannell R, Wood P, Rusin J, Bailey N, Halliwell C, Davies L, Sommerville M, Dobie A, Mitchell D, McDonagh M (1996) Shoreline clean up during the Sea Empress Incident: the role of surf washing (clay- M. oil flocculation), dispersants and bioremediation. Proceedings of the 19th Arctic and Marine Oilspill Program (AMOP) Technical Seminar, June 12–14, 1996, Calgary, Alberta, Canada, pp 1521–1540

Ma X, Cogswell A, Li Z, Lee K (2008) Particle size analysis of dispersed oil and oil-mineral aggregates with an automated epifluorescence microscopy system. Environ Technol 29:739–748

Mari XS, Passow U, Migon C, Burd AB, Legendre L (2017) Transparent exopolymer particles: effects on carbon cycling in the ocean. Prog Oceanogr 151:13–37

Mason OU, Scott NM, Gonzalez A, Robbins-Pianka A, Baelum J, Kimbrel J, Bouskill NJ, Prestat E, Borglin S, Joyner DC, Fortney JL, Jurelevicius D, Stringfellow WT, Alvarez-Cohen L, Hazen TC, Knight R, Gilbert JA, Jansson JK (2014) Metagenomics reveals sediment microbial community response to Deepwater Horizon oil spill. Int Soc Microbiol/Ecol J 8:1464–1475

McGenity T, Folwell B, McKew B, Sanni G (2012) Marine crude-oil biodegradation: a central role for interspecies interactions. Aquat Biosyst 16:10. https://doi.org/10.1186/2046-9063-8-10

Menon VB, Nagarajan R, Wasan DT (1987) Separation of fine particles from non-aqueous media: free energy analysis and oil loss estimation. Sep Sci Technol 22:2295–2322

Mitra S, Kimmel DG, Snyder J, Scalise K, McGlaughon BD, Roman MR, Jahn GL, Pierson JJ, Brandt SB, Montoya JP, Rosenbauer RJ, Lorenson TD, Wong FL, Campbell PL (2012) Macondo-1 well oil-derived polycyclic aromatic hydrocarbons in mesozooplankton from the northern Gulf of Mexico. Geophys Res Lett 39:L01605. https://doi.org/10.1029/2011GL049505

Montagna PA, Baguley JG, Cooksey C, Hartwell I, Hyde LJ, Hyland JL, Kalke RD, Kracker LM, Reuscher M, Rhodes ACE (2013) Deep-sea benthic footprint of the Deepwater Horizon blowout. PLoS One 8(8):e70540. https://doi.org/10.1371/journal.pone.0070540

Nishizawa S, Fukuda M, Inoue N (1954) Photographic study of suspended matter and plankton in the sea. Bull Fac Fish, Hokkaido Univ 5:36–40

Niu H, Lee K (2013) Study the transport of oil-mineral-aggregates (OMAs) in marine environment and assessment of their potential risks to benthic organisms. Int J Environ Pollut 52:32–51. https://doi.org/10.1504/IJEP.2013.056356

O'Connor BS, Muller-Karger FE, Nero RW, Hu C, Peebles EB (2016) The role of Mississippi River discharge in offshore phytoplankton blooming in the northeastern Gulf of Mexico during August 2010. Remote Sens Environ 173:133–144

Omotoso OE, Munoz VA, Mikula RJ (2002) Mechanisms of crude oil–mineral interactions. Spill Sci Technol Bull 8:45–54

OSAT (2010) Summary report for sub-sea and sub-surface oil and dispersant detection: sampling and monitoring, Operational Science Advisory Team (OSAT) US. Department of Homeland Security, New Orleans, LA, pp 131

Owens EH, Lee K (2003) Interaction of oil and mineral fines on shorelines: review and assessment. Mar Pollut Bull 47:397–405

Passow U, Sweet J, Francis S, Xu C, Dissanayake AL, Lin J, Santschi PH, Quigg A (2019) Incorporation of oil into diatom aggregates. Mar Ecol Prog Ser 612:65–86. https://doi.org/10.3354/meps12881

Passow U, Sweet J, Quigg A (2017) How the dispersant Corexit impacts the formation of sinking marine oil snow. Mar Pollut Bull 125:139–145

Passow U, Hetland R (2016) What happened to all of the oil? Oceanography 29:88–95

Passow U, Ziervogel K (2016) Marine snow sedimented oil released during the Deepwater Horizon spill. Oceanography 29:118–125

Passow U (2016) Formation of rapidly-sinking, oil-associated marine snow. Deep-Sea Res II 129:232. https://doi.org/10.1016/j.dsr2.2014.10.001

Passow U, Ziervogel K, Asper V, Dierks A (2012) Marine snow formation in the aftermath of the Deepwater Horizon oil spill in the Gulf of Mexico. Environ Res Lett 7:11. https://doi.org/10.1088/1748-9326/7/3/035301

Passow U (2002) Transparent exopolymer particles (TEP) in aquatic environments. Prog Oceanogr 55:287–333

Passow U, Alldredge AL (1995) Aggregation of a diatom bloom in a mesocosm: the role of transparent exopolymer particles (TEP). Deep-Sea Res II 42:99–109

Passow U, Alldredge AL (1994) Distribution, size, and bacterial colonization of transparent exopolymer particles (TEP) in the ocean. Mar Ecol Prog Ser 113:185–198

Passow U, Alldredge AL, Logan BE (1994) The role of particulate carbohydrate exudates in the flocculation of diatom blooms. Deep-Sea Res Oceanogr Abstr 41:335–357

Patton JS, Rigler MW, Boehm PD, Fiest DL (1981) Ixtoc I oil spill: flaking of surface mousse in the Gulf of Mexico. Nature 290:235–238

Payne JR, Clayton JR, Kirstein BE (2003) Oil/suspended particulate material interactions and sedimentation. Spill Sci Technol Bull 8:201–221

Pernice MC, Irene Forn I, Gomes A, Lara E, Alonso-Sáez L, Arrieta JM, del Carmen GF, Hernando-Morales V, Mackenzie R, Mestre M, Sintes E, Teira E, Valencia J, Varela MM, Vaqué D, Duarte CM, Gasol JM, Massana R (2015) Global distribution of planktonic heterotrophic protists in the deep ocean. ISME J 9:782–792

Ploug H, Iversen MH, Fischer G (2008) Ballast, sinking velocity and apparent diffusivity within marine snow and fecal pellets: implications and substrate turnover by attached bacteria. Limnol Oceanogr 53:1878–1886

Pruppacher HR, Klett JD (2010) Microphysics of clouds and precipitation, 2nd edn. Springer, New York

Quigg A, Passow U, Chin W-C, Xu C, Doyle S, Bretherton L, Kamalanathan M, Williams AK, Sylvan JB, Finkel ZV, Knap AH, Schwehr KA, Zhang S, Sun L, Wade TL, Obeid W, Hatcher PG, Santschi PH (2016) The role of microbial exopolymers in determining the fate of oil and chemical dispersants in the ocean. Limnol Oceanogr Lett 1:3–26

Quigg A, Irwin AJ, Finkel ZV (2011) Evolutionary imprint of endosymbiosis of elemental stoichiometry: testing inheritance hypotheses. Proc R Soc: Biol Sci 278:526–534

Quigg A, Finkel ZV, Irwin AJ, Reinfelder JR, Rosenthal Y, Ho T-Y, Schofield O, Morel FMM, Falkowski PG (2003) The evolutionary inheritance of elemental stoichiometry in marine phytoplankton. Nature 425:291–294

Remsen A, Daly K, Kramer K (2015) Plankton and particle response during the Deepwater Horizon oil spill: observations from the SIPPER imaging system. NOAA NRDA Report

Reuscher MG, Baguley JG, Conrad-Forrest N, Cooksey C, Hyland JL, Lewis C, Montagna PA, Ricker RW, Rohal M, Washburn T (2017) Temporal patterns of the Deepwater Horizon impacts on the benthic infauna of the northern Gulf of Mexico continental slope. PLoS One 12(6):e0179923. https://doi.org/10.1371/journalpone.0179923

Riley GA (1963) Organic aggregates in seawater and the dynamics of their formation and utilization. Limnol Oceanogr 8:372–381

Romero IC, Toro-Farmer G, Diercks A-R, Schwing P, Muller-Karger F, Murawski S, Hollander DJ (2017) Large-scale deposition of weathered oil in the Gulf of Mexico following a deep-water oil spill. Environ Pollut 228:179–189

Romero IC, Schwing PT, Brooks GR, Larson RA, Hastings DW, Ellis G, Goddard EA, Hollander DJ (2015) Hydrocarbons in deep-sea sediments following the 2010 Deepwater Horizon blowout in the northeast Gulf of Mexico. PLoS One 10(5):e0128371

Salazar G, Cornejo-Castillo FM, Borrull E, Díez-Vives C, Lara E, Vaqué D, Arrieta JM, Duarte CM, Gasol JM, Acinas SG (2015) Particle-association lifestyle is a phylogenetically conserved trait in bathypelagic prokaryotes. Mol Ecol 24:5692–5706

Schwing PT, O'Malley BJ, Hollander DJ (2018) Resilience of Benthic Foraminifera in the Northern Gulf of Mexico following the Deepwater Horizon event (2011–2015). Ecol Indic 84:753–764. https://doi.org/10.1016/j.ecolind.2017.09.044

Schwing PT, Brooks GR, Larson RA, Holmes CW, O'Malley BJ, Hollander DJ (2017) Constraining the spatial extent of the Marine Oil Snow Sedimentation and Accumulation (MOSSFA) following the DWH event using a $^{210}Pb_{xs}$ inventory approach. Environ Sci Technol 51:5962–5968. https://doi.org/10.1021/acs.est.7b00450

Schwing PT, Romero IC, Brooks GR, Hastings DW, Larson RA, Hollander DJ (2015) A decline in Deep-Sea benthic foraminifera following the Deepwater Horizon event in the Northeastern Gulf of Mexico. PLOSone 10(3):e0120565. https://doi.org/10.1371/journal.pone.0120565

Silver M (2015) Marine snow: a brief historical sketch. Limnol Oceanogr Bull 24:5–10

Silver MW, Gowing MM (1991) The "particle" flux: origins and biological components. Prog Oceanogr 26:75–113

Silver MW, Shanks AL, Trent JD (1978) Marine snow: microplankton habitat and source of small-scale patchiness in pelagic populations. Science 201:371–373

Sterling MC, Bonner JS, Ernest ANS, Page CA, Autenrieth RL (2005) Application of fractal flocculation and vertical transport model to aquatic sol– sediment systems. Water Res 39:1818–1830

Sterling MC, Bonner JS, Ernest AN, Page CA, Autenrieth RL (2004) Characterizing aquatic sediment–oil aggregates using in situ instruments. Mar Pollut Bull 48:533–542

Stoffyn-Egli P, Lee K (2003) Formation and characterization of oil-mineral aggregates. Spill Sci Technol Bull 8:31–44

Stout SA, Payne JR (2016a) Macondo oil in deep-sea sediments: part 1 – sub-sea weathering of oil deposited on the seafloor. Mar Pollut Bull 108:365–380

Stout SA, Payne JR (2016b) Chemical composition of floating and sunken in-situ burn residues from the Deepwater Horizon oil spill. Mar Pollut Bull 111:186–202

Stout SA, German CR (2015) Characterization and Flux of Marine Oil Snow in the Viosca Knoll (Lophelia Reef) Area Due to the Deepwater Horizon Oil Spill Newfields, Rockland Massachusetts

Suja LD, Summers S, Gutierrez T (2017) Role of EPS, dispersant and nutrients on the microbial response and MOS formation in the subarctic Northeast Atlantic. Front Microbiol 8:676. https://doi.org/10.3389/fmicb.2017.00676

Sun J, Khelifa A, Zhao C, Zhao D, Wang Z (2014) Laboratory investigation of oil–suspended particulate matter aggregation under different mixing conditions. Sci Total Environ 473:742–749

Sun J, Zhao D, Zhao C, Liu F, Zheng X (2013) Investigation of the kinetics of oil–suspended particulate matter aggregation. Mar Pollut Bull 76:250–257. https://doi.org/10.1016/j.marpolbul.2013.08.030

Sun J, Khelifa A, Zheng X, Wang Z, So LL, Wong S, Yang C, Fieldhouse B (2010) A laboratory study on the kinetics of the formation of oil-suspended particulate matter aggregates using the NIST-1941b sediment. Mar Pollut Bull 60:1701–1707

Sun J, Zheng XL (2009) A review of oil-suspended particulate matter aggregation natural process of cleansing spilled oil in the aquatic environment. J Environ Monit 11:1801–1809

Teal JM, Howarth RW (1984) Oil spill studies: a review of ecological effects. Environ Manag 8:27–44

Tsujita T (1953) A preliminary study on naturally occurring suspended organic matter in waters adjacent to Japan. (in Japanese, with a summary in English). J Oceanogr Soc Jpn 8:1–14

Valentine DL, Fisher GB, Bagby SC, Nelson RK, Reddy CM, Sylva SP, Woo MA (2014) Fallout plume of submerged oil from Deepwater Horizon. Proc Natl Acad Sci U S A 111:15906–15911. https://doi.org/10.1073/pnas.1414873111

Verdugo P, Alldredge AL, Azam F, Kirchman DL, Passow U, Santschi PH (2004) The oceanic gel phase: a bridge in the DOM–POM continuum. Mar Chem 92:67–85

Volk T, Hoffert MI (1985) Ocean carbon pumps: analysis of relative strengths and efficiencies in ocean-driven atmospheric CO2 changes. In: Sundquist ET, Broecker WS (eds) The carbon cycle and atmospheric CO: natural variations archean to present. American Geophysical Union, Washington, DC, pp 99–110

Volkman JK, Tanoue E (2002) Chemical and biological studies of particulate organic matter in the ocean. J Oceanogr 58:265–279

Vonk SM, Hollander DJ, Murk ATJ (2015) Was the extreme and wide-spread marine oil-snow sedimentation and flocculent accumulation (MOSSFA) event during the Deepwater Horizon blow-out unique? Mar Pollut Bull 100:5–12

Wang W, Zheng Y, Lee K (2013) Chemical dispersion of oil with mineral fines in a low temperature environment. Mar Pollut Bull 72:205–212

Wang W, Zheng Y, Li Z, Lee K (2011) PIV investigation of oil–mineral interaction for an oil spill application. Chem Eng J 170:241–249

Washburn TW, Reuscher MG, Montagna PA, Cooksey C, Hyland JL (2017) Macrobenthic community structure in the deep Gulf of Mexico one year after the Deepwater Horizon blowout. Deep Sea Res Part 1 Oceanogr Res Pap 127:21–30. https://doi.org/10.1016/j.dsr.2017.06.001

Weise AM, Nalewajko C, Lee K (1999) Oil-mineral fine interactions facilitate oil biodegradation in seawater. Environ Technol 20:811–824

Wincele DE, Wrenn BA, Venosa AD (2004) Sedimentation of oil-mineral aggregates for remediation of vegetable oil spills. J Environ Eng 130:50–58

Wirth M, Passow U, Jeschek J, Hand I, Schulz-Bull DE (2018) Partitioning of oil compounds into marine oil snow: insights into prevailing mechanisms and dispersant effects. Marine Chemistry 206:62. https://doi.org/10.1016/j.marchem.2018.09.007

Wood PA, Lunel T, Daniel F, Swannell R, Lee K, Stoffyn-Egli P (1998) Influence of oil and mineral characteristics on oil-mineral interaction. In: Artic and marine oil spill program technical seminar. Ministry of Supply and Services, Canada, pp 51–78

Xavier M, Passow U, Migon C, Burd AB, Legendre L (2017) Transparent exopolymer particles: effects on carbon cycling in the ocean. Prog Oceanogr 151:13–37

Yan B, Passow U, Chanton J, Nöthig E-M, Asper V, Sweet J, Pitiranggon M, Diercks A, Pak D (2016) Sustained deposition of contaminants from the Deepwater Horizon oil spill. Proc Natl Acad Sci U S A:E3332–E3340. https://doi.org/10.1073/pnas.1513156113

Yang T, Nigro LM, Gutierrez T, D'ambrosio L, Joye SB, Highsmith R, Teske A (2016a) Pulsed blooms and persistent oil-degrading bacterial populations in the water column during and after the Deepwater Horizon blowout. Deep-Sea Res II Top Stud Oceanogr 129:282–291

Yang T, Speare K, McKay L, MacGregor BJ, Joye SB, Teske A (2016b) Distinct bacterial communities in surficial seafloor sediments following the 2010 Deepwater Horizon blowout. Front Microbiol 7:1384. https://doi.org/10.3389/fmicb.2016.01384. eCollection 2016

Zetsche E-M, Ploug H (2015) Particles in aquatic environments: from invisible exopolymers to sinking aggregates. Mar Chem 175:1–4

Zhang H, Khatibi M, Zheng Y, Lee K, Li Z, Mullin JV (2010) Investigation of OMA formation and the effect of minerals. Mar Pollut Bull 60:1433–1441

Zhao L, Boufadel MC, Katz J, Haspel G, Lee K, King T, Robison B (2017) A new mechanism of sediment attachment to oil in turbulent flows: projectile particles. Environ Sci Technol 51:11020–11028. https://doi.org/10.1021/acs.est.7b02032

Zhao L, Boufadel MC, Geng X, Lee K, King T, Robinson B, Fitzpatrick F (2016) A-DROP: a predictive model for the formation of oil particle aggregates (OPA). Mar Pollut Bull 106:245–259

Zhao L, Torlapati J, Boufadel MC, King T, Robinson B, Lee K (2014) VDROP: a comprehensive model for droplet formation of oils and gases in liquids-Incorporation of the interfacial tension and droplet viscosity. Chem Eng J 253:93–106

Ziervogel K, Joye SB, Arnosti C (2016) Microbial enzymatic activity and secondary production in sediments affected by the sedimentation pulse following the Deepwater Horizon oil spill. Deep-Sea Res II Top Stud Oceanogr 129:241–248

Ziervogel K, Joye SB, Arnosti C (2014) Microbial enzymatic activity and secondary production in sediments affected by the sedimentation event of oily-particulate matter from the Deepwater Horizon oil spill. Deep-Sea Res II Top Stud Oceanogr 129:241–248

Chapter 13
The Sedimentary Record of MOSSFA Events in the Gulf of Mexico: A Comparison of the *Deepwater Horizon* (2010) and Ixtoc 1 (1979) Oil Spills

Patrick T. Schwing, David J. Hollander, Gregg R. Brooks, Rebekka A. Larson, David W. Hastings, Jeffrey P. Chanton, Sara A. Lincoln, Jagoš R. Radović, and Alette Langenhoff

Abstract Marine Oil Snow Sedimentation and Flocculent Accumulation (MOSSFA) refers to the process of formation, sinking, and seafloor deposition of oil-contaminated marine snow and oil-mineral aggregates. MOSSFA was well documented in the northern Gulf of Mexico (GoM) in the aftermath of the *Deepwater Horizon* (DWH 2010) and likely occurred in the southern GoM during Ixtoc 1

P. T. Schwing (✉) · D. J. Hollander
University of South Florida, College of Marine Science, St. Petersburg, FL, USA
e-mail: pschwing@mail.usf.edu; davidh@usf.edu

G. R. Brooks · D. W. Hastings
Eckerd College, St. Petersburg, FL, USA
e-mail: brooksgr@eckerd.edu; hastindw@eckerd.edu

R. A. Larson
University of South Florida, College of Marine Science, St. Petersburg, FL, USA

Eckerd College, St. Petersburg, FL, USA
e-mail: larsonra@eckerd.edu

J. P. Chanton
Florida State University, Department of Earth, Ocean and Atmospheric Science,
Tallahassee, FL, USA
e-mail: jchanton@fsu.edu

S. A. Lincoln
The Pennsylvania State University, Department of Geosciences, University Park, PA, USA
e-mail: slincoln@psu.edu

J. R. Radović
University of Calgary, PRG, Department of Geoscience, Calgary, AB, Canada
e-mail: jagos.radovic@ucalgary.ca

A. Langenhoff
Wageningen University & Research, Sub-department of Environmental Technology,
Wageningen, The Netherlands
e-mail: alette.langenhoff@wur.nl

S. A. Murawski et al. (eds.), *Deep Oil Spills*,
https://doi.org/10.1007/978-3-030-11605-7_13

(1979–1980). This chapter introduces Part IV: Oil Spill Records in Deep Sea Sediments and addresses a series of questions regarding MOSSFA in the sedimentary record:

- What were the characteristics of MOSSFA sedimentary inputs?
- What was the extent of MOSSFA on the seafloor?
- What postdepositional processes took place as a result of MOSSFA?
- Can MOSSFA be preserved in the sedimentary record?

MOSSFA sedimentary inputs were comprised of three main components (biogenic, lithogenic, and petrogenic), many of which were surface derived. MOSSFA resulted in a four- to ten fold increase in bulk sediment accumulation rates, a two- to three fold increase in oil-derived hydrocarbon concentrations, a two- to three-order of magnitude increase in Corexit 9500A dispersant concentration, and two- to three fold increases in surface-derived biotic material (e.g., planktic foraminifera, diatoms). Estimates of the total spatial extent of MOSSFA on the seafloor of the northern GoM range from 1030 to 35,425 km^2, accounting for between 3.7% and 14.4% of the total petroleum released during DWH. Increased microbial respiration of organic carbon caused depleted surface sediment oxygen, intensifying reducing conditions up to 3 years following DWH. Multiple proxies provided evidence of multi-year preservation of both oil residue in the sediments associated with DWH, MOSSFA, and the sedimentary event in the geologic record. Despite confounding factors in the southern GoM including regional events (e.g., volcanoes, hurricanes) and complex hydrocarbon backgrounds (e.g., natural seeps, oil, and gas infrastructure), multiple sedimentary proxies have provided evidence of degraded Ixtoc 1 oil-residue input and the MOSSFA sedimentary event preserved in the geologic record greater than 35 years after Ixtoc 1. Federal and international policies can be benefitted by incorporating MOSSFA with regard to response strategies, weighing the ecological trade-off between oiled coastlines and offshore benthic environments.

Keywords MOSSFA · *Deepwater Horizon* · Ixtoc 1 · Seafloor · Oil residue · Sedimentation

13.1   Introduction

The amount of oil released from the *Deepwater Horizon* (DWH) subsurface marine petroleum well blowout to the northern Gulf of Mexico (GoM) is somewhat controversial. McNutt et al. (2012) calculated that the cumulative release was approximately 4.9 million barrels, of which 0.8 million barrels were collected. Joye et al. (2011) figured the amount released to the environment was between 4.5 and 6.3 million barrels. Griffiths (2012) estimated that the cumulative discharge was 5.4 million barrels of which 4.6 million barrels entered the northern GoM. A court settlement (US District Court 2015) set the estimate at 4 million barrels of cumulative release with 3.2 million barrels released to the northern GoM following response efforts. The release of petroleum into the northern GoM resulted in the occurrence of Marine Oil Snow Sedimentation and Flocculent Accumulation (MOSSFA) (Quigg et al. 2020; Passow et al. 2012; Ziervogel et al. 2012; Passow 2014; Brooks et al. 2015; Romero et al. 2015; Daly et al. 2016; Schwing et al. 2017a; Romero et al. 2017). There are several regionally and temporally dependent factors that affect the formation of MOSSFA including three primary components: (1) riverine influences (nutrients, clay minerals, salinity), (2) marine biota (microbial, algal, stress response formation of exopolymeric substances), and (3) petroleum (petrogenic and pyrogenic sources). Components 1 and 2 are exacerbated in deltaic systems with strong gradients and high organic carbon production and deposition rates (Quigg et al. 2020).

The Ixtoc 1 subsurface marine petroleum blowout released 3 million barrels of petroleum from 50 meters water depth into the southern GoM over the course of 290 days from 1979 to 1980 (Jernelov and Linden 1981). While in situ water column and seafloor measurements were spatially limited (primarily surrounding the wellhead) at the time of the spill, all of the necessary contributing factors for MOSSFA formation were present. Vonk et al. (2015) performed a meta-analysis of all major accidental oil releases and found that MOSSFA likely occurred during Ixtoc 1 based on the presence of microbial communities, algae, river-derived particulate matter (Gracia et al. 2013), and dispersant usage.

During and directly following DWH in 2010, a series of sediment cores was collected throughout the northern GoM to produce a multi-proxy record of MOSSFA inputs, postdepositional processes, and seafloor spatial extent (Fig. 13.1). These collections continued annually from 2010 to 2017. Sediment cores were also collected from 2015 to 2016 throughout the southern GoM in open marine and coastal environments to assess the presence and long-term preservation of MOSSFA related to Ixtoc 1. All cores were extruded at 2–5 mm scale (Schwing et al. 2016) to properly characterize both oil spill events at sufficient temporal resolution (sub-annual to sub-decadal). This chapter introduces Part IV: *Oil Spill Records in Deep Sea Sediments* and addresses a series of questions regarding MOSSFA in the sedimentary record:

- What were the characteristics of MOSSFA sedimentary inputs?
- What was the extent of MOSSFA on the seafloor?
- What postdepositional processes took place as a result of MOSSFA?
- Can MOSSFA be preserved in the sedimentary record?

13.2   What Were the Characteristics of MOSSFA Sedimentary Inputs?

The initial evidence of MOSSFA on the seafloor in the northern GoM, following DWH, was an increase in sediment accumulation rates. Brooks et al. (2015) found that mass accumulation rates increased by four- to tenfold in the northern GoM relative to pre-DWH accumulation rates using a novel short-lived radioisotope application of excess ^{234}Th (^{234}Th$_{xs}$). This approach can only be utilized in cases

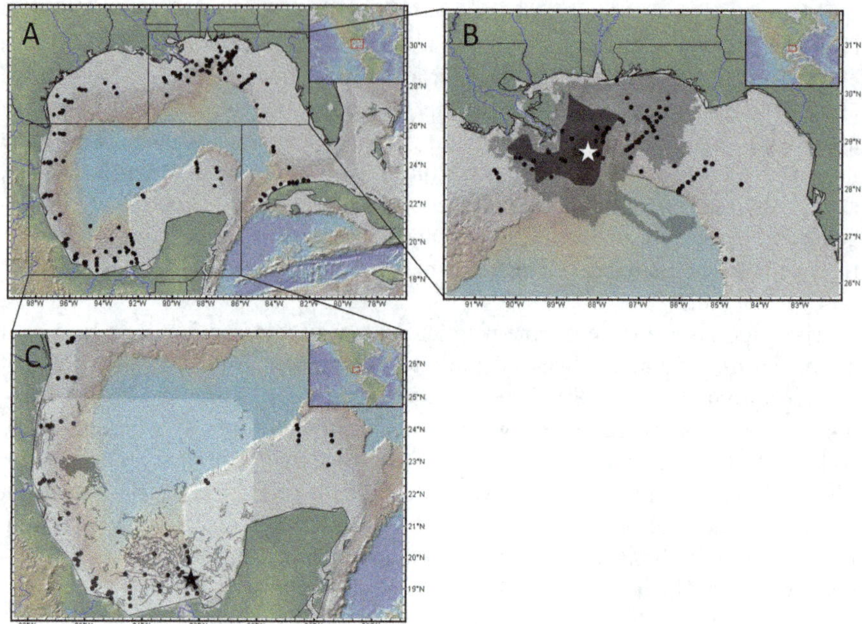

Fig. 13.1 Map of (**a**) all coring and surface sediment sampling locations (black circles) collected throughout the Gulf of Mexico from 2010 to 2017 as part of the C-IMAGE research effort including enlarged portions of the (**b**) northern Gulf of Mexico and the (**c**) southern Gulf of Mexico. In the enlarged northern Gulf of Mexico map (**b**), the dark gray shading is the surface extent of dispersant application (Environmental Response Management Application, ERMA), the light gray shading is the surface oil extent (ERMA), and the white star is the location of the *Deepwater Horizon* wellhead. In the enlarged southern Gulf of Mexico map (**c**), the light gray-shaded areas are the surface oil extent (Sun et al. 2015), and the black star is the location of the Ixtoc 1 wellhead

where there are high accumulation rates, fine-grained sediments, and no bioturbation present, which were satisfied by MOSSFA. Records based on ^{210}Pb (^{210}Pb$_{xs}$) also corroborated this finding (Brooks et al. 2015; Schwing et al. 2017a; Larson et al. 2018; Larson et al. 2020). The MOSSFA layer, in which the "pulse" of increased mass accumulation rates was identified, was primarily comprised of water column surface and near-surface-derived components. Fridrik et al. (2016) documented a fourfold increase in planktic foraminiferal accumulation rates in the post-DWH layer relative to pre-DWH (Fig. 13.2). Biogenic silica percentage and accumulation rate, which are indicative of surface diatom productivity, both increased significantly in the MOSSFA layer (Yan et al. 2016; Lee et al. 2018). This was corroborated by the three- to tenfold increase in the concentration of diatom 16S RNA sequences in the MOSSFA layer (Brooks et al. 2015). In addition to biological evidence of the surface-derived nature of MOSSFA sedimentary inputs, there was a 2–3 order of magnitude increase in the concentration of the primary component (DOSS) of the Corexit 9500A dispersant in post-DWH surface sediments (White et al. 2014) and a two- to three fold increase in high molecular weight

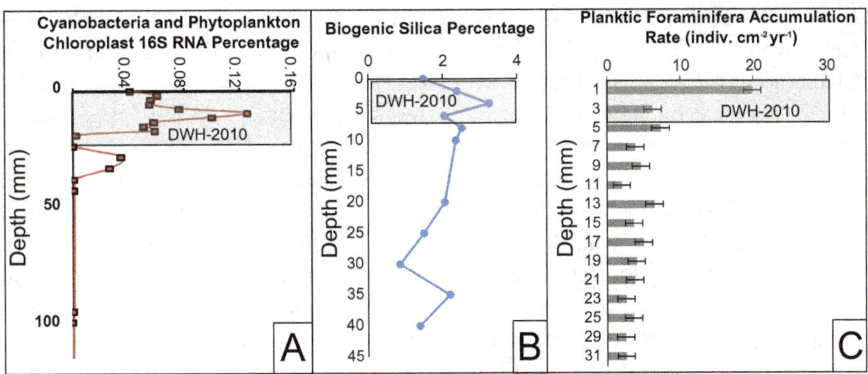

Fig. 13.2 Sedimentary records of surface-derived MOSSFA inputs with depth including (**a**) cyanobacteria and phytoplankton chloroplast 16 s RNA. (Modified from Brooks et al. 2015), (**b**) biogenic silica percentage. (Modified from Lee et al. 2018), and (**c**) planktic foraminifera accumulation rate (indiv. cm^{-2} yr^{-1}; modified from Fridrik et al. 2016). The gray-shaded regions are the depth intervals corresponding to 2010 according to short-lived radioisotope chronologies (Brooks et al. 2015)

polycyclic aromatic hydrocarbons (PAHs), which are indicative of surface burning response tactics (Romero et al. 2015). Surface sediments were radiocarbon depleted relative to deeper sediments, consistent with deposition of petrocarbon (Chanton et al. 2015). The characteristics and magnitude of deposition measured in the seafloor sedimentary records are consistent with those measured in sediment traps in the northern GoM (Yan et al. 2016; Stout and German 2017; Giering et al. 2018).

13.3 What Was the Extent of MOSSFA on the Seafloor?

Multiple approaches were employed to determine the spatial extent of MOSSFA on the deep-sea areas of the seafloor in the northern GoM following DWH. Overall, estimates from these various approaches indicate MOSSFA deposition over spatial extents in the deep sea that ranged from 1030 to 35,425 km^2 (Valentine et al. 2014; Chanton et al. 2015; Romero et al. 2015; Passow and Ziervogel 2016; Romero et al. 2017; Schwing et al. 2017a; Stout and German 2017) comprising between 3.7% and 14.4% of the total petroleum volume released that was not recovered or burned (3.19 million barrels, US District Court 2015). Using hopane and total PAH concentrations in surface sediments (relative to downcore sediments), Stout and German (2017) estimated that the deep-sea surface sediment spatial extent ranged from 1030 to 1910 km^2 and the sunken surface oil's spatial extent was 7600 km^2, which comprised 6.8–7.2% of the total petroleum released. Valentine et al. (2014) also utilized hopane concentrations in surface sediments to constrain an area of 3200 km^2. Both studies found that oil residues were primarily deposited proximal to the DWH wellhead with distribution extending toward the DeSoto Canyon to the northeast and toward the Mississippi Valley (i.e., Mississippi Canyon) to the southwest. Chanton et al. (2015) used radiocarbon as a tracer for petrocarbon deposition by identifying areas of depleted radiocarbon in surface sediments. Using this approach, they calculated that the seafloor extent of petrocarbon was 8400 km^2, comprising up to 14.4% of the total petroleum released and distributed primarily to the southwest of the DWH wellhead in the Mississippi Valley along with depletion apparent at sites in the DeSoto Canyon. Romero et al. (2017) used a suite of organic geochemical tracers to identify the deep-seafloor extent of oil residue (32,648 km^2) and found that deposition was primarily near the DWH wellhead and in the Mississippi Valley, comprising up to 3.7% of the total petroleum released that was not recovered or burned. Finally, Schwing et al. (2017a) utilized a short-lived radioisotope (excess ^{210}Pb) approach to constrain the spatial extent of increased sediment flux, which resulted in spatial seafloor coverage between 12,805 km^2 and 35,425 km^2. Again, the primary deep-sea areas that were impacted by increased sediment flux extended from the DWH wellhead southwest to the Mississippi Valley and northeast to the Desoto Canyon. The highest seafloor coverage estimates from these various studies (32,648–35,425 km^2) were consistent with the spatial extent of daily surface petroleum coverage 39,600 km^2 and suggest that MOSSFA affected broad sections of the continental slope and shelf where surface oil occurred (NRDA 2015; Stout et al. 2015). Despite the variability in spatial estimates between the various approaches, they all suggest that MOSSFA was focused in certain areas of the deep GoM, specifically surrounding the DWH wellhead, in the Mississippi Valley and the DeSoto Canyon. Bathymetric features such as the Mississippi Valley and the DeSoto Canyon likely acted as focusing factors for MOSSFA. Focusing of MOSSFA in these areas is also consistent with benthic biological impacts documented following DWH and should be taken into account during future oil spills (Montagna et al. 2013; Baguley et al. 2015: Schwing et al. 2015; Stout and German 2015; Schwing et al. 2017b, 2018a; Montagna and Girard 2020; Schwing and Machain Castillo 2020).

13.4 What Postdepositional Processes Took Place as a Result of MOSSFA?

Following DWH, MOSSFA in the northern GoM, there were several postdepositional sedimentary processes that occurred. These processes were largely due to the increased deposition of organic carbon (Brooks et al. 2015; Romero et al. 2015; Schwing et al. 2015; Yan et al. 2016; Chanton et al. 2018; Giering et al. 2018). Sedimentary total organic carbon (TOC) accumulation rates at sites up to 120 km away from the DWH wellhead increased by two orders of magnitude during DWH relative to downcore, pre-DWH measurements (Brooks et al. 2015; Romero et al. 2015). Yan et al. (2016) found an increase of up to an order of magnitude in organic carbon sedimentation rates in sediment traps during and directly following DWH compared to 2011, and ^{14}C depletion of this particulate organic carbon (POC) was observed for 3 years after the spill (Chanton et al. 2018). The increased TOC accumulation rates (including oil residue) delivered to the seafloor by MOSSFA caused an increase in the abundance of oil-degrading microbes (Kimes et al. 2013; Mason et al. 2014). Experimental evidence suggests that marine snow may have impeded the degradation of oil residue by up to 40% and that other MOSSFA biogenic organic material was preferentially degraded prior to oil-residue components (Rahsepar unpublished; Langenhoff et al. 2020). The increased respiration of organic carbon caused depleted surface sediment oxygen and intensification of reducing conditions up to 3 years following DWH (Hastings et al. 2016, 2020). The intensification of reducing conditions likely limited bioturbation, which may have in turn prolonged the reducing conditions in the surface sediment. There is evidence that bioturbation was present at the majority of the study sites throughout the northern GoM beginning in 2013 (Larson et al. 2018; Schwing et al. 2018a; Larson et al. 2020). The return of bioturbation in 2013 was consistent with the return to steady state of redox-sensitive metals (Fig. 13.3) from 2013 to 2016 (Hastings et al. 2016, 2020).

13.5   Can MOSSFA Be Preserved in the Sedimentary Record?

The preservation of oil residue in the sediments associated with DWH and the sedimentary event in the geologic record has been demonstrated using multiple tracers (Larson et al. 2020; Romero et al. 2020; Radović et al. unpublished; Larson et al. 2018). There was a preserved stratigraphic transition in grain size (finer or coarser depending on location) for the period of 2010–2011 in the majority of time series stations collected throughout the northern GoM relative to pre-DWH and post-DWH sedimentary intervals surrounding DWH (Larson et al. 2018; see Larson et al. 2020). Short-lived radioisotope profiles, specifically excess ^{210}Pb, were also able resolve the preservation of DWH, which manifested as a plateau in activity and was consistent with increased sedimentation (Larson et al. 2018, 2020). Radović et al. (unpublished) utilized polar lipid biomarkers, pigments, hydrocarbons, and stable isotopes of benthic foraminifera (Schwing et al. 2018b: Schwing and Machain Castillo 2020), to demonstrate the multi-year preservation of DWH in MOSSFA. The most recalcitrant DWH petroleum-derived compounds are likely to persist in the sedimentary record on the order of a decade if only degradation is taken into account (Romero et al. 2020; Gros et al. 2014; Turner et al. 2014; Romero et al. 2017). However, oxygenated weathering products derived from saturated and aromatic DWH compounds may persist longer (Aeppli et al. 2014; White et al. 2014). Regardless of degradation and weathering, downslope transport of sediments must also be taken into account, as discussed in Larson et al. (2020) and Romero et al. (2020).

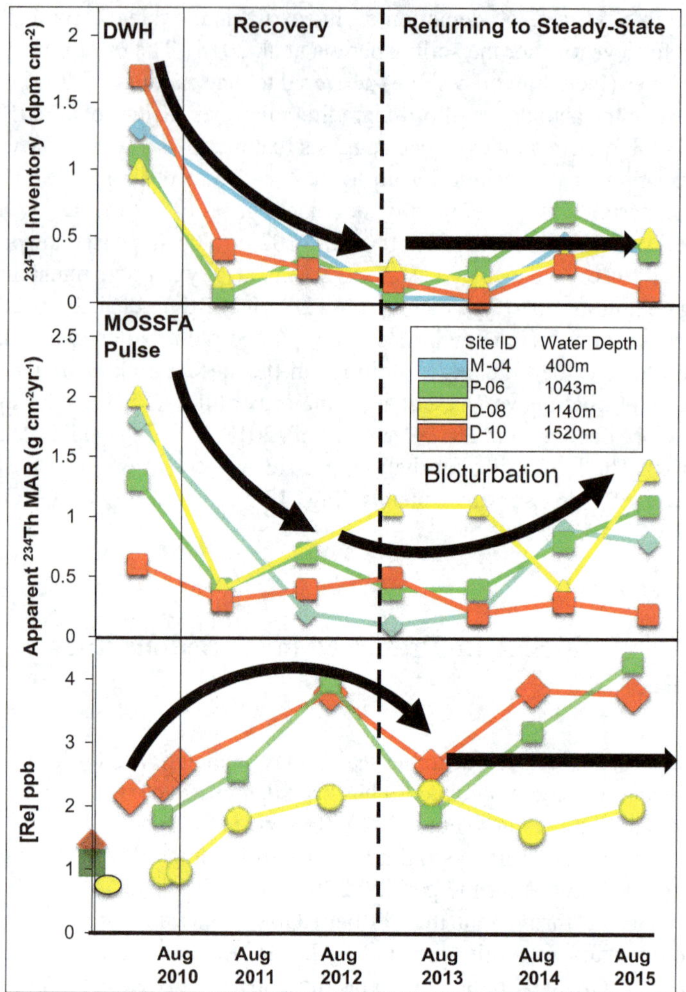

Fig. 13.3 Stacked plots of ^{234}Th inventory (disintegrations per minute per cm^2; modified from Larson et al. 2018), apparent ^{234}Th accumulation rate (g cm^{-2} yr^{-1}; modified from Larson et al. 2018), and rhenium concentration (modified from Hastings et al. 2016) from 2010 to 2015 for four time series sites in the northern GoM (M-04, blue; P-06, green; D-08, yellow; D-10, red). These plots detail the three postdepositional process periods of (1) MOSSFA flux, (2) recovery, and (3) return to steady state

To test the preservation potential of the various MOSSFA constituents (lithogenic, biogenic, petrogenic), sedimentary records from the southern GoM were assessed for any remaining evidence of Ixtoc 1 in 1979–1980. Preliminary visual inspection provided evidence that a complex laminated section with a preserved redox record (solid-state manganese) was preserved during the 1979–1980 interval in the sedimentary record (Fig. 13.4). These laminations and visible redox boundary were similar to those observed following DWH in the northern GoM.

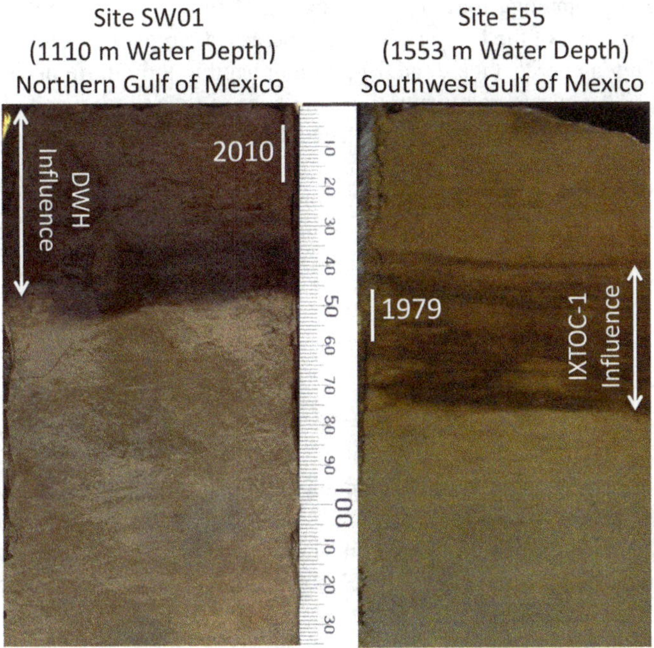

Fig. 13.4 Sediment core photographs comparing records (site SW01, collected at 1110 m water depth, photo credit: R. Larson) from the northern GoM and the southern GoM (site E55, collected at 1553 m water depth, photo credit: M. Machain). The white arrows are the depth interval influenced by the *Deepwater Horizon* event in the northern Gulf of Mexico and by Ixtoc 1 in the southern Gulf of Mexico, respectively. The white lines are the depth interval corresponding to the year 2010 in the northern GoM and 1979 in the southern GoM, respectively, determined by short-lived radioisotope chronology (Brooks et al. 2015; Sanchez-Cabeza 2016). Influences seen above and below the actual date of each event are due to carbon loading, redox processes, and redeposition

There were complex hydrocarbon backgrounds throughout many of the records in the southern GoM, which confounded fingerprinting petroleum inputs (Lincoln et al. unpublished; see Lincoln et al. 2020). At sites up to 80 km west of the Ixtoc 1 wellhead, terpane biomarkers, alkylated PAHs, and dibenzothiophenes provided evidence of Ixtoc 1 oil-residue input greater than 35 years after the spill (Lincoln et al. unpublished, 2020). Lincoln et al. (unpublished) and Radović et al. (unpublished) have been able to fingerprint hydrocarbons in deepwater sediments and relate them to Ixtoc 1, based on the available reservoir geochemistry data of Campeche basin petroleum systems but also by analyzing the reference Ixtoc 1 oil sample. Fourier transform ion cyclotron resonance mass spectrometry (FTICR-MS) was used to assess the transformation and preservation trends in heavy, non-GC amenable oil fractions (Radović et al. 2020). The analysis of putative Ixtoc 1-impacted stratigraphic horizon reveals abundant content sulfur containing residues, including poly-sulfur and oxy-sulfur species, likely decadal-scale residues of spilled oil. Using the same analytical tool, similar oxygen containing products of biodegradation have been discovered in reservoir settings providing another useful biogeochemical analogy for our understanding of subsurface oil spill fate. At sites up to 250 km west of the Ixtoc 1 wellhead, increased terpenoid accumulation rates, decreased retene percentages, and depletion in benthic foraminiferal calcite during the 1979–1980 interval relative to the remainder of the record were all consistent with MOSSFA (Schwing et al. unpublished; Schwing and Machain Castillo 2020). However, there were many variables that confound MOSSFA preservation in the southern GoM. These include other regional events such as volcanoes and hurricanes, other petroleum sources such as natural seeps and persistent oil and gas infrastructure, and highly variable bathymetry (e.g., Campeche canyon).

13.6 Conclusions

Considering the impact of MOSSFA on the GoM, these processes should be a new consideration for real-time oil spill response. In advance of future oil spills, numerical models need to be developed to predict the spatial and temporal formation of MOSSFA. MOSSFA should also be taken into account for response strategy decisions including freshwater discharge (clay particles and nutrients), surface burning, and dispersant application. MOSSFA could also potentially be used as concentrating mechanism of mineral-oil-biota aggregates as a response strategy. These factors are important for oil cleanup efforts and oil budget calculations.

These response decisions must also weigh the ecological trade-off between oiled coastlines and offshore benthic environments. Enhancing MOSSFA will likely result in increased impact in the offshore benthic environment, where MOSSFA constituents have proven to be recalcitrant. However, this may offer a way to predict the area of benthic ecosystem impact. If MOSSFA is transported to the continental slope and buried, then this may minimize impact and be the best alternative to remove oil from biologically active areas. Ultimately, federal and international policies should be discussed and modified with regard to the implications of MOSSFA prior to future oil spills.

Acknowledgments This research was made possible by a from the Gulf of Mexico Research Initiative/C-IMAGE. Data are publicly available through the Gulf of Mexico Research Initiative Information and Data Cooperative (GRIIDC) at https://data.gulfresearchinitiative.org/ (doi: 10.7266/n7-repn-q515, 10.7266/N7F18WRF, 10.7266/n7-jm62-5b12, 10.7266/n7-1k5d-4z68, 10.7266/n7-ynsd-gk73).

References

Aeppli C, Nelson RK, Radovic JR, Carmichael CA, Valentine DL, Reddy CM (2014) Recalcitrance and degradation of petroleum biomarkers upon abiotic and biotic natural weathering of Deepwater Horizon oil. Environ Sci Technol 48:6726–6734

Baguley J, Montagna P, Cooksey C, Hyland J, Bang H, Morrison C, Ricci M (2015) Community response of deep-sea soft-sediment metazoan meiofauna to the Deepwater Horizon blowout and oil spill. Mar Ecol Prog Ser 528:127–140. https://doi.org/10.3354/meps11290

Brooks GR, Larson RA, Schwing PT, Romero I, Moore C, Reichart GJ, Jilbert T, Chanton JP, Hastings DW, Overholt WA, Marks KP, Kostka JE, Holmes CW, Hollander D (2015) Sediment pulse in the NE Gulf of Mexico following the 2010 DWH blowout. PLoS One 10(7):e0132341. https://doi.org/10.1371/journal.pone.0132341

Chanton J, Zhao T, Rosenheim BE, Joye S, Bosman S, Brunner C, Yeager KM, Diercks AR, Hollander D (2015) Using natural abundance radiocarbon to trace the flux of Petrocarbon to the seafloor following the Deepwater Horizon oil spill. Environ Sci Technol 49(2):847–854. https://doi.org/10.1021/es5046524

Chanton JP, Giering SL, Bosman S, Rogers K, Sweet J, Asper V, Diercks AR, Passow U (2018) Isotope composition of sinking particles: oil effects, recovery and baselines in the Gulf of Mexico, 2010–2015. Elementa: Sci Anthrop https://doi.org/10.1525/elementa.298

Daly KL, Passow U, Chanton J, Hollander D (2016) Assessing the impacts of oil associated marine snow formation and sedimentation during and after the Deepwater Horizon oil spill. Anthropocene 12:18–33. https://doi.org/10.1016/j.ancene.2016.01.006

Draft Programmatic Environmental Impact Statement; Natural Resource Damage Assessment, National Ocean Service, National Oceanic and Atmospheric Administration (2015). http://www.Gulfspillrestoration.noaa.gov/wp-content/uploads/Chapter-4_Injury-to-Natural-Resources.pdf

ERMA (Environmental Response Management Application). ERMA Deepwater Gulf Response web application. http://gomex.erma.noaa.gov/

Fridrik EE, Schwing PT, Ramirez H, Larson RA, Brooks GR, O'Malley BJ, Hollander DJ (2016) Comparative records of Planktic Foraminiferal mass accumulation rates following the DWH and IXTOC events. In: Proceedings of the Gulf of Mexico oil spill and ecosystem science conference, Tampa

Giering SLC, Yan B, Sweet J, Asper V, Diercks A, Chanton J, Passow U (2018) The ecosystem baseline for particle flux in the northern Gulf of Mexico. Elementa: Sci Anthrop 6(1):6. https://doi.org/10.1525/elementa.264

Gracia A, Enciso Sánchez G, Alexander Valdés HM (2013) Composición y volúmen de contaminantes de las descargas costeras al Golfo de México. In: Botello AV, Rendón von Osten J, Benítez J, Gold-Boucht G (eds) Golfo de México. Contaminación e impacto ambiental: diagnóstico y tendencias. uac, unam-icmyl, cinvestav, Unidad Mérida

Griffiths SK (2012) Oil release from Macondo well MC252 following the Deepwater Horizon accident. Environ Sci Technol 46:5616–5622. https://doi.org/10.1021/es204569t

Gros J, Reddy CM, Aeppi C, Nelson RK, Carmichael CA, Smauel Arey J (2014) Resolving biodegradation patterns of persistent saturated hydrocarbons in weathered oil samples from the Deepwater Horizon disaster. Environ Sci Technol 48:1628–1637

Hastings DW, Schwing PT, Brooks GR, Larson RA, Morford JL, Roeder T, Quinn KA, Bartlett T, Romero IC, Hollander DJ (2016) Changes in sediment redox conditions following the BP DWH Blowout event. Deep-Sea Res II 2015. https://doi.org/10.1016/j.dsr2.2014.12.009

Hastings DW, Bartlett T, Brooks GR, Larson RA, Quinn KA, Razionale D, Schwing PT, Bernal LHP, Ruiz-Fernández AC, Sánchez-Cabeza J-A, Hollander DJ (2020) Changes in redox conditions of surface sediments following the *Deepwater Horizon* and Ixtoc 1 events (Chap. 16). In: Murawski SA, Ainsworth C, Gilbert S, Hollander D, Paris CB, Schlüter M, Wetzel D (eds) Deep oil spills: facts, fate, effects. Springer, Cham

Jernelöv A, Lindén O (1981) Ixtoc I: a case study of the world's largest oil spill. Ambio 10:299–306

Joye SB, MacDonald IR, Leifer I, Asper V (2011) Magnitude and oxidation potential of hydrocarbon gases released from the BP oil well blowout. Nat Geosci 4:160–165

Kimes NE, Callaghan AV, Aktas DF, Smith WL, Sunner J, Golding BT, Drozdowska M, Hazen TC, Suflita JM, Morris PJ (2013) Metagenomic analysis and metabolite profiling of deep-sea sediments from the Gulf of Mexico following the Deepwater Horizon oil spill. Front Microbiol, (MAR). https://doi.org/10.3389/fmicb.2013.00050

Langenhoff AM, Rahsepar S, van Eenennaam JS, Radović JR, Oldenburg TBP, Foekema E, Murk AJ (2020) Effect of marine snow on microbial oil degradation (Chap. 18). In: Murawski SA, Ainsworth C, Gilbert S, Hollander D, Paris CB, Schlüter M, Wetzel D (eds) Deep oil spills: facts, fate, effects. Springer, Cham

Larson RA, Brooks GR, Schwing PT, Diercks AR, Holmes CW, Chanton JP, Diaz-Asencio M, Hollander DJ (2020) Characterization of the sedimentation associated with the *Deepwater Horizon* blowout: depositional pulse, initial response, and stabilization (Chap. 14). In: Murawski SA, Ainsworth C, Gilbert S, Hollander D, Paris CB, Schlüter M, Wetzel D (eds) Deep oil spills: facts, fate, effects. Springer, Cham

Larson RA, Brooks GR, Schwing PT, Carter S, Hollander DJ (2018) High resolution investigation of event-driven sedimentation: Northeastern Gulf of Mexico. Anthropocene, 24, 40–50, 10.1016/j.ancene.2018.11.002.

Lee JJ, Schwing PT, Romero IC, Brooks GR, Larson RA, Overholt WA, Kostka JE, Hollander DJ (2018) Elevated rates of biogenic silica deposition during the Deepwater Horizon oil spill, Gulf of Mexico. In: Proceedings of the Gulf of Mexico oil spill and ecosystem science conference, New Orleans

Lincoln SA, Radović JR, Gracia A, Jaggi A, Oldenburg TBP, Larter SR, Freeman KH (2020) Molecular legacy of the 1979 Ixtoc 1 oil spill in deep-sea sediments of the southern Gulf of Mexico (Chap. 19). In: Murawski SA, Ainsworth C, Gilbert S, Hollander D, Paris CB, Schlüter M, Wetzel D (eds) Deep oil spills: facts, fate, effects. Springer, Cham

Mason OU, Scott NM, Gonzalez A, Robbins-Pianka A, Bælum J, Kimbrel J, Bouskill NJ, Prestat E, Borglin S, Joyner DC, Fortney JL, Jurelevicius D, Stringfellow WT, Alvarex-Cohen L, Hazen TC, Knight R, Gilbert JA, Jansson JK (2014) Metagenomics reveals sediment microbial community response to Deepwater Horizon oil spill. ISME J 8(7):1464–1475. https://doi.org/10.1038/ismej.2013.254

McNutt MK, Camilli R, Crone TJ, Guthrie GD, Hsieh PA, Ryerson TB, Savas O, Shaffer F (2012) Review of flow rate estimates of the Deepwater Horizon oil spill. Proc Natl Acad Sci 109:20260–20267. https://doi.org/10.1073/pnas.1112139108

Montagna PA, Girard F (2020) Deep-sea benthic faunal impacts and community evolution before, during and after the *Deepwater Horizon* event (Chap. 22). In: Murawski SA, Ainsworth C, Gilbert S, Hollander D, Paris CB, Schlüter M, Wetzel D (eds) Deep oil spills: facts, fate, effects. Springer, Cham

Montagna PA, Baguley JG, Cooksey C, Hartwell I, Hyde LJ, Hyland JL, Kalke RD, Kracker LM, Reuscher M, Rhodes ACE (2013) Deep-Sea benthic footprint of the Deepwater Horizon blowout. PLoS One 8(8):e70540. https://doi.org/10.1371/journal.pone.0070540

Passow U (2014) Formation of rapidly-sinking, oil-associated marine snow. Deep-Sea Res II 129:232–240. https://doi.org/10.1016/j.dsr2.2014.10.001

Passow U, Ziervogel K (2016) Marine snow sedimented oil released during the Deepwater Horizon spill. Oceanography 29(3):118–125. https://doi.org/10.5670/oceanog.2016.76

Passow U, Ziervogel K, Asper V, Diercks A (2012) Marine snow formation in the aftermath of the Deepwater Horizon oil spill in the Gulf of Mexico. Environ Res Lett 7:035301

Quigg A, Passow U, Daly KL, Burd A, Hollander DJ, Schwing PT, Lee K (2020) (Marine oil snow sedimentation and flocculent accumulation) events: past and present (Chap. 12). In: Murawski SA, Ainsworth C, Gilbert S, Hollander D, Paris CB, Schlüter M, Wetzel D (eds) Deep oil spills: facts, fate, effects. Springer, Cham

Radović JR, Jaggi A, Silva RC, Snowdon R, Waggoner DC, Hatcher PG, Larter SR, Oldenburg TBP (2020) Applications of FTICR-MS in oil spill studies (Chap. 15). In: Murawski SA, Ainsworth C, Gilbert S, Hollander D, Paris CB, Schlüter M, Wetzel D (eds) Deep oil spills: facts, fate, effects. Springer, Cham

Romero IC, Schwing PT, Brooks GR, Larson RA, Hastings DW, Ellis G, Goddard EA, Hollander DJ (2015) Hydrocarbons in Deep-Sea sediments following the 2010 Deepwater Horizon blowout in the Northeast Gulf of Mexico. PLoS One 10(5):e0128371. https://doi.org/10.1371/journal.pone.0128371

Romero IC, Toro-Farmer G, Diercks AR, Schwing PT, Muller-Karger F, Murawski S, Hollander DJ (2017) Large scale deposition of weathered oil in the Gulf of Mexico following a Deepwater oil spill. Environ Pollut 228:179–189. https://doi.org/10.1016/j.envpol.2017.05.019

Romero IC, Chanton JP, Rosenheim BE, Radović JR, Schwing PT, Hollander DJ, Larter SR, Oldenburg TBP (2020) Long-term preservation of oil spill events in sediments: the case for the *Deepwater Horizon* spill in the northern Gulf of Mexico (Chap. 17). In: Murawski SA, Ainsworth C, Gilbert S, Hollander D, Paris CB, Schlüter M, Wetzel D (eds) Deep oil spills: facts, fate, effects. Springer, Cham

Sanchez-Cabeza JA (2016) Recent Sedimentation in the Southern Gulf of Mexico. In: Proceedings of the Gulf of Mexico oil spill and ecosystem science conference, Tampa

Schwing PT, Machain-Castillo ML (2020) Impact and resilience of benthic foraminifera in the aftermath of the *Deepwater Horizon* and Ixtoc 1 oil spills (Chap. 23). In: Murawski SA, Ainsworth C, Gilbert S, Hollander D, Paris CB, Schlüter M, Wetzel D (eds) Deep oil spills: facts, fate, effects. Springer, Cham

Schwing PT, Romero IC, Brooks GR, Hastings DW, Larson RA, Hollander DJ (2015) A decline in Deep-Sea benthic foraminifera following the Deepwater Horizon event in the Northeastern Gulf of Mexico. PLoS One 10(3):e0120565. https://doi.org/10.1371/journal.pone.0120565

Schwing PT, Romero IC, Larson RA, O'Malley BJ, Fridrik EE, Goddard EA, Brooks GR, Hastings DW, Rosenheim BE, Hollander DJ, Grant G, Mulhollan J (2016) Sediment core extrusion method at millimeter resolution using a calibrated, threaded-rod. J Vis Exp 114

Schwing PT, Brooks GR, Larson RA, Holmes CW, O'Malley BJ, Hollander DJ (2017a) Constraining the spatial extent of the Marine Oil Snow Sedimentation and Accumulation (MOSSFA) following the DWH event using a $^{210}Pb_{xs}$ inventory approach. Environ Sci Technol 51:5962–5968. https://doi.org/10.1021/acs.est.7b00450

Schwing PT, O'Malley BJ, Romero IC, Martinez-Colon M, Hastings DW, Glabach MA, Hladky EM, Greco A, Hollander DJ (2017b) Characterizing the variability of benthic foraminifera in the Northeastern Gulf of Mexico following the Deepwater Horizon event (2010–2012). Environ Sci Pollut Res 24:2754. https://doi.org/10.1007/s11356-016-7996-z

Schwing PT, O'Malley BJ, Hollander DJ (2018a) Resilience of benthic foraminifera in the northern Gulf of Mexico following the Deepwater Horizon event (2011–2015). Ecol Indic 84:753–764. https://doi.org/10.1016/j.ecolind.2017.09.044

Schwing PT, Chanton JP, Romero IC, Hollander DJ, Goddard EA, Brooks GR, Larson RA (2018b) Tracing the incorporation of petroleum carbon into benthic foraminiferal calcite following the Deepwater Horizon event. Environ Pollut 237:424–429. https://doi.org/10.1016/j.envpol.2018.02.066

Stout AS, German CR (2015) Characterization and flux of marine oil snow into the Visca knoll (Lophelia reef) area due to the Deepwater Horizon oil spill. US Department of the Interior, Deepwater Horizon Response & Restoration, Administrative Record, 34. http://pub-dwhdatadiver.orr.noaa.gov/dwh-ar-documents/946/DWH-AR0039084.pdf

Stout SA, German CR (2017) Characterization and flux of marine oil snow settling toward the sea floor in the northern Gulf of Mexico during the Deepwater Horizon incident : evidence for input from surface oil and impact on shallow shelf sediments. Mar Pollut Bull 129:695–713. https://doi.org/10.1016/j.marpolbul.2017.10.059

Stout AS, Rouhani S, Liu B, Oehrig J (2015) Spatial extent ("Footprint") and volume of Macondo oil found on the deep-sea floor following the Deepwater Horizon oil spill. US Department of the Interior, Deepwater Horizon Response & Restoration, Administrative Record, DWH-AR0260244, 29, http://pub-dwhdatadiver.orr.noaa.gov/dwh-ar-documents/946/DWH-AR0260244.pdf

Sun S, Hu C, Tunnell JW (2015) Surface oil footprint and trajectory of the Ixtoc-I oil spill determined from Landsat/MSS and CZCS observations. Mar Pollut Bull 101(2):632–641. https://doi.org/10.1016/j.marpolbul.2015.10.036

Turner RE, Overton EB, Meyer BM, Miles MS, McClenachan G, Hooper-Bui L, Engel AS, Swenson EM, Lee JM, Milan CS, Gao H (2014) Distribution and recovery trajectory of Macondo (Mississippi canyon 252) oil in Louisiana coastal wetlands. Mar Pollut Bull 87:57–67

U.S. District Court Findings of Facts and Conclusions of Law –Phase 2 Trial. Case 2: 10-md-02179-cjb-ss, pp 1e44. Document 14021 Filed Jan. 15, 2015. http://www.laed.uscourts.gov/sites/default/files/OilSpill/Orders/1152015FindingsPhaseTwo.pdf

Valentine DL, Fisher GB, Bagby SC, Nelson RK, Reddy CM, Sylva SP (2014) Fallout plume of submerged oil from Deepwater Horizon. Proc Natl Acad Sci 111:15906–15911. https://doi.org/10.1073/pnas.1414873111

Vonk SM, Hollander DJ, Murk ATJ (2015) Was the extreme and wide-spread marine oil-snow sedimentation and flocculent accumulation (MOSSFA) event during the Deepwater Horizon blowout unique? Mar Pollut Bull 100(1):5–12. https://doi.org/10.1016/j.marpolbul.2015.08.023

White HK, Lyons SL, Harrison SJ, Findley DM, Liu Y, Kujawinski EB (2014) Long-term persistence of dispersants following the Deepwater Horizon oil spill. Environ Sci Techn Let 1(7):295–299

Yan B, Passow U, Chanton JP, Nothig EM, Asper V, Sweet J, Pitianggon M, Diercks A, Pak D (2016) Sustained deposition of contaminants from the Deepwater Horizon spill. Proc Natl Acad Sci 113:E3332. https://doi.org/10.1073/pnas.1513156113

Ziervogel K, Mckay L, Rhodes B, Osburn CL, Dickson-Brown J, Arnosti C, Teske A (2012) Microbial activities and dissolved organic matter dynamics in oil-contaminated surface seawater from the Deepwater Horizon oil spill site. PLoS One 7(4):e34816

Chapter 14
Characterization of the Sedimentation Associated with the *Deepwater Horizon* Blowout: Depositional Pulse, Initial Response, and Stabilization

Rebekka A. Larson, Gregg R. Brooks, Patrick T. Schwing, Arne R. Diercks, Charles W. Holmes, Jeffrey P. Chanton, Misael Diaz-Asencio, and David J. Hollander

Abstract The *Deepwater Horizon* (DWH) blowout led to a depositional pulse in the northeast Gulf of Mexico in the Fall of 2010 associated with an observed Marine Oil Snow Sedimentation and Flocculent Accumulation (MOSSFA) event. A time series (2010–2016) of annually collected sediment cores at four sites characterize the sedimentary response to the event, post-event, and stabilization/recovery. The depositional pulse (2010–2011) was characterized by high sedimentation rates with

R. A. Larson (✉)
Eckerd College, Department of Marine Science, St. Petersburg, FL, USA

University of South Florida, College of Marine Science, St. Petersburg, FL, USA
e-mail: larsonra@eckerd.edu; ralarso2@mail.usf.edu

G. R. Brooks
Eckerd College, Department of Marine Science, St. Petersburg, FL, USA
e-mail: brooksgr@eckerd.edu

P. T. Schwing · D. J. Hollander
University of South Florida, College of Marine Science, St. Petersburg, FL, USA
e-mail: pschwing@mail.usf.edu; davidh@usf.edu

A. R. Diercks
University of Southern Mississippi, School of Ocean Science and Engineering, Stennis Space Center, MS, USA
e-mail: Arne.Diercks@usm.edu

C. W. Holmes
Environchron, Tallahassee, FL, USA
e-mail: cwholmes@environchron.com

J. P. Chanton
Florida State University, Department of Earth, Ocean and Atmospheric Science, Tallahassee, FL, USA
e-mail: jchanton@fsu.edu

M. Diaz-Asencio
Ensenada Center for Scientific Research and Higher Education, Ensenada, BC, Mexico

© Springer Nature Switzerland AG 2020
S. A. Murawski et al. (eds.), *Deep Oil Spills*,
https://doi.org/10.1007/978-3-030-11605-7_14

little to no bioturbation and large excursions in % silt. The lack of changes in sediment composition indicate that the same sediment sources dominated during the event, but the rates of sedimentation increased. In the years following the event (2011–2012), sedimentation rates were lower, and bioturbation was absent, and the initial excursions in % silt began to become undetectable in the sedimentary record. Between 2013 and 2016, a spatially and temporally variable return of bioturbation was detected at most sites. Sedimentation rates at all sites remained low, but increases in $^{234}Th_{xs}$ apparent mass accumulation rates indicated a return of bioturbation and potential stabilization and/or recovery of the sedimentary system. The deepest site (~1500 m) did not have any indication of bioturbation as of the 2016 collections, which may reflect a lack of recovery or that bioturbation was never present. In 2012, $^{210}Pb_{xs}$ age dating began to resolve the depositional pulse suggesting it may be applied to determine changes in the pulse deposit over time, and/or its preservation in the sedimentary record. Factors that may influence preservation include burial, bioturbation, degradation of the pulse signature, and remobilization of pulse sediments.

Keywords Sediment · Chronology · MOSSFA · Short-lived radioisotopes · Sedimentation

14.1 Introduction

The *Deepwater Horizon* (DWH) blowout resulted in multiple sequences of events leading to the deposition of sediment and oil-contaminated sediment to the deep-sea benthos and subsequent integration into the sedimentary system. Beginning in April 2010, the DWH blowout led to the release of oil and gas at a water depth of ~1500 m in the northeastern Gulf of Mexico (NEGoM) for a duration of 87 days (Passow and Hetland 2016). The released oil formed a rising plume from the leak at the seafloor to the sea surface, a subsurface plume at ~1000 m water depth (Diercks et al. 2010; Joye et al. 2011), and a sea surface oil slick (Thibodeaux et al. 2011). A variety of strategies to stop the release of oil and mitigate impacts were utilized to contain, remove, and influence the distribution and degradation (biotic and abiotic) of released oil (US Coast Guard 2010; BOEM 2011). This included the addition of dispersants (Corexit) at the wellhead as well as at sea surface slicks (Yan et al. 2016), skimming and burning at the sea surface, water release from the Mississippi River, and the addition of drilling mud at the wellhead (Liu et al. 2018).

The DWH blowout occurred in a region that is sedimentologically complex (Fig. 14.1a). West of the DeSoto Canyon seafloor sediments are dominantly siliciclastic, associated with Mississippi River discharge. East of the DeSoto Canyon seafloor sediments are dominantly carbonate, associated with the west Florida Platform and low river influence (Balsam and Beeson 2003; Holmes 1976; Harbison 1968). This resulted in the potential impacts of oil released in both siliciclastic and carbonate sedimentary regimes.

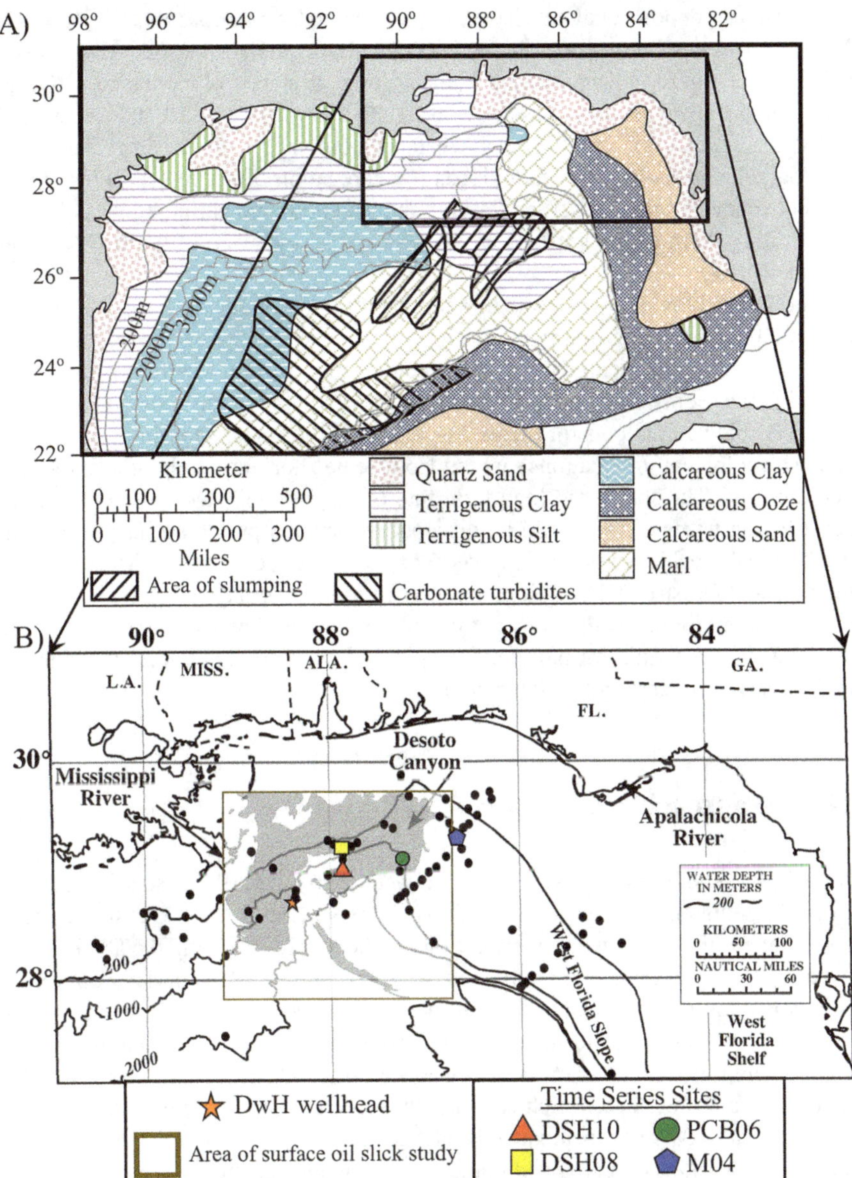

Fig. 14.1 Maps of the Gulf of Mexico (GoM) showing the physiographic features: (**a**) Map indicating the study area in the NEGoM with sedimentologic regions. (Modified from Brooks et al. 2020; Balsam and Beeson 2003; Holmes 1976), (**b**) map of NEGoM showing location of the DWH wellhead (orange star), surface sediment slick (gray shading in brown box, from Garcia-Pineda et al. 2013) and C-IMAGE sediment core locations (black circles) and time-series sites (DSH10 = red triangle, DSH08 = yellow square, PCB06 = green circle, and M04 = blue pentagon). (Modified from Brooks et al. 2015)

A primary depositional mechanism for oiled sedimentation was the observed Marine Oil Snow Sedimentation and Flocculent accumulation (MOSSFA), which consisted of surface oil interacting with biology, primarily phytoplankton (Quigg et al. 2020; Daly et al. 2016; Passow et al. 2016; Ziervogel et al. 2012; Passow et al. 2012). The interaction of this sticky substance combined with oil, dispersants, and clay particles in the water column leads to aggregation of marine oil snow, which lost buoyancy and rapidly settled through the water column, stripping particles and transporting them to the seafloor as MOSSFA (Passow et al. 2012; Brooks et al. 2015).

The DWH blowout event was of short duration (geologically), on the scale of months. Understanding the manifestation of this event in the sedimentary system and impacts to the benthos required rapid response and adaptive approaches. Studying the sedimentary system before, during, and after an oil spill event, such as the DWH blowout, allows for the assessment of benthic sedimentological, biological, and ecological implications ranging from acute/short-term to long-term/permanent impacts. The short- and long-term fate of oil-contaminated sediments can be described in terms of (1) the initial distribution patterns providing insight into the depositional mechanism(s) and source(s) of sedimented oil, such as MOSSFA, (2) the potential redistribution of oil-contaminated sediments and secondary deposition in benthic environments that may not have been initially impacted, and (3) the ultimate fate and potential for burial and sequestration of oil-contaminated sediments in deep-sea sediments and by the benthos.

14.2 Approach/Methods

A total of 179 sediment cores collected between 2010 and 2016 at 80 sites to investigate the sedimentary impacts of the DWH blowout (Fig. 14.1b). Due to the "real-time" nature of the sediment investigations immediately following the DWH event, it was expected that impacted sediments would be at the surface of the seafloor and core collection with an intact sediment water interface was of critical importance. Therefore, cores were collected using an Ocean Instruments MC-800 multicorer as it delicately collects sediment cores up to ~60 cm in length while preserving the sediment water interface, which is generally sufficient for capturing sedimentation over the past ~100 years to adequately assess baseline sedimentation patterns. Also, the MC-800 collects 8 cores simultaneously allowing for interdisciplinary studies (sedimentology, biology, chemistry, etc.) to fully characterize the benthic response and evolution during and following the event.

14.2.1 Time-Series Approach/High-Resolution Sampling

A time series of annually collected cores from the same sites allowed for the detection of the event in the sedimentary record, the annual-scale evolution of the sedimentary deposit over the subsequent years, and the potential burial and preservation in the sedimentary system. In the absence of any baseline information, such as for the DWH blowout, a time-series approach provides a post-event baseline. This approach also helps identify how indicators of oil-contaminated sediments may change over time in the sedimentary record. It also can assist in determining the preservation potential of oil-contaminated sediments including influences of burial, bioturbation/mixing, postdepositional alteration, and resuspension/erosion.

Both the siliciclastic and carbonate sedimentary regimes are represented in the time series with sites DSH08 and DSH10 located on the siliciclastic side of the Desoto Canyon and sites PCB06 and M04 located on the carbonate side of the DeSoto Canyon (Fig. 14.1). High-resolution sub-sampling of cores provided high temporal resolution to define pre-event baselines (downcore), the event, and post-event sedimentation. Sub-sampling of cores at 2 mm resolution by extrusion methods described by Schwing et al. (2016) provided the highest temporal resolution possible for analyses.

14.2.2 Chronometers: Timing of Deposition

Defining the timing of sedimentation is critical to correlate the deviations in the sedimentary record with the DWH event. The most commonly used chronometers for recent deposition are short-lived radioisotopes including excess ^{234}Th (^{234}Th$_{xs}$), excess ^{210}Pb (^{210}Pb$_{xs}$), ^{137}Cs, and ^{7}Be (Swarzenski 2014; Holmes 1998; Appleby 2001). Ideally, multiple chronometers would be utilized as they each have strengths and limitations. Sediment cores collected from the deep-sea oceanic setting (all analyzed to date) did not contain detectable ^{137}Cs or ^{7}Be, so they are not viable in this environment as chronometers or for corroborating other age dating techniques. Therefore, ^{234}Th$_{xs}$ and ^{210}Pb$_{xs}$ are the primary chronometers for defining sedimentation on monthly (^{234}Th$_{xs}$) and annual/decadal (^{210}Pb$_{xs}$) time scales. Sediment core samples were analyzed for short-lived radioisotopes by gamma spectrometry on Series HPGe (high-purity germanium) Coaxial Planar Photon Detectors for total ^{210}Pb (46.5 keV), ^{214}Pb (295 keV and 351 keV), ^{214}Bi (609 keV), and ^{234}Th (63 keV) in order to determine ^{234}Th$_{xs}$ and ^{210}Pb$_{xs}$ activities expressed as disintegrations per minute per gram (dpm/g) using methodology described by Brooks et al. (2015) and Larson et al. (2018).

The short duration of the sedimentation pulse required a high temporal resolution chronometer, such as ^{234}Th$_{xs}$, to identify the most recent sedimentation that would be associated with the event. Uranium-238 is soluble in seawater and behaves conservatively with salinity (Chen et al. 1986), and the decay product, ^{234}Th$_{xs}$, strongly adsorbs onto particles producing a flux of $^{234}Th_{xs}$ to the seafloor associated with sedimentation (Winkler and Rosner 2000). $^{234}Th_{xs}$ has a short half-life of ~24 days and is usually only detectable in surficial sediments representing sedimentation and/or bioturbation. Due to the short (~24 day) half-life, $^{234}Th_{xs}$ has traditionally been used to determine the depth of bioturbation (McClintic et al. 2008; Pope et al. 1996; Yeager et al. 2004), but under certain conditions (high sedimentation, high sampling resolution, stratigraphic integrity), it may be used as a chronological tool (Brooks et al. 2015). Activities of $^{234}Th_{xs}$ were decay corrected (DC $^{234}Th_{xs}$) for activity lost due to decay between the time of core collection and sample analysis, and data are expressed as Inventories and as mass accumulation rates (MAR) (Brooks et al. 2015; Larson et al. 2018). $^{234}Th_{xs}$ Inventory is the sum of activity in a core, which is an indicator of the flux of $^{234}Th_{xs}$ to the seafloor associated with sedimentation and is not influenced by bioturbation (Brooks et al. 2015). $^{234}Th_{xs}$ MAR is a function of the $^{234}Th_{xs}$ activity in a sediment core, as well as the depth downcore of the $^{234}Th_{xs}$ profile, and can be influenced by both sedimentation and bioturbation (downcore mixing deepening the $^{234}Th_{xs}$ profile). By utilizing the $^{234}Th_{xs}$ Inventory as a measure of sedimentation, the $^{234}Th_{xs}$ MAR can be used as an indicator of bioturbation/mixing in surficial sediments. Specifically, high Inventory with a high MAR indicates high sedimentation, and low Inventory with a low MAR indicates low sedimentation. The combination of a low Inventory with a high MAR indicates low sedimentation with the influence of bioturbation/mixing. Under these conditions, MAR is not an accurate indicator of sedimentation/accumulation due to the influence of bioturbation and is referred to as apparent MAR (A-MAR).

With a 22.3-year half-life, $^{210}Pb_{xs}$ provides age control and sedimentation rates over the past ~120 years but is less sensitive to monthly time scales (Chisté and Bé 2007; Appleby 2001). Therefore, $^{210}Pb_{xs}$ age dating is not as useful immediately following an event but becomes more useful as the primary dating tool in years to decades following a sedimentation event. The constant rate of supply (CRS) model, which is appropriate under conditions of varying accumulation rates, is used to provide $^{210}Pb_{xs}$ age dating for pre-event downcore baselines and to detect the sedimentation pulse in subsequent years (2012–2016) (Appleby 2001; Binford 1990; Appleby and Oldfield 1983).

14.2.3   Sediment Texture and Composition

Sediment texture and composition analyses were conducted on extruded samples and included grain size by methodology described by Folk (1965) and calcium carbonate content (%CaCO$_3$) by methodology by Milliman (1974) to determine variability in sediment sources and depositional mechanisms. Grain size is presented as % silt as this was the most indicative of changes in sediment texture (Larson et al. 2018). There are multiple indicators of oil-contaminated sediments and the benthic impacts including organic and inorganic geochemical indicators and biological responses. Organic geochemical fingerprinting is utilized to identify oil-contaminated sediments and specifically isolate DWH oil contamination and is described by Romero et al. (2020). Inorganic impacts and sedimentary signatures can include shifts in redox geochemistry as described by Hastings et al. (2020). Biological indicators of oil-contaminated sedimentation and benthic impacts also include benthic foraminifera by Schwing et al. (2020) and macrofaunal by Montagna et al. (2020).

14.3 Sedimentary Response: Depositional Pulse (2010–2011)

The sedimentary response of the DWH event was manifested in sediment cores collected in *the* Fall of 2010 through early 2011 as increased rates of sedimentation and subtle changes in sedimentology at the time-series sites (DSH08, DSH10, PCB06, M04). This depositional pulse associated with MOSSFA was characterized by high $^{234}Th_{xs}$ Inventories and high $^{234}Th_{xs}$ MAR, indicating high sedimentation, over the surficial 10–20 mm (Fig. 14.2). Downcore $^{210}Pb_{xs}$ geochronologies recorded MAR an order of magnitude lower than surficial rates based on $^{234}Th_{xs}$ (Brooks et al. 2015). However, rates calculated by different methods representing different time frames cannot be directly compared due to the differences in time scales involved, which is commonly referred to as the Sadler effect (Sadler 1981). Where $^{234}Th_{xs}$ represents monthly time-scale sedimentation, $^{210}Pb_{xs}$ is more reflective of annual- to decadal-scale sediment accumulation. This leads to a consistently higher estimation of MAR by $^{234}Th_{xs}$ as compared to $^{210}Pb_{xs}$ (Sadler 1981) and provides evidence to support the importance of comparisons using the same chronometer. As there were no pre-event $^{234}Th_{xs}$ data available for direct comparison, the high $^{234}Th_{xs}$ Inventory and MAR measured in 2010 and early 2011, associated with the depositional pulse, initially could not be confirmed (Brooks et al. 2015). The continued measurement of $^{234}Th_{xs}$ Inventory and MAR over the 6-year time series allowed for the confirmation of the depositional pulse and evaluation of sedimentation (Inventory and MAR) and bioturbation (A-MAR) as the sedimentary system evolved following the DWH event.

There were no major systematic changes in sedimentological parameters over the 10–20 mm-thick pulse layer as compared to underlying sediments, the exception to this being silt content (% silt), which exhibited major excursions in the surfaces of the time-series cores collected in late 2010 and early 2011 (Fig. 14.3). The % silt increased dramatically in some cores and decreased dramatically in others as compared to consistent downcore (pre-event) values, suggesting a deviation in sedimentation processes during 2010 and 2011 concurrent with the depositional pulse (Brooks et al. 2015). Below the surficial 10–20 mm (i.e., the depositional pulse layer), silt content did not vary, indicating reproducibility of high-resolution analyses and little spatial heterogeneity and suggesting a relatively stable sedimentological regime for >100 years preceding the depositional pulse (Larson et al. 2018). Composition, as represented by carbonate content (%carbonate), showed no systematic or detectable changes within the surficial 10–20 mm of the time-series cores

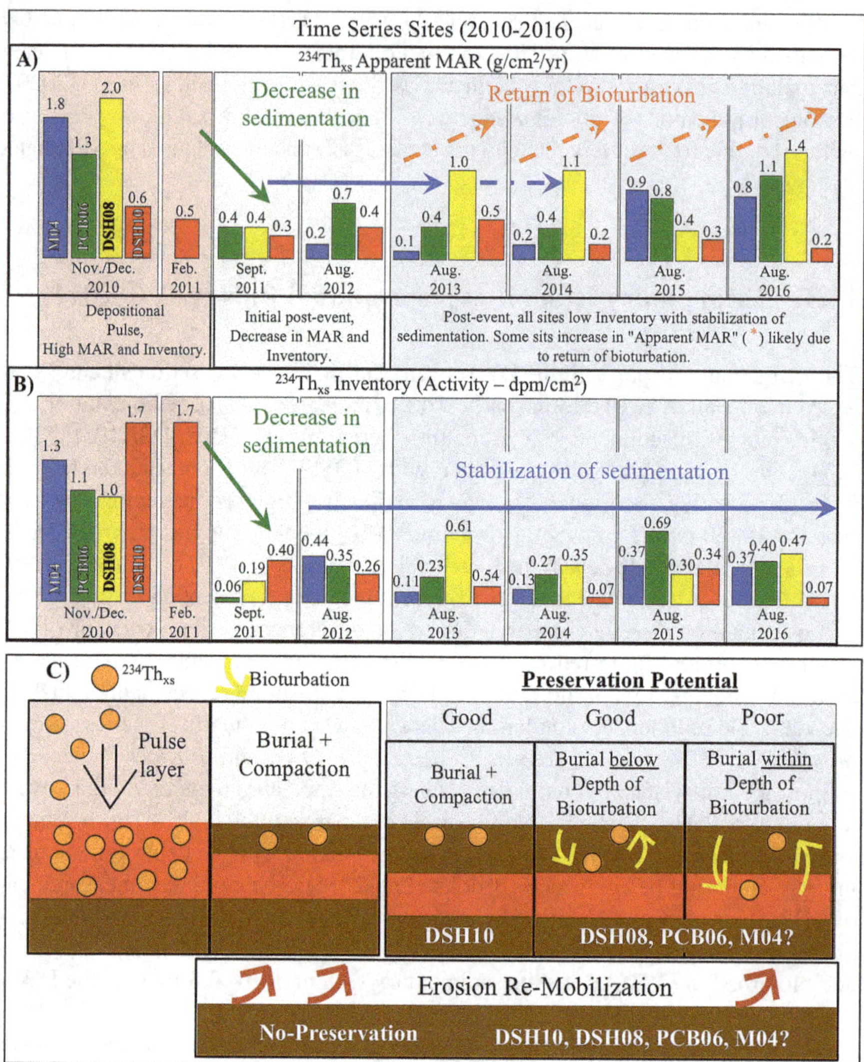

Fig. 14.2 2010–2016 time-series sites (**a**) ^{234}Th$_{xs}$ A-MAR, (**b**) ^{234}Th$_{xs}$ Inventory, showing the evolution of the sedimentation pulse (2010–Feb 2011) indicated by higher ^{234}Th$_{xs}$ Inventory and MAR, initial post pulse impact and response (2011–2012) with decrease in ^{234}Th$_{xs}$ Inventory and MAR, and stabilization with lower ^{234}Th$_{xs}$ Inventory with variable return of bioturbation/mixing (higher ^{234}Th$_{xs}$ A-MAR) at all sites except DSH10. (**c**) Diagram showing the influences on preservation potential of the pulse layer in the sedimentary record, with highest potential associated with burial and no bioturbation (DSH10) and poorest with bioturbation reaching the depth downcore of the pulse layer and remobilization of the pulse layer leading to no preservation at the sites of initial deposition (2010–2011). (Modified from Larson et al. 2018)

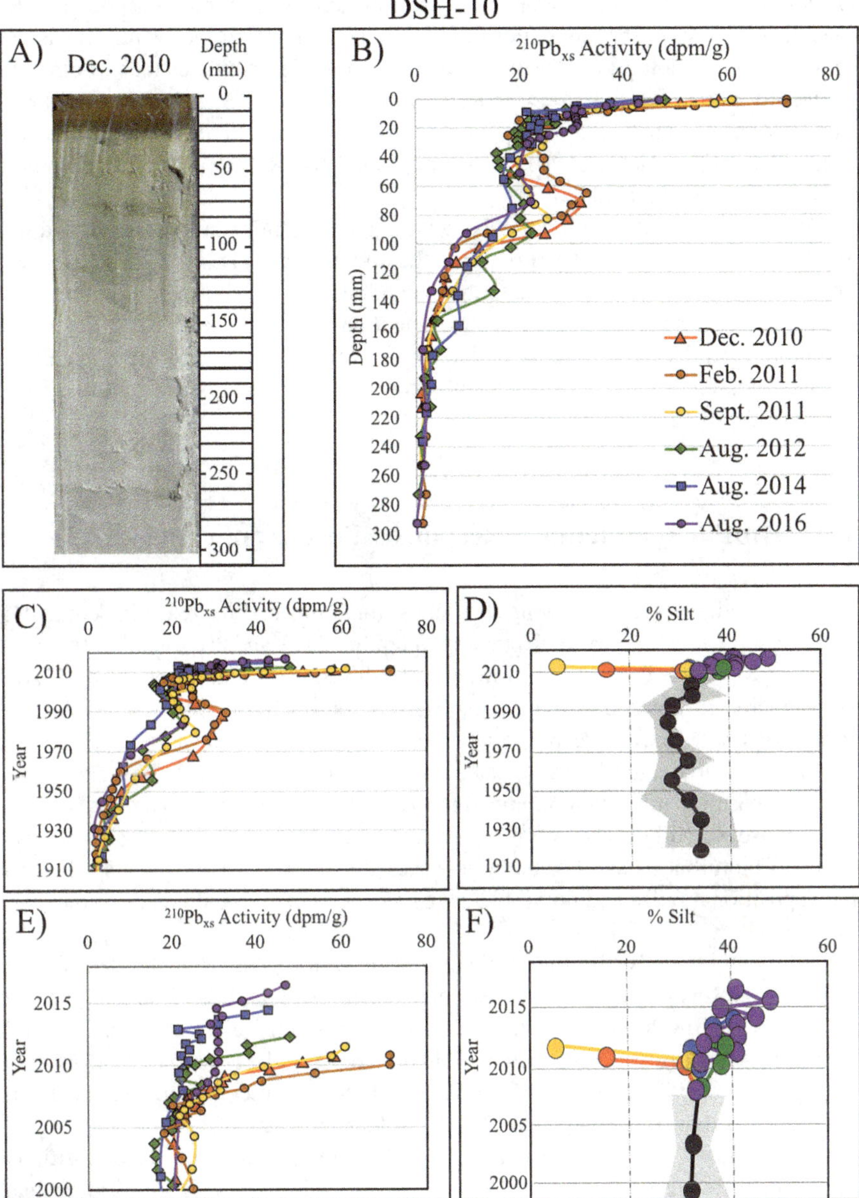

Fig. 14.3 Core data for site DSH10. (**a**) Core photograph for December 2010 collection. (**b**) ^{210}Pb$_{xs}$ profiles vs depth for all time-series collections. Note the reproducibility of profiles. (**c**) ^{210}Pb$_{xs}$ profiles vs year (CRS model). (**d**) % Silt profiles vs year for time series with averaged values and standard deviation shown as black line/circles and gray shading. (Modified from Larson et al. 2018). (**e**) ^{210}Pb$_{xs}$ profiles vs year since 2000 for time series showing evolution of ^{210}Pb$_{xs}$ signature of the depositional pulse. Note 2014 and 2016 cores have intervals of similar activity, which is possibly the detection of the depositional pulse. (**f**) % Silt profiles vs year since 2000 with decreases in % Silt in 2010 and 2011 cores associated with the depositional pulse, which is not detected in subsequent years. (Modified from Larson et al. 2018)

(Brooks et al. 2015). This indicates no detectable changes in sediment source(s) for the siliciclastic or carbonate regions of the study area during the event. This is due to MOSSFA stripping the overlying water column of particulates that would have eventually settled to the seafloor in each of the two sediment regimes, so composition may not be expected to change measurably even though the rate of sedimentation increased (Brooks et al. 2015; Chanton et al. 2014; Romero et al. 2015; Romero et al. 2017; Schwing et al. 2017).

Detection of the depositional pulse within a 4–5-month period that resulted in a 10–20 mm thick sediment layer required the unconventional use of $^{234}Th_{xs}$ as a geochronological tool. This was due to the exceptionally high (2 mm)-resolution sampling, the high sedimentation rate for detection of $^{234}Th_{xs}$ profiles to determine Inventories and MAR, as well as the interruption of bioturbation by the event allowing for detection of the depositional pulse by $^{234}Th_{xs}$ and % silt (Brooks et al. 2015) (Figs. 14.2 and 14.3).

14.4 Initial Sedimentary Response: Post-event (2011–2012)

In the immediate years following the depositional pulse, a decrease in sedimentation was indicated by cores collected in September 2011 and 2012 at all of the time-series sites. This is expressed as lower $^{234}Th_{xs}$ Inventories and lower $^{234}Th_{xs}$ MARs as compared to the depositional pulse (Larson et al. 2018) (Fig. 14.2). As these Inventories and MARs are based upon the same chronometer ($^{234}Th_{xs}$), they are directly comparable to those associated with the depositional pulse. This also serves as initial baseline information indicating that the sedimentation rates of the depositional pulse were high. The low $^{234}Th_{xs}$ Inventory and low $^{234}Th_{xs}$ MAR also indicate a lack of bioturbation at all of the time-series sites for the first few years following the depositional pulse and possible lack of recovery of the benthic ecosystem over this period.

Sedimentologically, the subtle diagnostic indicator (% silt) of the depositional pulse was not as detectable in cores collected in 2011 and 2012 as compared to those collected in 2010 (Fig. 14.3). Most cores exhibited a stabilization in % silt toward pre-event (downcore) values. This may be an indication that % silt is not a strong diagnostic indicator of MOSSFA in sedimentary records, potentially due to challenges in detecting changes in % silt, and/or preservation (Fig. 14.2).

Time-series cores collected in August 2012 showed that the 2010 depositional pulse began to be resolved by $^{210}Pb_{xs}$ (half-life ~22.3 years) in little more than 2 years following the event (Fig. 14.3). $^{210}Pb_{xs}$ profiles began to reflect several adjacent depth intervals with similar $^{210}Pb_{xs}$ activities (dpm/g), indicating intervals were of approximately the same age and deposited at the same time (Larson et al. 2018). In the surface intervals of cores collected in 2010, the $^{210}Pb_{xs}$ profile is more exponential in shape, lacking intervals with similar activities (dpm/g) illustrating the ineffectiveness of $^{210}Pb_{xs}$ to resolve the depositional pulse.

14.5   Stabilization/Recovery: Post-event (2013–2016)

Beginning ~3 years after the DWH blowout and associated depositional pulse, there were indications of variability in the evolution of the sedimentary system as it stabilized/recovered. At the time-series sites collected between 2013 and 2016, lower $^{234}Th_{xs}$ Inventories did remain consistent with the postdepositional pulse Inventories for 2011 and 2012 indicating continued lower rates of sedimentation (Fig. 14.2). The $^{234}Th_{xs}$ MARs exhibited specific increases at different collection years. Site DSH08 exhibited an apparent increase in $^{234}Th_{xs}$ MAR (A-MAR) beginning in August 2013, while sites M04 and PCB06 recorded apparent increases in $^{234}Th_{xs}$ MAR (A-MAR) beginning in August 2015. These A-MAR are not supported by the relatively stable $^{234}Th_{xs}$ Inventories from late 2011 through August 2016 (Larson et al. 2018) (Fig. 14.2) and therefore are not a reflection of increased sedimentation rates, but are likely due to the re-establishment of bioturbation. Bioturbation mixes $^{234}Th_{xs}$ downward, deepening the $^{234}Th_{xs}$ profile downcore and falsely increasing MARs (Larson et al. 2018). It is also unknown the reason(s) that bioturbation was re-established at specific sites at different times. Site DSH10 was the exception as $^{234}Th_{xs}$ MARs remained low through 2016 indicating low sedimentation rates and a lack of bioturbation (Larson et al. 2018) (Fig. 14.2). Whether bioturbation was never present at DSH10 (pre-event) or if this site remains in transition and/or has not recovered from the depositional pulse is unknown. The lack of bioturbation at site DSH10 increases the preservation potential of the depositional pulse at site DSH10 (Fig. 14.2). Alternatively, the return of bioturbation at sites DSH08, PCB06, and M04 decreases the preservation potential of the depositional pulse depending on if the depositional pulse layer is buried below the depth of bioturbation (Fig. 14.2).

The silt content continues to become more consistent/stable and similar to pre-event values in 2013–2016 cores as compared to the 2010–2011 event signature. This reinforces the time-sensitive nature of using % silt as an indicator of MOSSFA as it does not appear to be well preserved and/or detectable in the sedimentary record. In contrast, $^{210}Pb_{xs}$ dating continues to progressively resolve the depositional pulse beginning in the 2012 collections through the 2016 collections (Larson et al. 2018) (Fig. 14.3) with sequential intervals of similar activity capped by increases in activity from post-event deposition (Larson et al. 2018).

14.6   Preservation Potential in the Sedimentary Record

From a sedimentological perspective, the potential for preservation of deposited oil-contaminated sediments associated with the DWH event, and their detection in the sedimentary record, is dependent on a variety of factors including (1) bioturbation/mixing, (2) burial, (3) degradation of the signature, (4) sediment compaction, and (5) remobilization, transport, and secondary deposition (Fig. 14.2). The presence of bioturbation and mixing will decrease the potential for preservation/

detection in the sedimentary record by smearing and diluting the signature to below detection. If there is a lack of bioturbation immediately following a deposition event, it is possible for burial by subsequent sedimentation to increase the preservation potential as long as the depositional layer is deeper (downcore) than the depth of bioturbation (Fig. 14.2). Burial of the impacted layer by subsequent sedimentation will lead to the highest potential for preservation and detection in the sedimentary record (Fig. 14.2). Areas with higher sedimentation rates generally have a higher probability of burial. Some of the indicators of oiled sedimentation are susceptible to degradation and will become more difficult to detect over time (Romero et al. 2020). With subsequent burial, compaction will lead to thinning of the depositional layer, making it more difficult to detect and requiring high-resolution sampling. The remobilization of impacted sediments can lead to a lack of preservation at the location of the initial deposition, as well as modification of the geographic extent of oil-contaminated sediments on the seafloor. This often is a function of resuspension and downslope transport to deeper water with the potential for focusing contaminated sediments in areas of high deposition, and/or distribution over a broader area. Research shows that seafloor morphology affects resuspension (Diercks et al. 2018; Turnewitsch et al. 2017) and redistribution (Durrieu De Madron et al. 2017), as well as subsequent settling and deposition of material (Turnewitsch et al. 2013, 2004) focusing erosion and deposition on specific areas on the seafloor. This provides a mechanism for uneven sediment deposition and distribution within a given time frame.

14.7 Critical Approaches/Methods

The investigation of the DWH blowout in the sedimentary system was developed using a variety of approaches that maximized the detection of sedimentary impacts without pre-existing knowledge of the evolution and processes influencing sedimentation associated with a deepwater oil blowout. This included the rapid collection of sediment cores, an annual time series to characterize the event and its evolution in the sedimentary system over subsequent years, high-resolution sampling to maximize temporal resolution, and a multidisciplinary approach to determine diagnostic as well as corroborative indicators of oil-contaminated sediment deposition.

14.7.1 Rapid Response and Collection of Cores

The rapid collection of sediment cores following the initiation of the blowout event allowed for investigation of the immediate sedimentary response and benthic impacts and the use of time-sensitive indicators to define sedimentation associated with the event. Time-sensitive indicators included short-lived chronometers ($^{234}Th_{xs}$)

to identify recent sedimentation, as well as sedimentological, chemical, and biological indicators that are susceptible to degradation, rapid modification/change, and/or low preservation potential in the sedimentary record. Also, as mentioned, impacted sedimentation was expected to be at the very sediment surface, and this allowed for targeted analysis for rapid determination of oil-contaminated sedimentation.

14.7.2 Time Series

The continued collection of sediment cores over the subsequent years following DWH provided the ability to define how the sedimentary signature was distinctly different from non-event sedimentation and how the sedimentary signature evolved over the following months to years. This was particularly critical to be able to use the same chronometer (^{234}Th$_{xs}$) for direct comparison of sedimentation rates (event vs post-event) (Sadler 1981). The time series defined the duration of the event and timing of the post-event response and if/when the benthic system stabilizes and/or recovers following such an event. It also provided the opportunity to investigate how the sedimentary system, including the signature of oil-contaminated sediments, evolved over subsequent years and the preservation potential of oil-contaminated sediments in the sedimentary record.

14.7.3 Sampling Resolution

Due to the short duration (months) of the blowout event and subsequent rapid collection of cores, the appropriate sub-sampling resolution to detect sedimentation on a monthly time scale was also of critical importance. The utilization of high-resolution sampling (2 mm sub-sampling) of cores allowed for the highest temporal resolution to detect the sedimentary signature of the blowout event with multiple sample intervals that represented the depositional pulse. It also allowed for the utilization of ^{234}Th$_{xs}$ as an indicator of short-term (months) sedimentation rates of the depositional pulse, as well as a direct comparison to post-event rates. This provided post-event baselines to define the deviation in sedimentation rates associated with the depositional pulse. The continued use of 2 mm sampling resolution was critical for determining the post-event sedimentation rates using ^{234}Th$_{xs}$ as this indicator was often only detectable in the upper 2–10 mm. Coarser sampling resolution would not have been able to define post-event sedimentation rates using ^{234}Th$_{xs}$.

Factors to consider with respect to sub-sampling resolution for detection of an event in the sedimentary record are the thickness of the depositional event and the required sensitivity of the analytical measurements used to detect the sedimentary signature of the event. Coarser sampling resolution may be appropriate for depositional events of greater thickness, often associated with long duration and/or high magnitude/sedimentary responses. The thinner the depositional event the finer the sampling resolution as over sampling (i.e., more than just the depositional unit) will lead to dilution of event sedimentation with non-event sedimentation. Dilution of the sedimentary signature of an event can lead to changing interpretations of impact and/or making it more difficult to detect in the sedimentary record. This is particularly important for more subtle indicators of a sedimentation event, and increased sampling resolution would increase the potential for detection as well as the number of analyses performed providing a more robust record. Also, of consideration is the required sampling resolution to obtain comparable baseline sedimentological data on similar time scales (months/years) for direct comparison to determine deviations, from baseline, of sedimentation events. Often, baseline sedimentation is at lower rates as compared to events.

14.7.4 *MultiDisciplinary Approach*

The simultaneous collection of up to eight cores allowed for a multidisciplinary approach for investigating DWH blowout impacts in the sedimentary record. This provided a more robust definition of the sedimentary signature and detection of the event. As most diagnostic indicators were subtle, the combination of multiple lines of evidence for the presence of oil-contaminated sediments was extremely valuable for defining the impacts of spatial deposition. It also assisted in determining the sediment sources (i.e., surface waters, etc.), depositional mechanisms, and biological/ecological impacts.

14.8 Conclusions

The DWH blowout event led to the formation of MOSSFA and a depositional pulse to the deep-sea benthos in the NEGoM in the Fall of 2010. A time series of sediment cores from four sites collected annually between 2010 and 2016 characterizes the event, post-event response, and stabilization of the sedimentary system with respect to sedimentation rates, sedimentology, and preservation potential. All sites collected in 2010 and early 2011 had large excursions in % silt and high $^{234}Th_{xs}$ Inventories and MAR indicating a depositional pulse with high sedimentation rates associated with the observed MOSSFA event. There were no distinctive changes in sediment composition associated with the depositional pulse, as the MOSSFA event stripped existing particles from the water column and did not significantly change the source(s) of sediment in either the siliciclastic or carbonated dominated regions. In the following years, 2011–2012, $^{234}Th_{xs}$ Inventories and MAR were lower (at all sites) indicating lower sedimentation rates and a lack of bioturbation following the event. Over this period the initial deviations in % silt began to become undetectable in the sedimentary record. Beginning in 2013 and continuing through 2016, there was a site-specific return of bioturbation and stabilization of the sedimentary system. Over the late 2011–2016 period, all sites had low $^{234}Th_{xs}$ Inventories indicating consistently low sedimentation rates and stabilization of sedimentation patterns. The return of bioturbation, as indicated by higher $^{234}Th_{xs}$ Apparent MAR with no supporting increase in $^{234}Th_{xs}$ Inventory, was spatially and temporally variable with site DSH08 beginning in 2013, sites PCB06 and M04 in 2015. The absence of bioturbation at site DSH10 over the 2010–2016 period may indicate a lack of recovery at this site or that bioturbation was never present (pre-event, event, and post-event). The return of bioturbation decreases the preservation potential of the depositional pulse with site DSH10 having the highest preservation potential. This is assuming remobilization of the depositional pulse did not occur at these sites. Beginning in 2012, $^{210}Pb_{xs}$ age dating for cores appeared to resolve the depositional pulse and may be used to assist in determining the burial and preservation of oil-contaminated sediments and their detection in the sedimentary record. The combination of the rapid collection of cores following the DWH event, high-resolution sampling, and time series of the annual collection of cores allowed for the detection of the event (2010–early 2011) and evolution of the sedimentary system over the post-event impact (2011–2012) and stabilization (2013–2016).

Dedication This chapter is dedicated to Charles "Chuck" Holmes who was a mentor, colleague, and friend. In his career, he advanced geologic studies in the Gulf of Mexico and beyond, as well as the methods and applications of short-lived radioisotope geochronology. Without his guidance and knowledge, the use of short-lived radioisotopes for investigating sediment records at high resolution in this study and others would not have been as far-reaching.

Funding Information This research was made possible by grants from The Gulf of Mexico Research Initiative through its consortia: The Center for the Integrated Modeling and Analysis of the Gulf Ecosystem (C-IMAGE) and Sea to Coast Connectivity in the Eastern Gulf of Mexico (Deep-C). Data are publicly available through the Gulf of Mexico Research Initiative Information & Data Cooperative (GRIIDC) at https://data.gulfresearchinitiative.org (doi: 10.7266/N7FJ2F94; 10.7266/N79S1PJZ; 10.7266/N7610XTJ).

References

Appleby PG (2001) Chapter 9: Chronostratigraphic techniques in recent sediments. In: Last WM, Smol JP (eds) Tracking environmental change using lake sediments volume 1. Kluwer Academic Publishers, The Netherlands

Appleby PG, Oldfield F (1983) The assessment of ^{210}Pb data from sites with varying sediment accumulation rates. Hydrobiologia 103:29–35

Balsam WL, Beeson JP (2003) Sea-floor sediment distribution in the Gulf of Mexico. Deep Sea Res Part I Oceanogr Res Pap 50(12):1421–1444

Binford MW (1990) Calculation and uncertainty analysis of ^{210}Pb dates for PIRLA project Lake sediment cores. J Paleolimnol 3:253–267

BOEM (2011) Report Regarding the Causes of the April 20, 2010 Macondo Well Blowout. https://www.bsee.gov/sites/bsee.gov/files/reports/blowout-prevention/DWHfinaldoi-volumeii.pdf

Brooks GR, Larson RA, Schwing PT, Romero I, Moore C, Reichart GJ, Jilbert T, Chanton JP, Hastings DW, Overholt WA, Marks KP, Kostka JE, Holmes CW, Hollander D (2015)

Sedimentation pulse in the NE Gulf of Mexico following the 2010 DWH blowout. PLoS One 10(7):1–24

Brooks GR, Larson RA, Schwing PT, Diercks AR, Armenteros-Almanza M, Diaz-Asencio M, Martinez-Suarez A, Sánchez-Cabeza JA, Ruiz-Fernandez AC, Herguera Garcia JC, Perez-Bernal LH, Hollander DJ (2020) Gulf of Mexico (GoM) bottom sediments and depositional processes: a baseline for future oil spills (Chap. 5). In: Murawski SA, Ainsworth C, Gilbert S, Hollander D, Paris CB, Schlüter M, Wetzel D (eds) Scenarios and responses to future deep oil spills: fighting the next war. Springer, Cham

Chanton J, Zhao T, Rosenheim BE, Joye S, Bosman S, Bruner C, Yeager KM, Diercks AR, Hollander D (2014) Using natural abundance radiocarbon to trace the flux of Petrocarbon to the seafloor following the Deepwater horizon oil spill. Environ Sci Technol 49(2):847–854. https://doi.org/10.1021/es5046524

Chen JH, Lawrence ER, Wasserburg GJ (1986) ^{238}U,^{234}U and ^{232}Th in seawater. Earth Planet Sci Lett 80:241–251. https://doi.org/10.1016/0012-821X(86)90108-1

Chisté V, Bé MM (2007) Table of radionuclides ^{210}Pb – comments on evaluation of decay data, LNHB – bureau international des Poids et Mesures, vol 8. Sevres Cedex, France. http://www.nucleide.org/DDEP_WG/DDEPdata.htm

Daly KL, Passow U, Chanton J, Hollander DJ (2016) Assessing the impacts of oil-associated marine snow formation and sedimentation during and after the Deepwater Horizon oil spill. Anthropocene 13:18–33. https://doi.org/10.1016/j.ancene.2016.01.006

Diercks AR, Highsmith RC, Asper VL, Joung D, Zhou Z, Guo L, Shiller AM, Joye SB, Teske AP, Guinasso NL Jr, Wade TL, Lohrenz SE (2010) Characterization of subsurface polycyclic aromatic hydrocarbons at the Deepwater horizon site. Geophys Res Lett 37:L20602. https://doi.org/10.1029/2010GL045046

Diercks AR, Dike C, Asper VL, DiMarco SF, Chanton JP, Passow U (2018) Scales of seafloor sediment resuspension in the northern Gulf of Mexico. Elem Sci Anth 6:32. https://doi.org/10.1525/elementa.285

Durrieu De Madron X, Ramondenc S, Berline L, Houpert L, Bosse A, Martini S, Guidi L, Conan P, Curtil C, Delsaut N, Kunesch S, Ghiglione JF, Marsaleix P, Pujo-Pay M, Séverin T, Testor P, Tamburini C (2017) Deep sediment resuspension and thick nepheloid layer generation by open-ocean convection. J Geophys Res Oceans 122:2291–2318. https://doi.org/10.1002/2016JC012062

Folk RL (1965) Petrology of sedimentary rocks. Hemphillis, Austin

Garcia-Pineda O, MacDonald I, Hu C, Svejkovsky J, Hess M, Dukhovskoy D, Morey SL (2013) Detection of floating oil anomalies from the Deepwater horizon oil spill with synthetic aperture radar. Oceanography 26:124–137

Harbison RN (1968) Geology of DeSoto canyon. J Geophys Res 73:5175–5185

Hastings DW, Bartlett R, Brooks R, Larson RA, Quinn KA, Razionale D, Schwing PT, Pérez Bernal LH, Ruiz-Fernández AC, Sánchez-Cabeza JA, Hollander DJ (2020) Changes in Redox conditions of surface sediments following the BP Deepwater horizon and Ixtoc 1 events (Chap. 16). In: Murawski SA, Ainsworth C, Gilbert S, Hollander D, Paris CB, Schlüter M, Wetzel D (eds) Deep oil spills: facts, fate and effects. Springer, Cham

Holmes CW (1976) Distribution, regional variation, and geochemical coherence of selected elements in the sediments of the central Gulf of Mexico. Geological Survey Professional Paper 928. US Government Printing Office, Washington, D.C.

Holmes CW (1998) Short-lived isotopic chronometers – a means of measuring decadal sedimentary dynamics. U.S. Geological Survey, Dept. of the Interior. Fact Sheet FS-073-98

Joye SB, MacDonald LR, Leifer I, Asper V (2011) Magnitude and oxidation potential of hydrocarbon gases released from the BP oil well blowout. Nat Geosci 4:160–164

Larson RA, Brooks GR, Schwing PT, Carter S, Hollander DJ (2018) High resolution investigation of event-driven sedimentation: northeastern Gulf of Mexico. Anthropocene 24:40–50. https://doi.org/10.1016/j.ancene.2018.11.002

Liu G, Bracco A, Passow U (2018) The influence of mesoscale and submesoscale circulation on sinking particles in the northern Gulf of Mexico. Elementa: science of the Anthropocene 6(1):36. https://doi.org/10.1525/elementa.292

McClintic MA, DeMaster DJ, Thomas CJ, Smith CR (2008) Testing the FOODBANCS hypothesis: seasonal variations in near-bottom particle flux, bioturbation intensity, and deposit feeding based on [234]Th measurements. Deep Sea Res Part II: Topical Studies in Oceanography 55(22–23):2425–2437

Milliman JD (1974) Marine carbonates. Springer-Verlag, New York

Montagna PA, Girard F (2020) Deep-sea benthic faunal impacts and community evolution before, during and after the Deepwater Horizon event (Chap. 22). In: Murawski SA, Ainsworth C, Gilbert S, Hollander D, Paris CB, Schlüter M, Wetzel D (eds) Deep oil spills: facts, fate and effects. Springer, Cham

Passow U (2016) Formation of rapidly-sinking, oil-associated marine snow. Deep Sea Res Part II: Topical Studies in Oceanography,. The Gulf of Mexico Ecosystem – before, during and after the Macondo Blowout 129:232–240. https://doi.org/10.1016/j.dsr2.2014.10.001

Passow U, Hetland R (2016) What happened to all of the oil? Oceanography 29:88–95. https://doi.org/10.5670/oceanog.2016.73

Passow U, Ziervogel K, Asper VL, Diercks AR (2012) Marine snow formation in the aftermath of the Deepwater horizon oil spill in the Gulf of Mexico. Environ Res Lett 7:035301. https://doi.org/10.1088/1748-9326/7/3/035301

Pope RH, Demaster JD, Smith CR, Seltmann H (1996) Rapid bioturbation in equatorial Pacific sediments: evidence from excess [234]Th measurements. Deep-Sea Res II Top Stud Oceanogr 43:1339–1364

Quigg A, Passow U, Hollander DJ, Daly KL, Burd A, Schwing PT, Lee K (2020) Formation and sinking of MOSSFA (Marine Oil Snow Sedimentation and Flocculent Accumulation) Events: past and present (Chap. 12). In: Murawski SA, Ainsworth C, Gilbert S, Hollander D, Paris CB, Schlüter M, Wetzel D (eds) Deep oil spills: facts, fate and effects. Springer, Cham

Romero IC, Schwing PT, Brooks GR, Larson RA, Hastings DW, Ellis GE, Goddard EA, Hollander DJ (2015) Hydrocarbons in Deep-Sea sediments following the 2010 Deepwater horizon blowout in the Northeast Gulf of Mexico. PLoS One 10(5):e0128371

Romero IC, Toro-Farmer G, Diercks AR, Schwing PT, Muller-Karger F, Murawski S, Hollander DJ (2017) Large scale deposition of weathered oil in the Gulf of Mexico following a Deepwater oil spill. Environ Pollut 228:179–189. https://doi.org/10.1016/j.envpol.2017.05.019

Romero IC, Chanton JP, Rosenheim BE, Radović J, Schwing PT, Hollander DJ, Larter SR, Oldenburg TBP (2020) Long-term preservation of oil spill events in sediments: the case for the Deepwater Horizon Spill in the Northern Gulf of Mexico (Chap. 17). In: Murawski SA, Ainsworth C, Gilbert S, Hollander D, Paris CB, Schlüter M, Wetzel D (eds) Deep oil spills: facts, fate and effects. Springer, Cham

Sadler P (1981) Sedimentation rates and the completeness of stratigraphic sections. J Geol 89:569–584

Schwing PT, Machain Castillo ML (2020) Impact and resilience of benthic foraminifera in the aftermath of the Deepwater Horizon and Ixtoc 1 oil spills (Chap. 23). In: Murawski SA, Ainsworth C, Gilbert S, Hollander D, Paris CB, Schlüter M, Wetzel D (eds) Deep oil spills: facts, fate and effects. Springer, Cham

Schwing PT, Romero IC, Larson RA, O'Malley BJ, Fridrik EE, Goddard EA, Brooks GR, Hastings DW, Rosenheim BE, Hollander DJ, Grant G, Mulhollan J (2016) Sediment Core extrusion method at millimeter resolution using a calibrated, threaded-rod. J Vis Exp 114:e54363. https://doi.org/10.3791/54363

Schwing PT, Brooks GR, Larson RA, Holmes CW, O'Malley BJ, Hollander DJ (2017) Constraining the spatial extent of the marine oil snow sedimentation and accumulation (MOSSFA) following the DWH event using a [210]Pb$_{xs}$ inventory approach. Environ Sci Technol 51:5962–5968. https://doi.org/10.1021/acs.est.7b00450

Swarzenski PW (2014) ^{210}Pb dating. In: Encyclopedia of scientific dating methods. Springer Science and Business Media, Dordrecht. https://doi.org/10.1007/978-94-007-6326-5_236-1

Thibodeaux LJ, Valsaraj KT, John VJ, Papadopoulos KD, Pratt LR, Pesika NS (2011) Marine oil fate: knowledge gaps, basic research, and development needs: a perspective based on the Deepwater horizon spill. Environ Eng Sci 28(2):87–93. https://doi.org/10.1089/eex.2010.0276

Turnewitsch R, Reyss JL, Chapman DC, Thomson J, Lampitt RS (2004) Evidence for a sedimentary fingerprint of an asymmetric flow field surrounding a short seamount. Earth Planet Sci Lett 222:1023–1036. https://doi.org/10.1016/j.epsl.2004.03.042

Turnewitsch R, Falahat S, Nycander J, Dale A, Scott RB, Furnival D (2013) Deep-sea fluid and sediment dynamics—influence of hill- to seamount-scale seafloor topography. Earth Sci Rev 127:203–241. https://doi.org/10.1016/j.earscirev.2013.10.005

Turnewitsch R, Dale A, Lahajnar N, Lampitt RS, Sakamoto K (2017) Can neap-spring tidal cycles modulate biogeochemical fluxes in the abyssal near-seafloor water column? Prog Oceanogr 154:1–24. https://doi.org/10.1016/j.pocean.2017.04.006

US Coast Guard (2010) Report of investigation into the circumstances surrounding the explosion, fire, sinking and loss of eleven crew members aboard the Mobile offshore drilling unit Deepwater horizon in the Gulf of Mexico April 20–22, 2010. Vol I MISLE Activity Number: 3721503. https://www.bsee.gov/newsroom/library/deepwater-horizon-reading-room/joint-investigation-team-report

Winkler R, Rosner G (2000) Seasonal and long-term variation in ^{210}Pb concentration in air, atmoshpere deposition rate and total deposition velocity in South Germany. Sci Total Environ 263:57–68

Yan B, Passow U, Chanton JP, Nöthig EM, Asper V, Sweet J, Pitiranggon M, Diercks AR, Pak D (2016) Sustained deposition of contaminants from the Deepwater horizon spill. PNAS 113:24E3332–24E3340. https://doi.org/10.1073/pnas.1513156113

Yeager KM, Stantschi PH, Owe GT (2004) Sediment accumulation and radionuclide inventories ([239,240]Pu, ^{210}Pb and [234]Th) in the northern Gulf of Mexico, as influenced by organic matter and macrofaunal density. Mar Chem 91:1–14

Ziervogel K, McKay L, Rhodes B, Osburn CL, Dickson-Brown J, Arnosti C, Teske A (2012) Microbial activities and dissolved organic matter dynamics in oil-contaminated surface seawater from the Deepwater horizon oil spill site. PLoS One 7(4):e34816

Chapter 15
Applications of FTICR-MS in Oil Spill Studies

Jagoš R. Radović, Aprami Jaggi, Renzo C. Silva, Ryan Snowdon, Derek C. Waggoner, Patrick G. Hatcher, Stephen R. Larter, and Thomas B. P. Oldenburg

Abstract During the past decade, Fourier transform ion cyclotron resonance mass spectrometry (FTICR-MS) has been established as a technique of choice for the comprehensive chemical assessment of some of the most complex organic mixtures, such as petroleum, or dissolved organic matter. In the aftermath of the *Deepwater Horizon* (DWH) blowout, FTICR-MS demonstrated its applicability for the characterization of oil spill residues produced by abiotic weathering, such as photooxidation, and/or microbial processes and interactions, for example, marine oil snow aggregation. Such residues are abundant in high molecular weight, polar, and heteroatom-bearing chemical species, which cannot be analyzed by the typical oil spill forensics tools such as gas chromatography. Therefore, the expansion of the analytical window afforded by FTICR-MS is crucial for the monitoring and understanding of long-term oil spill fate. Furthermore, capability of FTICR-MS to characterize non-hydrocarbon petroleum fractions will be very important in the case of potential future spills of heavy, unconventional oils, such as bitumen.

Keywords FTICR-MS · Oil spills · *Deepwater Horizon* · Biomarkers · MOSSFA

15.1 Introduction

Since the inception of organic (bio)geochemistry, fundamental discoveries in this area have been, to a large degree, driven by the advances in the instrumental analytical technologies. Gas chromatography, mass spectrometry, carbon isotopic

J. R. Radović (✉) · A. Jaggi · R. C. Silva · R. Snowdon · S. R. Larter · T. B. P. Oldenburg
University of Calgary, PRG, Department of Geoscience, Calgary, AB, Canada
e-mail: Jagos.Radovic@ucalgary.ca; aprami.jaggi@ucalgary.ca; rcsilva@ucalgary.ca; rwsnowdo@ucalgary.ca; slarter@ucalgary.ca; toldenbu@ucalgary.ca

D. C. Waggoner · P. G. Hatcher
Old Dominion University, Norfolk, VA, USA
e-mail: dwagg002@odu.edu; PHatcher@odu.edu

© Springer Nature Switzerland AG 2020
S. A. Murawski et al. (eds.), *Deep Oil Spills*,
https://doi.org/10.1007/978-3-030-11605-7_15

measurements, Rock-Eval pyrolysis, and hydrous pyrolysis, to name but a few, made possible a more comprehensive analysis and a much better understanding of the origin and fate of organic matter (OM) in the environment (Hunt et al. 2002; Kvenvolden 2002).

Similarly, Fourier transform ion cyclotron resonance mass spectrometry (FTICR-MS), a technology pioneered by Comisarow and Marshall (1974), at the University of British Columbia, entered the field of complex organic mixtures analysis at the turn of the century, with the first practical applications of ultrahigh-resolution mass spectrometry ($m/\Delta m50\% > 100{,}000$, where $\Delta m50\%$ is a mass spectral peak width at half-maximum peak height), reported later on (Rodgers et al. 2001). Some recent improvements of this technique included the altering of the ICR cell design (Nikolaev et al. 2011) and/or increasing the magnetic field strength applied to the ion contents in the ICR cell (Hendrickson et al. 2015), resulting in an unmatched mass resolution, e.g., 12 million at m/z 675 (Popov et al. 2014), and unambiguous molecular identification.

Because of this, currently FTICR-MS is the technique of choice for characterization of the most complex mixtures of organic species, such as petroleum and petroleum residues (petroleomics), cellular constituents (proteomics, metabolomics, and lipidomics), dissolved organic matter (DOM), and, more recently, sedimentary OM (Han et al. 2008; Rodgers and Marshall 2007; Riedel and Dittmar 2014; Radović et al. 2016a; Wörmer et al. 2014).

The use of FTICR-MS has expanded the analytical window of complex mixture analysis due to its diverse and selective ionization modes (electrospray, photoionization, laser desorption, etc.), broad range of spectral detection, and high mass resolution, enabling it to access thousands of high molecular weight, nonvolatile, thermally unstable, and/or highly polar acidic and basic species, which are typically not GC-amenable, Fig. 15.1 (McKenna et al. 2013). Some limitations of this technique include the inability of the method to resolve all the possible isomers of a compound (since they have the same exact m/z), and its component quantitation capabilities. However, a correlation between the relative abundances of equivalent species in FTICR-MS and GC-MS analyses has been demonstrated, suggesting some rudimentary quantitative capabilities for FTICR-MS (Oldenburg et al. 2014).

As to the FTICR-MS applications in oil spill studies, in the years following the *Deepwater Horizon* (DWH) blowout, FTICR-MS made contributions to the characterization of spilled Macondo well (MW) oil, and its residues (McKenna et al. 2013; Ruddy et al. 2014). Furthermore, it has been used to assess the oil-microbial aggregates (Hatcher et al. 2018; Wozniak et al. 2018), and the baseline organic geochemical compositions of GoM waters and sediments (Radović et al. 2016a, b; Jaggi 2018). Finally, it has provided chemical characterization of potential future sources of oil spills, such as heavy oils and oil sands bitumens (Radović et al. 2018). In this chapter, we will highlight some of these applications.

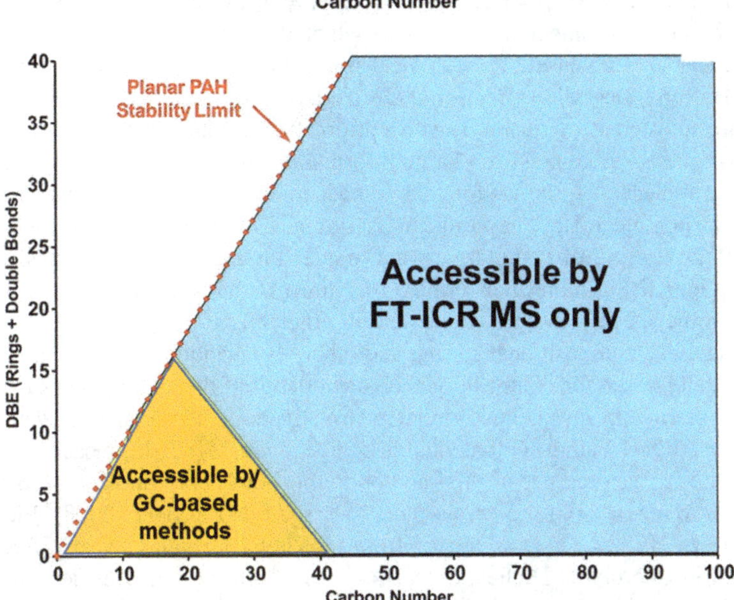

Fig. 15.1 Analytical space (defined by analytes' carbon number and double bond equivalents) which can be accessed by gas chromatography-based methods and by FTICR-MS. (Adapted with permission from McKenna et al. (2013). Copyright (2013) American Chemical Society)

15.2   FTICR-MS Basics

FTICR-MS can be paired with several soft ionization source mechanisms, each of which is used to focus analysis on different types of chemical compounds. For example, APPI (atmospheric pressure photoionization) is most effective on compounds with aromatic rings and sulfur-containing compounds, as these have electrons that can be easily removed with a UV photon or charged indirectly via a dopant. Alternatively, ESI (electrospray ionization) tends to be most useful for analysis of acidic or basic compounds, by adding or removing charged protons, depending on the ion polarity selected for detection. The molecular mass range that FTICR-MS analysis can encompass is very broad, thus readily accessing the petroleum compounds which are not GC-MS amenable due to their high molecular weight, (Fig. 15.1).

This analytical technique functions by driving ions to orbit within a homogeneous magnetic field executing an ion cyclotron motion, the orbital frequency of which is directly proportional to the strength of the magnetic field applied and inversely proportional to the m/z ratio of the ion. When trapped in an ICR cell and excited, the ions induce an image current on detector plates. Image currents are recorded in the time domain and then are transformed to the frequency domain via a Fourier transform. Mass spectra are obtained by converting signal frequency to ion m/z values, using the cyclotron equation $\upsilon(Hz) = 1.535611 \cdot 10^7\, B_0 \cdot (m/z)^{-1}$, where B_0 is the magnetic field strength of the instrument, in Tesla (Marshall et al. 1998).

In order to compensate for non-ideal interactions such as space charge effects and other ion suppression effects, the spectrum is calibrated, and the masses of each peak are adjusted using standards with a priori chemical information or known compounds. The accurate mass to charge information for each ion signal is then converted to molecular formulas for each ion accurate mass assessed. This is possible due to the nonintegral unique atomic masses of most elements (unique mass defects).

Data processing of complex, organic matter ultrahigh-resolution mass spectra yields a long list of molecular formula assignments and intensities related to peaks found in the spectra. In order to be able to efficiently interpret and compare such "big" datasets, compositional sorting and other visualization strategies are needed, which could reveal the compositional characteristics of the investigated sample set. Typical petroleum geochemistry-related investigations use sequential data layers based on the heteroatom content, double bond equivalent (DBE, a measure of unsaturation due to the presence of rings and/or double bonds in the molecular structure), and carbon number (C#) to create plots (Marshall and Rodgers 2008). Elemental ratios (e.g., H/C vs. O/C) calculated from molecular formulas are also frequently used in organic matter studies, especially in the form of van Krevelen plots (Wu et al. 2004). While detailed explanation of FTICR-MS terminology is out of this chapter's scope, more information can be found in other publications (Marshall and Rodgers 2008; McKenna et al. 2013; Oldenburg et al. 2014, 2017). Again, the use of software packages, either developed in-house or commercially available, makes the investigation of the several layers of data easier.

15.3 Characterization of Source Oils and Weathered Oil Residues Using FTICR-MS

The oil released during the DWH spill was a sweet, light Louisiana crude, rich in volatile and semi-volatile saturated and aromatic hydrocarbon compounds, readily amenable for standard gas chromatography analyses (Overton et al. 2016). Notwithstanding, when characterized with FTICR-MS, a plethora of additional, high molecular weight, and/or non-hydrocarbon polar oil constituents of MW oil also becomes accessible. McKenna et al. (2013) have used atmospheric pressure photoionization (APPI) and electrospray ionization (ESI) to inventory more than 16,000 unique molecular monoisotopic elemental compositions for the acidic, basic, and nonpolar components of the source MW oil. These included carbazoles and benzocarbazoles, indoles, carboxylic (naphthenic) acids, pyridines, and quinolines (McKenna et al. 2013).

However, weathering processes such as photooxidation and biodegradation have transformed the parent oil compounds in the spilled MW oil during the months and years after the release, to a complex mixture of oxygen-containing products, which escape the detection window of conventional gas chromatography tools (Aeppli et al. 2012, 2014; Radović et al. 2014; Ward et al. 2018). In this case, the capability of FTICR-MS to access the pool of nonvolatile compounds in the MW oil residues has been crucial and was leveraged, in combination with two-dimensional gas chromatography, to characterize and identify the weathering products of MW oil as ketones, carboxylic acids, and higher numbered (>3) oxygen-containing species (Ruddy et al. 2014). Figure 15.2 illustrates the compound class distribution of the source MW oil, and the weathered coastal residues, containing the oxidized products of parent oil compounds.

Recently, with unconventional oil resource development and pipeline projects, heavy oils and bitumens, such as the ones found in Western Canada, have come into focus as potential sources of future petroleum spills (NAS 2016). In-reservoir biodegradation, to which these oils were subjected to over long geological timescales, depleted saturated and aromatic hydrocarbons and enriched non-GC-amenable alkyl-substituted aromatic hydrocarbons and non-hydrocarbons containing sulfur and nitrogen. In order to characterize these chemical species, FTICR-MS is an indispensable tool (Radović et al. 2018; Oldenburg et al. 2017). For example, Fig. 15.3 shows the distribution of the main compound classes in a bitumen sample from the Peace River oil sands deposit (Western Canada), obtained using FTICR-MS

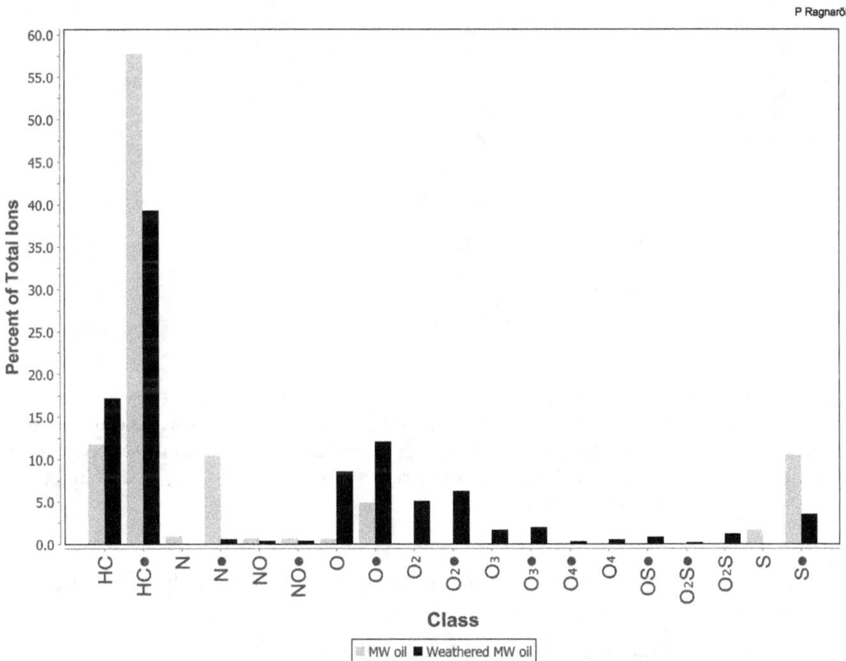

Fig. 15.2 Distribution of different compound classes in the source Macondo well oil, and its weathered residues, as obtained by FTICR-MS analysis in positive-ion APPI mode. Note the depletion of parent oil compounds, e.g., hydrocarbons (HC•), and the appearance of oxidized compound classes, products of oil transformations (e.g., photooxidation, biodegradation). The dot denotes radical ion, one of the possible ionization products in APPI (see Oldenburg et al. (2014) for a more detailed explanation of ionization mechanisms in FTICR-MS)

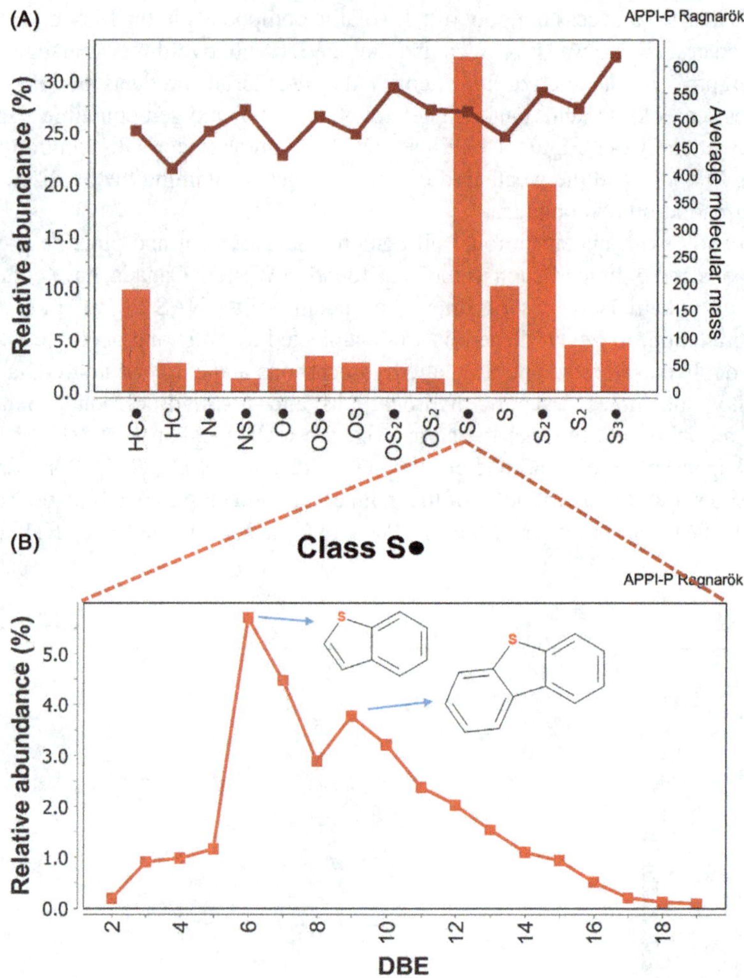

Fig. 15.3 (**a**) Compound class composition of heavy, biodegraded oil sand bitumen from the Peace River deposit. Note the predominance of non-hydrocarbon species, such as the high molecular weight sulfur heteroaromatic compounds (**b**), which, in a case of a spill, could not be characterized efficiently using conventional, gas chromatography-based methods. (Reprinted from Radović et al. (2018), with permission from Elsevier)

in positive-ion APPI (APPI-P) mode (Radović et al. 2018). Notably, the majority of chemical species in this sample belong to non-hydrocarbon compounds, such as the alkylated sulfur containing benzothiophenic (DBE 6) and dibenzothiophenic (DBE 9) analogs, with averaged molecular mass higher than 500 Da. Given the fact that typical GC-MS-based methods can detect only a limited range of thiophenic species with molecular masses up to ~250 Da (Andersson and Schade 2004), the importance of FTICR-MS for source and spill characterization of heavy, unconventional oils becomes evident.

15.4 FTICR-MS Characterization of Dissolved Organic Matter and Its Relevance for Oil Spill Assessments

Aquatic dissolved organic matter (DOM) comprises a cocktail of hundreds of thousands of different compounds, which can be classified into four broad groups of compounds based on their origin: (i) biomolecules released, exuded, excreted, or leached from the living and decaying biota; (ii) biomolecules released from the detritus of marine organisms; (iii) biomolecules percolating from surrounding waters, atmospheric deposition, sediments, and terrigenous sources; and (iv) molecules from anthropogenic input (such as oil spills, agricultural and industrial runoffs) (Gonsior et al. 2011; Mopper et al. 2007). These molecules are commonly polyfunctional, heterogeneous species, have disperse molecular weights, and are dissolved in natural marine waters at low concentrations in a ~0.6 M ionic strength inorganic salt solution, which further adds to the chemical complexity and analytical challenges. The characterization of aquatic DOM with FTICR-MS allows the complexity of the range of species in DOM to be identified by their molecular formulas, which demonstrates not only the large complexity and variability of this carbon reservoir but also its response to environmental processes that affect its molecular-level composition (Stubbins et al. 2010), which, in turn, can be leveraged for oil spill assessment.

The physicochemical properties of crude oils change upon their release in the aquatic environment, due to microbial degradation and weathering processes such as evaporation, aqueous dissolution, volatilization, biodegradation, etc. (Atlas and Hazen 2011; Radović et al. 2012), which selectively remove the low molecular weight petroleum compounds, leaving the relatively heavier compounds, like those found in the asphaltene and resin fractions of crude oil, in the marine environment for years (Howard et al. 2005). Prior to the advent of ultrahigh-resolution mass spectrometry, the understanding of the oil spill signature in the marine environment was limited to the GC-amenable parent oil compounds (Aeppli et al. 2012; Blumer et al. 1973; Harayama et al. 1999; Radović et al. 2014). The capability of FTICR-MS to analyze high molecular weight, polyfunctional, and polar compounds makes it ideal to determine the fate of oil and its transformation products in the aquatic DOM.

The studies using FTICR-MS to characterize aquatic DOM reveal FTICR-MS spectra enriched in multi-oxygenated species ($C_xH_yO_z$) (Koch et al. 2007; Sleighter and Hatcher 2008). The aquatic DOM spectra display a similarity in terms of the broad compound class distribution, attributed to the constant diagenetic transformations in marine ecosystems due to microbial degradation and photooxidation (Jaggi 2018; Mentges et al. 2017). With the use of ultrahigh-resolution mass spectrometry, Liu and Kujawinski (2015) identified molecular fingerprints for oil input in the marine environment in the form of preferential partitioning of compounds with higher heteroatom to carbon ratios from the oil to the water phase. These water-soluble oil-derived organics exhibit smaller molecular size than the natural marine solid phase extracted DOM, as determined by Seidel et al. (2016).

Further, FTICR-MS has also been utilized as a valuable tool in tracking the fate of dispersants applied following oil spills in the water column. Kujawinski et al. (2011)

used FTICR-MS and liquid chromatography with tandem mass spectrometry to identify and quantify the anionic surfactant DOSS (dioctyl sodium sulfosuccinate), a key component in the dispersant applied during the DWH oil spill. The group was able to detect the DOSS signature in the oil plume that was formed at 1000–1200 m water depth near the wellhead, implying that applied dispersant stayed in the plume without appreciably degrading for 64 days after the dispersant application ceased. Further, the Seidel et al. (2016) group tracked dispersant fate by characterizing the biodegradation markers in the aquatic DOM. The addition of dispersants to the water alters the metabolic pathways of organic matter biodegradation enabling the tracking of dispersant using dispersant-derived metabolite markers, such as sulfur-containing species, likely derivatives of DOSS surfactant (Seidel et al. 2016).

Overall, the advent of ultrahigh-resolution mass spectrometry has revolutionized the assessment of oil spill signature in the water column.

15.5 FTICR-MS Characterization of Marine Sediments and Its Relevance for Oil Spill Assessments

Recent sediments are a unique record of present and past biogeochemical processes and conditions in a given marine system, which reflects source OM inputs, both biogenic and anthropogenic. Therefore, the knowledge of the composition of sedimentary organic species is essential to understand the "normal," background state of a given environment, as well as to identify the impacts of major perturbations, such as oil spills. Given the complexity and chemical diversity of organic markers which can be found in sediments, nontargeted, broad-range analytical tools, such as FTICR-MS, are particularly useful.

For example, very commonly studied microbial markers in sediments are glycerol dialkyl glycerol tetraethers (GDGTs), lipid membrane constituents of Archaea, and some bacteria (Schouten et al. 2013). Due to the fact that the distribution of specific GDGT species changes with environmental conditions such as temperature, nutrients, or pH (Schouten et al. 2013), these species are extensively used as environmental monitoring proxies with wide application in paleoclimate studies. Radović et al. (2016b) developed a nontargeted APPI-P FTICR-MS method, abbreviated "RADAR" (Rapid Analyte Detection and Reconnaissance), to explore the compositional complexity of GDGT analogs present in the lipid extracts of the Gulf of Mexico recent sediments. In this study, the complete series of core GDGT species (0 to 8 alicyclic rings), including a completely resolved GDGT-4 peak, could be identified. Mono- and dihydroxy analogs, as well as glycerol dialkanol diethers, were also reported. The usefulness of the nontargeted analytical approach provided by FTICR-MS was clearly evident when the authors offered the putative identification of dihydroxy-GDGT species, as well as several other $C_{83-87}H_xO_{4-7}$ previously unknown GDGT analogs. Furthermore, commonly used sea surface temperature indices, such as CCaT, TEX^{L}86, and the methane index (Schouten et al. 2013), were determined based on the monoisotopic intensity of appropriate peaks detected in the spectra, illustrating that FTICR-MS measurements could provide a rapid and reproducible assessment of environmental variability at a given deepwater site with minimal sample workup. Radović and coworkers also expanded the range of putatively identified biomarkers to include pheopigments, branched GDGTs, and other, more tentatively identified structures, but with clear biomarker potential, Fig. 15.4 (Radović et al. 2016a). From a spill forensics perspective, distributions and excursions of these

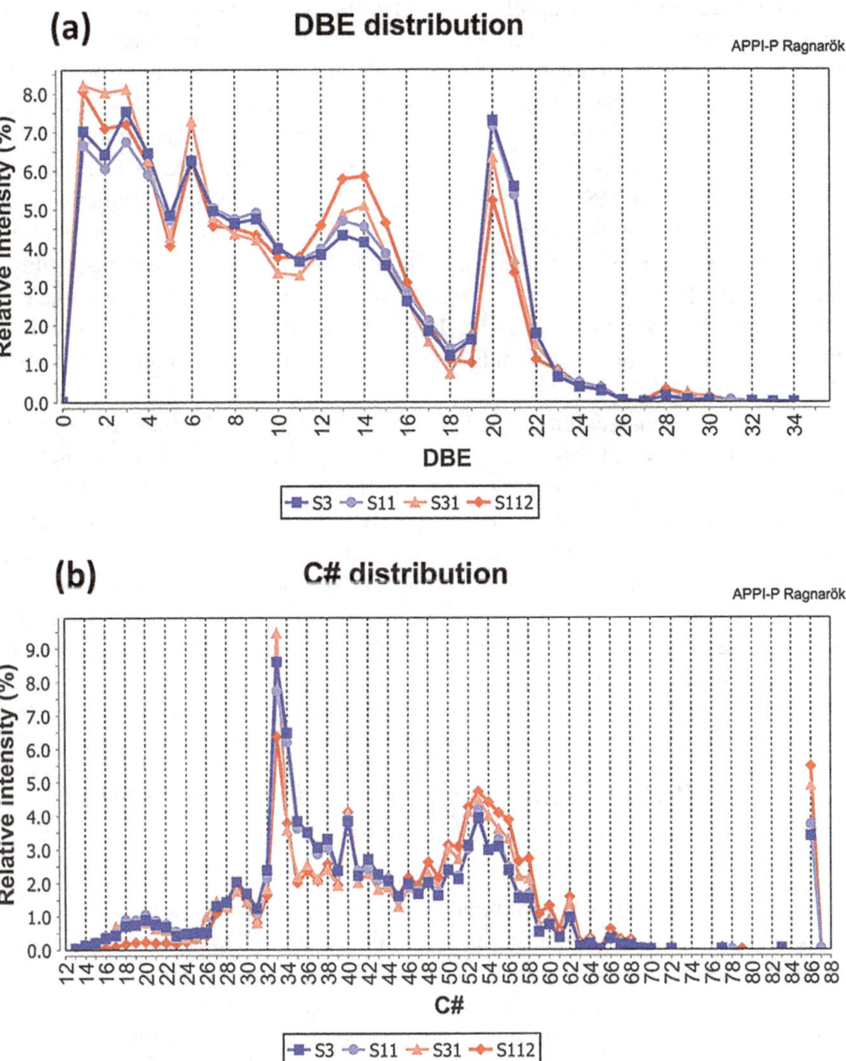

Fig. 15.4 (**a**) Double bond equivalent (DBE) and (**b**) carbon number distributions of the molecular species detected in a sediment core from the northern GoM, which are related to characteristic microbial biomarkers (e.g., Archaea, phytoplankton). Distributions of these biomarkers are useful proxies of "normal," background microbiome dynamics, but also of the perturbations related to oil spills, such as the MOSFFA (Marine Oil Snow Sedimentation and Flocculent Accumulation) event. The average depth of the sediment layers is from 3 mm (S3) to 112 mm (S112). (Reprinted from Radović et al. (2016a). Copyright (2016) John Wiley & Sons, Ltd.)

biomarkers in stratigraphic horizons corresponding to the DWH blowout can be related to a hypothesized MOSFFA (Marine Oil Snow Sedimentation and Flocculent Accumulation) event, which rapidly deposited large amount of mineral particles, spilled MW oil, and microbial biomass to the seafloor (Daly et al. 2016).

More recently, Jaggi (2018) used FTICR-MS to analyze water-extractable organic matter (WEOM) from the northern GoM (Gulf of Mexico) sediment, revealing a spectrum enriched with nitrogen-containing compounds, containing up to six nitrogen heteroatoms, likely related to proteinaceous cell constituents. The sediment WEOM, when compared to the aquatic DOM, showed lower carbon number and DBE values in the dominant species, likely representing the low molecular weight intermediates from a series of hydrolytic, fermentative, and eventually respiratory processes taking place in the sediments. While the aquatic DOM creates homogenous FTICR-MS spectra due to the slower transformations, dilutions, and longer residence times of species in the water column, WEOM of recent sediments shows a higher variability in the composition between different locations, owing to the relatively faster deposition of fresh biomass. Thus, in general, the chemical character of both water-extractable and lipid species in the sediments better reflects the input signatures of different sources and likely acts as a better alternative for oil spill assessments.

Noticeably, recent sediment analyses using ultrahigh-resolution mass spectrometry are still scarce. In this sense, nontargeted approaches, such as the RADAR mode FTICR-MS approach previously explained, are potential tools to rapidly expand the inventory of biomarkers and early diagenetic degradation products of biological materials in recently deposited organic matter, where complex mechanisms are involved and multiproxy nontargeted approaches seem to be a better fit. Such approaches show great promise for both pre-spill baseline characterization, as well as for assessment of planktonic and benthic responses of microbial communities to petrogenic inputs.

15.5.1 FTICR-MS Characterization of Marine Oil Snow Associations Generated by Oil Spills

In addition to previously known removal mechanisms for oil released into oceans, one important discovery resulting from the DWH oil spill was the realization that MOSSFA represents a significant fate for oil released into systems such as the Gulf of Mexico (Daly et al. 2016; Quigg et al. 2016). MOSSFA is believed to occur when sticky extracellular polymeric substances (EPS), excreted by microbial communities in response to oil presence, bring together oil, mineral, and biological materials to form marine oil snow (MOS) particles that eventually settle out of the water column and contribute to sedimentary material (Quigg et al. 2016). Studies conducted following the DWH spill suggest between 4% and 31% of oil released during the spill may have been entrained in MOS and contributed to sediments (Chanton et al. 2015; Valentine et al. 2014). FTICR-MS makes it possible to determine what oil components are associated with MOS and track the fate of that oil over time in complex mixtures.

Tracking the chemical changes of oil that become associated with MOS in the open ocean is particularly challenging as one would have to be on site during the event and be able to collect the MOS from the water column in a very gentle fashion. An alternative approach is to collect MOS from laboratory mesocosms that simulate the process of MOS formation (Quigg et al. 2016). Using this mesocosm approach, it is possible to compare marine snow formed by biota in clean seawater (control) to marine snow formed in seawater saturated with crude oil or water-accommodated fraction (WAF) (Wade et al. 2017).The mesocosms produce MOS in sufficient quantity that oil associated with the sinking particles can be extracted using methylene chloride from MOS isolated by filtration onto glass fiber filters (0.7 microns pore size). Once extracted, the oil associated with the marine snow could be examined by ESI-FTICR-MS (Wozniak et al. 2018) and ^{13}C NMR (Hatcher et al. 2018).

Another simulation methodology, described by Passow et al. (2012), involves the production of marine snow in roller tanks, done in conjunction with mesocosm experiments described above. These roller tanks are designed to simulate settling in a deep oceanic system where marine snow settles through the water column for an extended period of time. The particles formed in such systems settle to the bottom of the tanks when the experiment is terminated by stopping the rollers. In the case described here, the settling occurred for a period of 4 days after the initial set-up. Once terminated, the marine snow was allowed to settle to the bottom of the tank where it was then collected. Solvent-soluble oil that had been associated with the particles was extracted using methylene chloride, subjected to ESI-FTICR-MS analysis, and compared to extracts from natural particles without oil (control), as well as to the initial Macondo surrogate oil.

By examining the extracts from Macondo surrogate oil and MOS from WAF tanks after 4 days of settling, FTICR-MS is able to illustrate the transformations that the oil associated with MOS undergoes during settling. ESI-FTICR-MS shows that the polar compounds of the initial surrogate oil are primarily molecules containing a small number of oxygens (CHO_1, CHO_2, CHO_3, CHO_1N, CHO_2N). Association of oil with sinking particles over the course of just 4 days, Fig. 15.5, can significantly alter the oil, mainly by increasing the oxygen content of molecules as is typically observed for long-term (>100 days) weathering of oil (Aeppli et al. 2012; Chen et al. 2016). The distribution of CHO molecules in the initial oil is dominated by CHO_2, with DBE distributions centered between 3 and 5. Following association with MOS for 4 days, oil shows a change to CHO_x where x is mainly 3–5 oxygens, having DBE values distributed between 5 and 7 with a pronounced shift toward higher DBE. The H/C ratio for molecules can also be used as an indicator for alteration, where H/C ratios in the initial oil are generally between 0.75 and 2.0, typical for mostly aliphatic and alicyclic structures. The 4-day oil displays H/C ratios extending to lower values (H/C 0.5), suggestive of increasing proportions of aromatic structures when combined with the shift toward higher DBE. The alterations shown by FTICR-MS are consistent with studies investigating the

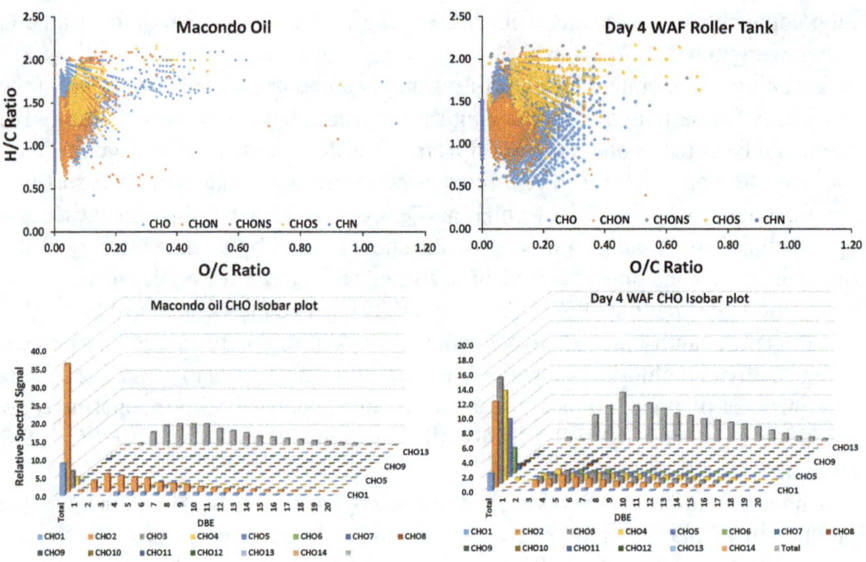

Fig. 15.5 Van Krevelen plots and isobar plots for the Macondo surrogate oil (left) and oil associated with sinking particles after 4 days of roller tank settling (right)

microbes associated with sinking particles (Doyle et al. 2018), where rapid colonization of the MOS occurred as it provided an ideal substrate, rich in readily degradable organic matter (exopolymeric exudates and oil).

Another important change observed involves N-containing molecules. A significant increasing amount of oxygenation is observed for molecules containing both one and two nitrogen atoms. The CHON molecules grow in abundance during the 4-day settling, and they shift toward higher O/C ratio as they become more oxygenated. This is analogous to the transformations observed for CHO molecules.

15.6 Conclusions and Future Directions

FTICR-MS provides a rapid, high-resolution screening approach to examine both natural diagenetic processes for organic matter in sediments and water columns and expands the analytical window of typical spill forensics tools such as gas chromatography, to include high molecular weight, polar, and heteroatom-bearing species. In this way, it facilitates the detection and monitoring of more complex geochemical signals of petroleum spills, related to products of abiotic weathering (e.g., photooxidation) or microbially mediated processes and interactions (e.g., MOSSFA). Coupled with other state-of-the-art instrumentation such as liquid chromatography, nuclear magnetic resonance (NMR), and especially the development of capabilities for making natural abundance isotope measurements on components identified in

FT-MS spectra (Eiler et al. 2017), FTICR-MS could provide revolutionary capabilities in tracing and quantifying the incorporation of petroleum carbon into the near-surface biogeosphere.

Acknowledgments This research was made possible in part by a grant from the Gulf of Mexico Research Initiative through the Center for the Integrated Modeling and Analysis of the Gulf Ecosystem (C-IMAGE) and in part from CFI, NSERC, the University of Calgary, and the Canada Research Chairs. Data are publicly available through the Gulf of Mexico Research Initiative Information & Data Cooperative (GRIIDC) at http://data.gulfresearchinitiative.org (doi: 10.7266/N71R6NGQ).

References

Aeppli C, Carmichael CA, Nelson RK, Lemkau KL, Graham WM, Redmond MC, Valentine DL, Reddy CM (2012) Oil weathering after the Deepwater horizon disaster led to the formation of oxygenated residues. Environ Sci Technol 46(16):8799–8807. https://doi.org/10.1021/es3015138

Aeppli C, Nelson RK, Radović JR, Carmichael CA, Valentine DL, Reddy CM (2014) Recalcitrance and degradation of petroleum biomarkers upon abiotic and biotic natural weathering of Deepwater horizon oil. Environ Sci Technol 48(12):6726–6734. https://doi.org/10.1021/es500825q

Andersson JT, Schade T (2004) Higher alkylated dibenzothiophenes in some crude oils and hydrodesulfurized fuels. ACS Division of Fuel Chemistry, Preprints 49

Atlas RM, Hazen TC (2011) Oil biodegradation and bioremediation: a tale of the two worst spills in U.S. history. Environ Sci Technol 45(16):6709–6715. https://doi.org/10.1021/es2013227

Blumer M, Ehrhardt M, Jones JII (1973) The environmental fate of stranded crude oil. Deep-Sea Res Oceanogr Abstr 20(3):239–259. https://doi.org/10.1016/0011-7471(73)90014-4

Chanton J, Zhao T, Rosenheim BE, Joye S, Bosman S, Brunner C, Yeager KM, Diercks AR, Hollander D (2015) Using natural abundance radiocarbon to trace the flux of Petrocarbon to the seafloor following the Deepwater horizon oil spill. Environ Sci Technol 49(2):847–854. https://doi.org/10.1021/es5046524

Chen H, Hou A, Corilo YE, Lin Q, Lu J, Mendelssohn IA, Zhang R, Rodgers RP, McKenna AM (2016) 4 years after the Deepwater horizon spill: molecular transformation of Macondo well oil in Louisiana salt marsh sediments revealed by FT-ICR mass spectrometry. Environ Sci Technol 50(17):9061–9069. https://doi.org/10.1021/acs.est.6b01156

Comisarow MB, Marshall AG (1974) Fourier transform ion cyclotron resonance spectroscopy. Chem Phys Lett 25:282–283. https://doi.org/10.1016/0009-2614(74)89137-2

Daly KL, Passow U, Chanton J, Hollander D (2016) Assessing the impacts of oil-associated marine snow formation and sedimentation during and after the Deepwater horizon oil spill. Anthropocene 13:18–33. https://doi.org/10.1016/j.ancene.2016.01.006

Doyle SM, Whitaker EA, De Pascuale V, Wade TL, Knap AH, Santschi PH, Quigg A, Sylvan JB (2018) Rapid formation of microbe-oil aggregates and changes in community composition in coastal surface water following exposure to oil and the dispersant Corexit. Front Microbiol 9(689). https://doi.org/10.3389/fmicb.2018.00689

Eiler J, Cesar J, Chimiak L, Dallas B, Grice K, Griep-Raming J, Juchelka D, Kitchen N, Lloyd M, Makarov A, Robins R, Schwieters J (2017) Analysis of molecular isotopic structures at high precision and accuracy by Orbitrap mass spectrometry. Int J Mass Spectrom 422:126–142. https://doi.org/10.1016/j.ijms.2017.10.002

Gonsior M, Zwartjes M, Cooper WJ, Song W, Ishida KP, Tseng LY, Jeung MK, Rosso D, Hertkorn N, Schmitt-Kopplin P (2011) Molecular characterization of effluent organic matter identified by ultrahigh resolution mass spectrometry. Water Res 45(9):2943–2953. https://doi.org/10.1016/j.watres.2011.03.016

Han J, Danell RM, Patel JR, Gumerov DR, Scarlett CO, Speir JP, Parker CE, Rusyn I, Zeisel S, Borchers CH (2008) Towards high-throughput metabolomics using ultrahigh-field Fourier transform ion cyclotron resonance mass spectrometry. Metabolomics 4(2):128–140. https://doi.org/10.1007/s11306-008-0104-8

Harayama S, Kishira H, Kasai Y, Shutsubo K (1999) Petroleum biodegradation in marine environments. J Mol Microbiol Biotechnol 1(1):63–70

Hatcher PG, Obeid W, Wozniak AS, Xu C, Zhang S, Santschi PH, Quigg A (2018) Identifying oil/marine snow associations in mesocosm simulations of the Deepwater Horizon oil spill event using solid-state 13C NMR spectroscopy. Mar Pollut Bull 126:159–165. https://doi.org/10.1016/j.marpolbul.2017.11.004

Hendrickson CL, Quinn JP, Kaiser NK, Smith DF, Blakney GT, Chen T, Marshall AG, Weisbrod CR, Beu SC (2015) 21 tesla Fourier transform ion cyclotron resonance mass spectrometer: a National Resource for ultrahigh resolution mass analysis. J Am Soc Mass Spectrom 26(9):1626–1632. https://doi.org/10.1007/s13361-015-1182-2

Howard P, Meylan W, Aronson D, Stiteler W, Tunkel J, Comber M, Parkerton TF (2005) A new biodegradation prediction model specific to petroleum hydrocarbons. Environ Toxicol Chem 24(8):1847–1860

Hunt JM, Philp RP, Kvenvolden KA (2002) Early developments in petroleum geochemistry. Org Geochem 33(9):1025–1052. https://doi.org/10.1016/S0146-6380(02)00056-6

Jaggi A (2018) Dissolved organic matter in marine environments: a study of the origin, lability and molecular composition. University of Calgary, AB, Canada; https://prism.ucalgary.ca/handle/1880/106479; http://dx.doi.org/10.11575/PRISM/31770

Koch BP, Dittmar T, Witt M, Kattner G (2007) Fundamentals of molecular formula assignment to ultrahigh resolution mass data of natural organic matter. Anal Chem 79(4):1758–1763. https://doi.org/10.1021/ac061949s

Kujawinski EB, Kido Soule MC, Valentine DL, Boysen AK, Longnecker K, Redmond MC (2011) Fate of dispersants associated with the Deepwater horizon oil spill. Environ Sci Technol 45(4):1298–1306. https://doi.org/10.1021/es103838p

Kvenvolden KA (2002) History of the recognition of organic geochemistry in geoscience. Org Geochem 33(4):517–521. https://doi.org/10.1016/S0146-6380(01)00172-3

Liu Y, Kujawinski EB (2015) Chemical composition and potential environmental impacts of water-soluble polar crude oil components inferred from ESI FT-ICR MS. PLoS One 10(9):e0136376. https://doi.org/10.1371/journal.pone.0136376

Marshall AG, Rodgers RP (2008) Petroleomics: chemistry of the underworld. Proc Natl Acad Sci 105(47):18090–18095. https://doi.org/10.1073/pnas.0805069105

Marshall AG, Hendrickson CL, Jackson GS (1998) Fourier transform ion cyclotron resonance mass spectrometry: a primer. Mass Spectrom Rev 17(1):1–35. https://doi.org/10.1002/(SICI)1098-2787(1998)17:1<1::AID-MAS1>3.0.CO;2-K

McKenna AM, Nelson RK, Reddy CM, Savory JJ, Kaiser NK, Fitzsimmons JE, Marshall AG, Rodgers RP (2013) Expansion of the analytical window for oil spill characterization by ultrahigh resolution mass spectrometry: beyond gas chromatography. Environ Sci Technol 47(13):7530–7539. https://doi.org/10.1021/es305284t

Mopper K, Stubbins A, Ritchie JD, Bialk HM, Hatcher PG (2007) Advanced instrumental approaches for characterization of marine dissolved organic matter: extraction techniques, mass spectrometry, and nuclear magnetic resonance spectroscopy. Chem Rev 107(2):419–442. https://doi.org/10.1021/cr050359b

Mentges A, Feenders C, Seibt M, Blasius B, Dittmar T (2017) Functional molecular diversity of marine dissolved organic matter is reduced during degradation. Front Mar Sci, 4, p.194.

NAS (2016) Spills of Diluted Bitumen from pipelines: a comparative study of environmental fate, effects, and response. National Academies of Sciences, Engineering, and Medicine, Washington, D.C.

Nikolaev EN, Boldin IA, Jertz R, Baykut G (2011) Initial experimental characterization of a new ultra-high resolution FTICR cell with dynamic harmonization. J Am Soc Mass Spectrom 22:1125–1133. https://doi.org/10.1007/s13361-011-0125-9

Oldenburg TBP, Brown M, Bennett B, Larter SR (2014) The impact of thermal maturity level on the composition of crude oils, assessed using ultra-high resolution mass spectrometry. Org Geochem 75:151–168. https://doi.org/10.1016/j.orggeochem.2014.07.002

Oldenburg TBP, Jones M, Huang H, Bennett B, Shafiee NS, Head I, Larter SR (2017) The controls on the composition of biodegraded oils in the deep subsurface – part 4. Destruction and production of high molecular weight non-hydrocarbon species and destruction of aromatic hydrocarbons during progressive in-reservoir biodegradation. Org Geochem 114:57–80. https://doi.org/10.1016/j.orggeochem.2017.09.003

Overton EB, Wade TL, Radović JR, Meyer BM, Miles MS, Larter SR (2016) Chemical composition of macondo and other crude oils and compositional alterations during oil spills. Oceanography 29(3):50–63. https://doi.org/10.5670/oceanog.2016.62

Passow U, Ziervogel K, Asper V, Diercks A (2012) Marine snow formation in the aftermath of the Deepwater horizon oil spill in the Gulf of Mexico. Environ Res Lett 7(3):035301

Popov IA, Nagornov K, Vladimirov GN, Kostyukevich YI, Nikolaev EN (2014) Twelve million resolving power on 4.7 T Fourier transform ion cyclotron resonance instrument with dynamically harmonized cell—observation of fine structure in peptide mass spectra. J Am Soc Mass Spectrom 25:790–799. https://doi.org/10.1007/s13361-014-0846-7

Quigg A, Passow U, Chin W-C, Xu C, Doyle S, Bretherton L, Kamalanathan M, Williams AK, Sylvan JB, Finkel ZV, Knap AH, Schwehr KA, Zhang S, Sun L, Wade TL, Obeid W, Hatcher PG, Santschi PH (2016) The role of microbial exopolymers in determining the fate of oil and chemical dispersants in the ocean. Limnol Oceanogr Lett 1(1):3–26. https://doi.org/10.1002/lol2.10030

Radović JR, Domínguez C, Laffont K, Díez S, Readman JW, Albaigés J, Bayona JM (2012) Compositional properties characterizing commonly transported oils and controlling their fate in the marine environment. J Environ Monit 14(12):3220–3229. https://doi.org/10.1039/c2em30385j

Radović JR, Aeppli C, Nelson RK, Jimenez N, Reddy CM, Bayona JM, Albaigés J (2014) Assessment of photochemical processes in marine oil spill fingerprinting. Mar Pollut Bull 79(1–2):268–277. https://doi.org/10.1016/j.marpolbul.2013.11.029

Radović JR, Silva RC, Snowdon R, Brown M, Larter SR, Oldenburg TBP (2016a) A rapid method to assess a broad inventory of organic species in marine sediments using ultra-high resolution mass spectrometry. Rapid Commun Mass Spectrom 30(8):1273

Radović JR, Silva RC, Snowdon R, Larter SR, Oldenburg TBP (2016b) Rapid screening of glycerol ether lipid biomarkers in recent marine sediment using atmospheric pressure photoionization in positive mode Fourier transform ion cyclotron resonance mass spectrometry. Anal Chem 88(2):1128–1137. https://doi.org/10.1021/acs.analchem.5b02571

Radović JR, Oldenburg TBP, Larter SR (2018) Chapter 19 – Environmental assessment of spills related to oil exploitation in Canada's Oil Sands region. In: Wang Z, Stout SA (eds) Oil spill environmental forensics case studies. Heinemann, Butterworth, pp 401–417. https://doi.org/10.1016/B978-0-12-804434-6.00019-7

Riedel T, Dittmar T (2014) A method detection limit for the analysis of natural organic matter via Fourier transform ion cyclotron resonance mass spectrometry. Anal Chem 86(16):8376–8382. https://doi.org/10.1021/ac501946m

Rodgers RP, Marshall AG (2007) Petroleomics: advanced characterization of petroleum-derived materials by Fourier transform ion cyclotron resonance mass spectrometry (FT-ICR MS). In: Mullins OC, Sheu EY, Hammami A, Marshall AG (eds) Asphaltenes, heavy oils, and Petroleomics. Springer, New York, pp 63–93. https://doi.org/10.1007/0-387-68903-6_3

Rodgers RP, Hendrickson CL, Emmett MR, Marshall AG, Greaney MA, Qian K (2001) Molecular characterization of Petroporphyrins in crude oil by electrospray ionization Fourier transform ion cyclotron resonance mass spectrometry. Can J Chem 79:546–551. https://doi.org/10.1139/cjc-79-5/6-546

Ruddy BM, Huettel M, Kostka JE, Lobodin VV, Bythell BJ, McKenna AM, Aeppli C, Reddy CM, Nelson RK, Marshall AG, Rodgers RP (2014) Targeted Petroleomics: analytical investigation of Macondo well oil oxidation products from Pensacola Beach. Energy Fuel 28(6):4043–4050. https://doi.org/10.1021/ef500427n

Schouten S, Hopmans EC, Sinninghe Damsté JS (2013) The organic geochemistry of glycerol dialkyl glycerol tetraether lipids: a review. Org Geochem 54:19–61. https://doi.org/10.1016/j.orggeochem.2012.09.006

Seidel M, Kleindienst S, Dittmar T, Joye SB, Medeiros PM (2016) Biodegradation of crude oil and dispersants in deep seawater from the Gulf of Mexico: insights from ultra-high resolution mass spectrometry. Deep-Sea Res II Top Stud Oceanogr 129:108–118. https://doi.org/10.1016/j.dsr2.2015.05.012

Sleighter RL, Hatcher PG (2008) Molecular characterization of dissolved organic matter (DOM) along a river to ocean transect of the lower Chesapeake Bay by ultrahigh resolution electrospray ionization Fourier transform ion cyclotron resonance mass spectrometry. Mar Chem 110(3):140–152. https://doi.org/10.1016/j.marchem.2008.04.008

Stubbins A, Spencer RGM, Chen H, Hatcher PG, Mopper K, Hernes PJ, Mwamba VL, Mangangu AM, Wabakanghanzi JN, Six J (2010) Illuminated darkness: molecular signatures of Congo River dissolved organic matter and its photochemical alteration as revealed by ultrahigh precision mass spectrometry. Limnology and Oceanography 55(4):1467–1477. https://doi.org/10.4319/lo.2010.55.4.1467

Valentine DL, Fisher GB, Bagby SC, Nelson RK, Reddy CM, Sylva SP, Woo MA (2014) Fallout plume of submerged oil from *Deepwater Horizon*. Proc Natl Acad Sci 111(45):15906–15911. https://doi.org/10.1073/pnas.1414873111

Wade TL, Morales-McDevitt M, Bera G, Shi D, Sweet S, Wang B, Gold-Bouchot G, Quigg A, Knap AH (2017) A method for the production of large volumes of WAF and CEWAF for dosing mesocosms to understand marine oil snow formation. Heliyon 3(10):e00419. https://doi.org/10.1016/j.heliyon.2017.e00419

Ward CP, Sharpless CM, Valentine DL, French-McCay DP, Aeppli C, White HK, Rodgers RP, Gosselin KM, Nelson RK, Reddy CM (2018) Partial photochemical oxidation was a dominant fate of Deepwater horizon surface oil. Environ Sci Technol 52(4):1797–1805. https://doi.org/10.1021/acs.est.7b05948

Wörmer L, Elvert M, Fuchser J, Lipp JS, Buttigieg PL, Zabel M, Hinrichs K-U (2014) Ultrahigh-resolution paleoenvironmental records via direct laser-based analysis of lipid biomarkers in sediment core samples. Proc Natl Acad Sci 111(44):15669–15674. https://doi.org/10.1073/pnas.1405237111

Wozniak AS, Prem PM, Obeid W, Quigg A, Xu C, Santschi PH, Schwehr KA, Hatcher PG (2018) Rapid degradation of oil in mesocosm simulations of marine oil snow events. Environ Sci Technol https://doi.org/10.1021/acs.est.8b06532

Wu Z, Rodgers RP, Marshall AG (2004) Two- and three-dimensional van Krevelen diagrams: a graphical analysis complementary to the Kendrick mass plot for sorting elemental compositions of complex organic mixtures based on ultrahigh-resolution broadband Fourier transform ion cyclotron resonance mass measurements. Anal Chem 76(9):2511–2516. https://doi.org/10.1021/ac0355449

Chapter 16
Changes in Redox Conditions of Surface Sediments Following the Deepwater Horizon and Ixtoc 1 Events

David W. Hastings, Thea Bartlett, Gregg R. Brooks, Rebekka A. Larson, Kelly A. Quinn, Daniel Razionale, Patrick T. Schwing, Libia Hascibe Pérez Bernal, Ana Carolina Ruiz-Fernández, Joan-Albert Sánchez-Cabeza, and David J. Hollander

Abstract Following the blowout of the Macondo well, a sedimentation pulse resulted in significant changes in sedimentary redox conditions. This is demonstrated by downcore and temporal changes in the concentration of redox-sensitive metals: Mn and Re. Sediment cores collected in the NE Gulf of Mexico reveal increased sedimentation after the *Deepwater Horizon* (DWH) blowout. The formation of mucous-rich marine snow in surface waters and subsequent rapid deposition to sediments is the likely cause. Respiration of this material resulted in decreased pore-water oxygen and a shoaled redoxcline, resulting in two distinct Mn peaks in sediments following the event, one typically in the top 5–7 mm, with the other at 20–30 mm. Cores near the wellhead reveal this nonsteady-state behavior for 3–5 years after the event. A time series reveals that bulk sediment Re increased 3–4 times compared to the pre-impact baseline value for 2–3 years indicating sediments are increasingly more reducing. Three years after the blowout, subsurface Re reaches a plateau suggesting a return to steady-state conditions. In select sites where

D. W. Hastings (✉) · G. R. Brooks · D. Razionale
Eckerd College, St. Petersburg, FL, USA
e-mail: hastindw@eckerd.edu; brooksgr@eckerd.edu; dprazion@eckerd.edu

T. Bartlett · K. A. Quinn · P. T. Schwing · D. J. Hollander
College of Marine Science, University of South Florida, St. Petersburg, FL, USA
e-mail: theabartlett@mail.usf.edu; kaquinn2@usf.edu; pschwing@mail.usf.edu; davidh@usf.edu

R. A. Larson
Eckerd College, St. Petersburg, FL, USA

College of Marine Science, University of South Florida, St. Petersburg, FL, USA
e-mail: larsonra@eckerd.edu

L. H. P. Bernal · A. C. Ruiz-Fernández · J.-A. Sánchez-Cabeza
Instituto de Ciencias del Mar y Limnología, Universidad Nacional Autónoma de México, Mexico City, Mexico
e-mail: lbernal@ola.icmyl.unam.mx; caro@ola.icmyl.unam.mx; jasanchez@cmarl.unam.mx

© Springer Nature Switzerland AG 2020
S. A. Murawski et al. (eds.), *Deep Oil Spills*,
https://doi.org/10.1007/978-3-030-11605-7_16

benthic foraminifera were counted, an assemblage-wide decrease is coincident with reducing conditions, demonstrating the important consequences of changing redox conditions on benthic ecosystems.

Another major submarine blowout in the southern Gulf of Mexico (Ixtoc 1; 1979–1980) released a large volume of crude oil below the surface. We observe multiple Mn oxide peaks associated with a shoaling redoxcline and Re maxima associated with more reducing conditions. Nonsteady-state behavior at sites near DWH and Ixtoc 1 is consistent with a MOSSFA (marine oil snow sedimentation and flocculent accumulation) event at both locations.

Keywords Oil spill · Gulf of Mexico · Deepwater Horizon · Redox · Trace metal · Rhenium · Manganese

16.1 Introduction

The oil spill associated with the 2010 blowout of the Macondo Well at the *Deepwater Horizon* oil rig released more oil than any other marine oil spill in history from an ultra-deep well at 1500 m. The exact amount is somewhat controversial; Joye et al. (2011) estimated the amount released was between 4.5 and 6.3 million barrels, while the US District Court (2015) set the amount at four million barrels released, which is considered a legal compromise. More dispersant was used in cleanup operations than ever before (about 2.1 million gallons). Jack Davis referred to it as "the worst oil spill in history" in his recent Pulitzer Prize winning book on the history of the Gulf of Mexico (Davis 2017).

Another remarkable aspect was the unexpected sedimentation event during and after the blowout. A massive pulse in sedimentation over an extensive area in the NE Gulf of Mexico (NGoM) occurred over a 4- to 5-month period during and after the oil spill (Daly et al. 2016; Passow 2016). This mass deposition was likely due to the formation of mucous-rich marine snow in surface waters and subsequent rapid deposition to marine sediments (Daly et al. 2016; Passow 2016). Mass accumulation rates following the event were far greater than before the spill (Brooks et al. 2015; Larson et al. 2020). In spite of its lower density, oil from the spill was transported to the seafloor via rapidly sinking, oil-mineral aggregates. This oil-associated marine snow became a focus of attention and resulted in the description of a new phenomenon termed marine oil snow sedimentation and flocculent accumulation (MOSSFA), fully explained in Chaps. 12 (Quigg et al. 2020) and 13 (Schwing et al. 2020).

The spatial extent of MOSSFA in the deep Gulf of Mexico was on the order of 35,000 km^2 (Romero et al. 2017). Several mechanisms have been proposed to explain the formation of the MOSSFA event, including coagulation of phytoplankton and/or suspended matter with the oil droplets; and production of mucosoid material from the degraders of the oil, which grew and multiplied rapidly following the event (Passow et al. 2012). The microbial production of sticky transparent exopolymeric particles (TEP) enhances the aggregation process, as would the release of exopolymeric substances (EPS) by phytoplankton (Passow 2000; Verdugo and Santschi 2010). Marine snow may have formed within subsurface oil intrusions as well as at the surface (Ziervogel et al. 2012; Passow et al. 2012; Passow 2014). Consistent with this pulse, a visually distinctive, brown, fine-grained surface layer in the top 1–2 cm with dark brown or black bands is seen in sediment cores affected by the blowout.

What is the consequence of this pulse of organic-rich material once it reached the seafloor? Our hypothesis is that it resulted in increased respiration of organic carbon in the sediments, which in turn decreased oxygen concentration in sediment pore waters. Since pore water oxygen was not measured, we use changes in the relative concentration of two redox-sensitive elements, manganese (Mn) and rhenium (Re), to constrain changes in the redox state of marine sediments following the blowout.

A brief review of the chemistry of the select redox-sensitive metals (Mn, Re) is useful to allow interpretation of our data (e.g., Algeo and Rowe 2012; Morford and Emerson 1999; Tribovillard et al. 2006; Crusius and Thomson 2000). The redox chemistry of Mn in marine sediments is central to understanding redox conditions, due to the importance of Mn cycling across the redoxcline in reducing environments. Mn exists primarily in two oxidation states: Mn(IV) and Mn(II). Mn(IV) oxides are readily reduced to aqueous Mn(II) in reducing environments. Soluble Mn(II) diffuses upward, and where pore water oxygen is present, it is then oxidized to Mn(IV) oxide completing the Mn redox cycle (e.g., Froelich et al. 1979; Burdige and Gieskes 1983; Gobeil et al. 1997). A distinct peak in bulk Mn typically marks the top of the redoxcline. Recent findings of abundant pore water Mn(III) in hemiplegic sediments require a revision of this classic redox model to include one electron transfer reactions for the Mn cycle (Madison et al. 2013).

Rhenium is mobile under oxic conditions and precipitates under mildly reducing and anoxic conditions (Crusius et al. 1996). It is ideally suited as a redox tracer since its detrital concentration is very low relative to authigenic deposition; oxic marine sediments have low concentrations of 0.4–0.6 ppb (Boyko et al. 1986; Koide et al. 1986). Re is enriched in reducing sediments because dissolved $Re(VII)O_4^-$ is reduced and precipitates in the solid phase, most likely as $Re(IV)O_2$ (Crusius et al. 1996). Re enrichment occurs under both anoxic (Colodner et al. 1993) and suboxic conditions, below the zone of Fe reduction and prior to sulfate reduction (Crusius et al. 1996; Morford et al. 2005).

In summary, oxic sediments are typically characterized by elevated Mn values, while more reducing and anoxic sediments have relatively low Mn content and elevated Re.

16.2   Analytical Approach

As part of the C-IMAGE I and II and Deep-C consortia, we collected sediment cores at over 250 stations in the Gulf of Mexico (GoM) from August 2010 to December 2017 (Larson et al. 2020). In this chapter, we present data from select cores with an emphasis on the temporal evolution (Fig. 16.1). Cores were collected

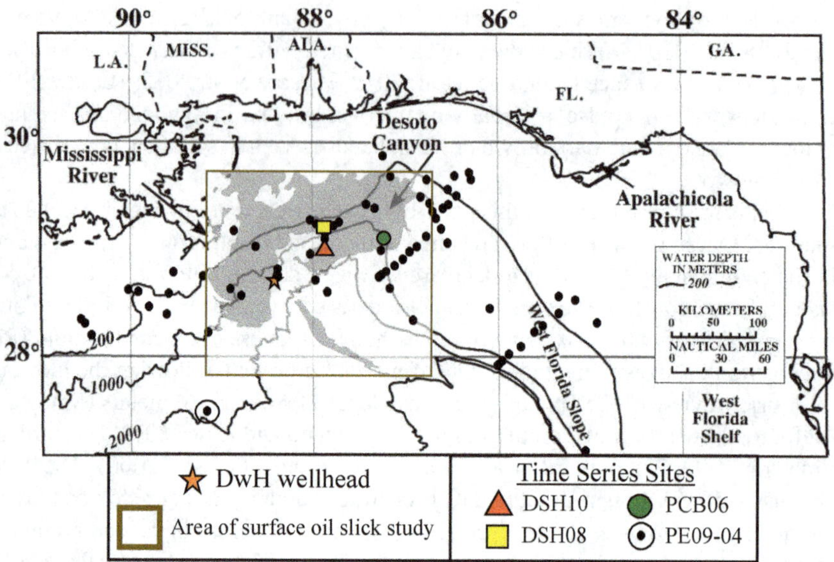

Fig. 16.1 Map of sites. (Modified from Brooks et al. 2015)

without disturbing the sediment-water interface and carefully extruded at very high resolution: 2 mm resolution in the top 20 mm and 5 mm resolution below 20 mm (Schwing et al. 2016).

Several sediment cores were used as a pre-event control, including a box core beneath a long-term sediment trap mooring in the NGoM collected in January 2009 (PE09–04 MC1;1132 m; 27° 31.508 N, 90° 10.126 W).

Subsamples were freeze dried and microwave digested in closed Teflon® digestion vessels with concentrated trace metal grade HNO_3 at high temperature (165 °C) and high pressure (~25 bar) for 15 minutes according to standard methods (US EPA method 3051a) (EPA 1994). The digest was diluted, filtered, and spiked with an internal standard to correct for instrumental drift. Since hydrofluoric acid was not used in the digestion method, the digest did not include refractory components such as aluminosilicates, but did include authigenic phases, crude oil, organic phases, FeMn oxides, and carbonates. Digested samples were analyzed by inductively coupled plasma mass spectrometry (ICP-MS).

16.3   Results and Discussion

16.3.1   *Pre-impact Geochemistry*

It is well established that the downward flux of organic carbon to the sediments from overlying surface waters, and the subsequent oxidation of that carbon, is the driving force for early diagenesis – transformation of organic carbon – in marine sediments, resulting in the reduction of a series of electron acceptors, oxygen, nitrate, Mn oxides, Fe oxides, and sulfate, in order of decreasing free energy per mole organic carbon (e.g., Froelich et al. 1979; Emerson et al. 1980). After oxygen is nearly depleted in these sediments, nitrate is reduced, followed by reduction of solid Mn(IV) oxides to aqueous Mn(II). The dissolved Mn(II) diffuses upward, is re-oxidized to Mn(IV) by pore water oxygen, and is then trapped within the sediment. Under steady-state conditions, this leads to the formation of a single Mn peak, which typically defines the depth of the redoxcline (e.g., Burdige and Gieskes 1983). The redox state of surface sediments and the depth of the redoxcline is controlled by the flux of organic carbon to sediments, bottom water oxygen concentration, and how labile the carbon is (e.g., Emerson et al. 1985).

We first establish natural, pre-event levels of redox-sensitive metals using sediment cores in the NGoM. Site PE 09-04 shows a characteristic single Mn oxide peak at 77 mm, with enrichment of Re below that at 85–100 mm (Fig. 16.2). This is consistent with typical Mn profiles in slope sediments where Mn is relatively constant in oxic surface sediments, increases to a single and substantial Mn oxide peak just above the manganese redox boundary, and then decreases to baseline levels as pore water oxygen decreases (e.g., Burdige and Gieskes 1983; Burdige 1993). At this site, surface Re is low, ~0.5 ppb in the oxic surficial sediments with Re enrichment below the redoxcline; this Re enrichment is an excellent indicator of more reducing sediments. Three other sites in the NGoM were sampled to establish pre-event conditions and are similar (Hastings et al. 2016).

This single Mn peak typical of continental slope sediments is observed in other cores we sampled before the event from Fisk Basin and Garrison Basin at 80 and 165 mm, respectively (Hastings et al. 2016). The Mn peak is considerably shallower in pre-event cores near the wellhead (e.g., PE-1031-6) since they are closer to the Mississippi River, with its high nutrient load and resulting high-productivity and high-sedimentation rate. As before, Re increases below the Mn peak, consistent with Re enrichment in mildly reducing sediments.

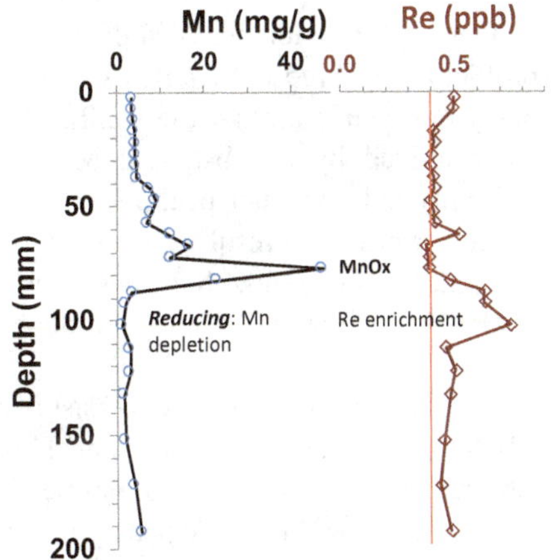

Fig. 16.2 Pre-impact redox-sensitive metal profiles at site PE09-04. A single Mn peak is typical and defines the depth of the redoxcline. Re enrichment is just below the redoxcline, indicating reducing sediments

16.3.2 Post-impact: Organic Geochemistry

Following the blowout event, the MOSSFA event and accumulation of marine snow resulted in a substantial sedimentation pulse. In addition to an increase in sedimentation rate, the quality of the organic carbon associated with the sediment pulse and the oil-associated marine snow was likely more labile and would be remineralized more quickly than the relatively low-quality partially degraded fecal pellets that make up most of the organic carbon rain. "High-quality," fresh diatom-derived organic carbon is remineralized >300% faster than "low-quality" carbon from fecal pellets, showing that different organic matter makes a significant difference in respiration rates and residence time of carbon on the ocean floor (Mayor et al. 2012). At least 50% of the recently deposited aliphatic hydrocarbons in each of the three times series cores degraded between the December 2010 and February 2011 sampling (Romero et al. 2017, 2020).

16.3.3 Post-impact: Mn Geochemistry

Our approach is to describe general trends and present downcore results that are representative of elemental profiles at different stations, rather than present data from all the downcore metal profiles from the different sites we sampled. One of the most interesting changes in the geochemistry of surface sediments after the event is the occurrence of multiple Mn oxide peaks. For example, two distinct Mn peaks are present at 5 and 20 mm at DSH10 (1520 m) in August 2010 with a modest Re enrichment of 0.1–0.2 ppb coincident with the Mn minimum between the two peaks (Fig. 16.3a). Below the second Mn peak, Re increases to a distinct relative maximum of 2.1 ppb. The DSH10 profile sampled two years later is another illustration of the double Mn peaks. As before, there is a large Mn peak close to the surface (5 mm), and a smaller Mn peak at 20 mm (Fig. 16.3b). There is also a Re peak at the base of the second Mn peak at 29 mm, just below the Mn redox boundary.

Double Mn oxide peaks in the top 40 mm are observed in virtually every depth profile at the three time series stations (DSH10, DSH08, and PCB06) as well as many other sites around the wellhead sampled after the event. The Mn profile is characterized by a substantial Mn peak close to the sediment-water interface (~5 mm) and a smaller peak, deeper in the sediment, about 20–35 mm deep. The shallower peak is a result of the shoaled redoxcline following the large sedimentation pulse associated with the event with the relic Mn peak beneath it. See Hastings et al. (2016) for more results and depth profiles demonstrating these multiple Mn peaks.

Do these double Mn peaks persist over time? We sampled these three sites annually for 8 years following the event. This time series reveal that double Mn peaks or a broad peak spanning 10–20 mm persists for 3–5 years following the event. Over time, the shallow Mn oxide peak gradually decreases in magnitude, while the deeper

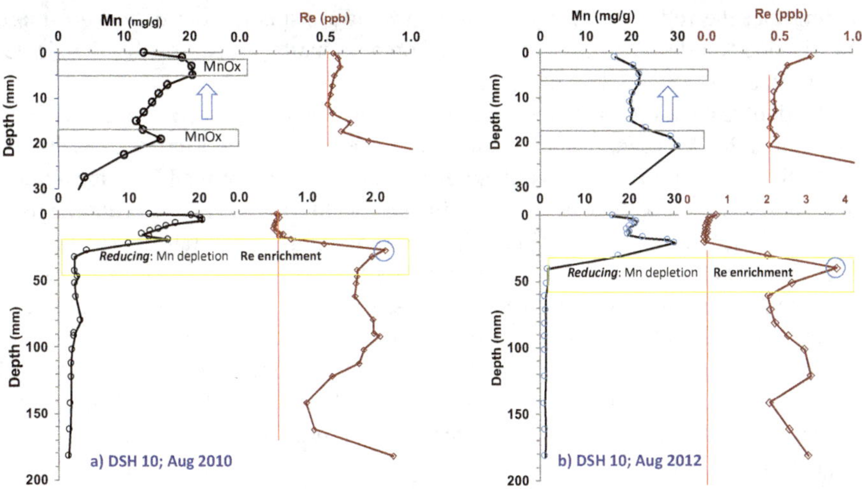

Fig. 16.3 Redox-sensitive metal profiles: (**a**) DSH10, Aug 2010; (**b**) DSH10, Aug 2012. Depth scale is expanded in the top pane. Note shoaling of Mn peak from ~20 to 5 mm in both profiles. The increasing subsurface Re (circled in red) over the 2-year interval between samples indicates more reducing conditions. This subsurface Re maximum is presented in Fig. 16.3 over 8 years

peak increases. The most recent profile at DSH10 reveals a single Mn peak at 20–30 mm, similar to the pre-event profile.

In the December 2010 profile at DSH08, the surface Mn peak is 16 mg/g but decreases over the next several years of sampling. By 2017, the surface Mn peak decreased to 6 mg/g Mn. The unusual double Mn peaks are indicative of nonsteady-state behavior, and the pulse of reactive carbon fades over the next 3–5 years reverting back to the canonical single Mn peak. This is the same time frame required for the diversity of benthic foraminifera to return to steady-state conditions (Schwing et al. 2018a) and for the incorporation of oil-derived $\delta^{13}C$ in the foraminiferal tests to return to steady state (Schwing et al. 2018b).

16.3.3.1   Post-impact: Explanation of Double Mn Peak

The sedimentation pulse mixed with fresh organic carbon would result in increased microbial respiration with consumption of oxygen and a shoaling or shallowing, of the redoxcline in the sediments. As the redox boundary migrates upward, a new, shallower Mn oxide peak would be expected to form, leaving behind a relic Mn peak at depth. The observation of two or more Mn peaks, and the concept of nonsteady-state diagenesis, was explored to describe paleoredox variations and changes in paleoproductivity (Finney et al. 1988), in the modern carbon cycle (Kuzyk et al. 2011), and in turbidite sequences in the North East Atlantic (Colley et al. 1984; Thomson et al. 1993). Pulsed sedimentation to the seafloor is not uncommon (e.g., Honjo et al. 1982; Gobeil et al. 1997; Gobeil et al. 2001). It is worth noting that the redox boundary can migrate upward quickly but downward more slowly due to relatively fast oxidation of organic matter in contrast to the slower diffusion of oxygen (Gobeil et al. 2001).

No double peak for Mn is observed at select sites including DSH08 and PCB06 in September 2011 and PCB06 in August 2012. The deeper relic Mn peak is subtle at DSH08 in August 2012. This is likely due to the reduction of the Mn oxide associated with the relic peak under more reducing conditions. There is a wide range in reported kinetics for the reduction of Mn oxide with first-order rate constants ranging from 0.002 d^{-1} for pelagic sediments (Burdige and Gieskes 1983) to 8–60 d^{-1} for coastal environments (Aller 1980). A pulse of organic carbon to the sea floor can trigger reduction of Mn and Fe oxides within several days with a first-order rate constant of 0.159 d^{-1} (Magen et al. 2011), demonstrating that Mn oxides in surficial sediments can be rapidly reduced. By August 2013 the relic Mn oxide peak is no longer present in the majority of the sites; by August 2016 only one Mn oxide peak is present at all sites.

16.3.4 Post-impact: Re Geochemistry

Re is the clearest indicator of reducing conditions, since detrital Re is very low, has no known role in biogeochemical processes, and is not associated with cycling of Mn oxides (Schaller et al. 2000; Morford et al. 2005). Before the event, downcore Re is effectively constant at 0.5 ppb in surface sediments at PE09-04 (Fig. 16.2). Below the Mn oxide peak, Re increases as sediments become more reducing with depth.

Several features of the Re profiles provide insight into the evolution of reducing conditions following the impact. A modest Re enrichment of 0.1–0.3 ppb typically occurs between the two Mn peaks, and coincident with the Mn minimum, characteristically at 15–20 mm between August 2010 and February 2011 (Fig. 16.3a). This infers reducing conditions relatively shallow in the sediment. This is consistent with the massive pulse of organic-rich material to surface sediments during and following the MOSSFA event, which is rapidly respired, leading to rapid consumption of pore water oxygen, reducing conditions, and a shoaled redoxcline. A distinct subsurface Re peak at 30–40 mm at DSH10 (Figs. 16.3a and 16.3b) and at 50–70 mm at PCB06 is consistent with substantially more reducing conditions following the sediment pulse as organic carbon degrades relatively rapidly in surface sediments.

16.3.4.1 Post-impact: Evolution of Re Enrichment

Three sites were sampled 9–10 times between August 2010 and August 2017, allowing us to determine how the event impacted surficial sediments and to constrain the temporal evolution of reducing conditions. We are able to best describe the temporal evolution of reducing conditions by measuring the change of Re over time.

We examined the change in Re for three stations for which an 8-year time series exists. Since Re changes substantially with depth, as well as over time, choosing a consistent depth interval is important. For DSH10 and PCB06, we chose the distinct Re peak centered at 40 mm and 60 mm, respectively. For DSH08 where the Re maximum is less distinct, we chose Re at 50 mm, which is where Re peaks occurred at the two other sites. We determined the pre-impact value for Re based on the suite of cores (PE1031) collected at different water depths just after the blowout occurred at the wellhead, and before substantial sedimentation or deposition of oil. These pre-impact Re values are indicated on the y-axis on Fig. 16.4.

At each site, subsurface authigenic Re concentrations increased substantially over 2 years following the blowout, increasing 3–4 times compared to the pre-impact value for Re (Fig. 16.4). The increase in Re demonstrates a clear change in sedimentary redox conditions after the event and continuing over the next 2 years. Re enrichment suggests persistent reducing conditions due to continued remineralization of organic carbon. The pulse of organic carbon must be incorporated into the sediment – not simply at the sediment-water interface – for it to have an effect on the oxygen content within the sediment (Gobeil et al. 1997). Since accumulation of sediment from the pulse was at least 20 mm, this criterion is satisfied.

Three years after the blowout and the first sampling, subsurface Re values no longer increase, but reach a plateau with some variability over time. This is the same time frame for benthic foraminifera to return to steady state (Schwing et al. 2018b). Re values in August 2013 are lower for both DSH10 and PCB06, indicating a possible return toward pre-impact conditions or integration of the subsurface Re peak into adjacent depths. Subsurface Re does not increase for the next 5 years, from 2013 to 2017.

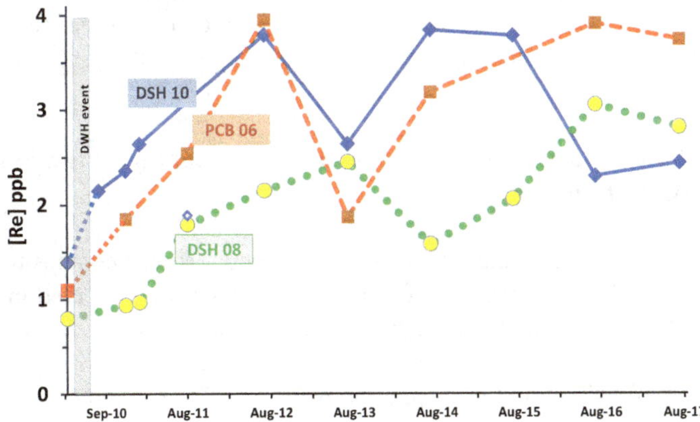

Fig. 16.4 Temporal evolution of Re enrichment following the DWH blowout. Re associated with the subsurface, Re peak is plotted for the same three stations over 8 years of sampling. Estimates of pre-event Re are indicated on the y-axis. Re increased 3–4 times following the DWH event as a result of more reducing conditions

16.3.5 Post-impact: Ecological Consequences – Benthic Foraminifera

Downcore changes in Mn and Re reveal clear evidence of changes in redox conditions following the event. What is the ecological impact of such changes on the benthic community? Given that redox changes occurred at the mm scale, benthic foraminifera are potentially sensitive indicators of changes in reducing conditions. We have documented changes in density of benthic foraminifera at two sites, DSH08 and PCB06, in December 2010 and February 2011 (Schwing et al. 2015; Schwing and Machain Castillo 2020). Where Mn and Re indicate significant reducing conditions by Mn depletion and Re enrichment, there is a clear decrease in density of the dominant genera of benthic foraminifera (Hastings et al. 2016; Schwing et al. 2015).

At site DSH08 in December 2010, a 40–60% reduction in both *Bulimina* spp. and *Uvigerina* spp., the most abundant genera of benthic foraminifera, occurs between 13 and 17 mm. This is coincident with the Mn minimum and a modest (0.20 ppb; 35%) Re enrichment (Hastings et al. 2016). At the surface, a significant decrease in density of benthic foraminifera occurs, while a modest (0.16 ppb; 30%) Re increase and Mn minimum occurs, consistent with more reducing conditions. On the subsequent sampling date, February 2011, a dramatic reduction in the most abundant genera, *Bulimina* spp. and *Uvigerina* spp., occurs at 13 mm, coincident with the Mn minimum and a 0.14 ppb increase in Re (Hastings et al. 2016). A similarly large reduction in *Brizalina* spp. and *Bolivina* spp. also occurs at this depth (Schwing et al. 2015). Site PCB06 in December 2010 and February 2011 reveals a similar relationship between redox-sensitive metals and benthic foraminiferal density: a significant and substantial decrease in density of the three dominant genera occurs when both Mn and Re indicate reducing conditions. While reducing conditions can explain these changes, polycyclic aromatic hydrocarbons (PAHs) and other toxic organic compounds found in crude oil may be another reason for the changes in foraminiferal density.

16.3.5.1 Post-impact: Ecological Consequences – Benthic Foraminifera (C-13 Depletion)

A modest but persistent depletion in the stable carbon isotope composition of benthic foraminiferal calcite extracted from impacted sites was observed in 2010 and early 2011 (Schwing et al. 2018b; Schwing and Machain Castillo 2020). The depletion was likely caused by intensifying reducing conditions brought about by increased microbial respiration of marine oil snow. Benthic foraminiferal carbon isotope composition took about 3 years to return to steady state, which corresponds to less intense reducing conditions, and the steady state achieved in Re and Mn concentrations we observe during the same time frame.

16.3.5.2   Post-impact: Ecological Consequences – Benthic Foraminiferal Resilience

Following the decrease in abundance after the DWH event, the diversity of benthic foraminifera at impacted sites increased from 2011 to 2013 and plateaued from 2013 to 2015 (Schwing et al. 2018a; Schwing and Machain Castillo 2020). This is the same time frame as the return to steady-state conditions seen in redox-sensitive elements Re and Mn.

16.3.6   Thirty Years After Ixtoc 1: What Do Redox-Sensitive Metals Reveal?

Thirty years before the *Deepwater Horizon* event, another major submarine blowout of 130 million gallons of crude oil occurred in the Bay of Campeche of the Gulf of Mexico in 1979–1980. The events were similar in many aspects (Schwing et al. 2020), including the massive accidental release of crude oil in the water column, followed by the application of million gallons of dispersant. The three decades between the two blowouts allow us to compare the two events, assess the fate of the oil, and estimate the possible long-term impacts and implications of the DWH spill.

This earlier spill allows us to test the hypothesis that the large sedimentation pulse which was a consequence of the 2010 blowout is not unique to the DWH event and also occurred following the blowout at Ixtoc 1. In July 2015, we sampled sites near the wellhead where the blowout occurred, and west and northwest of the wellhead where the crude oil drifted. The same sampling approach and analytical methods were used to determine redox-sensitive metals in sediments. Pore water oxygen was measured in situ with an optode fiber microsensor.

At the shallowest site located at the wellhead (Ixtoc 1, 60 m depth), Mn is reduced at the surface, with an oxygen penetration depth of 1 mm, virtually at the sediment-water interface (Fig. 16.5a). This is expected given the relatively shallow water depth and high-sedimentation rate. No Mn(IV) oxide exists. At IXW 500 (1200 m water depth), IXN 500 (1240 m), and IXNW 750 (2043 m) multiple Mn oxide peaks are present (Fig. 16.5b, c, d), which is consistent with a shoaled redoxcline following a MOSSFA event similar to the one observed and documented after the DWH blowout.

These Mn profiles document nonsteady-state behavior in both NGoM and sites near Ixtoc 1, consistent with the hypothesis that MOSSFA events, with a large pulse of organic-rich material, are not unique to the DWH event and may occur in many spills of this type.

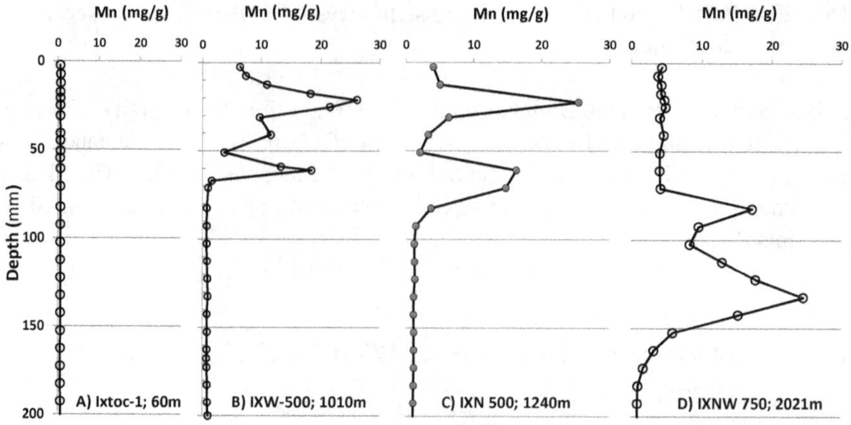

Fig. 16.5 Mn profiles at sites close to site of Ixtoc 1 blowout in the SGoM in increasing water depth. Note double MnOx peak at each site, except for the shallow site at the wellhead

16.4 Conclusions

Following the blowout, a substantial marine snow episode was recorded in NGoM sediments which led to a large sedimentation pulse. Subsequent respiration of the organic carbon associated with the marine snow resulted in reducing conditions, as evidenced by downcore changes in redox-sensitive metals Mn and Re.

After the event, double Mn peaks are present in surface sediments consistent with the nonsteady-state behavior associated with a shoaled redoxcline, precipitation of a new Mn peak, and the deeper relic Mn peak. These double Mn peaks result from the pulse of labile organic carbon to surface sediments. This is a result of the oil-associated marine snow, first documented in this oil spill, referred to as MOSSFA. Double Mn peaks are seen at depth at numerous sites in the SGoM near the Ixtoc 1 wellhead, the site of a massive oil spill in 1979. This is consistent with a MOSSFA event during the Ixtoc 1 spill, suggesting that oil-associated marine snow is not unique to the DWH event.

Re in subsurface sediments increases for 2–3 years after the event at all time series sites, demonstrating that sediments became more reducing. From 2013 to 2017, subsurface Re values remain high and relatively constant, suggesting a return to steady-state conditions. Benthic foraminifera indices affected by the event, including abundance, carbon isotopic composition, and diversity returned to steady-state conditions after 2–3 years, the same time frame required for sedimentary Re to return to steady-state values. These results suggest an important consequence of changing redox conditions in sediments to biotic communities following the DWH event.

Acknowledgments Many thanks to the numerous Eckerd College undergraduate students who helped in the laboratory and at sea including Brigid Carr, Shannon Hammaker, Chloe Holzinger, Farley Miller, Claire Miller, and Corday Selden. Grateful acknowledgments to Alan Shiller, who provided important insight at a critical time. We are grateful to the exceptional crew of the R/V Weatherbird II for their skilled help at sea collecting samples and staying safe during the field operations.

This research was made possible by funding from The Gulf of Mexico Research Initiative to the Center for the Integrated Modeling and Analysis of the Gulf Ecosystem (C-IMAGE) Consortium Deep and the Deep Sea to Coast Connectivity in the Eastern Gulf of Mexico (Deep-C) Consortium. We also acknowledge partial funding for summer student support from Eckerd College NSSRP program. The complete data set, including all elements determined by ICP-MS, can be accessed at the GRIIDC website: https://data.gulfresearchinitiative.org/ (doi: 10.7266/ N7DN43JM; 10.7266/N7C24TD; 10.7266/N7RX9914).

References

Algeo TJ, Rowe H (2012) Paleoceanographic applications of trace-metal concentration data. Chem Geol 324-325:6–18. https://doi.org/10.1016/j.chemgeo.2011.09.002

Aller RC (1980) Diagenetic processes near the sediment-water interface of Long Island sound. II. Fe and Mn. In: Barry S (ed) Advances in geophysics, vol 22. Elsevier, pp 351–415. https://doi.org/10.1016/S0065-2687(08)60068-0

Boyko T, Baturin G, Miller A (1986) Rhenium in recent ocean sediments. Geochem Int 23:38–47

Brooks GR, Larson RA, Schwing PT, Romero I, Moore C, Reichart G-J, Jilbert T, Chanton JP, Hastings DW, Overholt WA, Marks KP, Kostka JE, Holmes CW, Hollander D (2015) Sedimentation Pulse in the NE Gulf of Mexico following the 2010 DWH Blowout. PLoS One 10(7):e0132341. https://doi.org/10.1371/journal.pone.0132341

Bruce P. Finney, Mitchell W. Lyle, G. Ross Heath, (1988) Sedimentation at MANOP Site H (eastern equatorial Pacific) over the past 400,000 years: Climatically induced redox variations and their effects on transition metal cycling.Paleoceanography 3 (2):169–189

Burdige DJ (1993) The biogeochemistry of manganese and iron reduction in marine sediments. Earth Sci Rev 35(3):249–284. https://doi.org/10.1016/0012-8252(93)90040-E

Burdige DJ, Gieskes JM (1983) A pore water/solid phase diagenetic model for manganese in marine sediments. Am J Sci 283(1):29–47. https://doi.org/10.2475/ajs.283.1.29

Colley S, Thomson J, Wilson TRS, Higgs NC (1984) Post-depositional migration of elements during diagenesis in brown clays and turbidite sequences in the Northeast Atlantic. Geochim Cosmochim Acta 48:1223–1236

Colodner D, Sachs J, Ravizza G, Turekian K, Edmond J, Boyle E (1993) The geochemical cycle of rhenium: a reconnaissance. Earth Planet Sci Lett 117:205–221

Crusius J, Calvert S, Pedersen T, Sage D (1996) Rhenium and molybdenum enrichments in sediments as indicators of oxic, suboxic and sulfidic conditions of deposition. Earth Planet Sci Lett 145:65–78

Crusius J, Thomson J (2000) Comparative behavior of authigenic Re, U, and Mo during reoxidation and subsequent long-term burial in marine sediments. Geochim Cosmochim Acta 64(13):2233–2242. https://doi.org/10.1016/S0016-7037(99)00433-0

Daly KL, Passow U, Chanton J, Hollander D (2016) Assessing the impacts of oil-associated marine snow formation and sedimentation during and after the Deepwater Horizon oil spill. Anthropocene 13:18–33. https://doi.org/10.1016/j.ancene.2016.01.006

Davis JE (2017) The Gulf: the making of an American Sea. Liveright Publishing, New York

Emerson S, Jahnke R, Bender M, Froelich P, Klinkhammer G, Bowser C, Setlock G (1980) Early diagenesis in sediments from the eastern equatorial Pacific. I. Pore water nutrient and carbonate profiles. Earth Planet Sci Lett 49:57–80

Emerson SE, Fischer K, Reimers C, Heggie D (1985) Organic carbon dynamics and preservation in deep-sea sediments. Deep-Sea Res 32:1–22

EPA (1994) Microwave assisted acid digestion of sediments, sludges, soils, and oils, Method 3051. U.S. Government Printing Office, Washington, DC

Froelich PN, Klinkhammer GP, Bender ML, Leudtke N, Heath GR, Cullen D, Dauphin P, Hammond D, Hartman B (1979) Early oxidation of organic matter in pelagic sediments of the eastern Equatorial Atlantic: suboxic diagenesis. Geochim Cosmochim Acta 43:1075–1090

Finney BP, Lyle MW, Heath GR (1988) Sedimentation at MANOP Site H (eastern equatorial Pacific) over the past 400,000 years: climatically induced redox variations and their effects on transition metal cycling. Paleoceanography 3(2):169–189

Gobeil C, Macdonald RW, Sundby B (1997) Diagenetic separation of cadmium and manganese in suboxic continental margin sediments. Geochim Cosmochim Acta 61(21):4647–4654. https://doi.org/10.1016/S0016-7037(97)00255-X

Gobeil C, Sundby B, Macdonald RW, Smith JN (2001) Recent change in organic carbon flux to Arctic Ocean deep basins: evidence from acid volatile sulfide, manganese and rhenium discord in sediments. Geophys Res Lett 28(9):1743–1746. https://doi.org/10.1029/2000GL012491

Hastings DW, Schwing PT, Brooks GR, Larson RA, Morford JL, Roeder T, Quinn KA, Bartlett T, Romero IC, Hollander DJ (2016) Changes in sediment redox conditions following the BP DWH blowout event. Deep-Sea Res II Top Stud Oceanogr 129:167–178. https://doi.org/10.1016/j.dsr2.2014.12.009

Honjo S, Manganini SJ, Cole JJ (1982) Sedimentation of biogenic matter in the deep ocean. Deep-Sea Res 29(5):609–625. https://doi.org/10.1016/0198-0149(82)90079-6

Joye SB, MacDonald IR, Leifer I, Asper V (2011) Magnitude and oxidation potential of hydrocarbon gases released from the BP oil well blowout. Nat Geosci 4(3):160–164. http://www.nature.com/ngeo/journal/v4/n3/abs/ngeo1067.html#supplementary-information

Koide M, Hodge VF, Yang JS, Stallard M, Goldberg EG, Calhoun J, Bertine KK (1986) Some comparative marine chemistries of rhenium, gold, silver and molybdenum. Appl Geochem 1(6):705–714. https://doi.org/10.1016/0883-2927(86)90092-2

Kuzyk ZZA, Macdonald RW, Stern GA, Gobeil C (2011) Inferences about the modern organic carbon cycle from diagenesis of redox-sensitive elements in Hudson Bay. J Mar Syst 88(3):451–462. https://doi.org/10.1016/j.jmarsys.2010.11.001

Larson RA, Brooks GR, Schwing PT, Diercks AR, Holmes CW, Chanton JP, Diaz-Asencio M, Hollander DJ (2020) Characterization of the sedimentation associated with the Deepwater Horizon blowout: depositional pulse, initial response, and stabilization (Chap. 14). In: Murawski SA, Ainsworth C, Gilbert S, Hollander D, Paris CB, Schlüter M, Wetzel D (eds) Scenarios and responses to future deep oil spills – fighting the next war. Springer, Cham

Madison AS, Tebo BM, Mucci A, Sundby B, Luther GW (2013) Abundant Porewater Mn(III) is a major component of the sedimentary redox system. Science 341(6148):875–878. https://doi.org/10.1126/science.1241396

Magen C, Mucci A, Sundby B (2011) Reduction rates of sedimentary Mn and Fe oxides: an incubation experiment with Arctic Ocean sediments. Aquat Geochem 17(4–5):629–643. https://doi.org/10.1007/s10498-010-9117-9

Mayor DJ, Thornton B, Hay S, Zuur AF, Nicol GW, McWilliam JM, Witte UF (2012) Resource quality affects carbon cycling in deep-sea sediments. ISME J 6(9):1740–1748. https://doi.org/10.1038/ismej.2012.14

Morford JL, Emerson SE (1999) The geochemistry of redox sensitive trace metals in sediments. Geochim Cosmochim Acta 63(11–12):1735–1750. https://doi.org/10.1016/S0016-7037(99)00126-X

Morford JL, Emerson SR, Breckel EJ, Kim SH (2005) Diagenesis of oxyanions (V, U, Re, and Mo) in pore waters and sediments from a continental margin. Geochim Cosmochim Acta 69(21):5021–5032

Passow U (2000) Formation of transparent exopolymer particles, TEP, from dissolved precursor material. Mar Ecol Prog Ser 192:1–11

Passow U (2014) Formation of rapidly-sinking, oil-associated marine snow. Deep-Sea Res II Top Stud Oceanogr 129:232. https://doi.org/10.1016/j.dsr2.2014.10.001

Passow U (2016) Formation of rapidly-sinking, oil-associated marine snow. Deep-Sea Res II Top Stud Oceanogr 129:232–240. https://doi.org/10.1016/j.dsr2.2014.10.001

Passow U, Ziervogel K, Asper V, Diercks A (2012) Marine snow formation in the aftermath of the Deepwater Horizon oil spill in the Gulf of Mexico. Environ Res Lett 7(3):035301

Quigg A, Passow U, Hollander DJ, Daly KL, Burd A, Lee K (2020) Formation and sinking of MOSSFA (marine oil snow sedimentation and flocculent accumulation) events: past and present (Chap. 12). In: Murawski SA, Ainsworth C, Gilbert S, Hollander D, Paris CB, Schlüter M, Wetzel D (eds) Deep oil spills – facts, fate and effects. Springer, Cham

Romero IC, Chanton JP, Rosenheim BE, Radović J, Schwing PT, Hollander DJ, Larter SR, Oldenburg TBP (2020) Long-term preservation of oil spill events in sediments: the case for the Deepwater Horizon Spill in the Northern Gulf of Mexico (Chap. 17). In: Murawski SA, Ainsworth C, Gilbert S, Hollander D, Paris CB, Schlüter M, Wetzel D (eds) Deep oil spills – facts, fate and effects. Springer, Cham

Romero IC, Toro-Farmer G, Diercks A-R, Schwing P, Muller-Karger F, Murawski S, Hollander DJ (2017) Large-scale deposition of weathered oil in the Gulf of Mexico following a deep-water oil spill. Environ Pollut 228:179–189. https://doi.org/10.1016/j.envpol.2017.05.019

Schaller T, Morford J, Emerson SR, Feely RA (2000) Oxyanions in metalliferous sediments: tracers for paleoseawater metal concentrations? Geochim Cosmochim Acta 64(13):2243–2254. https://doi.org/10.1016/S0016-7037(99)00443-3

Schwing PT, Chanton JP, Romero IC, Hollander DJ, Goddard EA, Brooks GR, Larson RA (2018a) Tracing the incorporation of carbon into benthic foraminiferal calcite following the Deepwater Horizon event. Environ Pollut 237:424–429. https://doi.org/10.1016/j.envpol.2018.02.066

Schwing PT, O'Malley BJ, Hollander DJ (2018b) Resilience of benthic foraminifera in the Northern Gulf of Mexico following the Deepwater Horizon event (2011–2015). Ecol Indic 84:753–764. https://doi.org/10.1016/j.ecolind.2017.09.044

Schwing PT, Hollander DJ, Brooks GR, Larson RA, Hastings DW, Chanton JP, Lincoln SA, Radović JR, Langenhoff A (2020) The sedimentary record of MOSSFA events in the Gulf of Mexico: a comparison of the Deepwater Horizon (2010) and Ixtoc 1 (1979) oil spills (Chap. 13). In: Murawski SA, Ainsworth C, Gilbert S, Hollander D, Paris CB, Schlüter M, Wetzel D (eds) Deep oil spills – facts, fate and effects. Springer, Cham

Schwing PT, Machain Castillo ML (2020) Impact and resilience of benthic foraminifera in the aftermath of the Deepwater Horizon and Ixtoc 1 oil spills (Chap. 23). In: Murawski SA, Ainsworth C, Gilbert S, Hollander D, Paris CB, Schlüter M, Wetzel D (eds) Deep oil spills – facts, fate and effects. Springer, Cham

Schwing PT, Romero IC, Brooks GR, Hastings DW, Larson RA, Hollander DJ (2015) A decline in benthic foraminifera following the Deepwater Horizon event in the Northeastern Gulf of Mexico. PLoS One 10(3):e0128505. https://doi.org/10.1371/journal.pone.0120565

Schwing PT, Romero IC, Larson RA, O'Malley BJ, Fridrik EE, Goddard EA, Brooks GR, Hastings DW, Rosenheim BE, Hollander DJ, Grant G, Mulhollan J (2016) Sediment core extrusion method at millimeter resolution using a calibrated, threaded-rod. J Vis Exp (114):54363. https://doi.org/10.3791/54363

Thomson J, Higgs NC, Croudace IW, Colley S, Hydes DJ (1993) Redox zonation of elements at an oxic/post-oxic boundary in deep-sea sediments. Geochim Cosmochim Acta 57:579–595

Tribovillard N, Algeo TJ, Lyons T, Riboulleau A (2006) Trace metals as paleoredox and paleo-productivity proxies: an update. Chem Geol 232(1,Äì2):12–32. https://doi.org/10.1016/j.chemgeo.2006.02.012

U.S. District Court Findings of Facts and Conclusions of Law –Phase 2 Trial. Case 2: 10-md-02179-cjb-ss, pp 1e44. Document 14021 Filed Jan. 15, 2015. http://www.laed.uscourts.gov/sites/default/files/OilSpill/Orders/1152015FindingsPhaseTwo.pdf

Verdugo P, Santschi PH (2010) Polymer dynamics of DOC networks and gel formation in seawater. Deep-Sea Res II Top Stud Oceanogr 57(16):1486–1493. https://doi.org/10.1016/j.dsr2.2010.03.002

Ziervogel K, McKay L, Rhodes B, Osburn CL, Dickson-Brown J, Arnosti C, Teske A (2012) Microbial activities and dissolved organic matter dynamics in oil-contaminated surface seawater from the Deepwater Horizon oil spill site. PLoS One 7(4):e34816. https://doi.org/10.1371/journal.pone.0034816

Chapter 17
Long-Term Preservation of Oil Spill Events in Sediments: The Case for the *Deepwater Horizon* Oil Spill in the Northern Gulf of Mexico

Isabel C. Romero, Jeffrey P. Chanton, Brad E. Roseheim, Jagoš R. Radović, Patrick T. Schwing, David J. Hollander, Stephen R. Larter, and Thomas B. P. Oldenburg

Abstract Geochemical studies can provide a record of environmental changes and biogeochemical processes in sedimentary systems. Analytical methods are in need of high-throughput procedures targeting recalcitrant and multiple chemical species for delineating ecological patterns and ecosystem health. The goal of this chapter is to summarize the analytical methods, recalcitrant molecules and trans-formed organic material used in previous studies as chemical indicators of the impact and fate of *Deepwater Horizon* (DWH) oil residues in sediments. Further monitoring of recalcitrant molecules and transformed material will help to elucidate the long-term fate of the DWH weathered oil in sedimentary environments of the Gulf of Mexico (GoM).

Keywords Deep-sea sediments · GC/MS/MS-MRM · Ramped pyrolysis · Stable isotopes · FTICR-MS

I. C. Romero (✉) · B. E. Roseheim · P. T. Schwing · D. J. Hollander
University of South Florida, College of Marine Science, St. Petersburg, FL, USA
e-mail: isabelromero@mail.usf.edu; brosenheim@usf.edu; pschwing@mail.usf.edu; davidh@usf.edu

J. P. Chanton
Florida State University, Department of Earth, Ocean and Atmospheric Science, Tallahassee, FL, USA
e-mail: jchanton@fsu.edu

J. R. Radović · S. R. Larter · T. B. P. Oldenburg
University of Calgary, PRG, Department of Geoscience, Calgary, AB, Canada
e-mail: Jagos.Radovic@ucalgary.ca; slarter@ucalgary.ca; toldenbu@ucalgary.ca

© Springer Nature Switzerland AG 2020
S. A. Murawski et al. (eds.), *Deep Oil Spills*,
https://doi.org/10.1007/978-3-030-11605-7_17

17.1 Introduction

Marine sediments can provide a valuable record of natural and human-induced historical events that have affected the water column. For example, after the *Deepwater Horizon* (DWH) oil spill in the Gulf of Mexico (GoM), analyses of sediment and water column samples demonstrated multiple processes for weathered oil deposition on the seafloor, as oil-associated marine snow (MOS), oil-mineral aggregates (OMAs), and dissolved compounds (Ryerson et al. 2012; Passow et al. 2012; Daly et al. 2016; Romero et al. 2017). Oil-associated deposition was observed to occur mostly as a combination of sticky material of weathered oil mixed with plankton, bacteria (MOS), and/or sediment particles (OMAs), referred to as MOSSFA (Marine Oil Snow Sedimentation and Flocculent Accumulation; Daly et al. 2016). Also, based on sedimentological, geochemical, and biological measurements, deposition of weathered oil onto the seafloor seems to have occurred over a ~5-month period after the DWH oil spill (Brooks et al. 2015). Furthermore, results from studies using sediment traps indicate deposition of oil-based petrocarbon for as long as 3 years following the spill (Yan et al. 2016). As a consequence, this short-term transport of weathered oil to depth led to the deposition of an unprecedented amount of oil residues over ~110,000 km^2 of the seafloor (~0.8–1.9 million barrels), from coastal (including bays and estuaries) to deep-sea areas (down to 2600 m water column depth) (Romero et al. 2017). Specifically for the deep sea, several studies using different analytical methods have quantified similar amounts of deposited weathered oil (~0.1–0.4 million barrels) (Valentine et al. 2014; Chanton et al. 2015; Stout et al. 2016; Romero et al. 2017). The accumulation of oil residues on surface sediments can be still found years after the spill on beaches (Hayworth et al. 2015; Yin et al. 2015; White et al. 2016) and deep-sea sediments (Stout et al. 2016). However, changes in the chemical composition of the deposited oil residues have been observed with biodegradation and transformation processes as the primary drivers of oil weathering in the environment (Aeppli et al. 2014; Stout et al. 2016). In addition, chemical analyses have indicated the persistence of molecular signatures, recalcitrant to weathering processes, serving as long-term indicators of the deposition and accumulation of oil residues on sediments (Aeppli et al. 2014; Rosenheim et al. 2014; Chanton et al. 2015; Stout et al. 2016; Romero et al. 2017). The goal of this chapter is to summarize the chemical methods targeting recalcitrant molecules and transformed organic material in sediments as indicators of the DWH event. Monitoring of these recalcitrant molecules and transformed material will help to elucidate the long-term fate of the DWH weathered oil in sedimentary environments of the GoM.

17.2 Markers of Preservation

17.2.1 *Hydrocarbons*

Organic molecules formed by carbon and hydrogen (hydrocarbons) are found widely in nature (e.g., plants, animal tissues, oil). In environmental geochemistry, and specifically in chemical fingerprinting of spilled oils, studies have often focused on the quantification of saturates (straight- and branched-chain molecules, cycloalkanes, terpanes, steranes), aromatics (e.g., 2–6-ring polycyclic aromatic hydrocarbons (PAHs) and their homologues) and, more recently, oxidation products of weathered oil (Ruddy et al. 2014). Sediment samples and spilled oils are dominated by saturates such as n-alkanes (carbon range C8–C41) and isoprenoids (pristine and phytane). The ratio of these saturates in sediments are frequently used to distinguish between natural and oil sources and the degree of degradation of the organic matter (Wang et al. 2006, 2016). Other saturates such as terpanes and steranes are used as source-specific markers for oil identification due to its stability during weathering in the environment (e.g., hopanes C27–C32; Aeppli et al. 2014; Romero et al. 2015; Wang et al. 2016). Some exceptions have been found, where specific biomarkers (e.g., hopanes C33–C35) have shown to change due to particular weathering processes (Aeppli et al. 2014). In this case, using compounds that are more recalcitrant (e.g., compared to n-alkanes) enables quantifying not only the degradation levels but also the active processes at advanced stages of weathering (Aeppli et al. 2014). Aromatic compounds, such as PAHs, are used to identify natural (e.g., terrestrial plants, degradation products), pyrogenic (e.g., incomplete combustion of coal, oil, and wood), and petrogenic (crude and weathered oil, seeps) sources in environmental samples (Wang et al. 2006, 2016). Moreover, some PAHs present toxic, mutagenic, and/or carcinogenic properties (Louvado et al. 2015), critical for understanding the ecotoxicological nature and health of natural systems. Due to the complex mixture of hydrocarbons in crude oil and residues, it is critical to include in geochemical studies, a wide range of chemical compounds to better understand the ecosystem health and weathering of organics in the environment after an oil spill. Due to the advances in gas chromatography, now coupled with tandem mass spectrometry and operated in multiple reaction monitoring mode (GC/MS/MS-MRM), it is possible to identify multiple target compounds (e.g., n-alkanes, terpanes, steranes, PAHs) from complex matrices in one analytical run, thereby enabling the identification and quantification of compounds at very low concentrations with no interferences. The application of these techniques over time and space allows characterization of preservation and weathering processes that affect hydrocarbons deposited in the marine and coastal environments.

In coastal environments, oil slicks were transported and deposited during the DWH oil spill in 2010. Several studies indicated a severe decrease (up to ~80%) in the abundance of low-molecular-weight (LMW) compounds after 19–24 months of the spill in areas like beaches and wetlands (Aeppli et al. 2014; Gros et al. 2014; Turner et al. 2014). Different weathering processes have been identified like biodegradation of LMW *n*-alkanes (<C29) and PAHs (2–3 ring), photooxidation of triaromatic steroids (TAS: C20, C28), and preferential dissolution of parent to alkyl LMW PAHs (e.g., phenanthrenes). Also preferential loss of homohopanes with longer alkyl chains (>C31) was observed (up to ~50%) related to microbial degradation (Aeppli et al. 2014).

In the deep-sea area, severe oil weathering occurred during vertical and lateral transport in the water column before deposition. Specifically, within 8 km of the DWH site, hydrocarbon analyses indicate weathering increased with distance from the DWH site with preferential loss of *n*-alkanes (<C25; ~90%) and PAHs (2–3 ring; ~95%) and less (~20–80%) of TAS (C26–C28), homohopanes (C34–C35), and steranes (C27–C29) (Stout and Payne 2016). Similarly, severe weathering was observed for LMW compounds (*n*-alkanes and PAHs) in sediments collected in 2010 at the DeSoto Canyon (~56 km from the DWH site; Romero et al. 2015, 2017). The composition of hydrocarbons (GC-amenable fraction) in the DeSoto Canyon was dominated by high-molecular-weight (HMW) *n*-alkanes (>C24; 67%), followed by LMW *n*-alkanes (9%), and LMW PAHs (6%) (Fig. 17.1). This composition of hydrocarbons remained relatively unchanged 3 years after the spill, even though significant reductions in concentration were observed for homohopanes (~67%) and LMW compounds (LMW PAHs, ~66%; LMW *n*-alkanes, 65%) and to a lesser extent for HMW *n*-alkanes (~43%) and HMW PAHs (~12%). Similarly, Stout et al. (2016) found a significant reduction (~80–90%) of hopane (17α(H),21β(H)-hopane) and PAHs (TPAH$_{50}$) in an area around the DWH site in 2014.

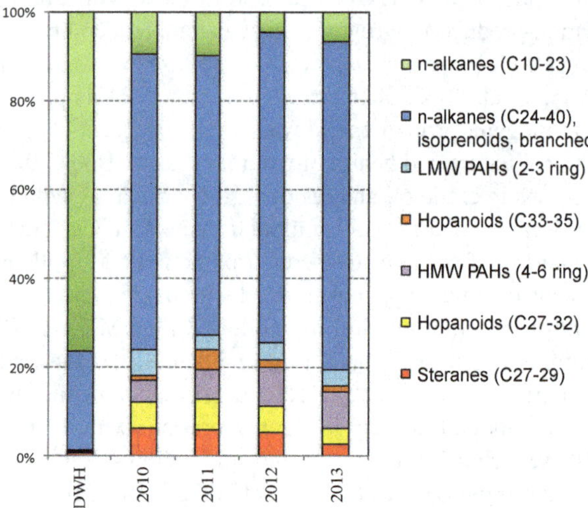

Fig. 17.1 Relative abundance (%) of hydrocarbon compound groups in deep-sea sediments from the DeSoto Canyon. Data shown for the reference oil standard (DWH: NIST 2779) and the sediment layer deposited after the DWH oil spill (2010) that was buried in the years after the spill (2011–2013). (Modified figure from Romero et al. (2017). Data are publicly available through the Gulf of Mexico Research Initiative Information and Data Cooperative (GRIIDC) at https://data.gulfresearchinitiative.org (doi: https://doi.org/10.7266/N7ZG6Q71 and doi: https://doi.org/10.7266/N7N29V16))

The significant reduction of both LMW and recalcitrant compounds over an extensive area indicates that not only biodegradation but also redistribution (downslope transport) processes may have affected the fate of the hydrocarbons deposited after the spill in deep-sea sediments. If only biodegradation of hydrocarbon compounds is taken into account, results from in situ observations and degradation models support the hypothesis of a long-term persistence (~ decade) from coastal to deep-sea environments of the most recalcitrant oil-derived hydrocarbons deposited in 2010–2011 (Gros et al. 2014; Turner et al. 2014; Romero et al. 2017). In addition, the formation of oxygenated products from weathering of both aromatic and saturated compounds suggest an even longer persistence of DWH-derived compounds in the environment (Aeppli et al. 2012; Hall et al. 2013; Lewan et al. 2014; White et al. 2016). However, the fact that also a significant reduction was observed for the most recalcitrant compounds (e.g., hopanoids) indicates potential downslope transport of bottom sediments. This is supported by a large-scale assessment of hydrocarbons in the GoM (Romero et al. 2017) showing that (1) the deep-sea area serves as a repository system for hydrocarbons in the northern GoM and (2) a gradient of oil-derived hydrocarbons toward deeper depths was observed (>2400 m) along bottom drainage paths. Also small-scale resuspension events in the GoM may occur more frequently than previously thought, perhaps as frequently as daily (Diercks et al. 2018). Therefore, biodegradation as well as redistribution of surface sediments containing oil residues should be considered as critical processes for modeling the long-term fate of oil spills at depth.

17.2.2 Bulk Carbon Isotopes

Regardless of chemical change, isotopic signatures from oil residues can be maintained in the sediment throughout the weathering process as long as the end result is not dissolved inorganic carbon (DIC). Oil from Miocene source rock, such as the light, sweet Louisiana crude oil that leaked after the DWH, has two important properties: (1) the stable carbon isotope composition is lighter than marine organic matter, and (2) it is essentially devoid of radiocarbon (^{14}C). Stable isotopes for sedimentary organic carbon (δ^{13}C) are reported relative to VPDB, an international standard carbonate reference material as $\delta^{13}C = (R_{sam}/R_{std} - 1) \times 1000$, where $R = {}^{13}C/{}^{12}C$, R_{sam} refers to the ratio of heavy to light isotopes in a sample, and R_{std} refers to the ratio in the international standard. Radiocarbon values can be reported according to the Δ notation as described by Stuiver and Pollach (1977). The Δ notation normalizes the radiocarbon content of any sample to a common δ^{13}C value (−25‰) and a common time interval. This ^{13}C correction means that mass-dependent isotope effects (isotopic fractionation) that occur in nature are removed from comparisons between ^{14}C values that are made using this notation. Another benefit of this notation is that it is a linear scale starting at −1000‰ when a sample has no radiocarbon content (McNichol and Aluwihare 2007). Modern carbon prior to atmospheric weapon testing had a value of 0‰ on this scale, but in the 1960s additional radiocarbon was produced in the atmosphere due to nuclear testing and warfare. This additional radiocarbon is slowly mixing into the earth's pool of surface carbon so that contemporary atmospheric CO_2 at the time of the oil spill was about +30 to 40‰ (Graven et al. 2012). At present (2018), the value of atmospheric CO_2 has been further reduced, and because of continual and growing fossil fuel combustion and the addition of that CO_2 to the atmosphere, organic matter fixed in 2050 may be indistinguishable from organic carbon fixed in the year 1050 (Graven 2015).

Oil and methane from fossil carbon reserves have characteristic isotopic values that allow them to be followed in the environment and in food webs (Chanton et al. 2012; Wilson et al. 2016). These compounds formed millions of years ago and have been locked away in the subsurface so they are radiocarbon-free with an isotopic signature of -1000‰ on the $\Delta^{14}C$ scale. Carbon fixed at the time of the oil spill by photosynthetic processes had a value of about +40‰ (Chanton et al. 2012, 2018). Oil and methane are also depleted in $\delta^{13}C$ (-27‰ to roughly -60‰, respectively) relative to modern photosynthetic marine production (-20 to -22‰). Away from seep sites, sedimentary organic matter in the Gulf has a $\delta^{13}C$ value that reflects marine production and a surface ^{14}C value of -200‰ $\pm$ 29‰ (Chanton et al. 2012, 2015).

Significant transformations can accompany oil weathering and biodegradation. The hydrocarbons can be oxygenated (Aeppli et al. 2012, 2014), while saturated and lighter hydrocarbons are lost due to evaporation and biodegradation. The oxygenated hydrocarbons are apparently rather persistent in the environment (Aeppli et al. 2018). The products of the transformed petroleum in the environment may be defined generically as "petrocarbon." Despite these reactions, the fossil-based carbon retains its distinctive ^{14}C-depleted isotopic composition relative to background carbon values allowing its detection in the environment. The larger range for $\Delta^{14}C$ relative to the $\delta^{13}C$ variation makes ^{14}C a significant tracer for finding evidence of petrocarbon in the sediment after the release of oil. The evidence for oil or methane contamination in the sediments consists of a surficial $\Delta^{14}C$-depleted layer overlying relatively $\Delta^{14}C$-enriched background layers (Fig. 17.2, Chanton et al. 2015). Sediments in the surface-most layer were as depleted as -500‰ in 2010 following the oil spill. Over time, ^{14}C depletion at some sites was ameliorated, returning toward background values of -200‰ but sediments as depleted as -420‰ were still observed in surface layers closer to the oil spill site in 2015. Both the distance of a site from the source of oil contamination and the quantity of that contamination influence the recovery rate of an oil-contaminated site (Bagby et al. 2016; Stout and Payne 2016; Roger et al. unpublished). Greater levels of petroleum contamination may reduce oil biodegradation rates, potentially due to larger particle sizes, which fall through the oxic water column more rapidly (Bagby et al. 2016). Less exposure to the oxic water column results in less weathering to the oil in transit, and in this manner both distance and particle size influence the state of the hydrocarbons upon their arrival on the suboxic sea floor. Consistent with this idea, elevated levels of PAHs were found at distances less than 35 km from the wellhead 3 years after the blowout; however, sediment farther than 35 km had returned to background levels (Adhikari et al. 2016). Biodegradation of hydrocarbon is more rapid in the oxygenated water column prior to deposition on the seafloor, so the more the oil traveled in the water column, the more it was degraded (Stout and Payne 2016).

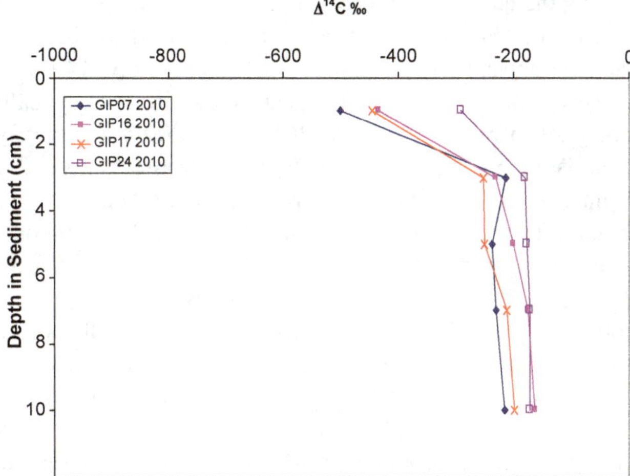

Fig. 17.2 $\Delta^{14}C$ values ‰ as a function of depth at several sites SW of the wellhead in 2010. Radiocarbon depletion in surface sediments is indicative of petrocarbon input, while underlying sediments are classified as background or pre-spill values, -200 ± 29‰. Locations were as follows: GIP 17, 28.6373, −88.5188; GIP 16, 28.7231, −88.4096; GIP 24, 28.7706, 88.3812; GIP 07, 28.2397, 89.1207. (Modified figure from Chanton et al. (2015). Data are publicly available through the Gulf of Mexico Research Initiative Information and Data Cooperative (GRIIDC) at https:// data.gulfresearchinitiative.org (UDI: R1.x138.078:0024))

With regard to the seafloor of the Gulf of Mexico in the years following the DWH oil spill, we suggest that the area surrounding the wellhead is returning back to baseline levels, with the area closest to the spill being the last area to return to background levels. Thus one can envision oil contamination spreading outward from the source in 2010, in a somewhat concentric fashion, and in the years following, oil contamination attenuating, retreating in an inverse manner relative to which it spreads.

17.2.3 Ramped Pyrolysis

As hydrocarbons and their degradation products are weathered in the environment, the carbon isotopic signature remains the same with the exception of carbon gained or lost during these processes. Whereas certain GC-amenable compounds can be extracted from the organic matter in seafloor sediments that contain hydrocarbon and degradation derivatives, there are myriad compounds and compound classes that are not extractable or GC amenable. Pseudo-bulk isotope techniques can be employed in these cases to ascertain and quantify the amount of petrocarbon in the complex mixture of organic matter. One such technique that has been applied to GoM oil spill studies after the DWH oil spill is Ramped PyrOx (RPO) carbon isotope analysis. This technique takes a minimally treated sample and heats it under a He atmosphere, gradually raising the temperature by 5 K/min. Degradation products are swept from the reaction chamber in flowing He gas and introduced to an oxidative chamber (~10% O_2 and CuO wire) for conversion to CO_2. The evolved CO_2, which happens over ~500 K between 200 °C and 700 °C, can be trapped cryogenically and purified on a standard gas separation line (Rosenheim et al. 2008). Ultimately, this method can provide an isotope composition spectrum of carbon released at different finite temperature intervals on a continuous temperature ramp. Thus it is a powerful tool in recognizing and quantifying petrocarbon in non-extractable and non-GC-amenable portions of the bulk sample (Pendergraft et al. 2013; Pendergraft and Rosenheim 2014; Adhikari et al. 2016). These studies have shown that immediately after the oil spill, compounds such as PAHs were negatively correlated with the amount of radiocarbon (Pendergraft et al. 2013). Several years after the spill and at deep-sea sites far afield from the wellhead, PAHs returned to background levels as radiocarbon levels approached natural abundance (Adhikari et al. 2016). However, in coastal sites, a significant portion of radiocarbon-depleted petrocarbon was observed through time shifting to higher temperature intervals of RPO analysis (Pendergraft and Rosenheim 2014), indicating that weathering was increasing the stability of the compounds and compound classes containing petrocarbon.

17.2.4   Foraminifera $\delta^{13}C$

Foraminifera are single-celled protists that are globally abundant in both planktonic and benthic habitats (Sen Gupta 1999). Many produce a calcite shell (test), while others produce a calcite matrix to which they adhere sediment grains (agglutinate) (Sen Gupta 1999). Benthic foraminifera density and diversity indices have proven to be valuable indicators of benthic impact and habitat suitability (Romero et al. 2015; Schwing et al. 2015, 2017). Benthic foraminifera are of particular interest in the context of marine oil spills considering the potential impact that the marine oil-snow sedimentation and flocculent accumulations (MOSSFA) had on the seafloor after the DWH oil spill (Brooks et al. 2015; Romero et al. 2015; Hastings et al. 2016). MOSSFA is the combination of biological, lithic, and oil residues, which caused rapid settling of flocculent material and deposition on the seafloor during and following the DWH (Passow et al. 2012; Brooks et al. 2015; Daly et al. 2016; Romero et al. 2017; Chap. 12 this volume). On the seafloor, it is likely that the deposited marine oil snow (MOS) rich in organic matter changed the quality of the organic carbon, increased microbial respiration, and produced lower pore-water oxygen concentrations (Hastings et al. 2016).

The stable carbon isotopic composition of benthic foraminiferal calcite ($\delta^{13}C_{CaCO3}$) has been shown previously to predominantly reflect the isotopic composition of dissolved inorganic carbon (DIC) in bottom or pore waters and changes in food sources (e.g., microbial biomass) and organic matter fluxes (TOC) in sediments (Torres et al. 2003; Hill et al. 2004; Nomaki et al. 2006; Zarriess and MacKensen 2011; Theodor et al. 2016). The stable isotope composition of benthic foraminifera tests ($\delta^{13}C_{CaCO3}$) in sediment cores collected in the DeSoto Canyon

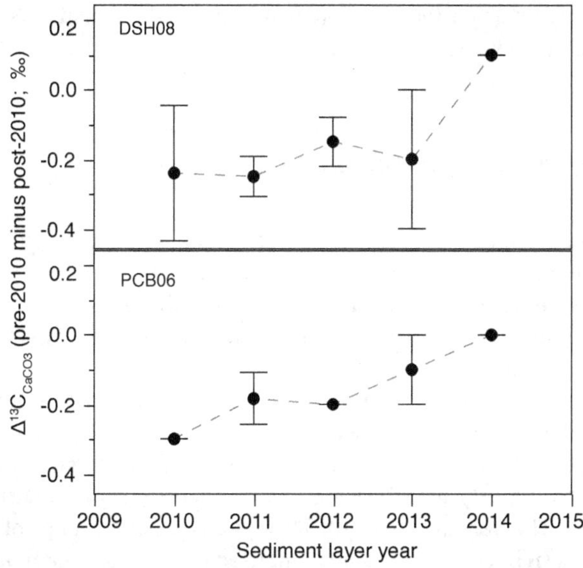

Fig. 17.3 Mean $\Delta^{13}C_{CaCO3}$ (pre-2010 minus post-2010; ‰) values in sediments from two sites in the DeSoto Canyon (DSH08 and PCB06). Sediment intervals are denoted by the year of deposition from cores collected between 2010 and 2014. Chronology in sediments was determined by short-lived radioisotope chronology ($^{210}Pb_{xs}$ and $^{234}Th_{xs}$). Data are shown as $\delta^{13}C_{CaCO3}$ mean ± 1SD (Schwing et al. 2018). (Data are publicly available through the Gulf of Mexico Research Initiative Information and Data Cooperative (GRIIDC) at https://data.gulfresearchinitiative.org (DOI: https://doi.org/10.7266/N79021PB, https://doi.org/10.7266/N7S180HN, https://doi.org/10.7266/N7HH6H32))

after the DWH oil spill showed to record the MOSSFA event, likely due to changes in sediment accumulation rates (Schwing et al. 2018). From 2010 to 2012, the $\delta^{13}C_{CaCO3}$ signature of benthic foraminifera on the surface of the seafloor was depleted beyond any natural variations recorded in the sediments for the years before the DWH oil spill. Small but consistent depletions in $\delta^{13}C_{CaCO3}$ (0.2–0.4‰) were documented (Fig. 17.3). The $\delta^{13}C_{CaCO3}$ depletion occurred in the sediment layers where other tracers indicated the presence of enhanced sedimentation rates and oil residues from the DWH oil spill (Brooks et al. 2015; Romero et al. 2015). Moreover, the $\delta^{13}C_{CaCO3}$ records from sediment cores collected in 2014 show that the sediment layer (at 1 cm below the surface) corresponding to 2010 (DWH oil spill) was depleted similar to the surface layer of the cores collected from 2010–2012 (DWH oil spill). However, the $\delta^{13}C_{CaCO3}$ at the surface of the 2014 sediment core was the same as prior to the DWH oil spill, indicating recovery of the system (Fig. 17.3). In addition, results indicate that the $\delta^{13}C_{CaCO3}$ of benthic foraminifera have preserved evidence of the MOSSFA event in the seafloor sediment record for 4 years. Therefore, benthic foraminifera calcite offers a modest but recalcitrant record of the MOSSFA event and subsequent burial. In certain areas with high seasonal variability in TOC flux, it may be difficult to distinguish a MOSSFA event from other organic fluxes as the depletion in $\delta^{13}C_{CaCO3}$ documented during and after the DWH oil spill was on the same order of magnitude as natural/seasonal fluxes elsewhere (Zarriess and MacKensen 2011; Theodor et al. 2016).

17.2.5 Microbial Lipid Biomarkers

Lipid biomarkers found in recent marine sediments are valuable biogeochemical proxies which abundance and distributions can be related to their source microorganisms and various marine system variables which control the composition of marine microbiome – nutrients, redox state, water temperature of the surface waters, and nutrients to name a few.

In the past decade, a specific class of lipid biomarkers, known as glycerol dialkyl glycerol tetraethers (GDGTs), has been extensively studied as (paleo)environmental proxies (Tierney 2012; Schouten et al. 2013a). GDGTs are microbial membrane lipids found ubiquitously in different environmental settings – marine, lacustrine, and terrestrial (Schouten et al. 2013b). There are two main groups of GDGTs: isoprenoid GDGTs (iGDGTs), which are comprised of two C_{40} isoprenoid units with a varying number (0 to 8) of alicyclic moieties, bounded via glycerol groups in a head-to-head configuration, and branched GDGTs (bGDGTs), which are comprised of 4 to 6 methyl groups attached to the ether-bounded C_{28} alkyl chains and with up to two cyclopentyl moieties formed by internal cyclization. In general, the main sources of iGDGTs are marine (planktonic archaea), while bGDGTs are predominantly from terrestrial bacteria, which is why they are used to calculate the BIT index (branched and isoprenoid tetraether index), a proxy to assess marine vs. terrestrial inputs to a given marine sediment (Damsté et al. 2002; Hopmans et al. 2004). Typically, concentrations of GDGT species are determined by normal-phase liquid chromatography (NP-HPLC) coupled to mass spectrometry (MS) detectors using atmospheric pressure chemical ionization (APCI) (Schouten et al. 2013a). Methods have also been developed for analysis of the intact ether lipids (i.e., still bound to polar head groups), via LC separation using aqueous eluents, followed by tandem MS with electrospray ionization (ESI) interface (Sturt et al. 2004), and simultaneous detection of both intact and core ether lipids, by reversed-phase LC (Zhu et al. 2013). A useful complementary tool to the abovementioned methods is the ultrahigh-resolution Fourier transform mass spectrometry (FTICR-MS) because of its broad range and high resolution of mass detection and diverse ionization modes. FTICR-MS is a more comprehensive analytical method, targeting multiple chemical species present in complex environmental matrices, after minimal sample preparation, ideal for exploratory, non-targeted detection of various polar species, including lipids.

More recently a simplified APPI-P (atmospheric pressure photoionization in positive mode) FTICR-MS workflow was developed (RADAR, Rapid Analyte Detection and Reconnaissance) and applied to analyze iGDGTs, bGDGTs, and other lipid biomarkers, such as pigments, in northern GoM sediments collected in 2014 (Radović et al. 2016a, b). Interestingly, fractional abundances (f) of iGDGTs (i.e., abundance of each GDGT species relative to total GDGT abundance) in the collected sediment cores follow the distribution typically observed in warmer water environments such as GoM (i.e., GDGT-5 > GDGT-0 > > GDGT-1 ~ GDGT-2 > > GDGT-3; Pearson and Ingalls 2013; Schouten et al. 2002), demonstrating a basic comparability of results obtained using FTICR-MS and traditional HPLC-MS methods for GDGTs. Despite generally similar trends of GDGT species in the analyzed northern GoM sediment cores, the variability of their ratios (i.e., proxy indices) records subtle changes related to the perturbations of the marine system at a given location. A recent study has observed an excursion of GDGT-derived BIT proxy and stable carbon isotope signature of benthic foraminifera in a sediment core collected in the DeSoto Canyon in 2014 (Fig. 17.4). Excursion of BIT to lower values (0.11) at 9 mm downcore, is due to the higher abundance of the iGDGT-5 peak (crenarchaeol and its isomers). Similar BIT changes in the northern GoM driven by crenarchaeol concentrations have been observed previously and attributed to seasonably variable delivery of marine planktonic OM to sediments (Smith et al. 2012). A geochronological model sets this horizon within the DWH timeframe, putatively suggesting that this observation might be reflecting MOSSFA-driven deposition of planktonic archaeal biomass. This is corroborated by the depletion of foraminiferal

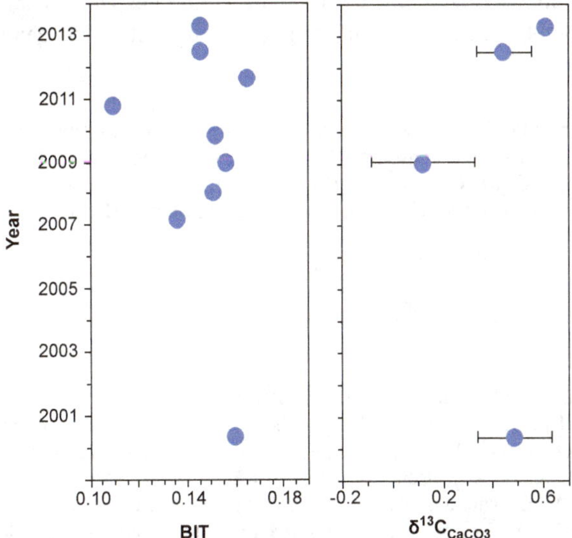

Fig. 17.4 Branched and isoprenoid tetraether (BIT) index and stable carbon isotope signature of benthic foraminifera in a sediment core collected at the DSH08 site in 2014. Note the negative excursions of the two proxies, a putative record of coeval deposition of planktonic biomass and petrocarbon via MOSSFA. For each sediment interval, the average reconstructed year is shown, based on the short-lived radioisotope geochronology model. Associated error of the model is approx. ±1 year. (Data are publicly available through the Gulf of Mexico Research Initiative Information and Data Cooperative (GRIIDC) at https://data.gulfresearchinitiative.org (DOI: https://doi.org/10.7266/N71R6NGQ, https://doi.org/10.7266/N7S180HN))

carbonate isotope signature (Fig. 17.4) which is in agreement with the depletion of pre-DWH versus post-DWH sedimentary organic carbon $\delta^{13}C$, documented by Rosenheim et al. (2016) and similar observations made on benthic foraminifera collected at other sites in the NGoM (Schwing et al. 2018).

17.3 Summary

Environmental studies are in need of rapid, high-throughput analytical methods, capable of multi-species screening, which would enable multiproxy approaches to defining ecological systems and health. In this respect, novel analytical strategies which use high-resolution mass spectrometry showed high promise to resolve a complex inventory of chemical species, over a broad mass range. Simplified sample preparation significantly reduces the time for sample preprocessing and preserves the complexity of the sediment extract. A multiproxy biogeochemical approach provides more robust stratigraphic (paleo)environmental reconstructions and can also be successfully applied for the studies of anthropogenic perturbations of the natural environment, such as major oil spills. The response to the DWH blowout disaster also saw development and application of new techniques which can be added to the cadre of rapid-response tools oceanographers and geochemists can apply to future events.

Funding Information This research was made possible by grants from The Gulf of Mexico Research Initiative through its consortia: The Center for the Integrated Modeling and Analysis of the Gulf Ecosystem (C-IMAGE), Ecosystem Impacts of Oil and Gas Inputs to the Gulf (ECOGIG), and Deep Sea to Coast Connectivity in the Eastern Gulf of Mexico (Deep-C).

References

Adhikari PL, Maiti K, Overton EB, Rosenheim BE, Marx BD (2016) Distributions and accumulation rates of polycyclic aromatic hydrocarbons in the northern Gulf of Mexico sediments. Environ Pollut 212:413–423. https://doi.org/10.1016/j.envpol.2016.01.064

Aeppli C, Carmichael CA, Nelson RK, Lemkau KL, Graham WM, Redmond MC, Valentine DL, Reddy CM, Graham M, Redmond MC, Valentine DL, Reddy CM (2012) Oil weathering after the Deepwater Horizon disaster led to the formation of oxygenated residues. Environ Sci Technol 46:8799–8807. https://doi.org/10.1021/es3015138

Aeppli C, Nelson RK, Radović JR, Carmichael CA, Valentine DL, Reddy CM (2014) Recalcitrance and degradation of petroleum biomarkers upon abiotic and biotic natural weathering of Deepwater Horizon oil. Environ Sci Technol 48:6726–6734. https://doi.org/10.1021/es500825q

Aeppli C, Swarthout RF, O'Neil GW, Katz SD, Nabi D, Ward CP, Nelson RK, Sharpless CM, Reddy CM (2018) How persistent and bioavailable are oxygenated Deepwater Horizon oil transformation products? Environ Sci Technol 52:7250–7258. https://doi.org/10.1021/acs.est.8b01001

Bagby SC, Reddy CM, Aeppli C, Fisher GB, Valentine DL (2016) Persistence and biodegradation of oil at the ocean floor following Deepwater Horizon. Proc Natl Acad Sci. 201610110. https://doi.org/10.1073/pnas.1610110114

Brooks GR, Larson RA, Schwing PT, Romero I, Moore C, Reichart G-J, Jilbert T, Chanton JP, Hastings DW, Overholt WA, Marks KP, Kostka JE, Holmes CW, Hollander D (2015) Sedimentation pulse in the NE Gulf of Mexico following the 2010 DWH blowout. PLoS One 10. https://doi.org/10.1371/journal.pone.0132341

Chanton JP, Cherrier J, Wilson RM, Sarkodee-Adoo J, Bosman S, Mickle A, Graham WM (2012) Radiocarbon evidence that carbon from the Deepwater Horizon spill entered the planktonic food web of the Gulf of Mexico. Environ Res Lett 7:045303. https://doi.org/10.1088/1748-9326/7/4/045303

Chanton J, Zhao T, Rosenheim BE, Joye S, Bosman S, Brunner C, Yeager KM, Diercks AR, Hollander D (2015) Using natural abundance radiocarbon to trace the flux of petrocarbon to the seafloor following the Deepwater Horizon oil spill. Environ Sci Technol 49:847–854. https://doi.org/10.1021/es5046524

Chanton JP, Giering SLC, Bosman SH, Rogers KL, Sweet J, Asper VL, Diercks AR, Passow U (2018) Isotopic composition of sinking particles: oil effects, recovery and baselines in the Gulf of Mexico, 2010–2015. Elementa Sci Anthropocene 6:43. https://doi.org/10.1525/elementa.298

Daly KL, Passow U, Chanton J, Hollander D (2016) Assessing the impacts of oil-associated marine snow formation and sedimentation during and after the Deepwater Horizon oil spill. Anthropocene:1–16. https://doi.org/10.1016/j.ancene.2016.01.006

Damsté JSS, Schouten S, Hopmans EC, van Duin ACT, Geenevasen JAJ (2002) Crenarchaeol. J Lipid Res 43:1641–1651. https://doi.org/10.1194/jlr.M200148-JLR200

Diercks A, Dike C, Asper VL, Dimarco SF, Jeffrey P (2018) Scales of seafloor sediment resuspension in the northern Gulf of Mexico. Elementa Sci Anthropocene 1:1–28. https://doi.org/10.1525/elementa.285

Graven HD (2015) Impact of fossil fuel emissions on atmospheric radiocarbon and various applications of radiocarbon over this century. Proc Natl Acad Sci 112:9542–9545. https://doi.org/10.1073/pnas.1504467112

Graven HD, Guilderson TP, Keeling RF (2012) Observations of radiocarbon in CO2 at la Jolla, California, USA 1992-2007: analysis of the long-term trend. J Geophys Res Atmos 117:1–14. https://doi.org/10.1029/2011JD016533

Gros J, Reddy CM, Aeppli C, Nelson RK, Carmichael C, Arey JS (2014) Resolving biodegradation patterns of persistent saturated hydrocarbons in weathered oil samples from the Deepwater Horizon disaster. Environ Sci Technol 48:1628–1637. https://doi.org/10.1021/es4042836

Hall GJ, Frysinger GS, Aeppli C, Carmichael CA, Gros J, Lemkau KL, Nelson RK, Reddy CM (2013) Oxygenated weathering products of Deepwater Horizon oil come from surprising precursors. Mar Pollut Bull 75:140–149. https://doi.org/10.1016/j.marpolbul.2013.07.048

Hastings DW, Schwing PT, Brooks GR, Larson RA, Morford JL, Roeder T, Quinn KA, Bartlett T, Romero IC, Hollander DJ (2016) Changes in sediment redox conditions following the BP DWH blowout event. Deep-Sea Res II Top Stud Oceanogr 129:167–178. https://doi.org/10.1016/j.dsr2.2014.12.009

Hayworth JS, Prabakhar Clement T, John GF, Yin F (2015) Fate of Deepwater Horizon oil in Alabama's beach system: understanding physical evolution processes based on observational data. Mar Pollut Bull 90:95–105. https://doi.org/10.1016/j.marpolbul.2014.11.016

Hill TM, Kennett JP, Valentine DL (2004) Isotopic evidence for the incorporation of methane-derived carbon into foraminifera from modern methane seeps, Hydrate Ridge, Northeast Pacific. Geochim Cosmochim Acta 68:4619–4627. https://doi.org/10.1016/j.gca.2004.07.012

Hopmans EC, Weijers JWH, Schefuß E, Herfort L, Sinninghe Damsté JS, Schouten S (2004) A novel proxy for terrestrial organic matter in sediments based on branched and isoprenoid tetraether lipids. Earth Planet Sci Lett 224:107–116. https://doi.org/10.1016/j.epsl.2004.05.012

Lewan MD, Warden A, Dias RF, Lowry ZK, Hannah TL, Lillis PG, Kokaly RF, Hoefen TM, Swayze GA, Mills CT, Harris SH, Plumlee GS (2014) Asphaltene content and composition as a measure of Deepwater Horizon oil spill losses within the first 80 days. Org Geochem 75:54–60. https://doi.org/10.1016/j.orggeochem.2014.06.004

Louvado A, Gomes NCM, Simões MMQ, Almeida A, Cleary DFR, Cunha A (2015) Polycyclic aromatic hydrocarbons in deep sea sediments : microbe – pollutant interactions in a remote environment. Sci Total Environ 526:312–328. https://doi.org/10.1016/j.scitotenv.2015.04.048

McNichol AP, Aluwihare LI (2007) The power of radiocarbon in biogeochemical studies of the marine carbon cycle: insights from studies of dissolved and particulate organic carbon (DOC and POC). Chem Rev 107:443–466. https://doi.org/10.1021/cr050374g

Nomaki H, Heinz P, Nakatsuka T, Shimanaga M, Ohkouchi N, Ogawa NO, Kogure K, Ikemoto E, Kitazato H (2006) Different ingestion patterns of 13C-labeled bacteria and algae by deep-sea benthic foraminifera. Mar Ecol Prog Ser 310:95–108. https://doi.org/10.3354/meps310095

Passow U, Ziervogel K, Asper V, Diercks A (2012) Marine snow formation in the aftermath of the Deepwater Horizon oil spill in the Gulf of Mexico. Environ Res Lett 7:035301. https://doi.org/10.1088/1748-9326/7/3/035301

Pearson A, Ingalls AE (2013) Assessing the use of archaeal lipids as marine environmental proxies. Annu Rev Earth Planet Sci 41:359–384

Pendergraft MA, Rosenheim BE (2014) Varying relative degradation rates of oil in different forms and environments revealed by ramped pyrolysis. Environ Sci Technol 48:10966–10974. https://doi.org/10.1021/es501354c

Pendergraft MA, Dincer Z, Sericano JL, Wade TL, Kolasinski J, Rosenheim BE (2013) Linking ramped pyrolysis isotope data to oil content through PAH analysis. Environ Res Lett 8(4):1–11. https://doi.org/10.1088/1748-9326/8/4/044038

Radović JR, Silva RC, Snowdon R, Larter SR, Oldenburg TBP (2016a) Rapid screening of glycerol ether lipid biomarkers in recent marine sediment using atmospheric pressure photoionization in positive mode fourier transform ion cyclotron resonance mass spectrometry. Anal Chem 88:1128–1137. https://doi.org/10.1021/acs.analchem.5b02571

Radović JR, Silva RC, Snowdon RW, Brown M, Larter S, Oldenburg TBP (2016b) A rapid method to assess a broad inventory of organic species in marine sediments using ultra-high resolution mass spectrometry. Rapid Commun Mass Spectrom 30:1273–1282. https://doi.org/10.1002/rcm.7556

Romero IC, Schwing PT, Brooks GR, Larson RA, Hastings DW, Flower BP, Goddard EA, Hollander DJ (2015) Hydrocarbons in deep-sea sediments following the 2010 Deepwater Horizon Blowout in the Northeast Gulf of Mexico. PLoS One 10:1–23. https://doi.org/10.1371/journal.pone.0128371

Romero IC, Toro-Farmer G, Diercks A, Schwing P, Muller-Karger F, Murawski S, Hollander DJ (2017) Large-scale deposition of weathered oil in the Gulf of Mexico following a deep-water oil spill. Environ Pollut 228:179–189. https://doi.org/10.1016/j.envpol.2017.05.019

Rosenheim BE, Day MB, Domack E, Schrum H, Benthien A, Hayes JM (2008) Antarctic sediment chronology by programmed-temperature pyrolysis: methodology and data treatment. Geochem Geophys Geosyst 9:1–16. https://doi.org/10.1029/2007GC001816

Rosenheim BE, Pendergraft MA, Flowers GC, Carney R, Sericano JL, Amer RM, Chanton J, Dincer Z, Wade TL (2014) Employing extant stable carbon isotope data in Gulf of Mexico sedimentary organic matter for oil spill studies. Deep-Sea Res II Top Stud Oceanogr Elsevier:1–10. https://doi.org/10.1016/j.dsr2.2014.03.020

Rosenheim BE, Pendergraft MA, Flowers GC, Carney R, Sericano JL, Amer RM, Chanton J, Dincer Z, Wade TL (2016) Employing extant stable carbon isotope data in Gulf of Mexico sedimentary organic matter for oil spill studies. Deep-Sea Res II Top Stud Oceanogr 129:249–258. https://doi.org/10.1016/j.dsr2.2014.03.020

Ruddy BM, Huettel M, Kostka JE, Lobodin VV, Bythell BJ, McKenna AM, Aeppli C, Reddy CM, Nelson RK, Marshall AG, Rodgers RP (2014) Targeted petroleomics: analytical investigation of Macondo well oil oxidation products from Pensacola Beach. Energy Fuel 28:4043–4050. https://doi.org/10.1021/ef500427n

Ryerson TB, Camilli R, Kessler JD, Kujawinski EB, Reddy CM, Valentine DL, Atlas E, Blake DR, de Gouw J, Meinardi S, Parrish DD, Peischl J, Seewald JS, Warneke C (2012) Chemical data quantify Deepwater Horizon hydrocarbon flow rate and environmental distribution. Proc Natl Acad Sci U S A 109:20246–20253. https://doi.org/10.1073/pnas.1110564109

Schouten S, Hopmans EC, Rosell-Melé A, Pearson A, Adam P, Bauersachs T, Bard E, Bernasconi SM, Bianchi TS, Brocks JJ, Carlson LT, Castañeda IS, Derenne S, Selver AD, Dutta K, Eglinton T, Fosse C, Galy V, Grice K, Hinrichs KU, Huang Y, Huguet A, Huguet C, Hurley S, Ingalls A, Jia G, Keely B, Knappy C, Kondo M, Krishnan S, Lincoln S, Lipp J, Mangelsdorf K, Martínez-García A, Ménot G, Mets A, Mollenhauer G, Ohkouchi N, Ossebaar J, Pagani M, Pancost RD, Pearson EJ, Peterse F, Reichart GJ, Schaeffer P, Schmitt G, Schwark L, Shah SR, Smith RW, Smittenberg RH, Summons RE, Takano Y, Talbot HM, Taylor KWR, Tarozo R, Uchida M, Van Dongen BE, Van Mooy BAS, Wang J, Warren C, Weijers JWH, Werne JP, Woltering M, Xie S, Yamamoto M, Yang H, Zhang CL, Zhang Y, Zhao M, Damsté JSS (2013a) An interlaboratory study of TEX86 and BIT analysis of sediments, extracts, and standard mixtures. Geochem Geophys Geosyst 14:5263–5285. https://doi.org/10.1002/2013GC004904

Schouten S, Hopmans EC, Schefuß E, Damsté JSS (2002) Distributional variations in marine crenarchaeotal membrane lipids: A new tool for reconstructing ancient sea water temperatures? Earth Planet Sci Lett 204:265–274

Schouten S, Hopmans EC, Sinninghe Damsté JS (2013b) The organic geochemistry of glycerol dialkyl glycerol tetraether lipids: a review. Org Geochem 54:19–61. https://doi.org/10.1016/j.orggeochem.2012.09.006

Schwing PT, Romero IC, Brooks GR, Hastings DW, Larson RA, Hollander DJ (2015) A decline in benthic foraminifera following the Deepwater Horizon event in the Northeastern Gulf of Mexico. PLoS One 10:1–22. https://doi.org/10.7266/N79021PB.Funding

Schwing PT, O'Malley BJ, Romero IC, Martínez-Colón M, Hastings DW, Glabach MA, Hladky EM, Greco A, Hollander DJ (2017) Characterizing the variability of benthic foraminifera in the northeastern Gulf of Mexico following the Deepwater Horizon event (2010–2012). Environ Sci Pollut Res 24. https://doi.org/10.1007/s11356-016-7996-z

Schwing PT, Chanton JP, Romero IC, Hollander DJ, Goddard EA, Brooks GR, Larson RA (2018) Tracing the incorporation of carbon into benthic foraminiferal calcite following the Deepwater Horizon event. Environ Pollut 237:424–429. https://doi.org/10.1016/j.envpol.2018.02.066

Sen Gupta BK (1999) Modern Foraminifera. Kluwer Academic Publishers, Dordrecht. Chicago (Author-Date, 15th ed). 385p

Smith RW, Bianchi TS, Li X (2012) A re-evaluation of the use of branched GDGTs as terrestrial biomarkers: implications for the BIT Index. Geochim Cosmochim Acta 80:14–29. https://doi.org/10.1016/j.gca.2011.11.025

Stout SA, Payne JR (2016) Macondo oil in deep-sea sediments: part 1 – sub-sea weathering of oil deposited on the seafloor. Mar Pollut Bull 111(1–2):365–380. https://doi.org/10.1016/j.marpolbul.2016.07.036

Stout SA, Rouhani S, Liu B, Oehrig J, Ricker RW, Baker G, Lewis C (2016) Assessing the footprint and volume of oil deposited in deep-sea sediments following the Deepwater Horizon oil spill. Mar Pollut Bull 114:327–342. https://doi.org/10.1016/j.marpolbul.2016.09.046

Stuiver M, Pollach HA (1977) Discussion: reporting of ^{14}C data. Radiocarbon 19:355–363

Sturt HF, Summons RE, Smith K, Elvert M, Hinrichs KU (2004) Intact polar membrane lipids in prokaryotes and sediments deciphered by high-performance liquid chromatography/electrospray ionization multistage mass spectrometry - New biomarkers for biogeochemistry and microbial ecology. Rapid Commun Mass Spectrom 18:617–628. https://doi.org/10.1002/rcm.1378

Theodor M, Schmiedl G, Mackensen A (2016) Stable isotope composition of deep-sea benthic foraminifera under contrasting trophic conditions in the western Mediterranean Sea. Mar Micropaleontol 124:16–28. https://doi.org/10.1016/j.marmicro.2016.02.001

Tierney JE (2012) GDGT Thermometry: lipid tools for reconstructing paleotemperatures. Paleontol Soc Papers 18:115–131. https://doi.org/10.1017/s1089332600002588

Torres ME, Mix AC, Kinports K, Haley B, Klinkhammer GP, McManus J, de Angelis MA (2003) Is methane venting at the seafloor recorded by δ 13 C of benthic foraminifera shells? Paleoceanography 18. https://doi.org/10.1029/2002PA000824

Turner RE, Overton EB, Meyer BM, Miles MS, Hooper-Bui L (2014) Changes in the concentration and relative abundance of alkanes and PAHs from the Deepwater Horizon oiling of coastal marshes. Mar Pollut Bull 86:291–297. https://doi.org/10.1016/j.marpolbul.2014.07.003

Valentine DL, Fisher GB, Bagby SC, Nelson RK, Reddy CM, Sylva SP, Woo M (2014) Fallout plume of submerged oil from Deepwater Horizon. Proc Natl Acad Sci 111(45):15906–15911. https://doi.org/10.1073/pnas.1414873111

Wang Z, Stout S, Fingas M (2006) Forensic fingerprinting of biomarkers for oil spill characterization and source identification. Environ Forensic 7:105–146. https://doi.org/10.1080/15275920600667104

Wang Z, Yang C, Yang Z, Brown CE, Hollebone BP, Stout SA (2016) Petroleum biomarker fingerprinting for oil spill characterization and source identification BT. In: Standard handbook oil spill environmental forensics, 2nd edn. Elsevier Inc, pp 131–254

White HK, Wang CH, Williams PL, Findley DM, Thurston AM, Simister RL, Aeppli C, Nelson RK, Reddy CM (2016) Long-term weathering and continued oxidation of oil residues from the Deepwater Horizon spill. Mar Pollut Bull 113:380–386. https://doi.org/10.1016/j.marpolbul.2016.10.029

Wilson RM, Cherrier J, Sarkodee-Adoo J, Bosman S, Mickle A, Chanton JP (2016) Tracing the intrusion of fossil carbon into coastal Louisiana macrofauna using natural ^{14}C and ^{13}C abundances. Deep-Sea Res II Top Stud Oceanogr 129:89–95. https://doi.org/10.1016/j.dsr2.2015.05.014

Yan B, Passow U, Chanton JP, Nöthig E-M, Asper V, Sweet J, Pitiranggon M, Diercks A, Pak D (2016) Sustained deposition of contaminants from the Deepwater Horizon spill. Proc Natl Acad Sci 113:E3332–E3340. https://doi.org/10.1073/pnas.1513156113

Yin F, John GF, Hayworth JS, Clement TP (2015) Long-term monitoring data to describe the fate of polycyclic aromatic hydrocarbons in Deepwater Horizon oil submerged off Alabama's beaches. Sci Total Environ 508:46–56. https://doi.org/10.1016/j.scitotenv.2014.10.105

Zarriess M, MacKensen A (2011) Testing the impact of seasonal phytodetritus deposition on ^{13}C of epibenthic foraminifer *Cibicidoides wuellerstorfi*: a 31,000 year high-resolution record from the northwest African continental slope. Paleoceanography 26(2):1–8. https://doi.org/10.1029/2010PA001944

Zhu C, Lipp JS, Wörmer L, Becker KW, Schröder J, Hinrichs KU (2013) Comprehensive glycerol ether lipid fingerprints through a novel reversed phase liquid chromatography-mass spectrometry protocol. Org Geochem 65:53–62. https://doi.org/10.1016/j.orggeochem.2013.09.012

Chapter 18
Effect of Marine Snow on Microbial Oil Degradation

Alette A. M. Langenhoff, Shokouh Rahsepar, Justine S. van Eenennaam, Jagoš R. Radović, Thomas B. P. Oldenburg, Edwin Foekema, and AlberTinka J. Murk

Abstract In the aftermath of an oil spill, a possible response is the addition of chemical dispersants to prevent further spreading of the spilled oil on the ocean surface. The main objective is to enhance the formation of smaller oil droplets by reducing the interfacial tension between oil and water, thus dispersing the oil into the water column. The resulting solubilized oil with microdroplets along with the associated toxic compounds will be swiftly incorporated into the seawater. The formation of smaller oil droplets and the dispersant enhanced solubilized oil will increase its availability for bacteria and thus the biodegradability. Subsequently, the number and activity of oil-degrading bacteria increases, and more oil will be degraded in a shorter period of time (Kessler et al., Science 331:312–315, 2011).

A. A. M. Langenhoff (✉) · S. Rahsepar
Wageningen University & Research, Department of Environmental Technology, Wageningen, The Netherlands
e-mail: Alette.langenhoff@wur.nl; Shokouh.rahsepar@wur.nl

J. S. van Eenennaam
Wageningen University & Research, Department of Environmental Technology, Wageningen, The Netherlands

Regulatory Affairs Department, Safety Assessment, Charles River Laboratories, 's Hertogenbosch, The Netherlands

J. R. Radović · T. B. P. Oldenburg
University of Calgary, PRG, Department of Geoscience, Calgary, AB, Canada
e-mail: jagos.radovic@ucalgary.ca

E. Foekema
Wageningen University & Research, Marine Animal Ecology Group, Wageningen, The Netherlands

Wageningen Marine Research, Den Helder, The Netherlands
e-mail: Edwin.foekema@wur.nl

A. J. Murk
Wageningen University & Research, Marine Animal Ecology Group, Wageningen, The Netherlands
e-mail: Tinka.murk@wur.nl

© Springer Nature Switzerland AG 2020
S. A. Murawski et al. (eds.), *Deep Oil Spills*,
https://doi.org/10.1007/978-3-030-11605-7_18

"""

However, during the immediate release of the dispersed oil, volatile hydrocarbons including some of the more toxic compounds of benzene, toluene, ethylbenzene, and xylenes (BTEX) can inhibit the oil degradation (Sherry et al., Front Microbiol 5:131, 2014).

Depending on the oceanic conditions, the addition of chemical dispersants can result in excessive formation of marine snow. It has been shown that the application of dispersants during phytoplankton blooms can trigger the formation of marine snow to which the sticky dispersed oil can bind. In the presence of mineral particles, oiled snow complexes are being formed that become negatively buoyant and sink to the ocean floor. As a result, oiled marine snow accumulates on the ocean floor where biodegradation is inhibited due to oxygen depletion.

The abovementioned two mechanisms of inhibition of oil biodegradation upon application of oil spill dispersants will be discussed in this chapter.

Keywords Biodegradation · Marine snow · Oil spill · Sediment · MOSSFA · *Deepwater Horizon* oil spill · Dispersants · Enhanced dissolution

18.1 Introduction

After an oil spill, various responses are possible, stopping the flow, containing the oil, and removal of oil from the sea surface (Fingas 2011). Methods to remove and prevent spreading of the spilled oil involve skimming, in situ burning, or the addition of chemical dispersants. Dispersant application was one of the responses during the *Deepwater Horizon* (DWH) oil spill, when ~640 million liters (4 million barrels) of oil were released into the northern Gulf of Mexico (GoM) over a period of almost 3 months. Around 8 million liters of the chemical dispersants Corexit 9527 and Corexit 9500A were added to the water surface or injected into the wellhead at a depth of 1500 m (Kujawinski et al. 2011; Brooks et al. 2015), the first time dispersants were used at both spots. Chemical dispersants reduce the interfacial tension between oil and water, stabilize the smaller oil droplets that are formed, and stimulate the dissolution of the more hydrophilic and toxic oil compounds (Brooks et al. 2015). Smaller droplet sizes provide a larger surface area for oil-degrading bacteria (Lessard and DeMarco 2000), thus increasing the bioavailability of oil. Subsequently, this can enhance the biodegradability of the dispersed oil. As a result, the number and activity of oil-degrading bacteria can increase, and more oil will be degraded in a shorter period of time (Kessler et al. 2011). These are the underlying mechanisms by which chemical dispersants enhance oil biodegradation.

There was an additional effect observed during the DWH oil spill following the application of chemical dispersants. The application of dispersants in the northern GoM triggered the formation of flocculent material and sinking of marine snow (biological debris) during the spring phytoplankton bloom, as a biological stress response to dispersants and oil (Passow et al. 2012; Van Eenennaam et al. 2016; Ziervogel et al. 2012, 2016). This has been reported previously (Vonk et al. 2015).

Marine snow is generally defined as rapidly sinking composite particles (>0.5 mm) that consist of many particles including bacteria, phytoplankton, and feces (Daly et al. 2016; Passow et al. 2012, 2014). Microbes (e.g., bacteria, phytoplankton, and bacterioplankton) secrete sticky extracellular polymeric substances (EPS) to absorb nutrients, to create a protective microenvironment, and to act as a sink of excess fixed carbon (Bullerjahn and Post 2014). The resulting formed marine snow can also have an effect on the degradation of oil during an oil spill.

This chapter will discuss the effect of chemical dispersants and marine snow on oil biodegradation, using both weathered and non-weathered oil.

18.2 Effect of Chemical Dispersants on Oil Biodegradation

To date, contradicting results are published in the literature regarding the effect of chemical dispersants on oil biodegradation, as both stimulation and inhibition have been described. Various mechanisms play a role and interfere with each other, such as (1) oxygen competition between oil- and dispersant-degrading bacteria; (2) enhanced bioavailability of oil due to the dispersion; and (3) decreased biodegradation of oil due to acute release of toxic oil components by chemical dispersants. These three mechanisms are discussed in the next paragraphs.

18.2.1 Oxygen Competition Between Oil- and Dispersant-Degrading Bacteria

The added chemical dispersants themselves can be degraded by bacteria present in the ocean, thus competing with oil-degrading bacteria for the available oxygen. The two chemical dispersants that were used during the DWH oil spill (Corexit 9500A and 9527) are composed of various solvents and nonpolar surfactants including petroleum distillates, propylene glycol, 2-butoxyethanol, Tween 80, Tween 85, Span 80, dioctyl sulfosuccinate (DOSS), and dipropylene glycol butyl ether (Glover et al. 2014). DOSS is the most commonly studied chemical of this mixture, and a combination of biodegradation and hydrolysis is responsible for its breakdown (Campo et al. 2013).

When comparing the oxygen consumption for the degradation of dispersants and oil, results show that oxygen depletion was much faster in mixtures of oil and seawater, compared to mixtures of dispersant and seawater. This shows that aerobic degradation occurs much faster in oil compared to Corexit (Fig. 18.1).

Limited oxygen consumption (less than 1%) was observed during Corexit degradation, compared to the amount used for oil degradation after 7 days of incubation, dropping from 20% to 8% in the headspace. These data suggest that biodegradation of Corexit will not result in a significant competing effect with oil-degrading bacteria. Similar observations were made regarding the degradation of DOSS, one of the

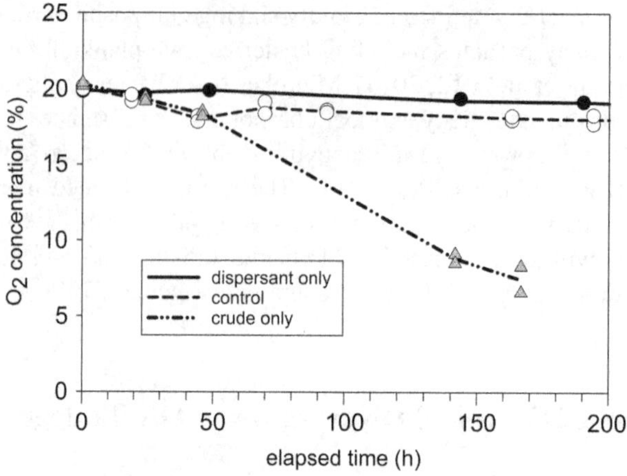

Fig. 18.1 Oxygen concentration during degradation of oil or Corexit 9500A in artificial seawater in the dark. The control contained no oil and no dispersant. Practical setup as described in Rahsepar et al. (2016)

key components of Corexit 9500A that was documented to have undergone negligible rates of biodegradation in the affected waters of the deepwater hydrocarbon plume of the DHW area (Kujawinski et al. 2011).

A more relevant process for the removal of Corexit than biodegradation in the water phase is photolysis with high degradation rates, suggesting that this process may play a significant role in the overall fate of Corexit in the ocean (Glover et al. 2014).

18.2.2 Enhanced Bioavailability and Biodegradation of Oil with Dispersant

A number of studies show the positive effect of dispersants on oil biodegradation by mixed bacterial communities. Generally, dispersants enhance the biodegradation of oil, through higher bioavailability of the dispersed oil (Kujawinski et al. 2011).

Hazen et al. (2010) reported the stimulation of deep-sea indigenous *Gammaproteobacteria* by the dispersed hydrocarbon plume from the DWH blowout; *Gammaproteobacteria* are closely related to known petroleum degraders. In that study, 16 distinct taxa classified as *Gammaproteobacteria* were found to be significantly enriched in the plume by using bacterial taxa microarray analysis. They conclude that the deep sea has a potential for intrinsic bioremediation of the dispersed oil plume by aerobic bacteria in the deepwater column without substantial oxygen depletion.

A recent study by Tremblay et al. (2017) demonstrated that the addition of dispersants to weathered crude oil increased degradation rates and favored the abundance and expression of oil-degrading genes in near-surface seawater from the vicinity of crude oil and natural gas production facilities off eastern Canada. This again shows the stimulating effect of chemical dispersants on oil degradation.

A less pronounced effect was shown by microbial hydrocarbon enrichments grown at 5 and 25 °C. Only slight increases in biodegradation were observed in the presence of Corexit 9500A at both temperatures (Techtmann et al. 2017), with oil biodegradation patterns consistent with those reported in the literature (i.e., aliphatics were degraded faster than aromatics).

18.2.3 Decreased Biodegradation of Oil Due to Dispersion by Chemical Dispersants

The stimulating effect of chemical dispersants is often related to the enhanced dissolution of oil; a better bioavailability leads to a better biodegradation. However, these smaller droplets also increase the dissolution of toxic oil compounds into the water phase, such as BTEX.

Lindstrom and Braddock (2002) compared biodegradation of fresh, weathered, or dispersed oil with and without the chemical dispersant Corexit 9500 and showed that aromatic compounds and alkanes were most rapidly mineralized in fresh oil, followed by weathered oil, and finally dispersed oil. However, when looking at individual components of crude oil, variable effects of the presence of dispersed and non-dispersed compounds were observed (Lindstrom and Braddock 2002). However, especially with light, fresh oil, the acutely released toxic compounds can in turn result in a toxic effect on oil-degrading bacteria, which might inhibit oil-degrading bacteria.

This toxic effect was shown in a recent study by Rahsepar et al. (2016) using pure bacterial cultures, fresh and weathered Macondo oil, and Corexit 9500A. When the alkane degrader *Rhodococcus qingshengii* TUHH-12 was introduced to a Corexit and fresh oil mixture, biodegradation inhibition was observed to occur for a period of at least 50 days (the incubation time), whereas when added to Corexit the weathered oil mixture, biodegradation was only inhibited for 10 days (Fig. 18.2).

The difference in inhibition may be attributed to the difference in chemical composition between crude and weathered oil. Chemical analyses showed that crude oil had high concentrations of BTEX compounds ($\approx$2000 µg/g), whereas weathered oil contains negligible amounts of BTEX compounds (0.25 µg/g oil), and less light aliphatic compounds (Rahsepar et al. 2016). The addition of Corexit 9500A to the fresh oil results in higher concentrations of BTEX in the dispersed small oil droplets and in the water phase (Prince 2015). Those higher concentrations of BTEX compounds in the water column inhibited the activity of the bacteria, thus explaining the inhibitory effect observed for *R. qingshengii* TUHH-12. Since weathered oil is devoid of toxic monoaromatic compounds, this inhibitory effect of high concentration of dissolved BTEX for *R. qingshengii* TUHH-12 did not occur in chemically dispersed weathered oil. In follow-up experiments, both a BTEX-degrading culture

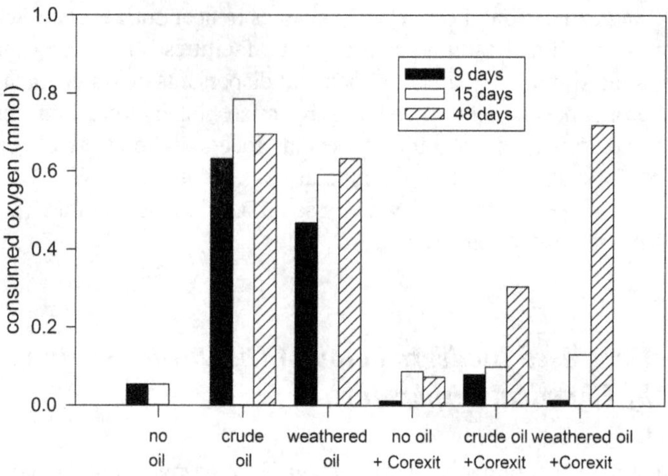

Fig. 18.2 Cumulative oxygen consumption by *R. qingshengii* TUHH-12 as measure of biodegradation of crude or weathered oil with and without Corexit 9500A. (Data from Rahsepar et al. 2016)

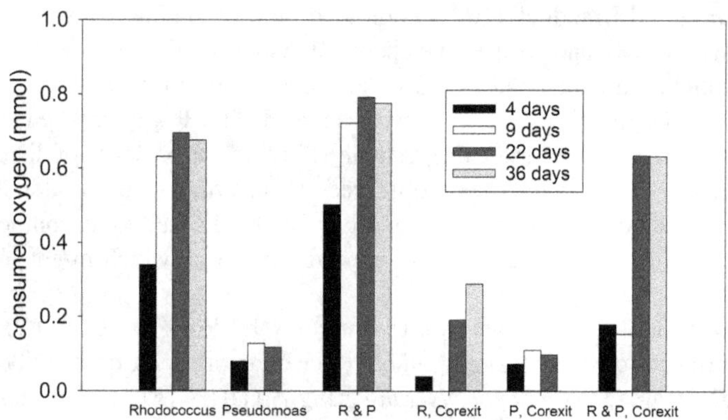

Fig. 18.3 Cumulative oxygen consumption by *R. qingshengii* TUHH-12 and/or *P. putida* F1 as measure of biodegradation of crude oil with and without Corexit. (Data from Rahsepar et al. 2016)

P. putida F1 and an *n*-alkane-degrading culture *R. qingshengii* TUHH-12 were used for the degradation of crude oil (Fig. 18.3).

Using both cultures resulted in an immediate degradation of BTEX compounds, followed by degradation of *n*-alkanes. The inhibition of chemically dispersed crude oil lasted only 10 days due to the fast biodegradation of the dissolved light aromatic (BTEX) compounds by *P. putida* F1.

Finally, hydrocarbon-degrading bacteria can also be inhibited by high concentrations of the chemical dispersant itself, as demonstrated by Hamdan and Fulmer (2011). They studied the composition of the microbial community in oil that had been freshly deposited on a beach in Louisiana as a result of the DWH oil spill and showed a decrease in microbial activity and viability of *Acinetobacter* and

Marinobacter in the presence of Corexit 9500A. *Marinobacter* was most sensitive to the dispersant, with nearly 100% reduction in viability and production after exposure to Corexit in concentrations likely to be encountered during the response to the spill (1–10 g/L).

These data suggest that the use of dispersants has the potential to reduce the capacity of the environment to bioremediate spills, as hydrocarbon-degrading bacteria can be inhibited by chemical dispersants itself or by the BTEX compounds released from fresh oil.

18.3 Biodegradation of Oil in the Presence of Marine Snow in Water Phase and Benthic Zone

In the presence of high concentrations of phytoplankton, dispersant application can induce the formation of marine snow (Van Eenennaam et al. 2016). During the DWH oil spill, an unusual formation of marine snow was observed after the addition of chemical dispersants to combat the oil during a phytoplankton bloom (Brooks et al. 2015). Dispersed oil is very sticky and binds to the marine snow. Furthermore, in the presence of mineral particles, oiled snow complexes are formed. During the spill also a large amount of suspended solids was present due to the flushing of the Mississippi River as a spill response (Bianchi et al. 2011; Hu et al. 2011; O'Connor 2013). When mineral particles bind to oiled marine snow particles, these complexes become negatively buoyant and sink.

Studies have shown that microbes tend to produce more EPS as a "stress" reaction or as a natural dispersant to degrade the dispersed oil (Passow et al. 2012). During oil spills, production of EPS could enhance the breakup of oil slicks into droplets to induce the degradation of oil (Sohm et al. 2011). On the other hand, more EPS production can enhance the aggregation of oil complexes, microorganisms, and other smaller particles into marine snow that settles in massive amounts on the seabed, a phenomenon known as Marine Oil Snow Sedimentation and Flocculent Accumulation (MOSSFA). This oiled marine snow will reduce the bioavailability of oil, and thus biodegradation will be low. Furthermore, these particles can sink to the sea floor by gravitational settling (Sohm et al. 2011; Passow et al. 2012, Schwing et al. 2020), thus further reducing oil biodegradation.

Few studies describe the effect of marine snow on bacterial activity, as marine snow in sufficient and reproducible quantities is not available for testing. However, alginate, kaolin, and algae mix particles representing artificial marine snow can be used as an approximation of real marine snow (Rahsepar et al. 2017).

Using this artificial marine snow in batch experiments, it was shown that artificial marine snow enhanced oil biodegradation in the water phase in the abundance of dissolved oxygen (Rahsepar et al. 2017). As described in Sect. 18.2, oil biodegradation was inhibited in the presence of Corexit 9500A, due to the increased BTEX concentrations. However, in the presence of artificial marine snow, this inhibition was lower, probably because the toxic BTEX compounds were bound to the marine snow, thus reducing their concentrations in the water phase. Furthermore, it was hypothesized that an inner aggregate environment is created, where mass transfer of the oil to the bacteria is enhanced. This will lead to the release of degradable oil compounds and shortens the diffusion distances between the oil phase and the oil-degrading bacteria (Rahsepar et al. 2017).

The influence of artificial marine snow on oil degradation was further tested in small-scale aquarium experiments with artificial seawater and a sediment layer by comparing the oxygen consumption in the presence and absence of artificial marine snow and/or oil during continuous aeration of the aquaria (Fig. 18.4).

Although air was bubbled in the water, the oxygen concentration in the water 10 cm above the sediment dropped from 9 mg/L to 4.4 mg/L after 6 days of incubation in the presence of artificial marine snow, whereas the water phase remained saturated with oxygen when only oil was present. This shows an enhanced oxygen consumption due to artificial marine snow degradation (Rahsepar *unpublished*). The carbohydrate compounds of marine snow have oxidized groups in their molecular structures which are readily available for further enzymatic conversions and mineralization (Alldredge 1998; Bochdansky et al. 2010), thus depleting the oxygen in the water phase. As can be seen in Fig. 18.5, the presence of marine snow reduces the degradation of oil alkanes by 40%.

Relative to marine snow, oil is more recalcitrant because of its stably reduced aliphatic or aromatic molecular structures that need to be activated for further mineralization, i.e., through oxygen inclusion by oxygenases (Chikere et al. 2011). When high concentrations of artificial marine snow or other easily biodegradable carbon sources are present, aerobic biodegradation of artificial marine snow is preferred following the biodegradation of oil. Oil degradation is then further hampered by oxygen depletion especially during the first 6 days of incubation (Fig. 18.4).

As biodegradation of marine snow reduces the oxygen concentration, this can result in depletion of oxygen and the formation of an anaerobic layer in the sediment at the ocean sea floor. As anaerobic degradation of oil is generally a slower process (see also Bubenheim et al. 2020), this will further reduce the overall

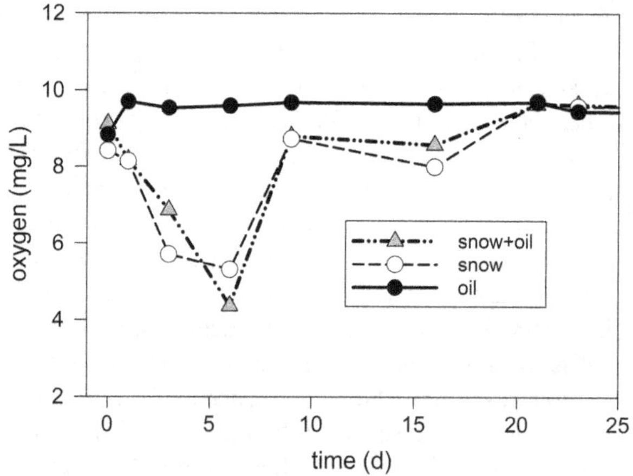

Fig. 18.4 Dissolved oxygen concentration at 10 cm above the sediment layer in the water phase of the aquarium. (Data from Rahsepar unpublished, data can be accessed through GRIIDC at https://data.gulfresearchinitiative.org/data/R4.x267.179:0017)

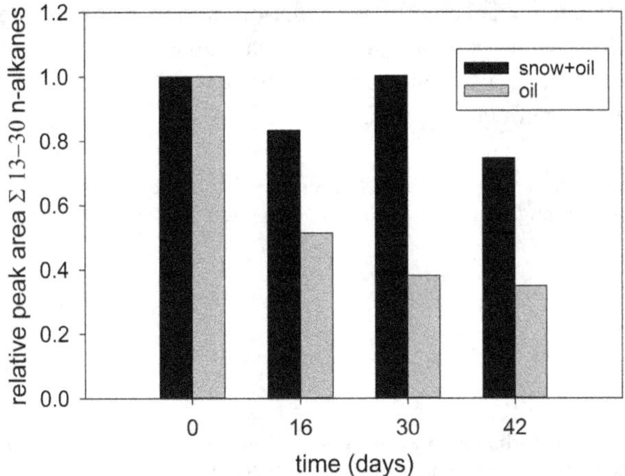

Fig. 18.5 Relative peak area of C30-hopane normalized *n*-alkanes (C13–C30) and isoprenoid alkanes in oil with and without the presence of marine snow. (Data from Rahsepar, data can be accessed through GRIIDC at https://data.gulfresearchinitiative.org/data/R4.x267.179:0017)

oil biodegradation rate in the benthic zone. The effects of oiled artificial marine snow on the benthic macroinvertebrates *Corophium volutator*, *Hydrobia ulvae*, and *Macoma balthica* are described in Foekema et al. (2020).

18.4 Conclusions

While dispersants are used to treat oil spills, there is an ongoing debate regarding their effects on the fate of oil, oil biodegradation, and ultimately their cumulative effects on the oil-degrading microbial communities. The scientific and technical understanding of the physicochemical interactions taking place between dispersed oil, dispersant, marine snow, and how they subsequently affect biological activities is not (yet) complete. The wide range of results in literature are most likely related to the varied chemical composition of the oils used (crude oil, weathered oil, or fresh oil), concentrations, the concentration of the added dispersants, and the characteristics of the microbial population.

In summary, the following general conclusions can be drawn based on the available literature about the effects of dispersants and (artificial) marine snow on oil biodegradation:

- Chemical dispersants inhibit the degradation of oil, and this effect can be minimized when both alkane- and BTEX-degrading bacteria are present.
- Release of the more toxic BTEX components of crude oil when chemical dispersants are added is toxic for bacteria and hampers biodegradation. This toxicity effect is less relevant for weathered oil.

- Marine snow can enhance oil biodegradation in the water phase in the abundant presence of dissolved oxygen; however, because the biodegradation of the marine snow occurs more rapidly, thus reducing the oygen concentration, the amount of oil degradation is effectively limited.

Funding Information This research was made possible by grants from the Gulf of Mexico Research Initiative through the Center for the Integrated Modeling and Analysis of the Gulf Ecosystem (C-IMAGE).

References

Alldredge A (1998) The carbon, nitrogen and mass content of marine snow as a function of aggregate size. Deep-Sea Res I Oceanogr Res Pap 45:529–541

Bianchi TS, Cook RL, Perdue EM, Kolic PE, Green N, Zhang Y, Smith RW, Kolker AS, Ameen A, King G, Ojwang LM, Schneider CL, Normand AE, Hetland R (2011) Impacts of diverted freshwater on dissolved organic matter and microbial communities in Barataria Bay, Louisiana, U.S.A. Mar Environ Res 72(5):248–257

Bochdansky AB, van Aken HM, Herndl GJ (2010) Role of macroscopic particles in deep-sea oxygen consumption. Proc Natl Acad Sci 107:8287–8291

Brooks GR, Larson RA, Schwing PT, Romero I, Moore C, Reichart G-J, Jilbert T, Chanton JP, Hastings DW, Overholt WA, Marks KP, Kostka JE, Holmes CW, Hollander D (2015) Sedimentation pulse in the NE Gulf of Mexico following the 2010 DwH blowout. PLoS One 10(7):e0132341. https://doi.org/10.1371/journal.pone.0132341

Bubenheim P, Hackbusch S, Joye SB, Kostka JE, Larter SR, Liese A, Lincoln SA, Marietou A, Müller R, Noirungsee N, Oldenburg TBP, Radović JR, Viamonte J (2020) Biodegradation of hydrocarbons in deep water and sediments (Chap. 7). In: Murawski SA, Ainsworth C, Gilbert S, Hollander D, Paris CB, Schlüter M, Wetzel D (eds) Deep oil spills – facts, fate and effects. Springer, Cham

Bullerjahn GS, Post AF (2014) Physiology and molecular biology of aquatic cyanobacteria. Front Microbiol 5:359

Campo P, Venosa AD, Suidan MT (2013) Biodegradability of Corexit 9500 and dispersed south Louisiana crude oil at 5 and 25 °C. Environ Sci Technol 47:1960–1967

Chikere CB, Okpokwasili GC, Chikere BO (2011) Monitoring of microbial hydrocarbon remediation in the soil. 3 Biotech 1(3):117–138. https://doi.org/10.1007/s13205-011-0014-8

Daly KL, Passow U, Chanton J, Hollander D (2016) Assessing the impacts of oil-associated marine snow formation and sedimentation during and after the Deepwater Horizon oil spill. Anthropocene 13:18–33. https://doi.org/10.1016/j.ancene.2016.01.006

Fingas M (2011) Introduction to oil spill contingency planning and response initiation. In: Fingan M (ed) Oil spill science and technology. Gulf Professional Publishing, Boston, pp 1027–1031

Foekema EM, van Eenennaam JS, Hollander DJ, Langenhoff AM, Oldenburg TBP, Radović JR, Rohal M, Romero IC, Schwing PT, Murk AJ (2020) Biodegration of hydrocarbons in deep water and sediments (Chap. 17). In: Murawski SA, Ainsworth C, Gilbert S, Hollander D, Paris CB, Schlüter M, Wetzel D (eds) Scenarios and responses to future deep oil spills – fighting the next war. Springer, Cham

Glover CM, Mezyk SP, Linden KG, Rosario-Ortiza FL (2014) Photochemical degradation of Corexit components in ocean water. Chemosphere 111(9):596–602. https://doi.org/10.1016/j.chemosphere.2014.05.012

Hamdan LJ, Fulmer PA (2011) Effects of COREXIT® EC9500A on bacteria from a beach oiled by the Deepwater Horizon spill. Aquat Microb Ecol 63(2):101–109. https://doi.org/10.3354/ame01482

Hazen TC, Dubinsky EA, DeSantis TZ et al (2010) Deep sea oil plume enriches indigenous oil degrading bacteria. Science 330(6001):204–208. https://doi.org/10.1126/science.1195979

Hu C, Weisberg RH, Liu Y, Zheng L, Daly KL, English DC, Zhao J, Vargo GA (2011) Did the northeastern Gulf of Mexico become greener after the Deepwater Horizon oil spill? Geophys Res Lett 38(9):L09601

Kessler JD, Valentine DL, Redmond MC, Du M, Chan EW, Mendes SD, Quiroz EW, Villanueva CJ, Shusta SS, Werra LM, Yvon-Lewis SA, Weber TC (2011) A persistent oxygen anomaly reveals the fate of spilled methane in the deep Gulf of Mexico. Science 331:312–315

Kujawinski EB, Kido Soule MC, Valentine DL, Boysen AK, Longnecker K, Redmond MC (2011) Fate of dispersants associated with the Deepwater Horizon oil spill. Environ Sci Technol 45(4):1298–1306. https://doi.org/10.1021/es103838

Lessard RR, DeMarco G (2000) The significance of oil spill dispersants. Spill Sci Technol Bull 6:59–68

Lindstrom JE, Braddock JF (2002) Biodegradation of petroleum hydrocarbons at low temperature in the presence of the dispersant Corexit 9500. Mar Pollut Bull 44(8):739–747. https://doi.org/10.1016/S0025-326X(02)00050-4

O'Connor B (2013) Impacts of the anomalous Mississippi river discharge and diversions on Phytoplankton Blooming in Northeastern Gulf of Mexico. MSc, University of South Florida

Passow U, Ziervogel K, Asper V, Diercks A (2012) Marine snow formation in the aftermath of the Deepwater Horizon oil spill in the Gulf of Mexico. Environ Res Lett 7:035301

Passow U, De La Rocha CL, Fairfield C, Schmidt K (2014) Aggregation as a function of and mineral particles. Limnol Oceanogr 59(2):532–547. https://doi.org/10.4319/lo.2014.59.2.0532

Prince RC (2015) Oil spill dispersants: boon or bane? Environ Sci Technol 49(11):6376–6384. https://doi.org/10.1021/acs.est.5b00961

Rahsepar S, Smit MPJ, Murk AJ, Rijnaarts HHM, Langenhoff AAM (2016) Biodegradation of chemically dispersed crude and weathered oil. Mar Pollut Bull 108:113–119

Rahsepar S, Langenhoff AAM, Smit MPJ, Van Eenennaam J, Murk A, Rijnaarts HHM (2017) Oil biodegradation: interactions of artificial marine snow, clay particles, oil and Corexit. Mar Pollut Bull 125(1–2):186–191. https://doi.org/10.1016/j.marpolbul.2017.08.021

Schwing PT, Hollander DJ, Brooks GR, Larson RA, Hastings DW, Chanton JP, Lincoln SA, Radović JR, Langenhoff AL (2020) The sedimentary record of MOSSFA events in the Gulf of Mexico: a comparison of the Deepwater Horizon (2010) and Ixtoc 1 (1979) oil spills (Chap. 13). In: Murawski SA, Ainsworth C, Gilbert S, Hollander D, Paris CB, Schlüter M, Wetzel D (eds) Deep oil spills – facts, fate and effects. Springer, Cham

Sohm JA, Edwards BR, Wilson BG, Webb EA (2011) Constitutive Extracellular Polysaccharide (EPS) production by specific isolates of Crocosphaera watsonii. Front. Microbiol. 2:229. https://doi.org/10.3389/fmicb.2011.00229

Techtmann SM, Zhuang M, Campo P, Holder E, Elk M, Hazen TC, Conmy R, Santo Domingo JW (2017) Corexit 9500 enhances oil biodegradation and changes active bacterial community structure of oil-enriched microcosms. Appl Environ Microbiol 83(10):e03462–e03416. https://doi.org/10.1128/AEM.03462-16

Tremblay J, Yergeau E, Fortin N, Cobanli S, Elias M, King TL, Lee K, Greer CW (2017) Chemical dispersants enhance the activity of oil- and gas condensate-degrading marine bacteria. ISME J 11(12):2793–2808. https://doi.org/10.1038/ismej.2017.129

Van Eenennaam JS, Wei Y, Grolle KCF, Foekema EM, Murk AJ (2016) Oil spill dispersants induce formation of marine snow by phytoplankton-associated bacteria. Mar Pollut Bull 104:294–302

Vonk SM, Hollander DJ, Murk AJ (2015) Was the extreme and wide-spread marine oil-snow sedimentation and flocculent accumulation (MOSSFA) event during the Deepwater Horizon blowout unique? Mar Pollut Bull 100:5–12. https://doi.org/10.1016/j.marpolbul.2015.08.023.

Ziervogel K, McKay L, Rhodes B, Osburn CL, Dickson-Brown J, Arnosti C, Teske A (2012) Microbial activities and dissolved organic matter dynamics in oil-contaminated surface seawater from the Deepwater Horizon oil spill site. PLoS One 7:e34816

Ziervogel K, Joye SB, Arnosti C (2016) Microbial enzymatic activity and secondary production in sediments affected by the sedimentation pulse following the Deepwater Horizon oil spill. Deep Sea Res II Trop Stud Oceanogr 129:241–248

Chapter 19
Molecular Legacy of the 1979 Ixtoc 1 Oil Spill in Deep-Sea Sediments of the Southern Gulf of Mexico

Sara A. Lincoln, Jagoš R. Radović, Adolfo Gracia, Aprami Jaggi, Thomas B. P. Oldenburg, Stephen R. Larter, and Katherine H. Freeman

Abstract The 2010 *Deepwater Horizon* blowout in the northern Gulf of Mexico (GoM) was the first major oil spill to impact the deep sea. This lack of precedent creates challenges for predicting the long-term fate of spilled oil constituents in the deep GoM. Depletion of oil residues in oxygenated seafloor sediments is thought to be relatively rapid, but as impacted horizons are buried by sediments and oxygen is depleted, biodegradation rates are likely to decrease. The sedimentary record of the 1979 Ixtoc 1 oil spill (southern GoM), the second largest marine oil spill to date, affords a window for forecasting the fate of constituents of spilled oil in the northern GoM over multi-decadal time scales.

This chapter summarizes evidence for the persistence of traces of Ixtoc 1 oil in sediments underlying deep (>500 m) waters of the southern GoM over ~4 decades.

Recognizing the complex hydrocarbon background of the region, which is impacted by natural and anthropogenic sources of petroleum, we focused on metrics with the greatest potential for assessing Ixtoc 1 input. Ratios of polycyclic aromatic hydrocarbons, sulfur heterocycles, and petroleum biomarkers, and molecular compound classes revealed by Fourier transform ion cyclotron resonance-mass spectrometry provide evidence that recalcitrant signatures of Ixtoc 1 oil can still be detected in Bay of Campeche sediments, although they are heavily overprinted by the regional hydrocarbon background. We discuss the spatial heterogeneity of

S. A. Lincoln (✉) · K. H. Freeman
The Pennsylvania State University, Department of Geosciences, University Park, PA, USA
e-mail: slincoln@psu.edu; khf4@psu.edu

J. R. Radović · A. Jaggi · T. B. P. Oldenburg · S. R. Larter
University of Calgary, PRG, Department of Geoscience, Calgary, AB, Canada
e-mail: jagos.radovic@ucalgary.ca; aprami.jaggi@ucalgary.ca; toldenbu@ucalgary.ca; slarter@ucalgary.ca

A. Gracia
Universidad Nacional Autónoma de México, Instituto Ciencias del Mar y Limnología
Apartado, Mexico City, DF, Mexico
e-mail: gracia@unam.mx

© Springer Nature Switzerland AG 2020
S. A. Murawski et al. (eds.), *Deep Oil Spills*,
https://doi.org/10.1007/978-3-030-11605-7_19

deep-sea oil deposition and conditions that favor its preservation or degradation in the context of both Ixtoc 1 and *Deepwater Horizon* spills.

Keywords Oil spill forensics · Tithonian oil families · Marine sediments · Ixtoc 1 · Bay of Campeche · Deepwater Horizon

19.1 Introduction

The 2010 *Deepwater Horizon* (DWH) blowout was the first major deep-sea oil spill in history. The lack of precedent hindered prediction of the fate of residues of spilled oil in seafloor sediments in the northern Gulf of Mexico (GoM). Early reports have found the half-lives of Macondo well (MW) oil constituents in sediments to be quite short (e.g., Bagby et al. 2017), but as impacted horizons are buried by sediment, degradation rates of residual petrocarbon are likely to decrease due to oxygen depletion. The longer-term fate and persistence of residues of MW oil in seafloor sediment remain uncertain.

The 1979 Ixtoc 1 oil spill in the southern GoM provides a window for forecasting the fate of MW oil over multi-decadal time scales. Approximately 500,000 metric tons of oil were released over a 10-month period following the Ixtoc 1 blowout (PCEESC 1980; CSLP 1980), creating the second largest marine oil spill to date. Although the blowout occurred in shallow waters (56 m depth) of the Bay of Campeche, the oil slick extended to deep waters (Sun et al. 2015), and residues were transported as far as the Texas coast (Farrington 1983).

This chapter summarizes evidence for the persistence of traces of Ixtoc oil in deep-sea sediments of the southern GoM over ~4 decades. Recognizing the complex hydrocarbon background of the region, which is impacted by both natural and anthropogenic sources of petroleum, we focused on metrics with the greatest potential for assessing Ixtoc 1 input. Polycyclic aromatic hydrocarbons (PAHs), sulfur heterocycles, and petroleum biomarkers measured using conventional gas chromatography-mass spectrometry as well as molecular compound classes revealed by Fourier transform ion cyclotron resonance-mass spectrometry (FTICR-MS) indicated that signatures of Ixtoc 1 oil can still be detected in Bay of Campeche sediments underlying >500 m of water (Lincoln et al. 2017, Radović et al. 2017). Finally, we discuss the spatial heterogeneity of deep-sea oil deposition and conditions that may favor its preservation or degradation in the context of both Ixtoc 1 and *Deepwater Horizon* spills.

19.2 The Ixtoc 1 Oil Spill

The subsea blowout of the Petróleos Mexicanos (PEMEX) exploratory well Ixtoc 1 released ~500,000 metric tons of oil into the GoM in 1979 (CSLP 1980; Jernelöv and Lindén 1981). It was the second largest marine oil spill in history, surpassed only by the 2010 *Deepwater Horizon* (DWH) spill in the northern GoM.

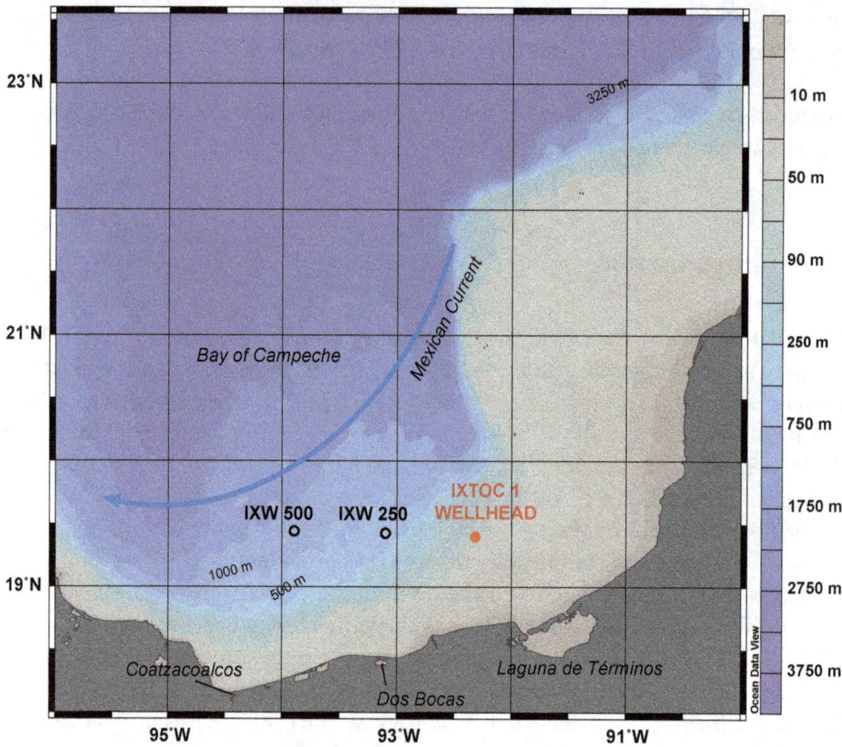

Fig. 19.1 Map of the Bay of Campeche, showing the location of the Ixtoc 1 wellhead and two sites where sediment cores discussed in this chapter were collected: *IXW-250*, 80 km from wellhead, water depth 583 m, and *IXW-500*, 164 km from wellhead, water depth 1010 m (Lincoln et al. 2017)

Initial attempts to plug the damaged well and divert oil via relief wells were unsuccessful, and the release of oil from the Ixtoc 1 well continued from June 3, 1979, until March 23, 1980 (Jernelöv and Lindén 1981). The wellhead (Fig. 19.1) was in shallow (56 m) water of the continental shelf ~80 km northwest of Ciudad del Carmen, Campeche. Much of the Mexican GoM coast was rapidly affected, and the Gulf Loop Current transported oil as far as the Texas coast (Farrington 1983). Recent research efforts to assess coastal impacts in the southern GoM are described in Radović et al. (2020).

There have been far fewer investigations of the extent and impacts of the Ixtoc 1 spill than of the DWH spill, and the deep sediments of the southern GoM are particularly understudied in this respect. Recent reconstructions of the Ixtoc 1 surface oil slick from satellite data show that it extended well beyond the continental shelf, overlying waters of depths up to ~3500 m (Sun et al. 2015), suggesting that the impact of the spill was not limited to coastal regions. Deep sediments potentially received Ixtoc 1 oil residues via marine oil snow processes in overlying waters ("MOSSFA," Quigg et al. 2020; Schwing et al. 2020) as well as via offshore transport of oiled coastal sediments to depocenters.

The limited information available from reports documenting the sedimentary footprint of the Ixtoc 1 spill immediately following the blowout (e.g., Boehm and Fiest 1980) precludes multi-decadal time-series approaches to understanding the fate of the oil in the deep southern GoM. Forensic investigations using conventional gas chromatography-mass spectrometry (Lincoln et al. 2017) and ultrahigh-resolution mass spectrometry (Radović et al. 2017) in tandem with sediment geochronology, however, have provided a means of assessing whether Ixtoc 1 oil residues reached deep sediments, evaluating the fidelity and longevity of diagnostic fingerprints, and provide insight into the preservation and transformation of petroleum compounds in the oceans over multi-decadal time scales.

19.3   Sources of Petroleum Constituents in the Southern Gulf of Mexico

The southern Gulf of Mexico is a dynamic marine system with multiple natural and anthropogenic petroleum inputs, many of which have the potential to confound oil spill forensic efforts. Oil seeps are numerous and broadly distributed in the region (Holguin-Quiñones et al. 2005; MacDonald et al. 2015; Sun et al. 2015), and rivers transport petroleum hydrocarbons from land to coastal sediments (e.g., Ruiz-Fernández et al. 2016). Terrestrial sources of hydrocarbons include onshore seeps, sediments, urban runoff, and pollution from petroleum refineries and infrastructure in the increasingly industrialized watersheds of the southern GoM. The petroleum industry has a long history on the GoM coast, with small refineries active in Tuxpan and Papantla as early as 1876 and 1880 (Alvarez de la Borda 2006). Rapid development of the prolific offshore oil fields in the Bay of Campeche in the 1970s increased the risk of direct crude oil release into southern GoM waters (García-Cuéllar et al. 2004). Industry infrastructure and operations may be both episodic and chronic sources of contamination; for instance, all produced waters from Bay of Campeche oil fields are discharged from the port of Dos Bocas (Fig. 19.1), and sediments near it contain biomarker ratios consistent with a mature petroleum signal (Schifter et al. 2015). Storm activity, coastal erosion, and resuspension of petrocarbon from sites previously impacted by spills or natural sources also have the potential to complicate efforts to identify Ixtoc 1 impact horizons in sediments.

Finally, although the Ixtoc 1 spill was the largest spill in the southern GoM, it was not an isolated incident, and accidental industrial releases of hydrocarbons such as the one caused by the Abkatun platform explosion (O'Malley et al. 2016) continue to occur.

19.4   Previous Assessments of Hydrocarbon Contamination in Southern Gulf of Mexico Sediments

Early-response cruises attempted to assess oil contamination in sediments near the Ixtoc 1 wellhead in 1979 (Farrington 1983). Based on total petroleum hydrocarbon (TPH) data, Boehm and Fiest (1980) estimated that 1–3% of the oil was deposited in sediments within a 30 km radius of the Ixtoc 1 wellhead. This estimate was thought to be conservative due to restricted sampling and, in particular, a lack of samples from sediments beneath the course of the oil as it was carried westward by the Gulf Loop Current.

TPH is a broad and non-specific measure of petrocarbon contamination, and no universal analytical method for measuring it exists; commonly, organic extracts of sediments are analyzed by gas chromatography-flame ionized detection, but methods may differ in terms of solvents used, sample purification, and chromatographic method. Nevertheless, TPH has proven useful in assessing contamination shortly after an oil spill. An early time-series study documented an increase in TPH near the Ixtoc 1 wellhead from 1978 to 1980, followed by a decline to pre-spill values by 1981 (Botello and Villanueva 1985). In cases with weaker spatial and temporal connections between source and spill, however, TPH is less useful because it provides little information about the composition or provenance of the hydrocarbons detected. Further, the analytical window of TPH is relatively narrow, only enabling detection of hydrocarbons that are amenable to gas chromatography and excluding large and polar residues of petroleum and its weathering products. The rapid attenuation of TPH observed after the Ixtoc 1 spill (Botello and Villanueva 1985) was likely the result of combined removal mechanisms (e.g., microbial respiration and abiotic degradation) and transformation processes that created polar moieties and insoluble residues not readily extracted or measured using standard TPH protocols. Following the DWH spill, such transformation products have been well studied using Fourier transform ion cyclotron resonance-mass spectrometry (FTICR-MS) and thin-layer chromatography-flame ionized detection (Aeppli et al. 2012; Radović et al. 2014). Evidence for similar transformation of Ixtoc 1 hydrocarbons in southern GoM environments is presented in Sect. 19.5.3 and in Radović et al. (2020).

Polycyclic aromatic hydrocarbons (PAHs) have greater forensic potential than TPH, but they may be rapidly biodegraded; for instance, Bagby et al. (2017) estimated half-lives of alkylated phenanthrenes to be <6 years in northern GoM sediments containing Macondo well oil residues from the DWH blowout. Only PAH with >4 rings had half-lives >10 years, and alkylated dibenzothiophene half-lives were as short as ~160 days. PAH concentrations, like those of more labile oil compounds such as n-alkanes, are thus likely to be of limited utility in tracing spilled oil over decadal time scales. Ratios of specific compounds that degrade at similar rates, however, can remain useful even as their concentrations decline.

Boehm and Fiest (1980) reported a predominance of petrogenic over pyrogenic PAHs and abundant dibenzothiophenes (DBT) in sediments containing putative Ixtoc 1 oil. Ratios of alkylated phenanthrenes (three-ringed PAHs with two or three methyl groups) and alkylated dibenzothiophenes were considered most diagnostic after the spill (Boehm et al. 1983). These metrics provided evidence of Ixtoc 1 oil on the seafloor 50 km from the wellhead and helped to identify Ixtoc 1 oil residues in coastal Texas sediments. Recently, Lincoln et al. (2017) used the same ratios to screen southern GoM seafloor sediments for residual traces of Ixtoc 1 petroleum input.

PAH and TPH measurements were also used in subsequent investigations of hydrocarbon contamination in sediments near zones of petroleum production in the Bay of Campeche (e.g., Marchand et al. 1982; Botello et al. 1991; Alpuche-Gual and Gold-Bouchot, 2014) and its coastal regions (e.g., Vazquez et al. 1991; Gold-

Bouchot et al. 1997; Carmona Uristegui 2015; Schifter et al. 2015; Ruiz-Fernández et al. 2016). Chronic PAH sources can be significant: Gracia et al. (2016) estimated that PAH input from Laguna de Términos (Fig. 19.1) inlets were as much as 0.1 metric tons/day from 2004 to 2010. Information about hydrocarbon contamination in deep-sea sediments is more limited. PAH and TPH measurements were monitored by PEMEX studies in the deep southern GoM during 2005–2010 and reported in internal documents (Gracia 2012). Recently investigations of sediment hydrocarbons were carried in the southern Gulf of Mexico in an area encompassing 150,000 km^2 in a 187–3800 m depth range and related to bacterial diversity (Godoy-Lozano et al. 2018).

Oil spill fingerprinting employs a broad suite of petroleum biomarkers, reviewed in Stout and Wang (2016). The molecule 17α(H),21β(H) C$_{30}$ hopane (referred to hereafter as "hopane") is one of the most commonly used markers. Hopane, a pentacyclic triterpane biomarker that is a diagenetic product of many bacterial membrane lipids, is a common constituent of crude oils. It has been widely used as a tracer for petroleum input to the environment, including northern GoM sediments impacted by the DWH spill (e.g., Valentine et al. 2014). Hopane has been considered recalcitrant (highly resistant to degradation) and likely to persist in the environment over long time scales, but several recent studies have found it may be less conservative than previously thought (Huesemann et al. 2010; Bagby et al. 2017). Nevertheless, it is likely to remain a useful member of the broad array of oil spill forensic biomarkers, all of which have individual limitations and are most powerful when employed as a suite.

Reportedly low concentrations of pentacyclic triterpanes in Ixtoc 1 oil, together with the pervasive hopane background in southern GoM sediments, precluded the use of hopanes in forensic studies just after the spill (Boehm and Fiest 1980). Few reports of hopanes or other hydrocarbon biomarkers in Bay of Campeche sediments have been published in the decades following the Ixtoc 1 spill. Available data are from surficial sediment samples collected in 1998 or later, likely to have been too recently deposited to contain any primary record of Ixtoc 1 oil input. They can, however, provide insight into the nature and extent of petroleum inputs to the region. Carmona Uristegui (2015) reported moderate hopane contamination (56–96 ng/g) of coastal sediments near the Terminal Marítima Dos Bocas, Tabasco, but did not attribute it to a specific source or discuss terpane biomarker distributions. In a study of sediments near the mouth of the Coatzacoalcos River (Fig. 19.1), a major tributary of the southern GoM and site of multiple refineries and chemical plants, Farrán et al. (1987) reported a uniform hydrocarbon pollution signature, with n-alkane, hopane, and sterane distributions showing no clear spatial trends. Scholz-Böttcher et al. (2008, 2009) analyzed biomarkers in seafloor sediments collected during a PEMEX environmental monitoring program and compared them to asphalt seep, crude oil, and drill cuttings from the region. They found complex hydrocarbon backgrounds, with sediments containing varying proportions of a mature petroleum component, recent organic matter, and a low-to-moderate maturity component thought to be derived from drill cuttings disposed of offshore. The seafloor surface sediment focus of this study meant that it was unlikely to have detected a primary, autochthonous Ixtoc 1 spill record, which would have been buried by several decades of sedimentation.

19.5   Evidence of Residual Ixtoc 1 Oil-Derived Compounds in Southern GoM Sediments >35 Years After the Spill

In 2015 the Gulf of Mexico Research Initiative-funded Center for the Integrated Modeling and Analysis of the Gulf Ecosystem (C-IMAGE) research consortium undertook a sampling campaign to evaluate the potential presence and impacts of Ixtoc 1 in the southern GoM. Radović et al. (2020) discusses evidence for Ixtoc 1 oil input to coastal environments; here, we focus exclusively on evidence for preservation in the deep sea. Sediment cores were retrieved from the seafloor at sites across the Bay of Campeche using a multicorer deployed from the B/O Justo Sierra. Site selection was guided by historical satellite data and observations of oiling during and after the spill (Sun et al. 2015). Tandem organic geochemical and geochronological analyses were used to search for compositional and temporal matches between sediment hydrocarbons and putative Ixtoc 1 inputs (Lincoln et al. 2017; Radović et al. 2017). The results of these analyses are summarized below after a brief explanation of the geochronological and chemostratigraphic approaches that enabled age estimates for oil spill event horizons.

19.5.1   Temporal Constraints from Geochronology and PCB Chronostratigraphy

Environmental oil fingerprinting can be difficult and contentious even when undertaken shortly after a spill, particularly if oils have experienced extensive weathering; the challenge is only compounded as sedimentation buries any potential record beneath the seafloor. Northern GoM studies have either made the implicit assumption that core tops (0–0.5 to 0–2 cm below seafloor) should contain any sedimentary record of DWH-derived petrocarbon <5 years after the spill (e.g., Valentine et al. 2014; Chanton et al. 2014), which seems reasonable given knowledge of sedimentation rates in the region, or have taken a higher-resolution event stratigraphy approach, sampling cores at depth intervals of as little as 2 mm and employing geochronological techniques to more tightly constrain horizons containing potential spill-related records (e.g., Romero et al. 2015; Brooks et al. 2015; Schwing et al. 2016). Geochronology becomes more critical as sediments deposited during and after spills are buried and compacted over time. Short-lived radioisotopes (specifically, the excess ^{210}Pb approach) have been used to resolve the DWH event (Larson et al. 2020) and to generate age models for sediment cores retrieved from the southern GoM. At the site with the strongest terpane biomarker evidence for Ixtoc 1 input (described in Sect. 19.5.2.1), ^{210}Pb dating placed the putative event horizon at a depth of 6–7 cm below seafloor (Lincoln et al. 2017).

Polychlorinated biphenyl compound (PCB) concentration profiles provided additional chronostratigraphic constraints in the southern GoM (Lincoln et al. 2017). First synthesized in 1881 and widely manufactured from the 1930s through the 1970s for use in electrical transformers and capacitors and paper processing (Robertson and Hansen 2001), PCBs are classified as persistent organic pollutants. They have been detected in sediments around the world. PCB use began to taper in the 1960s, their production was banned by the United States in 1979, and their manufacture was banned globally in 2001 (UNEP 2001). These once ubiquitous compounds thus have the potential to mark a sharp time window in undisturbed marine sediments, serving as a useful tool for recent chronostratigraphy (e.g., Turetsky et al. 2004). At IXW-250 (Fig. 19.1), the southern GoM site with strongest biomarker evidence for Ixtoc 1 input (Sect. 19.5.2.1), a sharp decline to very low PCB concentrations occurred above a horizon with a ^{210}Pb date of 1984 (Fig. 19.2). This general concordance between ^{210}Pb and PCBs, two independent chronostratigraphic records, increases confidence in implicating Ixtoc 1 input as the cause of shifts observed in biomarkers and other molecular geochemical profiles at this site (Lincoln et al. 2017).

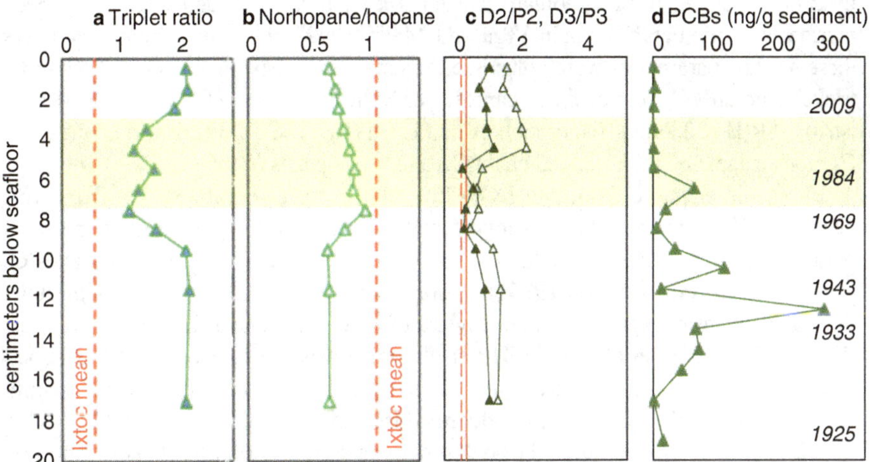

Fig. 19.2 Depth profiles of parameters indicating putative Ixtoc 1 input in sediment cores from site IXW-250; the yellow bar marks Ixtoc 1 impact horizons based on the following: (**a**) The triplet ratio of C_{26} R + C_{26} S tricyclic terpanes/C_{24} tetracyclic terpane (Magoon 1985), in green, shows a decrease from background values (~2) toward mean values detected near the wellhead (dashed red line). (**b**) The norhopane/hopane ratio increases, approaching the Ixtoc 1 value (dashed red line). (**c**) The ratios of dialkylated dibenzothiophenes to dialkylated phenanthrenes (D2/P2; solid triangle symbols) approach the Ixtoc 1 mean (solid red line), and the ratios of trialkylated dibenzothiophenes to trialkylated phenanthrenes (D3/P3; open triangle symbols) approach the Ixtoc 1 mean (dashed red line). (**d**) Total polychlorinated biphenyl (PCB) concentrations (green symbols) and ^{210}Pb dates (years on right) provide independent age constraints: PCBs decline from peak concentration at 12–13 cm below seafloor to near zero at ~6 cm below seafloor, likely corresponding to a rapid decline in the environment following their 1979 ban, consistent with a ^{210}Pb date of the mid-1980s for this horizon. (Data replotted from Lincoln et al. 2017)

19.5.2 Conventional Petroleum Fingerprinting

19.5.2.1 Terpane Signatures

All of the 2015 southern GoM sediment cores analyzed by Lincoln et al. (2017) contained a complex hydrocarbon background, consistent with observations of Boehm and Fiest (1980) and Scholz-Böttcher et al. (2008, 2009). Hopane was abundant and widespread; concentrations ranged from 21 to 711 ng/g sediment, with highest levels near the Ixtoc 1 wellhead.

Acknowledging the high hydrocarbon background that likely contains petroleum compounds derived from multiple oil sources, Lincoln et al. (2017) focused on biomarker parameters with the highest potential for differentiating Ixtoc 1 oil residues in sediments from other oil families in the Bay of Campeche. Terpane signatures measured in sediments near the Ixtoc 1 site were most consistent with those of southern GoM family 2b oils derived from Tithonian (late Jurassic) source rocks deposited in restricted marine environments (Guzmán-Vega and Mello 1999). For fingerprinting purposes, these terpane parameters were considered to have the highest diagnostic potential: C_{23} tricyclic terpane, C_{24} tetracyclic terpane, C_{26} tricyclic terpanes (with R and S stereochemistry), 18R(H)-22,29,30-trisnorneohopane (Ts), 17R(H)-22,29,30-trisnorhopane (Tm), C_{29} norhopane, and the extended chain C_{34} and C_{35} hopanes (R and S).

The sediment core from site IXW-250 (583 m water depth, 80 km from wellhead; Fig. 19.1) contained the strongest evidence of Ixtoc 1 input based on the above parameters. In particular, the "triplet ratio" of C_{26} R + C_{26} S tricyclic terpanes/C_{24} tetracyclic terpane (Magoon 1985) and norhopane/hopane ratios showed significant deviations from background values toward those detected near the Ixtoc 1 wellhead (Fig. 19.2; Lincoln et al. 2017). The triplet ratio was lowest overall at the Ixtoc 1 site, ranging from 0.5 to 0.6; at other sites it was higher and more variable within cores. At IXW-250, the triplet ratio declined from background values of ~2 to as low as 1.2 at a depth of potential Ixtoc 1 input (based on geochronology). Similarly, norhopane/hopane ratios were typically >1 at the Ixtoc 1 site, consistent with southern GoM family 2b oils (Guzmán-Vega and Mello 1999), and nearly always <1 at the other sites. At IXW-250, however, this ratio reached a maximum of 1.03 at 7.5 cm, suggestive of Ixtoc 1 oil input (Fig. 19.2).

Principal component analysis focusing on data from analyses for which southern GoM oil family data were available for comparison (Guzmán-Vega and Mello 1999) enabled further investigation of potential Ixtoc 1 oil input to the IXW-250 sediment core: samples from 5 to 9 cmbsf plotted most closely to the family 2b oils (Bubenheim et al. 2020). The relative contributions of specific Ixtoc 1 biomarkers to the pool detected in IXW-250 horizons containing evidence of putative impact (3–9 cmbsf) were estimated using a simple mixing model requiring the assumption of a compositionally consistent background signal at the site (Lincoln et al. 2017). Using this approach with the triplet ratio, an estimated 63% of the C_{24} tetracyclic and C_{26} tricyclic terpanes in the impacted horizon may be derived from Ixtoc 1 oil. Applying the same model to norhopane/hopane, Ixtoc 1 input of norhopane and hopane appears to have contributed 76% of these compounds present at IXW-250.

19.5.2.2 Alkylated Dibenzothiophenes and Phenanthrenes

Ratios of dimethylated phenanthrenes to dimethylated dibenzothiophenes (P2/D2) and of trimethylated phenanthrenes to trimethylated dibenzothiophenes (P3/D3) were considered most diagnostic in fingerprinting Ixtoc 1 oil in the GoM immediately after the spill (Boehm et al. 1983). Because P2/D2 and P3/D3 ratios have been demonstrated to be useful for oil source correlation until 98% depletion of these compounds occurs (Douglas et al. 1996), they were among the most promising tools for identifying Ixtoc 1 input decades after the spill.

P2/D2 and P3/D3 ratios in the top 9 cm of a sediment core collected near the Ixtoc 1 site in 2015 (Lincoln et al. 2017) were consistent with those reported in Ixtoc 1 oil (Boehm and Fiest 1980). At IXW-250, these ratios deviated from core background values between 6 and 9 cmbsf (Fig. 19.2). In that depth range, both P2/D2 and P3/D3 declined, and P3/D3 fell to <1. The two ratios are closest in this horizon and diverge above it. Both the decline in the P2/D2 and P3/D3 and the decreased difference between them (a pattern contrasting with that of sediments at depths above 6 cmbsf) are consistent with a contribution from Ixtoc 1 oil.

Significant differences in the P2/D2 and P3/D3 ratios between sites and their utility as an indicator of Ixtoc 1 oil in the IXW-250 sediment core support their apparent robustness as a metric of contaminant-source correlation over multi-decadal time scales.

19.5.3 Fourier Transform Ion Cyclotron Resonance-Mass Spectrometry

Alteration of oil residues due to biotic and abiotic weathering can transform oil compounds, particularly aromatics, to oxygen-containing residues (Aeppli et al. 2012, 2014; Radović et al. 2014; Ruddy et al. 2014), many of which are not detectable with standard GC-MS fingerprinting protocols. Fourier transform ion cyclotron resonance-mass spectrometry (FTICR-MS) is a useful complement to conventional GC-MS in that it extends the analytical window for petroleum analysis to include thousands of non-GC-amenable, high-molecular-weight, polar heteroatom-containing compounds, due to its diverse ionization modes, broad mass range, and ultrahigh mass resolution (McKenna et al. 2013). Following its demonstrated utility in characterizing the source Macondo well oil and the weathered residues collected in the aftermath of the DWH spill (McKenna et al. 2013; Ruddy et al. 2014), Radović et al. (2017) employed FTICR-MS to assess Ixtoc 1 input to southern GoM sediments, comparing signals detected in sediments with those of a reference sample of Ixtoc 1 crude oil obtained from PEMEX.

FTICR-MS analyses focused on one sediment core in which striking visual evidence provided the first clue of petrogenic input: the organic solvent extract of sediment 70–72 mm below seafloor at site IXW-500 (Fig. 19.1, 1010 m water depth) had a pronounced oily appearance (Radović et al. 2017). Comparison of FTICR-MS spectra from this horizon, seafloor surface sediment (0–5 mm below seafloor), more deeply buried sediment (200–210 mm below seafloor), and the Ixtoc 1 reference oil revealed important differences. Most notably, both the visibly oiled horizon and the Ixtoc 1 reference oil spectra showed a characteristic pseudo-Gaussian distribution peaking at m/z 400–500, a feature typical of FTICR-MS spectra of petroleum (McKenna et al. 2013). Spectra of the 0–5 mm below seafloor horizon, by contrast, had an irregular distribution indicative of recent organic matter derived from the water column. Both the oiled horizon (70–72 mm) and the surficial sediment (0–5 mm) showed distinctive peaks at high molecular weight corresponding to tetraether lipids (glycerol dialkyl glycerol tetraethers, or GDGTS; Radović et al. 2015) synthesized by water column and sediment archaea and bacteria, indicating that the petrogenic signal overprinted the background biogenic organic composition.

Molecular formula assignment revealed that the 70–72 mm horizon contained a higher abundance of hydrocarbon-, nitrogen-, and sulfur-containing compound classes than either surficial or deeper horizons (Fig. 19.3) and that it had a higher double bond equivalent (DBE) distribution. Increased DBE values are consistent with an enrichment in condensed aromatic species, as has been observed during

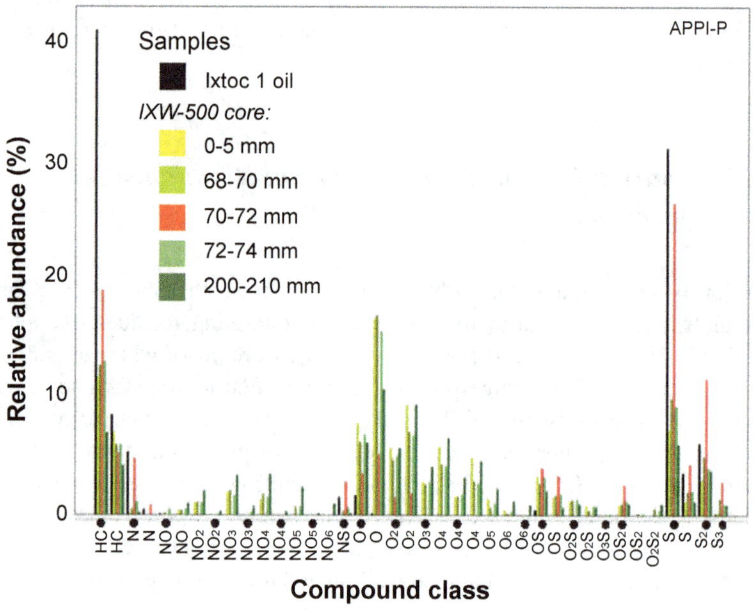

Fig. 19.3 Distribution and relative abundance of compound classes in extracts from core IXW-500 and the reference Ixtoc 1 oil (black). The sediment horizon 70–72 mm below seafloor (red) contained pronounced hydrocarbon (HC)-, nitrogen (N)-, and sulfur (S1–S3)-containing compound classes and more closely matched reference oil in these respects than it did sediments from the core top. (Data from Radović et al. Data are publicly available through the Gulf of Mexico Research Initiative Information & Data Cooperative (GRIIDC) at https://data.gulfresearchinitiative.org (https://doi.org/10.7266/n7-dya3-dj37))

weathering (Aeppli et al. 2014). Further evidence of weathered oil in the 70–72 mm below seafloor horizon was provided by two important differences in compound class distribution, as compared with the Ixtoc 1 reference oil: (1) a lower abundance of HCrad and Srad compound classes (defined in Oldenburg et al. 2014), suggesting preferential removal of lighter, abundant hydrocarbon ends and some sulfur heterocycles from sediments, and (2) a higher abundance of S_2rad and S_3rad compounds, likely reflecting the corresponding enrichment of highly sulfurized, recalcitrant compounds over the decades since the spill (Radović et al. 2017).

19.6 Summary and Discussion

Long-established oil fingerprinting parameters such as terpane and PAH ratios have provided evidence that traces of recalcitrant Ixtoc 1 oil compounds persist in seafloor sediments underlying deep waters (>500 m) of the southern Gulf of Mexico (Lincoln et al. 2017), and more recently developed, non-targeted, ultrahigh-resolution FTICR-MS techniques have revealed the presence of weathered petroleum residues that nonetheless share spectral and compound class characteristics with the Ixtoc 1 reference oil (Radović et al. 2017).

In this chapter we have presented the strongest evidence for residual Ixtoc 1 input to sediments that we detected in the southern GoM. We note that several other sediment cores collected to the west of the Ixtoc 1 wellhead showed no evidence of a signal that could be correlated with Ixtoc 1, possibly due to overprinting by the intense hydrocarbon background (Lincoln et al. 2017). However, deep-sea deposition of oil spill-derived hydrocarbons may simply be very heterogeneous and spotty, particularly if the primary mode of delivery to sediments is via sinking particles formed during MOSSFA events (Schwing et al. 2020). A single oiled particle may be adequate to impart a strong, diagnostic petroleum fingerprint, while nearby sediments are left unimpacted.

Considering the short half-life estimates of MW oil-derived petroleum compounds in primarily oxic northern GoM sediments underlying deep waters (e.g., Bagby et al. 2017), preservation of Ixtoc 1 compounds in southern GoM sediments >35 years since the spill may seem surprising. The preservation/biodegradation balance, however, is largely controlled by factors such as oxygen exposure time and sedimentation rates (Bubenheim et al. 2020); as sediments become anoxic, biodegradation slows and the likelihood that recalcitrant residues will remain increases. We conclude that, under optimal preservation conditions, a significant fraction of refractory oil spill-derived petroleum compounds can remain in the deep sea for decades, overprinting background signals and contributions from recent organic matter. A molecular legacy of the 2010 DWH oil spill will likely be detectable in deep GoM sediments in the year 2045 and beyond.

Funding Information This research was made possible by grants from the Gulf of Mexico Research Initiative through the Center for the Integrated Modeling and Analysis of the Gulf Ecosystem (C-IMAGE).

References

Aeppli C, Carmichael CA, Nelson RK, Lemkau KL, Graham WM, Redmond MC, Valentine DL, Reddy CM (2012) Oil weathering after the Deepwater Horizon disaster led to the formation of oxygenated residues. Environ Sci Technol 46(16):8799–8807

Aeppli C, Nelson RK, Radović JR, Carmichael CA, Valentine DL, Reddy CM (2014) Recalcitrance and degradation of petroleum biomarkers upon abiotic and biotic natural weathering of Deepwater Horizon oil. Environ Sci Technol 48(12):6726–6734

Alpuche-Gual L, Gold-Bouchot G (2014) Hidrocarburos totales en sedimentos cercanos a plataformas de exploración y de extracción de petróleo en la sonda de Campeche, p 383–398. In: Botello AV, Rendón von Osten J, Benítez JA, Gold-Bouchot G (eds) Golfo de México. Contaminación e impacto ambiental: Diagnóstico y tendencias. UAC, UNAM-ICMYL, CINVESTAV-Unidad Mérida. p 1176

Alvarez de la Borda J (2006) Cronica del petroleo en Mexico de 1863 a nuestros dias. PEMEX, México DF

Bagby SC, Reddy CM, Aeppli C, Fisher GB (2017) Persistence and biodegradation of oil at the ocean floor following Deepwater Horizon. Proc Natl Acad U S A 114:E9–E18

Boehm PD, Fiest DL (1980) Aspects of the transport of petroleum hydrocarbons to the offshore benthos during the IXTOC-I Blowout in the Bay of Campeche. In: N.O.A.A. (ed) Proceedings of a symposium on preliminary results from the September, 1979 RESEARCHER/PIERCE IXTOC-I cruise, Key Biscayne, Florida, June 9–10, 1980. Publications Office, NOAA/RD/MP3, Office of Marine Pollution Assessment, N.O.A.A., U.S. Dept. of Commerce, Boulder

Boehm PD, Fiest DL, Kaplan I, Mankiewicz P, Lewbel GS (1983) A natural resources damage assessment study: the Ixtoc I blowout. In: Proceedings of the 1983 Oil Spill Conference, American Petroleum Institute, Washington, D.C.

Botello AV, Villanueva S (1985) Vigilancia de los hidrocarburos fósiles en sistemas costeros del Golfo de México y áreas adyacentes. I. Sonda de Campeche. An Inst Cienc Mar Limnol 14:1–13

Botello AV, Gonzalez C, Diaz G (1991) Pollution by petroleum hydrocarbons in sediments from continental shelf of Tabasco, Mexico. Bull Environ Contam Toxicol 47:565–571

Brooks GR, Larson RA, Schwing PT, Romero I, Moore C, Reichart GJ, Jilbert T, Chanton JP, Hastings DW, Overholt WA, Marks KP, Kostka JE, Homes CW, Hollander D (2015) Sedimentation pulse in the NE Gulf of Mexico following the 2010 DWH blowout. PLoS One 10(7):e0132341. https://doi.org/10.1371/journal.pone.0132341

Bubenheim P, Hackbusch S, Joye S, Kostka J, Larter SR, Liese A, Lincoln SA, Marietou A, Müller R, Noirungsee N, Oldenburg TBP, Radović J, Viamonte J (2020) Biodegradation of petroleum hydrocarbons in the deep sea (Chap. 7). In: Murawski SA, Ainsworth C, Gilbert S, Hollander D, Paris CB, Schlüter M, Wetzel D (eds) Deep oil spills: facts, fate and effects. Springer, Cham

Carmona Uristegui MA (2015) Análisis de la contaminación marina por la industria petrolera en la región costera de Dos Bocas, Tabasco, México. Thesis, Escuela Superior de Ingeniería Química e Industrias Extractivas, Instituto Mexicano del Petróleo

Center for Short-Lived Phenomena (CSLP; Smithsonian Institution) (1980) Oil spill intelligence report, special report: Ixtoc 1, p 2–35

Chanton J, Zhao T, Rosenheim BE, Joye S, Bosman S, Brunner C, Yeager KM, Diercks AR, Hollander D (2014) Using natural abundance radiocarbon to trace the flux of petrocarbon to the seafloor following the Deepwater Horizon oil spill. Environ Sci Technol 49(2):847–854

Douglas GS, Bence AE, Prince RP, McMillen SJ, Butler EL (1996) Environmental stability of selected petroleum hydrocarbon source and weathering ratios. Environ Sci Technol 30:2332–2339

Farrán A, Grimalt J, Albaigés J, Botello AV, Macko SA (1987) Assessment of petroleum pollution in a Mexican river by molecular markers and carbon isotope ratios. Mar Pollut Bull 18:284–289

Farrington JW (1983) NOAA Ship RESEARCHER/Contract Vessel PIERCE Cruise to IXTOC 1 oil spill: overview and integrative data assessment and interpretation. Office of Marine Pollution

Assessment, National Oceanographic and Atmospheric Administration, United States Division of Commerce, Boulder, CO

García-Cuéllar JA, Arreguín-Sánchez F, Vázquez SH, Lluch-Cota DB (2004) Impacto ecológico de la industria petrolera en la Sonda de Campeche, México, tras tres décadas de actividad: Una revision. Interciencia 29:311–319

Godoy-Lozano E, Escobar-Zepeda A, Raggi L, Merino E, Gutiérrez-Rios RM, Juárez K, Segovia L, Licea-Navarro AF, Gracia A, Sánchez-Flores A, Pardo-López L (2018) Bacterial diversity and geochemical landscape in the southwestern Gulf of Mexico. Front Microbiol 9:2528. https://doi.org/10.3389/fmicb.2018.02528

Gold-Bouchot G, Zavala-Coral M, Zapata-Pérez O, Ceja-Moreno V (1997) Hydrocarbon concentrations in Oysters (*Crassostrea virginica*) and recent sediments from three coastal lagoons in Tabasco, Mexico. Bull Environ Contam Toxicol 59:430–437

Gracia A (2012) Campaña Oceanográfica (SGM-2010). Informe Final. Gerencia de Seguridad Industrial, Protección Ambiental y Calidad Región Marina Noreste, PEMEX - EXPLORACIÓN – PRODUCCIÓN. Instituto de Ciencias del Mar y Limnología, UNAM, México

Gracia A, Alexander-Valdés HM, Ortega-Tenorio PL, Frausto-Castillo JA (2016) Source and distribution of polycyclic hydrocarbons in the Ixtoc 1 spill area. In: Proceedings of the Gulf of Mexico oil spill and ecosystem science conference, Tampa

Guzmán-Vega MA, Mello MR (1999) Origin of oil in the Sureste Basin. AAPG Bull 83:1068–1095

Holguin-Quiñones N, Brooks JM, Román-Ramos JR, Bernard, BB, Lara- Rodriguez J, Zumberge JE, Medrano-Morales L, Rosenfeld J, de Faragó Botella M, Maldonado-Villalón R, Martínez-Pontvianne G (2005) Estudio regional de manifestaciones superficiales de aceite y gas en el sur del Golfo de México. Boletín de la Asociación Mexicana de Geólogos Petroleros LII, 20–41

Huesemann MH, Hausmann TS, Fortman TJ (2010) Biodegradation of hopane prevents use as conservative biomarker during bioremediation of PAHs in petroleum contaminated soils. Biorem J 7:111–117

Jagoš R. Radović, Renzo C. Silva, Ryan Snowdon, Stephen R. Larter, Thomas B. P. Oldenburg, (2015) Rapid Screening of Glycerol Ether Lipid Biomarkers in Recent Marine Sediment Using Atmospheric Pressure Photoionization in Positive Mode Fourier Transform Ion Cyclotron Resonance Mass Spectrometry. Analytical Chemistry 88 (2):1128–1137

Jernelöv A, Lindén O (1981) Ixtoc I: a case study of the world's largest oil spill. Ambio 10:299–306

Larson RA, Brooks GR, Schwing PT, Diercks AR, Holmes CW, Chanton JP, Diaz-Ascencio M, Hollander DJ (2020) Characterization of the sedimentation associated with the Deepwater Horizon blowout: depositional pulse, initial response, and stabilization (Chap. 14). In: Murawski SA, Ainsworth C, Gilbert S, Hollander D, Paris CB, Schlüter M, Wetzel D (eds) Deep oil spills: facts, fate and effects. Springer, Cham

Lincoln SA, Radović JR, Larson RA, Romero I, Schwing PT, Bosman S, Brooks GR, Chanton JP, Oldenburg TBP, Hollander DJ, Freeman KH (2017) Tracing oil from the 1979 Ixtoc blowout in southern Gulf of Mexico seafloor sediments. Proceedings of the International meeting on organic geochemistry, Florence

MacDonald IR, Garcia-Pineda O, Beet A, Daneshgar Asl S, Feng L, Graettinger G, French-McCay D, Holmes J, Hu C, Huffer F, Leiffer I, Mueller-Karger F, Solow A, Silva M, Swayze GA (2015) Natural and unnatural oil slicks in the Gulf of Mexico. J Geophys Res Oceans 120:8364–8380

Magoon L (1985) Alaska north slope oil-rock correlation study: analysis of north slope crude. American Association of Petroleum Geologists, Tulsa, OK

Marchand M, Monfort J, Cortés-Rubio A (1982) Distribution of hydrocarbons in water and marine sediments after the Amoco Cadiz and Ixtoc-I oil spills. In: Keith LH (ed) Energy and environmental chemistry. Ann Arbor Science, Ann Arbor, pp 161–183

McKenna AM, Nelson RK, Reddy CM, Savory JJ, Kaiser NK, Fitzsimmons JE, Marshall AG, Rodgers RP (2013) Expansion of the analytical window for oil spill characterization by ultrahigh resolution mass spectrometry: beyond gas chromatography. Environ Sci Technol 47(13):7530–7539

O'Malley BJ, Schwing PT, Hollander DJ (2016) Comparison of the 2015 Abkatun (Mexico) and 2013 Hercules-265 (USA) blowout events in the southern and northern Gulf of Mexico using benthic foraminiferal species richness as an environmental proxy for contamination. In: Proceedings of the Gulf of Mexico oil spill and ecosystem science conference, Tampa

Oldenburg TBP, Brown M, Bennett B, Larter SR (2014) The impact of thermal maturity on the composition of crude oils, assessed using ultra-high resolution mass spectrometry. Org Geochem 75:151–168

Programa Coordinado de Estudios Ecológicos en la Sonda de Campeche (PCEESC) (1980) Informe de los trabajos realizados para el control del pozo Ixtoc 1, el derrame de petróleo y determinación de sus efectos sobre el ambient marino. Instituto Mexicano del Petróleo, México City

Quigg A, Passow U, Hollander DJ, Daly K, Burd A, Lee K (2020) Marine Oil Snow Sedimentation and Flocculent Accumulation (MOSSFA) events: learning from the past to predict the future (Chap. 12). In: Murawski SA, Ainsworth C, Gilbert S, Hollander D, Paris CB, Schlüter M, Wetzel D (eds) Deep oil spills: facts, fate and effects. Springer, Cham

Radović JR, Aeppli C, Nelson RK, Jimenez N, Reddy CM, Bayona JM, Albaigés J (2014) Assessment of photochemical processes in marine oil spill fingerprinting. Mar Pollut Bull 79(1–2):268–277

Radović JR, Silva RC, Snowdon R, Larter SR, Oldenburg TBP (2015) Rapid screening of glycerol ether lipid biomarkers in recent marine sediment using atmospheric pressure photoionization in positive mode Fourier transform ion cyclotron resonance mass spectrometry. Anal Chem 88(2):1128–1137

Radović J, Lincoln S, Silva R, Jaggi A, Romero I, Schwing P, Larson R, Brooks G, Freeman K, Hollander D, Larter S, Oldenburg T (2017) The 1979 Ixtoc-I oil spill revisited. In: Proceedings of the Goldschmidt conference, Paris

Radović JR, Romero IC, Oldenburg TBP, Larter SR, Tunnell JW (2020) 40 years of weathering of coastal oil residues in the southern Gulf of Mexico (Chap. 20). In: Murawski SA, Ainsworth C, Gilbert S, Hollander D, Paris CB, Schlüter M, Wetzel D (eds) Deep oil spills: facts, fate and effects. Springer, Cham

Robertson LW, Hansen LG (2001) PCBs: recent advances in environmental toxicology and health effects. University Press of Kentucky, Lexington

Romero IC, Schwing PT, Brooks GR, Larson RA, Hastings DW, Ellis G, Goddard EA, Hollander DJ (2015) Hydrocarbons in deep-sea sediments following the 2010 Deepwater Horizon blowout in the northeast Gulf of Mexico. PLoS One 10(5):e0128371

Ruddy BM, Huettel M, Kostka JE, Lobodin VV, Bythell BJ, McKenna AM, Aeppli C, Reddy CM, Nelson RK, Marshall AG, Rodgers RP (2014) Targeted petroleomics: analytical investigation of Macondo well oil oxidation products from Pensacola Beach. Energy Fuel 28(6):4043–4050

Ruiz-Fernández AC, Portela JMB, Sericano JL, Sanchez-Cabeza J-A, Espinosa LF, Cardoso-Mohedano JG, Pérez-Bernal LH, Tinoco JAG (2016) Coexisting sea-based and land-based sources of contamination by PAHs in the continental shelf sediments of Coatzacoalcos River discharge area (Gulf of Mexico). Chemosphere 144:591–598

Schifter I, González-Macías C, Salazar-Coria L, Sánchez-Reyna G, González-Lozano C (2015) Long-term effects of discharges of produced water the marine environment from petroleum-related activities at Sonda de Campeche, Gulf of México. Environ Monit Assess 187:723

Scholz-Böttcher BM, Ahlf S, Vázquez-Gutiérrez F, Rullkötter J (2008) Sources of hydrocarbon pollution in surface sediments of the Campeche Sound, Gulf of Mexico, revealed by biomarker analysis. Org Geochem 39:1104–1108

Scholz-Böttcher BM, Ahlf S, Vázquez-Gutiérrez F, Rullkötter J (2009) Natural vs. anthropogenic sources of hydrocarbons as revealed through biomarker analysis: a case study in the southern Gulf of Mexico. Bol Soc Geol Mex 61:1–10

Schwing PT, Romero IC, Larson RA, O'Malley BJ, Fridrik EE, Goddard EA, Brooks GR, Hastings DW, Rosenheim E, Hollander DJ, Grant G, Mulhollan J (2016) Sediment core extrusion method at millimeter resolution using a calibrated, threaded-rod. J Vis Exp 114:e54363

Schwing PT, Hollander DJ, Brooks G, Larson RA, Hastings DW, Chanton JP, Lincoln SA, Radović J, Langenhoff A (2020) The sedimentary record of MOSSFA events in the Gulf of Mexico: a comparison of the Deepwater Horizon (2010) and Ixtoc 1 (1979) oil spills (Chap. 13). In: Murawski SA, Ainsworth C, Gilbert S, Hollander D, Paris CB, Schlüter M, Wetzel D (eds) Deep oil spills: facts, fate and effects. Springer, Cham

Stout SA, Wang Z (eds) (2016) Standard handbook of oil spill environmental forensics: fingerprinting and source identification. Elsevier, Amsterdam

Sun SJ, Hu CM, Tunnel JW (2015) Surface oil footprint and trajectory of the Ixtoc-I oil spill determined from Landsat/MSS and CZSC observations. Mar Pollut Bull 101:632–641

Turetsky MR, Manning SW, Wieder RK (2004) Dating recent peat deposits. Wetlands 24:324–256

United Nations Environment Programme (UNEP) (2001) Stockholm convention on persistent organic pollutants, Stockholm. https://treaties.un.org/doc/Treaties/2001/05/20010522%20 12-55%20PM/Ch_XXVII_15p.pdf

Valentine DL, Fisher GB, Bagby SC, Nelson RK, Reddy CM, Sylva SP, Woo MA (2014) Fallout plume of submerged oil from Deepwater Horizon. Proc Natl Acad Sci U S A 111(45): 15906–15911

Vazquez F, Sanchez M, Alexander H, Delgado D (1991) Distribution of Ni, V, and petroleum hydrocarbons in recent sediments from the Veracruz coast, Mexico. Bull Environ Contam Toxicol 46:774–781

Chapter 20
40 Years of Weathering of Coastal Oil Residues in the Southern Gulf of Mexico

Jagoš R. Radović, Isabel C. Romero, Thomas B. P. Oldenburg,
Stephen R. Larter, and John W. Tunnell Jr.

Abstract The oil spill from the Ixtoc 1 well in 1979 in the southern Gulf of Mexico (sGoM) was in many aspects very similar to the *Deepwater Horizon* (DWH) blowout offshore Louisiana 30 years later (2010), most importantly because of the subsurface nature of the oil release, the amount of oil released, and the extensive environmental distribution of the spilled oil, including coastal impacts. Because of that, the Ixtoc 1 spill can serve as an excellent analog to study and model the long-term oil weathering processes in coastal environments. In 2016, a research expedition sponsored by the Gulf of Mexico Research Initiative (GoMRI) visited many coastal sites in the sGoM, previously known to be impacted by the Ixtoc 1 spill, and collected oil residues. The residues were analyzed using targeted (GC-MS/MS) and non-targeted (FTICR-MS) approaches in order to assess their origin and the nature of weathering transformation products. The initial results suggest multi-decadal preservation potential of Ixtoc 1 spill residues in certain low-energy environments, such as coastal mangrove forests. These results provide valuable input for the modelling of long-term fate and impacts of the DWH spill.

Keywords Ixtoc 1 · Southern GoM · FTICR-MS · GC-MS/MS · Oil spill · Coast · Biomarkers · Fingerprinting · Weathering

J. R. Radović (✉) · T. B. P. Oldenburg · S. R. Larter
University of Calgary, PRG, Department of Geoscience, Calgary, AB, Canada
e-mail: Jagos.Radovic@ucalgary.ca; toldenbu@ucalgary.ca; slarter@ucalgary.ca

I. C. Romero
University of South Florida, College of Marine Science, St. Petersburg, FL, USA
e-mail: isabelromero@mail.usf.edu

J. W. Tunnell Jr. (Deceased)
Texas A&M University-Corpus Christi, Harte Research Institute for Gulf of Mexico Studies,
Corpus Christi, TX, USA
e-mail: Wes.Tunnell@tamucc.edu

© Springer Nature Switzerland AG 2020
S. A. Murawski et al. (eds.), *Deep Oil Spills*,
https://doi.org/10.1007/978-3-030-11605-7_20

20.1 Introduction

The southern Gulf of Mexico (sGoM) was the scene of the second largest accidental marine oil spill in history, overshadowed only by the 2010 *Deepwater Horizon* (DWH) disaster which happened in the northern part of the Gulf (nGoM). On June 3, 1979, a blowout on the Ixtoc 1 exploratory platform in the Bay of Campeche (lat, 19° 24′ 30.00″ N; long, −92° 19′ 30.00″ W, Fig. 20.1) started an uncontrolled release of oil from the wellhead in ~50 m deep water, which continued until March 23, 1980.

In total, during the Ixtoc 1 spill, approximately 3 million barrels of oil and associated gas was introduced to the sGoM marine system, out of which, by some estimates, 50% evaporated into the atmosphere, 25% sank to the bottom, and 12% was degraded biologically and photochemically (Jernelöv and Lindén 1981). Some of the remaining 13% was mechanically removed or burned; however, a significant portion of spilled oil caused oiling in coastal environments in Mexico and even Texas. In many aspects, including the coastal oiling, Ixtoc 1 was very similar to the DWH, where it has been estimated that approximately 10% of the oil released after

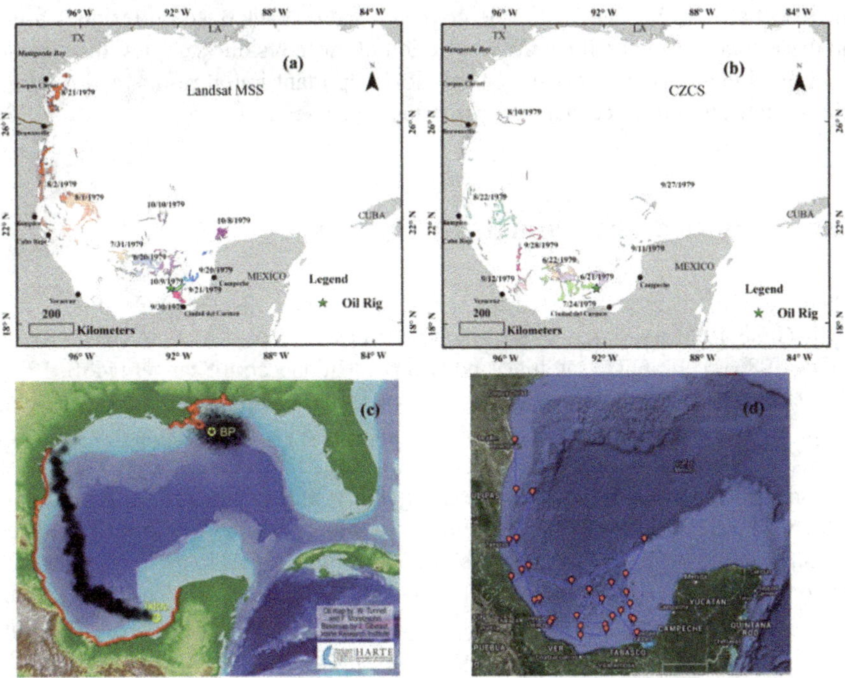

Fig. 20.1 (**a, b**) Surface oil slicks in the aftermath of the 1979–1980 Ixtoc 1 spill, reconstructed from the available historic satellite data. Colors differentiate the extent of slicks on a given date. (**c**) Ixtoc 1 spill trajectory in comparison to the surface expression of the 2010 Deepwater Horizon blowout. (**d**) Reconstructed Ixtoc 1 spill trajectory was used to plan field research campaigns, including offshore and coastal sampling, organized by the GoMRI-sponsored C-IMAGE consortium. (Reprinted from Sun et al. (2015), with permission from Elsevier)

the DWH formed surface oil slicks that eventually impacted nGoM shores (Ryerson et al. 2012). Weathering of beached DWH oil and coastal residues was driven by photooxidation, in particular shortly after the spill (Radović et al. 2014; Ward et al. 2018), and in later stages also by biodegradation (Aeppli et al. 2012, 2014). These weathering processes transformed the parent oil compounds to oxygen-containing derivatives (Aeppli et al. 2012; Radović et al. 2014; Ruddy et al. 2014), which seem to be continually produced and preserved/accumulated over periods of several years after the spill (White et al. 2016; Romero et al. 2017; Stout and Payne 2016).

Due to the post-spill fate similarities of the two GoM spills, Ixtoc 1 can serve as a good analog to help develop a predictive model of future long-term weathering effects and the ultimate fate of the DWH oil residues currently found along the nGoM coastline. However, at the time of the Ixtoc 1 spill, and in the following years, only limited research efforts were invested into monitoring of the post-spill fate and impacts of coastal residues (Lizarraga et al. 1984; Lizárraga-Partida et al. 1982; Soto et al. 1981, 2014). In order to fill in this research gap, in 2016, a group of researchers from the Gulf of Mexico Research Initiative (GoMRI)-funded C-IMAGE II consortium revisited the sites in the sGoM, which were affected by the Ixtoc 1 spill, according to the previous data and field observations by J.W. Tunnell in Sun et al. (2015) (Fig. 20.1), and collected and characterized oil residues found in these areas. Using an illustrative selection of those residue samples, this chapter will provide a general overview of the most important initial results and observations related to multi-decadal oil weathering processes.

20.2 Affected Areas and Sampling Sites

Sun et al. (2015) used available historic satellite data collected during the Ixtoc 1 spill (June 1979 to March 1980), namely, Coastal Zone Color Scanner (CZCS, 1978–1986 on Nimbus-7 satellite) and Landsat Multispectral Scanner (MSS, 1972–1999 on Landsat 1–5 satellites) datasets, to identify and quantify surface oil slicks from the Ixtoc 1 spill (Fig. 20.1). Their results suggest that the Ixtoc 1 sourced oil slicks most frequently occurred within 200 km north and west of the spill site. Most of them moved along the northwestern trajectory, reaching as far north as Corpus Christi, Texas. Later in the spill event (after mid-September 1979), oil was also detected in the southern and southeastern portion of the GoM, arriving all the way up to the offshore areas north of the Yucatan Peninsula. Satellite-derived reconstructions were corroborated by numerous field observations, both from the literature (Gundlach et al. 1981) and as recorded during the field trips conducted by John W. Tunnel in the aftermath of the spill. At many of these locations which were revisited in 2016, weathered oil residues could be found (Fig. 20.2).

Oil residues found at these locations were quite diverse, ranging from extremely weathered oil adhered to the exposed supratidal rocky shores, e.g., at Punta Delgado and Montepio sampling sites, to relatively preserved oil which was found buried in low-energy mangrove environments along the western coast of the Yucatan Peninsula (e.g., at the Isla Arenas site) (Fig. 20.2).

Fig. 20.2 Photos of the sites in the sGoM where oil residues were found during field trips by John W. Tunnel and the researchers from the GoMRI-funded C-IMAGE II consortium. (Courtesy John W. Tunnel and C-IMAGE II)

20.3 Chemical Composition of Ixtoc 1 Oil and Coastal Oil Residues

20.3.1 GC-MS/MS-MRM

Gas chromatography, coupled with tandem mass spectrometry (GC-MS/MS), operated in multiple reaction monitoring mode (MRM), is an analytical method that selectively characterizes multiple target compounds in complex matrices, such as oil residues in marine sediments. In GC-MS/MS-MRM, high selectivity of target compounds is achieved by filtering a precursor ion of interest to further produce product ions, after collision with a gas molecule and detecting only generated product ions with specific mass-to-charge ratios (m/z). This high ion selectivity enables the identification and quantification of compounds at very low concentrations with no interferences, giving an advantage to GC-MS/MS-MRM, over other targeted analytical methods. In petroleum forensic geochemistry, characterization of specific compounds is critical in order to accurately identify oil sources, quantify levels in the environment including in biological samples, and detect major weathering processes and molecular transformations that can be linked to toxicity and environmental health assessment studies. Specifically, for the DWH oil spill, GC-MS/MS-MRM analysis of biomarkers (e.g., hopanoids) helped identify for the first time the DWH oil residues in deep-sea sediments that were strongly weathered during deposition processes such as direct impingement of deep plumes on the continental slope and sinking of oiled marine snow (Romero et al. 2015). This study was further integrated into a multidisciplinary analytical approach that helped explain a large sedimentation pulse that occurred in 2010 as a product of the DWH spill (Brooks et al. 2015). Moreover, oil residues were found after 3 years in the same deep-sea areas with little post-deposition biodegradation (Romero et al. 2017). Other studies have also used GC-MS/MS-MRM for monitoring of toxic compounds (e.g., polycyclic aromatic hydrocarbons (PAHs)) in complex matrices such as sediments from submerged coastal and deep-sea areas (Adhikari et al. 2017; Yin et al. 2015), seafood (fishes, shrimp, oysters, and crabs) from areas affected by the DWH spill (Xia et al. 2012), and mesopelagic fishes over a long term in the nGoM (Romero et al. 2018). Altogether these studies have demonstrated a widespread occurrence of the DWH oil residues in the nGoM and a persistence of oil residues in the environment years after the spill with little human risk, but with potential adverse ecological effects across multiple levels of biological organization.

Similarly, GC-MS/MS-MRM analysis of biomarkers in the sGoM has been used to assess the source and weathering status of tar samples collected from mangrove prop roots (mangrove trees aerial network of roots; see Fig. 20.2). Recalcitrant ratios (e.g., 18α(H)-22,29,30-trisnorneohopane (Ts)/17α(H)-22,29,30-trisnorhopane (Tm); Ts/17α(H),21β(H)-hopane (HC30); 30,31-bishomohopane-22S/HC30) suggest Ixtoc 1 oil as the source of the tar samples. The effect of weathering over 37 years after the Ixtoc 1 spill on these samples was calculated as the percentage remaining of each hydrocarbon compound group (n-alkanes, PAHs, hopanes, steranes, and triaromatic steroids (TAS)) normalized to HC30 (recalcitrant internal marker; see Prince et al. (1994) and Garrett et al. (1998)) and compared to the normalized value in the reference Ixtoc 1 oil (% remaining = [compound$_a$/HC30]$_{sample}$/[compound$_a$/HC30]$_{Ixtoc\ 1\ oil}$). Results indicate remarkable persistence of homohopanes C27–C32 (~100% remaining), followed by TAS (~90% remaining) and high molecular weight PAHs (~60% remaining). Other compound groups showed lower relative amounts remaining in the tar samples, like homohopanes C33–C35 (~45%), steranes C27 (~30%), n-alkanes (~10%), and low molecular weight PAHs (~5%). In these coastal samples, high enrichment, relative to Ixtoc 1 crude oil, was found for steranes C28–C29 compounds (140–200%). Trends observed in homohopanes C27–C32 are similar to those found in previous studies 2–3 years after the DWH spill (Aeppli et al. 2014). However, differences in weathering patterns were found in the tar samples from the mangrove forests relative to previous studies after the DWH spill, showing low % remaining of steranes C27 and high % remaining of TAS and high molecular weight PAHs (Fig. 20.3).

20.3.2 FTICR-MS

Contrary to the GC-MS/MS, Fourier transform ion cyclotron mass spectrometry (FTICR-MS) is a non-targeted instrumental technique with a broad range of species detection and ultrahigh mass resolution, making it a tool for exploratory qualitative

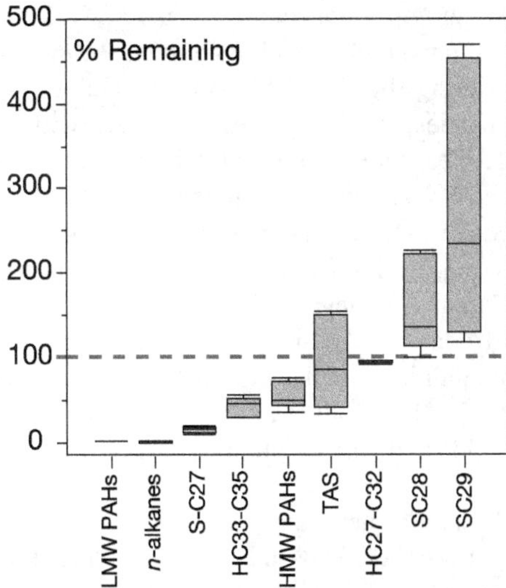

Fig. 20.3 Percent of hydrocarbon compounds remaining 37 years after the Ixtoc 1 spill in the oil residues collected in the coastal mangrove forest in the southern Gulf of Mexico. LMW PAHs, 2–3 rings, including alkylated homologues; *n*-alkanes, C_{10-40}; S-C27, 5α(H),14α(H),17α(H)-20-cholestane (S/R), 5α(H),14β(H),17β(H)-20-cholestane (S/R); HC33–C35, 17α(H),21β(H)-22R-trishomohopane, tetrakishomohopane, pentakishomohopane; HMW PAHs, 4–6 rings, including alkylated homologues; TAS, triaromatic steroid C26–C28; HC27–C32, Tm, Ts, 30-norhopane + 18a(H)-30-norneohopane, 18a(H)-30-norneohopane-normoretane, 17α(H),21β(H)-hopane, 17β(H),21α(H)-hopane-moretane, 17α(H),21β(H)-22R-homohopane, bishomohopane; SC28, 24-methyl-5α(H),14α(H),17α(H)-20-cholestane (S/R), 24-methyl-5α(H),14β(H),17β(H) 20 cholestane (S/R); SC29, 24-ethyl-5α(H),14α(H),17α(H)-20-cholestane (S/R), 24-ethyl-5α(H),14β(H),17β(H)-20-cholestane (S/R). Graph shows shaded boxes as the interquartile ranges with horizontal lines indicating median values and whiskers representing the 10th and 90th percentiles

screening of very complex samples, such as petroleum (Rodgers and Marshall 2007). Its advantage, in comparison to more conventional gas chromatography-based methods used in oil spill research, is the capability to ionize and detect high molecular weight, nonvolatile, and heteroatom (nitrogen, sulfur, and oxygen, plus trace metals)-containing chemical species. Data processing of complex mixtures of diverse organic species (e.g., crude oil), using ultrahigh-resolution mass spectra, yields a long list of molecular formula assignments and intensities related to mass spectral peaks found in the spectra. In order to be able to efficiently interpret and compare such "big" datasets, compositional sorting and other visualization strategies are needed, which could reveal the compositional characteristics of the investigated sample set. Typical petroleum geochemistry-related investigations use sequential data layers based on the heteroatom content, DBE, and carbon number (C#) to create plots. For more details on FTICR-MS and its applications in oil spill science, see Radović et al. (2019).

FTICR-MS played an important role in the post-DWH research – it expanded the analytical window for the chemical assessment of the source Macondo well (MW) oil, released during the DWH blowout, to include thousands of neutral, basic, and acidic species, typically not analyzed with traditional methods (McKenna et al. 2013); but more importantly, it helped to characterize weathered coastal residues of the DWH spill. In combination with other tools such as thin-layer chromatography, infrared spectroscopy, and comprehensive two-dimensional gas chromatography, FTICR-MS confirmed that weathering processes, such as photooxidation and biodegradation, were indeed responsible for the conversion of parent oil compounds, aromatic species, and saturated hydrocarbon species, to oxygen-containing weathering products (Aeppli et al. 2012; Radović et al. 2014; Ruddy et al. 2014; Hall et al. 2013). The identified products included ketone, carboxylic acid, and higher numbered (>3) oxygen-containing species (Ruddy et al. 2014).

Similarly, APPI-P (atmospheric pressure photoionization in positive mode) FTICR-MS analyses of weathered coastal residues collected in the sGoM show the presence of numerous oxidized classes of compounds, likely weathering transformation products. This is particularly evident in the samples collected in the low-energy, protected mangrove areas, for example, at the Isla Arenas site (Fig. 20.2), which are abundant in oxygen-bearing compound classes, such as O_1-O_4, akin to "oxyhydrocarbons" observed in the nGoM coastal residues (Aeppli et al. 2012; Ruddy et al. 2014), and various types of O_xS_x classes (Fig. 20.4), most likely weathering products of parent sulfur species presented in the spilled oil. Here we have to note that the parent Ixtoc 1 oil has appreciably higher sulfur content in comparison to the MW oil spilled during the DWH blowout, which is a characteristic feature of most of the petroleum systems of the Campeche Sound basin, where the Ixtoc 1 well was drilled (Atwood 1980; Guzman-Vega and Mello 1999; Santamaría-Orozco et al. 1998). To illustrate, sulfur-to-carbon ratio (S/C), as calculated from the APPI-P FTICR-MS data, is approximately four times higher in the Ixtoc 1 source oil than in the MW oil (S/C of 0.0142 and 0.0035, respectively). In comparison, in the weathered oil residue collected in the Isla Arenas mangroves and putatively related to Ixtoc 1 oil, S/C ratio is 0.0097; under a very simplified assumption of direct conversion of parent sulfur compounds to weathering products, this would signify that during 37 years, approximately 30% of sulfur species initially present in the Ixtoc 1 oil have been transformed to O_xS_x, and other products. On the other hand, in the oil residues found at supratidal rocky shores in the sGoM, such as at the Punta Delgado or Montepio sites (Fig. 20.2), the most prominent chemical feature, as detected by the APPI-P FTICR-MS, is the pronounced abundance of sulfur-containing species, including classes of compounds with multiple sulfur heteroatoms (up to four S atoms), and also the presence of O_xS_x compound classes (Fig. 20.4). For example, S/C ratio of heavily weathered tar residues collected at the Punta Delgado and Montepio shores is, respectively, ~1.2 and 2.3 times higher than in the reference Ixtoc 1 oil. Potential implications of these observations are discussed in the section below.

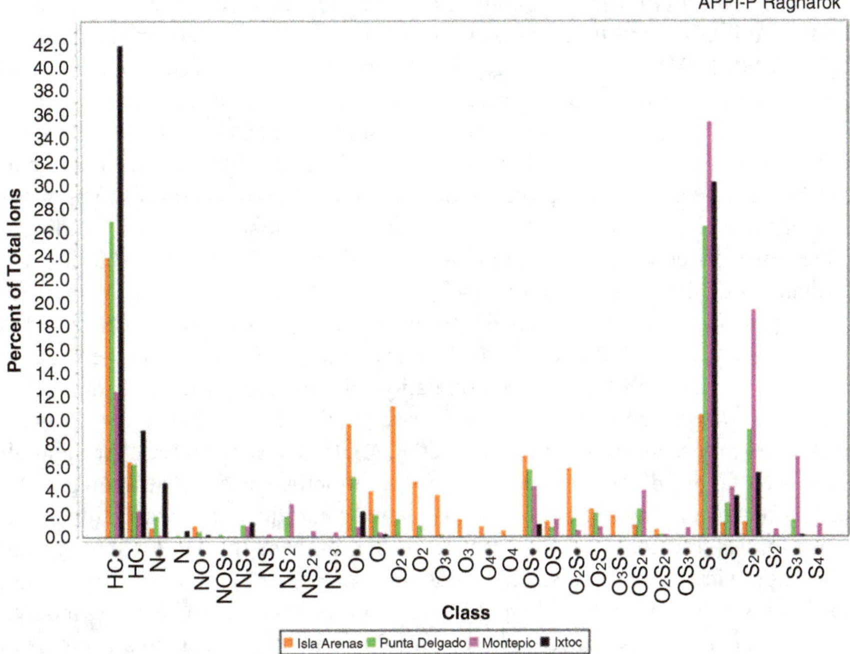

Fig. 20.4 APPI-P FTICR-MS-derived compound class distribution of the source Ixtoc 1 oil and the weathered petroleum residues found buried in the mangrove forests of Isla Arenas and in the heavily degraded, tar-like samples collected at the Punta Delgado and Montepio rocky shores. Note the presence of multiple oxygen-containing compound classes in the Isla Arenas sample and the relative enrichment of sulfur-containing classes in the samples from rocky shores

20.3.3 Decadal-Scale Weathering Processes and Analogies with the Weathering of Post-DWH Coastal Oil Residues

As demonstrated in various shallow systems (e.g., sandy beaches), time-series studies of weathered oil residues collected in the nGoM, the fate of the MW oil, released after the DWH blowout, most likely included pronounced photooxidation of the oil, in the days and weeks following the spill (Radović et al. 2014; Ward et al. 2018), followed by coupling to biotic oil transformations, on longer timescales, ~months to years (Aeppli et al. 2014), with abiotically oxidized metabolites possibly being good substrates for subsequent microbial degradation and/or prone to water washing (Harriman et al. 2017; Liu and Kujawinski 2015).

Based on the initial results from our study of coastal oil residues collected in the sGoM, we can potentially hypothesize how oil weathering might proceed in a multi-decadal timeframe. It seems that in low-energy environments, such as the mangrove forests of Yucatan Peninsula, which host residues related to the 1979 Ixtoc 1 spill as demonstrated by analyses of recalcitrant biomarkers (Sect. 20.3.1 GC-MS/MS-MRM), there is an apparent preservation of partially weathered oil. The samples from these sites, particularly the ones buried in the shallow subsurface, still had a characteristic oily appearance and smell and are relatively enriched in hydrocarbon content (e.g., homohopanes C27–C32). They also contain abundant oxygen-bearing species, likely oxidized metabolites of parent oil compounds, and an indication of incomplete oil weathering. Potential spill history leading to oil residues of such chemistry might have included initial biotic and abiotic transformations of the Ixtoc 1 surface slick, in the immediate aftermath of the spill, which produced "oxyhydrocarbon" products, analogous to the DWH spill fate (Aeppli et al. 2012; Ruddy et al. 2014; Ward et al. 2018). Based on the apparent relative preservation of TAS, photooxidation of surfaced Ixtoc 1 oil might have been less pronounced than in the DWH case (Radović et al. 2014). According to slick trajectory reconstructions by Sun et al. (2015), surface oil weathering might have been taking place over a period of ~3.5 months since the beginning of the spill, after which, by mid-September 1979 and later, the slicks reached the mangrove forests of the western Yucatan Peninsula. In these shaded, low-energy, and often hypoxic environments, oil degradation processes would slow down, and partially weathered Ixtoc 1 oil would have a good likelihood to be preserved over prolonged periods. As indicated by our results, this oil still contains appreciable amount of parent high molecular weight (HMW) PAHs, which can pose a chronic source of toxicity.

If we try to draw an analogy to the DWH case, it could be reasonably argued that the MW oil buried in the environments similar to sGoM mangrove forests, such as the salt marshes of Louisiana (Zengel et al. 2015; Nixon et al. 2016), has the potential to be preserved over multi-decadal timescales – an occurrence already reported in some historic spill cases, such as the 1969 West Falmouth spill (Reddy et al. 2002).

On the contrary, GC-MS/MS-MRM analysis of oil residues found exposed on supratidal rocks at sites such as Punta Delgado, Isla En Medio, and Montepio indicates possible modern oil sources. A similar conclusion could be drawn from the higher sulfur content (as detected by FTICR-MS), observed in these samples, compared to the mangrove tar samples. Contributions from concurrent, contemporary oil releases are not an unlikely possibility given the extent of oil exploration and the number of natural oil seep sites in the area (Garcia et al. 2009). On the other hand, this might also point to a very advanced stage of weathering, relatively enriching the content of sulfur species, including the occurrence of multiple sulfurized compound classes, possibly indicating S-crosslinking (natural vulcanization). Sulfur heteroatom oil fractions have been previously shown to be more recalcitrant to photooxidation (Radović et al. 2014) and biodegradation (Larter and Head 2014; Oldenburg et al. 2017).

20.4   Conclusions and Future Research Directions

In this chapter we discuss initial observations related to the analysis of oil residues found in coastal sGoM environments known to be impacted by the historic Ixtoc 1 spill, using an approach which combines targeted fingerprinting of GC-amenable biomarkers (GC-MS/MS-MRM) with non-targeted exploratory screening of heavy molecular weight, heteroatom-containing compound classes, using ultrahigh-resolution FTICR-MS. Biomarker ratios in the oil residues from the prop roots of mangrove trees of the western Yucatan Peninsula show a reasonably good match to the source Ixtoc 1 oil, given the time elapsed since the spill, with apparent decadal recalcitrance of C27–C32 homohopanes, triaromatic steroids, and high molecular weight PAHs, a trend which is somewhat similar to the observations made in the coastal residues of the DWH spill. Further, the samples also show the presence of oxygen-bearing compound classes, as indicated by FTICR-MS, likely indicating the preservation of partially weathered oil in these low-energy environments, a good analog for the prediction of DWH residues' fate in the similar, marshy areas of the nGoM.

On the contrary, heavily weathered samples from the exposed supratidal rocky shorelines show enrichment of sulfur-bearing compound classes, including polysul-furized species. Possible contributors to these residues may include the modern exploration of oil in the area and/or natural seeps. Regardless of the origin, the composition of these samples might reveal pathways of sulfur compound preservation during prolonged and extreme weathering. Relatively high sulfur content found in these oil residues is an interesting feature, which should be explored further using complementary techniques such as comprehensive two-dimensional gas chromatography or sulfur isotope analysis to improve and strengthen the conventional source apportionment using biomarker fingerprinting ratios.

Finally, an exact molecular characterization of the range of oil transformation products found in the samples, using tools such as LC-MS or NMR, coupled to environmental and toxicological modelling and biological transect studies in the affected areas, is warranted for better understanding of long-term risks of oil spill impacts in coastal areas.

Acknowledgments   †Dr. John "Wes" Tunnell passed away during the preparation of this chapter. Dr. Tunnell was a marine ecology and biology professor at Texas A&M University-Corpus Christi and an early orchestrator of the Harte Research Institute (HRI) for Gulf of Mexico studies. Within C-IMAGE, Dr. Tunnell provided a foundation to expand our oil spill studies to include the entire Gulf of Mexico studying the decade-long impacts of the 1979 Ixtoc 1 oil spill to compare with the recent Deepwater Horizon oil spill.

This research was made possible in part by a grant from the Gulf of Mexico Research Initiative, C-IMAGE, and in part thanks to support from CFI, NSERC, the University of Calgary, and the Canada Research Chairs. Data are publicly available through the Gulf of Mexico Research Initiative Information and Data Cooperative (GRIIDC) at https://data.gulfresearchinitiative.org/data/R6.x805.000:0064 [doi:10.7266/n7-zeda-tw26].

References

Adhikari PL, Wong RL, Overton EB (2017) Application of enhanced gas chromatography/ triple quadrupole mass spectrometry for monitoring petroleum weathering and forensic source fingerprinting in samples impacted by the Deepwater Horizon oil spill. Chemosphere 184:939–950. https://doi.org/10.1016/j.chemosphere.2017.06.077

Aeppli C, Carmichael CA, Nelson RK, Lemkau KL, Graham WM, Redmond MC, Valentine DL, Reddy CM (2012) Oil weathering after the Deepwater Horizon disaster led to the formation of oxygenated residues. Environ Sci Technol 46(16):8799–8807. https://doi.org/10.1021/es3015138

Aeppli C, Nelson RK, Radović JR, Carmichael CA, Valentine DL, Reddy CM (2014) Recalcitrance and degradation of petroleum biomarkers upon abiotic and biotic natural weathering of Deepwater Horizon oil. Environ Sci Technol 48(12):6726–6734. https://doi.org/10.1021/es500825q

Atwood D (1980) Proceedings of a symposium on preliminary results from the September 1979 Researcher/Pierce IXTOC-1 cruise. National Oceanic and Atmospheric Administration, Boulder, CO, USA

Brooks GR, Larson RA, Schwing PT, Romero I, Moore C, Reichart G-J, Jilbert T, Chanton JP, Hastings DW, Overholt WA, Marks KP, Kostka JE, Holmes CW, Hollander D (2015) Sedimentation pulse in the NE Gulf of Mexico following the 2010 DWH blowout. PLoS One 10(7):e0132341. https://doi.org/10.1371/journal.pone.0132341

Garcia O, MacDonald I, Zimmer B, Shedd W, Frye M (2009) Satellite SAR inventory of Gulf of Mexico oil seeps and shallow gas hydrates. In: EGU general assembly conference abstracts, p 11633

Garrett RM, Pickering IJ, Haith CE, Prince RC (1998) Photooxidation of crude oils. Environ Sci Technol 32(23):3719–3723. https://doi.org/10.1021/es980201r

Gundlach ER, Finkelstein KJ, Sadd JL (1981) Impact and persistence of Ixtoc 1 oil on the South Texas coast. Int Oil Spill Conf Proc 1981(1):477–485. https://doi.org/10.7901/2169-3358-1981-1-477

Guzman-Vega MA, Mello MR (1999) Origin of oil in the Sureste Basin, Mexico. AAPG Bull 83(7):1068–1094

Hall GJ, Frysinger GS, Aeppli C, Carmichael CA, Gros J, Lemkau KL, Nelson RK, Reddy CM (2013) Oxygenated weathering products of Deepwater Horizon oil come from surprising precursors. Mar Pollut Bull 75(1–2):140–149

Harriman B, Zito P, Podgorski DC, Tarr MA, Suflita JM (2017) Impact of Photooxidation and biodegradation on the fate of oil spilled during the Deepwater Horizon incident: advanced stages of weathering. Environ Sci Technol. https://doi.org/10.1021/acs.est.7b01278

Jernelöv A, Lindén O (1981) Ixtoc I: a case study of the world's largest oil spill. Ambio:299–306

Larter SR, Head IM (2014) Oil sands and heavy oil: origin and exploitation. Elements 10(4):277–283. https://doi.org/10.2113/gselements.10.4.277

Liu Y, Kujawinski EB (2015) Chemical composition and potential environmental impacts of water-soluble polar crude oil components inferred from ESI FT-ICR MS. PLoS One 10(9):e0136376. https://doi.org/10.1371/journal.pone.0136376

Lizarraga P, Munoz R, Porras A, Izquierdo V, Wongchang I (1984) Taxonomy and distribution of hydrocarbonoclastic bacteria from the Ixtoc-1 area. In: 2. Colloque International de Bacteriologie Marine, Brest, pp 1–5

Lizárraga-Partida ML, Rodríguez-Santiago H, Romero-Jarero JM (1982) Effects of the Ixtoc I blowout on heterotrophic bacteria. Mar Pollut Bull 13(2):67–70. https://doi.org/10.1016/0025-326X(82)90445-3

McKenna AM, Nelson RK, Reddy CM, Savory JJ, Kaiser NK, Fitzsimmons JE, Marshall AG, Rodgers RP (2013) Expansion of the analytical window for oil spill characterization by ultrahigh resolution mass spectrometry: beyond gas chromatography. Environ Sci Technol 47(13):7530–7539. https://doi.org/10.1021/es305284t

Nixon Z, Zengel S, Baker M, Steinhoff M, Fricano G, Rouhani S, Michel J (2016) Shoreline oiling from the Deepwater Horizon oil spill. Mar Pollut Bull 107(1):170–178. https://doi.org/10.1016/j.marpolbul.2016.04.003

Oldenburg TBP, Jones M, Huang H, Bennett B, Shafiee NS, Head I, Larter SR (2017) The controls on the composition of biodegraded oils in the deep subsurface – part 4. Destruction and production of high molecular weight non-hydrocarbon species and destruction of aromatic hydrocarbons during progressive in-reservoir biodegradation. Org Geochem 114:57–80. https://doi.org/10.1016/j.orggeochem.2017.09.003

Prince RC, Elmendorf DL, Lute JR, Hsu CS, Haith CE, Senius JD, Dechert GJ, Douglas GS, Butler EL (1994) 17.alpha.(H)-21.beta.(H)-hopane as a conserved internal marker for estimating the biodegradation of crude oil. Environ Sci Technol 28(1):142–145. https://doi.org/10.1021/es00050a019

Radović JR, Aeppli C, Nelson RK, Jimenez N, Reddy CM, Bayona JM, Albaigés J (2014) Assessment of photochemical processes in marine oil spill fingerprinting. Mar Pollut Bull 79(1–2):268–277. https://doi.org/10.1016/j.marpolbul.2013.11.029

Radović JR, Jaggi A, Silva AC, Snowdon R, Waggoner DC, Hatcher PG, Larter SR, Oldenburg TBP (2019) Applications of FTICR-MS in oil spill studies (Chap. 15). In: Murawski SA, Ainsworth C, Gilbert S, Hollander D, Paris CB, Schlüter M, Wetzel D (eds) Deep oil spills – facts, fate and effects. Springer, Cham

Reddy CM, Eglinton TI, Hounshell A, White HK, Xu L, Gaines RB, Frysinger GS (2002) The West Falmouth oil spill after thirty years: the persistence of petroleum hydrocarbons in marsh sediments. Environ Sci Technol 36(22):4754–4760. https://doi.org/10.1021/es020656n

Rodgers RP, Marshall AG (2007) Petroleomics: advanced characterization of petroleum-derived materials by Fourier transform ion cyclotron resonance mass spectrometry (FT-ICR MS). In: Mullins OC, Sheu EY, Hammami A, Marshall AG (eds) Asphaltenes, heavy oils, and Petroleomics. Springer New York, New York, pp 63–93. https://doi.org/10.1007/0-387-68903-6_3

Romero IC, Schwing PT, Brooks GR, Larson RA, Hastings DW, Ellis G, Goddard EA, Hollander DJ (2015) Hydrocarbons in Deep-Sea sediments following the 2010 Deepwater Horizon blowout in the Northeast Gulf of Mexico. PLoS One 10(5):e0128371. https://doi.org/10.1371/journal.pone.0128371

Romero IC, Toro-Farmer G, Diercks A-R, Schwing P, Muller-Karger F, Murawski S, Hollander DJ (2017) Large-scale deposition of weathered oil in the Gulf of Mexico following a deep-water oil spill. Environ Pollut 228(Supplement C):179–189. https://doi.org/10.1016/j.envpol.2017.05.019

Romero IC, Sutton T, Carr B, Quintana-Rizzo E, Ross SW, Hollander DJ, Torres JJ (2018) Decadal assessment of polycyclic aromatic hydrocarbons in mesopelagic fishes from the Gulf of Mexico reveals exposure to oil-derived sources. Environ Sci Technol 52:10985. https://doi.org/10.1021/acs.est.8b02243

Ruddy BM, Huettel M, Kostka JE, Lobodin VV, Bythell BJ, McKenna AM, Aeppli C, Reddy CM, Nelson RK, Marshall AG, Rodgers RP (2014) Targeted Petroleomics: analytical investigation of Macondo well oil oxidation products from Pensacola Beach. Energy Fuel 28(6):4043–4050. https://doi.org/10.1021/ef500427n

Ryerson TB, Camilli R, Kessler JD, Kujawinski EB, Reddy CM, Valentine DL, Atlas E, Blake DR, de Gouw J, Meinardi S, Parrish DD, Peischl J, Seewald JS, Warneke C (2012) Chemical data quantify Deepwater Horizon hydrocarbon flow rate and environmental distribution. Proc Natl Acad Sci 109(50):20246–20253. https://doi.org/10.1073/pnas.1110564109

Santamaria-Orozco D, Horsfield B, di Primio R, Welte DH (1998) Influence of maturity on distributions of benzo- and dibenzothiophenes in Tithonian source rocks and crude oils, Sonda de Campeche, Mexico. Org Geochem 28(7):423–439. https://doi.org/10.1016/S0146-6380(98)00009-6

Soto L, García A, Botello AV (1981) Study of the penaeid shrimp population in relation to petroleum hydrocarbons in Campeche Bank

Soto LA, Botello AV, Licea-Durán S, Lizárraga-Partida ML, Yáñez-Arancibia A (2014) The environmental legacy of the Ixtoc 1 oil spill in Campeche Sound, southwestern Gulf of Mexico. Front Mar Sci 1(57). https://doi.org/10.3389/fmars.2014.00057

Stout SA, Payne JR (2016) Macondo oil in deep-sea sediments: part 1 – sub-sea weathering of oil deposited on the seafloor. Mar Pollut Bull 111(1):365–380. https://doi.org/10.1016/j.marpolbul.2016.07.036

Sun S, Hu C, Tunnell JW (2015) Surface oil footprint and trajectory of the Ixtoc 1 oil spill determined from Landsat/MSS and CZCS observations. Mar Pollut Bull 101(2):632–641. https://doi.org/10.1016/j.marpolbul.2015.10.036

Ward CP, Sharpless CM, Valentine DL, French-McCay DP, Aeppli C, White HK, Rodgers RP, Gosselin KM, Nelson RK, Reddy CM (2018) Partial photochemical oxidation was a dominant fate of Deepwater Horizon surface oil. Environ Sci Technol 52(4):1797–1805. https://doi.org/10.1021/acs.est.7b05948

White HK, Wang CH, Williams PL, Findley DM, Thurston AM, Simister RL, Aeppli C, Nelson RK, Reddy CM (2016) Long-term weathering and continued oxidation of oil residues from the Deepwater horizon spill. Mar Pollut Bull. in press. https://doi.org/10.1016/j.marpolbul.2016.10.029

Xia K, Hagood G, Childers C, Atkins J, Rogers B, Ware L, Armbrust K, Jewell J, Diaz D, Gatian N, Folmer H (2012) Polycyclic Aromatic Hydrocarbons (PAHs) in Mississippi seafood from areas affected by the Deepwater Horizon oil spill. Environ Sci Technol 46(10):5310–5318. https://doi.org/10.1021/es2042433

Yin F, John GF, Hayworth JS, Clement TP (2015) Long-term monitoring data to describe the fate of polycyclic aromatic hydrocarbons in Deepwater Horizon oil submerged off Alabama's beaches. Sci Total Environ 508:46–56. https://doi.org/10.1016/j.scitotenv.2014.10.105

Zengel S, Bernik BM, Rutherford N, Nixon Z, Michel J (2015) Heavily oiled salt marsh following the Deepwater Horizon oil spill, ecological comparisons of shoreline cleanup treatments and recovery. PLoS One 10(7):e0132324. https://doi.org/10.1371/journal.pone.0132324

Impacts of Deep Spills on Plankton, Fishes, and Protected Resources

Curtis Whitwam
Snapper School
Watercolor on Aquaboard
20" × 16"

Chapter 21
Overview of Ecological Impacts of Deep Spills: *Deepwater Horizon*

Christopher G. Lewis and Robert W. Ricker

Abstract This chapter provides a general introduction to Part V: Impacts of Deep Spills on Plankton, Fishes, and Protected Resources. As a general overview of ecological impacts of deep spills, it orients the reader to the deep sea environment and resources potentially at risk from petroleum releases at depth, and sets the stage for subsequent chapters presenting findings from investigations of specific biological resources. Much of what is known about impacts from deep spills stems from the significant body of research performed in the wake of the *Deepwater Horizon* oil spill. Therefore, this chapter provides: an explanation of the regulatory and legal context for much of the deep sea research conducted post-spill, an overarching ecosystem-level perspective of recorded impacts, and insights gained on the part of the authors from overseeing government-funded investigative work to assess injuries to deep sea natural resources in the context of the natural resource damage assessment process. Key findings indicate that pathways of resource injury were multiple and varied, exposure to oil was patchy, a broad range of organisms were harmed by (though some benefitted from) the spill, impact severity decreased with increasing distance from the well head, and subtle and longer term injuries were difficult to identify. Significant gaps in our understanding of the effects of deep-sea oil spills remain, and stem primarily from logistical, technical, and funding constraints on our ability to conduct deep sea research.

Keywords Natural resource damage assessment · NRDA · *Deepwater Horizon* oil spill · Responsible party · Trustees · Marine snow · Conceptual models

C. G. Lewis (✉)
Industrial Economics, Incorporated, Cambridge, MA, USA
e-mail: CLewis@indecon.com

R. W. Ricker
Formerly National Oceanic and Atmospheric Administration (NOAA), Office of Response and Restoration, Assessment and Restoration Division, Silver Spring, MD, USA

© Springer Nature Switzerland AG 2020
S. A. Murawski et al. (eds.), *Deep Oil Spills*,
https://doi.org/10.1007/978-3-030-11605-7_21

21.1 Introduction

The deep ocean represents one of the few places left on the planet where humans are still actively exploring and where innumerable insights have yet to be discovered. It is therefore not surprising that the ecological impacts of oil spills on natural resources of the deep ocean are only partially understood. Nevertheless, this chapter aims to: describe some of the known adverse effects of releases of oil on deep ocean resources, identify a broader range of potential spill-related impacts in deep-sea habitats, and give examples of challenges that exist for scientists who attempt damage assessment work in the deep ocean. While there have been a number of accidents over the course of modern history where oil spills have occurred at depth, none of these previous events were of the same scale as the 2010 *Deepwater Horizon* oil spill. In light of the pace and trajectory of offshore oil and gas development, supported by recent federal policies aimed at further increasing the pace of offshore oil and gas development (Federal Register 2017), it seems unlikely that the *Deepwater Horizon* spill will be the last deep spill of its kind. We therefore draw heavily from our experiences with the *Deepwater Horizon* oil spill in this chapter as we discuss the impacts of deep oil spills on marine resources, with the caveat that, although there are similarities in deep ocean habitats across the globe, and certain species have widespread abundance, the information presented herein may not apply uniformly to all deep ocean habitats. We also highlight that we do not discuss herein the impacts to surface and shoreline resources, as those impacts have been studied extensively in the broader oil spill literature, but rather focus on the impacts of deep oil spills to deep benthic resources.

One fortunate result of the *Deepwater Horizon* oil spill was the unprecedented magnitude of scientific investigation that followed. Subsequent chapters in this book will provide in detail specific results from this body of work. In order to set the stage for these more detailed discussions, this chapter: (1) explains the regulatory and legal context that directed tens of millions of dollars of deep-sea research and ultimately led to the conclusions reached by the natural resource Trustees related to the ecological impacts of the spill, (2) provides an overarching ecosystem-level perspective of impacts, and (3) shares some of the insights gained from overseeing and guiding the government-funded investigative work to assess injuries to deep-sea natural resources. This chapter draws from information presented originally in the US Government's Programmatic Damage Assessment and Restoration Plan, and readers are encouraged to review that plan's entirety for additional detail and context (Trustees 2016).

The ocean floor of the Gulf of Mexico is generally subdivided into three distinct depth zones: the continental shelf (0–200 m); the continental slope (200–800 m), which drops off more abruptly; and the deep benthos (below 800 m) (Fig. 21.1, Trustees 2016). Characteristic biological resources inhabit each of these diverse zones, due to differences in habitat (although some mobile resources move between them). The vast majority of the seafloor along the Gulf of Mexico consists of muddy sediment substrate, though in certain areas hard substrates are present (Love et al. 2013).

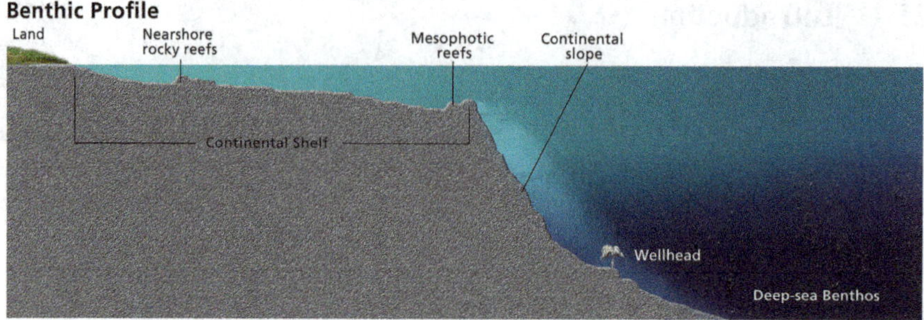

Fig. 21.1 The zones of benthos of the Gulf of Mexico. (Source: Trustees 2016; Kate Sweeney for NOAA)

The so-called hardbottom habitat is present in the form of nearshore rocky reefs on the continental shelf, mesophotic reefs situated between 60 and 90 m of depth at the edge of the continental shelf, and deep-sea hardground formed from upward migration of salt domes and presence of authigenic sediments, commonly calcium carbonates (Gittings et al. 1992; Love et al. 2013).

In the days and weeks that followed the initial blowout of the *Deepwater Horizon* oil spill, experts amassed in Houma, LA; Houston, TX; and elsewhere along the Gulf Coast to identify approaches and technologies to stem the release of oil, recover oil from the surface of the ocean, and prepare for the eventual landfall of oil slicks and tar balls along the Gulf Coast. Among them were separate teams of government-affiliated scientists directed by the natural resource Trustees who were beginning the natural resource damage assessment (NRDA) process. Natural resource trustees are federal, state, and tribal representatives appointed by the government to work on behalf of the public. NRDA process is codified in the Oil Pollution Act of 1990 and described in regulations promulgated by the National Oceanic and Atmospheric Administration at 15 C.F.R. Part 990. The goal of the NRDA process is "to make the environment and public whole for injuries to natural resources and services resulting from an incident involving a discharge or substantial threat of a discharge of oil (incident)" (15 C.F.R. 990.10). It is a compensatory, not punitive, process. As such, it is designed to identify direct compensation for losses, typically through damages (i.e., money paid by the responsible party) that sufficiently compensate the public for the loss of trust resources and their services. Responsible parties may also directly implement projects to restore identical or comparable types of resources lost as a result of the spill. Trust resources are public natural resources that are not owned by any private party but are held in trust by the government-designated Trustees. These include resources such as water, soil, plants, and wildlife. Additionally, the services these resources provide, such as provision of opportunities for wildlife viewing, noncommercial fishing, and various forms of recreation, are also compensable if harmed or lost. The language of the NRDA regulations highlights natural resource restoration as the ultimate goal of the NRDA process. To accomplish this, the regulations identify key steps Trustees should follow, which include documenting the release of oil and the pathways it travels to reach and expose natural resources, followed by approaches for identifying and quantifying the magnitude of injuries, and identifying and scaling appropriate restoration actions.

In light of this multistep process, it is important for Trustees to develop conceptual models at the outset of an oil spill to identify likely fate and transport mechanisms and potential resources at risk. This was particularly important in the case of the *Deepwater Horizon* oil spill, where it was quickly realized that oil and gas (and dispersant, applied to break up and disperse the oil, as well as other materials used to stem the flow from the broken wellhead) were being released at an extreme depth. A conceptual model can also be useful for identifying where assessment efforts should be expended, where data gaps might exist, and ultimately opportunities for restoration. In the case of the *Deepwater Horizon* oil spill, numerous conceptual models were developed by the Trustees, and these continued to evolve as additional information and data were gathered (Fig. 21.2). In particular, the scope of the spill forced the Trustees to consider atypical pathways and injuries—including to deep-sea resources that in some cases scientists didn't even have names for.

Given time, financial, and logistical constraints, not every potential resource at risk can be assessed. Trustees therefore must limit their focus to specific resources that are deemed "representative" to reflect the wider range of resources exposed to and potentially injured by the released oil. NRDA scientists try to focus on resources that adequately reflect the broad range of resources in the environment, such as keystone or dominant species, given the potentially litigious and public nature of NRDA. Trustees also sometimes focus on charismatic resources and/or resources in which injury can be readily demonstrated using existing scientific methods. Trustees,

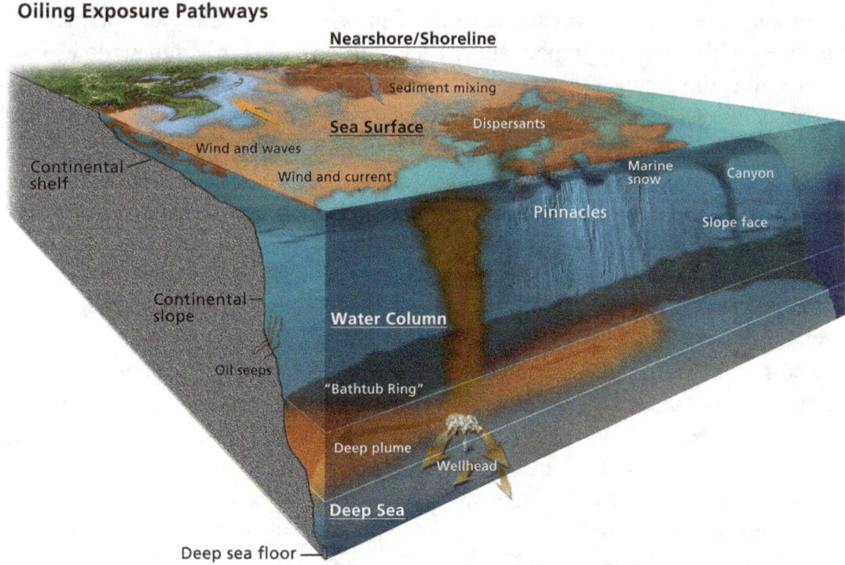

Fig. 21.2 Conceptual model for exposure of benthic resources from the Deepwater Horizon oil spill. (Source: Trustees 2016; Kate Sweeney for NOAA)

when demonstrating that injuries have occurred, must consider the baseline condition of the resources—that is, the condition the resources would have been in, in the absence of the release in question. Injuries are defined in the NRDA regulations as "an observable or measurable adverse change in a natural resource or impairment of a natural resource service" (15 C.F.R. 990.30). Trustees often rely on research, data, and information that have been collected prior to the spill and aim to leverage this information during the assessment. Consequently, representative resources typically include resources that have been studied previously. In the case of the deep benthos, in spite of challenges in this regard, these same considerations ultimately guided the selection of key resources assessed by the Trustees.

Given the unique nature of the releases of oil and dispersant at depth, the Trustees spent substantial effort characterizing: where the oil traveled subsurface, concentrations and changes in oil composition, microbial breakdown, and eventual deposition of the oil in the benthos if it did not remain on the water surface or strand on shorelines. Similarly, the Trustees spent a significant amount of time and energy simply trying to identify the presence or absence of resources within affected areas. Fortunately, the Trustees were also not the only entities conducting research post-spill. Teams of scientists, including academics and experts from various government agencies, nonprofit organizations, and industry, leveraged GoMRI (the Gulf of Mexico Research Initiative) and other sources of funding to spearhead investigations to understand the impacts of the spill on the deep sea. The Trustees devoted considerable time in monitoring, reviewing, and utilizing information from these independent efforts. In a number of cases, different lines of evidence generated by the NRDA and independent scientists were combined to assemble clearer understandings of environmental fate and transport processes or impacts. One key example of such cooperation was the pairing of information on elevated marine snow quantities within the geographic area of the spill observed in sediment cores taken on two oil spill response cruises directed by Dr. Paul Montagna (Fig. 21.3) with in situ observations and laboratory-based experiments by Dr. Uta Passow. The research and hypotheses promoted by Dr. David Hollander (Passow 2014; Passow et al. 2012; Stout and Passow 2015; Brooks et al. 2015) were critical to understanding and interpreting the pathways that oil traveled and where it deposited in the benthos. Together, disparate lines of evidence clearly painted a picture of marine snow as an initially overlooked but significant driver of impacts to the deep sea, with an estimated 6.9–7.7% of the 3.19 million barrels of oil spilled being deposited into the sediments surrounding the wellhead (Stout et al. 2017).

The Trustees ultimately must balance a number of factors, including the strength and volume of underlying research, legal strategy considerations, public interest, and regulatory constraints to assemble a defensible claim for natural resource restoration. This is because the Trustees bear the burden of demonstrating how the individual lines of evidence fit together and tell a compelling and consistent story about what happened as a result of the spill. Ideally, the story that emerges from the research performed during the NRDA will fit with the conceptual models; but if inconsistences arise, then the conceptual models may need to be changed, data sets scrutinized, or data gaps filled.

Fig. 21.3 Photos of sediment cores from the R/V Ocean Veritas Response Cruise showing cores unaffected (**a**) and affected (**b**) by marine snow. (Source: Trustees 2016; Photos provided by Jeff Baguley)

21.2   Overview of Impacts from the *Deepwater Horizon* Spill

One of the first steps taken by the Trustees was to convene and employ government and academic scientists with expertise in deep ocean resources. This included redirecting research cruises that were already previously funded and planned prior to the spill toward assessment-related activities (Boland et al. 2010). Some of these early research cruises helped to identify both readily apparent resource injuries and conversely resources that appeared to be unaffected. For example, the extensive *Lophelia* reefs that reside along the continental slope were found to be unaffected. By contrast, researchers reported soon after the release began overt apparent injuries to red crabs, deep-sea hardground corals, and mesophotic reef fish (Serata 2010). The Trustees quickly identified relevant experts in the field and existing data sets, and subsequently pursued certain lines of evidence of injury and discontinued research into others. For example, by late 2010, it became apparent that shallow coral reefs off the coast of Florida would fortunately remain unscathed as the predictive oil fate and transport models that initially forecasted the potential for oil to travel past Florida out to the Atlantic were not supported by field observations or documentation of substantial oil in these areas. Teams of shallow coral researchers who had worked furiously for six months to collect baseline samples and prepare impacts studies subsequently discontinued their work on the NRDA.

Other areas of the seafloor did not fare as well, and the Trustees assessing impacts to the deep ocean coalesced assessment efforts around four primary lines of evidence of injury, to: the mesophotic reefs along the continental shelf edge led by Dr. Ken Sulak of the United States Geological Survey, Dr. Peter Etnoyer of NOAA, and Dr. Ian MacDonald of Florida State University; deep-sea hardground corals led by Dr. Chuck Fisher from Pennsylvania State University and Dr. Erik Cordes from Temple University; red crabs led by Ms. Harriet Perry formerly from the University of Southern Mississippi and Tuck Hines from the Smithsonian; and benthic infauna within deep-sea sediments led by Dr. Paul Montagna from Texas A&M University Corpus Christi and Dr. Jeff Baguley from the University of Nevada, Reno. Over time, these disparate lines of evidence of injury were combined with other information from independent academics and other NRDA researchers working on the fate and transport of oil in the deep sea, including extensive work related to the fingerprinting of polycyclic aromatic hydrocarbons (PAHs) conducted by Dr. Scott Stout, into a broader ecosystem-level understanding of injury. What emerged was evidence of four concentric zones of readily apparent ecological impacts to over 2000 km^2 of the deep sea immediately surrounding the wellhead and a larger zone spanning approximately 9200 km^2 of uncertain impacts extending to the northeast, north, and west of the wellhead. In addition, significant losses of both resident corals and planktivorous fish were documented across approximately 10 km^2 of mesophotic reef habitat (Fig. 21.4; Trustees 2016).

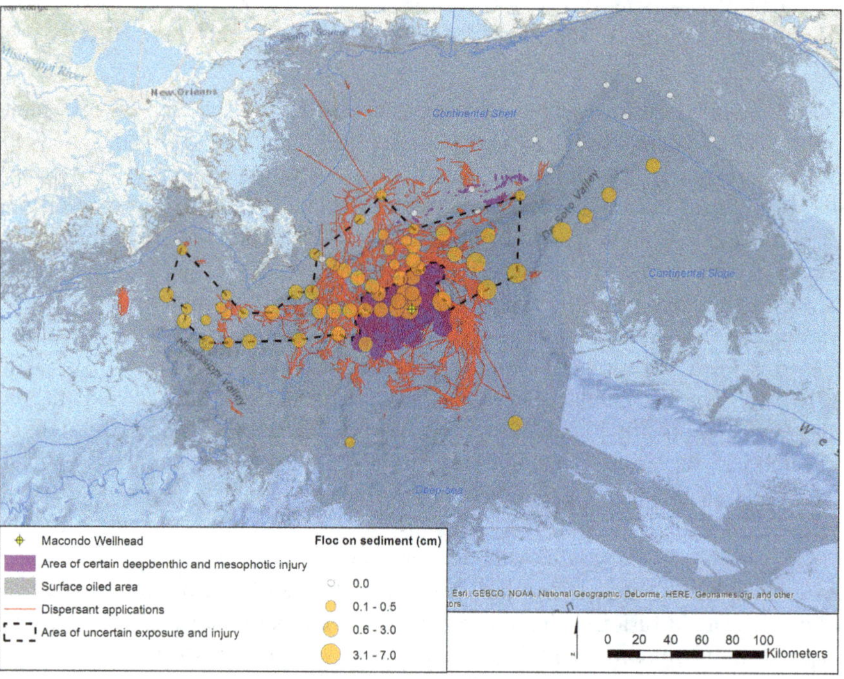

Fig. 21.4 Map of impacts from the *Deepwater Horizon* oil spill. (Source: Trustees 2016; Graphic by Cat Foley)

Specifically, the innermost zone covering 28 km^2 exhibited injuries including smothering and direct toxicity, resulting in a reduction by half in the diversity of sediment-dwelling infauna. The second and third concentric zones of 195 and 793 km^2, respectively, experienced mortality of hardground corals and reductions in sediment infauna diversity. The fourth concentric zone of 1275 km^2 was fouled by PAH concentrations in excess of toxicity values (i.e., site-specific LC20 and LC50 values for *Leptocheirus*) and experienced decreased abundance of benthic foraminifera (Trustees 2016).

This pattern of impacts demonstrated the following:

- *Pathways of injury were multiple and varied.* Based on experience from other spills, certain pathways were expected, including direct impacts from oil exposure in the nearfield environment and indirect injuries caused as a result of response efforts. However, the Trustees did not anticipate marine snow to be a significant pathway; yet studies showed this to be a significant transport mechanism for moving petroleum constituents from the surface and water column to the seafloor.
- *Exposure was patchy.* Likely due in part to currents present in the deep sea, exposure was patchy. This patchiness was apparent across multiple biological resources, including between and within hardground coral sites, as well as across the seafloor sediment, and even within various areas of the mesophotic reefs.
- *A broad range of organisms experienced harm, though some organisms benefitted.* For the most part, at least within the vicinity of the wellhead, adverse impacts were readily identifiable. While this could very well be related to the scope of the spill, it was somewhat surprising to see that in an environment that experiences natural oil seepage, the Trustees were able to identify overt injuries across a spectrum of organisms. Such findings hint at the potential for perturbations to propagate throughout the ecosystem. However, we also observed resiliency in certain organisms, and in some cases, certain species tended to benefit; for example, some of the more pollution-tolerant nematodes and polychaete worms increased in abundance in oil-impacted sediments.
- *The severity of impacts tended to decrease with increasing distance.* Not surprisingly, the Gulf of Mexico exhibited a certain capacity to dilute the impacts of the oil, particularly with increasing distance from the release point at the wellhead. As alluded to above, with distance, exposure became more patchy, and injuries became more difficult to identify, let alone directly link back to the spill. These difficulties were in part due to the scale of the area over which the Trustees were working to identify impacts.
- *Subtle and/or longer term impacts were difficult to identify.* Due primarily to a general dearth of scientific knowledge of many resources of the deep sea, the Trustees were unable to identify or quantify injuries that were not overt. That is, the Trustees focused on injuries that were obvious and easily tied to the release. However, some chronic impacts from the spill that were possible, or even likely, were insufficiently supported by data for the Trustees to claim damages for had the NRDA case gone to trial. Some spill-related injuries predicted to ecological

functions and connectivity between deep-sea habitats and communities also lacked adequate scientific study. In short, our limited understanding of deep benthic resources, especially regarding long-term trends, stability of communities, recruitment, and resiliency (or the ability of populations to recover after significant deepwater pollution incidents), limited our ability to identify and make claims for resource injuries.

21.3　Discussion

Although the *Deepwater Horizon* oil spill was a modern tragedy, the likes of which the public had never before experienced, the subsequent investigations that ensued, and continue at the time of this publication, represented a watershed moment in our understanding of the ecology of the Gulf of Mexico. This is evidenced by the range of literature that has been published over the past eight years since the release, which includes documentation of newly identified species, novel chemical techniques, advances in modeling, and an increased understanding of the fate and transport in, and impacts of oil on, the environment. The disaster also provided us an opportunity to redouble our efforts to ensure that oil and gas development is conducted in a safe manner for both workers and the environment.

Nevertheless, gaps in our understanding of effects of deep spills remain. These gaps stem primarily from the fact that the deep ocean is, relatively, inaccessible. The logistical challenges of studying the deep ocean are great, due primarily to the difficulties associated with accessing and working in deep water. In implementing studies of sediment infauna for the assessment, for example, it was highlighted that it took one hour simply to lower and raise the sediment coring equipment when working in 1500 m of water depth. This severely limited the overall number of sediment samples that could even be taken in the vicinity of the wellhead. Further, much of the historical research that has been done has been focused more on serving the oil and gas industry, as opposed to understanding the ecosystem. Biological surveys are often limited in geographic scope—many times restricted to areas where drilling is planned or oil extraction is underway. These limited surveys provide valuable information, but extrapolating to areas further removed creates unacceptable uncertainty for use in damage assessments and understanding potential impacts when habitats and communities are as patchy as they are in the deep sea. Clearly the work by the oil industry has yielded very helpful information, and the tools developed for deep-sea oil and gas development, ranging from remotely operated and autonomous underwater vehicles to undersea sonar and other imaging techniques, have served the academic research community immensely.

There are also real technical challenges associated with conducting valid scientific experiments under the harsh conditions present in the deep sea. For example, organisms adapted to the cold, darkness, and high pressure of the deep ocean are difficult to study ex situ, and conducting experiments at depth is challenged by the issues of inaccessibility noted above and all of the confounding factors associated with field work. While there are some promising new technologies that have the potential to transform and facilitate research in the deep sea, including consistent advances in remotely operated technologies and advanced biochemical techniques such as eDNA to document the presence of species, the most direct way to fill this gap in knowledge is continued funding for basic marine scientific research. As alluded to above, the NRDA framework necessarily limited the types of research the Trustees could, and were willing to, fund. The public and litigious setting dictated that only "tried and true" approaches, for the most part, were supported.

Despite the extensive funding that was committed for deep-sea research as a result of the *Deepwater Horizon* oil spill, one of the key data gaps that remained elusive was an understanding of what, if any, were the larger ecosystem-level effects. The NRDA approach taken is by its very nature deconstructive, whereby numerous component parts of the ecosystem were systematically examined with the goal of identifying and measuring concrete impacts. The Trustees were ultimately successful in documenting what became a laundry list of harms to individual resources. Everywhere we looked where oil was present in substantial quantity, we found change. Yet understanding the broader impacts of those changes was a challenge. While there were some early attempts to formulate and parameterize ecosystem-level models, this approach was ultimately abandoned. But questions of whether or not the Trustees (and the myriad other researchers working in the Gulf of Mexico) adequately captured the impacts of the spill remain. It is possible, and some may claim, that there were no ecosystem-level impacts. But it is more likely, given the interconnectedness of ecosystems, that we have simply failed to measure them. In particular, given the role of the deep ocean in providing ecological services related to energetic recycling, one is left to wonder how such a large input of organic material did not cause a measurable perturbation. As alluded to above, though, many researchers have also been surprised by the resiliency of the Gulf of Mexico and its component natural resources. Given the slow growth of many deep-sea organisms, however, it may be that continued research over time will yield additional information related to the chronic effects of the spill.

21.4 Conclusion

The release of oil (and associated response actions, including but not limited to the use of dispersant at depth) presented a unique threat to the deep-sea ecosystem. It was apparent from investigations following the *Deepwater Horizon* oil spill, the largest deep-sea spill to date, that despite natural seepage of oil in the Gulf of Mexico, such a spill had devastating effects on deep-sea natural resources. These include slow-growing and long-lived corals at various depths, the vast swathes of sediment and associated infauna and megafauna, as well as mobile fauna that reside in these habitats. Post-spill research not only identified unexpected pathways for resource exposure—in particular marine snow as a vector for delivery of hydrocarbons from the sea surface and water column to the seafloor—but also identified extensive impacts across multiple biological resources and habitats. The impacts were widespread, though severity decreased with increasing distance from the wellhead, and were increasingly patchy, likely due to variability in the fate and transport of toxic spill-related compounds. Several of the specific documented injuries to biological resources are detailed in the chapters that follow. Nevertheless, significant gaps remain in our understanding of the effects of deep-sea oil spills. These gaps stem primarily from logistical and technical, but also funding, constraints on our ability to conduct basic scientific research in the deep sea.

References

Boland G, Brewer G, Cordes E, Demopoulos A, Fisher C, German C, Sulak K (2010) Mississippi Canyon 252 Incident NRDA Tier 1 for deepwater communities: work plan and SOPs. June 27. Retrieved from https://www.fws.gov/doiddata/dwh-ar-documents/800/DWH-AR0016996.pdf

Brooks GR, Larson RA, Schwing PT, Romero I, Moore C, Reichart G-J, Jilbert T, Chanton JP, Hastings DW, Overholt WA, Marks KP, Kostka JE, Holmes C, Hollander D (2015) Sedimentation pulse in the NE Gulf of Mexico following the 2010 DWH blowout. PLoS One 10(7):24. https://doi.org/10.1371/journal.pone.0132341

Deepwater Horizon Natural Resource Damage Assessment Trustees (2016) Deepwater Horizon oil spill: final programmatic damage assessment and restoration plan and final programmatic environmental impact statement. Retrieved from http://www.gulfspillrestoration.noaa.gov/restoration-planning/gulf-plan

Federal Register (2017) Executive Order 13795 of April 28, 2017 Implementing an America-First Offshore Energy Strategy 82(84):20815–20818 May 3

Gittings SR, Bright TJ, Schroeder WW, Sager WW, Laswell SJ, Rezak R (1992) Invertebrate assemblages and ecological controls on topographic features in the Northeast Gulf of Mexico. Bull Mar Sci 50:435–455

Love M, Baldera A, Yeung C, Robbins C (2013) The Gulf of Mexico ecosystem: a coastal and marine atlas. Ocean Conservancy, Gulf Restoration Center, New Orleans

Passow U (2014) Formation of rapidly-sinking, oil-associated marine snow. Deep-Sea Res II Top Stud Oceanogr. https://doi.org/10.1016/j.dsr2.2014.10.001

Passow U, Ziervogel K, Asper V, Diercks A (2012) Marine snow formation in the aftermath of the Deepwater Horizon oil spill in the Gulf of Mexico. Environ Res Lett 7(035301). Retrieved from http://stacks.iop.org/1748-9326/7/i=3/a=035301

Serata B (2010) Are Red Crabs the Latest Victims of the Gulf Oil Disaster? National Wildlife Federation Blog. December 14. Retrieved from http://blog.nwf.org/2010/12/are-red-crabs-the-latest-victims-of-the-gulf-oil-disaster/

Stout SA, Passow U (2015) Character and sedimentation rate of "lingering" Macondo oil in the deep-sea up to one year after the Deepwater Horizon oil spill. DWH Natural Resource Exposure NRDA Technical Working Group Report. Seattle, WA. Retrieved from https://www.fws.gov/doiddata/dwh-ar-documents/946/DWH-AR0038895.pdf

Stout SA, Rouhani S, Liu B, Oehrig J, Ricker RW, Baker G, Lewis C (2017) Assessing the footprint and volume of oil deposited in deep-sea sediments following the Deepwater Horizon oil spill. Mar Pollut Bull 114(1):327–342. https://doi.org/10.1016/j.marpolbul.2016.09.046

Chapter 22
Deep-Sea Benthic Faunal Impacts and Community Evolution Before, During, and After the *Deepwater Horizon* Event

Paul A. Montagna and Fanny Girard

Abstract Oil from the Deepwater Horizon blowout reached the seafloor through deep-sea plumes and sedimentation of oil and oiled marine snow. This oil caused extensive damage over wide areas to both hard-bottom and soft-bottom communities. The most sensitive bioindicators were deep-sea planar octocorals for hard-bottoms and macrofauna and meiofauna diversity and taxa richness for soft-bottoms. Both hard-bottom and soft-bottom communities are very vulnerable to deep-sea oil spills. Deep-sea corals grow slowly and thus have extremely slow recovery rates. Four years after the spill, there was no recovery of the lost biodiversity of the macrofauna and meiofauna. Future research should be focused toward recovery and restoration. For hard-bottoms this could take the form of restoration projects. For soft-bottoms the restoration strategy could be "restoration in place" because fresh sediments, which fall to the seafloor continuously, can cap the contaminated sediments over time. Both strategies require monitoring to ensure desired outcomes are achieved.

Keywords Biodiversity · Deep-sea coral · Infauna · Macrofauna · Meiofauna · Oil spill · Sediments

P. A. Montagna (✉)
Texas A&M University-Corpus Christi, Harte Research Institute for Gulf of Mexico Studies, Corpus Christi, TX, USA
e-mail: Paul.Montagna@tamucc.edu

F. Girard
Institut Français de Recherche pour l'Exploitation de la Mer (Ifremer), Département Etude des Ecosystèmes Profonds, Laboratoire Environnement Profond, Centre de Brest, Plouzané, France
e-mail: Fanny.Girard@ifremer.fr

© Springer Nature Switzerland AG 2020
S. A. Murawski et al. (eds.), *Deep Oil Spills*,
https://doi.org/10.1007/978-3-030-11605-7_22

22.1 Introduction

The *Deepwater Horizon* (DWH) blowout and oil spill was a significant event. It is uncertain exactly how much oil was released from April through July 2010. One estimate is about 4.9–6.2 million barrels (Griffiths 2012), but the estimate is 3.19 million barrels as determined by court ruling (Deepwater Horizon Natural Resource Damage Assessment Trustees 2016). This is an immense amount of oil, more than enough to fuel a single household through the year 500,000 or power the world for 40 minutes. It is also $\approx$10 times greater than the combined leakage rate of all of the oil seeps in the Gulf of Mexico (GoM) (MacDonald et al. 2015). To contain the oil spill, about 3 million liters of dispersants were also released during the spill (Kujawinski et al. 2013). The dispersants are toxic as well (Montagna and Arismendez 2019; Almeda et al. 2014) and can increase the toxicity of the oil-derived compounds by making them more biologically available (Fern et al. 2015; Gong et al. 2014).

This was the first deepwater oil release, so there was no available information to predict impacts and consequences from this spill. Many assumed there would be no impact to the bottom because of the buoyancy of oil. The first sign of trouble was the report of deep-sea plumes at 1000–1300 m deep (Camilli et al. 2010; Joye et al. 2011), and the second sign was the sedimentation event that included oiled marine snow (Passow 2016). Eventually, it was estimated that as much as 35% of the oil was entrained in the deep-sea plumes (Ryerson et al. 2012) and that 4–31% of the oil was trapped in the deep sea, covering an area of 3200 km^2 (Valentine et al. 2014). This extensive coverage of a highly impacted area was a cause of concern for deep-sea benthic habitats.

Much research followed to develop an understanding of the impacts to the deep-sea benthos. Part of the motivation was to assess damage to the natural resources, part was scientific curiosity, and part was concern for the extensive habitats that support the oceanic food webs of the GoM. The two main habitats in the deep GoM are the soft-bottom and hard-bottom benthos. The soft-bottom is the dominant habitat making up about 95% of the seafloor bottom composed primarily of muddy bottom where the main faunal component is infauna that live in the mud. There are two dominant taxa of infauna, the small macrofauna (greater than 300 μm in size) and the very small meiofauna (between 43 μm and 300 μm in size). Many geological features in the GoM are hard and create outcroppings. These outcroppings have attached benthos such as corals. These organisms are also considered epibenthos organisms that live on the sediment surface. Here, a review is presented of the studies of the effect of the deepwater deposition of oil on the benthos, both soft-bottom and hard-bottom.

22.2 Hard-Bottom Communities: Deep-Sea Corals and Associated Fauna

Although much less widespread than sediment environments (Glover and Smith 2003), hard grounds are essential in the deep sea as they provide substrate to foundation species such as deep-sea corals, creating hot spots of biodiversity. Unlike their shallow-water counterparts, deepwater corals occur at all latitudes, commonly between 200 and 1000 m, and down to 6000 m for some species (Grasshoff 1981; Roberts et al. 2006; Watling et al. 2011). Because deepwater coral's diet is based on plankton or particles in suspension in the water column, they are usually found in areas of pronounced topographic relief, where currents are strong enough to provide them with food and prevent smothering by sediment (Bryan and Metaxas 2006; Roberts et al. 2006).

Whether they form reefs, like the well-studied scleractinian *Lophelia pertusa*, or dense aggregations, deep-sea corals enhance diversity by increasing the structural complexity of the seafloor (Buhl-Mortensen and Mortensen 2005; Jensen and Frederiksen 1992; Krieger and Wing 2002; Stone 2006). Numerous species use deepwater corals as habitat, feeding ground, shelter from predators, or nursery, resulting in coral ecosystems being associated with a higher diversity and abundance of megafaunal organism than the surrounding deep-sea environment (Baillon et al. 2012; Buhl-Mortensen and Mortensen 2005; Etnoyer and Warrenchuk 2007; Krieger and Wing 2002). If many studies have focused on characterizing the fauna associated with deepwater corals, the nature of the relationship between corals and their associates is still poorly understood (Buhl-Mortensen 2004). Of the approximately 1000 invertebrate species reported on deep-sea corals, about 100 have been characterized as symbionts. The large majority of these symbionts were either described as parasites or commensals, and only a few species appear to have a mutualistic relationship with corals (Buhl-Mortensen 2004; Girard et al. 2016).

The high biodiversity ecosystems sustained by deep-sea corals are also longlasting due to the high longevities generally recorded for deepwater corals. Several species can live for hundreds of years, and the oldest coral colony recorded is over 4000 years old (Andrews et al. 2002; Prouty et al. 2016; Roark et al. 2006). However, deep-sea corals often have slow growth and metabolic rates and are thus particularly vulnerable to the ever-increasing number of anthropogenic threats (Andrews et al. 2002; Cordes et al. 2001; Prouty et al. 2016).

22.2.1 Hard-Bottoms Before DWH

The presence of deep-sea corals in the GoM has been known for decades (Moore and Bullis 1960), but until more recently, few studies focused on these ecosystems. In order to address this knowledge gap and evaluate the potential environmental impacts from the oil and gas industry on hard-bottom communities, several multidisciplinary studies funded by the Bureau of Ocean Energy Management (BOEM) and the National Oceanic and Atmospheric Administration (NOAA) were conducted after 2007 (Boland et al. 2017). These studies significantly contributed to improving our knowledge of the biology and distribution of deepwater corals in the GoM.

Deep-sea corals occur throughout the GoM. In the deep GoM, they usually grow on authigenic carbonates that formed as a by-product of the microbial alteration of hydrocarbons coupled with anaerobic methane oxidation in the subsurface sediment at seeps (Cordes 2009). Newly formed seeps are generally characterized by the presence of microbial mats and rare megafauna. The precipitation of carbonates resulting from microbial activity eventually provides hard substrate to foundation species such as symbiont-containing vestimentiferan tubeworms and bathymodiolus mussels (Fisher et al. 2007). As seepage slows down, carbonates that are exposed due to the erosion of the surrounding sediment can be colonized by deepwater corals. Deep-sea corals have also been observed growing on man-made structures such as shipwrecks or oil platforms (Brooks et al. 2012; Larcom et al. 2014).

A total of 258 deep-sea coral species, predominantly octocorals (137 species) and hexacorals (84 scleractinian and 28 antipatharian species), have been documented in the northern GoM (Etnoyer and Cairns 2016). Species distribution is mostly dependent on depth. The reef-forming *Lophelia pertusa* is the most common coral species on the upper continental slope and also the most extensively studied (Cordes et al. 2008; CSA International 2007; Lessard-Pilon et al. 2010). Octocorals cover a greater depth range, down to at least 3000 m, with species assemblages being structured by depth (Quattrini et al. 2013).

As in the rest of the world, deep-sea corals enhance biodiversity in the deep GoM. For instance, studies show that *L. pertusa* reefs in the GoM are associated with diverse communities of associated fauna and that these communities are unique and differ from communities associated with vestimentiferan tubeworms aggregations, even when in close proximity (Cordes et al. 2008; Lessard-Pilon et al. 2010). Although the communities associated with octocorals in the GoM are less known, numerous species including ophiuroids have been observed in association with octocorals, and the documented presence of egg cases of a species of scyliorhinid catshark on the octocoral *Callogorgia delta* indicates their importance as nursery habitat (Etnoyer and Warrenchuk 2007; Girard et al. 2016).

22.2.2 *Hard-Bottoms During DWH (2010)*

To investigate potential impacts from the DWH oil spill on deep-sea corals, nine known coral sites were explored between 15 October and 1 November (3 months after the well was capped) during a Natural Resource Damage Assessment (NRDA) expedition funded by BOEM and NOAA. All these sites were located more than 20 km from the well between 290 m and 2600 m depth, and no visible evidence of impact was detected on coral communities there. However, a visibly impacted coral community was discovered on 2 November in lease block Mississippi Canyon (MC)

294, a site where 3D seismic reflectivity data suggested the presence of hard grounds (White et al. 2012b). At this site, located 13 km from the well at a depth of 1370 m, the majority of corals were, at least partially, covered in a brown flocculent material (floc) and displayed signs of stress such as tissue loss and excess mucus production. Analyses of the floc revealed that it contained traces of oil from the Macondo well as well as dioctyl sodium sulfosuccinate, a component of the dispersant deployed during the spill (White et al. 2012b, 2014). The composition of the floc, coupled with the location, timing of occurrence, and the fact that no visual damage has been observed on any other coral communities in the last decade, strongly links the damages observed at MC 294 to the DWH oil spill (White et al. 2012a).

The coral community at MC 294 was dominated by the octocoral species *Paramuricea biscaya* (Grasshoff 1977), although a few other species such as *Paragorgia regalis* (Nutting 1913) and *Swiftia* spp. were present and showed signs of impact. Toxicity experiments performed on a closely related species (*Paramuricea* sp. B3) indicated that if oil alone had little effect on coral's health, mixtures of oil and dispersant, as found in the floc, were particularly toxic to corals (DeLeo et al. 2015).

P. biscaya colonies were associated with hormathiid anemones and ophiuroids of the species *Asteroschema clavigerum* (Verrill 1894) (Hsing et al. 2013; White et al. 2012b). Ophiuroids were observed on the majority of coral colonies, and several individuals displayed an abnormal white color (*A. clavigerum* is generally tan to red in color). Additionally, several ophiuroids had loosely coiled or splayed out arms, an unusual behavior for this species (White et al. 2012b). These observations suggested an impact of the spill on ophiuroids associated with corals at MC 294.

22.2.3 Hard-Bottoms After DWH (2011–2014)

To determine whether other coral communities were impacted by the spill, surveys using towed cameras and autonomous underwater vehicles (AUVs) were conducted in 2011 in the vicinity of the Macondo well. These surveys led to the discovery of five previously unknown coral communities, and further exploration of these sites using remotely operated vehicles (ROVs) resulted in the identification of two additional impacted coral communities (Fisher et al. 2014a, b). These communities were located in lease blocks MC 297 and MC 344, 6 and 22 km from the Macondo well at 1560 and 1850 m depths, respectively (Fisher et al. 2014b). Both coral communities were dominated by *Paramuricea biscaya*, the same coral species present at MC 294. At these two sites, corals were no longer covered in floc but displayed the characteristic patchy impact and hydroid colonization also observed at MC 294 at this point in time (Fisher et al. 2014b).

Paramuricea biscaya has a planar morphology and is thus well suited for image analysis as all branches can be seen on the same image. Therefore, an image-based monitoring program was initiated in 2010 at MC 294 and 2011 at MC 297 and MC 344 in order to characterize the recovery of corals impacted by the spill. Over 300

Fig. 22.1 Impacted coral colony imaged every year between 2011 and 2017 (A to G). The state of all branches (healthy (green), non-healthy (red), hydroid-colonized (yellow), and broken (blue)) was digitized each year to quantify impact (H to N). (Reprinted Girard and Fisher (2018), Copyright (2018) with permission from Elsevier)

individual coral colonies were imaged every year between 2011 and 2017 at the three impacted sites as well as at two sites that were not affected by the DWH oil spill (AT 357 and GC 852, 183 and 325 km from the Macondo well at 1050 and 1400 m depths, respectively) (Girard and Fisher 2018; Hsing et al. 2013). High-resolution images were analyzed to characterize the long-term impact of the spill and evaluate the recovery potential of these deep-sea corals (Fig. 22.1).

Early results on the recovery of *P. biscaya* at MC 294 between 2010 and 2012 indicated that recovery was dependent on the initial level of total visible impact (Hsing et al. 2013). Indeed, apparent recovery from impact and hydroid colonization, which started in 2011 and had been increasing ever since, was negatively correlated with the proportion of the colony covered by floc in 2010. Although the median level of impact decreased between 2010 and 2012 and corals that were the least impacted (20% or less of the colony was impacted) appeared to have recovered by 2012, the majority of corals at MC 294 still showed signs of impact in 2012 (Hsing et al. 2013).

A later study that examined data collected between 2011 and 2017 at all sites found similar patterns (Girard and Fisher 2018). Recovery was slow and depended on the initial level of total visible impact. If the median level of total visible impact was steady after 2011, it remained higher than at the reference sites (AT 357 and GC 852) through 2017. Moreover, the initial level of total visible impact had a significant negative effect on both the ability of individual branches to remain healthy and their recovery from impact and hydroid colonization between consecutive years, even between 2016 and 2017. Impacted coral colonies were also more likely to lose branches than healthy colonies, and branch loss at both MC 294 and MC 344 was still higher than at the reference sites between 2016 and 2017. All these results indicate an ongoing, long-term non-acute effect of the DWH oil spill on deep-sea corals (Girard and Fisher 2018).

Results from the long-term monitoring of corals impacted by the spill showed that the fate of these corals was still uncertain 7 years after the spill. In order to plan for future monitoring, a study used a modeling approach to assess the recovery of impacted corals over the longer term (Girard et al. 2018). In this study, an impact-dependent matrix model was parameterized based on data collected as part of the long-term monitoring program. This model was then used to estimate the time necessary for every impacted coral colony to reach a state where all branches are visibly healthy (time to recovery) and branch loss. Results from the model indicated that, depending on their initial level of impact, the majority of impacted coral colonies will have recovered within 10 years. However, the most heavily impacted corals could take up to 30 years to reach a state where all branches appear healthy. Moreover, some of these impacted colonies are projected to lose a large number of branches, leading to a 10% reduction in coral biomass at some impacted sites by the time all corals will have visibly recovered. Based on these results, Girard et al. (2018) suggested that corals should be monitored every 2 years (sufficient as recovery is slow) until 2021 and then less frequently for a further 20 years to assess potential non-acute effects and follow the potential recovery of the most heavily impacted corals.

High-resolution images of individual coral colonies were also used to measure in situ growth rates Girard et al. 2019). Growth rates were estimated for both *Paramuricea biscaya* and *Paramuricea* sp. B3, providing important baseline information on the biology of these coral species. Additionally, the impact of the DWH oil spill on growth was characterized. No negative long-term impact of the spill was detected on growth. On the contrary, the level of total visible impact had a significant positive effect on growth after 2014 at two of the impacted sites (MC 294 and MC 297). However, this positive effect was not sufficient to compensate for the high levels of branch loss experienced by impacted corals as growth rates were extremely slow (average growth rates for healthy *P. biscaya* ranged from 0.14 to 1.2 cm/year/colony) (Girard et al. 2019). As a result, impacted coral colonies at MC 294 are expected to take over 50 years on average to grow back to their original size, and some impacted colonies could even take hundreds of years.

Results on the recovery and growth rates of *P. biscaya* not only suggested that this coral species has a low resilience to anthropogenic impact but also showed that its resilience differs between sites. Overall, corals at MC 297 and MC 294 tended to head toward recovery, while the health of corals at MC 344 tended to deteriorate (Girard and Fisher 2018). Moreover, healthy MC 344 corals had lower growth rates, and none of the impacted corals at this site visibly grew between 2011 and 2017 (Girard et al. 2019). These differences in coral's response to the spill show that historical and environmental variables inherent to each site can have a significant influence on the recovery potential of coral communities after impact.

No apparent long-term effect of the spill was observed on the ophiuroid *A. clavigerum*, the most common associate observed on *P. biscaya*. Several individuals displayed abnormal behaviors and colors in 2010 (White et al. 2012b), and some ophiuroids left their coral host between 2010 and 2011, which is also unusual as individuals from this species rarely move away from their host (Girard et al. 2016). However, these negative effects of the spill on ophiuroids did not persist after 2011. Using images collected as part of the long-term monitoring program, a study showed that *A. clavigerum* had a beneficial effect on coral recovery (Girard et al. 2016). Corals associated with ophiuroids were less impacted than corals without ophiuroid associates, and ophiuroids facilitated the recovery of branches in their vicinity from both impact and hydroids colonization. *A. clavigerum* likely removes particles depositing on coral branches and may also prevent the settlement of epibionts such as hydroids (Girard et al. 2016). To date, the potential effects of the DWH on other coral associates have not been characterized.

22.3 Soft-Bottom Habitats

About 70% of the Earth is covered by ocean, and about 65% is covered by deep-sea benthic habitat, making it one of the largest habitats on Earth. It is well known that deep-sea sediments are reservoirs of marine biodiversity in many places of the world. The organisms in the sediments are composed primarily of two groups of benthic invertebrates: the smaller meiofauna and larger macrofauna. The distinction between, and definition of, the two groups is said to be operational because it is based on size (Montagna et al. 2016). Macrofauna are retained on a 300 μm sieve, and meiofauna pass through a 300 μm sieve and are retained on a 45 μm sieve. However, there are also many taxonomic differences between the two groups. Meiofauna are phyletically diverse (Fig. 22.2), and many of the phyla only occur in marine sediment habitats. However, the meiofauna community is dominated by just two taxa: the Nematoda and the Harpacticoida (an order of Crustacea). While most phyla have representatives in the deep-sea macrofauna, just three phyla make up the bulk of organisms found as infauna: Polychaeta, Crustacea, and Mollusca (Fig. 22.3). These are the worms, shrimp-like organisms, and the hard-shelled clams and snails familiar to all.

The diversity of infauna that live in sediments has been used as a sensitive bioindicator in many ecological health studies and in many different contexts. The reason diversity is an important indicator of change is simple: during a disturbance or exposure to a toxic material, the sensitive species will be lost or disappear entirely, and the tolerant species will remain or even increase. Thus, the ecological model for disturbance detection is a loss of biodiversity and sometimes an increase in abundance of the "weedy" species. This is the basic approach used to assess the ecological health of all benthic communities and was applied to the GoM as well.

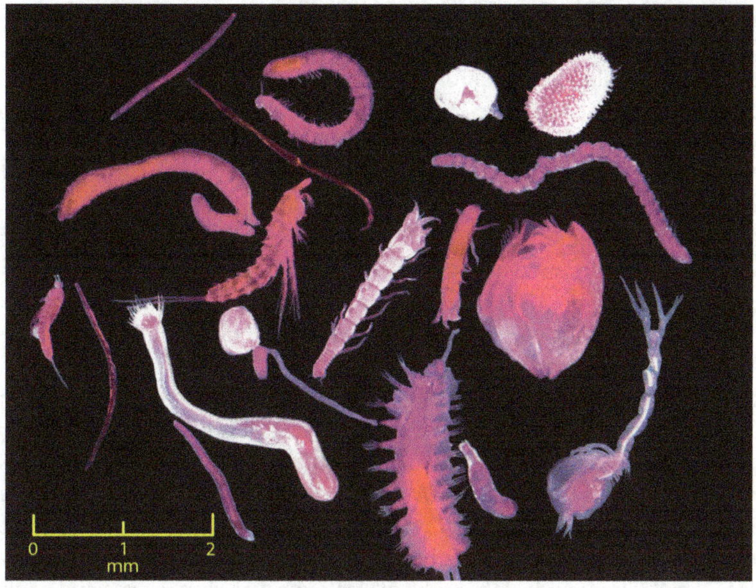

Fig. 22.2 Examples of common meiofauna taxa

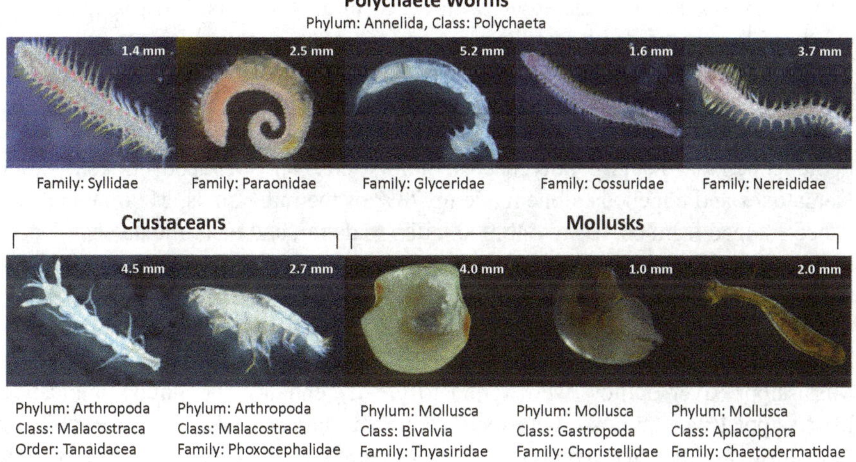

Fig. 22.3 Examples of macrofauna commonly found in the Gulf of Mexico deep-sea sediments

22.3.1 Soft-Bottom Habitats Before DWH

The US environmental laws and rules require an environmental impact statement (EIS) prior to leasing offshore lands for oil and gas exploration and development. This has led to a series of studies to describe the natural resources of the continental shelf and, more recently, the deep sea in the US exclusive economic zone. The US

Department of the Interior is responsible for these studies. In June 2010, the Minerals Management Service (MMS) was renamed the Bureau of Ocean Energy Management, Regulation and Enforcement (BOEMRE). The current BOEMRE and past MMS carried out two large-scale surveys of the deep GoM. The Northern Gulf of Mexico Continental Slope Study (NGOMCSS) was carried out between 1983 and 1985 (Gallaway 1988; Pequengnat et al. 1990). The Deep Gulf of Mexico Benthos (DGoMB) was carried out between 2000 and 2002 (Rowe and Kennicutt 2008, 2009). Both studies measured meiofauna and macrofauna infauna communities, megafauna, and a suite of other environmental physical-chemical variables in water depths of 300 m–3000 m using a transect design that covered the western, central, and eastern parts of the northern GoM. The DGoMB study also sampled the Sigsbee Abyssal Plain.

In the NGOMCSS study, 98% of the meiofauna was found to be made up by nematodes, harpacticoid copepods, nauplii, polychaetes, ostracods, and kinorynchs in that order. Nematodes and copepods alone made up 77% of the individuals. For macrofauna, 86% of the community was dominated by polychaetes, ostracods, bivalves, tanaids, bryozoans, and isopods, respectively. Meiofauna abundances ranged from 120,000 to 551,000 individuals/m^2, and macrofauna abundances ranged from 1560 to 8250 individuals/m^2. The macrofauna was very diverse and dominated by rarity because only 110 of 1212 species were represented by 10 or more individuals. However, very few of the species had existing names, and there was uncertainty about the taxonomy. The highest abundances for both groups were found in the central transect and decreased with depth in all transects. Most species were restricted to specific depth distributions, and there was more variability with depth than there was among transects. There was a break with peak abundances and diversity in depths ranging from 500 m to 1000 m.

In the DGoMB study, 99% of the meiofauna were made up of nematodes, harpacticoid copepods, nauplii, polychaetes, kinorynchs, and ostracods in that order. Nematodes and copepods alone made up 78% of the individuals. Meiofauna abundances ranged from 60,400 to 540,000 individuals/m^2, and macrofauna abundances ranged from 284 to 6400 individuals/m^2. But there was one outlier, station MT1 at the base of the Mississippi Trough that had an abundance of 20,700 individuals/m^2. The greater abundances in the Central Gulf resulted from the proximity to the Mississippi River outflow (Baguley et al. 2006a). Canyon features in proximity of Mississippi River outflow were found to greatly enhance meiofauna abundance. Mississippi River outflow alters local sediment characteristics and interacts with loop current eddies and dynamic slope topography to increase particulate organic matter flux in the northeastern region, thus creating areas of higher than normal meiofauna abundance. Meiofauna abundance was also related to water depth. Harpacticoida were the only meiofauna species identified to the lowest taxonomic level possible (Baguley et al. 2006b). Species richness declined linearly with increasing depth and was highest at 1200 m.

For macrofauna, 86% of the community was dominated by polychaetes, ostracods, bivalves, tanaids, bryozoans, and isopods in that order. Not all macrofauna species were identified to the lowest taxonomic level possible. Few of the deep-sea organisms are known to the species level, for example, only 40% (207 of 517) of the polychaete species and 25% (31 of 124) of the amphipod species found in the DGoMB study could be identified to the species level. The isopods contained more than 60 new to science (Wilson 2008). Species richness was greatest at depths around 1200–1500 m. The lowest richness was in the Sigsbee Abyssal Plain. There were no isopod diversity trends with longitude.

22.3.2 Soft-Bottom Habitats During DWH

The DWH oil spill occurred in the lease block named Macondo 252 and continued until 15 July 2010 when it was closed by a cap. Field missions to examine the deep seafloor for oil were conducted on the R/V Gyre and R/V Ocean Veritas in September through October 2010. Oil did end up on the bottom, and these contaminants pose risks to benthic fauna, particularly the infauna that lives within or in close association with bottom substrates and unable to avoid exposure due to their relatively sedentary existence. The study below describes the effects of the DWH oil spill on the macrofauna and meiofauna 2–3 months after the well was capped.

The study collected samples at 58 stations, and assessments of chemical, physical, and biological parameters were carried out using multivariate analyses to reduce the dataset to one new variable representing the oil spill footprint (Montagna et al. 2013). The new variable contained scores for each sample where high total petroleum hydrocarbon (TPH), polycyclic aromatic hydrocarbons (PAH), and barium (Ba) concentrations were inversely correlated to meiofauna and macrofauna diversity and the nematode/copepod (NC) ratio. The NC ratio is a sensitive bioindicator used to classify impacts of pollution and organic enrichment in the field (Shiells and Anderson 1985), mesocosm studies (Gee et al. 1985), and impacts of offshore drilling activities in the GoM (Peterson et al. 1996). Thus, the new variable is easy to interpret because as pollution goes up, bioindicator metrics go down. The most severe relative reduction of faunal abundance and diversity extended to 3 km from the wellhead in all directions covering an area about 24 km^2 (about 9 square miles; Fig. 22.4). Moderate impacts were observed up to 17 km toward the southwest and 8.5 km toward the northeast of the wellhead, covering an area of 148 km^2 (about 57 square miles). Benthic effects were correlated to TPH, PAHs, and Ba concentrations and distance to the wellhead, but not distance to hydrocarbon seeps. Thus, benthic effects are more likely due to the oil spill, and not natural hydrocarbon seepage.

A later analysis of this data focusing on the meiofauna alone at 66 stations confirmed that the NC ratio is a good indicator (Baguley et al. 2015). Nematode dominance increased significantly with increasing impacts, and NC spiked near the wellhead. Copepod abundance decreased where impacts were most severe. The increasing abundance and dominance by nematodes with increasing pollution could represent a balance between organic enrichment and toxicity. Using just meiofauna diversity and NC yielded a larger spatial footprint because stations that had

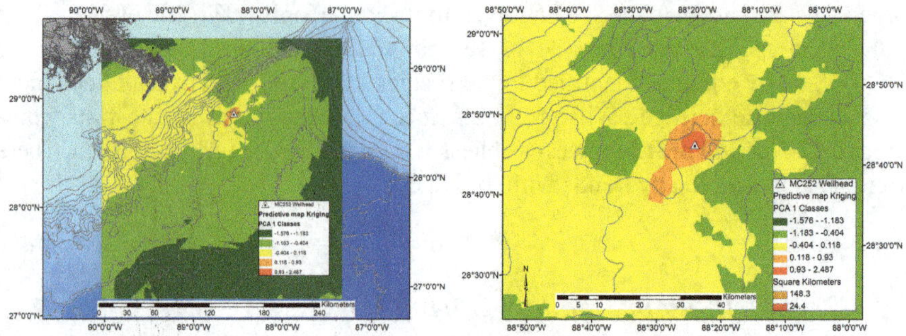

Fig. 22.4 Spatial extent of the Deepwater Horizon footprint on the seafloor (Montagna et al. 2013). Left: Interpolated area of deep-sea impact based on multivariate analysis of station scores (PC1). The interpolated area shown covers 70,166 km², of which 167 km² is considered moderately impacted and 24 km² is considered severely impacted. Right: Zoomed-in view of the interpolated area of deep-sea impact. The shape of the moderate impact area is asymmetrical, extending further to the southwest (about 17 km from the wellhead) than to the northeast (about 8.5 km from the wellhead). The diameter of the severely impacted zone is about 4 km

previously been identified as uncertain were now classified as impacted. The area of high impacts increased 80% from 172 km² to 310 km² compared to the earlier study.

There were also good macrofauna bioindicators of the oil spill because diversity decreased up to 10 km away from the wellhead (Washburn et al. 2016). Crustacean taxa were the most sensitive to the oil spill. There were six families of macrofauna that were tolerant of the presence of oil, and four were polychaetes. While polychaete taxa varied in their sensitivity, the family Dorvilleidae was the best indicator of effects close to the wellhead. Interestingly, high abundance or presence of Dorvilleidae is often associated with organic enrichment. Like the increase of nematodes, the increase of dorvillid polychaetes is an indicator of potential organic enrichment due to oil as well. Another interesting detail is that the loss of macrofauna diversity close to the wellhead was due to a loss in the surface sediments, because diversity in the lower 5–10 cm of sediments was similar in all stations. This is likely because the oil landed on the surface sediment and had yet to be buried or reworked by the macrofauna into the sediment. Thus toxicity was primarily at the surface.

Toxicity is evident in the benthic data, and sediment quality benchmarks were set using the macrofauna and meiofauna response to levels of TPH and PAHs in the sediment (Balthis et al. 2017). The low sample size contributes to uncertainty. The likelihood of impacts to benthic macrofauna and meiofauna communities is low (< 20%) at TPH concentrations of <606 mg/kg (ppm dry weight) and 700 mg/kg, respectively. Impacts are high (> 80%) at concentrations >2144 mg/kg and 2359 mg/kg, respectively. Impacts are intermediate at concentrations in between 700 mg/kg and 2144 mg/kg. For total PAH concentrations, the probability of impacts is low (<20%) at concentrations <4.0 mg/kg for both macrofauna and meiofauna; high (>80%) at concentrations >24 mg/kg and 25 mg/kg for macrofauna and meiofauna, respectively; and intermediate at concentrations in between. Dispersants can increase toxicity, but were not measured in 2010. Concentrations of dioctyl sulfosuccinate (DOSS), a principal component of Corexit 9500, were measured in June 2011, 11 months after the spill stopped, and they were very low, because 35% of the samples were below detection limits, and only one sample was as high as 84 ppb. The effect ranges found here for PAH were consistent with sediment quality guidelines developed for shallow estuarine and marine waters.

Sediments from a total of 179 stations were measured for sediment toxicity responses (Montagna and Arismendez 2019). Toxicity was assessed using the Microtox® system, where reductions of fluorescence of the bioluminescent bacterium, *Vibrio fischeri*, indicate corresponding increases in toxicity. Toxic responses were found as far away as 35 km from the spill site, and there were direct correlations between the presence of oil and biological and ecological effects of reduced macrofauna abundance and diversity. The Corexit dispersants were also highly toxic even when diluted 10,000 times, so a combination of dispersants and oil is responsible for the biological and ecological responses.

22.3.3 Soft-Bottom Habitats After DWH

In June 2011, 11 months after the well was capped, 32 stations from June 2010 were resampled (Montagna et al. 2017). The design compared 20 stations from the combined moderate and severe impact zones to 12 stations in the reference zones. There were no statistically significant differences in contaminant concentrations between the impact and non-impact zones from 2010 to 2011, which indicates contaminants persisted at the same levels after 1 year. Whereas meiofauna abundance and diversity exhibited some signs of recovery in 2011, there was evidence of persistent, statistically significant impacts to both macrofauna and meiofauna community structure. Macrofaunal taxa richness and diversity in 2011 were still 22.8% and 35.9% less, respectively, in the entire impact zone than in the surrounding non-impact area, and meiofaunal richness was 28.5% less in the entire impact zone than the surrounding area. The persistence of significant biodiversity losses and community structure change nearly 1 year after the wellhead was capped indicates that full recovery had yet to have occurred in 2011.

The macrofauna community response was examined for the 2011 sampling (Washburn et al. 2017). Richness, diversity, and evenness were severely impaired within a radius of approximately 1 km around the DWH wellhead. However, lower diversity than background was observed in several stations up to 29 km to the southwest of the wellhead. Abundance near the DWH wellhead increased in 2011 compared to 2010, and the highest abundances were found at stations within the 1 km radius of the wellhead. The increase was caused mostly by high abundance of opportunistic polychaetes of the family Dorvilleidae, genus *Ophryotrocha*. Diversity did not change in the impact zone near the DWH wellhead, but there was an increase of diversity at stations farther from the wellhead in the non-impact zone.

The meiofauna and macrofauna community responses were re-examined again in June 2014, approximately 4 years after the oil stopped flowing (Reucsher et al. 2017). The same 34 stations were sampled as in 2010 and 2011 from the impact zone and the non-impact zone. Even though chemical contaminant concentrations decreased slightly, they were still significantly different between the two zones. PAH concentrations averaged 218 ppb in the impact zone compared to 14 ppb in the non-impact zone. TPH concentrations averaged 1166 ppm in the impact zone compared to 102 ppm in the non-impact zone. While there was no difference between zones for meiofauna and macrofauna abundance, community diversity remained significantly lower in the impact zone. Meiofauna taxa richness over the three sampling periods averaged 8 taxa/sample in the impact zone, compared to 10 taxa/sample in the non-impact zone; and macrofauna richness averaged 25 taxa/sample in the impact zone compared to 30 taxa/sample in the non-impact zone. Oil originating from the DWH spill had a persistent negative impact on diversity of soft-bottom, deep-sea benthic communities. While there were signs of recovery for benthic metrics such as abundance, the low diversity and community structure differences indicate full recovery had yet to occur 4 years after the spill.

22.4 Summary and Conclusions

Oil from the DWH blowout reached the seafloor through deep-sea plumes and sedimentation of oil and oiled marine snow. This oil caused extensive damage to both hard-bottom and soft-bottom communities. The most sensitive bioindicators to detect the anthropogenic damage were deep-sea planar octocorals for hard-bottoms. For soft-bottom, the most sensitive indicators were increases in polychaetes (especially those in the family Dorvilleidae), macrofauna and meiofauna diversity and taxa richness, and the nematode/copepod ratio.

Four years after the spill, there was no recovery of the lost biodiversity of the macrofauna and meiofauna. Ironically, abundances of macrofauna and meiofauna can increase due to disturbance because the tolerant taxa, such as some polychaetes and nematodes, can increase to either take advantage of the open space vacated by sensitive taxa or by direct organic enrichment. While this response fits well into the classic toxicity-enrichment paradox, it is evidence of disturbance and that wide areas of the deep sea were impacted by the spill.

Both hard-bottom and soft-bottom communities appear to be very vulnerable to deep-sea oil spills. Deep-sea corals grow slowly and thus have extremely slow recovery rates, so it is necessary to prioritize a precautionary approach to protect them. Results from the long-term coral monitoring study following the DWH oil spill contributed to the recent designation of several coral sites as Habitat Areas of Particular Concern (HAPC) in the GoM.

Future research should be focused toward recovery and restoration. For hard-bottoms this could take the form of restoration projects. For soft-bottoms, the restoration strategy could be "restoration in place" because fresh sediments, which fall to the seafloor continuously, can cap the contaminated sediments over time. This would also require monitoring to ensure this strategy is viable.

Acknowledgments   The synthesis of this research was made possible by grants from the Gulf of Mexico Research Initiative through the Center for the Integrated Modeling and Analysis of the Gulf Ecosystem (C-IMAGE).

References

Almeda R, Bona S, Foster CR, Buskey EJ (2014) Dispersant Corexit 9500A and chemically dispersed crude oil decreases the growth rates of meroplanktonic barnacle nauplii (*Amphibalanus improvisus*) and tornaria larvae (*Schizocardium* sp.). Mar Environ Res 99:212–217. https://doi.org/10.1016/j.marenvres.2014.06.007

Andrews AH, Cordes EE, Mahoney MM, Munk K, Coale KH, Cailliet GM, Heifetz J (2002) Age, growth and radiometric age validation of a deep-sea, habitat-forming gorgonian (*Primnoa resedaeformis*) from the Gulf of Alaska. Hydrobiologia 471:101–110. https://doi.org/10.1023/A:1016501320206

Baguley JG, Montagna PA, Hyde LJ, Kalke RD, Rowe GT (2006a) Metazoan meiofauna abundance in relation to environmental variables in the northern Gulf of Mexico deep sea. Deep-Sea Res I 53:1344–1362. https://doi.org/10.1016/j.dsr2.2008.07.010

Baguley JG, Montagna PA, Lee W, Hyde LJ, Rowe GT (2006b) Spatial and bathymetric trends in Harpacticoida (Copepoda) community structure in the Northern Gulf of Mexico deep-sea. J Exp Mar Biol Ecol 330:327–341. https://doi.org/10.1016/j.jembe.2005.12.037

Baguley JG, Montagna PA, Cooksey C, Hyland JL, Bang HW, Morrison C, Kamikawa A, Bennetts P, Saiyo G, Parsons E, Herdener M, Ricci M (2015) Community response of deep-sea soft-sediment metazoan meiofauna to the Deepwater Horizon blowout and oil spill. Mar Ecol Prog Ser 528:127–140. https://doi.org/10.3354/meps11290

Baillon S, Hamel J-F, Wareham VE, Mercier A (2012) Deep cold-water corals as nurseries for fish larvae. Front Ecol Environ 10:351–356. https://doi.org/10.1890/120022

Balthis WL, Hyland JL, Cooksey C, Montagna PA, Baguley JG, Ricker RW, Lewis C (2017) Sediment quality benchmarks for assessing oil-related impacts to the deep-sea benthos. Integr Environ Assess Manag 13:840–851. https://doi.org/10.1002/ieam.1898/epdf

Boland GS, Etnoyer PJ, Fisher CR, Hickerson EL (2017) State of deep-sea coral and sponge ecosystems of the Gulf of Mexico Region. In: Hourigan TF, Etnoyer PJ, Cairns SD (eds) The state of deep-sea coral and sponge ecosystems of the United States. NOAA Technical Memorandum NMFS-OHC-4, Silver Spring, p 58. Available online: http://deepseacoraldata.noaa.gov/library

Brooks JM, Fisher C, Roberts H, Cordes E, Baums I, Bernard B, Brooke S, Church R, Demopoulos A, Etnoyer P, German C, Goehring E, Kellogg C, McDonald I, Morrison C, Nizinski M, Ross S, Shank T, Warren D, Welsh S, Wolff G (2012) Exploration and research of northern Gulf of Mexico deepwater natural and artificial hard-bottom habitats with emphasis on coral communities: reefs, rigs, and wrecks—"Lophelia II" Interim report. U.S. Dept. of the Interior, Bureau of Ocean Energy Management, Gulf of Mexico OCS Region, New Orleans. OCS Study BOEM 2012–106. p 126. Available online at https://www.boem.gov/ESPIS/5/5458.pdf

Bryan TL, Metaxas A (2006) Distribution of deep-water corals along the North American continental margins: Relationships with environmental factors. Deep-Sea Res I Oceanogr Res Pap 53:1865–1879. https://doi.org/10.1016/j.dsr.2006.09.006

Buhl-Mortensen L (2004) Symbiosis in deep-water corals. Symbiosis 37:33–61

Buhl-Mortensen L, Mortensen PB (2005) Distribution and diversity of species associated with deep-sea gorgonian corals off Atlantic Canada. In: Freiwald A, Roberts JM (eds) Cold-Water Corals and Ecosystems. Springer-V ed, Berlin, pp 849–879

Camilli R, Reddy CM, Yoerger DR, Van Mooy BAS, Jakuba MV, Kinsey JC, McIntyre CP, Sylva SP, Maloney JV (2010) Tracking hydrocarbon plume transport and biodegradation at Deepwater Horizon. Science 330:201–204. https://doi.org/10.1126/science.1195223

Cordes EE (2009) Macro-ecology of Gulf of Mexico cold seeps. Annu Rev Mar Sci 1:143–168. https://doi.org/10.1146/annurev.marine.010908.163912

Cordes EE, Nybakken JW, VanDykhuizen G (2001) Reproduction and growth of *Anthomastus ritteri* (Octocorallia: Alcyonacea) from Monterey Bay, California, USA. Mar Biol 138:491–501. https://doi.org/10.1007/s002270000470

Cordes EE, McGinley MP, Podowski EL, Becker EL, Lessard-Pilon S, Viada ST, Fisher CR (2008) Coral communities of the deep Gulf of Mexico. Deep-Sea Res I Oceanogr Res Pap 55:777–787. https://doi.org/10.1016/j.dsr.2008.03.005

CSA International I (2007) Characterization of Northern Gulf of Mexico deepwater hard-bottom communities with emphasis on Lophelia Coral. U.S. Department of the Interior, Minerals Management Service, Gulf of Mexico OCS Region, New Orleans, p 169. Available online at https://invertebrates.si.edu/boem/reports/LOPH_final.pdf

Deepwater Horizon Natural Resource Damage Assessment Trustees (2016) Deepwater Horizon oil spill: final programmatic damage assessment and restoration plan and final programmatic environmental impact statement. Available from: http://www.gulfspillrestoration.noaa.gov/restoration-planning/gulf-plan

DeLeo DM, Ruiz-Ramos DV, Baums IB, Cordes EE (2015) Response of deep-water corals to oil and chemical dispersant exposure. Deep-Sea Res II Top Stud Oceanogr 129:137–147. https://doi.org/10.1016/j.dsr2.2015.02.028

Etnoyer PJ, Cairns SD (2016) Deep-sea coral taxa in the US Gulf of Mexico: depth and geographical distribution. The state of deep-sea coral and sponge ecosystems of the United States. NOAA Technical Memorandum X: NOAA. Available online at https://deepseacoraldata.noaa.gov/library/2015-state-of-dsc-report-folder/NOAA_DSC-Species-List_GulfofMexico_Etnoyer-Cairns_2017.pdf

Etnoyer P, Warrenchuk J (2007) A catshark nursery in a deep gorgonian field in the Mississippi Canyon, Gulf of Mexico. Bull Mar Sci 81:553–559. http://www.ingentaconnect.com/contentone/umrsmas/bullmar/2007/00000081/00000003/art00019#expand/collapse

Fern R, Withers K, Zimba P, Wood T, Shoech L (2015) Toxicity of three dispersants alone and in combination with crude oil on blue crab *Callinectes sapidus* megalopae. Southeast Nat 14(4):G82–G92. https://www.eaglehill.us/SENAonline/articles/SENA-14-4/80-Fern.shtml

Fisher C, Roberts H, Cordes E, Bernard B (2007) Cold seeps and associated communities of the Gulf of Mexico. Oceanography 20:118–129. https://doi.org/10.5670/oceanog.2007.12

Fisher CR, Demopoulos AWJ, Cordes EE, Baums IB, White HK et al (2014a) Coral communities as indicators of ecosystem-level impacts of the Deepwater Horizon Spill. Bioscience 64:796–807. https://doi.org/10.1093/biosci/biu129

Fisher CR, Hsing P-Y, Kaiser CL, Yoerger DR, Roberts HH, Bourque JR (2014b) Footprint of Deepwater Horizon blowout impact to deep-water coral communities. Proc Natl Acad Sci 111:11744–11749. https://doi.org/10.1073/pnas.1403492111

Gallaway BJ (ed) (1988) Northern Gulf of Mexico continental slope study, final report: Year 4. Volume I: executive summary. Final report submitted to the Minerals Management Service, New Orleans. Contract No. 14-12-0001-30212. OCS Study/MMS 88-0052. p 69. Available online at https://invertebrates.si.edu/boem/reports/NGOMCS_88_ExecSum.pdf

Gee MJ, Warwick RM, Schaanning M, Berge JA, Ambrose WG (1985) Effects of organic enrichment on meiofaunal abundance and community structure insublittoral soft sediments. J Exp Mar Biol Ecol 91:247–262. https://doi.org/10.1016/0022-0981(85)90179-0

Girard F, Fisher CR (2018) Long-term impact of the Deepwater Horizon oil spill on deep-sea corals detected after seven years of monitoring. Biol Conserv 225:117–127. https://doi.org/10.1016/j.biocon.2018.06.028

Girard F, Fu B, Fisher CR (2016) Mutualistic symbiosis with ophiuroids limited the impact of the Deepwater Horizon oil spill on deep-sea octocorals. Mar Ecol Prog Ser 549:89–98. https://doi.org/10.3354/meps11697

Girard F, Shea K, Fisher CR (2018) Projecting the recovery of a long-lived deep-sea coral species after the Deepwater Horizon oil spill using state-structured models. J Appl Ecol 55(4):1812–1822. https://doi.org/10.1111/1365-2664.13141

Girard F, Cruz R, Glickman O, Harpster T, Fisher CR (2019) In situ growth of deep-sea octocorals after the deepwater horizon oil spill. Elem Sci Anth 7(1):12. https://doi.org/10.1525/elementa.349

Glover AG, Smith CR (2003) The deep-sea floor ecosystem: current status and prospects of anthropogenic change by the year 2025. Environ Conserv 30(3):219–241. https://doi.org/10.1017/S0376892903000225

Gong Y, Zhoa X, O'Reilly SE, Qian T, Zhoa D (2014) Effects of oil dispersant and oil on sorption and desorption of phenanthrene with Gulf Coast marine sediments. Environ Pollut 185:240–249. https://doi.org/10.1016/j.envpol.2013.10.031

Grasshoff M (1977) Die Gorgonarien des östlichen Nordatlantik und des Mittelmeeres III. Die Familie Paramuriceidae (Cnidaria, Anthozoa). Meteor Forsch Ergeb 27:5–76

Grasshoff M (1981) Die Gorgonaria, Pennatularia und Antipatharia des Tiefwassers der Biskaya (Cnidaria, Anthozoa). Ergebnisse der franzosischen Expeditionen Biogas, Polygas, Géomanche, Incal, Noratlante und Fahrten der Thalassa 1. Allgemeiner Teil. Bull Mus Natl Hist Nat 4:732–766

Griffiths SK (2012) Oil release from Macondo Well MC252 following the Deepwater Horizon accident. Environ Sci Technol 46:5616–5622. https://doi.org/10.1021/es204569t

Hsing P-Y, Fu B, Larcom E, Berlet SP, Shank TM, Govindarajan AF, Lukasiewiccz AJ, Dixson PM, Fisher CR (2013) Evidence of lasting impact of the Deepwater Horizon oil spill on a deep Gulf of Mexico coral community. Elementa: Sci Anthropocene 1:000012. https://doi.org/10.12952/journal.elementa.000012

Jensen A, Frederiksen R (1992) The fauna associated with the bank-forming deepwater coral *Lophelia pertusa* (Scleractinaria) on the Faroe shelf. Sarsia 77:53–69. https://doi.org/10.108 0/00364827.1992.10413492

Joye SB, MacDonald IR, Leifer I, Asper V (2011) Magnitude and oxidation potential of hydrocarbon gases released from the BP oil well blowout. Nat Geosci Lett 4:160–164. https://doi.org/10.1038/NGEO1067

Krieger KJ, Wing BL (2002) Megafauna associations with deepwater corals (*Primnoa* spp.) in the Gulf of Alaska. Hydrobiologia 471:83–90. https://doi.org/10.1023/a:1016597119297

Kujawinski EB, Kido Soule MC, Valentine DL, Boysen AK, Longnecker K, Redmond MC (2013) Fate of dispersants associated with the Deepwater Horizon oil spill. Environ Sci Technol 45:1298–1306. https://doi.org/10.1021/es103838p

Larcom EA, McKean DL, Brooks JM, Fisher CR (2014) Growth rates, densities, and distribution of *Lophelia pertusa* on artificial structures in the Gulf of Mexico. Deep-Sea Res I Oceanogr Res Pap 85:101–109. https://doi.org/10.1016/j.dsr.2013.12.00

Lessard-Pilon SA, Podowski EL, Cordes EE, Fisher CR (2010) Megafauna community composition associated with *Lophelia pertusa* colonies in the Gulf of Mexico. Deep-Sea Res II Top Stud Oceanogr 57:1882–1890. https://doi.org/10.1016/j.dsr2.2010.05.013

MacDonald IR, Garcia-Pineda O, Beet A, Daneshgar Asl S, Feng L, Graettinger G, French-McCay D, Holmes J, Hu C, Huffer F, Leifer I, Muller-Karger F, Solow A, Silva M, Swayze G (2015) Natural and unnatural oil slicks in the Gulf of Mexico. J Geophys Res Oceans 120:8364–8380. https://doi.org/10.1002/2015JC011062

Montagna PA, Arismendez SS (2019) Crude oil pollution II. Effects of the deepwater horizon contamination on sediment toxicity in the Gulf of Mexico. In: D'Mello JPF (ed) A handbook of environmental toxicology: human disorders and ecotoxicology. CAB International, Oxfordshire. In Press

Montagna P, Baguley JG, Cooksey C, Hartwell I, Hyde LJ, Hyland JL, Kalke RD, Kracker LM, Reuscher M, Rhodes ACE (2013) Deep-sea benthic footprint of the Deepwater Horizon blowout. PLoS One 8(8):e70540. https://doi.org/10.1371/journal.pone.0070540

Montagna PA, Baguley JG, Hsiang C-Y, Reuscher MG (2016) Comparison of sampling methods for deep-sea infauna. Limnol Oceanogr Methods 15:166–183. https://doi.org/10.1002/lom3.10150

Montagna PA, Baguley JG, Cooksey C, Hyland JL (2017) Persistent impacts to the deep soft-bottom benthos one year after the Deepwater Horizon event. Integr Environ Assess Manag 13:342–351. https://doi.org/10.1002/ieam.1791

Moore DR, Bullis JHR (1960) A deep-water coral reef in the Gulf of Mexico. Bull Mar Sci 10(1):125–128. https://www.ingentaconnect.com/content/umrsmas/bullmar/1960/00000010/00000001/art00008

Nutting CC (1913) Descriptions of the Alcyonaria collected by the U. S. Fisheries steamer Albatross, mainly in Japanese waters, during 1906. Proc US Natl Mus 43:1–104. Available online at https://library.si.edu/digital-library/book/proceedingsofuni431913unit

Passow U (2016) Formation of rapidly-sinking, oil-associated marine snow. Deep-Sea Res II 129:232–240. https://doi.org/10.1016/j.dsr2.2014.10.001

Pequengnat WE, Gallaway BJ, Pequegnat LH (1990) Aspects of the ecology of the deep-water fauna of the Gulf of Mexico. Am Zool 30:45–64. https://doi.org/10.1093/icb/30.1.45

Peterson CH, Kennicutt MC II, Green RH, Montagna P, Harper DH Jr, Powell EN, Roscigno PF (1996) Ecological consequences of environmental perturbations associated with offshore hydrocarbon production: a perspective from study of long term exposures in the Gulf of Mexico. Can J Fish Aquat Sci 53:2637–2654. https://doi.org/10.1139/f96-220

Prouty NG, Fisher CR, Demopoulos AWJ, Druffel ERM (2016) Growth rates and ages of deep-sea corals impacted by the Deepwater Horizon oil spill. Deep-Sea Res II Top Stud Oceanogr 129:196–212. https://doi.org/10.1016/j.dsr2.2014.10.021

Quattrini AM, Etnoyer PJ, Doughty C, English L, Falco R, Remon N, Rittinghouse M, Cordes EE (2013) A phylogenetic approach to octocoral community structure in the deep Gulf of Mexico. Deep-Sea Res II Top Stud Oceanogr 99:92–102. https://doi.org/10.1016/j.dsr2.2013.05.027

Reucsher MG, Baguley JG, Conrad-Forrest N, Cooksey C, Hyland JL, Lewis C, Montagna PA, Ricker RW, Rohal M, Washburn T (2017) Temporal patterns of the Deepwater Horizon impacts on the benthic infauna of the northern Gulf of Mexico continental slope. PLoS One 12(6):e0179923. https://doi.org/10.1371/journal.pone.0179923

Roark EB, Guilderson TP, Dunbar RB, Ingram BL (2006) Radiocarbon-based ages and growth rates of Hawaiian deep-sea corals. Mar Ecol Prog Ser 327:1–14. https://doi.org/10.3354/meps327001

Roberts JM, Wheeler AJ, Freiwald A (2006) Reefs of the deep: the biology and geology of cold-water coral ecosystems. Science 312:543–547. https://doi.org/10.1126/science.1119861

Rowe GT, Kennicutt MC II (eds) (2009) Northern Gulf of Mexico continental slope habitats and benthic ecology study: final report. U.S. Dept. of the Interior, Minerals Management. Service, Gulf of Mexico OCS Region, New Orleans. OCS Study MMS 2009-039. pp 456. Available online at https://invertebrates.si.edu/boem/reports/DGOMB_Final.pdf

Rowe GT, Kennicutt MC (2008) Introduction to the deep Gulf of Mexico Benthos program. Deep Sea Res II 55:2536–2540. https://doi.org/10.1016/j.dsr2.2008.09.002

Ryerson TB, Camilli R, Kessler JD, Kujawinski EB, Reddy CM, Valentine DL, Atlas E, Blake DR, de Gouw J, Meinardi S, Parrish DD, Peischl J, Seewald JS, Warneke C (2012) Chemical data quantify Deepwater Horizon hydrocarbon flow rate and environmental distribution. Proc Natl Acad Sci 109:20246–20253. www.pnas.org/cgi/doi/10.1073/pnas.1110564109

Shiells GM, Anderson KJ (1985) Pollution monitoring using the nematode/copepod ratio. A practical application. Mar Pollut Bull 16:62–68. https://doi.org/10.1016/0025-326X(85)90125-0

Stone RP (2006) Coral habitat in the Aleutian Islands of Alaska: depth distribution, fine-scale species associations, and fisheries interactions. Coral Reefs 25:229–238. https://doi.org/10.1007/s00338-006-0091-z

Valentine DL, Fisher GB, Bagby SC, Nelson RK, Reddy CM, Sylva SP, Woo MA (2014) Fallout plume of submerged oil from Deepwater Horizon. Proc Natl Acad Sci 111:15906–15911. www.pnas.org/cgi/doi/10.1073/pnas.1414873111

Verrill AE (1894) Descriptions of new species of starfishes and ophiurans, with a revision of certain species formerly described; mostly from the collections made by the United States Commission of Fish and Fisheries. Proc US Natl Mus 1000:245–297. https://doi.org/10.5479/si.00963801.1000.245

Washburn T, Rhodes ACE, Montagna PA (2016) Benthic taxa as potential indicators of a deep-sea oil spill. Ecol Indic 71:587–597. https://doi.org/10.1016/j.ecolind.2016.07.045

Washburn T, Reuscher MG, Montagna PA, Cooksey C, Hyland JL (2017) Macrobenthic community structure in the deep Gulf of Mexico one year after the Deepwater Horizon blowout. Deep-Sea Res I 127:21–30. https://doi.org/10.1016/j.dsr.2017.06.001

Watling L, France SC, Pante E, Simpson A (2011) Biology of deep-water octocorals. Adv Mar Biol 60:42–122. https://doi.org/10.1016/B978-0-12-385529-9.00002-0

White HK, Hsing P-Y, Cho W, Shank TM, Cordes EE, Quattrinni AM, Nelson RK, Camilli R, Demopoulos AWJ, German CR, Brooks JM, Robert HH, Shedd W, Reddy AM, Fisher CR (2012a) Reply to Boehm and Carragher: multiple lines of evidence link deep-water coral damage to Deepwater Horizon oil spill. Proc Natl Acad Sci 109:E2648. https://doi.org/10.1073/pnas.1210413109

White HK, Hsing PY, Cho W, Shank TM, Cordes EE, Quattrinni AM, Nelson RK, Camilli R, Demopoulos AWJ, German CR, Brooks JM, Robert HH, Shedd W, Reddy AM, Fisher CR (2012b) Impact of the Deepwater Horizon oil spill on a deep-water coral community in the Gulf of Mexico. Proc Natl Acad Sci 109:20303–20308. https://doi.org/10.1073/pnas.1118029109

White HK, Lyons SL, Harrison SJ, Findley DM, Liu Y, Kujawinski EB (2014) Long-term persistence of dispersants following the Deepwater Horizon oil spill. Environ Sci Technol Lett 1:295–299. https://doi.org/10.1021/ez500168r

Wilson GDF (2008) Local and regional species diversity of benthic Isopoda (Crustacea) in the deep Gulf of Mexico. Deep-Sea Res II 55:2634–2649. https://doi.org/10.1016/j.dsr2.2008.07.014

Chapter 23
Impact and Resilience of Benthic Foraminifera in the Aftermath of the Deepwater Horizon and Ixtoc 1 Oil Spills

Patrick T. Schwing and Maria Luisa Machain-Castillo

Abstract Benthic foraminifera, which are single-celled protists that primarily produce calcite shells, have been commonly used as bioindicators of anthropogenic and natural perturbations. Numerous surveys of benthic foraminifera conducted in the Gulf of Mexico (GoM) prior to any major oils spills allow for a fair assessment of impact, response, and resilience during and following the *Deepwater Horizon* (DHW) oil spill in the northern GoM (2010) and Ixtoc 1 oil spill in the southern (1979–1980). Initially, in the aftermath of DWH, there was an 80–93% decrease in benthic foraminifera density and a 30–40% decrease in species richness and heterogeneity in the northern GoM. From 2010 to 2012, there was a continuous depletion in benthic foraminifera calcite stable carbon isotopes related to increased deposition of petroleum carbon (PC). This depletion has subsequently been preserved in the sedimentary record. Following this period of impact, benthic foraminifera density and diversity reached a resilient state of equilibrium from 2013 to 2015, suggesting that the rate of resilience for the benthic habitat is on the order of 3 years following an event like the DWH. Secondly, the sedimentary records of benthic foraminifera were used to assess the impact, resilience, and subsequent preservation of the Ixtoc oil spill. A noticeable decrease in benthic foraminifera density as well as a depletion in the stable carbon isotopes of benthic foraminifera calcite occurred in the sedimentary interval corresponding to 1979–1980. These results have implications for determining the long-term preservation of oil spills, assessing PC mineralization and burial, and contributing to overall oil spill budgets. Overall, benthic foraminifera have proven to be valuable indicators of impact, response, and resilience of the benthos and can provide useful information concerning benthic habitat suitability following oil spills in the future.

P. T. Schwing (✉)
University of South Florida, College of Marine Science, St. Petersburg, FL, USA
e-mail: pschwing@mail.usf.edu

M. L. Machain-Castillo
Universidad Nacional Autónoma de México, Instituto de Ciencias del Mar y Limnología, Mexico City, Mexico
e-mail: machain@cmarl.unam.mx

© Springer Nature Switzerland AG 2020
S. A. Murawski et al. (eds.), *Deep Oil Spills*,
https://doi.org/10.1007/978-3-030-11605-7_23

Keywords Benthic foraminifera · Oil spill · Deepwater Horizon · Ixtoc 1 · Benthic impact

23.1 Background: Benthic Foraminifera Distribution in the Gulf of Mexico

Benthic foraminifera are single-celled protists, which are abundant throughout the global oceans from coastal to abyssal depths (Sen Gupta 1999). Benthic foraminifera have been utilized widely as bioindicators of anthropogenic environmental stressors such as coastal pollution (Nigam et al. 2006) and estuarine and open-marine petroleum contamination (Morvan et al. 2004; Mojtahid et al. 2006; Denoyelle et al. 2010; Brunner et al. 2013; Lei et al. 2015).

Many surveys and reviews of benthic foraminifera have been conducted throughout the Gulf of Mexico (GoM) (Phleger and Parker 1951; Parker 1954; Culver and Buzas 1983; Poag 1984; Denne and Sen Gupta 1991; Sen Gupta and Aharon 1994; Buzas et al. 2007; Lobegeier and Sen Gupta 2008; and Bernhard et al. 2008; Poag 2015). Phleger and Parker (1951) identified six benthic foraminifera depth facies in the northwestern GoM based on the overlap of species, with depth boundaries of approximately 100 m, 200 m, 500 m, 1000 m, and 2000 m water depth. They attributed these facies changes primarily to changes in the physical properties of the overlying water such as temperature (Phleger and Parker 1951). Throughout the entire GoM, Poag (2015) identified 45 facies based on the predominance of genera and 5 additional geographically isolated biofacies distinguished by characteristic genera. Species distribution patterns for benthic foraminifera in the southern GoM reveal their association with depth, water masses, substrate, and other environmental patterns (Diego-Casimiro 1989; Mata-Mendoza 1982, 1987; Machain-Castillo et al. 2010). Previous studies have also documented the benthic foraminifera assemblages associated with natural petroleum and methane seeps in the northern GoM (Sen Gupta and Aharon 1994; Lobegeier and Sen Gupta 2008). Important calcareous taxa near natural hydrocarbon seeps between 500 and 1000 m water depth included *Bolivina ordinaria* and *Gavelinopsis translucens*, *Eponides turgidus*, *Uvigerina peregrina*, *Nonionella iridea*, *Bolivina ordinaria*, *Bolivina lowmani*, *Bolivina albatrossi*, *Bulimina aculeata*, and *Epistominella exigua* (Sen Gupta and Aharon 1994; Lobegeier and Sen Gupta 2008). These resources provide context in which to assess variability and dominant taxa of benthic foraminifera following oil spills.

23.2 The *Deepwater Horizon* Oil Spill

23.2.1 *Introduction to the* Deepwater Horizon

The *Deepwater Horizon* (DWH) event released 3.19 million barrels of oil into the northern GoM from April 20 to July 12, 2010 (US District Court 2015). Following the DWH event, a phenomenon known as marine oil snow sedimentation and flocculent accumulation (MOSSFA) produced an order of magnitude increase in oiled-flocculent flux and sedimentation throughout a large portion of the northern GoM (Passow et al. 2012; Ziervogel et al. 2012; Brooks et al. 2015; Romero et al. 2015; Daly et al. 2016; Schwing et al. 2017a; Romero et al. 2017; Quigg et al. 2020). Increased oiled-flocculent flux following DWH produced a threefold increase in polycyclic aromatic hydrocarbon (PAH) concentrations (Romero et al. 2015) and enhanced reducing conditions (Hastings et al. 2016). Both the increase in PAH concentrations and in redox conditions impacted the surface 4–12 mm of sediment (Brooks et al. 2015).

23.2.2 *Impact*

Benthic foraminifera density declined by as much as 80–93% at sites up to 120 km northeast of the DWH wellhead in 2010 and early 2011 (Fig. 23.1; Schwing et al. 2015). Benthic foraminifera diversity (Shannon and Fisher's Alpha indices) also decreased 30–40% at these sites in 2010 and early 2011 (Schwing et al. 2017b). The decline in density (Schwing et al. 2015) and diversity (Schwing et al. 2017b) was likely caused by inhibited reproduction or mortality and associated with the

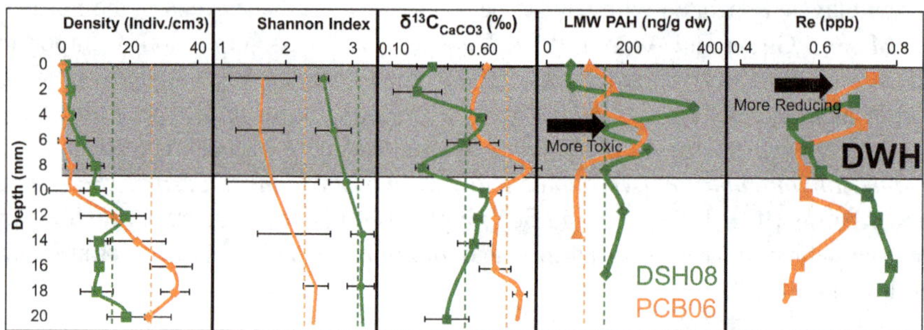

Fig. 23.1 Stacked depth plots of benthic foraminiferal density (individuals/cm³), Shannon index, and stable carbon isotopes ($\delta^{13}C_{CaCO3}$ ‰) along with environmental parameters including low molecular weight polycyclic aromatic hydrocarbons (LMW PAH, ng/g dry weight) and rhenium (Re) concentration. These plots demonstrate the concurrent decline in density, diversity, and depletion in stable carbon isotopes with increased petroleum toxicity and intensification of reducing conditions

intensification of reducing conditions and 2–3-fold increase in sedimentary PAH concentrations resulting from the MOSSFA event. Benthic foraminifera in the pre-DWH intervals (>10 mm) resembled assemblages from previous studies relating specific benthic foraminifera assemblages and to Caribbean Midwater mass (Denne and Sen Gupta 1991) and characterizing dominant species in continental slope sediments (Culver and Buzas 1983; Osterman 2003).

A modest (0.2–0.4‰) but persistent $\delta^{13}C_{CaCO3}$ depletion in *Cibicidoides* spp. (*Cibicidoides wuellerstorfi* and *Cibicidoides pachyderma*) extracted from DWH intervals of impacted sites was observed in 2010 and early 2011 (Fig. 23.1; Schwing et al. 2018a). The depletion was significantly beyond any natural (pre-DWH, background) variability and demonstrated that benthic foraminiferal calcite recorded the depositional event. The stable carbon composition of benthic foraminiferal calcite ($\delta^{13}C_{CaCO3}$) is affected by environmental changes to bottom water $\delta^{13}C$ of dissolved inorganic carbon (DIC), the carbon isotopic composition of food sources (e.g., microbial biomass, phytodetritus), and the flux of total organic carbon (TOC) (Rathburn et al. 2003; Torres 2003; Hill et al. 2004; Nomaki et al. 2006; Panieri 2006; Panieri and Sen Gupta 2008; Zariess and Mackensen 2011; Panieri et al. 2014; Theodor et al. 2016). Zariess and Mackensen (2011) found that seasonal increases in phytodetritus deposition caused depletion in *Cibicidoides wuellerstorfi* $\delta^{13}C_{CaCO3}$ from 0.1‰ to 0.4‰. The relationship between $\delta^{13}C_{CaCO3}$ and TOC flux is based on increased microbial respiration rates, which in turn depletes the $\delta^{13}C$ of dissolved inorganic carbon of bottom water. This relationship suggests that increased TOC flux could result in $\delta^{13}C_{CaCO3}$ depletion following the DWH event because of increased sedimentary mass accumulation rates (up to 4–10-fold; Brooks et al. 2015) and total organic carbon (TOC) flux (Romero et al. 2015; Schwing et al. 2015) caused by MOSSFA.

23.2.3 Response

Initial signs of benthic foraminiferal response were observed at two sites in the northern GoM in as short as 2 years after the DWH event, but the authors concluded that further work was needed to determine the full spatial extent and time frame of recovery (Schwing et al. 2017b). The response varied at each site. There was a protracted return to pre-DWH values with respect to density (at least 2 years) and a rapid (~1 year) return to pre-DWH diversity values at the site PCB06. However, there was a relatively fast return to pre-DWH values for density (~1 year) and a prolonged return to pre-DWH diversity values (at least 2 years) at site DSH08. Opportunistic taxa, which are able to tolerate high organic carbon deposition and petroleum concentration, were likely dominant in the areas with a rapid return to pre-DWH density values and a protracted return to pre-DWH diversity values. The benthic foraminifera response to the increased petroleum concentration and enhanced reducing conditions suggested that the benthic resilience, following an event such as the DWH, took on the order of 1–2 years to initiate (Schwing et al.

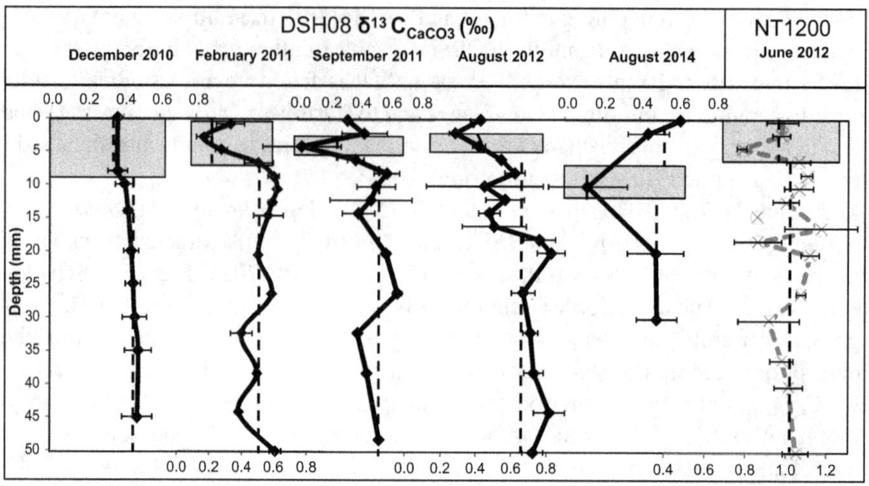

Fig. 23.2 Stacked depth plots of benthic foraminiferal stable carbon isotopes ($\delta^{13}C_{CaCO3}$ ‰). These plots demonstrate the concurrent depletion in stable carbon isotopes with the DWH event, continued surface depletion from 2010 to 2012, and burial of the DWH event as of 2014. The gray-shaded region represents the sediment increments dated as 2010–2011 (DWH) using short-lived radioisotope methods. The dashed lines signify the pre-DWH (down-core) and post-DWH mean values. (Reprinted from Schwing et al. (2018a), Copyright (2018), with permission from Elsevier)

2017b). Some of the metrics (e.g., density at PCB06) continued to be significantly below the densities found in previous studies conducted in the northern GoM before the DWH event until 2015.

The depletion (0.2–0.4‰) in $\delta^{13}C_{CaCO3}$ in *Cibicidoides* spp. (*Cibicidoides wuellerstorfi* and *Cibicidoides pachyderma*) persisted from 2010 to 2012 (Fig. 23.2; Schwing et al. 2018a). The longevity of the depletion in the $\delta^{13}C$ record suggested that not only did the benthic foraminifera record the change in organic matter caused by MOSSFA from 2010 to 2012 but that they were incorporating a depleted carbon source for up to 2 years at the sediment-water interface (Schwing et al. 2018a).

23.2.4 Resilience

Benthic foraminifera species richness and heterogeneity increased throughout the northern GoM from 2011 to 2015, which was consistent with resilience following the DWH event in 2010 (Fig. 23.3; Schwing et al. 2018b). Benthic foraminifera's highest coefficient of change for both species richness and heterogeneity occurred from 2011 to 2013. After 2013, both species richness and heterogeneity became more consistent, which implied that the northern GoM benthic resilience rate following DWH was approximately 3 years (Schwing et al. 2018b). As of 2014, $\delta^{13}C_{CaCO3}$ values were enriched in the surface intervals (0–4 mm depth) relative to the

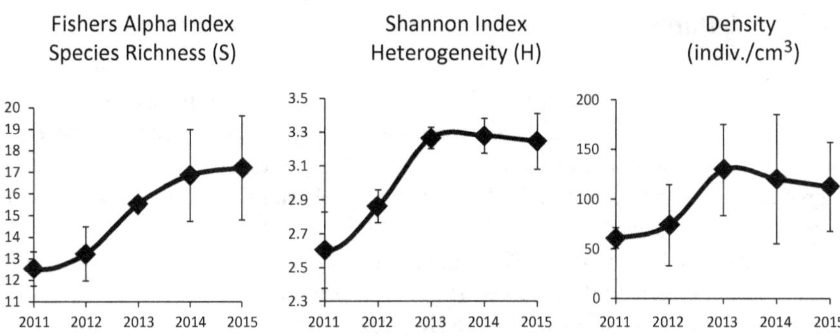

Fig. 23.3 Mean Fisher's alpha index, Shannon index, and density (individuals/cm³) for benthic foraminifera from seven sites in the northern Gulf of Mexico from 2011 to 2015 consistent with resilience rates of 3–5 years to achieve a steady state following the DWH event

DWH (2010) sediment interval (~10 mm depth). This suggested that the DWH interval, deposited in 2010, was buried by subsequent sedimentation of a sedimentary organic carbon source similar to the source prior to DWH (Schwing et al. 2018a).

23.3 The Ixtoc 1 Oil Spill

23.3.1 Event Background

The Ixtoc 1 oil spill occurred over a period of 290 days in 1979–1980 at 51 meters water depth (Jernelov and Linden 1981). The wellhead was located 80 km NW of Ciudad del Carmen, Mexico, and released a total of 475 million liters of oil into the southern GoM (Jernelov and Linden 1981). Approximately 50% of the spilled oil evaporated (NOAA Oil Budget Calculator 2011), and 120,000 metric tons of oil were deposited on the seafloor in the months following the event (Jernelov and Linden 1981).

MOSSFA has been well documented following DWH in the northern GoM in 2010 (Thibodeaux et al. 2011; Passow et al. 2012; Ziervogel et al. 2012; Paris and Hénaff 2012; Passow 2014; Brooks et al. 2015; Daly et al. 2016; Romero et al. 2017; Schwing et al. 2017a, Schwing et al. 2018b). Further, MOSSFA formation is predicated on the presence of petroleum, dispersants, bacteria, phytoplankton, and clay minerals (Daly et al. 2016). The area surrounding the Ixtoc 1 wellhead satisfies each of these criteria. Vonk et al. (2015) suggest that MOSSFA was likely to have occurred during Ixtoc 1 due to the particulate matter, algae, dispersants, and bacterial communities present. Gracia et al. (2013) documented the introduction of a significant amount of suspended solids into the GoM from surrounding rivers, especially during the summer, a high-precipitation season. The duration of Ixtoc 1 coincided with the high-precipitation season. The setting for Ixtoc 1 supports the occurrence of MOSSFA.

23.3.2 Impacts

Impacts on density and diversity of benthic foraminifera were measured on a transect, west from the Ixtoc 1 wellhead. The reduction in density and diversity observed in the southern GoM records during and following Ixtoc 1 was not to the same degree as the impacts observed in the northern GoM after DWH, which is likely due to preservation effects.

A decrease in density (42–63%) in the Ixtoc 1 layer (IL, dated using excess ^{210}Pb) relative to the pre-Ixtoc down-core mean was observed at all sites (Fig. 23.4). The largest decrease in density was in core IXW250. Density values were below pre-IL mean at, and after, the IL in core IXW250. Density decreased in core IXW500 at the IL, followed by a continuous increase for the next three increments, where it reached down-core mean values. However, density did not reach pre-IL values. After the decrease in density at the onset of the IL in core IXW750, there was an immediate increase to above pre-IL and down-core mean values (probably due to tolerant or opportunistic species that preyed on the available carbon). Therefore, the degree of impact in the area seems to be correlated with distance from the wellhead.

Benthic foraminiferal diversity exhibited a different behavior in the three sites. In the shallowest site (IXW250), there was a slight increase in species richness (from 14 to 17) and heterogeneity (from 3.0 to 3.5) at the onset of the event. These increases were possibly caused by an increase in organic carbon that initially provided more food for tolerant or opportunistic species. After this event, species richness slowly increased to down-core mean values. All of these variations were within the variability of the parameter in the core. There was a similar trend in diversity (S and H) in core IXW500 to that of density and a similar behavior to that observed in the northern GoM. Both indices decreased at the onset of the IL (H value was within the down-core variability of the parameter, but S was the lowest of all the studied sections) and increased immediately following the IL. Species richness was not affected at the IL in core IXW750 but increased abruptly following. In general, density and diversity values did not vary considerably throughout the cores. Machain-Castillo et al. (unpublished) suggest that the southern GoM has been under natural and/or anthropogenic oil seepage influence at least during the last 100 years and that multiple episodes of oil contamination were probable in the area.

There were three types of δ^{13}C records present in the southern GoM with respect to preservation of the Ixtoc 1 spill: (1) no depletion in $\delta^{13}C_{CaCO3}$ present in IL, (2) a depletion in $\delta^{13}C_{CaCO3}$ present in IL but also within natural variability (similar depletion present in pre-IL and post-IL), and (3) a depletion in $\delta^{13}C_{CaCO3}$ present in the IL but beyond natural variability (Fig. 23.5; Schwing et al. unpublished). The third category, which is most consistent with long-term preservation of Ixtoc 1, included cores IXW750 and IXN750.

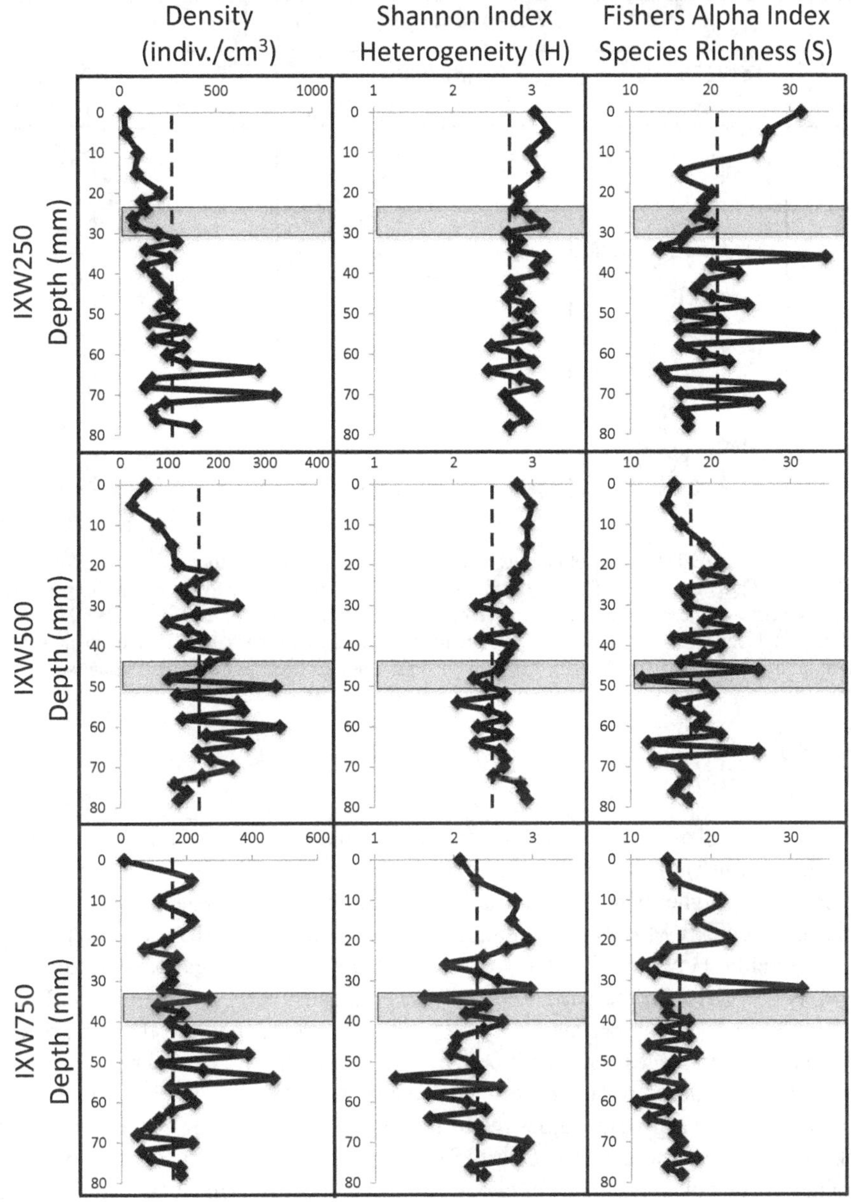

Fig. 23.4 Stacked depth plots of benthic foraminiferal density, Shannon index, and Fisher's alpha index for three sites in the southern Gulf of Mexico (IXW250, IXW500, and IXW750). The gray-shaded regions represent the Ixtoc I layer (IL), and the segmented lines represent the core mean

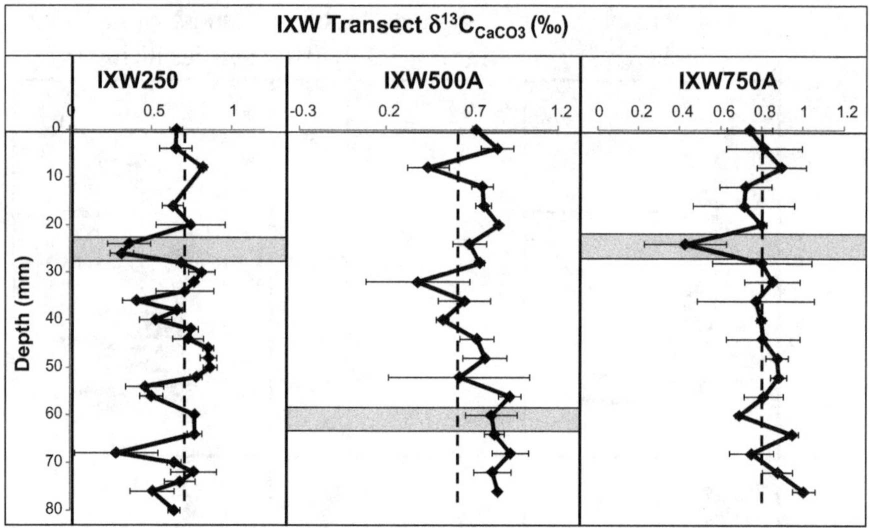

Fig. 23.5 Stacked depth plots of benthic foraminiferal stable carbon isotopes ($\delta^{13}C_{CaCO3}$ ‰). These plots depict the three types of $\delta^{13}C$ records present in the southern GoM with respect to preservation of Ixtoc 1. The gray-shaded region represents the sediment increments dated as 1979–1980 (Ixtoc 1) using short-lived radioisotope methods. The dashed lines signify the mean values throughout the entire core. These plots have been modified from Schwing et al. (unpublished)

23.3.3 Resilience

There was evidence of recovery in benthic foraminiferal density and diversity after the IL at different times in the three studied cores. In core IXW250, density values were clearly higher before the IL than after it, indicating there has not been a recovery in the density of benthic foraminifera at this site. Species richness did not vary considerably (between 14 and 33) and reached higher values (near pre-IL), approximately 15 years after the Ixtoc 1 oil spill. Heterogeneity decreased within the IL and remained near down-core mean values for approximately the next 10 years, reaching above mean values around 1990. In core IXW500, density did not recover (mean post-IL values are lower than pre-IL), as in core IXW250, but species richness and particularly heterogeneity values recovered to pre-IL values within 5 years after the IL similar to northern GoM records following DWH. However, species richness values were not constant throughout the core, and the trend was erratic. Core IXW750 was the only record where density recovered to higher than pre-IL values, which occurs almost 20 years after the IL. Species richness and heterogeneity increased immediately following the IL and then decreased below the core mean. The recovery time (higher than pre-IL values) for both records was on the order of 15–17 years. Therefore the recovery of density from Ixtoc 1 revealed that the effects were proportional to the distance to the wellhead. Diversity trends were not as clear. A larger suite of samples could reveal clearer trends.

Following the depletions in $\delta^{13}C_{CaCO3}$ assessed in the IL in the IXN750 and IXW750 records from the southern GoM, it took approximately 4–5 years for benthic foraminiferal $\delta^{13}C_{CaCO3}$ to return to the mean value for the entire record, signifying a return to a carbon source/flux similar to that prior to Ixtoc 1. This is consistent with the DWH records in the northern GoM which took 4 years (Schwing et al. 2018a).

23.4   Implications

23.4.1   Establishment of Baselines in Advance of Next Spill

Indices of benthic foraminifera diversity are robust proxies of benthic impacts, response, and ecosystem resilience and useful for establishing benthic resilience rates following oil spills. Metrics of benthic ecosystem resilience are essential for establishing current baselines (new normal) and providing context for quantitative characterization of the severity of benthic impact and recovery following submarine oil spills in the future.

23.4.2   Importance of Long-Term Time-Series Studies

These studies demonstrate the necessity of long-term time-series studies in the aftermath of events like DWH and Ixtoc 1. Longer-term (decadal) time-series collections and measurements will not only provide ongoing baselines but also provide context for seasonal, annual, and decadal changes to benthic communities. The variability and extent of natural (seasonal to decadal timescale) changes are essential to extricate a quantitative assessment of impact and recovery following any human perturbations such as oil spills.

23.4.3   Biomineralization and Preservation of Oil Events (MOSSFA)

Benthic foraminifera and the stable carbon isotopes of their tests provide a distinctive tool to assess the spatial extent of MOSSFA on the seafloor following oil spill events. Benthic foraminifera test carbon isotope composition is also likely a persistent recorder of marine oil snow sedimentation events (e.g., Ixtoc preservation). On a longer timescale (hundreds to thousands of years), the alteration of stable carbon isotope composition of benthic foraminifera during oil spills may have implications for long-term burial and biomineralization of petroleum carbon considering the recalcitrance of benthic foraminifera in the sediment. Further work is needed to fully characterize the extent of stable carbon isotope alteration during oil spills and the relative influence of petroleum carbon incorporation versus depleted bottom water dissolved inorganic carbon due to increased microbial respiration of MOSSFA material.

23.4.4 Connectivity to Other Trophic Levels (Microbial, Macrofaunal) and Benthic Habitat Suitability

Considering the utility of benthic foraminifera assemblage metrics (density, diversity, opportunistic species dominance, stable isotopes) as indicators of impact and resilience of the benthos following oil spills, these metrics could also be applied as benthic habitat suitability measures in support of demersal and benthic-dependent fisheries. To strengthen the understanding of benthic habitat suitability using benthic foraminifera assemblage metrics, future work is also necessary to document the connectivity of benthic foraminifera with other trophic levels (microbial, macrofaunal, demersal fishes). Development of a benthic foraminifera-based marine biotic index (Jorissen et al. 2018; Alve et al. 2016) for the Gulf of Mexico would be a valuable management tool for benthic habitat suitability assessment.

Acknowledgments This research was made possible in part by a grant from the Gulf of Mexico Research Initiative, C-IMAGE, DEEP-C, and in part by the British Petroleum/Florida Institute of Oceanography (BP/FIO)-Gulf Oil Spill Prevention, Response, and Recovery Grants Program. The authors also thank Bryan O'Malley, Xinantecatl A. Nava-Fernández, Alejandro Rodriguez-Ramírez, Laura E. Gómez-Lizárraga, Laura Almaraz-Ruiz, and Marysol Escorza Reyes for their assistance with laboratory analyses. Data are publicly available through the Gulf of Mexico Research Initiative Information and Data Cooperative (GRIIDC) at http://data.gulfresearchinitiative.org, https://doi.org/10.7266/N79021PB, https://doi.org/10.7266/N7CR5RDS, https://doi.org/10.7266/N70P0WZM, https://doi.org/10.7266/N7S180HN, https://doi.org/10.7266/n7-e90r-1v29, https://doi.org/10.7266/n7-repn-q515, https://doi.org/10.7266/n7-xh2e-et70.

References

Alve E, Korsun S, Schönfeld J, Dijkstra N, Golikova E, Hess S, Husum K, Panieri G (2016) Foram-AMBI: a sensitivity index based on benthic foraminiferal faunas from North-East Atlantic and Arctic fjords, continental shelves and slopes. Mar Micropaleontol 122:1–12. https://doi.org/10.1016/j.marmicro.2015.11.001

Bernhard JM, Sen Gupta BK, Baguley JG (2008) Benthic foraminifera living in Gulf of Mexico bathyal and abyssal sediments: community analysis and comparison to metazoan meiofaunal biomass and density. Deep-Sea Res II 55:2617–2626

Brooks GR, Larson RA, Schwing PT, Romero I, Moore C, Reichart GJ, Jilbert T, Chanton JP, Hastings DW, Overholt WA, Marks KP, Kostka JE, Holmes CW, Hollander D (2015) Sediment pulse in the NE Gulf of Mexico following the 2010 DWH blowout. PLoS One 10(7):e0132341. https://doi.org/10.1371/journal.pone.0132341

Brunner CA, Yeager KM, Hatch R, Simpson S, Keim J, Briggs KB, Louchouarn P (2013) Effects of oil from the 2010 Macondo well blowout on Marsh foraminifera of Mississippi and Louisiana, USA. Environ Sci Technol 47:9115–9123. https://doi.org/10.1021/es401943y

Buzas MA, Hayek LC, Culver SJ (2007) Community structure of benthic foraminifera in the Gulf of Mexico. Mar Micropaleontol 65:43–53

Culver SJ, Buzas MA (1983) Recent benthic foraminiferal provinces in the Gulf of Mexico. J Foraminifer Res 13:21–31

Daly KL, Passow U, Chanton J, Hollander D (2016) Assessing the impacts of oil associated marine snow formation and sedimentation during and after the Deepwater Horizon oil spill. Anthropocene 12:18–33. https://doi.org/10.1016/j.ancene.2016.01.006

Denne RA, Sen Gupta BK (1991) Association of bathyal foraminifera with water masses in the northwestern Gulf of Mexico. Mar Micropaleontol 17:173–193

Denoyelle M, Jorissen FJ, Martin D, Galgani F, Mine J (2010) Comparison of benthic foraminifera and macrofaunal indicators of the impact of oil-based drill mud disposal. Mar Pollut Bull 60(11):2007–2021. https://doi.org/10.1016/j.marpolbul.2010.07.024

Diego-Casimiro G (1989) Estudio preliminar de los foraminiferos bentonicos del area circundante a la IslaPerez, arrecife Alacran, Yucatan, Mexico. Secretaria de Marina. Direccion Investigaciones Oceanograficas, Boletin, pp 19–27

Gracia A, Enciso Sánchez G, Alexander Valdés HM (2013) Composición y volúmen de contaminantes de las descargas costeras al Golfo de México. In: Botello AV, Rendón von Osten J, Benítez J, Gold-Boucht G (eds) Golfo de México. Contaminación e impacto ambiental: diagnóstico y tendencias. uac, unam-icmyl, cinvestav-Unidad Mérida

Hastings DW, Schwing PT, Brooks GR, Larson RA, Morford JL, Roeder T, Quinn KA, Bartlett T, Romero IC, Hollander DJ (2016) Changes in sediment redox conditions following the BP DWH Blowout event. Deep-Sea Res II Top Stud Oceanogr 129:167–178. https://doi.org/10.1016/j.dsr2.2014.12.009

Hill TM, Kennett JP, Valentine DL (2004) Isotopic evidence for the incorporation of methane-derived carbon into foraminifera from modern methane seeps, Hydrate Ridge, Northeast Pacific. Geochim Cosmochim Acta 68(22):4619–4627. https://doi.org/10.1016/j.gca.2004.07.012

Jernelöv A, Lindén O (1981) Ixtoc I: a case study of the world's largest oil spill. Ambio 10:299–306

Jorissen F, Nardelli MP, Almogi-Labin A, Barras C, Bergamin L, Bicchi E, El Kateb A, Ferraro L, McGann M, Morigi C, Romano E, Sabbatini A, Schweizer M, Spezzaferri S (2018) Developing Foram-AMBI for biomonitoring in the Mediterranean: species assignments to ecological categories. Mar Micropaleontol 140:33–45. https://doi.org/10.1016/j.marmicro.2017.12.006

Lei YL, Li TG, Bi H, Cui WL, Song WP, Li JY, Li CC (2015) Responses of benthic foraminifera to the 2011 oil spill in the Bohai Sea, PR China. Mar Pollut Bull 96(1–2):245–260. https://doi.org/10.1016/j.marpolbul.2015.05.020

Lobegeier MK, Sen Gupta BK (2008) Foraminifera of hydrocarbon seep, Gulf of Mexico. J Foraminifer Res 38(2):93–116

Machain-Castillo ML, Gío-Argáez FR, Cuesta-Castillo LB, Alcala-Herera JA, Sen Gupta BK (2010) Last Glacial Maximum Deep water masses in southwestern Gulf of Mexico: clues from benthic foraminífera. No. Esp. "Paleoclimas del Cuaternario en ambientes tropicales y subtropicales". Bol Soc Geol Mex 62(3):453–467. https://doi.org/10.18268/BSGM2010v62n3a9

Mata ML (1987) Benthic foraminiferal assemblages from Mexican continental shelves. M.S. Thesis, Louisiana State University, Baton Rouge, Louisiana

Mata-Mendoza ML (1982) Foraminiferos Recientes de la Sonda de Campeche, Mexico. (DGO-DM- 20-78-04) Secretaria de Marina. Direccion de Investigaciones Oceanograficas 1:1–53

Mojtahid M, Jorissen F, Durrieu J, Galgani F, Howa H, Redois F, Camps R (2006) Benthic foraminifera as bio-indicators of drill cutting disposal in tropical east Atlantic outer shelf environments. Mar Micropaleontol 61:58–75

Morvan J, Le Cadre V, Jorissen FJ, Debenay JP (2004) Foraminifera as potential bio-indicators of the "Erika" oil spill in the Bay of Bourgneuf: field and experimental studies. Aquat Living Resour 17:317–322

National Oceanic and Atmospheric Association (2011) Deepwater Horizon oil budget calculator (Technical Documentation). Retrieved from: http://www.noaanews.noaa.gov/stories2010/PDFs/Oil BudgetCalc _Full_HQ-Print_111110.pdf

Nigam R, Saraswat R, Panchang R (2006) Application of foraminifers in ecotoxicology: retrospect, perspect and prospect. Environ Int 32:273–283

Nomaki H, Heinz P, Nakatsuka T, Shimanaga M, Ohkouchi N, Ogawa NO, Kogure K, Ikemoto E, Kitazato H (2006) Different ingestion patterns of ^{13}C-labelled bacteria and algae by deep-sea benthic foraminifera. Mar Ecol Prog Ser 310(95):1–8

Osterman LE (2003) Benthic foraminifers from the continental shelf and slope of the Gulf of Mexico: an indicator of shelf hypoxia. Estuar Coast Shelf Sci 58:17–35

Panieri G (2006) The effect of shallow marine hydrothermal vent activity on benthic foraminifera (Aeolian arc, Tyrrhenian Sea). J Foraminifer Res 36:1. https://doi.org/10.2113/36.1.3

Panieri G, Sen Gupta BK (2008) Benthic foraminifera of the Blake Ridge hydrate mound, Western North Atlantic Ocean. Mar Micropaleontol 66:91–102. https://doi.org/10.1016/j.marmicro.2007.08.002

Panieri G, James RH, Camerlenghi A, Westbrook GK, COnsolaro C, Cacho I, Cesari V, Sanchez Cerveza C (2014) Record of methane emissions from the West Svalbard continental margin during the last 23,500 years revealed by δ13C of benthic foraminifera. Glob Planet Chang 122:151–160. https://doi.org/10.1016/j.gloplacha.2014.08.014

Paris C, Hénaff M (2012) Evolution of the Macondo well blowout: simulating the effects of the circulation and synthetic dispersants on the subsea oil transport. Environ Sci Technol 46:13293–13302

Parker FL (1954) Distribution of foraminifera in the Northeastern Gulf of Mexico. Bull Mus Comp Zool 111:453–588

Passow U (2014) Formation of rapidly-sinking oil-associated marine snow. Deep-Sea Res II Top Stud Oceanogr 129. https://doi.org/10.1016/j.dsr2.2014.10.001i

Passow U, Ziervogel K, Asper V, Diercks A (2012) Marine snow formation in the aftermath of the Deepwater Horizon oil spill in the Gulf of Mexico. Environ Res Lett 7:035301

Phleger FB, Parker FL (1951) Gulf of Mexico foraminifera, Part 1 and 2. Geological Society of America Memoirs 46

Poag WC (1984) Distribution and ecology of deep-water benthic foraminifera in the Gulf of Mexico. Palaeogeogr Palaeoclimatol Palaeoecol 48:25–37

Poag WC (2015) Benthic foraminifera of the Gulf of Mexico: distribution, ecology, paleoecology. Texas A&M University Press, College Station

Quigg A, Passow U, Hollander DJ, Daly KL, Burd A, Lee K (2020) Marine oil snow sedimentation and flocculent accumulation (MOSSFA) events: learning from the past to predict the future (Chap. 12). In: Murawski SA, Ainsworth C, Gilbert S, Hollander D, Paris CB, Schlüter M, Wetzel D (eds) Deep oil spills – facts, fate and effects. Springer, Cham

Rathburn AE, Elena Pérez M, Martin JB, Day SA, Mahn C, Gieskes J, Ziebis W, Williams D, Bahls A (2003) Relationships between the distribution and stable isotopic composition of living benthic foraminifera and cold methane seep biogeochemistry in Monterey Bay, California. Geochem Geophys Geosyst 4:12. https://doi.org/10.1029/2003GC000595

Romero IC, Schwing PT, Brooks GR, Larson RA, Hastings DW, Ellis G, Goddard EA, Hollander DJ (2015) Hydrocarbons in deep-sea sediments following the 2010 Deepwater Horizon Blowout in the Northeast Gulf of Mexico. PLoS One 10(5):e0128371. https://doi.org/10.1371/journal.pone.0128371

Romero IC, Toro-Farmer G, Diercks AR, Schwing PT, Muller-Karger F, Murawski S, Hollander DJ (2017) Large scale deposition of weathered oil in the Gulf of Mexico following a deepwater oil spill. Environ Pollut 228:179–189. https://doi.org/10.1016/j.envpol.2017.05.019

Schwing PT, Romero IC, Brooks GR, Hastings DW, Larson RA, Hollander DJ (2015) A decline in deep-sea benthic foraminifera following the Deepwater Horizon event in the Northeastern Gulf of Mexico. PLoS One 10(3):e0120565. https://doi.org/10.1371/journal.pone.0120565

Schwing PT, Brooks GR, Larson RA, Holmes CW, O'Malley BJ, Hollander DJ (2017a) Constraining the spatial extent of the Marine Oil Snow Sedimentation and Accumulation (MOSSFA) following the DWH event using a $^{210}Pb_{xs}$ inventory approach. Environ Sci Technol 51:5962–5968. https://doi.org/10.1021/acs.est.7b00450

Schwing PT, O'Malley BJ, Romero IC, Martinez-Colon M, Hastings DW, Glabach MA, Hladky EM, Greco A, Hollander DJ (2017b) Characterizing the variability of benthic foraminifera in the Northeastern Gulf of Mexico following the Deepwater Horizon event (2010–2012). Environ Sci Pollut Res 24:2754. https://doi.org/10.1007/s11356-016-7996-z

Schwing PT, Chanton JP, Romero IC, Hollander DJ, Goddard EA, Brooks GR, Larson RA (2018a) Tracing the incorporation of petroleum carbon into benthic foraminiferal calcite following the Deepwater Horizon event. Environ Pollut 237:424–429. https://doi.org/10.1016/j.envpol.2018.02.066

Schwing PT, O'Malley BJ, Hollander DJ (2018b) Resilience of benthic foraminifera in the Northern Gulf of Mexico following the Deepwater Horizon event (2011–2015). Ecol Indic 84:753–764. https://doi.org/10.1016/j.ecolind.2017.09.044

Sen Gupta BK (ed) (1999) Modern foraminifera. Kluwer Academic Publishers, Great Britain

Sen Gupta BK, Aharon P (1994) Benthic foraminifera of bathyal hydrocarbon vents of the Gulf of Mexico: initial report on communities and stable isotopes. Geo-Mar Lett 14:88–96

Theodor M, Schmiedl G, Jorissen F, Mackensen A (2016) Stable carbon isotopes gradients in benthic foraminifera as proxy for organic fluxes in the Mediterranean Sea. Biogeosciences 13:6385–6404. https://doi.org/10.5194/bg-13-6385-2016

Thibodeaux LJ, Valsaraj KT, John VT, Papadopoulos KD, Pratt LR, Pesika NS (2011) Marine oil fate: knowledge gaps, basic research, and development needs; a perspective based on the Deepwater Horizon spill. Environ Eng Sci 28:87–93

Torres ME (2003) Is methane venting at the seafloor recorded by δ 13 C of benthic foraminifera shells? Paleoceanography 18(3):1–13. https://doi.org/10.1029/2002PA000824

U.S. District Court (2015) Findings of facts and conclusions of law – phase 2 trial. Case 2: 10-md-02179-cjb-ss, p 1e44. Document 14021 filed Jan. 15, 2015. http://www.laed.uscourts.gov/sites/default/files/OilSpill/Orders/1152015FindingsPhaseTwo.pdf

Vonk SM, Hollander DJ, Murk ATJ (2015) Was the extreme and wide-spread marine oil-snow sedimentation and flocculent accumulation (MOSSFA) event during the Deepwater Horizon blowout unique? Mar Pollut Bull 100(1):5–12. https://doi.org/10.1016/j.marpolbul.2015.08.023

Zariess M, Mackensen A (2011) Testing the impact of seasonal phytodetritus deposition on d13C of epibenthic foraminifer Cibicidoides wuellerstorfi: A 31,000 year high-resolution record from the northwest African continental slope. Paleoceanography 26:2202. https://doi.org/10.1029/2010PA001944

Ziervogel K, Mckay L, Rhodes B, Osburn CL, Dickson-Brown J, Arnosti C, Teske A (2012) Microbial activities and dissolved organic matter dynamics in oil-contaminated surface seawater from the Deepwater Horizon oil spill site. PLoS One 7(4):e34816

Chapter 24
Chronic Sub-lethal Effects Observed in Wild-Caught Fishes Following Two Major Oil Spills in the Gulf of Mexico: *Deepwater Horizon* and Ixtoc 1

Erin L. Pulster, Adolfo Gracia, Susan M. Snyder, Kristina Deak, Susan Fogelson, and Steven A. Murawski

Abstract During and subsequent to major oil spill events, considerable attention focuses on charismatic and economic megafauna – and especially fishes – and visual manifestations of impacts upon them. Beginning with a series of tanker accidents occurring in Europe and the USA in the 1970s–1990s, greater awareness of the potential for both acute and chronic sub-lethal impacts on fish populations has focused on exposure to polycyclic aromatic hydrocarbons (PAHs). The ambiguity of acute impacts observed during the *Deepwater Horizon* and Ixtoc 1 incidents has promoted considerable new research on alternative toxic endpoints that portend short- and long-term sub-lethal outcomes that influence the overall fitness of exposed populations. Laboratory-based exposure studies have traditionally focused on acute mortality-based endpoints (e.g., lethal concentrations at which 50% of the population dies = LC_{50}) and observed at test concentrations normally exceeding environmentally relevant concentrations in real-world spills. Consequently, using laboratory-based toxicity experiments can be problematic inferring impacts on wild fish populations. In this chapter we review historical and more recent information documenting changes in abundance, recruitment, habitat use, population dynamics,

E. L. Pulster (✉) · S. M. Snyder · S. A. Murawski
University of South Florida, College of Marine Science, St. Petersburg, FL, USA
e-mail: epulster@mail.usf.edu; ssnyder4@mail.usf.edu; smurawski@usf.edu

A. Gracia
Universidad Nacional Autónoma de México, Instituto de Ciencias del Mar y Limnología, México City, Mexico
e-mail: gracia@unam.mx

K. Deak
University of South Florida, College of Marine Science, St. Petersburg, FL, USA

Florida Fish and Wildlife Research Institute, St. Petersburg, FL, USA
e-mail: Kristina.Deak@MyFWC.com

S. Fogelson
Fishhead Labs, Stuart, FL, USA
e-mail: SFogelson@fishheadlabs.com

© Springer Nature Switzerland AG 2020
S. A. Murawski et al. (eds.), *Deep Oil Spills*,
https://doi.org/10.1007/978-3-030-11605-7_24

trophic changes, and various physiologically based sub-lethal effects on oil-exposed fishes and especially consider research undertaken following the *Deepwater Horizon* and Ixtoc 1 spills in the Gulf of Mexico.

Keywords PAHs · Oil-exposed fish · Sub-lethal effects · Population fitness · LC_{50}

24.1 Introduction

The 2010 *Deepwater Horizon* and the 1979–1980 Ixtoc 1 oil spills were the two largest anthropogenically induced oil releases to occur in the Gulf of Mexico (GoM) and the second and third largest global oil spills, respectively, in contaminant volume released (the largest being the Persian Gulf War). Both the DWH and Ixtoc 1 incidents were subsurface blowouts creating surface slicks, water column oil emulsions, and benthic oil deposits, all of which resulted in marine life encounters with oil. Spawning for most marine fishes in the GoM occurs during spring and summer months (Chancellor 2015; Chancellor et al. 2020), thus overlapping with both spills in the GoM. Yet, unlike the 1989 *Exxon Valdez* oil spill and other nearshore tanker accidents, in which acute impacts on wildlife were highly visible, the offshore, subsurface nature of both the DWH and Ixtoc 1 on fish and fisheries was less obvious and may have taken years to fully express in terms of their outcomes.

Crude oil and its derived petroleum compounds are comprised of ~97% hydrocarbons with the remaining 3% made up of minor elements including nitrogen, sulfur, and oxygen (NRC 2003). Polycyclic aromatic hydrocarbons (PAHs) comprise a small compositional percentage (0.2–7%) of crude oil, yet they are considered the most toxic components primarily due to their metabolites (James and Kleinow 1994; NRC 2003). The toxic potential of PAHs in fish exposed during oil spills have been well documented and reviewed extensively in the literature (e.g., Reynaud and Deschaux 2006; Beyer et al. 2016). Following DWH, a number of studies focusing on sub-lethal impacts examined a range of adverse acute and chronic effects, including skin lesions, cardiotoxicity, liver abnormalities, respiration changes, reproductive irregularities, histopathological changes, and also mortality (Murawski et al. 2014; Incardona et al. 2014; Brown-Peterson et al. 2015a, b; Klinger et al. 2015; Bentivegna et al. 2015; Hedgpeth and Griffitt 2016; Esbaugh et al. 2016; Olson et al. 2016; Grosell et al. 2020; Portnoy et al. 2020).

Toxicity testing has, in the past, concentrated on laboratory exposures using relatively high-dose concentrations (to elicit effects) and with "model" organisms, not usually found in actual spills. Consequently, using laboratory-based toxicity experiments can be problematic when making inferences about potential impacts in wild fish at environmentally relevant pollutant concentrations. Nevertheless, consistent sublethal effects have been observed between laboratory studies and field studies. This chapter explores sub-lethal effects observed primarily from sampling of naturally occurring fish populations in the GoM following two major oil spills, the *Deepwater Horizon* and the Ixtoc 1. In the case of the Ixtoc 1 spill, few fishes (perhaps a few long-lived groupers) are still alive from that spill 40 years ago, but post-spill data collected in the past 5 years allow interpretation of chronic contamination either from current hydrocarbon inputs or from previous spills sequestered in GoM sediments.

24.2 Larval Abundance and Growth Indices

The GoM is inhabited by over 1500 fish species (Murawski et al. 2016) including those that are reef-dwelling, demersal, pelagic, highly migratory, and estuarine-dependent, exhibiting a wide diversity of life-history strategies. Most marine fishes spawn pelagic, buoyant eggs that are fertilized externally and float in the water column near the sea surface (Kendall Jr. et al. 1983). These critical pelagic egg and larval phases, including species of commercial and recreational importance in the GoM, coincided with both the DWH and Ixtoc 1 spills (Chancellor 2015; Chancellor et al. 2020). Murawski et al. (2016) outlined three conditions that have to be met for oil spills to have significant population-level effects on fishes. Firstly, adverse effects include increased mortality and reduced fitness of animals as the result of oil exposure. Secondly, the observed environmental concentrations are sufficient to cause increased mortality and/or sub-lethal impairment. Lastly, a substantial fraction of the population must be exposed to critical levels of oil in the environment resulting in adverse effects (mortality and sub-lethal impairment). For the DWH and the Ixtoc 1 spills, all three conditions seem to have been met. For example, using broad-scale SEAMAP ichthyoplankton sampling data, Chancellor (2015) computed the overlap between the surface expression of the DWH spill and total fish larvae in the northern GoM (e.g., US waters). She estimated that ~27% of the total fish larvae, representing 110 taxa in the northern GoM, were at locations where they were potentially to be exposed to DWH oil (Chancellor 2015). The prolonged and widespread surface expression of the Ixtoc 1 spill also likely overlapped with shallow- and deepwater species over a broad stretch of the southwestern GoM (Sun et al. 2015). Planktonic early life-history stages are highly sensitive to fluctuations in many environmental conditions, such as light, temperature, salinity, oxygen, and anthropogenic pollution. Thus, adverse effects including decreased survival and recruitment from a large pollution event, such as an oil spill, would be highly plausible and anticipated.

Larval abundance and condition were evaluated pre- and post-DWH using field-caught larvae to assess the potential impacts of oil exposures. Ichthyoplankton surveys conducted before, during, and after *DWH* along the coast of Alabama investigated potential larval impacts on red snapper (*Lutjanus campechanus*) and Spanish mackerel (*Scomberomorus maculatus*) by evaluating abundance, condition indices (morphometrics and dry weight), environmental parameters (e.g., wind speed, water temperature, etc.), and changes in stable isotopes (Hernandez Jr et al. 2016; Ransom et al. 2016). There were no significant differences in their larval abundances in either the Spanish mackerel or the red snapper across years post-DWH. However, examining a wider spatial domain in the north central GoM,

Murawski et al. (2018a) documented declines in the abundance of large juvenile and adult red snapper following the spill in seven time series conducted by academic institutions and government (NOAA/NMFS) agencies (Fig. 24.1b). For most species, however, longline sampling in the north central GoM did not document species declines (Fig. 24.1a).

For large pelagic species (blackfin tuna, *Thunnus atlanticus*; blue marlin, *Makaira nigricans*; dolphinfish, *Coryphaena hippurus*; and sailfish, *Istiophorus platypterus*), larval abundance and occurrences were lower during the oil spill (June 2010) than for previous years (Rooker et al. 2013). Spanish mackerel larvae had relatively high condition indices during DWH (2010), and similar to 2007 and 2011,

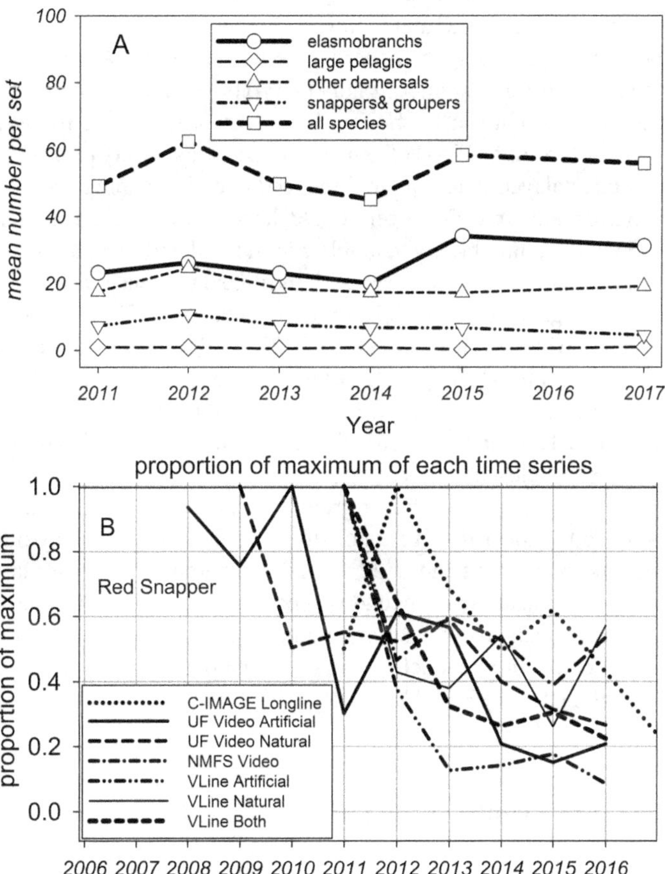

Fig. 24.1 (**a**) Catch per set of four species groups sampled in the north central Gulf of Mexico using demersal longline sampling, 2011–2017. (Data from Murawski et al. 2018a). (**b**) Catch per unit of effort of red snapper caught in seven fishery-independent sampling programs, 2008–2017. UF video = University of Florida video sampling on artificial and natural reefs, NMFS video = National Marine Fisheries Service video survey, VLine = vertical longline sampling on artificial and natural habitats and both combined. (From Murawski et al. 2018b)

whereas red snapper larvae were in poorer condition both during and after the spill (2010, 2011, and 2013) as compared to larvae collected in years prior (2007–2009; Ransom et al. 2016; Hernandez Jr et al. 2016). Studies are lacking on larval abundance of commercial species post Ixtoc 1; however, zooplankton biomass decreased four orders of magnitude post-spill, and changes in the structure and organization of zooplankton communities with a lower level of stability were detected (Guzmán del Próo et al. 1986).

Laboratory studies exposing both model species and non-model species at environmentally relevant water concentrations (ΣPAH 3–14 µg/L) demonstrate adverse cardiovascular responses of teleost embryos to petroleum-derived PAHs consistent with field observations (Incardona et al. 2014). Exposures <15 µg/L resulted in cardiotoxicity in large pelagic species (i.e., yellowfin tuna, *Thunnus albacares*; bluefin tuna, *Thunnus thynnus*; and yellowtail amberjack, *Seriola lalandi*), suggesting that the survival of embryos and/or larvae will be potentially reduced in GoM waters containing just a few micrograms of PAHs per liter. Moreover, Dubansky et al. (2013) exposed Gulf killifish (*Fundulus grandis*) embryos to field-collected sediments (~30–325 µg/g ΣPAH) from oil-impacted locations resulting in cardiovascular defects, delayed hatching, and reduced overall hatching success. As such, these environmentally relevant levels could have implications for many species within the GoM including highly valuable commercial and recreational species.

Laboratory exposure studies have also contributed evidence indicating oil exposures can result in growth declines (Kerambrun et al. 2012; Kerambrun et al. 2011; Brewton et al. 2013; Brown-Peterson et al. 2017). However, data on the growth impacts of oil on adult fish from field studies are generally lacking. To our knowledge, Herdter et al. (2017) were the first to evaluate the impacts of oil on the growth rates of adult fish. They analyzed growth rates using back-calculated length-at-age analysis from 822 field-caught red snapper otoliths collected post-DWH (2011–2013). This study revealed significant growth declines in the dominant age groups (4–6 years) of red snapper that were not strongly correlated with any other non-oil spill-related environmental factors (e.g., wind, temperature, river discharge). Given that size-at-age correlates with fecundity, changes in growth could have potential population-level effects by negatively impacting reproduction and recruitment, which is implicated in the declines in red snapper in years following DWH (Murawski et al. 2018a, b; Fig. 24.1).

24.3 Diet and Trophic Level Shifts

The direct oiling of marine resources (i.e., phytoplankton, zooplankton, fish, birds, and marine mammals) can result in adverse impacts on their abundances throughout the water column, consequently altering food web trophic structures. The EVOS resulted in delayed long-term trophic cascade effects that were in some cases devastating (Peterson et al. 2003; Harwell and Gentile 2006). The Gulf of Alaska kelp ecosystem, comprised of killer whales (*Orcinus orca*), sea otters (*Enhydra lutris*), sea urchins, kelp forests, fish, and invertebrates, was substantially changed post-EVOS.

Stable isotope ratios are frequently used as indicators of trophic position and diet compositions due to the differential fractionation of organic sources (i.e., marine vs. terrestrial). Therefore, stable isotopes have been used as a tool to reconstruct potential diet and/or trophic level shifts by marine species as a result of oil exposures. Isotopic ratios were evaluated in a number of species both pre- and post-*DWH* in order to identify potential diet or trophic level shifts (Graham et al. 2010; Tarnecki and Patterson 2015; Quintana-Rizzo et al. 2015; Ransom et al. 2016). Isotopic signatures from Spanish mackerel larvae appeared stable over time (2007–2011), yet monthly comparisons revealed a depletion of $\delta^{13}C$ during periods of peak oil coverage in coastal sites (June 2010) followed by an enrichment in $\delta^{13}C$ and $\delta^{15}N$ in later months (August 2010). This pattern of carbon depletion was consistent with those observed in mesozooplankton, suspended particulate samples, and mesopelagic fish and shrimp (Graham et al. 2010; Quintana-Rizzo et al. 2015). Depletion of carbon isotope values was suspected to be the result of trophic transfer of oil-derived carbon through the food web rather than strictly the presence of oil in the water column (Graham et al. 2010). Following the oil spill, nitrogen enrichment was observed in Spanish mackerel, red snapper, and mesopelagic fish and shrimp indicating dietary and/or trophic shifts (Quintana-Rizzo et al. 2015; Ransom et al. 2016; Tarnecki and Patterson 2015). Additional supporting evidence of dietary and trophic shifts were observed in red snapper through the additional analysis of $\delta^{34}S$ and gut contents (Tarnecki and Patterson 2015; Patterson III et al. 2020). Results indicated an increase in trophic level and a shift in diet from pelagic zooplankton prey to benthic decapods and fish which persisted weeks to months following DWH (Fig. 24.2). What is unknown is how this shift in elemental compounds of nitrogen, carbon, and other constituents impacted the food value (if at all) and implications for resource production.

24.4   Habitat Shifts

In general, fishes are highly mobile with daily movements and migrations that range from small to large scale depending on species and life stage. Fish use a suite of acoustic, olfactory, and visual sensory capabilities in order to detect, orient, and discriminate between habitats and settlement locations (Arvedlund and Takemura 2006; Arvedlund and Kavanagh 2009; Paris et al. 2013; Díaz-Gil et al. 2017). Cues from nutrients, natural degradation products (e.g., algal compounds), and anthropogenic chemicals, such as sunscreen, oil, fertilizers, etc., can result in habitat discrimination and altered selection (Hadfield and Paul 2011; Paris et al. 2013; Havel and Fuiman 2016; Gouraguine et al. 2017). Specifically, cues from anthropogenic chemicals have been demonstrated to alter fish behavior and increase habitat avoidance and are confound conspecific cues (Tierney et al. 2010; Kwan et al. 2015; Lecchini et al. 2017). Of particular concern are the variety of hydrocarbons and

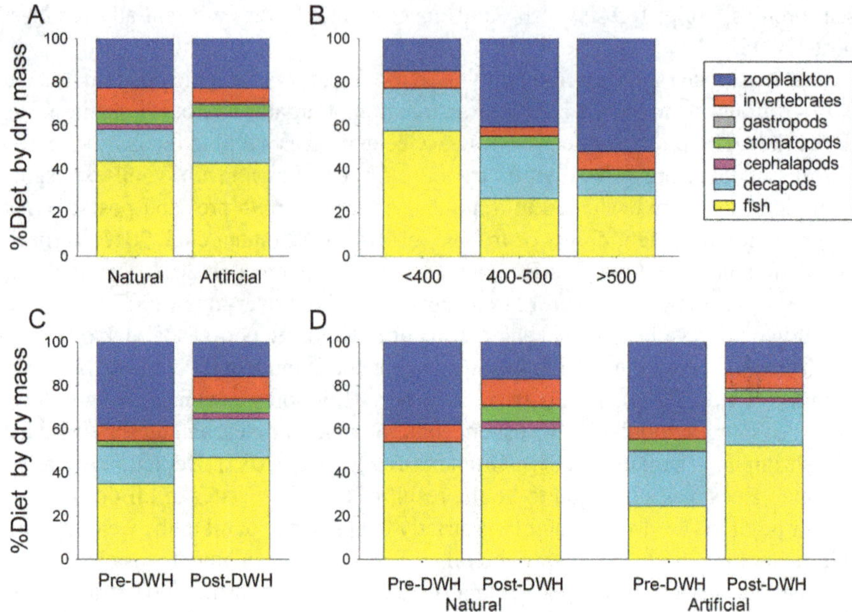

Fig. 24.2 Mean percent of red snapper diet by dry mass estimated among seven prey categories observed among stomach samples. Panels demonstrate the effect of (**a**) the habitat (natural versus artificial reefs), (**b**) the size class (total length in mm), (**c**) the timing of the *Deepwater Horizon* (DWH) oil spill, and (**d**) the interaction between habitat type and DWH timing on red snapper diet among samples collected in the northern Gulf of Mexico from June 2009 through August 2011. (Reproduced from Tarnecki and Patterson (2015))

surfactants that have been shown to elicit avoidance behavior in number of fish species (Hara and Thompson 1978; Maynard and Weber 1981; Dauble et al. 1985; Hidaka and Tatsukawa 1989; Ishida and Kobayashi 1995; Martin 2017; Lari and Pyle 2017; Reichert et al. 2017; Claireaux et al. 2018).

Post-DWH surveys conducted from Texas to Florida documented oiling along 2100 km of shoreline and more than 1100 km of coastal wetlands (Nixon et al. 2016). Ichthyoplankton surveys (2006–2010) from oil-impacted nearshore regions were used to assess potential habitat shifts and abundances of juvenile fish in seagrass nursery habitats in the northern GoM (Chandeleur Islands, LA to St. Joseph Bay, FL; Fodrie and Heck Jr 2011). Over 160,000 individual fishes representing 86 taxa were examined. Species composition of juvenile fish assemblages was unaltered, and there was little statistical evidence that potential exposures significantly impacted catch per unit effort (CPUE) during 2010. However, as compared to pre-spill catch rates, the fish characterized as of moderate (spring spawning, nearshore larvae) or high risk (spring-summer spawning, sparse larvae across continental shelf) exhibited decreases in CPUE following the spill in several of the bays assessed (Fodrie and Heck Jr 2011). Collectively, there were no significant post-spill shifts in community composition or juvenile cohorts in seagrass habitats, although the impacts of the fishery closures and the impacts of oil on the major predators of eggs and larvae could have contributed to the high catch rates in seagrass habitats in the northern GoM during 2010. Some of the wetland areas included in this study were considered minimally impacted (lightly oiled or no oil observed) based on oiling surveys (Nixon et al. 2016), which could also explain the lack of changes in juvenile fish assemblages and community composition in this area. A similar situation occurred in the SGoM where no significant impacts on demersal fish communities were reported as a result of the Ixtoc 1 oil spill (Yáñez-Arancibia 1986; Yáñez-Arancibia et al. 1982). Although the oil spill spread along the SGoM reaching Texas beaches potentially affecting extensive shorelines (1700–3000 km) (Tunnell Jr 2016; Sun et al. 2015), coastal areas were not uniformly heavy oiled (Anonymous 1980). The main nursery areas in the SGoM, like Términos Lagoon and nearby offshore areas in proximity to Ixtoc 1, were not heavily affected due to natural barriers from current patterns and freshwater plumes.

Habitat usage and fish assemblages in oiled and unoiled salt marshes in Louisiana in 2012 and 2013 were compared (Able et al. 2015). The study area spanned 90 km along the LA coast, including 22 sites within Terrebonne, Caminada, and Barataria Bays. There were no statistical differences in composition, abundance, size, or assemblage composition of fundulids, cyprinodontid, or poeciliids between oiled and unoiled sites in Barataria Bay. The inability to detect fish assemblage or abundance differences could be due to limitations related to statistical power, reference site locations (e.g., unknown oiling of sites considered "controls"), and timing of the fish collections. Conversely, a study conducted at similar sites within Barataria Bay found lower abundances of Gulf killifish at oil-impacted sites in Barataria Bay compared to reference or unoiled site located in Texas (Carr et al. 2018). Body weight, standard length, and total length were all statistically lower in fish collected at impacted sites versus the unimpacted, reference site. Remarkably, there were no sexually mature *F. grandis* greater than 12 g collected at any of the 11 sites sampled in Barataria Bay. The timing of these studies and the selection reference sites between these two studies are obviously important in interpreting results. The study by Carr et al. (2018) was conducted in 2010, whereas Able et al. (2015) conducted their study 2–3 years post-*DWH*. This time lag could have allowed for a period of recovery, particularly given the short life spans of the species involved. Also, Able et al. (2015) selected a reference location within Barataria Bay. Within the last three decades, at least nine major oil spills have occurred in coastal Louisiana, including Barataria Bay. Identifying reference sites or unoiled marshes within Barataria Bay might be challenging since Barataria Bay marsh sediments routinely receive input of PAHs from petrogenic, pyrogenic, and biogenic sources (DeLaune et al. 1979; Hester and Mendelssohn 2000; Dincer Kırman et al. 2016).

Time-series data and generalized additive models (GAM) were also used to characterize the patterns of abundance and habitat usage for large predatory pelagic fishes in the northern GoM for the years 2007–2010 (Rooker et al. 2013). This study examined the spatial extent of the *DWH* surface oil to determine whether habitat of blackfin tuna, blue marlin, dolphinfish, and sailfish was exposed to oil. The spatial distribution predications and impacts on suitable habitat in 2010, post-*DWH*, were highly variable depending on the species. For instance, the impact on high-quality habitat was much higher for blackfin tuna habitat (19%) compared to those critical areas for sailfish and blue marlin (<10%). This was largely due to the pattern of habitat usage, outer and upper shelf (i.e., blackfin tuna) versus the slope and abyssal regions (i.e., sailfish). Furthermore, electronic tagging revealed that the core usage areas of adult blue marlin shifted from the northwestern and north central GoM to a more limited western GoM usage following DWH. There was a twofold decrease in the usage of areas impacted by DWH in 2010 compared to previous years, possibly indicating they did not return to their spawning areas due to adverse environmental conditions, perhaps as a result of chemical cues, changes in prey distributions, or other factors (Rooker et al. 2013).

Significant declines and community structure shifts were observed in reef fish located on artificial reef systems in the northern GoM between 2009–2010 and 2011–2012 (Dahl et al. 2016). This decline ensued during the simultaneous occurrence of the lionfish (*Pterois* spp.) invasion and the *DWH* oil spill in the northern GoM. Experimental removal of lionfish had significant impacts on reef fish community structure, predominantly on the smaller lionfish prey fishes and obligate reef residents (e.g., damselfishes, cardinalfishes, wrasses, blennies, etc.). However, there were no substantial gains observed in the overall native reef fish abundance of most taxa, including the larger snappers and gray triggerfish (*Balistes capriscus*). The declines in these larger fishes are unlikely due to being directly consumed by lionfish and may be the result of direct oil spill mortality or emigration from oil spill impacted areas. It is also highly plausible that environmental disturbances and habitat degradation caused by the *DWH* increased the system susceptibility and vulnerability (e.g., compromised immune systems in fish), allowing for a relatively effortless lionfish invasion (Dahl et al. 2016).

In the SGoM, the Universidad Nacional Autónoma de Mexico (UNAM) conducted demersal fish surveys both pre-and post-Ixtoc 1. Results indicated that both the number of fish species collected and the diversity of species decreased significantly in areas surrounding the spill site (Amezcua-Linares et al. 2015). The number of fish species collected around the Ixtoc I spill area had decreased 38% between 1978 and 1980, pre- and post-spill, respectively. By 1982, 3 years post-spill, the number of fish species collected was similar to those collected in 1978. However, although the number of species rebounded, the pre- and post-spill assemblages were significantly different. In fact, three species (rough scad, *Trachurus lathami*; Atlantic thread herring, *Opisthonema oglinum*; and the blackcheek tonguefish, *Symphurus plaguisa*) that were abundant before the Ixtoc 1 spill were never caught again. By 1982, there was 50% decrease in commercial landings compared to 1981 and a significant composition change within Campeche Bay. Yet, the landings in 1982 were similar to pre-spill levels and higher than those in 1987 and 1989. These variations in landings could be due to environmental or seasonal fluctuations or changes in fishing patterns (Gracia et al. 2020). The reduction in landings was suspected to be directly related to larval stage mortalities, whereas the changes in composition that occurred immediately after the spill appeared to be related to habitat shifts and migrations to areas outside of the spill zone.

24.5   Immunotoxicity and Other Health Indices

Oil exposure has been linked to immunosuppression in a number of aquatic organisms, including marine mammals (De Guise et al. 2017; White et al. 2017) and fish (Reynaud and Deschaux 2006; Barron 2012; Bayha et al. 2017). Immunosuppression and immunotoxicity can result in a number of sub-lethal effects, such as enhanced or suppressed cell proliferation, decreased phagocytosis, and induced cell apoptosis, which may eventually result in disease, decreased resiliency, community structure changes, mortality, and ultimately population declines (Reynaud and Deschaux 2006; Kim et al. 2013; Lee et al. 2018). Sub-lethal negative health effects of PAH exposure on fishes are traditionally measured using biomarkers for immunotoxicity, oxidative stress, and genotoxicity (Ameur et al. 2015; Benedetti et al. 2015; Javed et al. 2017). However, biomarker responses can vary depending upon species, sex, reproductive status, size, and location of collection (Balfry and Iwama 2004; Fonseca et al. 2011; Wunderlich et al. 2015; Rehberger et al. 2017). Thus, for meaningful analysis of changes in health status of non-model species, reference intervals should first be established, which describe the response values for a biomarker in 95% of a representative population (Geffré et al. 2009).

Post-*DWH*, fish surveys conducted in 2011 and 2012 evaluated the disease status of over 7000 fishes (>100 species) collected from 150 sampling locations in the northern GoM (Murawski et al. 2014). In 2011, the incidence of skin lesions varied with species; however, the prevalence rates were higher in the demersal fishes such as tilefish (*Lopholatilus chamaeleonticeps*), snappers, and groupers compared to elasmobranchs (Fig. 24.3a). The frequencies of lesions ranged from ≤1% in elasmobranchs to ~5% in red snapper and upward of 10% in bottom dwelling tilefish. The overall frequency of skin lesions dropped significantly from 1.9% to 0.9% ($p = 0.019$) between 2011 and 2012. The hypothesis that DWH oil exposure was the cause of increased lesion frequency could not be rejected. Data collected subsequent to those published in the Murawski et al. (2014) paper (Fig. 24.3b) show further declines in lesion frequency to zero in 2015, with a slight uptick in 2017. Lesions were also observed in red drum (*Sciaenops ocellatus*) collected at several sites within Barataria Bay, Louisiana, where surveys documented moderate to heavy oiling following DWH (Harr et al. 2018). Of the red drum examined, 87% of fish assessed from impacted sites within Barataria Bay presented pathological changes in the skin. Moreover, lesions were induced in southern flounder (Fig. 24.4) exposed to oil and pathogens providing direct evidence that oil exposure induces immunosuppression, elevated bacteria levels, and increased incidences of bacterial infections (e.g., skin lesions; Bayha et al. 2017).

In addition, 18% of the red drum collected at impacted sites within Barataria Bay were also anemic compared to 0% of the fish collected at non-impacted sites (Harr et al. 2018). Bottlenose dolphins from this area also presented with anemia (Schwacke et al. 2014). Lowered hemoglobin concentrations in fish have previously been linked to oil exposures in a number of field and laboratory studies (Alkindi et al. 1996; Khan 2003; Bayha et al. 2017).

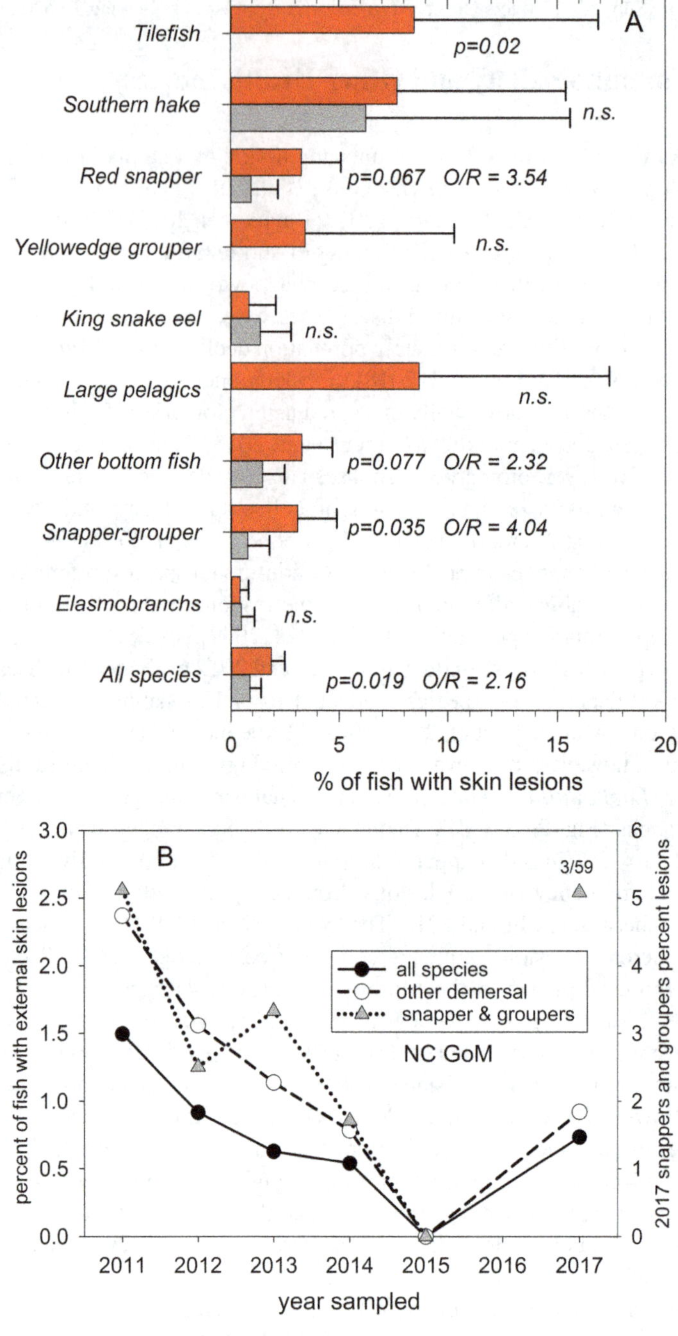

Fig. 24.3 (a) Mean percent (error bars show SEs) of fish within a species or species group with skin lesions examined in 26 repeated stations in 2011 (red) and 2012 (gray) in the NGM region. Abbreviations are as follows: O/R D the odds ratio of differences between years (2011/2012) and n.s. = not significant. (Reproduced from Murawski et al. (2014)). (b) Percent of fish with external skin lesions observed in the north central Gulf of Mexico, 2011–2017. Data are aggregated for three species groups. Data for snappers and groupers for 2017 are plotted as a symbol, given the low sample sizes examined (3 lesions observed from 59 fishes; Murawski et al. 2018a)

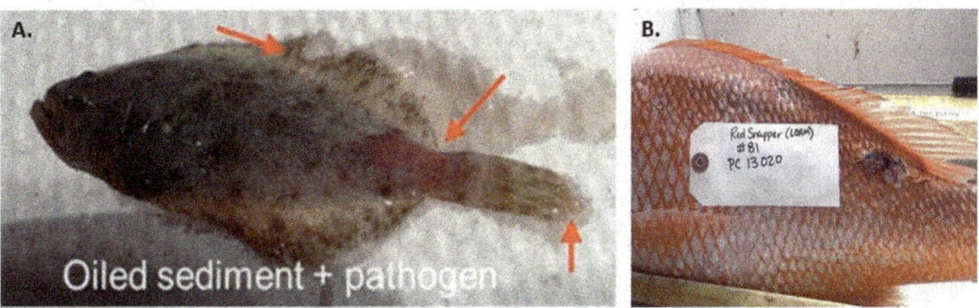

Fig. 24.4 (**a**) Oil-induced lesions in southern flounder exposure studies. Red arrows indicate observed lesions. (Reproduced with permission from PLOS One (Bayha et al. 2017)). (**b**) Lesions observed on field-caught red snappers in the north central region of the Gulf of Mexico. (Image courtesy of C-IMAGE)

From 2015 to 2017, golden tilefish ($n = 255$) and red snapper ($n = 125$) were collected throughout the GoM via demersal longline to (1) generate the first reference intervals for a variety of biomarkers for these species, (2) examine differences (and outliers) in responses around the Gulf, and (3) select suitable candidate specimens for a subsequent de novo golden tilefish transcriptome analysis (Deak et al. 2017). Preliminary reference intervals were successfully established in both species for biomarkers of immune status (hematocrit, leukocrit, differential white blood cell counts, and lysozyme), oxidative stress (sorbitol dehydrogenase, superoxide dismutase, and malondialdehyde), and genotoxicity (erythrocyte nuclear abnormalities, Deak (in prep), according to ASVCP guidelines (Friedrichs et al. 2012). While statistically significant differences were found in some variables by geographic location, sex, and size, partitioning into subgroups was unnecessary since the sample means did not meet the customary ¼ of the 95% confidence interval level for the entire reference population (Walton 2001).

Statistically significant differences in biomarker reference values were observed between species. The most pronounced of these differences occurred in elevated erythrocyte nuclear abnormality counts (genotoxicity) and sorbitol dehydrogenase (antioxidant) response in golden tilefish (Figs. 24.5 and 24.6). This is not surprising, given the disparate life histories between red snapper and golden tilefish, with the latter forming burrows in silt-clay sediments, thereby having a documented elevation of heavy metals and pollutant accumulation (Hall et al. 1978; Harris and Boyd 1990; Snyder et al. 2015). Differential biomarker responses were observed at stations Gulf-wide, particularly in expression of lysozyme, superoxide dismutase, sorbitol dehydrogenase, and nuclear abnormalities, further supporting their inclusion in health assessments (Deak et al. 2017).

A transcriptome analysis was performed by Deak et al. (2018) on 15 reproductively mature female goleden tilefishes, all caught during the same time period of the spawning cycle. Five fishes each were selected from three stations: one on the northern edge of the DeSoto Canyon (likely impacted by the DWH), one from the oil fields of Campeche Bay, and one station with no local oil platforms, off the

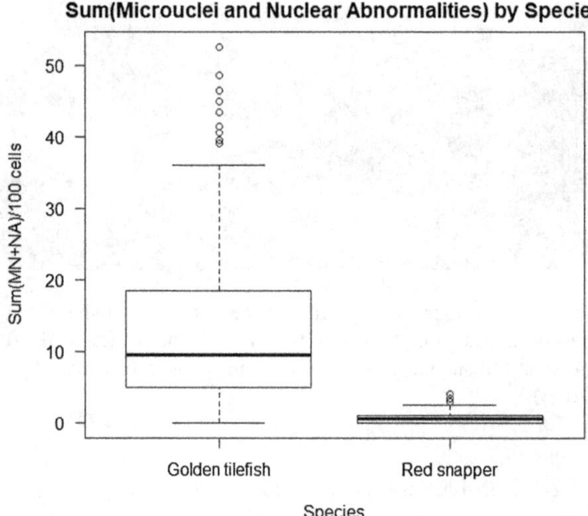

Fig. 24.5 Comparison between the sum of micronuclei and nuclear abnormalities in erythrocytes from golden tilefish ($n = 245$) and red snapper ($n = 116$) caught throughout the Gulf of Mexico in 2015–2017 (Deak et al. 2017, 2018)

Fig. 24.6 Comparison between plasma superoxide dismutase from golden tilefish ($n = 245$) and red snapper ($n = 116$) caught throughout the Gulf of Mexico in 2015–2017 (Deak et al. 2017, 2018)

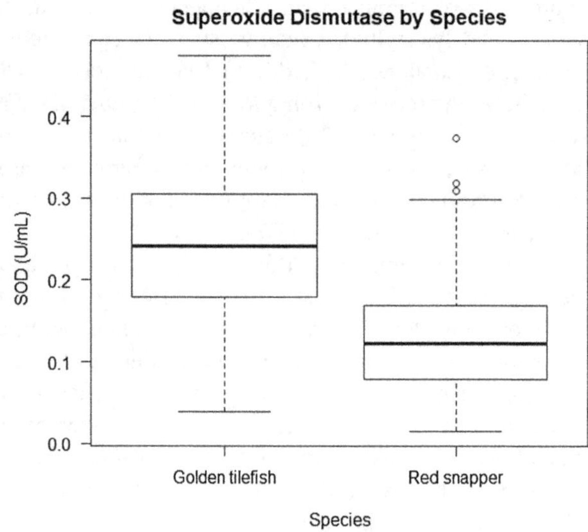

Yucatan Peninsula. After de novo assembly, over 1000 differentially expressed genes were observed between stations.

Fish from the DeSoto Canyon site had unique gene expressions for ontologies relating to reproduction, metabolism, immune response, and cellular repair. In particular, DeSoto Canyon fish had lower expression of multiple vitellogenin and zona pellucida genes compared to the other stations, suggesting reproductive impairment

(Brian et al. 2007). Furthermore, they had elevated glucose metabolism and glycolytic pathway activation, a common stress response in fishes (Naour et al. 2017). Their immune profile included elevated cytokines and chemokines, indicative of an ongoing inflammatory response (Wang and Secombes 2013). Multiple anti-apoptotic and antioxidant genes were engaged, as a potential protective mechanism against documented bile PAHs in the animals (Santana et al. 2018).

In addition, both oil field groups of fish (from DeSoto Canyon and Campeche Bay) displayed differential gene expression from the non-oil field fish taken off the Yucatan Peninsula. This was evidenced in the elevation of circadian rhythm genes, chemokines, and pro-apoptotic genes and decreased gene expression associated with blood coagulation and angiogenesis (Deak et al. 2018). This study underscores the utility of transcriptomics in elucidating molecular discrepancies found in populations after large contamination events.

24.6 Insights from Histopathology

Histopathological techniques to detect cellular alterations have been used in laboratory and environmental exposures and are considered important tools to detect direct toxic effects of chemical exposures (Schwaiger et al. 1997). Moreover, microscopic pathological studies have been used to differentiate between alterations that result from natural environmental stressors and secondary stress effects of anthropogenic pollution (Snieszko 1974; Teh et al. 1997; Schwaiger et al. 1997).

Planar PCBs, dioxins, and PAHs are all known to exert biological effects through aryl hydrocarbon receptor (AHR) signaling pathways evoking xenobiotic metabolism (i.e., cytochrome P450). The upregulation and expression of the CYP1A protein were detected in the gills, head kidney, and intestinal tissues from resident Gulf killifish collected in Barataria Bay post-*DWH* in 2010 and 1 year later (Whitehead et al. 2012; Dubansky et al. 2013, 2017). These CYP1A proteins were markedly elevated in the fish collected at oiled sites compared to those at unoiled locations. Additionally, body weights, germinal epithelium thickness, oocyte diameters, and gonadal somatic indices from histological measurements measured in Gulf killifish collected at impacted oil sites in Barataria Bay were all lower than those from non-impacted sites (Carr et al. 2018). Higher lesion prevalence in the gill, liver, stomach, and epithelial tissues of menhaden (*Brevoortia* sp.) was also observed in those collected in Barataria Bay compared to those collected along the north Atlantic coast (Bentivegna et al. 2015).

Offshore species were also been evaluated for histopathological alterations following the DWH and Ixtoc 1 events. Red snapper and golden tilefish were routinely sampled throughout the GoM over the course of 7 years (2011–2017; Murawski et al. 2014; Snyder et al. 2015). A variety of tissue samples were collected for PAH analysis and histopathology (Gracia et al. 2017, 2018; Snyder et al. 2019; Pulster et al. 2018). Skin lesions observed in these species were confirmed through histology (Fig. 24.7). A subset of red snapper hepatic tissues collected in the north central

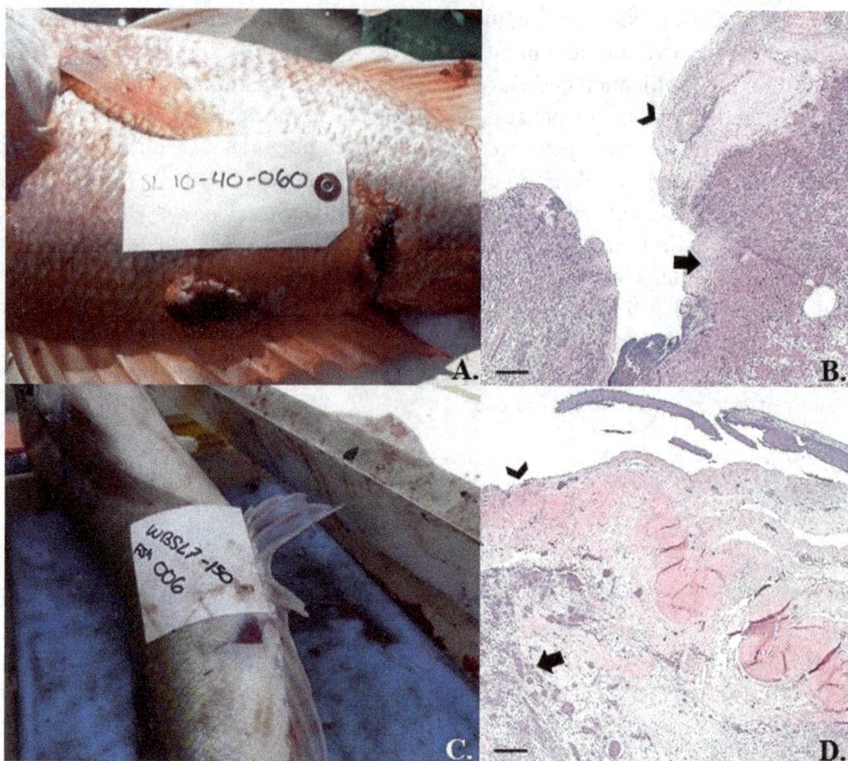

Fig. 24.7 Field photograph (**a**) and a photomicrograph (**b**) of skin ulceration observed on a red snapper collected in the northern Gulf of Mexico in 2013. Severe epidermal ulceration (arrowhead) with abundant, hemorrhage, fibrin, edema, necrotic cellular debris, inflammation, and granulation tissue expanding the exposed dermis (arrow). H&E, scale bar represents 500 µm. Field photograph (**c**) and photomicrograph (**d**) of a skin ulcer found on a golden tilefish collected in the northern Gulf of Mexico in 2012. Golden tilefish ulcerated epidermis (arrowhead), with extensive granulation tissue expanding the dermis and hypodermis (arrow). Inflammation extends deep into the body wall musculature. H&E, scale bar is 500 µm. (Photographs courtesy of C-IMAGE)

region of GoM (2012–2017) at sites located proximate to the DWH wellhead were analyzed for histopathological changes and disease manifestations (Pulster et al. 2018). Preliminary analysis of hepatic tissues, revealed that 99% exhibited morphological and/or pathological changes, including a spectrum of alterations such as vacuolar change (lipid and glycogen type), hepatocellular atrophy, lymphocytic inflammation, migrating larval nematodes, adult nematodes, and intra-biliary myxozoans. In the SGoM Ixtoc 1 spill area, red snapper livers also showed a high incidence (>80%) of a variety of hepatic parenchymal changes. Red snapper liver variances were found along the whole SGoM, and more than 85% of these pathologies could be related to hydrocarbons among other agents (Fig. 24.8). In contrast, 90% of fish gonad tissues were found to be healthy which might indicate the preservation of reproductive potential (Gracia et al. 2017). In the SGoM, an analysis of

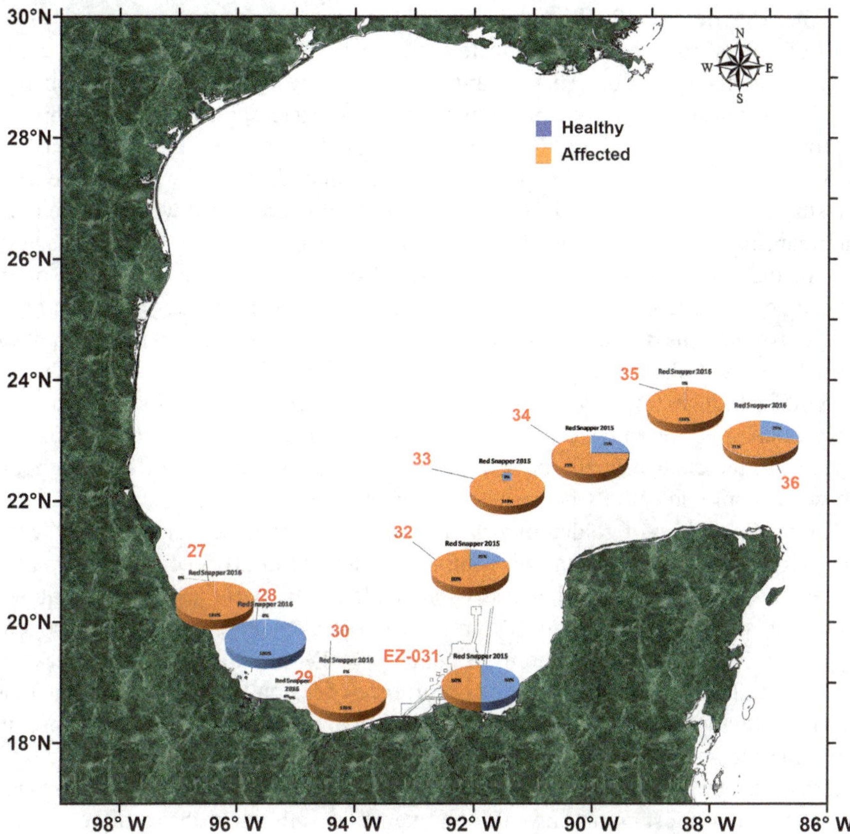

Fig. 24.8 Red snapper liver anomalies found along the southern Gulf of Mexico. More than 85% of these pathologies occur in areas where oil infrastructure or natural oil seeps occur (Gracia et al. 2017, 2018)

12 demersal deepwater fish species along a depth gradient of 10–800 m showed that pathological hepatic lesions were a frequent condition. The highest percentage of liver damage (~90%) was found in deepwater fish collected in a natural seep area, whereas fishes sampled around oil industry infrastructure near the Ixtoc 1 location presented with a lower percentage (55%) of hepatic lesions (Gracia et al. 2018). Although many of the observed hepatic changes have been previously associated with chemical exposures (e.g., PAHs, PCBs, heavy metals), parasites, wastewater, environmental stressors, and normal physiological processes, the simultaneous presence of multiple hepatic changes observed in red snapper (Pulster et al. 2018) combined with decreasing abundances of adult red snapper (Murawski et al. 2018a, b) could be indicative of chronic stress and potential disease progression.

A subset of golden tilefish hepatic tissues (n = 113) collected from the DeSoto Canyon region of the northern GoM (2012–2016) were analyzed for microscopic changes (Snyder et al. 2019). Similar to red snapper, preliminary microscopic analysis of golden tilefish tissues indicated 99% of individuals exhibited microscopic changes in hepatic tissue. These changes included glycogen-type vacuolar change (72%), increased occurrence of pigmented macrophage aggregates (55%), biliary fibrosis and duplication (37%), lipid-type vacuolar change (35%), granulomas (25%), hepatocellular atrophy (12%), foci of cellular alteration (9%) and one neoplasm (<1%). The occurrence of neoplasms typically associated with chronic contaminant exposure was rare ($n = 1$; hepatocellular adenoma); however, approximately 9% of individuals examined exhibited foci of cellular alteration, which are often considered "preneoplastic lesions." Preliminary data analysis suggests significant linear associations between hepatobiliary burdens of PAHs and hepatocellular atrophy, foci of cellular alteration, biliary fibrosis, and glycogen-type vacuolar change in golden tilefish from the DeSoto Canyon region (Fig. 24.9). However, further investigation is warranted as many of the observed changes could be normal for the species or the changes could be related to chronic stress, disease, or starvation, all of which could potentially be exacerbated by chronic toxicant exposure.

A smaller subset of golden tilefish hepatic tissues collected from offshore Texas in 2016 ($n = 26$) and the Bay of Campeche, Mexico, in 2015 ($n = 24$) were also analyzed for microscopic changes. Golden tilefish from these regions exhibited similar hepatic changes to those from the DeSoto Canyon region, with glycogen-type vacuolar change being the most common (92% of individuals from Mexico, 85% from Texas). The frequency of occurrence of lesion types was similar across the three regions, with a notable difference being higher frequency of granulomas in both the Mexico (88%) and Texas (65%) regions compared to the DeSoto Canyon (25%). The granulomas observed were likely the result of host response to a parasitic infection, most commonly of larval nematode migration through the tissues. Higher frequency of parasitic infection in these regions may be due to higher susceptibility of hosts in response to contamination in the environment.

24.7 Conclusions

While massive fish kills and similar evidence of severe impacts were not obvious from either the DWH or Ixtoc 1, there is nevertheless strong observational and correlative evidence of impacts on fish populations within the boundaries of observed oil distributions. Many sub-lethal effects were observed in field-caught specimens that were also apparent in controlled laboratory-based exposure trials. For example, elevated lesion frequencies in several species (Fig. 24.2; Murawski et al. 2014) were correlated with PAH exposure; subsequent laboratory trials were able to induce lesion occurrence when fishes were exposed to oil and elevated bacterial loads, but not to elevated pathogenic bacteria levels alone (Bayha et al. 2017). This is strong prima facie evidence for oil exposure being the proximal cause of skin lesions observed not only in scientific surveys but by fishers. This example demonstrates

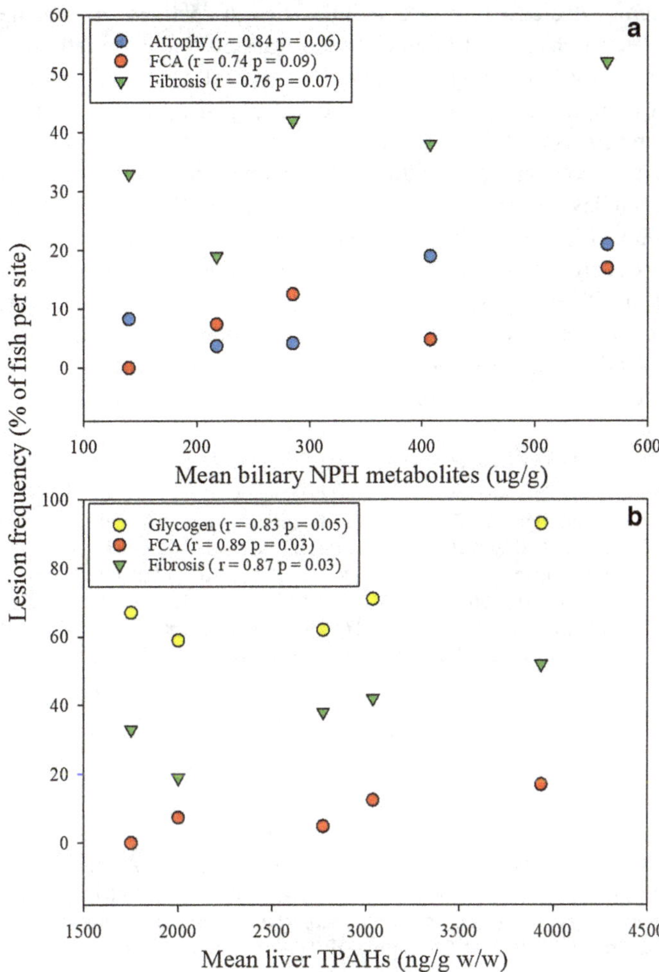

Fig. 24.9 (**a**) Linear correlations between average total PAHs and alkylated homologs (TPAH) in liver tissue and frequency of glycogen-type vacuolar change, FCA, and biliary fibrosis by station are strong and significant. (**b**) Correlations between average biliary naphthalene (NPH) metabolites and frequency of hepatocellular atrophy, FCA, and biliary fibrosis by station are strong but not significant at $\alpha = 0.05$ (Snyder et al. 2019)

the importance of combined field observations and experimental exposures to isolate causation when examining sub-lethal health effects of oil spills.

The underlying and critical question in evaluating oil spill effects *is* one of causation; do oil spills cause sub-lethal effects? Applying Hill's criteria (strength of association, consistency, specificity, temporality, biological gradient, plausibility, coherence, experiment, and analogy), typically used for evaluating causality in epidemiologic data, sub-lethal oil spill effects do occur in oil-exposed fishes, and they are potentially significant to short- and long-term population viability. Sub-lethal impacts such as decreased larval condition, growth declines, diet changes, habitat and trophic level shifts, and biomarker and transcriptomic alterations are evident from recent oil spill toxicological studies. While field-based evaluations of oil spill impacts are complicated by chronic exposure of organisms to other non-related contaminants and stressors, the effects of point-source release can be elucidated through the creation or comparison of biomarker reference intervals. In addition, recent computational developments have furthered the application of transcriptome analysis in non-model species with no sequenced genome, allowing researchers to document changes across thousands of genes at once, which will enhance data obtained from future health evaluations. The GoM is a unique and dynamic ecosystem that is highly productive and rich in resources, including both fisheries and the oil and gas industries. Understanding the interactions and impacts among these resources is essential for both of these economically vital resources and the entire health of the GoM ecosystem.

Acknowledgments This research was made possible by a grant from The Gulf of Mexico Research Initiative through the Center for the Integrated Modeling and Analysis of the Gulf Ecosystem (C-IMAGE I, II and III). Final data will be publicly available through the Gulf of Mexico Research Initiative Information & Data Cooperative (GRIIDC) at https://data.gulfresearchinitiative.org. (doi 10.7266/N7PG1PRZ, doi 10.7266/N7T43R3T, doi 10.7266/N7GT5K56, doi 10.7266/N7G73C4N, doi:10.7266/n7-g27a-x012, doi:10.7266/n7-nmsy-tq94, doi:10.7266/n7-460y-rz32, 10.7266/n7-5zt3-8c72, 10.7266/n7-t3g7-sk92, 10.7266/n7-qjz0-bt64, 10.7266/n7-5y5c-9348).

References

Able KW, López-Duarte PC, Fodrie FJ, Jensen OP, Martin CW, Roberts BJ, Valenti J, O'Connor K, Halbert SC (2015) Fish assemblages in Louisiana salt marshes: effects of the Macondo oil spill. Estuar Coasts 38:1385–1398. https://doi.org/10.1007/s12237-014-9890-6

Alkindi AYA, Brown JA, Waring CP, Collins JE (1996) Endocrine, osmoregulatory, respiratory and haematological parameters in flounder exposed to the water soluble fraction of crude oil. J Fish Biol 49:1291–1305

Ameur WB, El Megdiche Y, de Lapuente J, Barhoumi B, Trabelsi S, Ennaceur S, Camps L, Serret J, Ramos-López D, Gonzalez-Linares J, Touil S, Driss MR, Borràs M (2015) Oxidative stress, genotoxicity and histopathology biomarker responses in *Mugil cephalus* and *Dicentrarchus labrax* gill exposed to persistent pollutants. A field study in the Bizerte Lagoon: Tunisia. Chemosphere 135:67–74. https://doi.org/10.1016/j.chemosphere.2015.02.050

Amezcua-Linares F, Amezcua F, Gil-Manrique B (2015) Effects of the Ixtoc I oil spill on fish assemblages in the Southern Gulf of Mexico. In: Alford JB, Peterson MS, Green CC (eds) Impacts of oil spill disasters on marine habitats and fisheries in North America. CRC Press, Boca Raton, pp 209–236

Anonymous (1980) Informe de los trabajos realizados para el control del pozo IXTOC I, el combate, derrame de petróleo y determinación de sus efectos sobre el ambiente marino. Programa Coordinado de Estudios Ecológicos de la Sonda de Campeche. Instituto Mexicano del Petróleo, México, D.F.

Arvedlund M, Kavanagh K (2009) The senses and environmental cues used by marine larvae of fish and decapod crustaceans to find tropical coastal ecosystems. In: Nagelkerken I (ed)

Ecological connectivity among tropical coastal ecosystems. Springer Netherlands, Dordrecht, pp 135–184. https://doi.org/10.1007/978-90-481-2406-0_5

Arvedlund M, Takemura A (2006) The importance of chemical environmental cues for juvenile *Lethrinus nebulosus* Forsskål (Lethrinidae, Teleostei) when settling into their first benthic habitat. J Exp Mar Biol Ecol 338:112–122. https://doi.org/10.1016/j.jembe.2006.07.001

Balfry SK, Iwama GK (2004) Observations on the inherent variability of measuring lysozyme activity in coho salmon (*Oncorhynchus kisutch*). Comp Biochem Physiol B: Biochem Mol Biol 138:207–211. https://doi.org/10.1016/j.cbpc.2003.12.010

Barron MG (2012) Ecological impacts of the *Deepwater Horizon* oil spill: implications for immunotoxicity. Toxicol Pathol 40:315–320. https://doi.org/10.1177/0192623311428474

Bayha KM, Ortell N, Ryan CN, Griffitt KJ, Krasnec M, Sena J, Ramaraj T, Takeshita R, Mayer GD, Schilkey F, Griffitt RJ (2017) Crude oil impairs immune function and increases susceptibility to pathogenic bacteria in southern flounder. PLoS One 12:e0176559. https://doi.org/10.1371/journal.pone.0176559

Benedetti M, Giuliani ME, Regoli F (2015) Oxidative metabolism of chemical pollutants in marine organisms: molecular and biochemical biomarkers in environmental toxicology. Ann NY Acad Sci 1340:8–19. https://doi.org/10.1111/nyas.12698

Bentivegna CS, Cooper KR, Olson G, Pena EA, Millemann DR, Portier RJ (2015) Chemical and histological comparisons between *Brevoortia sp.* (menhaden) collected in fall 2010 from Barataria Bay, LA and Delaware Bay, NJ following the *Deepwater Horizon* (DWH) oil spill. Mar Environ Res 112(Part A):21–34. https://doi.org/10.1016/j.marenvres.2015.08.011

Beyer J, Trannum HC, Bakke T, Hodson PV, Collier TK (2016) Environmental effects of the *Deepwater Horizon* oil spill: a review. Mar Pollut Bull 110:28–51. https://doi.org/10.1016/j.marpolbul.2016.06.027

Brewton RA, Fulford R, Griffitt RJ (2013) Gene expression and growth as indicators of effects of the BP *Deepwater Horizon* oil spill on spotted seatrout (*Cynoscion nebulosus*). J Toxic Environ Health A 76:1198–1209. https://doi.org/10.1080/15287394.2013.848394

Brian JV, Harris CA, Scholze M, Kortenkamp A, Booy P, Lamoree M, Pojana G, Jonkers N, Marcomini A, Sumpter JP (2007) Evidence of estrogenic mixture effects on the reproductive performance of fish. Environ Sci Technol 41:337–344. https://doi.org/10.1021/es0617439

Brown-Peterson NJ, Brewton RA, Griffitt RJ, Fulford RS (2015a) Impacts of the *Deepwater Horizon* oil spill on the reproductive biology of spotted seatrout (*Cynoscion nebulosus*). In: Alford JB, Peterson MS, Green CC (eds) Impacts of oil spill disasters on marine habitats and fisheries in North America, CRC marine biology series. CRC Press, Boca Raton, pp 237–252

Brown-Peterson NJ, Krasnec M, Takeshita R, Ryan CN, Griffitt KJ, Lay C, Mayer GD, Bayha KM, Hawkins WE, Lipton I, Morris J, Griffitt RJ (2015b) A multiple endpoint analysis of the effects of chronic exposure to sediment contaminated with *Deepwater Horizon* oil on juvenile southern flounder and their associated microbiomes. Aquat Toxicol 165:197–209. https://doi.org/10.1016/j.aquatox.2015.06.001

Brown-Peterson NJ, Krasnec MO, Lay CR, Morris JM, Griffitt RJ (2017) Responses of juvenile southern flounder exposed to *Deepwater Horizon* oil-contaminated sediments. Environ Toxicol Chem 36:1067–1076. https://doi.org/10.1002/etc.3629

Carr DL, Smith EE, Thiyagarajah A, Cromie M, Crumly C, Davis A, Dong MJ, Garcia C, Heintzman L, Hopper T, Kouth K, Morris K, Ruehlen A, Snodgrass P, Vaughn K, Carr JA (2018) Assessment of gonadal and thyroid histology in Gulf killifish (*Fundulus grandis*) from Barataria Bay Louisiana one year after the *Deepwater Horizon* oil spill. Ecotoxicol Environ Saf 154:245–254. https://doi.org/10.1016/j.ecoenv.2018.01.013

Chancellor E (2015) Vulnerability of larval fish populations to oil well blowouts in the Northern Gulf of Mexico. M.S. Thesis, College of Marine Science, University of South Florida, Scholar Commons

Chancellor E, Murawski SA, Paris CB, Perruso L, Perlin N (2020) Comparative environmental sensitivity of offshore Gulf of Mexico waters potentially impacted by ultra-deep oil well blowouts. In: Murawski SA, Ainsworth C, Gilbert S, Hollander D, Paris CB, Schlüter M, Wetzel D (eds) Scenarios and responses to future deep oil spills – fighting the next war. Springer, Cham

Claireaux G, Queau P, Marras S, Le Floch S, Farrell AP, Nicolas-Kopec A, Lemaire P, Domenici P (2018) Avoidance threshold to oil water-soluble fraction by a juvenile marine teleost fish. Environ Toxicol Chem 37:854–859. https://doi.org/10.1002/etc.4019

Dahl KA, Patterson WF, Snyder RA (2016) Experimental assessment of lionfish removals to mitigate reef fish community shifts on northern Gulf of Mexico artificial reefs. Mar Ecol Prog Ser 558:207–221. https://doi.org/10.3354/meps11898

Dauble DD, Gray RH, Skalski JR, Lusty EW, Simmons MA (1985) Avoidance of a water-soluble fraction of coal liquid by fathead minnows. Trans Am Fish Soc 114:754–760. https://doi.org/10.1577/1548-8659(1985)114<754:aoawfo>2.0.co;2

de Guise S, Levin M, Gebhard E, Jasperse L, Hart LB, Smith CR, Venn-Watson S, Townsend F, Wells R, Balmer B, Zolman E, Rowles T, Schwacke L (2017) Changes in immune functions in bottlenose dolphins in the northern Gulf of Mexico associated with the *Deepwater Horizon* oil spill. Endanger Species Res 33:290–303. https://doi.org/10.3354/esr00814

Deak K, Dishaw L, Murawski S (in prep) Establishment of baseline reference intervals for golden tilefish and red snapper caught throughout the Gulf of Mexico

Deak K, Dishaw L, Murawski SA (2017) Gauging Gulf-wide golden tilefish health: how do demersal denizens of De Soto Canyon compare to their Cuban, Mexican, and US neighbors? Presentation at: American Fisheries Society, Tampa, Florida, 2017

Deak K, Dishaw L, Murawski SA (2018) Tracking oil's toxicological transgressions through tilefish transcriptomics. Presentation at: Gulf of Mexico oil spill and ecosystem science conference, New Orleans, Louisiana, 2018

DeLaune RD, Patrick WH Jr, Buresh RJ (1979) Effect of crude oil on a Louisiana *Spartina alterniflora* salt marsh. Environ Pollut (1970) 20:21–31. https://doi.org/10.1016/0013-9327(79)90050-8

Díaz-Gil C, Cotgrove L, Smee SL, Simón-Otegui D, Hinz H, Grau A, Palmer M, Catalán IA (2017) Anthropogenic chemical cues can alter the swimming behaviour of juvenile stages of a temperate fish. Mar Environ Res 125:34–41. https://doi.org/10.1016/j.marenvres.2016.11.009

Dincer Kırman Z, Sericano JL, Wade TL, Bianchi TS, Marcantonio F, Kolker AS (2016) Composition and depth distribution of hydrocarbons in Barataria Bay marsh sediments after the *Deepwater Horizon* oil spill. Environ Pollut 214:101–113. https://doi.org/10.1016/j.envpol.2016.03.071

Dubansky B, Whitehead A, Miller JT, Rice CD, Galvez F (2013) Multitissue molecular, genomic, and developmental effects of the *Deepwater Horizon* oil spill on resident Gulf killifish (*Fundulus grandis*). Environ Sci Technol 47:5074–5082. https://doi.org/10.1021/es400458p

Dubansky B, Rice CD, Barrois LF, Galvez F (2017) Biomarkers of aryl-hydrocarbon receptor activity in Gulf killifish (*Fundulus grandis*) from northern Gulf of Mexico marshes following the *Deepwater Horizon* oil spill. Arch Environ Contam Toxicol 73:63–75. https://doi.org/10.1007/s00244-017-0417-6

Esbaugh AJ, Mager EM, Stieglitz JD, Hoenig R, Brown TL, French BL, Linbo TL, Lay C, Forth H, Scholz NL, Incardona JP, Morris JM, Benetti DD, Grosell M (2016) The effects of weathering and chemical dispersion on *Deepwater Horizon* crude oil toxicity to mahi-mahi (*Coryphaena hippurus*) early life stages. Sci Total Environ 543(Part A):644–651. https://doi.org/10.1016/j.scitotenv.2015.11.068

Fodrie FJ, Heck KL Jr (2011) Response of coastal fishes to the Gulf of Mexico oil disaster. PLoS One 6(7):1–8. https://doi.org/10.1371/journal.pone.0021609

Fonseca VF, França S, Vasconcelos RP, Serafim A, Company R, Lopes B, Bebianno MJ, Cabral HN (2011) Short-term variability of multiple biomarker response in fish from estuaries: influence of environmental dynamics. Mar Environ Res 72:172–178. https://doi.org/10.1016/j.marenvres.2011.08.001

Friedrichs KR, Harr KE, Freeman KP, Szladovits B, Walton RM, Barnhart KF, Blanco-Chavez J (2012) ASVCP reference interval guidelines: determination of *de novo* reference intervals in veterinary species and other related topics. Vet Clin Pathol 41:441–453. https://doi.org/10.1111/vcp.12006

Geffré A, Friedrichs K, Harr K, Concordet D, Trumel C, Braun J-P (2009) Reference values: a review. Veterinary Clinical Pathology 38:288–298. https://doi.org/10.1111/j.1939-165X.2009.00179.x

Gouraguine A, Díaz-Gil C, Reñones O, Otegui DS, Palmer M, Hinz H, Catalán IA, Smith DJ, Moranta J (2017) Behavioural response to detection of chemical stimuli of predation, feeding and schooling in a temperate juvenile fish. J Exp Mar Biol Ecol 486:140–147. https://doi.org/10.1016/j.jembe.2016.10.003

Gracia A, Murawski SA, Alexander-Valdés HM, Snyder S, López-Durán IM, Pulster EL, Ortega-Tenorio P, Frausto-Castillo A (2017) Impact of PAH at fish sub-individual level and resiliency consequences. Presentation at: Gulf of Mexico oil spill & ecosystem science, New Orleans, LA, 2017

Gracia A, Murawski SA, Alexander-Valdés RM, Vázquez-Bader AR, Snyder S, LópezDurán IM, Ortega-Tenorio P, Pulster EL, Frausto-Castillo JA (2018) Fish stock resiliency to environmental PAH. Presentation at: Gulf of Mexico oil spill & ecosystem science conference, New Orleans, LA, February 6–9, 2018

Gracia A, Murawski SA, Vázquez-Bader AR (2020) Impacts of deep spills on fish and fisheries (Chap. 25). In: Murawski SA, Ainsworth C, Gilbert S, Hollander D, Paris CB, Schlüter M, Wetzel D (eds) Deep oil spills – facts, fate and effects. Springer, Cham

Graham WM, Condon RH, Carmichael RH, D'Ambra I, Patterson HK, Linn LJ, Hernandez FJ (2010) Oil carbon entered the coastal planktonic food web during the *Deepwater Horizon* oil spill. Environ Res Lett 5. https://doi.org/10.1088/1748-9326/5/4/045301

Grosell M, Griffit RJ, Sherwood TA, Wetzel DL (2020) Digging deeper than LC/EC50: non-traditional endpoints and non-model species in oil spill toxicology (Chap. 29). In: Murawski SA, Ainsworth C, Gilbert S, Hollander D, Paris CB, Schlüter M, Wetzel D (eds) Deep oil spills – facts, fate and effects. Springer, Cham

Guzmán del Próo SA, Chávez EA, Alatriste FM, de la Campa S, De la Cruz G, Gómez L, Guadarrama R, Guerra A, Mille S, Torruco D (1986) The impact of the Ixtoc-1 oil spill on zooplankton. J Plankton Res 8:557–581. https://doi.org/10.1093/plankt/8.3.557

Hadfield MG, Paul VJ (2011) Natural chemical cues for settlement and metamorphosis of marine-invertebrate larvae. In: McClintock JB (ed) Marine chemical ecology. CRC Press, Boca Raton

Hall RA, Zook EG, Meaburn GM (1978) National Marine Fisheries Service survey of trace elements in the fishery resource. NOAA, Washington, DC

Hara TJ, Thompson BE (1978) The reaction of whitefish, *Coregonus clupeaformis*, to the anionic detergent sodium lauryl sulphate and its effects on their olfactory responses. Water Res 12:893–897. https://doi.org/10.1016/0043-1354(78)90042-8

Harr KE, Deak K, Murawski SA, Reavill DR, Takeshita RA (2018) Generation of red drum (*Sciaenops ocellatus*) hematology reference intervals with a focus on identified outliers. Vet Clin Pathol 47:22–28. https://doi.org/10.1111/vcp.12569

Harris EK, Boyd JC (1990) On dividing reference data into subgroups to produce separate reference ranges. Clin Chem 36:265–270

Harwell MA, Gentile JH (2006) Ecological significance of residual exposures and effects from the *Exxon Valdez* oil spill. Integr Environ Assess Manag 2:204–246. https://doi.org/10.1002/ieam.5630020303

Havel LN, Fuiman LA (2016) Settlement-size larval red drum (*Sciaenops ocellatus*) respond to estuarine chemical cues. Estuar Coasts 39:560–570. https://doi.org/10.1007/s12237-015-0008-6

Hedgpeth BM, Griffitt RJ (2016) Simultaneous exposure to chronic hypoxia and dissolved polycyclic aromatic hydrocarbons results in reduced egg production and larval survival in the sheepshead minnow (*Cyprinodon variegatus*). Environ Toxicol Chem 35:645–651. https://doi.org/10.1002/etc.3207

Herdter ES, Chambers DP, Stallings CD, Murawski SA (2017) Did the *Deepwater Horizon* oil spill affect growth of red snapper in the Gulf of Mexico? Fish Res 191:60–68. https://doi.org/10.1016/j.fishres.2017.03.005

Hernandez FJ Jr, Filbrun JE, Fang J, Ransom JT (2016) Condition of larval red snapper (*Lutjanus campechanus*) relative to environmental variability and the *Deepwater Horizon* oil spill. Environ Res Lett 11:094019

Hester MW, Mendelssohn IA (2000) Long-term recovery of a Louisiana brackish marsh plant community from oil-spill impact: vegetation response and mitigating effects of marsh surface elevation. Mar Environ Res 49:233–254. https://doi.org/10.1016/S0141-1136(99)00071-9

Hidaka H, Tatsukawa R (1989) Avoidance by olfaction in a fish, medaka (*Oryzias latipes*), to aquatic contaminants. Environ Pollut 56:299–309. https://doi.org/10.1016/0269-7491(89)90075-4

Incardona JP, Gardner LD, Linbo TL, Brown TL, Esbaugh AJ, Mager EM, Stieglitz JD, French BL, Labenia JS, Laetz CA, Tagal M, Sloan CA, Elizur A, Benetti DD, Grosell M, Block BA, Scholz NL (2014) *Deepwater Horizon* crude oil impacts the developing hearts of large predatory pelagic fish. Proc Natl Acad Sci USA 111:E1510–E1518. https://doi.org/10.1073/pnas.1320950111

Ishida Y, Kobayashi H (1995) Avoidance behavior of carp to pesticides and decrease of the avoidance threshold by addition of sodium lauryl sulfate. Fish Sci 61:441–446. https://doi.org/10.2331/fishsci.61.441

James MO, Kleinow KM (1994) Trophic transfer of chemicals in the aquatic environment. In: Malins DC, Ostrander GK (eds) Aquatic toxicology. Molecular, biochemical and cellular perspectives. Lewis Publishers, Boca Raton, pp 1–36

Javed M, Ahmad MI, Usmani N, Ahmad M (2017) Multiple biomarker responses (serum biochemistry, oxidative stress, genotoxicity and histopathology) in *Channa punctatus* exposed to heavy metal loaded waste water. Sci Rep 7:1675. https://doi.org/10.1038/s41598-017-01749-6

Kendall AW Jr, Ahlstrom EH, Moser HG (1983) Early life history stages of fishes and their characters. In: Ontogeny and systematics of fishes. Library of Congress, La Jolla

Kerambrun E, Sanchez W, Henry F, Amara R (2011) Are biochemical biomarker responses related to physiological performance of juvenile sea bass (*Dicentrarchus labrax*) and turbot (*Scophthalmus maximus*) caged in a polluted harbour? Comp Biochem Physiol Part C: Toxicol Pharmacol 154:187–195. https://doi.org/10.1016/j.cbpc.2011.05.006

Kerambrun E, Henry F, Courcot L, Gevaert F, Amara R (2012) Biological responses of caged juvenile sea bass (*Dicentrarchus labrax*) and turbot (*Scophthalmus maximus*) in a polluted harbour. Ecol Indic 19:161–171. https://doi.org/10.1016/j.ecolind.2011.06.035

Khan RA (2003) Health of flatfish from localities in Placentia Bay, Newfoundland, contaminated with petroleum and PCBs. Arch Environ Contam Toxicol 44:485–492. https://doi.org/10.1007/s00244-002-2063-9

Kim HN, Park CI, Chae YS, Shim WJ, Kim M, Addison RF, Jung JH (2013) Acute toxic responses of the rockfish (*Sebastes schlegelii*) to Iranian heavy crude oil: feeding disrupts the biotransformation and innate immune systems. Fish Shellfish Immunol 35:357–365. https://doi.org/10.1016/j.fsi.2013.04.041

Klinger DH, Dale JJ, Machado BE, Incardona JP, Farwell CJ, Block BA (2015) Exposure to *Deepwater Horizon* weathered crude oil increases routine metabolic demand in chub mackerel, *Scomber japonicus*. Mar Pollut Bull 98:259–266. https://doi.org/10.1016/j.marpolbul.2015.06.039

Kwan CK, Sanford E, Long J (2015) Copper pollution increases the relative importance of predation risk in an aquatic food web. PLoS One 10:e0133329. https://doi.org/10.1371/journal.pone.0133329

Lari E, Pyle GG (2017) Rainbow trout (*Oncorhynchus mykiss*) detection, avoidance, and chemosensory effects of oil sands process-affected water. Environ Pollut 225:40–46. https://doi.org/10.1016/j.envpol.2017.03.041

Lecchini D, Dixson DL, Lecellier G, Roux N, Frédérich B, Besson M, Tanaka Y, Banaigs B, Nakamura Y (2017) Habitat selection by marine larvae in changing chemical environments. Mar Pollut Bull 114:210–217. https://doi.org/10.1016/j.marpolbul.2016.08.083

Lee EH, Kim M, Moon YS, Yim UH, Ha SY, Jeong CB, Lee JS, Jung JH (2018) Adverse effects and immune dysfunction in response to oral administration of weathered Iranian heavy crude oil in the rockfish *Sebastes schlegelii*. Aquat Toxicol 200:127–135. https://doi.org/10.1016/j. aquatox.2018.04.010

Martin CW (2017) Avoidance of oil contaminated sediments by estuarine fishes. Mar Ecol Prog Ser 576:125–134. https://doi.org/10.3354/meps12084

Maynard DJ, Weber DD (1981) Avoidance reactions of juvenile coho salmon (*Oncorhynchus kisutch*) to monocyclic aromatics. Can J Fish Aquat Sci 38:772–778. https://doi.org/10.1139/ f81-105

Murawski SA, Hogarth WT, Peebles EB, Barbeiri L (2014) Prevalence of external skin lesions and polycyclic aromatic hydrocarbon concentrations in Gulf of Mexico fishes, Post-*Deepwater Horizon*. Trans Am Fish Soc 143:1084–1097. https://doi.org/10.1080/00028487.2014.911205

Murawski SA, Fleeger JW, Patterson WF, Hu C, Daly K, Romero I, Toro-Farmer GA (2016) How did the *Deepwater Horizon* oil spill affect coastal and continental shelf ecosystems of the Gulf of Mexico? Oceanography 29:160–173

Murawski SA, Patterson W II, Campbell M (2018a) Has abundance of continental shelf fish species declined after *Deepwater Horizon*? Presentation at: Gulf of Mexico Oil Spill & Ecosystem Science Conference, New Orleans, LA, 2018

Murawski SA, Peebles EB, Gracia A, Tunnell JW Jr, Armenteros M (2018b) Comparative abundance, species composition, and demographics of continental shelf fish assemblages throughout the Gulf of Mexico. Mar Coast Fish: Dyn Manag Ecosys Sci 10:325–346

Naour S, Espinoza BM, Aedo JE, Zuloaga R, Maldonado J, Bastias-Molina M, Silva H, Meneses C, Gallardo-Escarate C, Molina A, Valdes JA (2017) Transcriptomic analysis of the hepatic response to stress in the red cusk-eel (*Genypterus chilensis*): insights into lipid metabolism, oxidative stress and liver steatosis. Plos One 12. https://doi.org/10.1371/journal. pone.0176447

National Research Council (NRC) (2003) Oil in the sea III: inputs, fates, and effects. The National Academies Press, Washington, DC. https://doi.org/10.17226/10388

Nixon Z, Zengel S, Baker M, Steinhoff M, Fricano G, Rouhani S, Michel J (2016) Shoreline oiling from the *Deepwater Horizon* oil spill. Mar Pollut Bull 107:170–178. https://doi.org/10.1016/j. marpolbul.2016.04.003

Olson GM, Meyer BM, Portier RJ (2016) Assessment of the toxic potential of polycyclic aromatic hydrocarbons (PAHs) affecting Gulf menhaden (*Brevoortia patronus*) harvested from waters impacted by the BP *Deepwater Horizon* Spill. Chemosphere 145:322–328. https://doi. org/10.1016/j.chemosphere.2015.11.087

Paris CB, Atema J, Irisson J-O, Kingsford M, Gerlach G, Guigand CM (2013) Reef odor: a wake up call for navigation in reef fish larvae. PLoS One 8:e72808. https://doi.org/10.1371/journal. pone.0072808

Patterson WF III, Chanton JP, Barnett B, Joseph H, Tarnecki JH (2020) The utility of stable and radio isotopes in fish tissues as biogeochemical tracers of marine oil spill food web effects. In: Murawski SA, Ainsworth C, Gilbert S, Hollander D, Paris CB, Schlüter M, Wetzel D (eds) Scenarios and responses to future deep oil spills – fighting the next war. Springer, Cham

Peterson CH, Rice SD, Short JW, Esler D, Bodkin JL, Ballachey BE, Irons DB (2003) Long-term ecosystem response to the *Exxon Valdez* oil spill. Science 302(5653):2082–2086. https://doi. org/10.1126/science.1084282

Portnoy DA, Fields AT, Greer JB, Schlenk D (2020) Genetics and oil: transcriptomics, epigenetics and population genomics as tools to understand animal responses to exposure across different time scales (Chap. 30). In: Murawski SA, Ainsworth C, Gilbert S, Hollander D, Paris CB, Schlüter M, Wetzel D (eds) Deep oil spills – facts, fate and effects. Springer, Cham

Pulster EL, Fogelson SB, Carr B, Murawski SA (2018) A spatiotemporal analysis of hepatic polycyclic aromatic hydrocarbon levels and pathological findings in red snapper (*Lutjanus campechanus*), post-*Deepwater Horizon*. Presentation at: Gulf of Mexico oil spill & ecosystem science conference, New Orleans, LA, 2018

Quintana-Rizzo E, Torres JJ, Ross SW, Romero I, Watson K, Goddard E, Hollander D (2015) Delta C-13 and delta N-15 in deep-living fishes and shrimps after the *Deepwater Horizon* oil spill, Gulf of Mexico. Mar Pollut Bull 94:241–250. https://doi.org/10.1016/j.marpolbul.2015.02.002

Ransom JT, Filbrun JE, Hernandez FJ Jr (2016) Condition of larval Spanish mackerel *Scomberomorus maculatus* in relation to the *Deepwater Horizon* oil spill. Mar Ecol Prog Ser 558:143–152. https://doi.org/10.3354/meps11880

Rehberger K, Werner I, Hitzfeld B, Segner H, Baumann L (2017) 20 years of fish immunotoxicology – what we know and where we are. Crit Rev Toxicol 47:516–542. https://doi.org/10.1080/10408444.2017.1288024

Reichert M, Blunt B, Gabruch T, Zerulla T, Ralph A, El-Din MG, Sutherland BR, Tierney KB (2017) Sensory and behavioral responses of a model fish to oil sands process-affected water with and without treatment. Environ Sci Technol 51:7128–7137. https://doi.org/10.1021/acs.est.7b01650

Reynaud S, Deschaux P (2006) The effects of polycyclic aromatic hydrocarbons on the immune system of fish: a review. Aquat Toxicol 77:229–238. https://doi.org/10.1016/j.aquatox.2005.10.018

Rooker JR, Kitchens LL, Dance MA, Wells RJD, Falterman B, Cornic M (2013) Spatial, temporal, and habitat-related variation in abundance of pelagic fishes in the Gulf of Mexico: potential implications of the *Deepwater Horizon* oil spill. PLoS One 8:e76080. https://doi.org/10.1371/journal.pone.0076080

Santana MS, Sandrini-Neto L, Filipak Neto F, Oliveira Ribeiro CA, Di Domenico M, Prodocimo MM (2018) Biomarker responses in fish exposed to polycyclic aromatic hydrocarbons (PAHs): systematic review and meta-analysis. Environ Pollut 242:449–461. https://doi.org/10.1016/j.envpol.2018.07.004

Schwacke LH, Smith CR, Townsend FI, Wells RS, Hart LB, Balmer BC, Collier TK, De Guise S, Fry MM, Guillette LJ, Lamb SV, Lane SM, McFee WE, Place NJ, Tumlin MC, Ylitalo GM, Zolman ES, Rowles TK (2014) Health of common bottlenose dolphins (*Tursiops truncatus*) in Barataria Bay, Louisiana, following the *Deepwater Horizon* oil spill. Environ Sci Technol 48:93–103. https://doi.org/10.1021/es403610f

Schwaiger J, Wanke R, Adam S, Pawert M, Honnen W, Triebskorn R (1997) The use of histopathological indicators to evaluate contaminant-related stress in fish. J Aquat Ecosyst Stress Recover 6:75–86. https://doi.org/10.1023/a:1008212000208

Snieszko SF (1974) The effects of environmental stress on outbreaks of infectious diseases of fishes. J Fish Biol 6:197–208. https://doi.org/10.1111/j.1095-8649.1974.tb04537.x

Snyder SM, Pulster EL, Wetzel DL, Murawski SA (2015) PAH exposure in Gulf of Mexico demersal fishes, Post-*Deepwater Horizon*. Environ Sci Technol 49:8786–8795. https://doi.org/10.1021/acs.est.5b01870

Snyder SM, Pulster EL, Fogelson SB, Murawski SA (2019) Hepatic accumulation of PAHs and prevalence of hepatic lesions in golden tilefish from the northern Gulf of Mexico. Presentation at: Gulf of Mexico Oil Spill & Ecosystem Science Conference, New Orleans, LA, 2018

Sun S, Hu C, Tunnell JW Jr (2015) Surface oil footprint and trajectory of the Ixtoc-I oil spill determined from Landsat/MSS and CZCS observations. Mar Pollut Bull 101:632–641. https://doi.org/10.1016/j.marpolbul.2015.10.036

Tarnecki JH, Patterson WF (2015) Changes in red snapper diet and trophic ecology following the *Deepwater Horizon* oil spill. Mar Coast Fish 7:135–147. https://doi.org/10.1080/19425120.2015.1020402

Teh SJ, Adams SM, Hinton DE (1997) Histopathologic biomarkers in feral freshwater fish populations exposed to different types of contaminant stress. Aquat Toxicol 37:51–70. https://doi.org/10.1016/S0166-445X(96)00808-9

Tierney KB, Baldwin DH, Hara TJ, Ross PS, Scholz NL, Kennedy CJ (2010) Olfactory toxicity in fishes. Aquat Toxicol 96:2–26. https://doi.org/10.1016/j.aquatox.2009.09.019

Tunnell JW Jr (2016) Ixtoc I vs *Deepwater Horizon*: a different day, a different time, but with similarities. Presentation at: Gulf of Mexico Oil Spill and Ecosystem Science, New Orleans, Louisiana, 2016

Walton RM (2001) Establishing reference intervals: health as a relative concept. Semin Avian Exotic Pet Med 10:66–71. https://doi.org/10.1053/S1055-937X(01)80026-8

Wang TH, Secombes CJ (2013) The cytokine networks of adaptive immunity in fish. Fish Shellfish Immunol 35(6):1703–1718. https://doi.org/10.1016/j.fsi.2013.08.030

White ND, Godard-Codding C, Webb SJ, Bossart GD, Fair PA (2017) Immunotoxic effects of in vitro exposure of dolphin lymphocytes to Louisiana sweet crude oil and Corexit. J Appl Toxicol 37(6):676–682. https://doi.org/10.1002/jat.3414

Whitehead A, Dubansky B, Bodinier C, Garcia TI, Miles S, Pilley C, Raghunathan V, Roach JL, Walker N, Walter RB, Rice CD, Galvez F (2012) Genomic and physiological footprint of the *Deepwater Horizon* oil spill on resident marsh fishes. Proc Natl Acad Sci USA 109:20298–20302. https://doi.org/10.1073/pnas.1109545108

Wunderlich AC, Silva RJ, Zica ÉOP, Rebelo MF, Parente TEM, Vidal-Martínez VM (2015) The influence of seasonality, fish size and reproductive status on EROD activity in *Plagioscion squamosissimus*: implications for biomonitoring of tropical/subtropical reservoirs. Ecol Indic 58:267–276. https://doi.org/10.1016/j.ecolind.2015.05.063

Yáñez-Arancibia A (1986) Ecología, impacto ambiental y recursos pesqueros: El caso del Ixtoc-1 y los peces. In: Ecología de la Zona Costera Análisis de Siete Tópicos. México, p 180

Yáñez-Arancibia A, Lara-Domínguez AL, Sánchez-Gil P, Álvarez H, Vargas I, Aguirre A (1982) Caracterización Ambient al del Sistema Ecológico y Análisis Comparativo de las Poblaciones de Peces de la Sonda de Campeche y de la LagunadeTérminos antes y Después del Derrame Petrolero del Pozo IXTOC-I (Informe Final). vol 4 partes. PC-EESC/UNAM/ICML(IF)

Chapter 25
Impacts of Deep Oil Spills on Fish and Fisheries

Adolfo Gracia, Steven A. Murawski, and Ana Rosa Vázquez-Bader

Abstract The Gulf of Mexico (GoM) total fishery production varies around one million metric tons per year. Fishery production is based on a diverse set of invertebrate and finfish species, including estuarine, continental shelf, and open-ocean species. The GoM has been subjected to two large oil spills in the Southern (Ixtoc 1, 1979–1980) and in the Northern GoM (*Deepwater Horizon* 2010) that caused serious concern about impacts on the abundance and seafood safety of fishery resources. Scientific evidence does not indicate a collapse or a clear impact on long-term fishery productivity related to either the Ixtoc 1 or DWH oil spills. Fishery landings in the Northern GoM returned quickly to pre-spill levels, and short-term declines could be attributed to the extensive fishery closure in the US exclusive economic zone. In the Southern GoM, fishery production post-Ixtoc 1 decreased dramatically over time attributed primarily to overharvesting of the main target species. Although no oil spill impact on the fishery resources was apparent at the population level, there is considerable evidence of impacts at the organismal and sub-individual levels, and there is concern how these effects could impact fishery resources in the long term. The responses of fish and shellfish populations are analyzed in relation to reproductive strategies, distribution of nursery grounds and critical habitats, exploitation status, oil spill distribution, and overall pollution levels. Fish and shellfish populations show a high capacity to withstand/recover from natural and anthropogenic impacts by taking advantage of favorable environmental conditions and by evolving life history strategies robust to seasonal and interannual variability. Stock resiliency is affected by several factors but mainly overharvesting that may reduce reproductive potential and compromise fishery resource resiliency in the eventual case of another large-scale oil spill disaster.

A. Gracia (✉) · A. R. Vázquez-Bader
Instituto de Ciencias del Mar y Limnología Universidad Nacional Autonóma de México, Ciudad de México, Mexico
e-mail: gracia@unam.mx; ana-rosav@hotmail.com

S. A. Murawski
University of South Florida, College of Marine Science, St. Petersburg, FL, USA
e-mail: smurawski@usf.edu

© Springer Nature Switzerland AG 2020
S. A. Murawski et al. (eds.), *Deep Oil Spills*,
https://doi.org/10.1007/978-3-030-11605-7_25

Keywords Gulf of Mexico fisheries · Red snapper · Bluefin tuna · Pink shrimp · White shrimp · Oil spill effects on fisheries

25.1   Introduction

The Gulf of Mexico (GoM) is one of the most productive and diverse large marine ecosystems of the world contributing significantly to the overall fishery production of the three bordering countries. Its high diversity of fishes and shellfishes has been the basis of important commercial and recreational fisheries in the diverse ecosystems around the GoM. Species importance varies among the different regions, but all rely upon inshore, coastal, benthic, demersal, and oceanic-pelagic species, including snappers, groupers, drums, croakers, shrimps, blue crabs, stone crabs, spiny lobster, octopus, oyster, menhaden, mackerels, dolphinfish, billfish, and tunas, among others.

Total annual fishery landings of the three countries comprising the GoM (Cuba, Mexico, USA) average approximately one million metric tons. Fishery production is dominated by a few species that account for most of the landing volume and value, including coastal and pelagic species, shelf demersal species, and benthic species for crustaceans including shrimps (brown, pink, and white shrimp), blue crabs, and mollusks including oysters. Shrimp fisheries are the most valuable in both Mexico and the US GoM. The three countries have had similar harvesting histories reaching historical maximum exploitation levels in the 1980s–1990s, followed by a general decline in landings afterward. Management regimes show differences in the three countries, although they share a common situation: most of the stocks are fully exploited and some of them are overfished. In US waters, the Gulf of Mexico Fishery Council's Management Plans list three stocks as overfished. In the Southern Gulf of Mexico, most of the principal stocks have been overfished since the beginning of the current century (Díaz de León et al. 2004), and a high proportion of these fishery resources are in decline. A similar situation has been reported in Cuba where 74% of the stocks are overexploited and 5% collapsed with an increasing trend in overexploitation (Baisre 2018). Total US landings in the Northern Gulf are twice that of Mexican fishery production, and the countries' landings are considerably greater than Cuba's GoM landings. Overall abundance in the Northern GoM is higher than the other two regions (Tunnell Jr 2017; Murawski et al. 2018a).

Besides its fishery importance, the large marine ecosystem of the Gulf of Mexico is also the basis of many important coincident economic activities. Among them the oil industry plays a principal role as there is a very large historical development of oil and gas exploitation facilities beginning in the 1930s. Related to this activity, the GoM has been subjected to two mega oil spills along its history, one in the Southern GoM (Ixtoc 1, 1979–1980) and more recently in the Northern GoM (*Deepwater*

Horizon 2010). These two mega oil spills (the largest submarine blowouts ever to occur globally) caused serious concern regarding potential impact on the abundance and quality of fishery resources. Were the stocks seriously affected by contamination? How do the stocks respond to these oil spills? What is the resiliency of fishery resources of the GoM? These are some of the questions that have arisen as a result of the previous two mega oil blowouts. This also begs the question of how yet another large blowout would affect fishery resources, depending on where, when, and under what circumstances it might occur. In this chapter we analyze the impact of these two mega oil spills on fisheries of the GoM and discuss resiliency potential and strategies to make these stocks more resilient in the face of eventual future oil spill disasters in the GoM.

25.2 The Oil Spills

The Ixtoc 1 well blowout in the Southern GoM was considered the world's largest oil spill in the world when it happened (Jernelöv and Lindén 1981). The exploratory well was located about 94 km NW of Ciudad del Carmen, Campeche, at a water depth of 54 m. The Ixtoc 1 oil spill lasted almost 10 months (3 June 1979 to 23 March 1980) resulting in 3.1 million barrels of oil being spilled into the SW GoM. It was estimated that after the efforts to collect the oil and processes like oil burning, evaporation, and biodegradation, 1.023 million barrels of oil were left in the sea (PCEESC 1980). Oil was dispersed by the dominant current pattern along a large area in the Southern GoM, and some oil eventually reached Texas beaches (Sun et al. 2015). A fortunate situation was that very little oil was transported into coastal lagoons which are the main nursery grounds of many commercially important species (e.g., shrimps). High river flow due to heavy rains and current patterns played a critical role in preventing oil from reaching the main nursery grounds including the Laguna de Términos.

The *Deepwater Horizon* (DWH) blowout 30 years after Ixtoc 1 proved to be larger in both volume and areal extent. The explosion of the DWH occurred on 20 April 2010, 78 km off the Louisiana coast, spilling a large quantity of crude oil into the Gulf of Mexico. During the 87 days of the spill (20 April to 15 July), approximately 4.1 million barrels of light crude oil leaked into the marine environment (McNutt et al. 2012). The accident occurred at a depth of 1500 meters. Part of the oil was removed from the marine environment through a combination of cleanup activities and natural processes including dissolution, evaporation, biodegradation, and dispersion. The oil spilled into the GoM was estimated to be 3.7 million barrels (McNutt et al. 2012; Ryerson et al. 2012) that dispersed in a large area.

25.3 GoM Fishery Landings

The extension and timing of both mega oil spills caused great concern regarding potential impact on fish and shellfish stocks and their population dynamic processes including spawning, larval survival, recruitment, and eventually fishery production.

During the Ixtoc 1 spill, there was no general closure in the SGoM; only in highly contaminated areas, was fishing banned or restricted. Mexican fishery production did not undergo a large decrease in total fish and shellfish landings nor exhibit signs of collapsed stocks after the blowout (Fig. 25.1). Amezcua-Linares et al. (2015) suggested that a 50% landing decrease recorded in Campeche state in 1982 was related to high larval mortality occurring at the time of the Ixtoc 1 blowout. However, this reduction did not continue after 1982, and landings increased in the following years. Landing reductions were higher in Veracruz state but also recovered quickly in the following year (Fig. 25.2). Nonetheless, landing variations, except Veracruz, were typically of the order observed pre-spill and likely due to factors such as environmental variability or changes in fishing patterns. A general impact was not evident post-spill; landings showed a steady increase and attained a historical maximum in 1994 with a long-term decreasing trend thereafter. Overharvesting seems to be the main factor responsible of this trend that has resulted in 94% of fishery resources of the SGoM being in fully or overexploited condition (DOF 2018). In recent years an upward overfishing has diminished somewhat although only a small percentage (6%) of fishery resource species are considered underexploited.

During the *Deepwater Horizon* oil spill, fishery closures were established in US federal waters due to concern for the integrity and safety of seafood. Closures increased in size as the oil surface expanded reaching 290,000 km² equivalent to

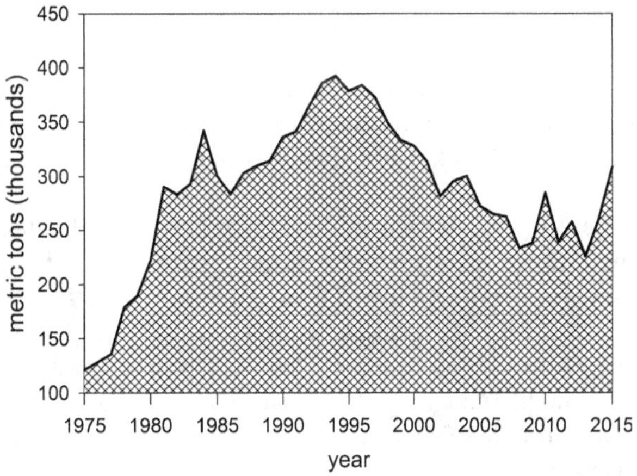

Fig. 25.1 Total fishery landings (thousands of metric tons) in the Mexican waters of the Gulf of Mexico, 1975–2014. (Source: CONAPESCA)

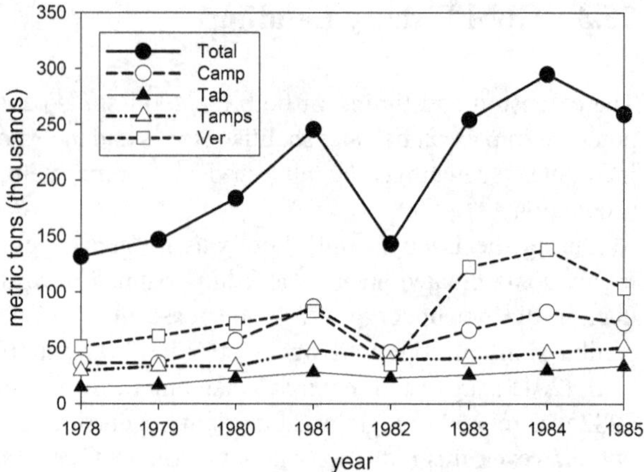

Fig. 25.2 Commercial landings of the Campeche (CAMP), Tabasco (TAB), Tamaulipas (TAMPS), and Veracruz (VER) states in the Gulf of Mexico, 1978–1985. (Source: CONAPESCA)

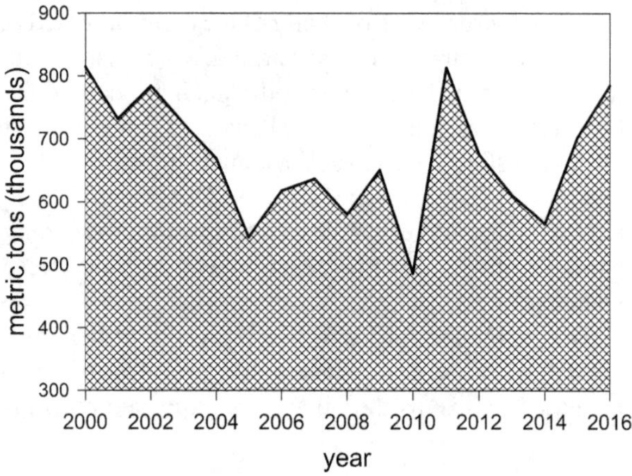

Fig. 25.3 US total fishery landings (thousands of metric tons) of the Gulf of Mexico, 2000–2016. (Source NMFS)

37% of Gulf waters (Ylitalo et al. 2012; Carroll et al. 2016). Fishery closures negatively affected total commercial landings that decreased 25%. Once closures were opened (by February of 2011), fishery landings returned quickly to pre-spill levels (Fig. 25.3).

Various studies focused on the evaluation of the oil spill impact on fishing resources in the affected area (Fodrie and Heck Jr 2011; Xia et al. 2012; Moddy et al. 2013; Rooker et al. 2013). Most of them pointed out that there was no collapse of fish or crustacean related to the *Deepwater Horizon* oil spill. Neither the US National Marine Fisheries Service reported any collapse or downward trend of fisheries related to the spill. Tunnell Jr (2014) conducted a review of the main commercial species of fish and shellfish, concluding that there were no adverse effects on fish and shellfish stocks of the Northern GoM, and he expected no negative effects in the future. Some stocks declined in the NGoM region after DWH (e.g., red snapper; Pulster et al. 2019), but most remained stable.

Scientific evidence did not indicate a collapse or a clear impact on fisheries (other than a near-term reduction due to closures in the NGoM) in the case of Ixtoc 1 or *DWH*. Although no dramatic oil spill impact on the fishery resources was registered at the population level, there is considerable evidence of impacts at the organismal and sub-individual levels (Incardona et al. 2014, Dubansky et al. 2013, Brown-Peterson et al. 2017; Pulster et al. 2019), and there is concern how these effects could impact fishery resources in the long term. Murawski et al. (2016) point out that although fishery landings rebounded in 1 or 2 years following the spill, there may be impacts at the population level that could affect stock productivity. These impacts could be differential depending on the life strategies of fish and shellfish. Species with a short life cycle may respond by either collapsing or recovering quickly depending on the impact at the population level. In longer-lived species, the impacts could manifest in the long term if spawning potential or growth rates are affected. In order to analyze these responses, we selected three representative species or groups of commercial/recreational importance (red snapper, tuna, and pink and white shrimp) with different habitats and ecological roles in the GoM (continental demersal platform, benthic, and pelagic-oceanic species) and compare their behaviors in response to the two mega oil blowouts.

25.4 Red Snapper

Red snapper is a traditionally important fishing resource in the GoM and is also sought as a valuable recreational species. Red snappers are bottom-dwelling and distributed throughout the GoM and Caribbean. They live on sandy, rocky, muddy bottoms and coral reefs. The red snapper can live for 50 years (Szedlmayer and Shipp 1994; Wilson and Nieland 2001) and mature at age of 2. A female can produce up to 55 million eggs throughout her life (SEDAR7 2005). Spawning occurs throughout the year, but mainly from April to September, with most reproduction occurring between June and August (Collins et al. 1995). After hatching, larvae remain in the pelagic environment from 26 to 30 days (Rooker et al. 2004; Gallaway et al. 2007), go through a metamorphosis, and settle on the seabed in reefs, in rocky habitats, or on shell remains in sand bottom. In the juvenile phase, under 2 years of age, red snappers are caught incidentally by shrimp trawlers. After 2 years of age, red snappers are subjected to fishing and remain so throughout their life (Gallaway et al. 2009).

During the Ixtoc 1 oil spill, Mexican red snapper landings did not collapse. Catches in 1979, 1980, and 1981 declined 7%, 35%, and 25%, respectively, in comparison to the average landings of the 1973–1978 pre-spill period (Fig. 25.4). Even though there was a 35% decrease in the immediate post-spill year, landings were similar to catches in 1976 and were in the range of observed fluctuations. Landings quickly returned to pre-spill levels in 1982 and increased steadily until 1993 when the historical maximum catch was attained. After the Ixtoc 1 oil spill, a long-term impact was not apparent in red snapper landings in the SGoM. The steep decreasing

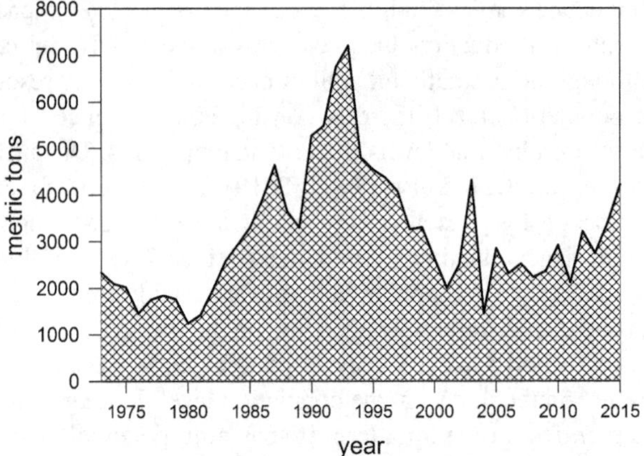

Fig. 25.4 Red snapper commercial landings (metrics tons) in the Mexican area of the Gulf of Mexico, 1973–2015. (Source: CONAPESCA)

Fig. 25.5 Red snapper commercial landings and recreational catch (metric tons) in the US area of Gulf of Mexico, 2000–2016. (Source: NMFS)

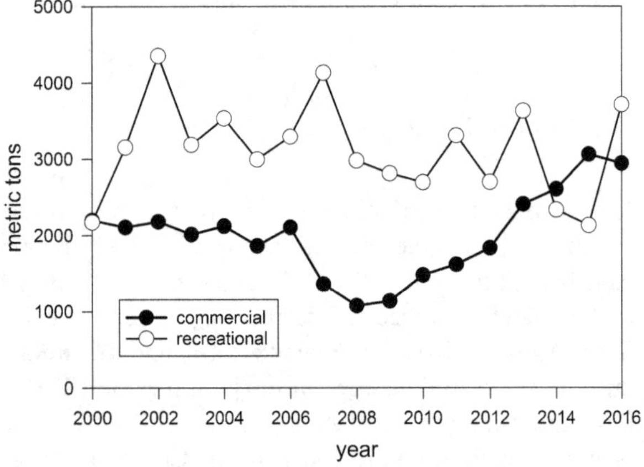

landing trend in catch levels after 1993 is attributable to overexploitation. Recently, the fishery has shown a slight landing increase; however, the fishery resource is considered to be in a fully to overexploited condition (DOF 2018).

Commercial landings of red snapper during the *Deepwater Horizon* oil spill did not exhibit a strong decline, and, despite the fishery closures, catches in 2010 were 25% greater than those in 2008 and 2009 pre-spill years. Moreover, a steady catch increase was apparent in post-spill years (Fig. 25.5). As well, red snapper recreational catch increased in recent years. Nonetheless, there is some evidence that the eastern GoM red snapper substock may have exhibited recruitment declines post-2010 (Murawski et al. 2018b; Pulster et al. 2019). This is contrary to the positive trend observed in the western GoM substock (SEDAR 2013) before 2010 that may be enhancing catches in the northeastern GoM. Szedlmayer and Mudrak (2014) conducted a field study to test possible reduced recruitment or year class failure of red snapper on reefs off the Alabama coast, finding no evidence of the oil spill affecting abundance of early life stages. They concluded that red snapper densities at age 0 (settlement on the seabed) were similar to previous years and that juveniles in 2011 were abundant on the reefs. In the eastern GoM substock, a sustained recruitment decline may have an impact on the spawning stock biomass in the long term. There is no conclusive evidence that this is due to the effect of the DWH oil spill or alternatively to the influence of environmental factors (*Deepwater Horizon* Natural Resource Damage Assessment Trustees 2016; Murawski et al. 2016). If environmental variability is responsible, the spawning stock may recover once normal condition is re-established due to contribution of different year classes.

25.5 Tunas

Tunas are highly migratory species that are under the management of the International Commission for the Conservation of Atlantic Tunas (ICCAT). In the GoM, yellowfin tuna is the principal target (both in Mexico and the USA), although incidental catches may include migratory species such as other tunas, including marlins, billfish, sharks, sea turtles, and other species of fish. Among the most important tunas is bluefin tuna because of its economic value and because it has been intensively overexploited, leading to near-collapse of the Eastern and Western Atlantic stocks (ICCAT 2007).

Tunas are long-lived species and have wide distribution in the pelagic environment of the GoM. Sexual maturity is reached between 2.5 and 3 years of age in yellowfin tuna (González et al. 2002), whereas bluefin tunas reach reproductive maturity at 8–12 years old. Yellowfin tunas spawn throughout the year, but intensity varies by region, while bluefin spawns over a narrower time span. Spawning occurs in the GoM in April to July for yellowfin tuna (Arocha et al. 2000) and April to June for bluefin tuna (ICCAT 2007). Both species have high fecundity with pelagic larvae subjected to the current pattern for dispersion. The NGoM was designated as the spawning area, and no fishing targeting on bluefin tuna is allowed in US waters of the GoM.

During the Ixtoc 1 oil spill, there was not a Mexican fishery for tuna, which was initiated after the oil spill in 1981. Atlantic landings of bluefin tuna show a decreasing trend since 2000 with an increasing trend in the last 10 years (Fig. 25.6).

The DWH oil spill overlapped the spawning area of bluefin tuna in the GoM, so there was considerable concern that the spill could affect larval survival. The oil spill could overlap the entire area, directly impacting the survival of young fish and, in the medium term, spawning stock biomass and productivity. However, subsequent studies showed that the potentially affected spawning area was less than 10% (Muhling et al. 2012). Hazen et al. (2016) suggested that although the oil overlapped with a small portion of predicted spawning area, the combined effect of environmental variation and bycatch mortality may result in a significant impact on

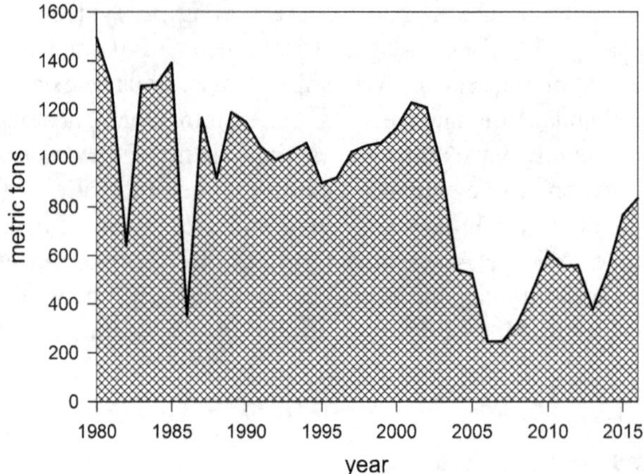

Fig. 25.6 Atlantic and Gulf of Mexico USA commercial landings of bluefin tuna (metric tons), 1980–2016. (Surce: NMFS)

Fig. 25.7 Catch per unit effort (CPUE) of bluefin tuna incidental catch of the Mexican tuna fleet, 1994–2015. (Source: ICCAT)

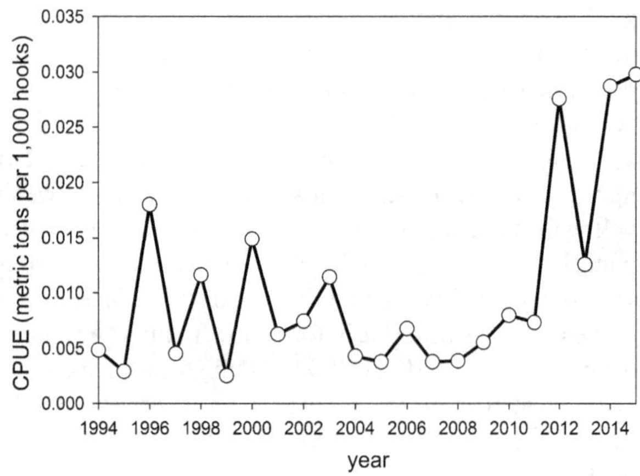

bluefin tuna. On the other hand, an ICCAT report (2015) showed that numbers of tuna larvae in 2011 and 2013 were the highest recorded in several decades which could be due to increasing spawner abundance. Coincidently Mexican fleet registered relatively high catch per unit effort of bluefin tuna in the 2012–2015 period (Fig. 25.7).

Tunas have a relatively long life with multiple age classes that contribute to the population spawning biomass in any given year. Therefore, an impact that is limited in time and space to a single annual class (2010) could have a relatively small effect on the total population spawning potential.

25.6 Shrimps

Shrimps are the most important fishery resources in the GoM due to their economic and social value. The main shrimp species throughout the GoM are brown, pink, and white shrimp. These species have been subjected both inshore and offshore to relatively high fishing effort since the last century by both artisanal and industrial fisheries, and many of their stocks are fully or overexploited in the GoM. The potential impact of the two mega oil spills on shrimp fisheries was of great concern during each event. The most important fisheries potentially impacted by the Ixtoc 1 and *DWH* oil spills were pink and white shrimp, because they are the most important shrimp fisheries in the oil spill areas.

Shrimps have an annual life cycle, spawning throughout much of the year, with two seasonal reproduction peaks that result in two main generation peaks in the year: late spring-early summer and autumn (Gracia et al. 1997). Sexual maturity is reached at an average of 8 months, and the species have relatively high fecundity. Spawning occurs offshore in the pelagic environment where larvae are dispersed by current patterns for approximately 2 weeks. Afterward, they enter estuaries and coastal lagoons as post-larvae, remain in these nursery areas for 3–4 months as juveniles, and then migrate to the marine environment (Gracia 1992). Shrimp juvenile abundance in the nursery grounds is critical for determining the recruitment to adult stocks offshore. Due to shrimp life history characteristics, the population renews annually.

The shrimp fishery in the Campeche Sound was the most important in GoM Mexican waters in the 1950–1980 period and was based mainly on pink and white shrimp. Shrimp stock exploitation reached maximum levels in 1970 (Fig. 25.8). Both stocks were fully exploited in the 1970s then started a steady production decline before Ixtoc 1 blowout that resulted in yields decreasing to around 10% of their maximum historical levels (Gracia and Vázquez-Bader 1999; Gracia 2004). This marked declining trend and fishery collapse was frequently attributed to

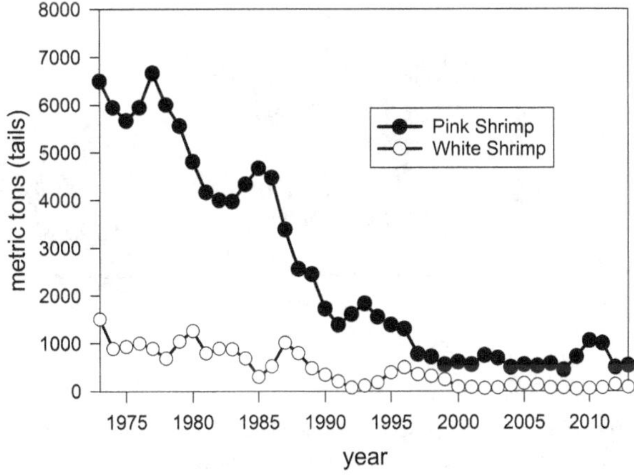

Fig. 25.8 Landings of pink and white shrimp in Mexican fisheries, 1973–2013. (Source: CONAPESCA)

long-term effects of Ixtoc 1 oil spill and/or combined impacts of chronic pollution related to oil industry activities in the Campeche Sound. However, in retrospect, overfishing seems to be the main driving factor responsible for the decline. White and pink shrimp spawners (age 8+ months) and recruits (age 4 months) pre- and post-spill [abundance estimated through virtual population analysis (Gracia 1991, 1995)] did not show an immediate effect of Ixtoc 1 for both stocks (Fig. 25.9). White and pink shrimp recruitment and spawner abundance variation seem to be more associated to environmental variability. However, a sustained declining trend was evident in recruit and spawner numbers of both stocks. Analysis of the

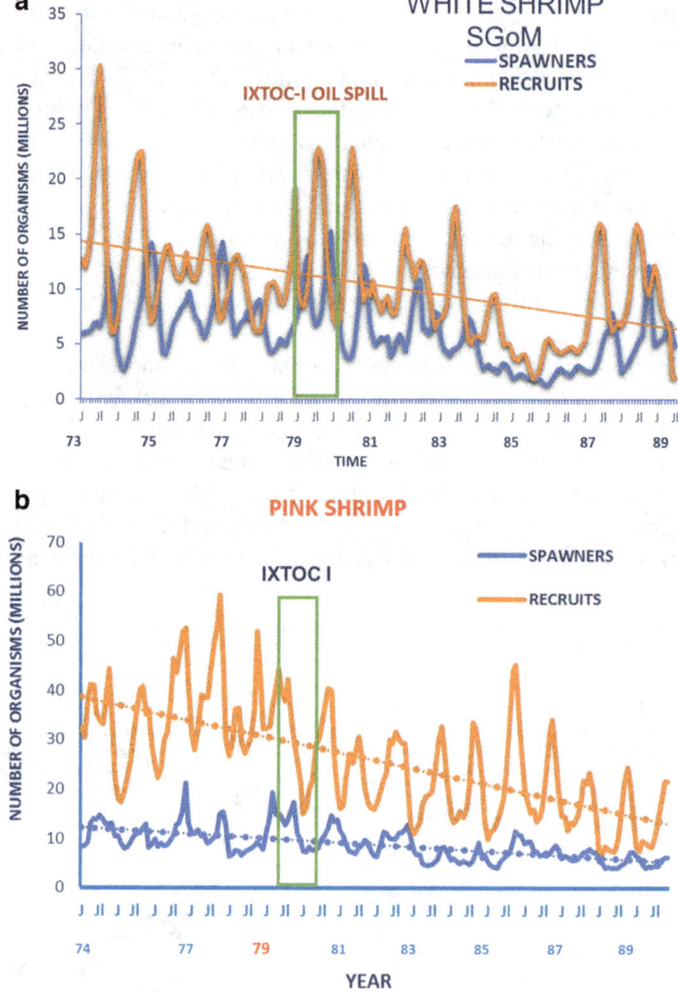

Fig. 25.9 Pre- and post-spill abundance of spawners and recruits of white (**a**) and pink shrimp (**b**) in the Ixtoc 1 area estimated through virtual population analysis (VPA), 1973–1989

stock-recruitment relationship of the two stocks arranged in biological years for white (June to May) and pink shrimp (October to September) showed no clear impact of Ixtoc 1 oil spill on the spawning stock and recruitment relationship (Fig. 25.10). Spawning stock and recruitment numbers of both species showed similar behavior until 1983. The stock-recruitment trajectories suggest that variations were within historical range before and after the Ixtoc 1 blowout. However, shrimp populations then shifted to a lower abundance stage after the oil spill. It seems that there was a sustained impact on the recruitment that affected the spawning stock biomass. The question is if this was a long-term impact of the Ixtoc 1 oil spill. Considering the annual renewal of shrimp populations is very unlikely and according to Gracia (1996) and Gracia and Vázquez-Bader (1998) is more related to overfishing due to emergence and development of intensive new artisanal fisheries acting

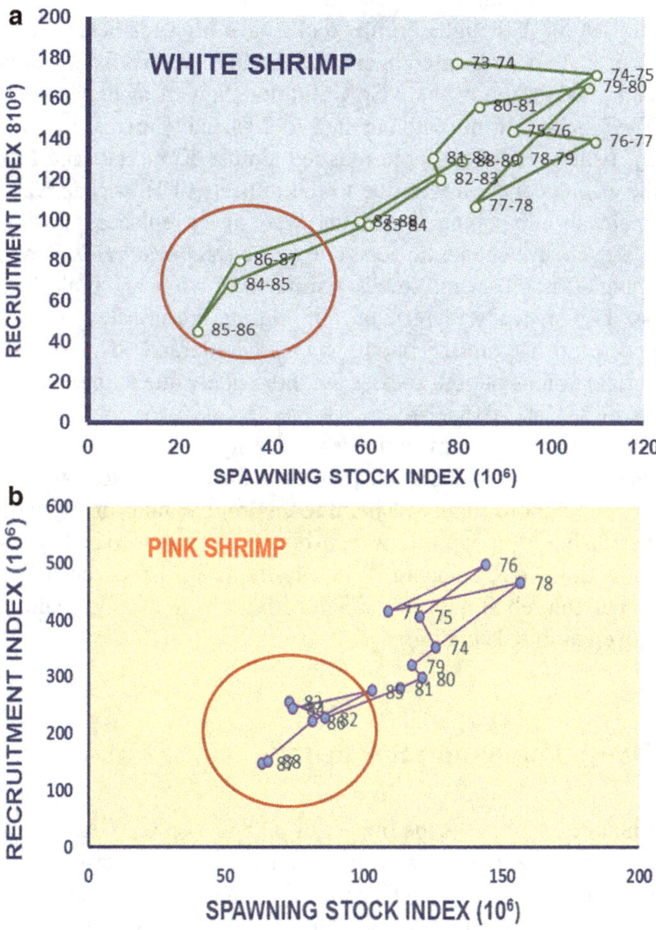

Fig. 25.10 Spawning stock-recruitment relationship (biological years) trajectories of white (**a**) and pink shrimp (**b**) in the Ixtoc 1 area. Time periods circled are post-Ixtoc 1

on white shrimp spawners offshore and on pink shrimp juveniles in the nursery areas in the early 1980s which added to the full developed offshore industrial fishing effort. Gracia (1996) stated that the increase of selective artisanal fishing pressure on white shrimp spawners led to recruitment overfishing affecting population renewal. In the case of pink shrimp, the new artisanal was directed to juvenile even under 2 months old (35 mm total length) in nursery grounds. Excessive fishing effort resulted in growth overfishing and then affected the spawning stock biomass, leading to recruitment overfishing.

Gracia (2004) stated that even though there could be many other factors affecting the shrimp stock productivity in the Campeche Sound, the accumulated fishing effort exerted on different shrimp life stages affected the shrimp reproductive potential and is the main factor driving shrimp fishery collapse. Signs of recoveries in white shrimp stock occurred in 1986–1987 (Fig. 25.9) and were associated with decreasing artisanal fishing effort combined with favorable environmental conditions (Gracia 1996). Although shrimp stocks have high resiliency, recurrent high fishing by artisanal effort resulted in chronic recruitment overfishing episodes.

White and pink shrimp in the NGoM did not show clear impacts on yield pre- and post-DWH. White shrimp catch in the NGoM steadily increased in recent years; coincidently, fishing effort has been reduced almost 50% (Hart and Nance 2013). Much of the effort reduction was due to the impacts of Hurricane Katrina, which destroyed many shrimp fishing vessels that were never replaced. Closing of almost 1/3 of the US exclusive economic zone due to the *Deepwater Horizon* oil spill also reduced fishing effort on shrimp stocks. Estimates of white and pink shrimp spawning biomass show a steady increasing trend during the last few years. As well as recruitment of white and pink shrimp has an increased trend. Spawning biomass and recruitment fluctuations of both species are more likely due to the effect of environmental variations (Hart 2015a, b).

Shrimp reproductive strategy with early maturity, spawning along the year, high fecundity, and two annual generations, which contribute to each other's spawning, make shrimp populations highly resilient to environmental or man-made stressors. In both fisheries, but in an opposite way, fishing effort seems to be the principal factor that is affecting stock behavior. While in the NGoM reduced fishing effort is allowing population rebuilding, in the SGoM high fishing effort maintains shrimp stock in an overexploited condition.

25.7 Fishery Population Resiliency

One of the main concerns was the impact of oil spills on reproduction and population in early stages. Murawski et al. (2016) point out three important conditions that affect significantly populations on its early life history: (1) exposure to oil results in increased mortality and reduced fitness of animals exposed, (2) oil concentrations in the environment are sufficient to cause increased mortality and sublethal impairment, and (3) a substantial fraction of the population is exposed to critical levels of oil in the environment.

These three conditions were not met in the cases of DWH and Ixtoc 1. Oil distributions in both spills were widespread but patchily distributed (Sun et al. 2015). Also, oil concentration and toxicity varied owing to dissolution, photodegradation, evaporation, biodegradation, and dispersion in the marine environment. In this way the matching of oil spill to main population processes in time and space was not large enough to have catastrophic impact. This is in contrast to other spills such as *Exxon Valdez*, which occurred in a more confined area but which overlapped with much of the distribution and critical spawning habitats of local fish stocks and other species that were affected by the spill (Rice et al. 1996).

Life history adaptations of fishery stocks confer them high resiliency to environmental stressors (either due to anthropogenic causes or natural environmental variation) as long as the life history strategy is not overwhelmed. Although the three species groups illustrated have different life histories—from annual to long lived—their reproductive strategies result in a high reproductive potential that allows them to opportunistically take advantage of favorable environmental windows. Sublethal effects that impair individual survival, growth, and reproduction could only have long-term impact if they undermine the overall fitness of the population in the evolutionary sense. If stocks are in healthy condition, effects could be mitigated by the cumulative population's reproductive potential. In the NGoM, the proportion of overfished species has reduced from 40% to less than 5% (Karnauskas et al. 2017), and during the Ixtoc 1 oil spill, most of the fishery stocks in the SGoM were not as overfished as they are now. Similarly, overfishing has reduced the resiliency potential of many stocks off Cuba (Baisre 2018). Stock resiliency could also be affected by other factors, but overharvesting is a key factor that affects reproductive potential and may compromise fishery resource conservation in the case of an eventual oil spill disaster.

Acknowledgments The synthesis of this research was made possible by grants from the Gulf of Mexico Research Initiative through the Center for the Integrated Modeling and Analysis of the Gulf Ecosystem (C-IMAGE).

References

Amezcua-Linares F, Amezcua F, Gil-Manrique B (2015) Effects of the Ixtoc I oil spill on fish assemblages in the Southern Gulf of Mexico. In: Alford JB, Peterson MS, Green CC (eds) Impacts of oil spill disasters on marine habitats and fisheries in North America. CRC Press, Boca Raton, pp 209–236

Arocha F, Lee DW, Marcano LA, Marcano JS (2000) Preliminary studies on the spawning of yellowfin tuna, *Thunnus albacares*, in the western Central Atlantic. Collect Vol Sci Pap ICCAT 52:538–551

Baisre JA (2018) An overview of Cuban commercial marine fisheries: the last 80 years. Bull Mar Sc. https://doi.org/10.5343/bms.2017.1015

Brown-Peterson NJ, Krasnec MO, Lay CR, Morris JM, Griffitt RJ (2017) Responses of juvenile southern flounder exposed *to Deepwater Horizon* oil-contaminated sediments. Environ Toxicol Chem 36:1067–1076

Carroll M, Genter B, Larkin S, Quigley K, Perlot N, Dehner L, Kroetz A (2016) An analysis of the impacts of the *Deepwater Horizon* oil spill on the Gulf of Mexico seafood industry. U. S. Dept. of the Interior. Bureau of Ocean Energy Management, Gulf of Mexico OCS Region, New Orleans, LA. OCS Study BOEM 2016-020

Collins LA, Johnson AG, Keim CP (1995) Spawning and annual fecundity of the Red Snapper (*Lutjanus campechanus*) from the northeastern Gulf of Mexico. In: Arreguín-Sánchez F, Munro JL, Balgos MC, Pauly D (eds) Biology, fisheries and culture of tropical groupers and snappers. ICLARM Conference Proceedings 48, pp 174–188

Deepwater Horizon Natural Resource Damage Assessment Trustees (2016) Final programmatic damage assessment and restoration plan and final programmatic environmental impact statement. http://www.gulfspillrestoration.noaa.gov/restoration-planning/gulf-plan

Diario Oficial de la Federación (DOF) (2018) Actualización de la Carta Nacional Pesquera. Secretaría de Agricultura, Ganadería, Desarrollo Rural, Pesca y Alimentación. Ciudad de México, México

Díaz de León A, Fernández I, Fernández-Méndez JI, Álvarez-Torres P, Ramírez-Flores O, López-Lemus LG (2004) Sustainability of fishing resources of the Gulf of Mexico. In: Caso M, Pisanty I, Ezcurra E (eds) Environmental diagnostics of the Gulf of Mexico. Ministry of Environment and Natural Resources, National Institute of Ecology, A. C., Harte Research Institute for the Gulf of México Studies, Vol. 2, pp 727–755

Dubansky B, Whitehead A, Miller JT, Rice CD, Gálvez F (2013) Multitissue molecular, genomic, and developmental effects of the Deepwater Horizon oil spill on resident gulf killifish (*Fundulus grandis*). Environ Sci Technol 47:5074–5082

Fodrie FJ, Heck KL Jr (2011) Response of coastal fishes to the Gulf of Mexico oil disaster. PLoS ONE 6:e21609. https://doi.org/10.1371/journal.pone.0021609

Gallaway BJ, Gazey WJ, Cole JG, Fechhelm RG (2007) Estimation of potential impacts from offshore liquefied natural gas terminals on Red Snapper and drum fisheries in the Gulf of Mexico: an alternative approach. Trans Am Fish Soc 136:655–677

Gallaway BJ, Zsedlmayer ST, Gazey WJ (2009) A life history review for Red Snapper in the Gulf of Mexico with an evaluation of the importance of offshore petroleum platforms and other artificial reefs. Rev Fish Sci 17:48–67

González-Ania LV, Ulloa PA, Arenas P (2002) Tuna fishery. In: Fishing in Veracruz and perspectives for development. SAGARPA, Ciudad de México, México, p 435

Gracia A (1991) Spawning stock-recruitment relationships of White Shrimp in the Southwestern Gulf of Mexico. Trans Am Fish Soc 120:519–527

Gracia A (1992) Explotación y manejo del recurso camarón. Ciencia y Desarrollo 18:82–95

Gracia A (1995) Impacto de la pesca artesanal sobre la producción de camarón rosado *Penaeus Farfantepenaeus duorarum* Burkenroad, 1939. Cienc Mar 21:343–359

Gracia A (1996) White shrimp, *Penaeus setiferus*, recruitment overfishing. Mar Freshw Res 47:59–63

Gracia A (2004) Utilization and conservation of the shrimp resource. In: Caso M, Pisanty I, Ezcurra E (eds) Environmental assessment of the Gulf of Mexico. Ministry of Environment and Natural Resources, National Institute of Ecology, A. C., Harte Research Institute for the Gulf of Mexico Studies, Vol. 2, pp 713–725

Gracia A, Vázquez-Bader AR, Arreguín-Sánchez F, Schultz-Ruiz L, Sánchez-Chávez J (1997) Ecology of Penaeid Shrimp in the Gulf of Mexico. In: Flores-Hernández D, Sánchez-Gil P, Seijo JC, Arreguín-Sánchez F (eds) Analysis and evaluation of critical fishing resources of the Gulf of Mexico. EPOMEX. Scientific Series, 7, pp 127–144

Gracia A, Vázquez-Bader AR (1998) The effects of artisanal fisheries on penaeid shrimp Stocks in the Gulf of México. In: Funk F, Quinn TJ, Heifetz J, Ianelli JN, Powers JE, Schweigert JF, Sullivan PJ, Zhang CI (eds) Proceedings of the International Symposium Fishery Stock Assessment Models for the 21st Century. University of Alaska, Sea Grant College, pp 977–998

Gracia A, Vázquez-Bader AR (1999) Shrimp fisheries in the South of the Gulf of Mexico. Present state and future management alternatives. In: Kumpf H, Steidinger D, Sherman K (eds) The Gulf of Mexico large marine ecosystem. Assessment, sustainability, and management. Blackwell Science, Berlin, pp 205–234

Hart RA (2015a) Stock assessment update for Pink Shrimp (*Farfantepenaeus duorarum*) in the U.S. Gulf of Mexico for 2014. NOAA Fisheries Southeast Science Center, Galveston Laboratory 18 pp.

Hart RA (2015b) Stock assessment update for White Shrimp (*Litopenaeus setiferus*) in the U. S. Gulf of Mexico for 2014. NOAA Fisheries Southeast Science Center, Galveston Laboratory 17 pp.

Hart RA, Nance JM (2013) Three decades of U.S. Gulf of Mexico White shrimp *Litopenaeus setiferus* catch commercial statistics. Mar Fish Rev 75:43–47

Hazen LE, Carlisle AB, Wilson SG, Gagnong JE, Castleton MR, Schallert RJ, Stokesbury MJW, Bograd SJ, Block BA (2016) Quantifying overlap between the *Deepwater Horizon* oil spill and predicted Bluefin Tuna spawning habitat in the Gulf of Mexico. Scientific Reports 6:33824. https://doi.org/10.10138/srep33824

International Commission for the Conservation of Atlantic Tuna (ICCAT) (2007) Report of the Standing Committee on Research and Statistics (SCRS). Madrid, Spain, October 1–5, 2007, 213 pp.

International Commission for the Conservation of Atlantic Tuna. Executive Summary BFTE. Web Mar. 6, 2015. https://www.iccat.int/Documents/SCRS/ExecSum/BFT_EN.pdf

Incardona JP, Gardner LD, Linbo TL, Brown TL, Esbaugh AJ, Mager EM, Stieglitz JD, French BL, Labenia JS, Laetz CA, Tagal M, Sloan CA, Elizur A, Benetti D, Grosell M, Block BA, Scholz NA (2014) Deepwater Horizon crude oil impacts the developing hearts of large predatory pelagic fish. Proc Natl Acad Sci USA, 111:E1510–E1518

Jernelöv A, Lindén O (1981) Ixtoc I: a case study of the world's largest oil spill. Ambio 10:299–306

Karnauskas, M, Kelble CR, Regan S, Quenée C, Allee R, Jepson M, Freitag A, Craig JK, Carollo C, Barbero L, Trifonova N, Hanisko D, Zapfe G (2017) Ecosystem status report update for the Gulf of Mexico. NOAA Technical Memorandum NMFS-SEFSC 706, 51 pp

McNutt M, Chu S, Lubchenco J, Hunter T, Dreyfus G, Murawski SA, Kennedy DM (2012) Applications of science and engineering to quantify and control the Deepwater Horizon oil spill. Proc Natl Acad Sci USA 109:20,222–20-228

Moddy RM, Cebrina J, Heck KL Jr (2013) Interannual recruitment dynamics for resident and transient marsh species: evidence for a lack of impact of the Macondo Oil Spill. PLos ONE 8(3):e58376. https://doi.org/10.1371/journal.pone.10058376

Muhling BA, Roffer MA, Lamkin JT, Ingram GW Jr, Upton MA, Gawlikowski G, Muller-Karger F, Habtes S, Richards WJ (2012) Overlap between Atlantic Bluefin Tuna spawning grounds and observed *Deepwater Horizon* surface oil in the northern Gulf of Mexico. Mar Pollut Bull 64:679–687

Murawski SA, Fleeger JW, Patterson WF, Hu C, Daly K, Romero I, Toro-Farmer G (2016) How did the Deepwater Horizon oil spill affect coastal and continental shelf ecosystems of the Gulf of Mexico? Oceanography 29(3):160–173

Murawski SA, Peebles EB, Gracia A, Tunnell JW Jr, Armenteros M (2018a) Comparative abundance, species composition and demographics of continental shelf fish assemblages throughout the Gulf of Mexico. Mar Coast Fish 10:325–346

Murawski SA, Patterson II W, Campbell M (2018b) Has abundance of continental shelf fish species declined after Deepwater Horizon? Presentation at: Gulf of Mexico Oil Spill & Ecosystem Science Conference, New Orleans, LA, 5–8 February 2018

Programa Coordinado de Estudios Ecológicos en la Sonda de Campeche (1980) Informe de los trabajos realizados para el control del pozo Ixtoc 1, el derrrame de petróleo y determinación de sus efectos sobre el ambiente marino. Instituto Mexicano del Petróleo, México City

Pulster EL, Gracia A, Snyder SM, Deak K, Folgelson S, Murawski SA (2019) Chronic sub-lethal effects observed in wild-caught fishes following two major oil spills in the Gulf of Mexico:

DWH and Ixtoc 1 (Chap. 24). In: Murawski SA, Ainsworth C, Gilbert S, Hollander D, Paris CB, Schlüter M, Wetzel D (eds) Deep oil spills: facts, fate and effects. Springer, Cham

Rice SD, Spies RB, Wolfe DA, Wright BA (1996) Proceedings of the *Exxon Valdez* oil spill symposium. American Fisheries Society Symposium 19, Bethesda, MD. 931 pp.

Rooker JR, Landry AM Jr, Geary BW, Harper JA (2004) Assessment of a shell bank and associated substrates as nursery habitat of post-settlement red snapper. Estuar Coast Shelf Sci 59:653–661

Rooker JR, Kitchens LL, Dance MA, Well RJD, Falterman B, Cornic M (2013) Spatial, temporal, and habitat-related variations in abundance of pelagic fishes in the Gulf of Mexico: potential implications of the *Deepwater Horizon* oil spill. PLoS ONE 8(10):e76080. https://doi.org/10.1371/journal.pone.0076080

Ryerson TB, Camilli R, Kessler JD, Kujawinski EB, Reddy CM, Valentine DL, Atlas E, Blake DR, de Gouw J, Meinardi S, Parrish DD, Peischi J, Seewald JS, Warneke C (2012) Chemical data quantify Deepwater Horizon hydrocarbon flow rate and environmental distribution. Proc Natl Acad Sci 109:20246–20253

SEDAR 7 (2005) Stock Assessment Report of SEDAR7. Gulf of Mexico Red Snapper, Charleston, SC, 480 pp

SEDAR 31 (2013) Stock assessment report: Gulf of Mexico Red Snapper. Southeast Data, Assessment, and Review

Sun S, Hu C, Tunnell JW Jr (2015) Surface oil footprint and trajectory of the Ixtoc-1 oil spill determined from Landsat /MSS and ZCS observations. Mar Pollut Bull 101:632–641

Szedlmayer ST, Shipp RL (1994) Movement and growth of red snapper, *Lutjanus campechanus*, from an artificial reef area in the Northeastern Gulf of Mexico. Bull Mar Sci 55:887–896

Szedlmayer ST, Mudrak PA (2014) Influence of age-1 conspecifics, sediment type, dissolved oxygen, and the *Deepwater Horizon* oil spill on recruitment of age-0 red snapper in the Northeast Gulf of Mexico during 2010 and 2011. N Am J Fish Manag 34:443–452

Tunnell JW Jr (2014) In: RE Oil Spill of the "*Deepwater Horizon*" Oil Rig in the Gulf of Mexico, on April 20, 210. Expert Report of Dr. John W. Tunnell, Jr. 155 pp

Tunnell JW Jr (2017) Shellfishes of the Gulf of Mexico. In: Ward HC (ed) Habitats and biota of the Gulf of Mexico: before the Deepwater Horizon oil spill, vol 1. Rice University, Houston, TX, pp 769–845

Wilson CA, Nieland DL (2001) Age and growth of red snapper, *Lutjanus campechanus*, from the northern Gulf of Mexico off Louisiana. Fish Bull 99:653–664

Xia K, Hagood G, Childers C, Atkins J, Rogers B, Ware L, Armbrust K, Jewell J, Díaz D, Gatian N, Folmer H (2012) Polycyclic aromatic hydrocarbons (PAHs) in Mississippi seafood from areas affected by the *Deepwater Horizon* oil spill. Environ Sci Technol 46:5310–5318

Ylitalo GM, Krahn MM, Dickhoff WW, Stein JE, Walker CC, Lassiter CL, Garrett ES, Desfosse LL, Mitchell KM, Noble BT, Wilson S, Beck NB, Benner RA, Koufopoulos PN, Dickey RW (2012) Federal seafood safety response to the Deepwater Horizon oil spill. Proc Natl Acad Sci 109:20274–20279

Chapter 26
Impacts of the Deepwater Horizon Oil Spill on Marine Mammals and Sea Turtles

Kaitlin E. Frasier, Alba Solsona-Berga, Lesley Stokes, and John A. Hildebrand

Abstract The Gulf of Mexico (GOM) is one of the most diverse ecosystems in the world (Fautin et al. PLoS One 5(8):e11914, 2010). Twenty-one species of marine mammals and five species of sea turtles were routinely identified in the region by the end of the twenty-first century (Waring et al. NOAA Tech Memo NMFS NE 231:361, 2015), a decrease from approximately 39 species prior to intensive exploitation (Darnell RM. The American sea: a natural history of the Gulf of Mexico. Texas A&M University Press, College Station, TX, 2015). Life histories of these megafauna species range from hyperlocal residence patterns of bottlenose dolphins to inter-ocean migrations of leatherback turtles. All species are subject to direct and indirect impacts associated with human activities. These impacts have intensified with major development and extraction efforts since the 1940s. The *Deepwater Horizon* (DWH) oil spill represents a new type of injury to this system: Unlike previous large oil spills, it not only exposed marine megafauna to surface slicks, it also involved an unprecedented release of dispersed oil into deep waters and pelagic habitats, where effects are difficult to observe and quantify. This chapter synthesizes the research conducted following the DWH oil spill to characterize acute and chronic offshore effects on oceanic marine mammals and sea turtles. Marine mammals and sea turtles were exposed to unprecedented amounts of oil and dispersants. Local declines in marine mammal presence observed using passive acoustic monitoring data suggest that the acute and chronic population-level impacts of this exposure were likely high and were underestimated based on coastal observations alone. These population declines may be related to reduced reproductive success as observed in nearshore proxies. Long-term monitoring of oceanic marine mammals is a focus of this chapter because impacts to these populations have not been extensively covered elsewhere. We provide an overview of impacts to sea turtles and

K. E. Frasier (✉) · A. Solsona-Berga · J. A. Hildebrand
University of California San Diego, Scripps Institution of Oceanography, Marine Physical Laboratory, La Jolla, CA, USA
e-mail: kfrasier@ucsd.edu; asolsonaberga@ucsd.edu; jhildebrand@ucsd.edu

L. Stokes
National Marine Fisheries Service, Southeast Fisheries Science Center, Miami, FL, USA
e-mail: lesley.stokes@noaa.gov

© Springer Nature Switzerland AG 2020
S. A. Murawski et al. (eds.), *Deep Oil Spills*,
https://doi.org/10.1007/978-3-030-11605-7_26

431

coastal marine mammals, but other more thorough resources are referenced for in depth reviews of these more widely covered species.

Keywords Marine mammal · Sperm whale · Beaked whale · Dolphin · Passive acoustic monitoring · PAM · Sea turtle · Loggerhead · Kemp's ridley · Megafauna · Bottlenose · Barataria Bay · Mammal · Odontocete · Bryde's whale · Spotted dolphin · *Stenella* · *Kogia* · Echolocation · Visual survey · Nesting · Entanglement · Ship strike · Noise · Airgun · UME · Unusual mortality event · Leatherback, Hawksbill · Stranding · Pinniped · HARP · Mississippi Canyon · Green Canyon · *Sargassum* · Green turtle · Trawl · Skimming · Risso's dolphin · Pilot whale · Tag · Aerial survey · Bycatch

26.1　Megafauna of the Gulf of Mexico

The Gulf of Mexico (GOM) supports 5 species of sea turtles and at least 21 species of marine mammals including 1 species of baleen whale, 19 species of toothed whales, and 1 species of manatee (Darnell 2015). Gulf marine mammal species fall into several ecological groups: Shallow-dwelling bottlenose dolphins (*Tursiops truncatus*) inhabit coastal waters including bays, sounds, and estuaries, as well as the broad continental shelf regions extending from the coast out to the shelf break. Atlantic spotted dolphins (*Stenella frontalis*) are also commonly found on the continental shelf. The majority of the marine mammal diversity in the GOM is found at or beyond the shelf break, often hundreds of kilometers offshore. Pelagic deep-diving species include sperm whales (*Physeter macrocephalus*), Gervais' and Cuvier's beaked whales (*Mesoplodon densirostris* and *Ziphius cavirostris*), and *Kogia* species, which execute long foraging dives to depths typically exceeding 200 meters. These species feed at depth, primarily on squid and do not typically exhibit diel foraging patterns. At least 13 species of pelagic, shallower diving delphinids (typical diving depths less than 200 m) are also found in the region. These species feed nocturnally on vertically migrating mesopelagic prey in the deep scattering layer. A single baleen whale species, the GOM Bryde's (*Balaenoptera edeni*) whale, is found in the northeastern GOM near Desoto Canyon (Soldevilla et al. 2017).

All marine mammal species currently known to the northern GOM are also found in other oceans; however, little is known about the migration ranges of Gulf populations or the degree to which they mix with populations in the southern GOM and broader Atlantic. The GOM Bryde's whale is thought to be a distinct and isolated subspecies (Rosel and Wilcox 2014). The GOM sperm whale population also appears to be resident in the area (Waring et al. 2009; Jochens et al. 2008). Sperm whale sightings in the GOM often consist of groups of females and juveniles; therefore, the region is thought to serve as a year-round nursing ground for sperm whales. Large solitary males, which are routinely observed in other oceans, are rarely encountered in the GOM, and tag data has shown that males may travel in and out of the Gulf (Jochens et al. 2008).

Leatherback (*Dermochelys coriacea*) and surface-pelagic juvenile loggerhead turtles (*Caretta caretta*) and green (*Chelonia mydas*), Kemp's ridley (*Lepidochelys kempii*), and Atlantic hawksbill (*Eretmochelys imbricata*) sea turtles are found in offshore waters in the GOM, while larger neritic juvenile and adult turtles are found in the continental shelf and nearshore/coastal waters; inshore areas host juvenile and adult Kemp's ridleys, loggerheads, and greens (Wallace et al. 2010, 2017). All five species are listed as endangered or threatened under the Endangered Species Act of 1973. Hatchlings emerge from nesting beaches and disperse into surface-pelagic developmental habitats in convergence zones, using *Sargassum* communities as a foraging resource that affords protection from predation and potential thermal benefits (Bolten 2003, Witherington et al. 2012, Mansfield et al. 2014). Foraging strategies differ by species, with the adult diet of green turtles dominated by seagrasses and algae, loggerheads feeding upon a broad range of pelagic and benthic invertebrates, hawksbills specializing primarily on sponges, Kemp's ridleys feeding mostly on crabs, and leatherbacks depending on cnidarians (see Bjorndal 1997 for a comprehensive review).

Loggerheads and to a lesser extent Kemp's ridleys and green turtles nest on northern GOM beaches in spring and summer months, although the Kemp's ridley's primary nesting beaches are found in the western Gulf in Tamaulipus and Veracruz, Mexico.

26.2   A Context of Chronic Impacts

The *Deepwater Horizon* (DWH) oil spill is one chapter in a long history of direct and indirect anthropogenic impacts on marine mammal and sea turtle populations in the GOM. The primary sources of stressors are summarized below.

26.2.1   Direct Harvest

Exploitation of GOM megafauna dates back to the Maya and Aztecs, who intensively harvested sea turtles and manatees (Lange 1971). Impacts were likely mainly limited to coastal zones until the late 1700s when the American whaling industry reached Gulf waters (Darnell 2015). Whalers primarily targeted sperm whales, with pilot whales as secondary targets, and reported that the waters of the mouth of the Mississippi River constituted one of the most profitable whaling grounds (Reeves et al. 2011). Reports of sightings and takes of "finback" whales taken in the region likely refer to the Gulf Bryde's whale, and these reports indicate that the range of this species included most of the Gulf (Reeves et al. 2011), although the current population appears to be restricted to a small region near Desoto Canyon (Soldevilla et al. 2017).

Sea turtles are threatened by direct harvest both in offshore habitats and on nesting beaches, from egg to adult stages. Adult green turtles were intensively harvested for their meat in the 1880s (Valverde and Holzwart 2017) when landings of that species alone are estimated at 4800 to 6000 animals per year, across the GOM and broader Caribbean (Darnell 2015). Adult loggerheads were harvested in Cuba through the mid-1990s (Gavilan 1998). Harvesting of sea turtles became illegal in the United States under the Endangered Species Act of 1973, and illegal poaching at sea is thought to be rare in the US Gulf of Mexico (NMFS 2011). However, active harvest may still occur outside of the US EEZ. An active illegal trade in hawksbill tortoiseshells persists (NOAA 2013). Poaching of sea turtle eggs from nesting sites continues in the United States and neighboring countries (NMFS 2008, 2011).

26.2.2　Shipping and Vessel Strikes

Commercial shipping has been a major industry in the GOM since the 1850s, when the port of New Orleans was the second largest in the country (Darnell 2015). In 2016, GOM ports accounted for nearly 50% of total tonnage transferred through American ports (US Army Corps of Engineers). Both marine mammals and sea turtles are at risk of vessel strikes, and these are likely highly underreported for pelagic species (Williams et al. 2011; Epperly et al. 1996).

26.2.3　Anthropogenic Noise

Offshore human activities also affect megafauna through elevated noise levels and pollution. Oil and gas development took hold in the GOM in 1947 (Darnell 2015), expanding rapidly thereafter. In 2018 over 50 thousand wells and 7 thousand drilling platforms were documented in the GOM (BOEM 2018). Seismic surveys using explosive sound sources (airguns) are used to map subsurface oil and gas deposits. These surveys are nearly continuous in the GOM, and they combine with shipping noise to make average low-frequency ambient noise levels very high in the GOM relative to levels in other ocean regions (Wiggins et al. 2016). Noise is considered a chronic stressor for marine mammals because these species rely on sound to interpret their environment and communicate with one another (e.g., Wright et al. 2007).

26.2.4　Debris Entanglement, Ingestion, and Bycatch

Commercial fishing efforts in the GOM expanded after WWII, adopting novel technologies including purse seines, longlines, trawls, and gillnets, which increased the occurrence of marine mammal and sea turtle entanglement in fishing gear (Lutcavage et al. 1997), as well bycatch rates and competition for prey species (Darnell 2015). Sea turtles and marine mammals are incidentally caught and killed in trawl, gillnet, hook-and-line, and longline fishing gear, and fishery bycatch is considered the most serious global threat to marine megafauna (Lewison et al. 2004; Wallace et al. 2010, 2013).

Marine mammals and turtles are affected by entanglement in gear and marine debris, with possible effects including injury and drowning (Walker and Coe 1989; Plotkin and Amos 1990). Loggerhead, green, hawksbill, and Kemp's ridley sea turtles take refuge as juveniles in *Sargassum* rafts and are particularly susceptible to entanglement in trash and ingestion of plastics (Witherington et al. 2012).

26.2.5 Oil and Gas Development

Oil and gas development in the GOM is linked to a variety of chronic impacts including occasional small- and large-scale spills (Asl et al. 2016; SERO n.d.), leaking infrastructure, chemical releases related to extraction activities (Neff 1990; Neff et al. 2011a, b), persistence of weathered oil and related compounds in the environment (Van Vleet and Pauly 1987; Botello et al. 1997; Van Vleet et al. 1984), increased vessel activity, and the construction and removal of offshore platforms (Gitschlag et al. 1995). Some amount of natural crude oil seepage also occurs from an estimated 914 natural seep zones in the GOM (MacDonald et al. 2015). Oil from small-scale releases and seeps weathers and spreads in the pelagic ecosystem, accumulating in offshore convergence zones. These zones, which aggregate drifts of *Sargassum* and other macroalgae species, act as vital habitats for surface-pelagic juvenile turtles, putting them at particular risk for exposure to oil accumulating in these zones (Witherington et al. 2012; Bolten 2003).

26.2.6 Nesting Beach Impacts

Turtle populations have susceptibilities related to their reliance on nesting beaches, which are impacted by coastal development, beach erosion, light pollution, dredging, beach re-nourishment programs and armoring, climate change and sea level rise, as well as native and exotic predators (Lutcavage et al. 1997). Current efforts to protect nesting beaches and rescue nests began in some areas as early as the 1950s, and successful nesting beach conservation efforts can result in rapid local population increases (Troëng and Rankin 2005; Hayes 2004).

26.2.7 Habitat and Environmental Degradation

Other major chronic impacts to both marine mammals and turtles include hypoxia in the Mississippi River outflow region, which affects prey quality and density in a previously rich foraging ground for GOM megafauna, as well as harmful algal blooms (Magaña et al. 2003). Direct impacts to turtle habitats including loss of nesting beaches, seagrass beds, and coral reefs, are primarily associated with coastal and continental shelf zones. The same suite of chronic impacts that affect GOM marine mammal and sea turtle health may also have some effect on the health and quality of their prey.

26.3 Quantifying Impacts on Pelagic Species/Stages and Habitats

Although occasional reports of various direct impacts to marine mammals and sea turtles including entanglements and ship strikes are reported, they are likely highly underreported (Williams et al. 2011; Epperly et al. 1996) because they occur offshore and may go undetected. Carcasses are unlikely to strand following pelagic deaths, particularly in regions or seasons where higher water temperatures accelerate decomposition (Nero et al. 2013). Equally difficult to quantify are the cumulative effects of chronic impacts such as pollution, noise, and prey depletion, since these effects likely accumulate gradually at sublethal levels over many years.

One possible indicator of chronic stress is the occurrence, intensity, and length of unusual mortality events (UMEs). A UME is an unexpected stranding event that represents a significant die-off in a marine mammal population. From 1990 to 2104, there were 12 UMEs in the GOM (Litz et al. 2014) with recovered carcass counts ranging from 31 to 1141 animals and lasting from 1 to 52 months (Resources NOoP 2018). Coastal bottlenose dolphins are the predominant species in the stranding record, likely because the bodies of this nearshore species are far more likely to reach beaches. The proximate cause of the majority of these UMEs is typically determined to be morbillivirus, biotoxins, and/or cold weather. However, the causes of some events remain unknown, as in the case of the largest, longest-lasting event that occurred from 2012 to 2014 with 1141 recovered carcasses. The occurrence of these events suggests population-level immunodeficiencies (Di Guardo and Mazzariol 2013) or other susceptibility in populations which are already experiencing chronic stress.

Turtle populations also experience mass stranding events such as cold stunning events (Milton and Lutz 2003). In January 2010, unusually cold temperatures resulted in a cold stunning event of unprecedented magnitude involving over 4500 sea turtles (primarily green turtles) in Florida (Avens et al. 2012). As in the case of marine mammals, chronic stress associated with anthropogenic activity may be acting to decrease overall resilience of turtle populations (Lamont et al. 2012).

One particularly observable anthropogenic impact is the loss of historic nesting beaches to coastal development projects. Female sea turtles generally exhibit high site fidelity to particular nesting beaches, and the availability of natural, undeveloped beaches is dwindling. It is unclear how many nesting beaches may have previously existed in the northern GOM. A 2006 study in the Caribbean estimated that 20% of historic (twentieth century) green and hawksbill nesting beaches in the Caribbean had been lost and that 50% of the remaining beaches were visited by fewer than ten nesting females (McClenachan et al. 2006). Beaches in the northern GOM today have minimal sea turtle nesting relative to the Atlantic regions (except for Kemp's ridleys), although previous nesting density is largely unknown (Hildebrand 1982; Valverde and Holzwart 2017). Currently, most nesting beaches in the US GOM are in Florida and Texas (see Valverde and Holzwart 2017 for a thorough account of nesting sites and habitats). However, most Kemp's ridleys found in US waters hatched on Mexican beaches; thus, conservation measures require an international perspective.

The majority of the existing data on population-level effects of oil spills on marine megafauna comes from shallow water spills and their effects on coastal populations. A long-term study of the effects of the Exxon Valdez spill on killer whale populations indicates that two pods suffered acute losses during the event (33% and 41% of their members) and had not returned to pre-spill numbers nearly 20 years later (Matkin et al. 2008). One of the two exposed pods appeared to be headed for extinction at the conclusion of the study (Matkin et al. 2008). Evidence suggests that lipophilic chemical contaminants are often offloaded from mother to calf in marine mammals, including bottlenose dolphins (Irwin 2005) and killer whales (Ylitalo et al. 2001), leading to higher levels of calf mortality.

26.3.1 Study Methods: Pre-DWH

Marine mammal and sea turtle populations in the offshore GOM have historically been quantified in the offshore GOM using shipboard and aerial visual surveys. Aerial continental shelf surveys began in 1979, initially conducted by the USFWS and then by NMFS. Offshore marine mammal surveys were conducted by NMFS in the spring or summer of 1990–1994, 1996–2001, 2003–2004, and 2009 (Waring et al. 2015; Mullin and Fulling 2004; Mullin 2007) (Fig. 26.1). Major survey efforts in the early to mid-1990s (GulfCet study, supported by the Minerals Management Service) focused on the continental slope region (100–2000 m) and recorded both marine mammals and turtles over eight surveys in all seasons (Würsig et al. 2000). Despite these survey efforts, few species had enough sightings to produce robust population size estimates, and none could be analyzed for long-term population trends due to low precision and infrequent assessments. As a result, little baseline information was available on population trends prior to the DWH spill. The only cetaceans with adequate population data were bottlenose dolphins resident in

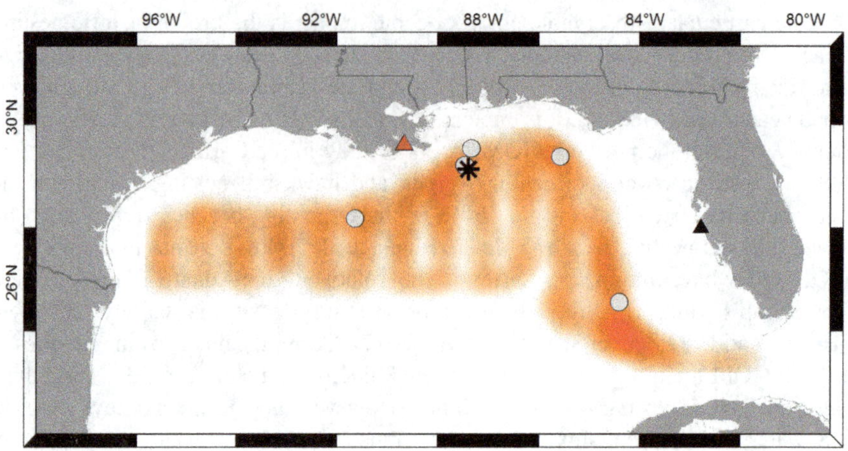

Fig. 26.1 Oceanic monitoring effort in the GOM. Colormap indicates high visual survey effort (red) to low visual survey effort (white) during deepwater (>200 m target depths) NOAA and NOAA-affiliate shipboard surveys 1992–2014. Gray dots indicate GOM HARP monitoring sites, 2010–2017. Asterisk is site of DWH, and triangles are locations of bottlenose dolphin studies in Barataria Bay (red) and Sarasota Bay (black)

selected bays and estuaries, where high-resolution monitoring was based on mark-recapture analysis (Wells 2014).

Offshore turtle surveys using aerial and shipboard methods have limitations because small turtles are difficult to detect and identify using these methods, although using satellite tags to monitor diving behavior to account for sightability may reduce uncertainty in estimates (Thomas et al. 2010; Seminoff et al. 2014). Beach counts of nesting females and clutch sizes are a reliable census method (Schroeder and Murphy 1999), but this approach only surveys adult females nesting in a given year (nesting cycles differ between species). Satellite systems and drones are now being adopted to survey sea turtles, and satellite tags are being used to track their movement over large distances (Rees et al. 2018).

26.3.2 Study Methods: Post-DWH

In response to the DWH event, numerous additional studies were initiated in the GOM to understand potential impacts from the spill. The longest-term marine mammal study was the GOM High-frequency Acoustic Recording Package (HARP) program based on passive acoustic monitoring of marine mammal sounds (Hildebrand et al. 2017); an 8 + −year broadband passive acoustic monitoring effort was initiated at five locations in the GOM in 2010 (Fig. 26.1). Three deepwater monitoring locations included a site in Mississippi Canyon near the DWH wellhead (site MC), an eastern site at Green Canyon outside of the DWH surface oil footprint (site GC), and a southern site outside of the oil footprint near the Dry Tortugas (site DT).

HARPs were maintained nearly continuously at these representative oil-exposed and oil-unexposed monitoring sites to detect marine mammal sound production as a proxy for animal presence across the region. Marine mammal species and genera were then distinguished in the long-term passive acoustic recordings based on the characteristics of their sounds. Species monitored include sperm whales, Cuvier's and Gervais' beaked whales, Risso's dolphin, and pygmy/dwarf sperm whales (*Kogia* spp.), delphinids in the genus *Stenella* (Atlantic spotted, pantropical spotted, spinner, striped, and Clymene dolphins), and blackfish (primarily short-finned pilot whales).

Bottlenose dolphin health assessments began in 2011 in Barataria Bay, Louisiana (Schwacke et al. 2013). Health metrics from resident bottlenose dolphins in Barataria Bay, which was heavily oiled by the DWH event (Michel et al. 2013), were compared to an unexposed reference population in Sarasota Bay, Florida (Fig. 26.1).

Shorter-term marine mammal studies following the DWH event included low-frequency passive acoustic monitoring for Bryde's whales on the west Florida shelf in 2010 and 2011 (Rice et al. 2014), satellite tagging efforts aimed at understanding sperm whale distributions (Mate unpublished data), and short-term passive acoustic monitoring near the DWH wellhead site (Ackleh et al. 2012). In the absence of long-term data, various population modeling efforts were also undertaken to try to estimate population-level effects and recovery times based on assumed vital rates (Farmer et al. 2018; Ackleh et al. 2018; Schwacke et al. 2017).

In an effort to mitigate the impact of the spill on sea turtle nesting beaches, 25,000 Kemp's ridley and loggerhead eggs were transported from beaches in the GOM to the Atlantic coast in Florida (Inkley et al. 2013). Following the DWH spill, transect searches were conducted in convergence zones within the spill area to rescue oceanic juvenile sea turtles and to document species composition and oiling status (McDonald et al. 2017). Aerial surveys were conducted on the continental shelf throughout the northern GOM to the 200-m isobath between April and September 2010, documenting the distribution and abundance of neritic sea turtles throughout the DWH spill area (Garrison 2015).

Estimates of the probabilities of oil exposure for sea turtles present within the area of the spill were generated from direct observations of surface-pelagic juvenile sea turtles (Stacy 2012) and satellite-derived surface oil distributions (Wallace et al. 2017). Abundance and source populations for impacted turtles were predicted using ocean circulation and particle tracking simulation models, estimating that 321,401 green, loggerhead, and Kemp's ridley turtles were likely within the spill site, originating primarily from Mexico and Costa Rica (Putman et al. 2015).

26.4 Acute Effects

We consider acute effects of the DWH oil spill on marine megafauna as those effects that caused immediate harm during the spill and response. This section focuses primarily on effects experienced in oceanic habitats, as coastal impacts have been reviewed extensively elsewhere. Primary acute effects include immediate injury and death from oil and chemical exposure, response activities, and strandings.

26.4.1 Oil Exposure: Marine Mammals

In an oil spill, marine megafauna can be exposed to oil and related compounds through surface slicks when breathing or resting at the air/sea interface (Trustees 2016) and through interaction with subsurface plumes during dives and foraging events. Oil compounds can be taken up through the skin, breathed into the lungs, or ingested with prey (Schwacke et al. 2013). Exposure studies conducted in the 1970s focused on polar species including pinnipeds, sea otters, and polar bears and found a range of effects including eye and skin lesions associated with continued exposure, uptake, distribution and accumulation of oil compounds into body tissues and fat reserves through oil ingestion (Englehardt 1977; Engelhardt 1982), and thermoregulation problems associated with oiling. Marine mammals and turtles were observed in oil-impacted areas during the DWH spill response while the well remained uncapped (Wilkin et al. 2017).

There is limited prior information on the effects of oil spills on marine mammal populations. Early oil spill studies noted that a wide range of marine mammal species including baleen whales, dolphins, and pinnipeds did not appear to avoid oil-contaminated waters (Goodale et al. 1981; Spooner 1967; Geraci 1990a, 1990b) despite the fact that captive bottlenose dolphins could detect and avoid oil in experimental settings (Geraci 1990a). In the case of the DWH oil spill, the acoustic record shows little evidence for near-term avoidance of the wellhead area by marine mammals (see Sect. 26.5).

Marine mammals were exposed to oil at the sea surface but also likely at depth. Deep-diving pelagic species including sperm whales and beaked whales forage at depths of 1000 m or more and likely interacted with the deep plume which formed at approximately 1100 m (Hildebrand et al. 2012). The deep plume formation is largely attributed to dispersant use (Kujawinski et al. 2011) and has not been reported in previous spills; therefore, it likely represents a new route of exposure for deep-diving cetaceans. A large amount of released oil did not reach the surface and likely was eventually deposited on the seafloor (Romero et al. 2017). GOM marine mammal species are not typically benthic foragers; however, bay, sound, and estuary bottlenose dolphins may use benthic hunting tactics (e.g., Lewis and Schroeder 2003; Rossbach and Herzing 1997), which could increase their exposure to deposited oil. In addition, GOM Bryde's whales appear to forage at or near the sea floor

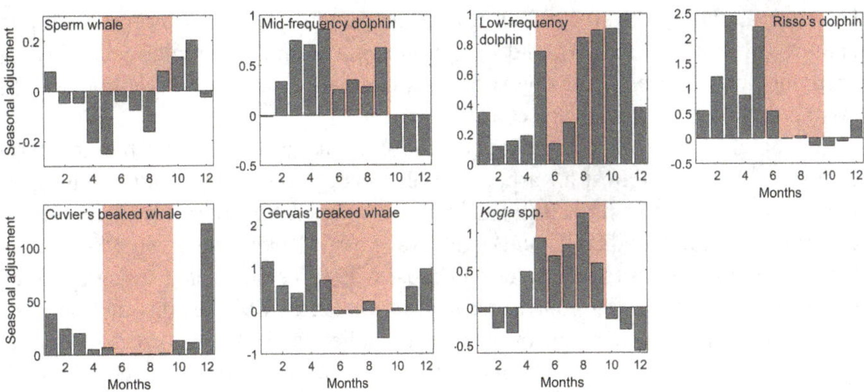

Fig. 26.2 Seasonal patterns in marine mammal presence at a passive acoustic monitoring site in Mississippi Canyon, located approximately 10 km from the DWH wellhead. The vertical axis indicates the factor by which seasonal presence varies relative to mean presence. Higher values indicate stronger seasonality

(Soldevilla et al. 2017) and, therefore, may also be at risk of exposure to oil deposited in sediments.

Differences in seasonal presence likely played a role in the extent to which marine mammal species were directly exposed to DWH oil and dispersants (Fig. 26.2). Sperm whales, Gervais' beaked whales, *Kogia* species, and mid-frequency dolphins (in the GOM this group primarily consists of species in the *Stenella* genus, categorized based on echolocation click peak frequencies) are found year-round in the region of the oil spill and were likely directly exposed to oil. In particular *Kogia* and mid-frequency delphinid species presence increases in the summer months, increasing the likelihood of exposure to DWH oil and dispersants. Risso's dolphins are seasonally present in spring through summer and therefore likely experienced direct exposure during the first months of the spill but less exposure as the summer progressed. In contrast, Cuvier's beaked whale presence is strongly seasonal near the wellhead with highest occurrence during winter months; therefore, these populations likely experienced minimal direct exposure during the spring spill.

26.4.2 *Oil Exposure: Sea Turtles*

Potential direct impacts to sea turtles from an oil spill differ depending on the life stage, but all stages are vulnerable to acute toxicity from volatile contaminants, exposure through inhalation and ingestion, physical impairment from heavy oiling, and a variety of physiological and clinicopathological impacts of exposure (see review in Shigenaka 2003). Sea turtles are unlikely to detect oil (Odell and MacMurray 1986), and in experimental conditions they showed no avoidance behavior (Lutcavage et al. 1995). They are continuously exposed by resurfacing to breathe (Milton et al. 2003), and pelagic juveniles are susceptible to floating tar accumulations in ocean convergence zones due to indiscriminate feeding patterns (Witherington 2002; Lutcavage et al. 1997).

In laboratory studies, juvenile loggerheads were adversely affected by short-term exposures to oil in almost all aspects of physiology (e.g., respiration, diving patterns, energy metabolism, salt gland function, oxygen transport, blood chemistry, and red and white blood cell count) (Lutcavage et al. 1995; Lutz et al. 1986). In sea turtles, oil clings to eyes and nares and causes skin to slough off leaving inflamed soft skin exposed to infection (Lutcavage et al. 1995). Skin lesions and necrosis were observed in leatherback oil exposure studies, and skin returned to normal appearance approximately 1 month after the turtles were removed from oil (Lutcavage et al. 1995). Following the Ixtoc 1 oil spill, necropsied sea turtles were found to have to ingested large amounts of oil, with indications that the ingestion was eventually lethal (Hall et al. 1983). Effects of oil ingestion in loggerheads dying from oil exposure in the Canary Islands include esophageal impaction, necrotizing dermatitis and gastroenteritis, and necrotizing hepatitis (Camacho et al. 2013).

During the DWH spill, live oiled turtles admitted for rehabilitation exhibited abnormalities including relatively severe metabolic and osmoregulatory derangements resulting from a combination of stress, exertion, exhaustion, and dehydration related to oiling, capture and transport (Stacy et al. 2017). Mortalities were examined for evidence of internal exposure to polycyclic aromatic hydrocarbons (PAHs) and dispersant component dioctyl sodium sulfosuccinate (DOSS) (Ylitalo et al. 2017). Visibly oiled turtles had higher concentrations of PAH than unoiled turtles, which may suggest low-level exposure from other sources, and DOSS levels were below the limit of quantitation in almost all samples (Ylitalo et al. 2017).

Transect searches conducted in convergence zones during rescue operations following the DWH spill documented 937 oceanic juvenile Kemp's ridley, green, loggerhead, and hawksbill turtles in the spill area, and 81% of those captured were visibly oiled (McDonald et al. 2017). Based on these observations, turtle density calculations, and spatial extent of the oil, the total number of pelagic-stage sea turtles exposed to DWH oil was estimated at 402,000, with 54,800 of these heavily oiled, although the majority of the dead turtles were believed to be unobserved and therefore unaccounted for in these estimates (McDonald et al. 2017). Researchers estimated an overall mortality of 30% for oceanic turtles within the footprint of the spill in addition to those presumed dead from heavy oiling (Mitchelmore et al. 2017). Dependence on floating *Sargassum* for shelter and food in convergence zones where oil and tar accumulate makes surface-pelagic turtles particularly vulnerable to ingesting oil and tar (Witherington 2002; Witherington et al. 2012). Stranding data indicated that sea turtle stranding rates were at record levels in 2010 and 2011, increasing as much as 5× after the spill (NMFS data).

Kemp's ridley's principal foraging habitat is in the northern GOM (Seney and Landry Jr 2008; Shaver et al. 2013). Stable isotope analyses conducted on nesting Kemp's showed that 51.5% of turtles sampled had evidence of oil exposure (Reich et al. 2017), indicating that the primary foraging grounds in the northern GOM were contaminated by oil and that Kemp's ridleys continued to forage in these areas after the spill. Loggerhead foraging sites characterized through satellite tracking demonstrated an overlap with the oil spill footprint, with 32% of tracked individuals taking up year-round residence in the northern GOM foraging habitats (Hart et al. 2014). Stable isotope analysis confirmed that loggerheads returned to the oiled area and did not change foraging patterns after the spill, increasing their risk of chronic exposure to oil and dispersants (Vander Zanden et al. 2016).

Declines in reproductive parameters of loggerheads in the northern GOM were reported (Lamont et al. 2012), although the decline could not be linked directly to the DWH spill. Observed declines in nesting may have been partly due to reduced prey availability and therefore an inhibited ability to allocate resources required for nesting. Colder temperatures in 2010 may have delayed or reduced nesting activity or suppressed the ability of turtles to reach breeding condition (Chaloupka et al. 2008; Lauritsen et al. 2017; Weishampel et al. 2010; Hawkes et al. 2007).

26.4.3 Response Activities

Surface skimming and burning of oil slicks during the DWH disaster response may have impacted an unquantified number juvenile turtles living in *Sargassum* (McDonald et al. 2017). Up to 23% of important *Sargassum* habitat was estimated as lost as a result of oil exposure (Trustees 2016).

Response activities related to cleanup efforts, such as mechanical beach cleaning of oiled sand with heavy machinery and the associated disturbance from noise and artificial lighting, impacted sea turtle nesting habitats in the northern GOM (Michel et al. 2013). Enhanced vessel activity and physical barriers (e.g., booms) in nearshore waters may have affected nesting activity as well (Lauritsen et al. 2017). Loggerhead nesting densities in 2010 in northwest Florida were 43.7% lower than expected based on previous data, and an estimated 250 loggerhead nests were lost due to DWH response activities on nesting beaches (Lauritsen et al. 2017).

26.4.4 Dispersants

In addition to being exposed to oil, marine mammals and sea turtles were also exposed to dispersants. Impacts of exposure to dispersants or dispersants in conjunction with oil are not well known, as there are few studies for marine mammals and sea turtles. Since oil itself is generally toxic and can be lethal, dispersants may improve short-term survival of marine megafauna by reducing formation of oil slicks, decreasing the probability of heavy oiling, and accelerating the initial degradation of released oil (Neff 1990).

The ramifications of the unprecedented release of high volumes of dispersant chemicals as part of the DWH spill response are widely unknown. Evidence for cytotoxicity and genotoxicity of Corexit 9527 and Corexit 9500, the two dispersant chemicals used during the DWH spill, to sperm whale skin cells has been demonstrated in a laboratory setting (Wise et al. 2014). These findings were consistent with cytotoxicity and cell survival studies using Corexit 9500 in human and rat cells (Bandele et al. 2012; Zheng et al. 2014); however, Corexit 9527 was found to be less cytotoxic to whale cells than reported for other species' cells. Cytotoxicity may lead to acute effects, while genotoxicity is expected to lead to delayed effects associated with genetic mutations in somatic and/or germ cells. Mutations in somatic cells from toxic exposure may be associated with cancer in exposed marine mammals (Gauthier et al. 1999), while mutations in germ cells are inherited by offspring.

Effects of dispersants on sea turtles are largely unknown, but dispersants have the potential to interfere with lung function, digestion, and salt gland function (Shigenaka 2003). In an exposure study that investigated the effects of crude oil, dispersant, and a crude oil/dispersant combination on loggerhead hatchlings, significant differences between treatment and nonexposed controls were detected in multiple blood chemistry parameters (Harms et al. 2014). Electrolyte imbalances and hydration challenges were worst in the combined oil/dispersant group, and the failure to gain weight was noted in dispersant and combined exposed hatchlings (Harms et al. 2014). Only one heavily oiled Kemp's ridley showed evidence of DOSS at detectable concentrations (Ylitalo et al. 2017). Recent studies have demonstrated that DOSS degrades more rapidly in surface conditions than under deepwater conditions (Campo et al. 2013; Batchu et al. 2014), suggesting that DOSS exposure was minimized in surface-pelagic turtles (Ylitalo et al. 2017).

26.4.5 Mortality Events

The 2010 marine mammal UME which began prior to the DWH spill complicated measurement of the fatalities from the DWH event itself. It is now thought that the UME was not caused by the spill but was aggravated and potentially prolonged and expanded by the event (Venn-Watson et al. 2015; Antonio et al. 2011). An exponential increase in sea turtle and cetacean mortality was reported beginning 38 days after the initial blowout (Antonio et al. 2011). The relationship between observed strandings and unobserved offshore mortality is difficult to assess, but it has been estimated that strandings accounted for at most 6.2% of the total dead marine mammals in the GOM following the DWH oil spill, depending upon the species (Williams et al. 2011). This study relied on highly uncertain population estimates and mortality rates but strongly suggests that stranded carcass counts are not an adequate means to estimate the total mortality. Similarly, sea turtle carcass stranding rates likely represent a fraction of total at-sea mortality, as carcasses are likely to sink prior to detection (Epperly et al. 1996). Winds, surface currents, and sea temperatures can bias stranding sites with respect to offshore source mortality locations (Nero et al. 2013).

26.5   Long-Term Effects

We consider long-term effects of the DWH oil spill on marine megafauna as those occurring after the initial response and cleanup period, extending months to years after the event.

26.5.1   Findings of Marine Mammal Passive Acoustic Studies

The GOM HARP project (Frasier et al. 2017; Hildebrand et al. 2015) provides the only long-term time series documenting marine mammal occurrence in oiled and unoiled oceanic habitats during and after the DWH oil spill. Data collected between 2010 and 2016 are discussed here. Mean weekly presence was calculated for each species (or species group) as the weekly average of time per day in which echolocation clicks were detected (Table 26.1). The seasonal component was removed from the weekly presence time series using a monthly seasonal trend decomposition procedure (Cleveland et al. 1990). Long-term trends in deseasoned mean weekly presence were then estimated for each site and species combination using a Theil-Sen regression (Table 26.2) with 5–95% confidence intervals obtained using a bootstrap method. The median slope across 500 pairs of points selected randomly with replacement within each time series was computed 100 times.

On average across the monitoring period, presence of sperm whales (Fig. 26.3) was substantial at the site adjacent to the wellhead (MC, 36.8% of 5-min time windows detected their presence; Table 26.2), slightly less at site GC (13.8% of time windows), and low at site DT (5.1% of time windows). Long-term trend estimates suggest a slow decline in mean presence of sperm whales at site MC (5 ± 1% annual reduction), between 2010 and 2016, and a greater decline at site GC (8% ±2% annual reduction). A possible slight increase in the presence of sperm whales was found at site DT (5 ± 5% annual increase); however, encounter rates were low and seasonally variable at this southern GOM location.

Table 26.1 Mean weekly marine mammal presence (as percentage) including [5th, 95th] percentiles at passive acoustic monitoring sites in the GOM HARP study, 2010–2016

Site	Sperm whale	*Kogia* spp.	Cuvier's BW	Gervais' BW	Risso's dolphin	*Stenella* delphinid	Blackfish delphinid
MC	36.8 [11.6, 64.2]	0.5 [0.0, 1.4]	0.1[a] [0, 0.6]	0.3[a] [0, 1.1]	1.3 [0, 5.7]	6.3 [0.5, 16.0]	0.6 [0, 2.0]
GC	13.8 [0.2, 38.4]	0.3 [0.0, 0.9]	0.1 [0, 0.5]	0.5 [0.0, 1.1]	0.2 [0, 1.1]	3.0 [0.2, 9.0]	0.4 [0, 1.8]
DT	5.1 [0, 19.5]	0.1 [0.0, 0.2]	3.6 [0.8, 7.2]	1.5 [0.3, 3.9]	4.5 [0, 23.4]	3.1 [0.0, 9.4]	0.4 [0, 2.1]

[a]Indicates subset from 2010 to 2013 was used to calculate the mean

Table 26.2 Estimated average annual percent change in marine mammal presence including [5th, 95th] confidence intervals at passive acoustic monitoring sites in the GOM HARP study, 2010–2016

Site	Sperm whale	*Kogia* spp.	Cuvier's BW	Gervais' BW	Risso's dolphin	*Stenella* delphinid	Blackfish delphinid
MC	−4.5 [−6.0, −3.3]	18.8 [12.9, 26.3]	5.4[a] [2.0, 8.7]	37.3[a] [24.3, 52.8]	8.7 [2.1, 19.4]	−1.6 [−3.5, 0.2]	−7.0 [−9.0, −4.6]
GC	−8.3 [−10.0, −6.4]	−15.5 [−16.3, −14.8]	1.0 [−3.0, 6.3]	4.1 [1.4, 7.3]	−5.0 [−9.1, 0.0]	−10.9 [−12.2, −9.7]	−11.1 [−13.5, −9.3]
DT	5.4 [−0.3, 13.6]	−9.0 [−11.8, −5.4]	−9.7 [−11.0, −8.9]	−8.1 [−9.1, −6.6]	4.2 [1.4, 8.1]	−13.0 [−14.2, −11.5]	1.3 [−4.1, 10.2]

[a]Indicates subset from 2010 to 2013 was used to calculate slope

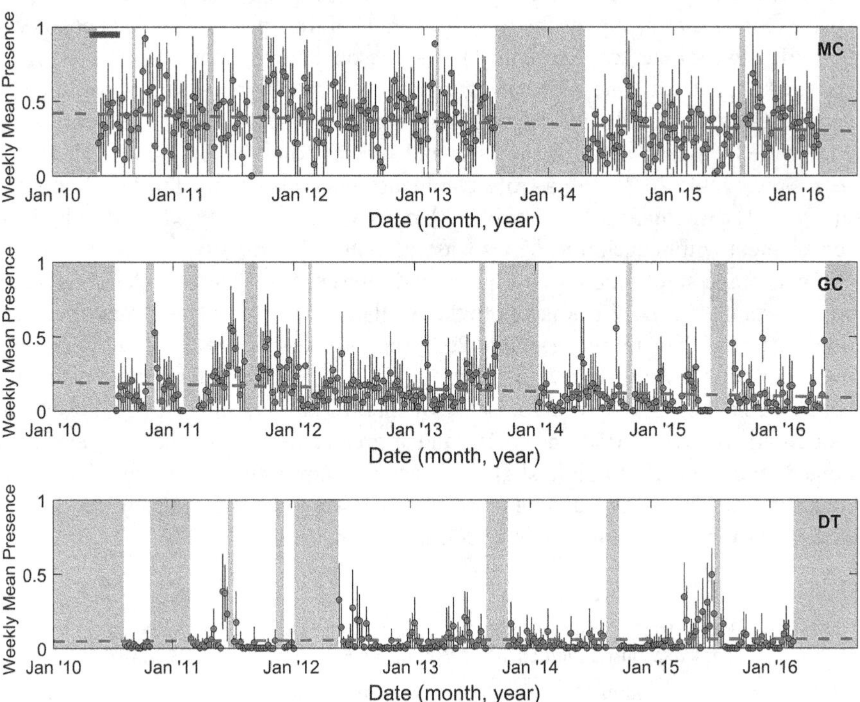

Fig. 26.3 Sperm whale weekly mean presence (open circles) as fraction of time present at passive acoustic monitoring sites from the GOM HARP study. Error bars indicate standard deviation within each week. Gray rectangles indicate periods without data. Red dashed line indicates estimated trend. Dark red bar on top plot indicates period during which the DWH well remained uncapped

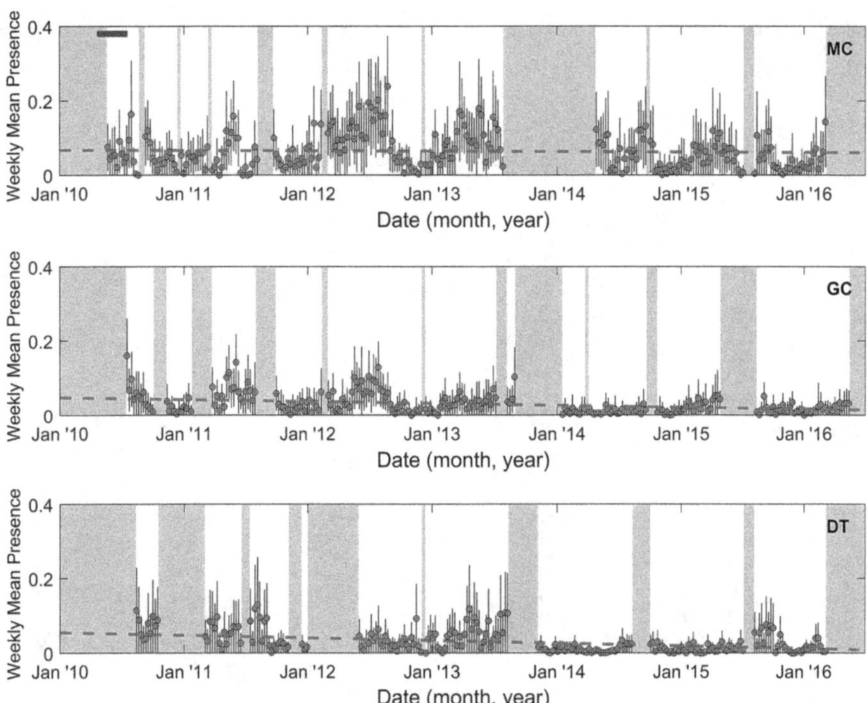

Fig. 26.4 Weekly mean *Stenella* sp. (mid-frequency delphinid) presence as fraction of time present at passive acoustic monitoring sites from the GOM HARP study. Markings as in Fig. 26.3

Stenella species and blackfish (presumably short-finned pilot whales) species were associated with mid- and low-frequency echolocation, respectively. *Stenella* had slightly higher presence at site MC (6.3% for *Stenella*) relative to other sites. Long-term declines in *Stenella* occurrence (Fig. 26.4) were observed at the sites GC (11 ± 1% annual reduction) and DT (13 ± 2% annual reduction) outside of the DWH surface slick footprint, but not at site MC where presence remained nearly constant (2 ± 2% annual reduction). However, relatively higher encounter rates in 2012 may be masking long-term decreases in *Stenella* delphinid presence at site MC. Blackfish presence was low overall (0.4–0.6%), with declines at sites MC and GC (7 ± 2% and 11 ± 2% annual reductions, respectively), but no significant change at site DT (1 ± 6% annual change) (Fig. 26.5). Risso's dolphin presence was low (0.2–4.5%) and strongly seasonal at all sites, and their presence increased slightly at site DT (4 ± 4% annual increase) and more strongly at site MC (9 ± 7% annual increase) (Fig. 26.6). A possible decline in Risso's dolphin presence was found at site GC (5 ± 5% annual reduction) where overall presence was low.

Beaked whale presence was highest at site DT for both Cuvier's (4%) and Gervais' (2%) beaked whale, and both were present year round (Figs. 26.7 and 26.8). Presence of both species declined at site DT (10 ± 1% annual reduction for Cuvier's; 8 ± 1% annual reduction for Gervais'), but remained constant or increased

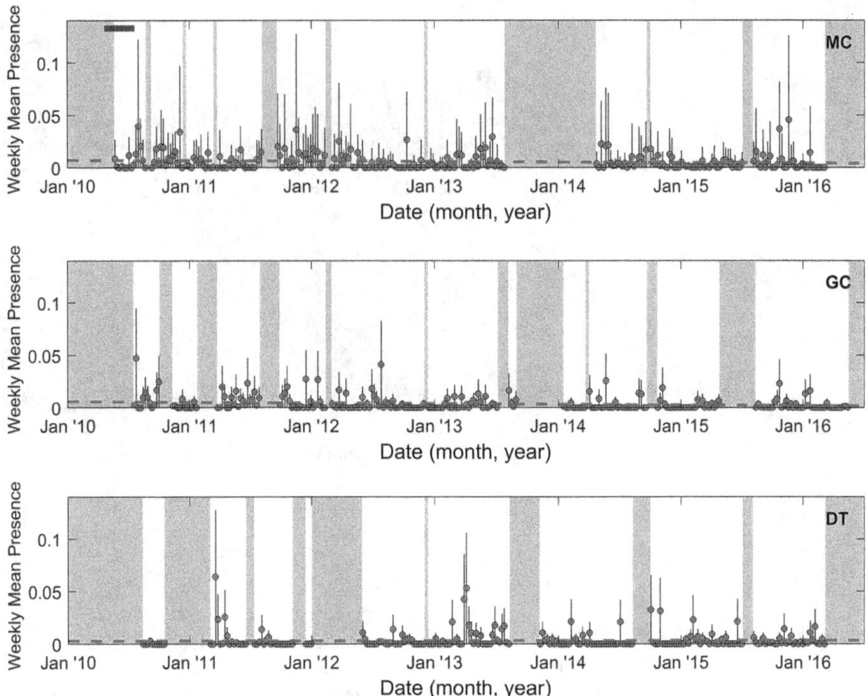

Fig. 26.5 Weekly mean blackfish (low-frequency echolocation) presence as fraction of time present at passive acoustic monitoring sites from the GOM HARP study. Markings as in Fig. 26.3

at site GC (1 ± 4% annual change for Cuvier's; 4 ± 3% annual increase for Gervais'). An increase in beaked whale presence at site MC is observed; however, analysis for beaked whale presence at this site occurred over a limited date range (2010–2013); therefore, trends are less robust.

Kogia spp. presence was relatively high (0.3–0.5% occurrence within a short detection range of <1 km) at site MC and GC (Fig. 26.9). Presence of *Kogia* spp. increased at site MC (19 ± 7% annual increase), but presence at site GC decreased strongly after 2013 resulting in a strongly negative long-term trend in mean presence at this site (15 ± 1% annual decrease). Presence also decreased at site DT (9 ± 3% annual decline), although overall encounter rates at that site were low (0.1%) throughout the monitoring period.

Population movements and declines may be convolved in the trends seen in GOM acoustic monitoring because of the limited number of monitoring locations in the HARP dataset. It is unclear to what degree changes in presence reflect population displacement around the GOM and beyond or rather indicate offshore mortality. Aspects of both processes may be influencing the long-term observed trends.

Population trends may be related to exposure: Based on seasonal trends and encounter rates during the oil spill at site MC, sperm whales, *Stenella*, and *Kogia*

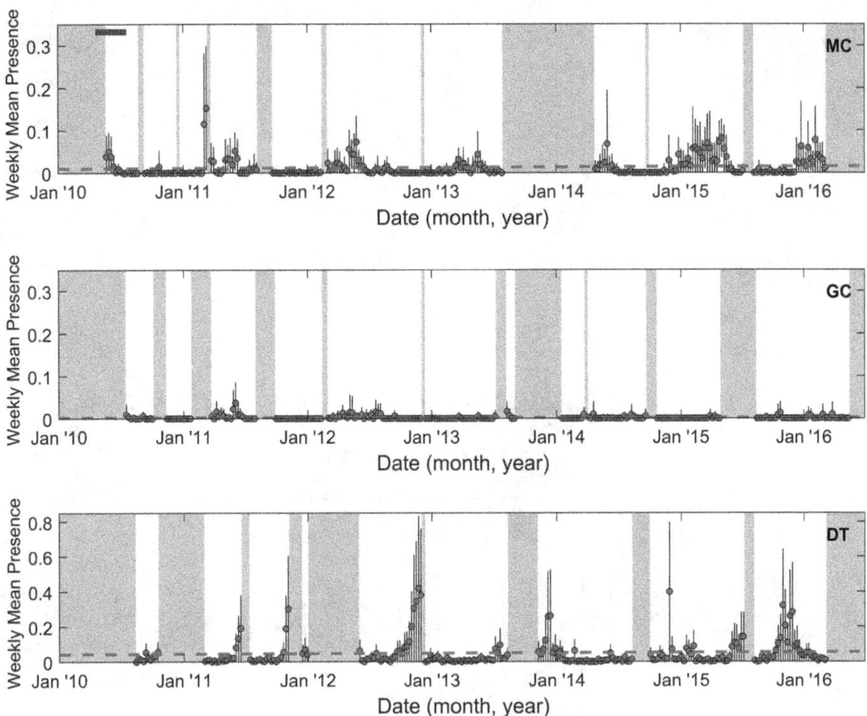

Fig. 26.6 Weekly mean Risso's dolphin presence as fraction of time present at passive acoustic monitoring sites from the GOM HARP study. Markings as in Fig. 26.3

species are most likely to have interacted with the DWH surface and subsurface footprints for extended periods of time in the spring and summer of 2010. *Stenella* and *Kogia* presence strongly declined at sites GC and DT, *Kogia* presence declined at site MC, and *Stenella* delphinids appear to have declined from 2012 to 2016 following a peak in 2012. Sperm whale presence declined steadily at GC and MC, and while possibly increasing at site DT, this site may not be part of core sperm whale habitat (Jochens et al. 2008) given overall low encounter rates there. Blackfish presence is highly variable at site MC; however, this group appears to have been present during the DWH event based on the acoustic record. Presence of blackfish delphinids has declined at sites MC and GC while remaining approximately constant at site DT. In contrast Risso's dolphins may not have been as strongly exposed to oil in 2010 due to the seasonality of their presence in the northern GOM, and Risso's dolphin encounter rates appear to be increasing at sites MC and DT while decreasing at site GC. Both beaked whale species appear to be declining at site DT, with limited change at site GC. Owing to the seasonality of Cuvier's beaked whales, only the Gervais' beaked whale appears to have been substantially exposed to oil during the DWH spill. The time series for beaked whale presence at site MC may be too short to robustly interpret long-term trends there.

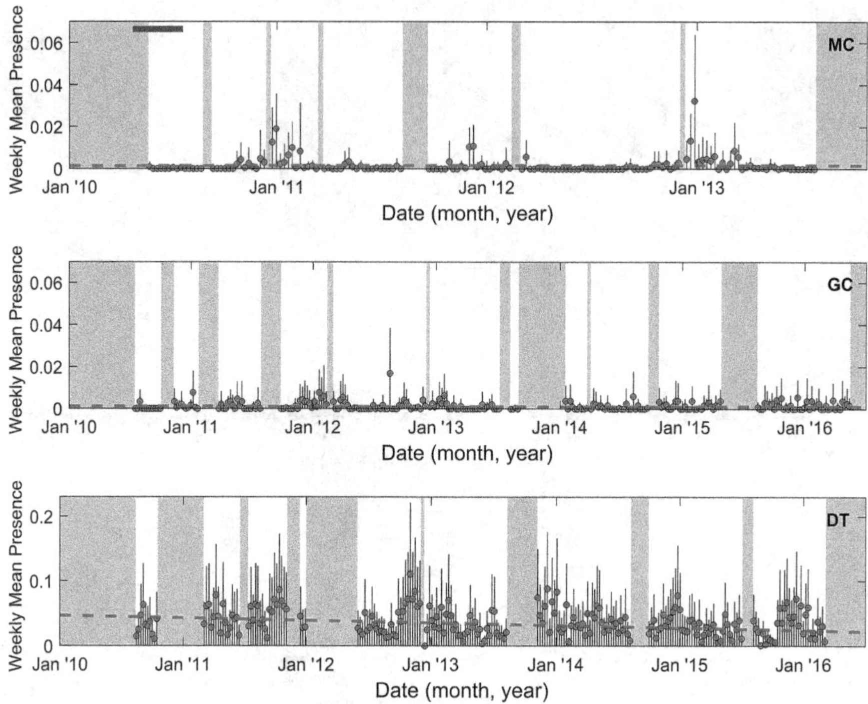

Fig. 26.7 Weekly mean Cuvier's beaked whale presence as fraction of time present at passive acoustic monitoring sites from the GOM HARP study. Markings as in Fig. 26.3

Population declines in the eastern and southern GOM may be unrelated to the DWH event, since some of the strongest declines are seen at the two sites outside the DWH oil footprint (GC and DT). However, seasonal cycles in the passive acoustic data suggest that these species' distributions shift over time, likely as animals seek out favorable conditions; therefore, many of these pelagic species may not be resident in specific areas throughout the year. The high productivity conditions created by the outflow of the Mississippi River have historically supported higher marine mammal densities than other regions of the GOM (Reeves et al. 2011), and populations may preferentially return to that region. Female sperm whales tagged in the MC region typically had long residence times in the area and appeared to use it as core habitat (Jochens et al. 2008). Declines at other less productive sites may indicate range contraction associated with population-level mortality (Rugh et al. 2010; Worm and Tittensor 2011), or might reflect population shifts in response to other drivers. A broader understanding of migratory patterns on a GOM-wide scale is needed to more confidently interpret site-level trends in the context of the broader GOM ecosystem.

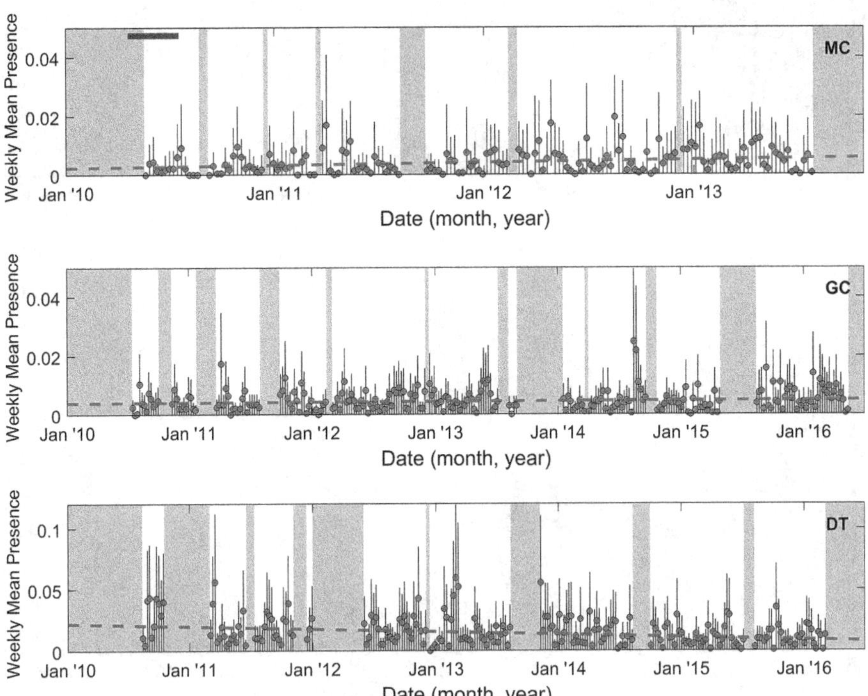

Fig. 26.8 Weekly mean Gervais' beaked whale presence as fraction of time present at passive acoustic monitoring sites from the GOM HARP study. Markings as in Fig. 26.3

26.5.2   Additional Marine Mammal Studies

Latent effects of exposure have been examined in the case study of the resident Barataria Bay bottlenose dolphin population. Over 5 years after heavy oiling of the bay during the DWH oil spill, successful calving rates were 20% compared to 83% for an unexposed reference population (Lane et al. 2015). It was unclear whether unsuccessful pregnancies were directly caused by oil exposure or were linked indirectly through poor maternal health (Schwacke et al. 2013). Similar reproductive failures occurring in offshore populations (e.g., Farmer et al. 2018) could explain the observed long-term declines in encounter rates at oceanic monitoring locations. Annual survival rates among adults were also lower (86.8%) than in comparable populations (95.1 to 96.2%; Lane et al. 2015). Bottlenose dolphins in Barataria Bay were five times more likely to have moderate to severe lung disease than a reference population (Schwacke et al. 2013).

A study comparing short-term (7–12-day) PAM recordings before and after the DWH spill at a site near the wellhead indicated possible declines in sperm whale occurrence (Ackleh et al. 2012), with an increase of 25 miles from the site. However, due to the high variability in sperm whale presence at fixed monitoring sites on

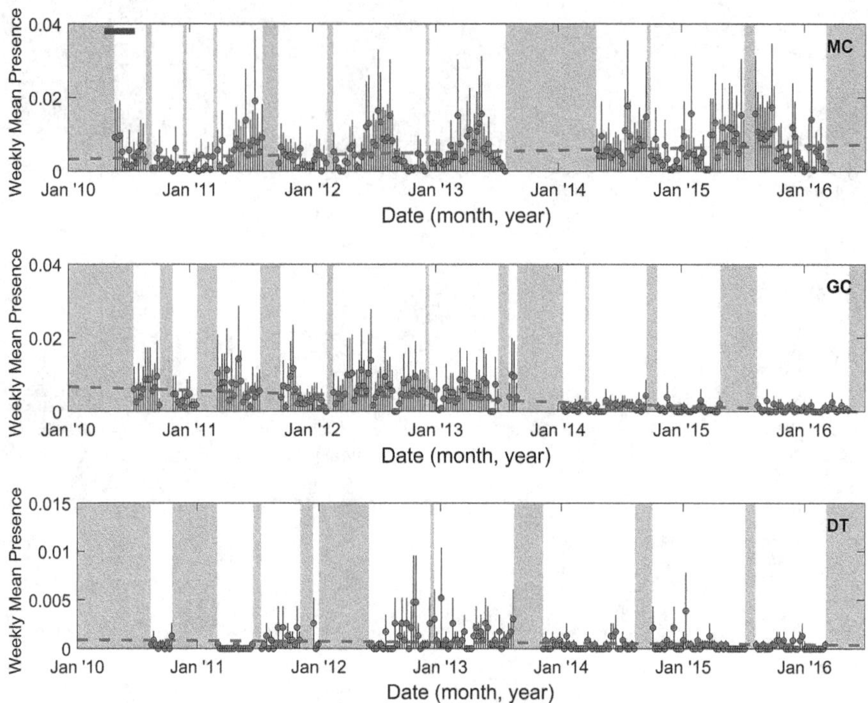

Fig. 26.9 Weekly mean *Kogia* spp. presence as fraction of time present at passive acoustic monitoring sites from the GOM HARP study. Markings as in Fig. 26.3

weekly timescales (Fig. 26.3), it is not possible to determine whether the difference between the two measurements reflects real change or normal variability.

26.5.3 Findings of Sea Turtle Studies

The long-term effects of oil exposure and the DWH oil spill on sea turtles are not well understood or quantified (Vander Zanden et al. 2016). The number of Kemp's ridley nests in Tamaulipas, Mexico, in 2010 was below predicted levels and has remained below expected levels every nesting season since (Dixon and Heppell 2015), but the reduction has not been definitively attributed to the DWH spill (Caillouet et al. 2016; Caillouet Jr 2014). Some have speculated that the large-scale oiling of *Sargassum* (Hu et al. 2016) and subsequent loss of developmental/foraging habitat for juvenile turtles may have long-term implications for population recovery.

In response to the DWH oil spill, stage-based spatial matrix models have been developed to simulate oil spills to assess the potential impact of oil spills on loggerhead populations, defining oceanic-stage survival followed by fecundity as the most sensitive parameters for eliciting changes in population growth (Leung et al. 2012). A geospatial assessment of cumulative stressors to evaluate where combined threats and impacts are greatest was conducted on a GOM-wide scale for Kemp's ridleys and loggerheads following the DWH oil spill (Love et al. 2017). This research showed a range of anthropogenic stressors including incidental bycatch in commercial and recreational fisheries and habitat degradation, and it demonstrated that few areas exist in their terrestrial or marine environment without cumulative impacts from multiple stressors (Love et al. 2017).

26.6   Remaining Knowledge Gaps

Efforts to assess the comprehensive immediate and long-term effects of the DWH oil spill on pelagic species are limited by a scarcity of pre-disaster baseline data (Bjorndal et al. 2011; Graham et al. 2011; Trustees 2016). Without the ability to compare pre-and post-spill measurements, many potential impacts are unquantifiable. Further, effects on these long-lived species may continue to play out over the coming decades (Schwacke et al. 2017). The assessment of cumulative impacts must be considered on an ecosystem level, as effects are based on direct mortality, degradation of habitat, quality and availability of prey resources, and sublethal impacts such as reduced foraging or reproductive potential (Love et al. 2017). Developing a better understanding of the spatiotemporal overlap of threats with the distribution and abundance of sea turtle populations will guide managers to develop geographically targeted management strategies to mitigate key stressors and restore injured resources (Love et al. 2017).

No comparable long-term data on marine mammal presence were collected in the period prior to the 2010 spill. At best, visual survey data give decadal-scale abundances that cannot be directly applied to understanding the impact of the spill. In addition, the GOM was not a pristine habitat prior to the 2010 spill; therefore, we cannot assume that pre-spill population levels were stable or attribute observed shifts to the DWH event with great confidence. Although a visual marine mammal survey was conducted prior to the spill in 2009 (Waring et al. 2013), it did not provide the kind of spatiotemporal resolution or precise abundance estimates needed to quantify acute impacts. Passive acoustic sensors were deployed 26 days after the initial blowout, so although they did record during the majority of the 152-day spill, recorders were not in place to capture pre-spill levels, and some immediate effects may have been missed. Further, passive acoustic sensors have limited detection ranges (Frasier et al. 2016), and additional research is needed to determine the spatial scale over which the observations from these monitoring locations can be extrapolated. Efforts to estimate chronic effects by any method have necessarily relied on uncertain assumptions regarding pre-spill population sizes, health, and distributions.

Much of the released oil is thought to have been deposited on the seafloor. Little is known about if and how marine megafauna might be interacting with deep water benthic oil, either directly or via the pelagic food web (see Pulster et al. 2020). Impacts of the spill on mesopelagic and bathypelagic prey availability remains unclear (Fisher et al. 2016). Different prey types likely have differing abilities to metabolize oil-derived compounds. In particular, cephalopods seem less capable of metabolizing polycyclic aromatic hydrocarbons (PAHs) and more likely to bioaccumulate heavy metals than fish (Reijnders et al. 2009). Trace metals are common in crude oil and may further concentrate in weathered oil (Gohlke et al. 2011). Deep-foraging, squid-eating cetaceans including sperm whales and beaked whales may be at higher risk of long-term exposure to oil-related pollutants through their prey. Toxicity of oil and oil-related compounds to marine mammals and sea turtles remains poorly understood.

Lastly, the unknown spatial ranges and movement patterns of most oceanic GOM marine mammal species and sparse habitat use, abundance, and distribution data for sea turtles result in broad uncertainty regarding exposure and long-term impacts of the spill and subsequent environmental pollution on these populations. It remains unclear to what degree observed animals are resident in or systematically return to affected habitats. Without coordinated, international GOM-wide monitoring efforts, it is not possible to determine whether local declines in encounter rates represent population shifts or population decreases.

26.7 Conclusion

The majority of research on the effects of oil spills on marine mammals and sea turtles has focused on nearshore species (coastal bottlenose dolphins, killer whales, and pinnipeds), coastal impacts (coastal strandings, sea turtle nesting beaches), and surface oiling. The DWH event was a large-volume oil spill that occurred offshore, with significant subsurface footprint, in poorly understood habitats, and with sparse baseline data. Long-term offshore monitoring suggests ongoing declines in marine mammal presence, which may be related to reduced reproductive success as observed in nearshore proxies. Oceanic species were most heavily and directly impacted by this spill, but discerning the immediate and long-term effects on oceanic populations requires piecing together a patchwork of sparse observations and studies. It is clear however that marine mammals and sea turtles were directly exposed to unprecedented amounts of oil and dispersants and that the acute and chronic population-level impacts of this exposure were likely high and underestimated based on coastal observations.

Funding Information This research was made possible by grants from the Gulf of Mexico Research Initiative through its consortium the Center for the Integrated Modeling and Analysis of the Gulf Ecosystem (C-IMAGE). Funding for HARP data collection and analysis was also provided by the Natural Resource Damage Assessment partners (20105138), the US Marine Mammal Commission (20104755/E4061753), the Southeast Fisheries Science Center under the Cooperative Institute for Marine Ecosystems and Climate (NA10OAR4320156) with support through Interagency Agreement #M11PG00041 between the Bureau of Offshore Energy Management,

Environmental Studies Program and the National Marine Fisheries Service, Southeast Fisheries Science Center. The analyses and opinions expressed are those of the authors and not necessarily those of the funding entities. The data used for this study are archived by the Gulf of Mexico Research Initiative at https://data.gulfresearchinitiative.org/data/R4.x267.180:0011 maintained by the Gulf Research Initiative Information and Data Cooperative.

References

Ackleh AS, Ioup GE, Ioup JW, Ma B, Newcomb JJ, Pal N, Sidorovskaia NA, Tiemann C (2012) Assessing the Deepwater Horizon oil spill impact on marine mammal population through acoustics: endangered sperm whales. J Acoust Soc Am 131(3):2306–2314

Ackleh AS, Caswell H, Chiquet RA, Tang T, Veprauskas A (2018) Sensitivity analysis of the recovery time for a population under the impact of an environmental disturbance. Nat Resour Model 32(1):e12166

Antonio FJ, Mendes RS, Thomaz SM (2011) Identifying and modeling patterns of tetrapod vertebrate mortality rates in the Gulf of Mexico oil spill. Aquat Toxicol 105(1–2):177–179

Asl SD, Amos J, Woods P, Garcia-Pineda O, MacDonald IR (2016) Chronic, anthropogenic hydrocarbon discharges in the Gulf of Mexico. Deep-Sea Res II Top Stud Oceanogr 129:187–195

Avens L, Goshe LR, Harms CA, Anderson ET, Hall AG, Cluse WM, Godfrey MH, Braun-McNeill J, Stacy B, Bailey R (2012) Population characteristics, age structure, and growth dynamics of neritic juvenile green turtles in the northeastern Gulf of Mexico. Mar Ecol Prog Ser 458:213–229

Bandele OJ, Santillo MF, Ferguson M, Wiesenfeld PL (2012) In vitro toxicity screening of chemical mixtures using HepG2/C3A cells. Food Chem Toxicol 50(5):1653–1659

Batchu SR, Ramirez CE, Gardinali PR (2014) Stability of dioctyl sulfosuccinate (DOSS) towards hydrolysis and photodegradation under simulated solar conditions. Sci Total Environ 481:260–265

Bjorndal KA, Bowen BW, Chaloupka M, Crowder LB, Heppell SS, Jones CM, Lutcavage ME, Policansky D, Solow AR, Witherington BE (2011) Better science needed for restoration in the Gulf of Mexico. Science 331(6017):537–538

BOEM Data Center (2018). https://www.data.boem.gov/Mapping/Files/platform.zip; https://www.data.boem.gov/Mapping/Files/wells.zip. Accessed 26th June 2018

Bolten AB (2003) Variation in sea turtle life history patterns: neritic vs. oceanic developmental stages. In: The biology of sea turtles, vol 2. CRC Press, Boca Raton, FL, pp 243–257

Botello AV, Villanueva FS, Diaz GG (1997) Petroleum pollution in the Gulf of Mexico and Caribbean Sea. Rev Environ Contam Toxicol. Springer 153:91–118

Caillouet CW, Gallaway BJ, Putman NF (2016) Kemp's ridley sea turtle saga and setback: novel analyses of cumulative hatchlings released and time-lagged annual nests in Tamaulipas, Mexico. Chelonian Conserv Biol 15(1):115–131

Caillouet CW Jr (2014) Interruption of the Kemp's ridley population's pre-2010 exponential growth in the Gulf of Mexico and its aftermath: one hypothesis. Mar Turt Newsl 143:1–7

Camacho M, Calabuig P, Luzardo OP, Boada LD, Zumbado M, Orós J (2013) Crude oil as a stranding cause among loggerhead sea turtles (Caretta caretta) in the Canary Islands, Spain (1998–2011). J Wildl Dis 49(3):637–640

Campo P, Venosa AD, Suidan MT (2013) Biodegradability of Corexit 9500 and dispersed South Louisiana crude oil at 5 and 25 C. Environ Sci Technol 47(4):1960–1967

Chaloupka M, Work TM, Balazs GH, Murakawa SK, Morris R (2008) Cause-specific temporal and spatial trends in green sea turtle strandings in the Hawaiian Archipelago (1982–2003). Mar Biol 154(5):887–898

Cleveland RB, Cleveland WS, McRae JE, Terpenning I (1990) STL: A seasonal-trend decomposition procedure based on loess. J Off Stat 6(1):3–33

Darnell RM (2015) The American sea: a natural history of the Gulf of Mexico. Texas A&M University Press, College Station, TX

Di Guardo G, Mazzariol S (2013) Dolphin Morbillivirus: a lethal but valuable infection model. Emerg Microbes Infect 2(11):e74

Dixon PM, Heppell SS (2015) Statistical analysis of Kemp's ridley nesting trends, vol ST_TR.10. DWH Sea Turtles NRDA Technical Working Group Report, Corpus Christi, TX

Engelhardt FR (1982) Hydrocarbon metabolism and cortisol balance in oil-exposed ringed seals, *Phoca hispida*. Comp Biochem Physiol C 72(1):133–136

Englehardt F (1977) Uptake and clearance of petroleum hydrocarbons in the ringed seal (*Phoca hispida*). J Fish Res Board Can 34:1143–1147

Epperly SP, Braun J, Chester AJ, Cross FA, Merriner JV, Tester PA, Churchill JH (1996) Beach strandings as an indicator of at-sea mortality of sea turtles. Bull Mar Sci 59(2):289–297

Farmer NA, Baker K, Zeddies DG, Denes SL, Noren DP, Garrison LP, Machernis A, Fougères EM, Zykov M (2018) Population consequences of disturbance by offshore oil and gas activity for endangered sperm whales (Physeter macrocephalus). Biol Conserv 227:189–204

Fautin D, Dalton P, Incze LS, Leong J-AC, Pautzke C, Rosenberg A, Sandifer P, Sedberry G, Tunnell JW Jr, Abbott I (2010) An overview of marine biodiversity in United States waters. PLoS One 5(8):e11914

Fuentes MMPB, Limpus CJ, Hamann M, Dawson J (2010) Potential impacts of projected sea-level rise on sea turtle rookeries. Aquat Conserv Mar Freshwat Ecosyst 20(2):132–139

Fisher CR, Montagna PA, Sutton TT (2016) How did the Deepwater Horizon oil spill impact deep-sea ecosystems? Oceanography 29(3):182–195

Frasier KE, Roch MA, Soldevilla MS, Wiggins SM, Garrison LP, Hildebrand JA (2017) Automated classification of dolphin echolocation click types from the Gulf of Mexico. PLoS Comput Biol 13(12):e1005823

Frasier KE, Wiggins SM, Harris D, Marques TA, Thomas L, Hildebrand JA (2016) Delphinid echolocation click detection probability on near-seafloor sensors. J Acoust Soc Am 140(3):1918–1930. https://doi.org/10.1121/1.4962279

Garrison L (2015) Estimated lethal and sub-lethal injuries of sea turtles on the continental shelf in the Northern Gulf of Mexico based on aerial survey abundance estimates and surface oiling. DWH Sea Turtles Technical Working Group Report

Gauthier J, Dubeau H, Rassart E, Jarman W, Wells R (1999) Biomarkers of DNA damage in marine mammals. Mutat Res Genet Toxicol Environ Mutagen 444(2):427–439

Gavilan FM (1998) Impact of regulatory measures on Cuban marine turtle fisheries. In: 18th International Sea Turtle Symposium, Mazatlán, Sinaloa, Mexico, pp 108–109

Geraci JR (1990a) Physiologic and toxic effects on cetaceans. In: Geraci JR, St. Aubin DJ (eds) Sea mammals and oil: confronting the risks. Academic Press, Inc., San Diego, CA, pp 167–192

Geraci JR (1990b) Physiologic and toxic effects on pinnipeds. In: Geraci JR, St. Aubin DJ (eds) Sea mammals and oil: confronting the risks. Academic Press, Inc., San Diego, CA, pp 107–127

Gitschlag GR, Herczeg BA, Barcak TR (1995) Observations of sea turtles and other marine life at the explosive removal of offshore oil and gas structures in the Gulf of Mexico. Gulf research reports 9

Gohlke JM, Doke D, Tipre M, Leader M, Fitzgerald T (2011) A review of seafood safety after the Deepwater Horizon blowout. Environ Health Perspect 119(8):1062

Goodale DR, Hyman MA, Winn HE, Edkel R, Tyrell M (1981) Cetacean responses in association with the Regal Sword oil spill. Cetacean and Turtle Assessment Program, University of Rhode Island, Annual Report 1979. U.S. Dept. of the Interior, Washington, DC

Graham B, Reilly WK, Beinecke F, Boesch DF, Garcia TD, Murray CA, Ulmer F (2011) Deep water: the Gulf oil disaster and the future of offshore drilling (report to the President). National Commission on the BP Deepwater Horizon Oil Spill and Offshore Drilling

Hall RJ, Belisle AA, Sileo L (1983) Residues of petroleum hydrocarbons in tissues of sea turtles exposed to the Ixtoc I oil spill. J Wildl Dis 19(2):106–109

Harms CA, McClellan-Green P, Godfrey M, Christiansen E, Broadhurst H, Godard-Codding C (2014) Clinical pathology effects of crude oil and dispersant on hatchling loggerhead sea turtles (Caretta caretta). Proceedings of the 45th Annual International Association for Aquatic Animal Medicine, Gold Coast, Australia, pp. 17–22

Hart KM, Lamont MM, Sartain AR, Fujisaki I (2014) Migration, foraging, and residency patterns for Northern Gulf loggerheads: implications of local threats and international movements. PLoS One 9(7):e103453

Hawkes LA, Broderick AC, Coyne MS, Godfrey MH, Godley BJ (2007) Only some like it hot—quantifying the environmental niche of the loggerhead sea turtle. Divers Distrib 13(4):447–457

Hays GC (2004) Good news for sea turtles. Trends Ecol Evol 19(7):349–351

Hildebrand HH (1982) A historical review of the status of sea turtle populations in the western Gulf of Mexico. In: Biology and conservation of sea turtles. Smithsonian Institution Press, Washington, DC, pp 447–453

Hildebrand J, Baumann-Pickering S, Frasier K, Tricky J, Merkens K, Wiggins SM, McDonald MA, Garrison LP, Harris D, Marques TA, Thomas L (2015) Passive acoustic monitoring of beaked whale densities in the Gulf of Mexico during and after the Deepwater Horizon oil spill. Nat Sci Rep 5:16343

Hildebrand JA, Armi L, Henkart PC (2012) Seismic imaging of the water-column deep layer associated with the Deepwater Horizon oil spill. Geophysics 77(2):EN11–EN16

Hildebrand JA, Frasier KE, Wiggins SM (2017) Trends in deep-diving whale populations in the Gulf of Mexico 2010 to 2016. In: Gulf of Mexico Oil Spill and Ecosystem Science Conference, New Orleans

Hu C, Hardy R, Ruder E, Geggel A, Feng L, Powers S, Hernandez F, Graettinger G, Bodnar J, McDonald T (2016) Sargassum coverage in the northeastern Gulf of Mexico during 2010 from Landsat and airborne observations: implications for the Deepwater Horizon oil spill impact assessment. Mar Pollut Bull 107(1):15–21

Inkley D, Gonzalez-Rothi Kronenthal S, McCormick L (2013) Restoring a degraded Gulf of Mexico: wildlife and wetlands three years into the gulf oil disaster. Report of the National Wildlife Federation. 15 pp

Irwin L-J (2005) Marine toxins: adverse health effects and biomonitoring with resident coastal dolphins. Aquat Mamm 31(2):195

Jochens A, Biggs D, Benoit-Bird K, Engelhaupt D, Gordon J, Hu C, Jaquet N, Johnson M, Leben R, Mate B, Miller P, Ortega-Ortiz J, Thode A, Tyack P, Würsig B (2008) Sperm whale seismic study in the Gulf of Mexico: synthesis report, Vol. OCS Study MMS 2008-006. US Dept. of Interior, Minerals Management Service, Gulf of Mexico OCS Region, New Orleans, LA

Kujawinski E, Soule M, Valentine D, Boysen A, Longnecker K, Redmond M (2011) Fate of dispersants associated with the Deepwater Horizon oil spill. Environ Sci Technol 45(4):1298–1306. https://doi.org/10.1021/es103838p

Lamont MM, Carthy RR, Fujisaki I (2012) Declining reproductive parameters highlight conservation needs of loggerhead turtles (Caretta caretta) in the northern Gulf of Mexico. Chelonian Conserv Biol 11(2):190–196

Lane SM, Smith CR, Mitchell J, Balmer BC, Barry KP, McDonald T, Mori CS, Rosel PE, Rowles TK, Speakman TR (2015) Reproductive outcome and survival of common bottlenose dolphins sampled in Barataria Bay, Louisiana, USA, following the Deepwater Horizon oil spill. Proc R Soc B 282(1818):20151944

Lange FW (1971) Marine resources: a viable subsistence alternative for the prehistoric lowland Maya1. Am Anthropol 73(3):619–639

Lauritsen AM, Dixon PM, Cacela D, Brost B, Hardy R, MacPherson SL, Meylan A, Wallace BP, Witherington B (2017) Impact of the Deepwater Horizon oil spill on loggerhead turtle Caretta caretta nest densities in northwest Florida. Endanger Species Res 33:83–93

Leung M-R, Marchand M, Stykel S, Huynh M, Flores JD (2012) Effect of localized oil spills on Atlantic loggerhead population dynamics. Open J Ecol 02(03):109–114

Lewis JS, Schroeder WW (2003) Mud plume feeding, a unique foraging behavior of the bottlenose dolphin in the Florida Keys. Gulf Mex Sci 21(1):9

Lewison R, Crowder L, Read A, Freeman S (2004) Understanding impacts of fisheries bycatch on marine megafauna. Trends Ecol Evol 19(11):598–604

Litz JA, Baran MA, Bowen-Stevens SR, Carmichael RH, Colegrove KM, Garrison LP, Fire SE, Fougeres EM, Hardy R, Holmes S et al (2014) Review of historical unusual mortality events (UMEs) in the Gulf of Mexico (1990- 2009): providing context for the multi-year northern Gulf of Mexico cetacean UME declared in 2010. Dis Aquat Org 112(2):161–175

Love M, Robbins C, Baldera A, Eastman S, Bolten A, Hardy R, Herren R, Metz T, Vander Zanden HB, Wallace B (2017) Restoration without borders: an assessment of cumulative stressors to guide large-scale, integrated restoration of sea turtles in the Gulf of Mexico. Ocean Conservancy report. https://doi.org/10.13140/RG.2.2.19832.55049

Lutcavage M, Lutz P, Bossart G, Hudson D (1995) Physiologic and clinicopathologic effects of crude oil on loggerhead sea turtles. Arch Environ Contam Toxicol 28(4):417–422

Lutcavage ME, Plotkin P, Witherington B, Lutz PL (1997) Human impacts on sea turtle survival. In: Lutz PL, Musick JA (eds) The biology of sea turtles, vol 1. CRC Press, Boca Raton, FL, pp 387–409

Lutz PL, Lutcavage M, Hudson D (1986) Physiological effects. Final Report. Study of the Effect of Oil on Marine Turtles. Minerals Management Service Contract Number 14-12-0001-30063, Florida Institute of Oceanography, St. Petersburg, FL

MacDonald IR, Garcia-Pineda O, Beet A, Asl SD, Feng L, Graettinger G, French-McCay D, Holmes J, Hu C, Huffer F (2015) Natural and unnatural oil slicks in the Gulf of Mexico. J Geophys Res Oceans 120(12):8364–8380

Magaña HA, Contreras C, Villareal TA (2003) A historical assessment of Karenia brevis in the western Gulf of Mexico. Harmful Algae 2(3):163–171

Mansfield KL, Wyneken J, Porter WP, Luo J (2014) First satellite tracks of neonate sea turtles redefine the 'lost years' oceanic niche. Proc R Soc B Biol Sci 281(1781):20133039–20133039

Matkin CO, Saulifis EL, Ellis GM, Olesiuk P, Rice SD (2008) Ongoing population-level impacts on killer whales Orcinus orca following the 'Exxon Valdez' oil spill in Prince William Sound, Alaska. Mar Ecol Prog Ser 356:269–281. https://doi.org/10.3354/meps07273

McClenachan L, Jackson JB, Newman MJ (2006) Conservation implications of historic sea turtle nesting beach loss. Front Ecol Environ 4(6):290–296

McDonald TL, Schroeder BA, Stacy BA, Wallace BP, Starcevich LA, Gorham J, Tumlin MC, Cacela D, Rissing M, McLamb DB (2017) Density and exposure of surface-pelagic juvenile sea turtles to Deepwater Horizon oil. Endanger Species Res 33:69–82

Michel J, Owens EH, Zengel S, Graham A, Nixon Z, Allard T, Holton W, Reimer PD, Lamarche A, White M et al (2013) Extent and degree of shoreline oiling: Deepwater Horizon oil spill, Gulf of Mexico, USA. PLoS One 8(6):e65087–e65087

Milton S, Lutz P, Shigenaka G (2003) Oil toxicity and impacts on sea turtles. In: Shigenaka G (ed) Oil and sea turtles: biology planning and response. NOAA, National Ocean Service, Office of Response and Restoration, Seattle, WA, pp 35–47

Milton SL, Lutz PL (2003) Physiological and genetic responses to environmental stress. In: The biology of sea turtles, vol 2

Mitchelmore C, Bishop C, Collier T (2017) Toxicological estimation of mortality of oceanic sea turtles oiled during the Deepwater Horizon oil spill. Endanger Species Res 33:39–50

Mullin KD (2007) Abundance of cetaceans in the oceanic Gulf of Mexico based on 2003-2004 ship surveys. Available from: NMFS, Southeast Fisheries Science Center, PO Drawer 1207

Mullin KD, Fulling GL (2004) Abundance of cetaceans in the oceanic northern Gulf of Mexico, 1996-2001. Mar Mamm Sci 20(4):787–807. https://doi.org/10.1111/j.1748-7692.2004.tb01193.x

Neff J, Lee K, DeBlois EM (2011a) Produced water: overview of composition, fates, and effects. In: Produced water. Springer, New York, pp 3–54

Neff J, Sauer TC, Hart AD (2011b) Bioaccumulation of hydrocarbons from produced water discharged to offshore waters of the US Gulf of Mexico. In: Produced water. Springer, New York, pp 441–477

Neff JM (1990) Composition and fate of petroleum and spill-treating agents in the marine environment. In: Sea mammals in oil: confronting the risks. Academic Press, Inc., San Diego, CA, pp 1–33

Nero RW, Cook M, Coleman AT, Solangi M, Hardy R (2013) Using an ocean model to predict likely drift tracks of sea turtle carcasses in the north central Gulf of Mexico. Endanger Species Res 21(3):191–203

NMFS U (2008) Recovery plan for the Northwest Atlantic population of the loggerhead sea turtle (Caretta caretta): Second revision. Office of Protected Resources, National Marine Fisheries Service, Silver Spring, MD, USA, January 325

NMFS U, SEMARNAT (Secretary of Environment and Natural Resources, Mexico) (2011) Bi-national recovery plan for the Kemp's ridley sea turtle (*Lepidochelys kempii*), 2nd Rev. National Marine Fisheries Service, Silver Spring, MD, USA

NOAA OoPR (2013) Hawksbill sea turtle (*Eretmochelys imbricata*) 5-Year Review: Summary and Evaluation. U.S. Department of Commerce U.S. Department of the Interior National Oceanic and Atmospheric Administration, Jacksonville, FL

Odell DK, MacMurray C (1986) Behavioural response to oil. In: Vargo S, Lutz PL, Odell DK, Van Vleet T, Bossart G (eds) Final report. Study of the effect of oil on marine turtles. Minerals Management Service Contract Number 14-12-0001-30063, Florida Institute of Oceanography, St. Petersburg, FL

Plotkin P, Amos AF (1990) Effects of anthropogenic debris on sea turtles in the northwestern Gulf of Mexico. In: Proceedings of the second international conference on marine debris. NOAA Technical Memorandum NMFS-SEFC-154, Honolulu, Hawaii, pp 736–743

Pulster E, Gracia A, Snyder SM, Deak K, Fogleson S, Murawski SA (2020) Chronic sublethal effects observed in wild caught fish following two major oil spills in the Gulf of Mexico: Deepwater Horizon and Ixtoc 1 (Chap. 24). In: Murawski SA, Ainsworth C, Gilbert S, Hollander D, Paris CB, Schlüter M, Wetzel D (eds) Deep oil spills – facts, fate and effects. Springer, Cham

Putman NF, Abreu-Grobois FA, Iturbe-Darkistade I, Putman EM, Richards PM, Verley P (2015) Deepwater Horizon oil spill impacts on sea turtles could span the Atlantic. Biol Lett 11(12):20150596

Rees AF, Avens L, Ballorain K, Bevan E, Broderick AC, Carthy RR, Christianen MJ, Duclos G, Heithaus MR, Johnston DW (2018) The potential of unmanned aerial systems for sea turtle research and conservation: a review and future directions. Endanger Species Res 35:81–100

Reeves RR, Lund JN, Smith TD, Josephson EA (2011) Insights from whaling logbooks on whales, dolphins, and whaling in the Gulf of Mexico. Gulf Mex Sci 29(1):4

Reich KJ, López-Castro MC, Shaver DJ, Iseton C, Hart KM, Hooper MJ, Schmitt CJ (2017) δ13C and δ15N in the endangered Kemp's ridley sea turtle Lepidochelys kempii after the Deepwater Horizon oil spill. Endanger Species Res 33:281–289

Reijnders PJH, Aguilar A, Borrell A (2009) Pollution and marine mammals. In: Perrin WF, Würsig B, Thewissen JGM (eds) Encyclopedia of marine mammals, 2nd edn. Academic Press, London, pp 890–898. https://doi.org/10.1016/B978-0-12-373553-9.00205-4

Resources NOoP (2018) 2010–2014 Cetacean Unusual Mortality Event in Northern Gulf of Mexico (Closed). https://www.fisheries.noaa.gov/national/marine-life-distress/2010-2014-cetacean-unusual-mortality-event-northern-gulf-mexico. Accessed Retrieved 19 Sept 2019

Rice AN, Palmer K, Tielens JT, Muirhead CA, Clark CW (2014) Potential Bryde's whale (Balaenoptera edeni) calls recorded in the northern Gulf of Mexico. J Acoust Soc Am 135(5):3066–3076

Romero IC, Toro-Farmer G, Diercks A-R, Schwing P, Muller-Karger F, Murawski S, Hollander DJ (2017) Large-scale deposition of weathered oil in the Gulf of Mexico following a deep-water oil spill. Environ Pollut 228:179–189

Rosel PE, Wilcox LA (2014) Genetic evidence reveals a unique lineage of Bryde's whales in the northern Gulf of Mexico. Endanger Species Res 25(1):19–34

Rossbach KA, Herzing DL (1997) Underwater observations of benthic-feeding bottlenose dolphins (Tursiops truncatus) near Grand Bahama Island, Bahamas. Mar Mamm Sci 13(3):498–504

Rugh DJ, Shelden KE, Hobbs RC (2010) Range contraction in a beluga whale population. Endanger Species Res 12(1):69–75

Schroeder B, Murphy S (1999) Population surveys (ground and aerial) on nesting beaches. In: Eckert K, Bjorndal K, Abreu-Grobois F, Donnelly M (eds) Research and management techniques for the conservation of sea turtles, vol 4. IUCN/SSC Marine Turtle Specialist Group Publication No. 4, p 45

Schwacke LH, Smith CR, Townsend FI, Wells RS, Hart LB, Balmer BC, Collier TK, De Guise S, Fry MM, Guillette LJ Jr (2013) Health of common bottlenose dolphins (Tursiops truncatus) in Barataria Bay, Louisiana, following the Deepwater Horizon oil spill. Environ Sci Technol 48(1):93–103

Schwacke LH, Thomas L, Wells RS, McFee WE, Hohn AA, Mullin KD, Zolman ES, Quigley BM, Rowles TK, Schwacke JH (2017) Quantifying injury to common bottlenose dolphins from the Deepwater Horizon oil spill using an age-, sex-and class-structured population model. Endanger Species Res 33:265–279

Seminoff JA, Eguchi T, Carretta J, Allen CD, Prosperi D, Rangel R, Gilpatrick JW Jr, Forney K, Peckham SH (2014) Loggerhead sea turtle abundance at a foraging hotspot in the eastern Pacific Ocean: implications for at-sea conservation. Endanger Species Res 24(3):207–220

Seney EE, Landry AM Jr (2008) Movements of Kemp's ridley sea turtles nesting on the upper Texas coast: implications for management. Endanger Species Res 4(1–2):73–84

SERO Other Significant Oil Spills in the Gulf of Mexico (n.d.) US Department of Commerce, National Oceanic and Atmospheric Administration, Office of Response and Restoration, Emergency Response Division

Shaver DJ, Hart KM, Fujisaki I, Rubio C, Sartain AR, Pena J, Burchfield PM, Gamez DG, Ortiz J (2013) Foraging area fidelity for Kemp's ridleys in the Gulf of Mexico. Ecol Evol 3(7):2002–2012

Shigenaka GE (2003) Oil and sea turtles: biology planning and response. NOAA, National Ocean Service, Office of Response and Restoration, Seattle, WA

Soldevilla MS, Hildebrand JA, Frasier KE, Dias LA, Martinez A, Mullin KD, Rosel PE, Garrison LP (2017) Spatial distribution and dive behavior of Gulf of Mexico Bryde's whales: potential risk of vessel strikes and fisheries interactions. Endanger. Species Res 32:533–550

Spooner M (1967) Biological effects of the Torrey Canyon disaster. J Devon Trust Nat Conserv:12–19

Stacy B (2012) Summary of findings for sea turtles documented by directed captures, stranding response, and incidental captures under response operations during the BP DWH MC252 oil spill. DWH sea turtles NRDA technical working group report. Prepared for NOAA Assessment and Restoration Division

Stacy N, Field C, Staggs L, MacLean R, Stacy B, Keene J, Cacela D, Pelton C, Cray C, Kelley M (2017) Clinicopathological findings in sea turtles assessed during the Deepwater Horizon oil spill response. Endanger Species Res 33:25–37

Thomas L, Buckland S, Rexstad E, Laake J, Strindberg S, Hedley S, Bishop J, Marques T, Burnham K (2010) Distance software: design and analysis of distance sampling surveys for estimating population size. J Appl Ecol 47(1):5–14. https://doi.org/10.1111/j.1365-2664.2009.01737.x

Troëng S, Rankin E (2005) Long-term conservation efforts contribute to positive green turtle Chelonia mydas nesting trend at Tortuguero, Costa Rica. Biol Conserv 121(1):111–116

Trustees DHNRDA (2016) Injury to Natural Resources. In: Final Programmatic Damage Assessment and Restoration (PDARP) Plan and Final Programmatic Environmental Impact Statement (PEIS). Retrieved from http://www.gulfspillrestoration.noaa.gov/restoration-planning/gulf-plan. p 516

Valverde RA, Holzwart KR (2017) Sea turtles of the Gulf of Mexico. In: Habitats and biota of the Gulf of Mexico: before the Deepwater Horizon oil spill. Springer, New York, pp 1189–1351

Van Vleet E, Sackett W, Reinhardt S, Mangini M (1984) Distribution, sources and fates of floating oil residues in the Eastern Gulf of Mexico. Mar Pollut Bull 15(3):106–110

Van Vleet ES, Pauly GG (1987) Characterization of oil residues scraped from stranded sea turtles from the Gulf of Mexico. Carib J Sci 23:77–83

Vander Zanden HB, Bolten AB, Tucker AD, Hart KM, Lamont MM, Fujisaki I, Reich KJ, Addison DS, Mansfield KL, Phillips KF (2016) Biomarkers reveal sea turtles remained in oiled areas following the Deepwater Horizon oil spill. Ecol Appl 26(7):2145–2155

Venn-Watson S, Garrison L, Litz J, Fougeres E, Mase B, Rappucci G, Stratton E, Carmichael R, Odell D, Shannon D (2015) Demographic clusters identified within the Northern Gulf of Mexico common bottlenose dolphin (Tursiops truncatus) Unusual Mortality Event: January 2010-June 2013. PLoS One 10(2):e0117248

Walker WA, Coe JM (1989) Survey of marine debris ingestion by odontocete cetaceans. In: Proceedings of the second international conference on marine debris, pp 2–7

Wallace BP, DiMatteo AD, Hurley BJ, Finkbeiner EM, Bolten AB, Chaloupka MY, Hutchinson BJ, Abreu-Grobois FA, Amorocho D, Bjorndal KA (2010) Regional management units for marine turtles: a novel framework for prioritizing conservation and research across multiple scales. PLoS One 5(12):e15465

Wallace BP, Kot CY, DiMatteo AD, Lee T, Crowder LB, Lewison RL (2013) Impacts of fisheries bycatch on marine turtle populations worldwide: toward conservation and research priorities. Ecosphere 4(3):art40

Wallace BP, Stacy BA, Rissing M, Cacela D, Garrison LP, Graettinger GD, Holmes JV, McDonald T, McLamb D, Schroeder B (2017) Estimating sea turtle exposures to Deepwater horizon oil. Endanger Species Res 33:51–67

Waring GT, Josephson E, Fairfield-Walsh CP, Maze-Foley K (2009) US Atlantic and Gulf of Mexico marine mammal stock assessments - 2008. NOAA Tech Memo NMFS NE 210(440):11.10

Waring GT, Josephson E, Maze-Foley K, Rosel PE (2013) U.S. Atlantic and Gulf of Mexico marine mammal stock assessments - 2012. NOAA Tech Memo NMFS NE 223:419

Waring GT, Josephson E, Maze-Foley K, Rosel PE (2015) US Atlantic and Gulf of Mexico marine mammal stock assessments - 2014. NOAA Tech Memo NMFS NE 231:361

Weishampel JF, Bagley DA, Ehrhart LM, Weishampel AC (2010) Nesting phenologies of two sympatric sea turtle species related to sea surface temperatures. Endanger Species Res 12(1):41–47

Wells RS (2014) Social structure and life history of bottlenose dolphins near Sarasota Bay, Florida: insights from four decades and five generations. In: Primates and cetaceans. Springer, Tokyo, pp 149–172

Wiggins SM, Hall JM, Thayre BJ, Hildebrand JA (2016) Gulf of Mexico low-frequency ocean soundscape impacted by airguns. J Acoust Soc Am 140(1):176–183

Wilkin SM, Rowles TK, Stratton E, Adimey N, Field CL, Wissmann S, Shigenaka G, Fougères E, Mase B, Network SRS (2017) Marine mammal response operations during the Deepwater Horizon oil spill. Endanger Species Res 33:107–118

Williams R, Gero S, Bejder L, Calambokidis J, Kraus SD, Lusseau D, Read AJ, Robbins J (2011) Underestimating the damage: interpreting cetacean carcass recoveries in the context of the Deepwater Horizon/BP incident. Conserv Lett 4(3):228–233

Wise CF, Wise JT, Wise SS, Thompson WD, Wise JP Jr, Wise JP Sr (2014) Chemical dispersants used in the Gulf of Mexico oil crisis are cytotoxic and genotoxic to sperm whale skin cells. Aquat Toxicol 152:335–340

Witherington B (2002) Ecology of neonate loggerhead turtles inhabiting lines of downwelling near a Gulf Stream front. Mar Biol 140(4):843–853

Witherington B, Hirama S, Hardy R (2012) Young sea turtles of the pelagic Sargassum-dominated drift community: habitat use, population density, and threats. Mar Ecol Prog Ser 463:1–22

Worm B, Tittensor DP (2011) Range contraction in large pelagic predators. Proc Natl Acad Sci 108(29):11942–11947

Wright AJ, Soto NA, Baldwin AL, Bateson M, Beale CM, Clark C, Deak T, Edwards EF, Fernández A, Godinho A (2007) Anthropogenic noise as a stressor in animals: a multidisciplinary perspective. Int J Comp Psychol 20(2)

Würsig B, Jefferson TA, Schmidly DJ (2000) The marine mammals of the Gulf of Mexico. Texas A&M University Press, College Station

Ylitalo GM, Collier TK, Anulacion BF, Juaire K, Boyer RH, da Silva DA, Keene JL, Stacy BA (2017) Determining oil and dispersant exposure in sea turtles from the northern Gulf of Mexico resulting from the Deepwater Horizon oil spill. Endanger Species Res 33:9–24

Ylitalo GM, Matkin CO, Buzitis J, Krahn MM, Jones LL, Rowles T, Stein JE (2001) Influence of life-history parameters on organochlorine concentrations in free-ranging killer whales (Orcinus orca) from Prince William Sound, AK. Sci Total Environ 281(1–3):183–203

Zheng M, Ahuja M, Bhattacharya D, Clement TP, Hayworth JS, Dhanasekaran M (2014) Evaluation of differential cytotoxic effects of the oil spill dispersant Corexit 9500. Life Sci 95(2):108–117

Part VI
Toxicology of Deep Oil Spills

Teri Navajo
Pain and Sorrow, Buried Deep
Acrylic, ink and gouache on Canvas
24" × 48"

Chapter 27
Ecotoxicology of Deep Ocean Spills

Mace G. Barron, Susan C. Chiasson, and Adriana C. Bejarano

Abstract Central issues in the ecotoxicology of oil spills in the subsea are the unknown sensitivity of deep ocean species and the complexity and diversity of habitats that may be affected in subsea, offshore, and coastal environments. This chapter reviews and discusses the unique aspects of assessing spill risks and impacts to deep ocean species and approaches for toxicity extrapolation across a diversity of ecological communities and environments. The focus of this chapter is on aquatic species, based primarily on the Gulf of Mexico (GoM) with information and experiences drawn from the *Deepwater Horizon* (DWH) oil spill.

Keywords Genomic · Species extrapolation · Toxicity · Sensitivity distribution · Chemosynthetic · Coral reef

27.1 Introduction

Central issues in the ecotoxicology of oil spills in the subsea are the unknown sensitivity of deep ocean species and the complexity and diversity of habitats that may be affected in subsea, offshore, and coastal environments. This chapter reviews and discusses the unique aspects of assessing spill risks and impacts to deep ocean species and approaches for toxicity extrapolation across a diversity of ecological

M. G. Barron (✉)
Environmental Protection Agency, Gulf Ecology Division, Gulf Breeze, FL, USA
e-mail: Barron.Mace@epa.gov

S. C. Chiasson
Loyola University, Biological Sciences, New Orleans, LA, USA
e-mail: scchiass@loyno.edu

A. C. Bejarano
Research Planning Inc., Columbia, SC, USA

Shell Oil Company, Shell Health - Americas, Houston, TX, USA
e-mail: abejarano@researchplanning.com; Adriana.Bejarano@shell.com

S. A. Murawski et al. (eds.), *Deep Oil Spills*,
https://doi.org/10.1007/978-3-030-11605-7_27

communities and environments. The focus of this chapter is on aquatic species, based primarily on the Gulf of Mexico (GoM) with information and experiences drawn from the *Deepwater Horizon* (DWH) oil spill.

27.2 Complexity of Potentially Affected Habitats

As observed during the DWH oil spill, a diversity of habitats may be affected from a deep ocean release as oil moves from the subsea to offshore and coastal environments (Carriger and Barron 2011; Barron 2012; Beyer et al. 2016). This section provides a brief overview of the complexity of potentially affected coastal and offshore habitats, and aspects of species sensitivity to hydrocarbons from subsea spills, with a focus on communities in the deep ocean (Fig. 27.1).

27.2.1 Deep Ocean

Some unique aspects of the deep ocean relevant to the sensitivity of species and aquatic communities to hydrocarbon exposure include the physical environment, the dynamics of a subsea oil spill, and exceptional habitats and ecological communities of the deep ocean (e.g., MMS 2000; Fisher et al. 2014). The deep ocean is found within the bathypelagic, abyssopelagic, and hadopelagic zones of the subsea at depths below 1000 meters, where there is limited to no light (Fig. 27.1). In the

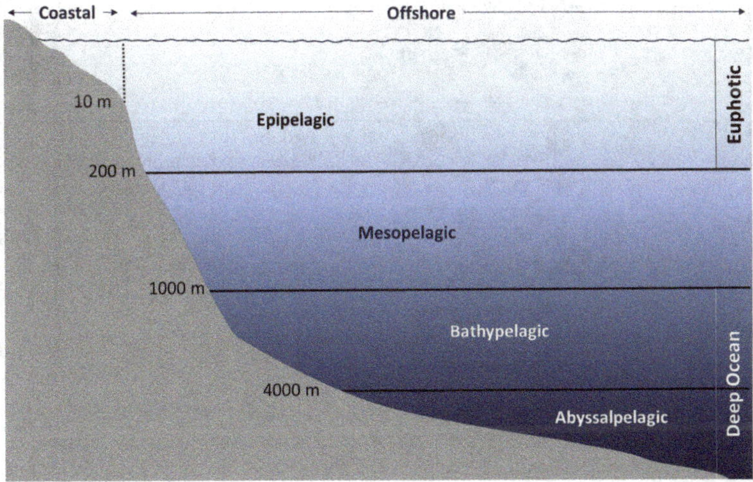

Fig. 27.1 Habitats and oceanographic zones potentially affected by deep ocean spills. (Adapted from French-McCay et al. (2018a))

GoM, oxygen levels in the deep ocean typically range between 3 and 5 mg/L, temperatures are a nearly uniform 4 °C, and pressures exceed 100 atmospheres. Salinity in the deep ocean is 36 parts per thousand except in localized areas around brine seeps where salinity can exceed 40 parts per thousand.

In contrast to oil spilled near the sea surface, oil released into the deep ocean results in an increased residence time of volatile hydrocarbons (gas) in the aqueous phase in the absence of atmospheric evaporation (Reddy et al. 2011). The residence time of hydrocarbons in the deep ocean is also affected by the decreased availability of dissolved oxygen for biodegradation (Camilli et al. 2010). During the DWH oil spill, a gas to oil ratio of 1600 cubic feet per petroleum barrel was measured, with methane as the predominant volatile hydrocarbon (less than five carbon atoms) comprising over 80 mole % of the gas hydrocarbon composition (Valentine et al. 2010; Reddy et al. 2011). Given the temperature and pressure conditions of the deep ocean, it is likely that hydrocarbons with a molecular weight greater than methane would predominate in the liquid oil phase (Fig. 27.2). Soluble hydrocarbons (gases, soluble aromatics) will primarily partition into the water column at the depth of oil release. In contrast, semi-soluble hydrocarbons are expected to dissolve to some extent at depth as the droplets rise through the water column, and relatively insoluble

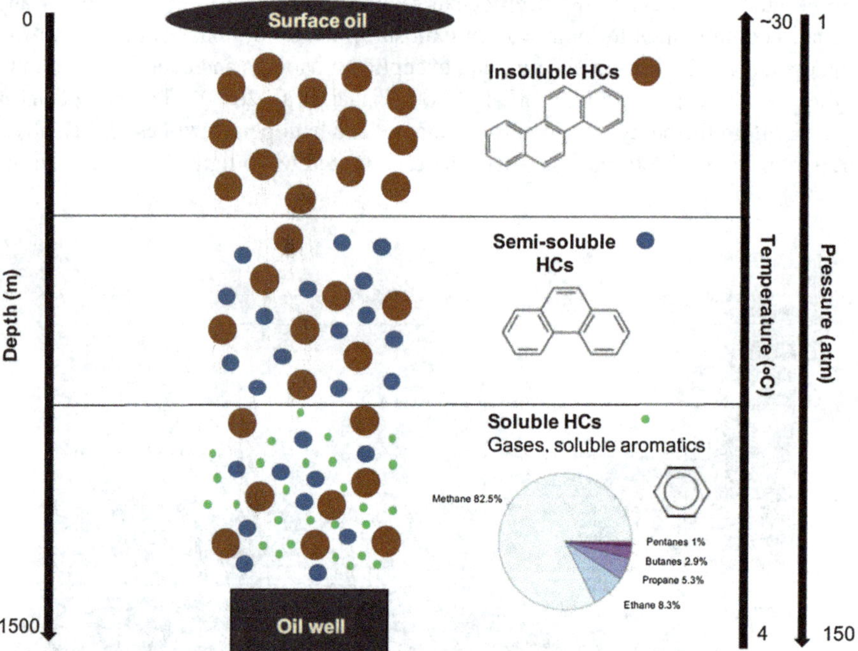

Fig. 27.2 Changes in the relative composition of gases and hydrocarbons (HC), temperature, and pressure with depth in the subsea. Gases and soluble HC dissolve primarily at the depth of the oil release, semi-soluble HCs dissolve as the droplets rise through the water column, and insoluble HCs rise to the ocean surface. (Oil behavior from French-McCay et al. 2018a; hydrocarbon proportions from Reddy et al. 2011)

hydrocarbons also rise to the surface or are deposited on the ocean floor (Reddy et al. 2011; French-McCay et al. 2018) (Fig. 27.2).

Deep ocean oil spills can impact a diversity of habitats in the subsea, including deepwater coral, bottom sediment, and chemosynthetic vent communities (Fisher et al. 2014). These habitats are home to a diversity of organisms including zooplankton, tube worms, shellfish, jellyfish, squid, and cartilaginous and teleost fish (Carney 1994; MMS 2000; Fisher et al. 2014; Beyer et al. 2016). Four types of chemosynthetic communities have been described in the Gulf of Mexico based on predominant taxa: (1) vestimentiferan tube worms, (2) mytilid mussels, (3) vesicomyid clams, and (4) infaunal clams (MMS 2000). Chemosynthetic vent and deepwater coral habitats are relatively rare in the deep ocean and can be several hundred years old (MMS 2000; Gerard et al. 2018; Gerard and Fisher 2018). Other benthic seafloor communities are composed of microbes (e.g., bacteria, protists), foraminifera and metazoa (e.g., harpacticoid copepods, nematodes), macrofauna (e.g., nematodes, polychaetes), and megafauna (e.g., sea stars, sea fans, anemones, isopods, demersal fishes; Rowe 2017). Invertebrate communities in bottom sediments have generally low abundance but high species diversity (Gallaway et al. 2001). Deep ocean communities have unknown sensitivities to hydrocarbons. Additionally, deepwater coral and chemosynthetic vent communities have unique habitat requirements, and many species are projected to have slow recoveries from a subsea oil spill because of low recruitment and growth rates in many deep ocean species (Gallaway et al. 2001; Gerard et al. 2018; Gerard and Fisher 2018).

27.2.2 Offshore

As observed during the DWH oil spill, oil released in the deep ocean will either be entrained as a subsurface plume of 10–50 μm droplets or move toward the surface as droplets greater than 50 μm (North et al. 2011; French-McCay et al. 2018b; Robinson et al. 2018). The offshore euphotic or epipelagic zone is characterized by depths to 200 meters with sufficient incident sunlight for photosynthesis (Fig. 27.1). Oil in the euphotic zone and at the sea surface will expose pelagic organisms including plankton, *Sargassum* communities, jellyfish, fish, sea turtles, marine mammals, and sea birds.

Surface oiling will substantially impact neuston and near-surface assemblages which contain early life stages of invertebrates and fish. As seen following the DWH oil spill, oil exposure of embryonic life stages of pelagic fish and other species to polycyclic aromatic hydrocarbon concentrations in the low μg/L range can lead to malformations and long-term consequences on swimming performance (Incardona et al. 2014; Beyer et al. 2016). Additionally, the limited attenuation of light can lead to both near-surface generation of photooxidation products of oil and photoenhanced toxicity of bioaccumulated oil residues at depths to 20 meters (Barron 2017).

Sargassum mats are a unique pelagic habitat, with estimated masses in the Gulf of Mexico greater than one million tons (Gower and King 2011). *Sargassum* mats host a complex community of marine organisms that may be vulnerable to oiling, mat sinking, and oxygen loss from oil and surface dispersant applications (Powers et al. 2013). However, the sensitivity of *Sargassum* communities to oil exposure from a deep ocean spill remains largely unknown.

Multiple species of cetaceans, sea turtles, and sea birds were severely impacted by the DWH oil spill from coating from surface oiling, ingestion and inhalation pathways, and impacts to dolphins continue to be documented (Barron 2012; Beyer et al. 2016; Smith et al. 2017). Prior to the DWH, marine mammals and sea turtles had been identified as particularly at risk from a large subsea spill, and recent simulation modeling of deep ocean oil releases also show the highest relative risks from surface oiling to these species (MMS 2000; Bock et al. 2018).

27.2.3   *Coastal Environments*

Coastal environments include nearshore surf zone habitats, estuaries, wetlands, and beaches, with depths greater than 10 m delineating offshore areas (Fig. 27.1) (CERC 1984). As seen in DWH, deep ocean spills can result in weathered oil impacting a diversity of coastal environments including oyster reef, seagrass, beach, tidal mud flat, mangrove, marsh, and wetland communities (Barron 2012; Beyer et al. 2016; Baker et al. 2017). For example, over 1700 km of shoreline were oiled during DWH, including beaches (51%) and saltwater marshes (45%), and oil remained on nearly 700 km of the shoreline 2 years after the spill (Michel et al. 2013). Sand beach habitat, sea grass, marsh vegetation, and subtidal oysters were severely injured from physical oiling, chemical toxicity, and response actions and cleanup activities during DWH (Baker et al. 2017).

Coastal environments provide nursery, refuge, and foraging habitat for a diversity of resident and migratory species of aquatic animals, including infaunal benthic invertebrates, snails, oysters and other shellfish, shrimp, crabs, and multiple species of fish (Beyer et al. 2016; Baker et al. 2017). Many coastal species such as oysters, shrimp, and fish are commercially important and may also be recreationally important and/or keystone species (e.g., Gulf menhaden, *Brevoortia patronus*) (Beyer et al. 2016; Baker et al. 2017; Bock et al. 2018).

Shallow water coral reef communities in proximity to dispersed oil may also be impacted from a deep ocean spill. The limited available research indicates that species and life stages of reef building corals are sensitive to oil and chemical dispersants. During DWH, species of sea fans appeared to be impacted in areas of persistent oil slicks (Beyer et al. 2016).

27.3 Determining Toxicity to Aquatic Species from Deep Ocean Spills

27.3.1 What Is Known about Deep Ocean Toxicity

Oil and gas exploration of the deep-sea environment have raised concerns about the potential impact of deep ocean well blowouts on resources inhabiting these remote areas of the ocean. Under a blowout spill scenario, oil exposures to deep-sea communities may occur through three pathways: (1) direct oil coating; (2) direct exposure to flocculated oil and oil residues deposited via marine snow; and (3) direct exposure to oil droplets and dissolved hydrocarbon fractions. However, understanding the relative contribution of each exposure mechanism requires careful experimental design that may not be easily achievable. The field of aquatic toxicology has primarily focused on understanding effects from oil spills under controlled laboratory exposures by means of water-accommodated fractions (WAF) with or without dispersants (chemically enhanced water-accommodated fractions; CEWAF), directly addressing the third exposure pathway.

The sensitivity of coastal aquatic species is relatively well understood compared to the species and aquatic communities of the deep ocean and offshore environments. In particular, the sensitivity of larval and juvenile life stages of estuarine organisms has been tested extensively in laboratory toxicity tests using a variety of oil products (see Mitchelmore et al. 2020). Over 40 coastal species were tested during DWH and shown to be sensitive to weathered oil and field samples of spill water (Echols et al. 2015; NOAA 2015). In recent years, several studies have emerged on the relative sensitivity of select deep ocean species to single hydrocarbons, oil products, and dispersants. Knap et al. (2017) conducted toxicity testing of a single hydrocarbon (1-methylnaphthalene) with the mysid shrimp (*Americamysis bahia*) and five deep ocean crustacean species. Results based on lethal concentrations and critical body burdens (CBB) indicated higher sensitivity of deep ocean species to this petroleum component compared to other species. However, only two of the five deep ocean crustaceans were more sensitive than the mysid shrimp, further supporting the possible use of this standard test species as a surrogate for deepwater species. Another study (McConville et al. 2018) evaluated the acute toxicity of several petroleum compounds (i.e., toluene, 2-methylnaphthalene, phenanthrene) and mixtures (i.e., Corexit 9500, CEWAF of Alaskan North Slope crude oil) to sablefish (*Anoplopoma fimbria*). Results showed that while this species has a sensitivity to Corexit 9500 and CEWAF within the range reported for other fish species, it appears to be more sensitive (based on CBB) than other aquatic species to single hydrocarbon compounds. Frometa et al. (2017) also found that the sensitivity of a deep ocean gorgonian octocoral (*Swiftia exserta*) to Corexit 9500 was within the range of previously tested aquatic species. Although limited, exposure data for aromatic hydrocarbons, petroleum products, dispersants (Knap et al. 2017; McConville et al. 2018), and even metals (Brown et al. 2017) suggest that the sensitivity of tested deep ocean species falls within the range of traditionally tested species. Thus, euphotic zone species including commonly used standard test species may serve as suitable surrogates for deep ocean species. However, there are number of challenges that need to be addressed before test species from shallow-water environments can be confidently used as surrogates for deep ocean species. Consequently, studies need to be expanded to confirm these generalizations, including determining the influence of low temperature and high pressure (Beyer et al. 2016; McConville et al. 2018).

27.3.2 Genomic Insights into Species Sensitivity

Impacts on gene expression following petroleum exposure in deep ocean communities are not well understood, making evaluation of some sublethal effects in these communities additionally complex. The protected status of sea turtles and marine mammals also limit genomic toxicity research endpoints on effects of petroleum exposure in these taxa. Fish and invertebrates that reside in the euphotic zone elicit different responses at the gene level to petroleum. Petroleum exposure can impact, for example, reproductive success and cardiac function and affect stress response sensitive genes (Whitehead et al. 2011; Incardona et al. 2014; Chiasson and Taylor 2017). Most taxa differentially express general stress markers in response to polycyclic aromatic hydrocarbon exposure (e.g., Cytochrome P450). Expression of more specific markers (e.g., indicators of endocrine disruption) is evident in fish but absent in, for example, crustaceans (see Portnoy et al. 2020). Genomic markers for vertebrate aquatic species are relatively well understood, whereas markers for invertebrate species are often extrapolated from vertebrate studies and may not reflect comparable responses between these taxa (Short et al. 2014). Early life stages are generally more sensitive to sublethal effects than adults, but comparative genomic sensitivity studies of oil exposure between larval and adult stages are scarce (Whitehead et al. 2011). Genomic fingerprints due to petroleum exposure can persist for decades in an aquatic population residing in a spill area without a significant population decline (Jewett et al. 2003).

27.4 Toxicity Extrapolation Approaches

27.4.1 Available Tools

In recent years, a number of tools and applications have been developed to help characterize what is known about the toxicity of crude oils. One data visualization tool that has gained acceptance in hazard assessments of chemicals, including crude oil and spill mitigating agents such as dispersants, is the species sensitivity distribution (SSD). SSDs describe the probability distribution of the sensitivity of an assemblage of taxa to a particular chemical by ranking toxicity data (i.e., geometric mean) for individual species from the most to least sensitive. From SSDs, hazard concentrations (HC) affecting a specified percentile of the assemblage are derived, serving as toxicity benchmark for hazard evaluations. Most commonly a fifth percentile HC (HC5), or concentration protective of 95% of the assemblage, is computed and employed in hazard and risk evaluations. However, the generation of SSDs requires both sufficient number of species and diversity of taxa (e.g., five to greater than ten aquatic species), which can be hindered by data availability and access to toxicity and chemistry information. Recent efforts by the response community to facilitate access to hazard and fate information allows for SSD development for petroleum and related chemicals. One of these tools is the Chemical Aquatic Fate and Effects (CAFE) database (NOAA/ER 2015; Bejarano et al. 2016). CAFE currently hosts aquatic toxicity data for many individual chemical compounds and complex mixtures of chemicals including select crude oils and dispersants.

Despite the availability of information repositories, insufficient toxicity data still limit the development of SSDs. However, mathematical relationships known as Interspecies Correlation Estimation (ICE) models have been developed to allow extrapolation of toxicity between a surrogate species to one or several predicted species. ICE models have been developed from toxicity data for chemicals with and without specific modes of toxic action (Willming et al. 2016; Raimondo et al. 2017), nonpolar narcotics (Bejarano and Barron 2016), and petroleum and dispersant products (Bejarano and Barron 2014). The use of ICE models can augment SSDs comprised of measured data, providing for more robust HC5 estimates (Awkerman et al. 2014).

27.4.2 Deep Ocean Species Sensitivity Distributions

To demonstrate the applicability of SSD and ICE toxicity extrapolation approaches, datasets for chemicals with at least one deep ocean species were used (Frometa et al. 2017; Knap et al. 2017; McConville et al. 2018). These data were augmented with either ICE predicted toxicity values (Bejarano and Barron 2014) based on hydrocarbon CBB or empirical data queried from CAFE. Data for Corexit 9500 were also extracted from a recent compilation of dispersant-only toxicity studies (Bejarano 2018). These datasets were used to generate SSDs along with HC5 estimates (Fig. 27.3). SSDs based on CBB for 1-methylnaphthalene and measured total petroleum hydrocarbon (TPH) concentrations from CEWAF prepared with fresh crude oil (Alaskan North Slope or Prudhoe Bay plus Corexit 9500) show that deep ocean species generally occur in the lower end of the sensitivity distribution. In the case of a larger dataset (Corexit 9500), sensitivities for two deep ocean species were in the upper portion of the distribution. In all cases, HC5s are protective of the most sensitive species in the SSD. Although preliminary, these results suggest that SSDs are useful in visualizing data for deep ocean species of concern. The available data for deep ocean species were near or above HC5 values indicating the potential protectiveness of HC5 estimates.

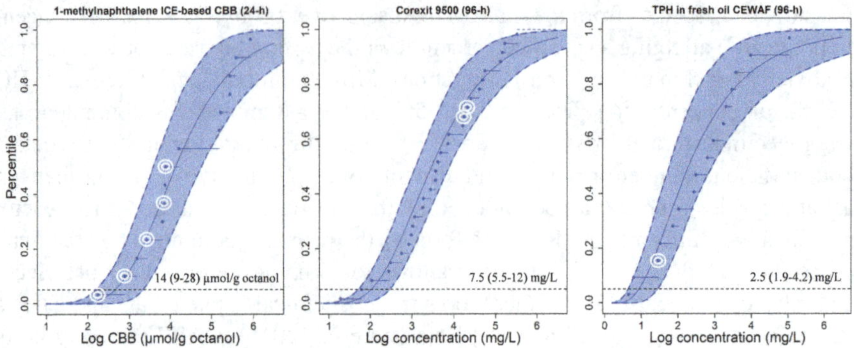

Fig. 27.3 SSDs and estimated HC5 values for 1-methylnaphthalene, Corexit 9500, and total petroleum hydrocarbons (TPH). SSDs were generated for 1-methylnaphthalene (left panel) using predicted data from ICE models; Corexit 9500 (center panel) using empirical data; and TPH (right panel) based on data for CEWAF prepared with fresh Alaska North Slope or Prudhoe Bay crude oil and Corexit 9500. Large circles are deep ocean species. (Frometa et al. 2017; Knap et al. 2017; McConville et al. 2018)

27.4.3 Limitations, Uncertainties, and Assumptions

There are many challenges in utilizing toxicity data and extrapolation tools to predict the sensitivity of deep ocean species and the aquatic communities inhabiting environments impacted from a subsea spill (Fig. 27.4). Limitations and uncertainties include (1) the unknown toxicity of untested species or species for which toxicity data are limited, (2) the uncertainties associated with the tools themselves (i.e., data quality, model domain of applicability, tool reliability), and (3) unique biotic (e.g., chemosynthesis, biolumi-nescence) and abiotic factors (e.g., temperature, pressure, oxygen, light) that may influence exposure and effects on deep ocean species. These challenges and limitations are not necessarily exclusive to the understanding of toxicological effects of oil spills on deep ocean species and are a common concern for spills in unique environments (e.g., polar regions, tropical habitats; Bejarano et al. 2017). To improve toxicity extrapolation to deep ocean species, multiple implicit assumptions remain to be evaluated. These include determining whether (1) standard test species can serve as surrogates for deep ocean species, (2) testing deep ocean species under standard laboratory conditions is representative of sensitivity in the deep ocean, and (3) use of extrapolation tools can adequately predict hazards for assemblages of deep ocean species (Fig. 27.4).

27.5 Lessons Learned, Recommendations, and Research Needs

The DWH spill highlighted the large data gaps in understanding the sensitivity of deep ocean species and the complexity and diversity of habitats that may be affected in subsea, offshore, and coastal environments (Carriger and Barron 2011; Barron

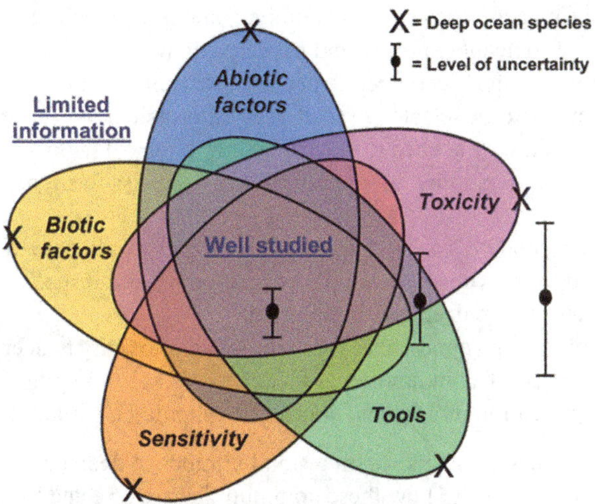

Fig. 27.4 Venn diagram of the challenges of predicting the effects of a subsea oil spill on deep ocean communities. Challenges (X symbol) and relative uncertainties (error bars) include limited information about how the unique biotic and abiotic factors of the deep ocean influence toxicity and species sensitivity; whether toxicity and species sensitivity can be extrapolated from experiments conducted under standard laboratory conditions; and uncertainty in available tools to generate reliable predictions based on well-studied species

2012; Beyer et al. 2016; Bock et al. 2018). Much of the information to support impact assessments was based on toxicity tests under controlled laboratory conditions, and relatively few studies focused on deep ocean species and communities. Toxicity testing protocols require careful consideration of habitat requirements and adaptations to the deep ocean environment (e.g., low temperature, high pressure) as these factors may influence the sensitivity and response of biological resources to petroleum. Additionally, temperature is known to influence the dissolution rate of hydrocarbon fractions, thus possibly resulting in delayed effects to exposed aquatic receptors.

To address knowledge gaps, comparative studies under relevant environmental conditions are recommended that include standard test species and closely related shallow- and deepwater species exposed to single hydrocarbon compounds, crude oil, and/or dispersants. Another aspect that is not well understood is the rate at which deep ocean communities recover from exposure to oil. Under the assumption that the physical constraints imposed by the deep ocean environment influence the growth rate and colonization of many subsea species, it is also likely that recovery would be slow. Some specific recommendations and research needed to advance the science of ecotoxicology of deep ocean spills include:

- Genetic markers for adverse effects in invertebrates are not well established, and more studies are needed to identify markers that are unique to invertebrates and not extrapolated from vertebrate studies.

- Omics data for deep ocean species is limited, and more studies are needed so that biomarkers of toxicant exposure and toxicity can be established.
- The toxicity data for deep ocean species is extremely limited, and additional studies are needed for selected taxa of deepwater fish and invertebrates.
- Additional studies are needed to determine how physical factors influence hydrocarbon toxicity, including low temperature and dissolved oxygen and high pressure.
- Understanding body burdens of deep ocean species would help determine if current knowledge and known sensitivity of commonly tested species also applies to aquatic species in the deep ocean habitats.
- Validation of existing approaches for extrapolating toxicity to deep ocean species and the diversity of communities affected by subsea oil spills is a continuing need, and application of genomic approaches should be explored.

Overall, comprehensive studies on a small number of deep ocean species would help reduce uncertainties of how these communities are affected by toxicity, biotic, and abiotic factors, their sensitivity in comparison to well-studied species, and improve the reliability of predictive tools. The role of ecotoxicological-based science in informing future response decisions and impact assessments continues to grow in importance, and incorporation of new approaches applicable to deep ocean species and habitats will be the next frontier.

References

Awkerman JS, Raimondo S, Jackson CR, Barron MG (2014) Augmenting aquatic species sensitivity distributions with interspecies toxicity estimation models. Environ Toxicol Chem 33:688–695

Baker MC, Steinhoff MA, Fricano GF (2017) Integrated effects of the Deepwater Horizon oil spill on nearshore ecosystems. Mar Ecol Prog Ser 576:219–234

Barron MG (2017) Photoenhanced toxicity of petroleum to aquatic invertebrates and fish. Arch Environ Contam Toxicol 73:40–46

Barron MG (2012) Ecological impacts of the Deepwater Horizon oil spill: implications for immunotoxicity. Toxicol Pathol 40:315–320

Bejarano AC, Barron MG (2014) Development and practical application of petroleum and dispersant interspecies correlation models for aquatic species. Environ Sci Technol 48(8):4564–4572

Bejarano AC, Barron MG (2016) Aqueous and tissue residue-based interspecies correlation estimation models provide conservative hazard estimates for aromatic compounds. Environ Toxicol Chem 35(1):56–64

Bejarano AC, Farr JK, Jenne P, Chu V, Hielscher A (2016) The chemical aquatic fate and effects database (CAFE), a tool that supports assessments of chemical spills in aquatic environments. Environ Toxicol Chem 35(6):1576–1586

Bejarano AC, Gardiner WW, Barron MG, Word JQ (2017) Relative sensitivity of Arctic species to physically and chemically dispersed oil determined from three hydrocarbon measures of aquatic toxicity. Mar Pollut Bull 122:316–322

Bejarano AC (2018) Critical review and analysis of aquatic toxicity data on oil spill dispersants. Environ Toxicol Chem 9999:1–13

Beyer J, Trannum HC, Bakke T, Hodson PV, Collier TK (2016) Environmental effects of the Deepwater Horizon oil spill: a review. Mar Pollut Bull 110:28–51

Bock M, Robinson H, Wenning R, French-McCay D, Rowe J, Hayward Walker A (2018) Comparative risk assessment of oil spill response options for a Deepwater oil well blowout: part II. Relative risk methodology. Mar Pollut Bull 133:984–1000

Brown A, Thatje S, Hauton C (2017) The effects of temperature and hydrostatic pressure on metal toxicity: insights into toxicity in the deep sea. Environ Sci Technol 51:10222–10231

Camilli R, Reddy CM, Yoerger DR, Van Mooy BAS, Jakuba MV, Kinsey JC, McIntyre CP, Sylva SP, Maloney JV (2010) Tracking hydrocarbon plume transport and biodegradation at Deepwater Horizon. Science 330:201–204

Carney RS (1994) Consideration of the oasis analogy for chemosynthetic communities at Gulf of Mexico hydrocarbon vents. Geo-Mar Lett 14:149–159

Carriger JF, Barron MG (2011) Minimizing risks from spilled oil to ecosystem services using influence diagrams: the Deepwater Horizon spill response. Environ Sci Technol 45:7631–7639

CERC (1984) Shore protection manual. Vol. 1. Coastal Engineering Research Center, Department of Army. Vicksburg, MS

Chiasson SC, Taylor CM (2017) Effects of crude oil and oil/dispersant mixture on growth and expression of vitellogenin and heat shock protein 90 in blue crab, *Callinectes sapidus*, juveniles. Mar Pollut Bull 119:128–132

Echols BS, Smith AJ, Gardinali PR, Rand GM (2015) Acute aquatic toxicity studies of Gulf of Mexico water samples collected following the Deepwater Horizon incident (May 12, 2010 to December 11, 2010). Chemosphere 120:131–137

Fisher CR, Demopoulos AWJ, Cordes EE, Baums IB, White HK, Bourque JR (2014) Coral communities as indicators of ecosystem-level impacts of the Deepwater Horizon spill. Bioscience 64:796–807

French-McCay DP, Crowley D, Rowe JJ, Bock M, Robinson H, Wenning R, Hayward Walker A, Joeckel J, Nedwed TJ, Parkerton TF (2018a) Comparative risk assessment of oil spill response options for a Deepwater oil well blowout: part 1. Oil spill modeling. Mar Pollut Bull 133:1001–1015

French-McCay DP, Horn M, Li Z, Jayko K, Spaulding ML, Crowley D, Mendelsohn D (2018b) Modeling distribution, fate, and concentrations of Deepwater Horizon oil in subsurface waters of the Gulf of Mexico. Chapter 31. In: Oil spill Environmental forensics case studies. Elsevier, pp 638–635

Frometa J, DeLorenzo ME, Pisarski EC, Etnoyer PJ (2017) Toxicity of oil and dispersant on the deep water gorgonian octocoral *Swiftia exserta*, with implications for the effects of the Deepwater Horizon oil spill. Mar Pollut Bull 122:91–99

Gallaway BJ, Cole JG, Martin LR (2001) The deep sea Gulf of Mexico: an overview and guide. U.S. Department of the Interior, Minerals Management Service, Gulf of Mexico OCS Region, New Orleans, LA. OCS Study MMS 2001-065. 27 pp.

Gerard F, Fisher CR (2018) Long-term impact of the Deepwater Horizon oil spill on deep-sea corals detected after seven years of monitoring. Biol Conserv 225:117–127

Gerard F, Shea K, Fisher CR (2018) Projecting the recovery of a long-lived deep-sea coral species after the Deepwater Horizon oil spill using state-structured models. J Appl Ecol 2018:1–11

Gower JFR, King SA (2011) Distribution of floating Sargassum in the Gulf of Mexico and the Atlantic Ocean mapped using MERIS. Int J Remote Sens 32:1917–1929

Incardona JP, Gardner LD, Linbo TL, Brown TL, Esbaugh AJ, Mager EM, Stieglitz JD, French BL, JS Labenia JS, Laetz CA, Tagal M, Sloan CA, Elizur A, Benetti DD, Grosell M, Block BA, Scholz NL (2014) *Deepwater Horizon* impacts the developing hearts of large predatory pelagic fish. Proc Natl Acad Sci 111:E1510–E1518

Jewett SC, Dean TA, Woodin BR, Hoberg MK, Stegeman JJ (2003) Exposure to hydrocarbons 10 years after the Exxon Valdez oi spill: evidence from cytochrome P4501A expression and biliary FACs in nearshore demersal fishes. Mar Environ Res 54:21–48

Knap A, Turner NR, Bera G, Renegar DA, Frank T, Sericano J, Riegl BM (2017) Short-term toxicity of 1-methylnaphthalene to *Americamysis bahia* and 5 deep-sea crustaceans. Environ Toxicol Chem 36:3415–3423

McConville MM, Roberts JP, Boulais M, Woodall B, Butler JD, Redma AD, Parkerton TF, Arnold WR, Guyomarch J, LeFloch S, Bytingsvik J, Camus L, Volety A, Brander SM (2018) The sensitivity of a deep-sea fish species (*Anoplopoma fimbria*) to oil-associated aromatic compounds, dispersant, and Alaskan North Slope crude oil. Environ Toxicol Chem 37:2210–2221

Michel J, Owens EH, Zengel S, Graham A, Nixon Z, Owens EH, Zengel S, Graham A, Nixon Z, Allard T, Holton W, Reimer PD, Lamarche A, White M, Rutherford N, Childs C, Mauseth G, Challenger G, Taylor E (2013) Extent and degree of shoreline oiling: Deepwater Horizon oil spill, Gulf of Mexico, USA. PLoS One 8(6):e65087

Mitchelmore C, Bejarano AC, Wetzel D (2020) A synthesis of DWH oil and dispersant aquatic standard laboratory acute and chronic toxicity studies (Chap. 28). In: Deep oil spills: facts, fate, effects, vol 1. Springer, Cham

MMS (2000) Gulf of Mexico Deepwater Operations and Activities. Environmental Assessment. OCS EIS/EA MMS 2000-001. Minerals management Service. New Orleans, May 2000. www.boem.gov/BOEM-Newsroom/Library/Publications/2000/2000-001.aspx

NOAA (2015) Deepwater Horizon Oil Spill Natural Resource Damage Assessment Comprehensive Toxicity Testing Program: Overview, Methods, and Results. National Oceanic and Atmospheric Administration. Seattle Washington. 805 pp. www.fws.gov/doiddata/dwh-ar-documents/952/DWH-AR0293761.pdf

NOAA/ERD (2015) Chemical Aquatic Fate and Effects (CAFE) Database. Version 1.1 [Computer Software]. National Oceanic and Atmospheric Administration, Emergency Response Division, Office of Response and Restoration, Seattle, WA. p. 40 + Appendices https://response.restoration.noaa.gov/oil-and-chemical-spills/chemical-spills/response-tools/cafe.html. [Internet]

North EW, Adams EE, Schlag Z, Sherwood CR, He R, Hoon Hyun K, Socolofsky SA (2011) Simulating oil droplet dispersal from the Deepwater Horizon spill with a Lagrangian approach. In: Monitoring and modeling the Deepwater Horizon oil spill: a record-breaking enterprise, Geophysical Monograph Series 195, vol 195. American Geophysical Union, pp 217–226

Portnoy D, Fields A, Greer J, Schlenk D (2020) Genetic and oil: transcriptomics, epigenetics and genomics as tools to understand animal responses to exposure across different time scales (Chap. 30). In: Deep oil spills: facts, fate, effects, vol 1. Springer, Cham

Powers SP, Hernandez FJ, Condon RH, Drymon JM, Free CM (2013) Novel pathways for injury from offshore oil spills: direct, sublethal and indirect effects of the Deepwater Horizon oil spill on pelagic Sargassum communities. PLoS One 8(9):e74802

Raimondo S, Jackson CR, Barron MG (2017) Web-based Interspecies Correlation Estimation (Web-ICE) for acute toxicity: user manual, Version 3.3. EPA/600/R-15/192; Office of Research and Development, U.S. Environmental Protection Agency: Gulf Breeze, FL. https://www3.epa.gov/ceampubl/fchain/webice/index.html. [Internet]

Reddy CM, Arey JS, Seewald JS, Sylva SP, Lemkau KL, Nelson RK, Carmichael CA, McIntyre CP, Fenwick J, Ventura GT, Van Mooy BAS, Camilli R (2011) Composition and fate of gas and pol released to the water column during the Deepwater Horizon oil spill. Proc Natl Acad Sci 109:20229–20234

Robinson H, Wenning R, Hayward Walker AH, Joeckel J, Nedwed TJ, Parkerton TF (2018) Comparative risk assessment of oil spill response options for a Deepwater oil well blowout: part 1. Oil spill modeling. Mar Pollut Bull 133:1001–1015

Rowe GT (2017) Offshore plankton and benthos of the Gulf of Mexico. Chapter 7. In: Ward CH (ed) Habitats and Biota of the Gulf of Mexico: before the Deepwater Horizon oil spill, New York, NY, pp 641–767

Short S, Yang G, Guler Y, Green Etxabe A, Kille P, Ford AT (2014) Crustacean intersexuality is feminization without demasculinization: implications for environmental toxicology. Environ Sci Technol 48:13520–13529

Smith CR, Rowles TK, Hart LB, Townsend FI, Wells RS, Zolman ES, Balmer BC, Quigley B, Ivancic M, McKercher W, Tumlin MC, Mullin KD, Adams JD, Wu Q, McFee W, Collier TK, Schwacke LH (2017) Slow recovery of Barataria Bay dolphin health following the Deepwater Horizon oil spill (2013–2014), with evidence of persistent lung disease and impaired stress response. Endanger Species Res 33:127–142

Valentine DL, Kessler JD, Redmond MC, Mendes SD, Heintz MB, Farwell C, Hu L, Kinnaman FS, Yvon-Lewis S, Du M, Chan EW, Garcia Tigreros F, Villanueva CJ (2010) Propane Respiration Jump-Starts Microbial Response to a Deep Oil Spill. Science 330:208–211

Whitehead A, Dubansky B, Bodinier C, Garcia TI, Miles S, Pilley C, Raghunathan V, Roach JL, Walker N, Ealter RB, Rice CD, Galvez F (2011) Genomic and physiological footprint of the Deepwater Horizon oil spill on resident marsh fishes. Proc Natl Acad Sci 109:20298–20302

Willming MM, Lilavois CR, Barron MG, Raimondo S (2016) Acute toxicity prediction to threatened and endangered species using Interspecies Correlation Estimation (ICE) models. Environ Sci Technol 50:10700–10707

Chapter 28
A Synthesis of DWH Oil: Chemical Dispersant and Chemically Dispersed Oil Aquatic Standard Laboratory Acute and Chronic Toxicity Studies

Carys L. Mitchelmore, Adriana C. Bejarano, and Dana L. Wetzel

Abstract In the event of an oil spill, like the *Deepwater Horizon* (DWH), a Natural Resource Damage Assessment (NRDA) is mandated to document observable or measurable adverse changes to a natural resource or predict a future impairment of a natural resource. The assessment of health and integrity of estuarine and marine aquatic resources following exposure to contaminants and/or other environmental stressors relies on using established standard protocols and metrics for evaluating chemical, biological, and toxicological status. One of the key tools available for assessing environmental damage is the use of realistic laboratory- and mesocosm-scale toxicity experiments to expose organisms to various concentrations of oil and/or chemical dispersant employing an array of biological endpoints. However, widely varying methods of laboratory testing have been used together with extrapolation procedures that are not fully reliable for environmental assessments of oil spills (NRC 2005; Coelho et al. 2013; Bejarano et al. 2014; Bejarano 2018). There is often a severe lack of understanding in how to conduct aquatic toxicity tests correctly with respect to the questions being asked (i.e., toxicological comparisons between different oil mixtures, different species or life stages, or specific parameters unique to a spill event). There are many examples of serious errors (i.e., under and overestimations of toxicity) resulting from using inadequate methods for experimental design, chemical analyses of exposure media, and data analysis/interpretation. Studies that evaluate biological responses and measureable adverse changes from

C. L. Mitchelmore (✉)
University of Maryland Center for Environmental Science, Chesapeake Biological Laboratory, Solomons, MD, USA
e-mail: mitchelmore@umces.edu

A. C. Bejarano
Research Planning Inc., Columbia, SC, USA

Shell Oil Company, Shell Health - Americas, Houston, TX, USA
e-mail: abejarano@researchplanning.com; Adriana.Bejarano@shell.com

D. L. Wetzel
Mote Marine Laboratory, Sarasota, FL, USA
e-mail: dana@mote.org

480

exposure to oil, dispersed oil, and dispersants must (1) follow rigorous protocols which are well supported in the scientific literature, (2) produce reproducible and verifiable results, and (3) have well-defined and accepted statistical criteria for interpreting as well as rejecting results (NRC 2005; Bejarano 2018).

This chapter summarizes the traditional laboratory toxicity tests that have been conducted on MC252 (and surrogate) oils and/or Corexit 9500, pertinent to the DWH oil spill for standardtest species and also nontraditional estuarine and marine species reporting lethal and sublethal (i.e., growth, reproduction) endpoints from both short-term acute and chronic toxicitytests (either published in peer reviewed literature or reported on NOAA DIVER).

Keywords NRDA · EPA · Standard toxicity testing · CROSERF · Corext 9500 · WAF · CEWAF · LC50 · EC50

28.1 Introduction

In the event of an oil spill, like the *Deepwater Horizon* (DWH), a Natural Resource Damage Assessment (NRDA) is mandated to document observable or measurable adverse changes to a natural resource or predict a future impairment of a natural resource. The assessment of health and integrity of estuarine and marine aquatic resources following exposure to contaminants and/or other environmental stressors relies on using established standard protocols and metrics for evaluating chemical, biological, and toxicological status. One of the key tools available for assessing environmental damage is the use of realistic laboratory- and mesocosm-scale toxicity experiments to expose organisms to various concentrations of oil and/or chemical dispersant employing an array of biological endpoints. However, widely varying methods of laboratory testing have been used together with extrapolation procedures that are not fully reliable for environmental assessments of oil spills (NRC 2005; Coelho et al. 2013; Bejarano et al. 2014; Bejarano 2018). There is often a severe lack of understanding in how to conduct aquatic toxicity tests correctly with respect to the questions being asked (i.e., toxicological comparisons between different oil mixtures, different species or life stages, or specific parameters unique to a spill event). There are many examples of serious errors (i.e., under and overestimations of toxicity) resulting from using inadequate methods for experimental design, chemical analyses of exposure media, and data analysis/interpretation. Studies that evaluate biological responses and measureable adverse changes from exposure to oil, dispersed oil, and dispersants must (1) follow rigorous protocols which are well supported in the scientific literature, (2) produce reproducible and verifiable results, and (3) have well-defined and accepted statistical criteria for interpreting as well as rejecting results (NRC 2005; Bejarano 2018).

It was clear nearly four decades ago that the critical issue in the interpretation and comparison of laboratory toxicity data for oil, dispersants, and dispersed oil was the lack of standardization of laboratory toxicity testing protocols (NRC 1989; Tjeerdema 1990). Therefore, in the mid-1990s, the Chemical Response to Oil Spills: Ecological Effects Research Forum (CROSERF) was established to provide guidelines to state, federal, industry, scientists, and consultants engaged in petroleum research, with respect to understanding the ecological and biological effects of oil spills, together with specific protocols and methods for carrying out oil and dispersant toxicity testing (Singer et al. 2000, 2001a, 2001b; Aurand and Coelho 2005). These standardized methods have been useful, not just for Natural Resource Damage Assessments (NRDAs) but also in response planning. Chemical and biological results from these standard toxicity tests have been incorporated in numerous models of oil fate and effects (Di Toro et al. 2000; French-McKay 2002; Bejarano et al. 2015; Bejarano 2018) and used for understanding chemical partitioning (Forth et al. 2017a,b; Sandoval et al. 2017; Redman and Parkerton 2015). However, despite the protocol recommendations from CROSERF, oil and dispersant toxicity experiments are still being conducted in ways that render the interpretation of the data nearly impossible or of very limited utility (NCR 2005; Coelho et al. 2013; Bejarano et al. 2014; Redman and Parkerton 2015; Bejarano 2018). For example, the nominal amounts of oil and/or dispersant (e.g., g oil/L water or a % of a nominal loading stock solution) are being reported as the actual exposure concentration rather than a concentration determined via using analytical instrumentation.

Over time, there have been suggestions for changes and updates to the original CROSERF protocols (NRC 2005), and many adjustments were used during the NOAA NRDA toxicity testing following the DWH oil spill (Carney et al. 2016). The chapter on *Modernizing Protocols for Aquatic Toxicity Testing of Oil and Dispersant* (Mitchelmore et al. (2020)) discusses this further in great detail. This chapter, however, summarizes the traditional laboratory toxicity tests that have been conducted on MC252 (and surrogate) oils and/or Corexit 9500, pertinent to the DWH oil spill for standard test species and also nontraditional estuarine and marine species reporting lethal and sublethal (i.e., growth, reproduction) endpoints from both short-term acute and chronic toxicity tests (either published in peer reviewed literature or reported on NOAA DIVER). Furthermore, all of these reported tests were conducted using a minimum set of testing criteria described in detail in Sect. 3 below. The focus of this chapter on traditional whole-organism toxicity testing complements *Digging Deeper than LC/EC50: Assessing pathways of exposure, nontraditional endpoints and non-model species in oil spill toxicology* (Grosell et al. 2020).

28.2 Overview of Aquatic Toxicity Testing Protocols

28.2.1 Types of Toxicity Tests

Traditional laboratory-based acute (short-term exposures, typically 24–96 hrs) and chronic (longer-term, often weeks to months) toxicity tests have routinely been used to determine the potential impacts of specific crude oils, chemical dispersants, and mixtures of both. The goal of toxicity testing is, of course, to assess toxicity using rigorously evaluated and widely accepted standard methods. By definition, and to ensure reproducibility, traditional standard toxicity testing requires that standardized species/life stages, test solutions, and environmental parameters are used,

Table 28.1 EPA standard acute and chronic toxicity testing for marine and estuarine organisms

Species	Duration	Life stage	Exposure	Endpoint
Cyprinodon variegatus	Chronic	Larval	7 days – static renewal	Survival and weight
Cyprinodon variegatus	Chronic	Embryo and larval	Fertilization to 4 dph – static renewal	Survival and deformities
Menidia beryllina	Chronic	Larval	7 days –static renewal	Survival and weight
Americamysis bahia	Chronic	Juvenile	7 days –static renewal	Survival, growth, and fecundity
Arbacia punctulata	Chronic	Sperm	1 hr. – static renewal	Fertilization
Champia parvula	Chronic	Sexually mature	48 hrs – static renewal	Sexual reproduction
Cyprinodon variegatus	Acute	Juvenile	24 hrs – static nonrenewal 48 hrs – static renewal or nonrenewal 96 hrs – static renewal or all flow-through	Mortality
Menidia beryllina	Acute	Juvenile	24 hrs – static nonrenewal 48 hrs – static renewal or nonrenewal 96 hrs – static renewal or all flow-through	Mortality
Menidia menidia	Acute	Juvenile	24 hrs – static nonrenewal 48 hrs – static renewal or nonrenewal 96 hrs – static renewal or all flow through	Mortality
Americamysis bahia	Acute	Juvenile	24 hrs – static nonrenewal 48 hrs – static renewal or nonrenewal 96 hrs – static renewal or all flow-through	Mortality

(EPA 2002, 2003)

regardless of where the testing is performed. Therefore, established standardized toxicity test protocols, such as those developed by the US Environmental Protection Agency (Table 28.1; EPA 2002, 2003), for determining the acute (mortality/death as the biological endpoint) and chronic toxicity (a variety of sublethal biological endpoints, often relating to growth and/or reproductive output) of contaminants to specific marine and estuarine species are commonly used in regulatory applications. The use of standard EPA test organisms allows a for the comparison of toxicity for different contaminants; however, the test species list is very restricted in scope and often does not reflect the species or life stages under a specific threat in a particular location, and the ecological application of standardized toxicity testing results can be quite limited. In some cases, because of these limitations, researchers have instead used more appropriate test organisms while still applying methods or modified methods based on these standard EPA toxicity test protocols.

Acute toxicity tests are often the most commonly employed assessments, as the measured endpoint is easily determined (mortality), but the levels of contaminant concentrations needed to produce mortality at the statistically required response rates are rarely found in the environment. But, even though high concentrations of contaminants may be unrealistic in the environment, the LC50 (lethal concentration that results in 50% mortality of the population) generated using standard toxicity protocols confirms the likely potential tipping point at which mortality will occur under standard toxicity testing conditions. This standard metric allows for assessments regarding the comparison of different oil types and/or dispersants and mixtures and can provide a ranking of species sensitivities to a range of toxicants, useful for many reasons. When an acute and highly concentrated contaminant spill occurs, and surrogate organism responses have already been documented, a prediction of mortality for the affected environment can be completed rapidly. Chronic and sublethal exposure studies use LC50 values to bracket the range of concentrations to target for test solution creation. These data are also useful in defining environmental risk, the combination of known exposure concentrations (i.e., measured concentrations in the field) and standard toxicological assessments.

In traditional acute toxicity studies, early life stages are generally preferred as they are often more sensitive to toxicant exposure, whereas either early life stages or mature adult organisms are used in chronic testing. The exposure regimes for both acute or chronic are either (1) static nonrenewal where the test organisms are exposed to the same test solution for the duration of the test (less than 24 hrs); (2) static renewal where solutions are replaced with fresh solutions at specified time intervals (i.e., every 48 hrs); or (3) flow-through, where there is a continuous passage of test solution or dilution water through a test chamber with no recycling. Flow-through regimes may employ either a single "spiked" exposure or a continuous concentration. The typical range of exposure duration for traditional toxicity testing is from 24 hrs to 21 days or more depending on the organism, life stage, and endpoint.

28.2.2   Preparation of Toxicity Test Exposure Media

Many of the test condition protocols for test solution makeup, environmental conditions (e.g., water quality, lighting regime, temperature), species, experimental design, endpoints, and analyses are detailed in the standard methods (EPA 2002, 2003). Normally, there are five test solution concentrations and a control required for single contaminant/mixture toxicity testing. Under ideal test designs, a response increases proportionally as a function of concentration to a maximum point (i.e., monotonic dose response) allowing the estimation of specific endpoint metrics (ECs and LCs) along the curve. The minimum control survival required to obtain a valid test is 80% or 90% but is test protocol specific. Statistical results for acute and chronic tests are calculated for lethal concentrations (LCs), sublethal effect concentrations (ECs), concentrations that result in 25% inhibition in biological measurement (IC25s), no observed effect concentrations (NOECs), and lowest observed effect concentrations (LOECs), using a variety of statistical approaches based on point estimates (e.g., Probit or Spearman-Karber) or hypothesis (comparison) methods.

Working with a complex mixture like crude oil with or without chemical dispersants poses many unique issues compared with single chemicals, with respect to test exposure solution makeup and true organismal chemical exposure for reporting of toxicity endpoints. These complexities and difficulties in preparing reproducible standard test solutions were part of the motivation behind the establishment of CROSERF which developed standardized robust oil toxicity testing methodology (Singer et al. 1992; Singer et al. 2000, 2001a, b; Aurand and Coelho 2005). Prior to DWH, various researchers and an NRC report discussed refinements to these protocols regarding the preparation and analysis of test media and interpretations from toxicity tests (Barron and Ka'aihua 2003; NRC 2005). After DWH, standard operating procedures (SOPs) to conduct toxicity tests were developed by BP and NOAA/trustees (Carney et al. 2016) based on CROSERF but included modifications, some made prior to DWH but some made applying the unique conditions of this spill. Deep-ocean, high-energy environments, which affect the environmental fate, transport, and physiochemical characteristics of the oil and dispersant, all had to be considered when designing toxicity exposure studies. For example, a high-energy WAF preparation method (HEWAF; Carney et al. 2016; Forth et al. 2017a, b) was designed to mimic the high-energy environment scenario occurring during the DWH event. Many of these issues, including other nuances of working with oil and dispersants and the pros and cons of various approaches (i.e., variable dilution/loading, sampling times, and type of analytical chemistry methods for exposure solutions), are discussed in *Modernizing Protocols for Aquatic Toxicity Testing of Oil and Dispersant* (see Mitchelmore et al. 2020) and summarized in Table 28.2.

28.2.3 Interpretation of Toxicity Test Results

There are inherent problems involved in the preparation of oil and water test solutions leading to widely varying bioavailable PAH concentrations limiting the utility and interpretation of the observed effects if detailed chemical characterizations of the exposure test solutions are not conducted. Using nominal oil concentrations or a limited characterization of the test media is sadly still common – and never appropriate. This approach provides no useful data and is an unethical use of laboratory organisms (vertebrate or invertebrate species). As described recently by Coelho et al. (2013) and Bejarano et al. (2014), toxicity studies are useless without a detailed understanding of true organismal exposure to oil.

The interpretation of toxicity tests is not just dependent on the composition of the analytes (i.e., PAHs) but in what phase these compounds are found. This is

Table 28.2 Considerations regarding the preparation of test media and the implementation and reporting requirements for oil, dispersant, and dispersed oil aquatic toxicity tests

Toxicity test type and procedures	
Specific test	Standard method, EPA/OECD/ASTM (state number and reference) or modified standard method (state all modifications)
	Acute or chronic
Species and life stages	Name (common/Latin), source (i.e., cultures, hatcheries, wild specimens, and locations), life stage, age, sex (if possible)
Test conditions	Water: freshwater, brackish, marine, and specific makeup, salts used for control/dilution water, volumes used/test chamber; report salinity
	Number of treatments, replicates, and negative and positive controls
	Test duration (e.g., 96 hrs or 7 days)
	Static nonrenewal, static renewal, constant exposure, spiked, flow-through
	Aeration/feeding regime
	Size of exposure chamber
	Water quality control and measurements
	Type of lighting (i.e., spectral quality and quantity) and day/night regime
	Water temperature
Biological endpoints	Endpoints measured (LC50, EC50, IC25, NOEC, LOEC, etc.)
	Time endpoints are measured (e.g., LT50 considerations)
	Utility of toxic units (TU) approach for WAFs (dissolved only test solutions)
Reporting and statistics	Specific statistical methods used
	Positive and negative control results
Test exposure preparation protocols	
Type of oil and/or chemical dispersant and water	Detail specific oil type used (e.g., CAS #)/source/weathering state
	Detail specific type of dispersant used
	Water used (freshwater, brackish, or marine), filtered/unfiltered water, salts used and their sources, salinity
Preparation of test concentrations	Variable loading or variable dilution approach (may depend on questions being asked and/or volume of test media required)
	Report oil/water and dispersant/oil ratio units used (e.g., mg/L)
	One or multiple preparations (e.g., for static renewal tests)
Test preparation method used	WAF, CEWAF, HEWAF, LEWAF
	Details on container used (total volume vs. test media volume, headspace, sealed vessels, etc.)
	Mixing energy, mixing time, and settlement time

(continued)

Table 28.2 (continued)

Test maintenance	Renewal time(s) and % of test media renewed, removal of dead organisms
	Aeration, feeding regime (timing and amount and adjustments based on organism death)
	Recording intervals for biological endpoints and water quality measures
Analytical: chemical characterization of exposure solutions	Recommended minimum of total PAH concentrations TPAH including alkyl homologues (MC 252 QAPP, 2011)
	Recommend analysis of VOCs (e.g., BTEX)
	For chemically (CEWAF) and physically dispersed oils (WAFs), recommend chemical analysis of both the dissolved and particulate phases (i.e., not just total exposure solution)
	Consider measurements for dispersant components and/or UV-Vis confirmation for dispersant-only concentrations
	For CEWAFs/WAFs, consider analysis of oil droplet concentration (volume, quantity) and size distribution
	Use of nominal concentrations of oil is unacceptable
Analytical: tissue residue	Measurement of TPAH in exposed organism tissue to relate internal exposure (body burden) to toxicity endpoints

particularly important when comparing the results between oil only solutions and chemically dispersed solutions. Oil components can partition between the fully dissolved phase and the particulate phases (oil droplets). If PAHs are not measured in the appropriate phase of the test solution for the questions asked, there will either be an over or underrepresentation of the toxicity observed depending on how a particular organism is exposed to the oil constituents. Some organisms will only be exposed to the dissolved fraction and others based on their feeding (i.e., indiscriminate feeding on oil droplets due to their size being similar to their prey items), or other physiologies are exposed to both the dissolved and oil droplets fractions. Therefore, determining oil partitioning between the particulate (oil droplets) and dissolved fractions, as well as the quantity and size distribution of oil droplets, is critically important to understand the environmental relevance of toxicity studies. Figure 28.1 demonstrates how the interpretation of toxicity can be misrepresented by the choice of analytical chemistry conducted on the test solutions.

28.3 Synthesis of DWH Aquatic Toxicity Data

For the specific purpose of this synthesis, study selection was based on a minimum set of criteria, which were as follows: (1) standard toxicity tests conducted using MC252 (all types) or surrogate (i.e., South Louisiana crude) oils; (2) toxicity data reported as measured hydrocarbons and based on the sum of total polycyclic aromatic hydrocarbons (TPAH or $\sum 50$PAH; MC 252 QAPP, 2011); studies reporting

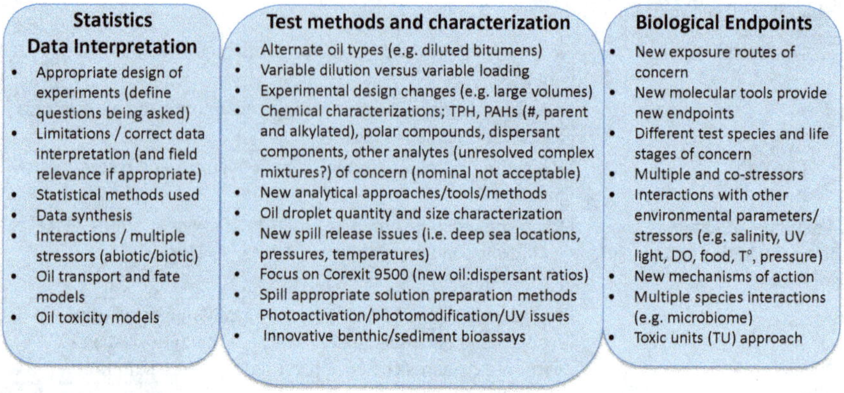

Statistics
Data Interpretation

- Appropriate design of experiments (define questions being asked)
- Limitations / correct data interpretation (and field relevance if appropriate)
- Statistical methods used
- Data synthesis
- Interactions / multiple stressors (abiotic/biotic)
- Oil transport and fate models
- Oil toxicity models

Test methods and characterization

- Alternate oil types (e.g. diluted bitumens)
- Variable dilution versus variable loading
- Experimental design changes (e.g. large volumes)
- Chemical characterizations; TPH, PAHs (#, parent and alkylated), polar compounds, dispersant components, other analytes (unresolved complex mixtures?) of concern (nominal not acceptable)
- New analytical approaches/tools/methods
- Oil droplet quantity and size characterization
- New spill release issues (i.e. deep sea locations, pressures, temperatures)
- Focus on Corexit 9500 (new oil:dispersant ratios)
- Spill appropriate solution preparation methods
  Photoactivation/photomodification/UV issues
- Innovative benthic/sediment bioassays

Biological Endpoints

- New exposure routes of concern
- New molecular tools provide new endpoints
- Different test species and life stages of concern
- Multiple and co-stressors
- Interactions with other environmental parameters/stressors (e.g. salinity, UV light, DO, food, T°, pressure)
- New mechanisms of action
- Multiple species interactions (e.g. microbiome)
- Toxic units (TU) approach

Fig. 28.1 Overview of changing protocols, developments, and considerations since CROSERF for oil toxicity studies

toxicity data on the basis of nominal concentrations were specifically excluded from this synthesis; and (3) biological responses with whole organisms reported as mortality (LC50) or sublethal endpoints (EC50), specifically growth (i.e., abnormalities, development, hatching failure) or reproduction (i.e., fertilization, viability). Only studies that used a standardized test procedure (i.e., acute or chronic), such as modified EPA and CROSERF tests, were included in this synthesis.

Fourteen studies met the criteria specified above; comprising a total of 381 aquatic toxicity data points (Table 28.3). This toxicity synthesis showed that the large majority of data (81%) were generated using standard acute toxicity exposure durations (24, 48, 72, 96 hrs), while exposure periods that deviated from these standard durations accounted for 19% of the data. Consistently, a larger effort focused on characterizing acute mortality (79% of all records), while less attention was generally given to chronic/sublethal endpoints (21%). Although toxicity data were generated using a large species diversity (32 aquatic species), most data (75%) were for a handful of test organisms (i.e., *Americamysis bahia, Callinectes sapidus, Coryphaena hippurus, Crassostrea virginica, Cyprinodon variegatus, Sciaenops ocellatus*). Similarly, most data (82%) were generated using early life stages (e.g., embryos, larvae), which are generally some of the most sensitive life stages. Toxicity test results vary over two orders of magnitude across aquatic species. These results are likely influenced by the preparation of the exposure media and experimental conditions, plus the inherent differences in sensitivity across test species. A related synthesis of dispersant-only toxicity data (Bejarano 2018) found that the large majority of data generated post-2010 were generated using non-US standard test species. Consistent with the findings above, most dispersant-only toxicity data were based on LC50s from standard test durations (24, 48, 72, 96 hrs) under constant static or static renewal test conditions. Toxicity data from chronic tests or from spiked flow-through exposures were generally uncommon in the DWH literature.

Table 28.3 Summary of LC50s and EC50s toxicity from standard acute and chronic whole-organism toxicity tests. Study selection was based on the following criteria: tested oil, MC252 or surrogate oil; hydrocarbon metric (TPAH or $\sum$50PAH- MC 252 AQAP, 2011); and endpoint, mortality, growth (i.e., abnormalities, development, hatching failure), or reproduction (i.e., fertilization, viability)

| Species | Physically dispersed oil | | | Chemically dispersed oil | | |
	Number of toxicity records	Concentration (mg/L)	References	Number of toxicity records	Concentration (mg/L)	Reference
LC50						
Crustaceans						
Acartia tonsa	3	0.05 ± 0.04	9	–	–	–
Americamysis bahia	19	0.10 ± 0.28	2, 5	22	0.02 ± 0.02	2, 5
Callinectes sapidus	10	0.15 ± 0.29	2, 9	6	0.04 ± 0.02	2, 9
Litopenaeus setiferus	2	0.41 ± 0.11	2, 9	2	0.11 ± 0.04	2
Litopenaeus vannamei	3	0.12 ± 0.03	2	–	–	–
Palaemonetes pugio	–	–	–	4	0.12 ± 0.01	9
Uca longisignalis	4	0.008 ± 0.005	2, 9	–	–	–
Fish						
Anchoa mitchilli	6	0.41 ± 0.64	2, 3	2	4.36 ± 0.56	3
Coryphaena hippurus	29	0.13 ± 0.19	2, 4, 9	13	0.11 ± 0.25	2, 4, 9
Cynoscion nebulosus	9	0.06 ± 0.12	2, 9	4	0.07 ± 0.11	2, 9
Cyprinodon variegatus	15	0.61 ± 0.65	2	22	0.18 ± 0.18	2
Fundulus grandis	–	–	–	2	0.36 ± 0.20	2
Lutjanus campechanus	2	0.04 ± 0.01	2	–	–	–
Menidia beryllina	4	0.04 ± 0.04	5	9	0.05 ± 0.04	5, 9
Paralichthys lethostigma	1	0.03	2	–	–	–
Rachycentron canadum	5	0.06 ± 0.03	9	–	–	–
Sciaenops ocellatus	25	0.31 ± 0.86	1, 2, 7, 9	5	0.08 ± 0.04	2, 9
Thunnus albacares	7	0.03 ± 0.05	2, 9	–	–	–
Thunnus maccoyii	5	0.004 ± 0.002	9	–	–	–
Mollusk						
Crassostrea virginica	18	0.05 ± 0.04	2, 8, 9, 11, 13, 14	20	0.08 ± 0.09	2, 8, 9, 11, 13, 14

(continued)

Table 28.3 (continued)

Species	Physically dispersed oil			Chemically dispersed oil		
	Number of toxicity records	Concentration (mg/L)	References	Number of toxicity records	Concentration (mg/L)	Reference
Crassostrea gigas	2	0.09 ± 0.11	11	2	0.09 ± 0.09	11
Mercenaria mercenaria	2	0.09 ± 0.10	11	2	0.14 ± 0.16	11
Mercenaria galloprovincialis	2	0.13 ± 0.01	11	2	0.34 ± 0.02	11
Other						
Strongylocentrotus purpuratus	3	0.09 ± 0.06	11	2	1.4 ± 0.50	11
Dendraster excentricus	2	0.10 ± 0.11	11	2	0.10 ± 0.07	11
EC50						
Fish						
Coryphaena hippurus	7	0.76 ± 0.74	9, 12	4	0.17 ± 0.02	2, 9
Cyprinodon variegatus	11	0.78 ± 0.80	2			
Fundulus grandis	2	0.58 ± 0.03	2	3	0.56 ± 0.26	2
Sciaenops ocellatus	5	0.67 ± 1.08	2, 7	1	0.11	2
Mollusk						
Crassostrea virginica	21	0.047 ± 0.71	2, 9, 13, 14	15	0.072 ± 0.05	2, 9, 13, 14
Phytoplankton						
Chaetoceros socialis	1	0.05	10	–	–	–
Chaetoceros sp.	–	–	–	3	0.11 ± 0.04	6
Ditylum brightwellii	1	0.03	10	–	–	–
Heterocapsa triquetra	1	0.01	10	–	–	–
Isochrysis galbana	–	–	–	3	0.11 ± 0.01	6
Pyrocystis lunula	1	0.01	10	–	–	–
Scrippsiella trochoidea	1	0.01	10	–	–	–

Data sources: [1]Bridges et al. 2018; [2]DIVER 2017; [3]Duffy et al. 2016; [4]Esbaugh et al. 2016; [5]Finch et al. 2017; [6]Garr et al. 2014; [7]Khursigara et al. 2017; [8]Laramore et al. 2014; [9]Morris et al. 2015; [10]Özhan et al. 2014; [11]Stefansson et al. 2016; [12]Sweet et al. 2017; [13]Vignier et al. 2015; [14]Vignier et al. 2016

28.4   Lessons Learned During DWH Toxicity Testing

Toxicity testing of physically and chemically dispersed oil, and chemical dispersants following the DWH oil spill, expanded substantially our knowledge on the biological responses of aquatic species to these complex chemical mixtures. Most commonly, toxicity tests focused on standard aquatic species across several taxa (e.g., mysid shrimp [*Americamysis bahia*]; inland silverside [*Menidia beryllina*]; EPA 2002) as standard testing protocols are well established including defined internal controls (i.e., control survival) on which to base the quality of test results. The overall assumption for using such species is that they provide conservative toxicity estimates that can also be protective of untested species and sensitive life stages. Although one of the drawbacks of using standard test species is that there are uncertainties about how their sensitivity compares to that of a species of concern (i.e., deep-sea and pelagic species), recent research supports the possible use of standard test species as surrogates for deepwater species and is discussed in detail in *Ecotoxicology of Deep Ocean Spills* (see Barron et al. 2020). However, for many species of concern, laboratory husbandry and standard methods specific for each species' requirements may not be available, which was a clear challenge during the DWH oil spill. Following the DWH oil spill, toxicity testing was performed using a range of aquatic species (i.e., several taxa, habitat preferences), resulting in toxicity test results for at least 20 different species, including standard test species. Test results using species that had not previously been tested offered new insights into their sensitivity relative to that of most commonly tested organisms. Based on the information generated to date, early life stages (e.g., fish embryos) are generally more sensitive to oil and dispersed oil than juveniles/adults of the same or related species. These findings are not necessarily new but provide a means of confirming earlier work. Similarly, the range of sensitivity of newly tested species appears to fall within the range of sensitivity of most commonly tested species (Table 28.3). The same finding applies to aquatic toxicity data from dispersant-only exposures (Bejarano 2018).

Since a single measure of aquatic toxicity does not provide sufficient information on which to infer protective thresholds, the concept of species sensitivity distributions (SSDs) has been recommended to facilitate evaluations of relative sensitivities under comparable exposure conditions. Several studies (e.g., Bejarano et al. 2014; Bejarano 2018; Barron et al. 2020) have shown that hazard concentrations from SSDs may be protective of a wide range to species, including sensitive species. This was recently demonstrated following a synthesis of dispersant-only toxicity data (Bejarano 2018). While such approach works relatively well for single chemicals or highly soluble chemical mixtures, there are a number of challenges when applied to complex hydrocarbon mixtures such as crude oil. Interlaboratory comparisons of species sensitivities based on toxicity test results with crude oil are difficult, unless standard protocols are carefully followed (Coelho et al. 2013; Redman and Parkerton 2015). As a result, hazard concentrations from SSDs may not be appropriate, unless toxicity data used in their development are from studies using comparable standardized procedures (i.e., exposure media preparation, analytical chemistry reporting, etc.). Hence, interlaboratory comparisons of test results would benefit from agreed-upon factors including preparation method, oil loading rate, mixing energy, dispersant-to-oil ratio, oil source, and weathering state, among others.

The large majority of aquatic toxicity data were generated via constant static or static renewal tests. While these types of tests may not necessarily reflect the dynamic field conditions that occur during an oil spill (i.e., water column mixing and dilution), these types of "controlled" conditions offer one key advantage: by allowing exposure concentrations to remain relatively constant over the test period, body burdens buildup to levels that lead to a measureable adverse effect (i.e., death). Thus, these types of data are useful in the development and validation of predictive models (McGrath et al. 2018) that could then be applied with confidence in the assessment of potential risks from oil spills. The challenge with these types of data and associated predictive models is that their application to more realistic field conditions could only be made via assumptions (i.e., extrapolation factors). These however should be supported toxicity data, which are currently largely absent from the scientific literature (i.e., static vs. dynamic tests).

28.5   Summary and Recommendations for Future Studies

Toxicity testing of oil and dispersants is essential in providing synthesis data to (1) improve NRDA, hazard assessments, and operational field decisions and (2) refine toxicity models to predict or forecast toxicity in new, understudied, or sensitive species for future oil spills. The DWH spill resulted in a massive expansion of laboratory toxicity studies in both standard and a diverse array of nonstandard species. These results supported NRDA (DIVER 2017) and provided data to populate new toxicity models, databases, and species sensitivity assessments (CAFE; NOAA/ ERD 2015). Many studies were of limited utility resulting from selecting inadequate methods for experimental design, chemical analyses of exposure media, and data analysis/interpretation. Thus, it is clear that there is a need to reevaluate and reiterate expected standards for conducting toxicity bioassays in the light of new knowledge and experiences gained following DWH. Even despite guidelines for conducting toxicity tests (CROSERF and modifications thereof; Singer et al. 2000; NRC 2005), many oil, dispersant, and dispersed oil studies were not of sufficient quality or lacked sufficient experimental details to be disregarded in a synthesis of DWH toxicity results (Coelho et al. 2013; Redman and Parkerton 2015; Bejarano et al. 2018). For example, in a recent synthesis of dispersant-only toxicity results, many studies were not included as they did not meet the minimal selection criteria, and a quarter of the studies included in that synthesis did not provide detailed information on which to assess study quality (Bejarano 2018). Table 28.2 summarizes some of the main points to consider when conducting oil and dispersant toxicity tests, and Fig. 28.2 highlights some of the main additional considerations post CROSERF which were included in many DWH investigations.

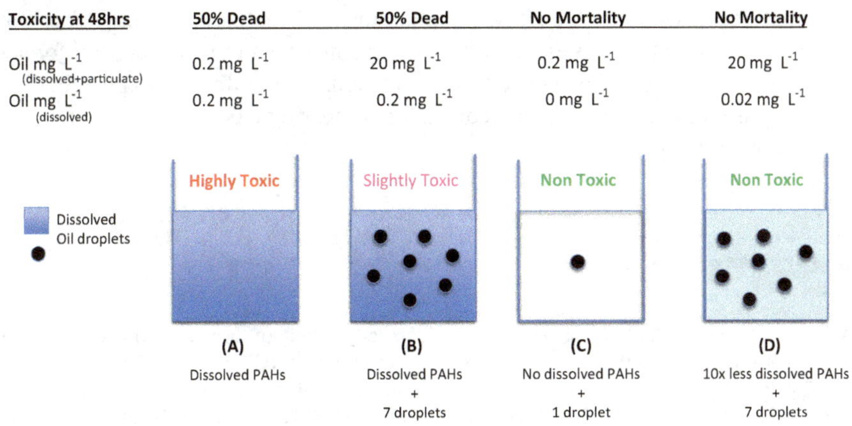

Fig. 28.2 Example of misinterpretation of toxicity test data using total PAH (or TPH) quantitation in a species where acute toxicity is via a narcosis mechanism of action related to the presence of dissolved PAHs (LC50 threshold of 0.2 mg L^{-1})

Moving forward, we recommend that standard protocols for conducting toxicity tests with oil and dispersants (CROSERF; Carney et al. 2016) should be updated to provide standard methods for future oil spill events that include the array of potential oil spill scenarios, oil types, and response options. Developments in analytical chemistry and molecular analyses have led to a far greater understanding of chemical exposure/fate and biological mechanisms of action and effect and need to be incorporated into these new standard protocols (also see- *Digging Deeper than LC/ EC50: Assessing pathways of exposure, non-traditional endpoints and non-model species in oil spill toxicology* (Grosell et al. 2020) and *Modernizing Protocols for Aquatic Toxicity Testing of Oil and Dispersant* (Vol. 2 Mitchelmore et al. Chap. 13).

Specific recommendations of this chapter include:

- Expansion of number of standard test species to include those more relevant to the spill site, including sensitive and understudied species.
- Conduct more environmentally relevant exposures i.e. longer-term, chronic, sublethal endpoints. Consideration of changing exposure characteristics (chemistry, etc.) over time.
- Specific minimum criteria required to conduct and interpret toxicity tests based on what they are being used for (e.g. minimum selection criteria for inclusion into toxicity models).
- Record toxicity data at multiple and earlier time points to assess additional endpoints, e.g. time to death (i.e. LT50).
- Studies that account for the relative contribution of oil droplets and truly dissolved fractions in exposure solutions.
- Minimum list of analytical chemistry to be completed of test exposure solutions dependent upon questions being asked and study design approach.
- Specific guidelines on laboratory lighting (i.e. timing and spectral quantity and quality [UV light]).

- Single species studies that allow for laboratory to field extrapolations. This may be accomplished by simultaneously running concurrent tests that control for and account for dilution (i.e., constant static vs. spiked flow-through exposures).
- Proper use of extrapolation procedures for comparing laboratory exposures with field-based measurements.

References

Aurand D, Coelho G (2005) Cooperative aquatic toxicity testing of dispersed oil and the chemical response to oil spills: Ecological Effects Research Forum (CROSERF). (Technical Report 07–03, Ecosystem Management & Associates, Inc. Lusby, 105 pages + Appendices)

Barron MG, Ka'aihue L (2003) Critical evaluation of CROSERF test methods for oil dispersant toxicity testing under subarctic conditions. Mar Pollut Bull 46:1191–1199

Barron MG, Chiasson SC, Bejarano AC (2020) Ecotoxicology of deep ocean spills (Chap. 27). In: Murawski SA, Ainsworth C, Gilbert S, Hollander D, Paris CB, Schlüter M, Wetzel D (eds) Deep oil spills: facts, fate, effects. Springer, Cham

Bejarano AC, Clark JR, Coelho GM (2014) Issues and challenges with oil toxicity data and implications for their use in decision making: a quantitative review. Environ Toxicol Chem 33(4):732–742

Bejarano AC, Farr JK, Jenne P, Chu V, Hielscher A (2015) The chemical aquatic fate and effects database (CAFÉ), a tool that supports assessments of chemical spills in aquatic environments. Environ Toxicol Chem 35(6):1576–1586

Bejarano AC (2018) Critical review and analysis of aquatic toxicity data on oil spill dispersants. Environ Toxicol Chem 9999:1–13. https://doi.org/10.1002/etc.4254

Bridges KN, Krasnec MO, Magnuson JT, Morris JM, Gielazyn ML, Chavez JR, Roberts AP (2018) Influence of variable ultraviolet radiation and oil exposure duration on survival of red drum (*Sciaenops ocellatus*) larvae. Environ Toxicol Chem 37(9):2372–2379. https://doi.org/10.1002/etc.4183

Carney MW, Forth HP, Krasnec MO, Takeshita R, Holmes JV, Morris JM (2016) Quality assurance project plan: Deepwater Horizon laboratory toxicity testing. DWH NRDA Toxicity Technical Working Group. Prepared for National Oceanic and Atmospheric Administration by Abt Associates, Boulder

Coelho G, Clark J, Aurand D (2013) Toxicity testing of dispersed oil requires adherence to standardized protocols to assess potential real world effects. Environ Pollut 177:185–188

Di Toro DM, McGrath JA, Hansen DJ (2000) Technical basis for narcotic chemicals and polycyclic aromatic hydrocarbon criteria. I. Water and tissue. J Environ Toxicol Chem 19:1951–1970

DIVER (2017) Web application: data integration visualization exploration and reporting application, National Oceanic and Atmospheric Administration. Retrieved: [September, 20, 2017], from https://www.diver.orr.noaa.gov

Duffy TA, Childress W, Portier R, Chesney EJ (2016) Responses of bay anchovy (*Anchoa mitchilli*) larvae under lethal and sublethal scenarios of crude oil exposure. Ecotoxicol Environ Saf 134:264–272

EPA (2002) Methods for measuring the acute toxicity of effluents and receiving waters to freshwater and marine organisms, fifth ed. EPA-821-R-02-012

EPA (2003) Short-term methods for estimating the chronic toxicity of effluents and receiving water to marine and estuarine organisms, third ed. EPA-821-R-02-014

Esbaugh AJ, Mager EM, Stieglitz JD, Hoenig R, Brown TL, French BL, Linbo TL, Lay C, Forth H, Scholz NL (2016) The effects of weathering and chemical dispersion on Deepwater Horizon crude oil toxicity to mahi-mahi (*Coryphaena hippurus*) early life stages. Sci Total Environ 543:644–651

Finch BE, Marzooghi S, Di Toro DM, Stubblefield WA (2017) Phototoxic potential of undispersed and dispersed fresh and weathered Macondo crude oils to Gulf of Mexico marine organisms. Environ Toxicol Chem 36:2640–2650

Forth HP, Mitchelmore CL, Morris JM, Lay CR, Lipton J (2017a) Characterization of dissolved and particulate phases of water accommodated fractions used to conduct aquatic toxicity testing in support of the Deepwater Horizon natural resource damage assessment. Environ Toxicol Chem 36:1460–1472

Forth HP, Mitchelmore CL, Morris JM, Lipton J (2017b) Characterization of oil and water accommodated fractions used to conduct aquatic toxicity testing in support of the Deepwater Horizon oil spill natural resource damage assessment. Environ Toxicol Chem 36:1450–1459

French-McKay DP (2002) Development and application of an oil toxicity and exposure model, (OilToxEx). J Environ Toxicol Chem 21:2080–2094

Garr AL, Laramore S, Krebs W (2014) Toxic effects of oil and dispersant on marine microalgae. Bull Environ Contam Toxicol 93:654–659

Grosell M, Griffitt RJ, Sherwood TA, Wetzel DL (2020) Digging deeper than LC/EC50: non-traditional endpoints and non-model species in oil spill toxicology (Chap. 29). In: Murawski SA, Ainsworth C, Gilbert S, Hollander D, Paris CB, Schlüter M, Wetzel D (eds) Deep oil spills: facts, fate, effects. Springer, Cham

Khursigara AJ, Perrichon P, Martinez Bautista N, Burggren WW, Esbaugh AJ (2017) Cardiac function and survival are affected by crude oil in larval red drum, *Sciaenops ocellatu*s. Sci Total Environ 579:797–804

Laramore S, Krebs W, Garr A (2014) Effects of Macondo Canyon 252 oil (naturally and chemically dispersed) on larval *Crassostrea virginica* (Gmelin, 1791). J Shellfish Res 33:709–718

McGrath JA, Fanelli CJ, Di Toro DM, Parkerton TF, Redman AD, Paumen ML, Comber M, Eadsforth CV, den Haan K (2018) Re-evaluation of target lipid model–derived HC5 predictions for hydrocarbons. Environ Toxicol Chem 37(6):1579–1593

Mitchelmore CL, Griffitt RJ, Coelho GM, Wetzel DL (2020) Modernizing protocols for aquatic toxicity testing of oil and dispersant (Chap. 14). In: Murawski SA, Ainsworth C, Gilbert S, Hollander D, Paris CB, Schlüter M, Wetzel D (eds) Scenarios and responses to future deep oil spills: fighting the next war. Springer, Cham

Morris J, Krasnec M, Carney M, Forth H, Lay C, Lipton I, McFadden A, Takeshita R, Cacela D, Holmes J (2015) Deepwater Horizon oil spill natural resource damage assessment comprehensive toxicity testing program: overview, methods, and results. DWH-AR0293761. Technical report. Prepared by Abt Associates, Boulder, CO and Chesapeake Biological Laboratory, Solomons, for National Oceanic and Atmospheric Administration Assessment and Restoration Division, Seattle, WA. Retrieved from https://www.fws.gov/doiddata/dwh-ar-documents/952/DWH-AR0302396.pdf

National Research Council (1989) Using oil spill dispersants on the sea. National Research Council, Committee on Effectiveness of Oil Spill Dispersants. National Academy Press, Washington, DC

National Research Council (2005) Oil spill dispersants; efficacy and effects. National Academies Press, Washington DC, p 400. https://doi.org/10.17226/11283

NOAA/ERD (2015) Chemical aquatic fate and effects (CAFE) database. Version 1.1 [computer software]. National Oceanic and Atmospheric Administration, Emergency Response Division, Office of Response and Restoration, Seattle. p. 40 + Appendices

Özhan K, Miles SM, Gao H, Bargu S (2014) Relative phytoplankton growth responses to physically and chemically dispersed South Louisiana sweet crude oil. Environ Monit Assess 186:3941–3956

Redman AD, Parkerton TF (2015) Guidance for improving comparability and relevance of oil toxicity tests. Mar Pollut Bull 98(1–2):156–170

Sandoval K, Ding Y, Gardinali P (2017) Characterization and environmental relevance of oil water preparations of fresh and weathered MC-252 Macondo oils used in toxicology testing. Sci Total Environ 576:118–128

Stefansson ES, Langdon CJ, Pargee SM, Blunt SM, Gage SJ, Stubblefield WA (2016) Acute effects of non-weathered and weathered crude oil and dispersant associated with the Deepwater Horizon incident on the development of marine bivalve and echinoderm larvae. Environ Toxicol Chem 35:2016–2028

Singer MM, Aurand D, Bragin GE, Clark JR, Coelho GM, Sowby ML, Tjeerdema RS (2000) Standardization of the preparation and quantitation of water-accommodated fractions of petroleum for toxicity testing. Mar Pollut Bull 40:1007–1016

Singer MM, Aurand D, Coelho G, Bragin GE, Clark JR, Jacobson S, Sowby ML, Tjeerdema RS (2001a) Making, measuring, and using water-accommodated fractions of petroleum for toxicity testing. In: Proceedings of the 2001 international oil spill conference. American Petroleum Institute, Washington, D.C., pp 1269–1274

Singer MM, Jacobson S, Tjeerdema RS, Sowby ML (2001b) Acute effects of fresh versus weathered oil to marine organisms: California findings. In: Proceedings of the 2001 international oil spill conference. American Petroleum Institute, Washington, D.C., pp 1263–1268

Sweet LE, Magnuson J, Garner TR, Alloy MM, Stieglitz JD, Benetti D, Grosell M, Roberts AP (2017) Exposure to ultraviolet radiation late in development increases the toxicity of oil to mahi-mahi (*Coryphaena hippurus*) embryos. Environ Toxicol Chem 36:1592–1598

Tjeerdema RS (1990) Oil spill cleanup agent research in California. Final report to the California Department of Fish and Game. Office of Oil Spill Prevention and Response, Sacramento, CA

Vignier J, Donaghy L, Soudant P, Chu FLE, Morris JM, Carney MW, Lay C, Krasnec M, Robert R, Volety AK (2015) Impacts of Deepwater Horizon oil and associated dispersant on early development of the eastern oyster *Crassostrea virginica*. Mar Pollut Bull 100:426–437

Vignier J, Soudant P, Chu F, Morris J, Carney M, Lay C, Krasnec M, Robert R, Volety A (2016) Lethal and sub-lethal effects of Deepwater Horizon slick oil and dispersant on oyster (*Crassostrea virginica*) larvae. Mar Environ Res 120:20–31

Chapter 29
Digging Deeper than LC/EC50: Nontraditional Endpoints and Non-model Species in Oil Spill Toxicology

Martin Grosell, Robert J. Griffitt, Tracy A. Sherwood, and Dana L. Wetzel

Abstract The response to the 2010 *Deepwater Horizon* (DWH) oil spill lead to a number of peer-reviewed publications examining the effects of the released oil and dispersant on fish species found in the northern Gulf of Mexico (GoM). Many of these papers, for very good reasons, focused on assessing toxicity by defining lethality through identification of dose-response curves that were specific to a given species, age class, exposure type, oil preparation method, and many other factors. Often those dose-response curves were used to predict LC or EC50 concentrations – amounts of oil that produced an effect on 50% of the exposed organisms. The advantage of this approach is obvious, in that it provides a single point estimate and variance of a concentration required to produce a given effect. This point estimate can then be compared across different exposure regimes to compare susceptibilities. Relevant LC/EC50 data is summarized and discussed in *A synthesis of DWH oil, chemical dispersant and chemically dispersed oil aquatic standard laboratory acute and chronic toxicity studies* (see Mitchelmore et al. A synthesis of Deepwater Horizon oil, chemical dispersant and chemically dispersed oil aquatic standard laboratory acute and chronic toxicity studies (Chap. 28). In: Murawski SA, Ainsworth C, Gilbert S, Hollander D, Paris CB, Schlüter M, Wetzel D (Eds.) Deep oil spills – facts, fate and effects. Springer; 2020). By constraining toxicity to this point estimate (often of lethality), however, researchers run the risk of missing effects that evince more subtle effects that do not manifest themselves as overt mortality in the short term. In the present chapter, we focus exclusively on fish and explore some of

M. Grosell (✉)
University of Miami, Department of Marine Biology and Ecology, Rosenstiel School of
Marine and Atmospheric Science, Miami, FL, USA
e-mail: mgrosell@rsmas.miami.edu

R. J. Griffitt
University of Southern Mississippi, Division of Coastal Sciences, School of Ocean Science
and Engineering, Ocean Springs, MS, USA
e-mail: Joe.Griffitt@usm.edu

T. A. Sherwood · D. L. Wetzel
Mote Marine Laboratory, Sarasota, FL, USA
e-mail: tsherwood@mote.org; dana@mote.org

© Springer Nature Switzerland AG 2020
S. A. Murawski et al. (eds.), *Deep Oil Spills*,
https://doi.org/10.1007/978-3-030-11605-7_29

497

these endpoints, many of which were successfully used in recent years to assess sublethal health impacts on marine fish as part of the response to the DWH spill. We compare what is known about differences in sensitivity, among species, and between age classes within species, examining both organismal and molecular endpoints. Developmental impacts on cardiac health, swim performance, and sensory systems have been widely studied. We discuss what is known about effects on fish immune and endocrine function, the microbiome of the intestine and gill, and intracellular effects such as altered gene expression, oxidative stress, and DNA damage. In conclusion, we attempt to compare the endpoints, assess the sensitivity and utility, and link molecular- and individual-level impacts to larger population and community-level effects.

Keywords ELS (early life stage) · Cardiotoxicity · CEWAF (chemically enhanced water accommodated fraction) · WAF (water accommodated fraction) · Toxicity testing · Microbiome · Immune function · RNA sequencing

29.1   Introduction

The response to the 2010 *Deepwater Horizon* (DWH) oil spill lead to a number of peer-reviewed publications examining the effects of the released oil and dispersant on fish species found in the northern Gulf of Mexico (GoM). Many of these papers, for very good reasons, focused on assessing toxicity by defining lethality through identification of dose-response curves that were specific to a given species, age class, exposure type, oil preparation method, and many other factors. Often those dose-response curves were used to predict LC or EC50 concentrations – amounts of oil that produced an effect on 50% of the exposed organisms. The advantage of this approach is obvious, in that it provides a single point estimate and variance of a concentration required to produce a given effect. This point estimate can then be compared across different exposure regimes to compare susceptibilities. Relevant LC/EC50 data is summarized and discussed in *A synthesis of DWH oil, chemical dispersant and chemically dispersed oil aquatic standard laboratory acute and chronic toxicity studies* (see Mitchelmore et al. 2020). By constraining toxicity to this point estimate (often of lethality), however, researchers run the risk of missing effects that evince more subtle effects that do not manifest themselves as overt mortality in the short term. In the present chapter, we focus exclusively on fish and explore some of these endpoints, many of which were successfully used in recent years to assess sublethal health impacts on marine fish as part of the response to the DWH spill. We compare what is known about differences in sensitivity, among species, and between age classes within species, examining both organismal and molecular endpoints. Developmental impacts on cardiac health, swim performance, and sensory systems have been widely studied. We discuss what is known about effects on fish immune and endocrine function, the microbiome of the intestine and gill, and intracellular effects such as altered gene expression, oxidative stress, and

DNA damage. In conclusion, we attempt to compare the endpoints, assess the sensitivity and utility, and link molecular- and individual-level impacts to larger population and community-level effects.

29.2   Test Organism Sensitivity

29.2.1   Species-Specific Differences in Sensitivity to Oil Exposure

Parallel studies, using identical WAF preparations and exposure conditions, of acute toxicity to larval fish have revealed substantial differences among species (McKim et al. 1985) underscoring the range of sensitivities in different species and the importance of selecting the optimal test organism. In these studies of acute toxicity, 7-day-old inland silverside (*Menidia beryllina*) was more sensitive than 3-day-old sheepshead minnow (*Cyprinodon variegatus*) and more sensitive than <1-day-old Gulf killifish (*Fundulus grandis*) (Finch et al. 2017). These apparent differences in sensitivity should be viewed in the context of the age of the test organisms. Age, and presumably size, confers a higher tolerance to acute exposures to oil as illustrated on studies of bay anchovy exposed to HEWAF preparations of Macondo oil. These 24-h exposure studies on bay anchovy established LC50 values of 9.71 µg/L TPAH for embryos, compared to 690 and 1610 µg/L TPAH for 5- and 21-day-old larvae, respectively (Duffy et al. 2016; O'Shaughnessy et al. 2018). Thus, the observation that older silverside are more sensitive than the newly hatched killifish further underscores the significant difference in sensitivity among organisms. Although other factors may be at play, sensitivity differences among species may be a reflection of differences in PAH metabolism. In a DWH oil intraperitoneal injection (IP) study assessing species-specific PAH metabolism differences of subadult red drum (*Sciaenops ocellatus*), Florida pompano (*Trachinotus carolinus*), and southern flounder (*Paralichthys lethostigma*), analysis of PAH bile metabolites indicated that pompano exhibited a much faster rate of PAH metabolism than red drum and southern flounder. Results suggest that pompano may be better able to depurate toxic PAH compounds (Pulster et al. 2017).

29.2.2   Differences in Sensitivity Among Life Stages

Early life stages are generally more sensitive to environmental insults than older conspecifics which form the foundation for acceptance of early life stage toxicity testing as a faster and more cost-effective alternative to full life-cycle tests. In most cases, early life stage testing accurately predicts outcomes of full life-cycle tests (McKim 1985). Size is undoubtedly an important factor determining differences in sensitivity within and among species, likely because size dictates metabolic demands and thereby needs for exchange of gasses and metabolic products as well as surface area/volume ratios (Bianchini et al. 2002; Grosell et al. 2002). The general relationship of smaller early life stage organisms being more sensitive than older and larger conspecifics holds true also for oil/PAH exposures as illustrated in the following. However, it is important to note that impacts to several endpoints, such as sensory function, aerobic scope, and reproduction, cannot be assessed in early life stages and that these endpoints in some cases appear highly sensitive (Stieglitz et al. 2016; Xu et al. 2018; Johansen and Esbaugh 2017) and very well may be of relevance for population-level impacts. Further, several studies have revealed that early life stage exposures may lead to effects that are not readily evident until later in life (Xu et al. 2018; Mager et al. 2014; Magnuson et al. 2018) underscoring the importance for studies also on later life stages.

29.3 Targets of Oil Toxicity

29.3.1 Cardiac Developmental and Functional Effects on ELS

Cardiac development and function in embryo and larval fish from the GoM have been demonstrated to be sensitive to crude oil exposure. The early life stages (ELS) of yellowfin tuna (*Thunnus albacares*) were found to display pericardial edema and reduced heart rate following 48 hrs of exposure to HEWAFs at EC50 concentrations of 2.5 and 7.5 μg/L$\sum$50PAH, respectively (Incardona et al. 2014). Similarly sensitive are mahi-mahi (*Coryphaena hippurus*) embryos and larvae with 48-hr EC50s ranging from >5.1 to 7.3 μg/L$\sum$50PAH for HEWAFs and 11.3 to 13.0 μg/L$\sum$50PAH for CEWAFs of slick, source, and weathered oil (Esbaugh et al. 2016). As with yellowfin tuna, mahi-mahi displayed pericardial edema but also presented an altered atrium-ventricle angle and evidence of reduced cardiac contractility (Edmunds et al. 2015), the later suggesting reduced stroke volume. The coastal red drum is no less sensitive with a pericardial edema 48-h EC50 of 2.4 μg/L$\sum$50PAH for HEWAF (Khursigara et al. 2017). For the first time in any oil-exposed fish ELS, red drum stroke volume was measured to reveal that this parameter is far more sensitive than heart rate with the two factors both contributing to cardiac output with a 48-h EC50 of 2.2 μg/L$\sum$50PAH for HEWAF (Khursigara et al. 2017). While the aryl hydrocarbon receptor (AhR) plays a role in the pericardial edema phenotype (Incardona et al. 2006, 2011) observed in ELS fish following oil exposure, AhR-independent effects have also been reported (Incardona et al. 2005). In mahi-mahi, as well as red drum, pathway analyses of RNA-seq experiments revealed cardiac Ca^{2+} homeostasis as a target for crude oil toxicity (Xu et al. 2016, 2017a, b). Impaired Ca^{2+} homeostasis in myocytes may certainly account for observations of reduced cardiac contractility and stroke volume (Edmunds et al. 2015; Khursigara et al. 2017) and has also been implicated in cardiac hypertrophy (Xu et al. 2017b).

29.3.2 Noncardiac Developmental Effects on ELS

In addition to cardiac impairments, several other biological systems have demonstrated sensitivity to oil exposure. Mahi-mahi embryos exhibit yolk sac edema when exposed to a low concentration (1.2 µg/L$\sum$50PAH) of HEWAF for 48 hrs (Pasparakis et al. 2016), and decrease in yolk sac depletion (nutrient uptake) was observed in embryos exposed to higher concentrations (14.9–26.6 µg/L$\sum$50PAH) of HEWAF for 24 hrs (Mager et al. 2014). Other embryonic developmental impairments observed in sheepshead minnows when exposed to 257 µg/L$\sum$50PAH of CEWAF for 48 hrs were retarded development and decreases in eye pigmentation and movement (Bosker et al. 2017). Both length and weight of larvae of several fish species appeared to be influenced by exposure to CEWAF or HEWAF. Bosker et al. (2017) observed a reduction in standard length sheepshead minnow larvae exposed for 48 hrs to 257 µg/L $\sum$50PAH CEWAF. A reduction in lenght was also seen in spotted seatrout exposed for 96 hrs (at much lower concentrations) of approximately 74 µg/L$\sum$50PAH exposure solutions, whether HEWAF or CEWAF (Brewton et al. 2013), whereas a marked reduction in weight was measured in bay anchovy exposed for 24 hrs to 0.24–1.23 µg/L $\sum$50PAH of CEWAF (Duffy et al. 2016). Data from RNA-seq pathway analysis of oil-exposed larval mahi-mahi indicated that several other biological systems were affected by a 48-h exposure to DWH source oil (4.6 µg/L$\sum$50PAH) or DWH slick oil (12 µg/L$\sum$50PAH) HEWAF (Xu et al. 2016). Particularly affected were genes involved in neurodegeneration of the central nervous system, eye degeneration, visual impairment, and abnormality of the vertebral column. These studies highlight some of the potential consequences of oil exposure on multiple biological systems leading to the likely increase in post-exposure mortality.

29.3.3 Cardiac Function and Swim Performance in Later Life Stages

While cardiotoxicity has long been a recognized effect of oil exposure in developing fish, it was not until recently that the developed heart of adult fish was also identified as a target for oil exposure. Yellowfin and bluefin tuna (*Thunnus thynnus*) myocytes exposed to crude oil solutions showed prolonged action potentials through blockage of the delayed rectifier potassium current and also decreased Ca^{2+} current and Ca^{2+} cycling leading to excitation-contraction uncoupling (Brette et al. 2014). Although it is unknown at present how blood plasma PAH concentrations relate to environmental concentrations, the effects of PAH exposure on isolated tuna myocytes were observed at low µg/L$\sum$50PAH concentrations in the bathing salines, suggesting that impacts to cardiac function may occur during environmentally realistic oil exposures. The first evidence of oil-induced impacts to cardiac function following exposures of adult GoM species came from a study exposing mahi-mahi to HEWAF dilutions of <10 µg/L$\sum$50PAH (Stieglitz et al. 2016). This study reported reduced maximal oxygen uptake, reduced aerobic scope, and a concomitant reduction in critical swimming speed (U_{crit}) with no apparent effect on gill histology after only 24 hrs of exposure in young adult mahi-mahi. Subsequent studies employing vascular blood flow probes inserted into anesthetized mahi-mahi revealed normal heart rate but greatly reduced stroke volume and thereby reduced cardiac output when exposed for 24 hrs to HEWAF at ~ 10 µg/L$\sum$50PAH prior to experimentation (Nelson et al. 2016). Similar exposures resulted in reduced U_{crit} for cobia (*Rachycentron canadum*) and a substantial reduction in stroke volume, although the later was largely compensated for by an elevated heart rate (Nelson et al. 2017). Further studies on cobia employing 24-h HEWAF exposures at ~ 22 µg/L$\sum$50PAH found reduced cardiac power output (Cox et al. 2017), although this effect could be mitigated by β-adrenergic stimulation alluding to possible compensations in intact animals. In addition to the cardiac impacts noted above for pelagic GoM species, the coastal red drum show reduced aerobic scope and U_{crit} following 24 hrs of acute exposure to HEWAF at 12 µg/L$\sum$50PAH. These impairments remained for 8 weeks after the 24-hr exposure attesting to the persistence of cardiac effects induced by oil exposure to adult GoM species (Johanson and Esbaugh 2017).

Interestingly, reduced swim performance has been observed in oil-exposed GoM species without a concomitant reduction on maximal oxygen uptake or aerobic scope. For red drum, a 24-h exposure to HEWAF at 4 µg/L$\sum$50PAH resulted in both reduced burst swimming speed (U_{burst}) and U_{crit} without any apparent reductions in aerobic scope (Johanson and Esbaugh 2017). Similarly, 48-hr HEWAF exposures to mahi-mahi embryos at 1.2 µg/L$\sum$50PAH followed by rearing in uncontaminated water for 30 days resulted in reduced U_{crit}, again without reductions in aerobic scope or maximal oxygen uptake (Mager et al. 2014). The observed effects on swim performance in absence of notable reductions on oxygen supply points to non-respiratory pathways for PAH impacts on swim performance. Three such possible nonexclusive pathways include (1) impaired neuromast function leading to reduced swimming efficiency (Johanson and Esbaugh 2017), (2) reduced efficiency of mitochondrial ATP production, and finally (3) impacted calcium cycling in skeletal muscle (Johanson and Esbaugh 2017) as seen for cardiomyocytes (Brette et al. 2014, Xu et al. 2016, 2017a, b).

29.3.4 *Impacts on Sensory Systems and Behavior Including Prey-Predator Interactions*

A recent study (Rowsey et al. 2019) on red drum determined that larvae exposed to HEWAF at ~25 µg/L$\sum$50PAH for 24 hrs, 21-day post fertilization (dpf), displayed increased thigmotaxis (time spent in open areas as opposed to in shelter) leading to an increased exploration of foraging area. Despite exploring larger foraging areas, oil-exposed red drum larvae (24 hrs at 35 dpf) caught less prey in a time-limited prey capture assay at ~55 µg/L$\sum$50PAH (Rowsey et al. 2019). Similar behavioral impacts have been observed for a number of larval great barrier reef fish species exposed to HEWAF for 24 hrs (Johansen et al. 2017). In mesocosm experiments, larval reef fish showed altered habitat preference, increased exploration of different habitats (more movement), and less tendency for shoaling (lower average group size) following 24 hrs of exposure to 5.7 µg/L$\sum$50PAH (Johansen et al. 2017). When exposed to predators under simulated natural conditions, these oil-induced behavioral changes in larval fish resulted in increased predation mortality (Johansen et al. 2017). While integrative observations of behavioral changes may reflect impaired sensory function, impaired central nervous system function, or both, RNA-seq studies on mahi-mahi as well as red drum have revealed that brief exposures during larval development may alter expression of genes related to sensory function. For mahi-mahi, 24–96 hrs of exposure to HEWAFs at concentrations as low as 7 µg/L$\sum$50PAH showed developmental impacts on the peripheral nervous system through exposure to oil. In particular, eye development and the development of eye function were among the pathways most impacted by oil exposure (Xu et al. 2016). Similar findings were reported for red drum larvae exposed to HEWAF at ~5 µg/L$\sum$50PAH for 24–72 hrs where pathway analyses pointed to likely impacts to the nervous system (Xu et al. 2017b). These observations of gene expression changes were paralleled by observations of reductions in eye and brain areas (Xu et al. 2017b). Functional implications of these gene expression and developmental impacts have been noted using an optomotor response to assess eye function. As a consequence of exposure during embryonic development, larval sheepshead minnow show reduced eye function following exposure to HEWAF at 159 µg/L$\sum$50PAH (and higher) (Magnuson et al. 2018). The same assay indicated that red rum larvae exposed during embryonic development were more sensitive, displaying reduced eye function following HEWAF exposure at 2.7 µg/L$\sum$50PAH (Magnuson et al. 2018). However, larval mahi-mahi exposed to HEWAF during embryonic development were even more sensitive, showing effects at the low levels of 0.67 µg/L$\sum$50PAH (Xu et al. 2018). With the exception of one study on great barrier reef species (Johansen et al. 2017), it is unclear to what extent impaired sensory function and altered central nervous system function may affect overall performance and survival under natural conditions. However, considering the effect thresholds for impaired eye function and the apparent ubiquity of this response across species, these impacts may well influence population level responses to oil exposure.

29.3.5 Impacts on Microbiomes

In-depth analysis of the microbiome (defined loosely as the suite of microorganisms that are associated with a host organism in different organs) offers an intriguing opportunity to examine subtle effects of exposure that may, nonetheless, have important effects on the host organism. Several studies clearly demonstrated that an incursion of oil from the DWH spill altered the microbial community structure in water and sediment (Hazen et al. 2010; Kostka et al. 2011; Mason et al. 2014), but only a handful of papers have examined this effect on marine fish species. Arias et al. (2013) and Larsen et al. (2015) examined the effects of oil exposure on the skin microbiome of the Gulf killifish and found no evidence for a linkage between oil exposure and skin microbiome shifts using culture-based approaches and ribosomal intergenic spacer analysis (RISA), although neither paper reported concentrations of PAHs in water or sediment where the samples were collected. In contrast, both Brown-Peterson et al. (2015) and Bayha et al. (2017) showed that laboratory exposure to oil-contaminated sediment (54 and 57 mg/kg $\sum$50PAH, respectively) altered the microbiome of both gills and intestines in southern flounder in ways that were clearly associated with the oil exposure, including a significant increase in the prevalence of bacteria linked to hydrocarbon degradation in upper gill, lower gill, and intestinal communities. Both studies used predictive metagenomics to show that KEGG (Kyoto Encyclopedia of Genes and Genomes) pathways involved in contaminant degradation increased in abundance following oil exposure and pathways involved in biosynthesis of metabolically important compounds were strongly decreased in abundance following exposure to oil-contaminated sediments.

The effect of oil exposure on the commensal microbiome is likely worthy of further study as there is a clearly demonstrated linkage between intestinal Ahr receptor ligands and the intestinal microbiome in vertebrates. In particular, Ahr activation has been demonstrated to play a strong role in mediating tetrachlorodibenzofuran contaminant-driven shifts in intestinal microbiome structure in mice (Zhang et al. 2018). The idea that expression of certain proteins affected by oil exposure (such as Ahr) could influence the structure of the host microbiome has interesting consequences for studying the interaction of host and contaminants.

29.3.6 *Impacts on Immune Function*

The immune system of fish, like that of mammals, is comprised of innate and adaptive immunity. Innate immunity is considered the first line of defense through physical barriers and nonspecific recognition of microbial molecular patterns, whereas adaptive immunity, the second line of defense, is comprised of humoral and cell-mediated immunity, in which specific antigen recognition leads to either B cell proliferation and antibody production or cytotoxic T cell proliferation. Both the innate and adaptive immune responses use a complex network of signaling molecules (cytokines, chemokines), specialized cells, and organs that work in concert to clear microbial pathogens and maintain immunological homeostasis (Rauta et al. 2012).

Oil exposure can have a direct effect on the immune system through biotransformation or detoxification systems such as the cytochrome P450 (CYP) enzyme system. These systems are shown to be interrelated and not exclusively independent. In fish, immune organs such as the head, kidney, and spleen have the capacity to detoxify xenobiotics indicating a biotransformation system within the immune system (Reynaud et al. 2008). In addition, in the liver, PAH activation of the AhR and PAH metabolism by the CYP enzymes along with Ca^{2+} mobilization can induce immunotoxicity (Reynaud and Deschaux 2006).

Three field studies of adult Gulf killifish collected in 2010 from DWH oil-contaminated sites of the northern GoM indicated modulations to both innate and adaptive immunities had occurred. Modulation of the adaptive immune system included notable decreases of circulating lymphocytes and enlargement of splenic melano-macrophage centers (MMCs) in relation to increased activity of ethoxyresorufin-O-deethylase (EROD), an established biomarker for PAH exposure in the environment (Omar-Ali et al. 2014). In addition, RNA-seq analysis of liver tissue indicated upregulation of adaptive immune genes, IgM, and tumor necrosis factor receptors 14 and 21 (TNFr14 and 21), relative to upregulation of genes associated with hypoxia and the CYP/AhR pathway (Garcia et al. 2012). Whereas microarray analysis of gill tissue demonstrated modulation of innate immunity with enrichment of genes involved in wound healing, inflammation, and the acute phase response, there was also a correlation with enrichment of CYP genes that have oil-associated expression (Dubansky et al. 2013).

RNA-seq and microarray analyses revealed immunosuppression in the livers of GoM fish exposed to either oil or dispersed oil. In juvenile red drum exposed for 3 days to a CEWAF of 1.73 μg/L$\sum$50PAH, differential gene expression analysis indicated the majority of immune genes were significantly downregulated. These included genes associated with granulocyte and macrophage function (23 genes), T cell receptor (4 genes), and B cell receptor (2 genes). This downregulation of immune genes further correlated to significant upregulation of CYP1A, GST, and UDP-glucuronosyltransferase genes involved in metabolism of xenobiotics (Wetzel DOI: 10.7266/NT8P5Z08). Altered immune processes were observed in adult sheepshead minnow exposed for 7 or 14 days to environmentally relevant HEWAF concentrations of 0.26–5.95 μg/L$\sum$50PAH and CEWAF concentrations of 0.35–1.10 μg/L$\sum$50PAH, with the most affected responses related to macrophage and granulocyte functions (Jones et al. 2017).

Following the DWH spill, offshore surveys of fish populations observed incidences of skin lesions in GoM fish especially in bottom-dwelling species near the DWH site (Murawski et al. 2014). The lesions were hypothesized to be caused by PAH suppression of the immune system allowing for opportunistic bacteria colonization. Bayhe et al. (2017) were able to demonstrate this phenomenon in juvenile southern flounder exposed to weathered DWH oil-contaminated sediments at a concentration of 57.40 mg/kg$\sum$50PAH for 7 days, followed by a 1-h bacterial challenge with *V. anguillarum*, resulting in body lesions similar to those reported in the field. The observed lesions were most likely due to a reduction of IgM gene expression in the spleen and kidney of oil-exposed southern flounder. IgM is the main immunoglobulin in fish that is essential for clearance of bacterial pathogens.

29.3.7 *Oxidative Stress and DNA Damage*

Oxidative stress occurs when there is an imbalance between the production of reactive oxygen species (ROS) and the organism's ability to detoxify reactive intermediates, such as those generated by metabolism of PAHs through cytochrome P450. Depending on the severity of oxidative stress, this imbalance between excessive production of ROS and antioxidant defenses can lead to lipid peroxidation (LPO), resulting from a suite of chain reactions triggered by radicals causing loss of membrane integrity and DNA damage. Potential damage to DNA can be caused by a variety of mechanistic endpoints, such as oxidized bases, apurinic/apyrimidinic sites, single- or double-strand breaks, and formation of DNA adducts, in which PAH metabolites intercalate into the DNA (Xue and Warshawsky 2005; Starostenko et al. 2017).

Monitoring organismal antioxidant defenses as markers for oxidative stress due to PAH exposure in fish is relatively new. The antioxidant enzymes, superoxide dismutase (SOD), and catalase (CAT) activities have been extensively used as biomarkers for a variety of contaminants; however direct links could not be established for SOD-CAT activities to a specific class of contaminants including PAHs (van der Oost et al. 2003), rendering these ineffective as markers for PAH-associated oxidative stress. Reduced glutathione (GSH), an important non-enzymatic antioxidant, acts as a scavenger for ROS (particularly H_2O_2) and PAH reactive intermediates to produce oxidized glutathione (GSSG), which is then recycled back to GSH for antioxidant defense. The relationship of PAH exposure and associated oxidative stress has been established using GSH/GSSG for antioxidant defense assessments. Other useful assessments of oxidative stress include measuring levels of malondialdehyde (MDA; a by-product of LPO), lipid hydroperoxides, and the total antioxidant capacity (TAC). In nuclear and mitochondrial DNA, 8-hydroxy-2-deoxyguanosine (8-OHdG) is one of the predominant forms of free radical-induced oxidative lesions and it has been widely used as a biomarker for oxidative stress leading to DNA oxidative damage.

In a series of controlled DWH surrogate oil exposures, targeted GoM fish species exhibited an overall decrease in antioxidants, increase of lipid peroxidation, and evidence of DNA damage. A correlation of $\sum$50PAH in the liver and decrease in antioxidant defense was observed in Florida pompano exposed to CEWAF concentrations of 14.7 µg/L$\sum$50PAH (Wetzel DOI: 10.7266/N7PZ5793). Red drum experienced a significant reduction in antioxidant defenses, increase in lipid peroxidation, a decrease in total antioxidant capacity, and a slight (not significant) increase in liver DNA oxidative damage (8-OHdG) when exposed to 485.07 mg/kg$\sum$50PAH and 688.94 mg/kg$\sum$50PAH DWH surrogate oil-contaminated feed for 14 days (Wetzel DOI: 10.7266/N7Q23XRS). Similarly, effects of oxidative stress were observed in southern flounder exposed to 47.50 mg/kg$\sum$50PAH DWH surrogate oiled sediment for 30 days (Wetzel DOI: 10.7266/N7WH2NGJ). In a laboratory study using adult Gulf killifish, genotoxicity was observed in the form of single-stranded DNA breaks, responsive genes for DNA damage, and apoptosis when exposed to weathered DWH surrogate oil WAF concentrations ranging from 300 µg/L to 3000 µg/L $\sum$50PAH. These exposure levels were chosen to simulate PAH concentrations found in field-collected sediments (284 to >8000 µg/L $\sum$50PAH) not long after the DWH spill (Pilcher et al. 2014).

29.3.8 Reproduction

Reproduction in fish is controlled by an organized endocrine system, the hypothalamic-pituitary-gonadal axis (HPG axis), which consists of the hypothalamus, pituitary gland, and gonadal glands. In oviparous females, the HPG axis is referred to as the HPG-liver axis given that many egg yolk and chorionic (e.g., vitellogenin and choriogenin) proteins are synthesized in the liver, necessary for oocyte growth and development.

There is evidence that oil exposure can impact reproductive fitness of fish through endocrine disruption and cytotoxic effects on germ cells (Iwanowicz and Blazer 2009; Zhang et al. 2016; Lee et al. 2017). However, there have been relatively few studies investigating the impacts on reproduction from the 2010 DWH oil spill. Microarray and RNA-seq analysis of gene expression in field-collected Gulf killifish showed a decrease of vitellogenin, choriogenin, and zona pellucida (glycoprotein layer surrounding egg) genes in the liver (Whitehead et al. 2012; Garcia et al. 2012) coinciding with upregulation of the CYP/AhR pathway genes. Though these studies illustrate alterations of genes involved in egg development, what remains unclear are the long-term consequences oil exposure may have on egg development and fecundity. In a full life-cycle (larvae through reproduction) study using sheepshead minnows exposed to oil-contaminated sediments for 19 weeks, Raimondo et al. (2016) were able to demonstrate a decrease in the first-generation fecundity of fish exposed to 358 and 751 mg/kg$\sum$50PAH contaminated sediments. While these studies may shed light on possible mechanisms of impairment, additional studies are warranted to determine the sublethal effects of the DWH oil spill on reproductive fitness and the ramifications those effects have on wild populations.

29.3.9 Transcriptomic Effects Not Yet "Anchored" by Phenotypic Observations

The use of global transcriptome profiling techniques has great utility for understanding subtle, sublethal effects of contaminant exposure on exposed species (Schirmer et al. 2010). The recent development and adoption of RNA-seq as a high throughput sequencing approach has allowed researchers to rapidly and accurately quantify the state of the transcriptome in non-model fish species following exposure to anthropogenic and environmental stressors (Vega-Retter et al. 2018). *Genetics and oil: Transcriptomics, epigenetics and genomics as tools to understand animal responses to exposure across different time scales* (see Portnoy et al. 2020) discusses what the field of transcriptomics is and what it can tell us about effects of exposure to oil.

Several studies have linked transcriptome changes following exposure to DWH oil with endpoints of ecological significance. Whitehead et al. (2012) showed differences in gene expression profiles of Gulf killifish in response to oil contamination from the 2010 DWH oil spill, particularly decreased physiological functions in osmoregulation, respiration, and excretion, which persisted for 2 months post-oiling. Oil exposure affected multiple molecular pathways in developing mahi-mahi, including E1/E2 signaling, steroid biosynthesis, and ribosome biogenesis pathways (Xu et al. 2016). Dubansky et al. (2013) showed that Gulf killifish collected from areas affected by DWH oil had significantly altered transcriptional patterns, including a number of genes and pathways involved in xenobiotic response or oil detoxification. Exposure to very low levels (<5 µg/L$\sum$PAH50) caused altered expression of many pathways in early life stage red drum in a manner that was dependent on both oil type and age of the embryo (Xu et al. 2017b), indicating that results from one exposure regime or age class, even within species, may not be fully informative of the transcriptional effects of other exposure regimes or ages. Exposure to oil-contaminated sediment altered liver transcriptional patterns in juvenile southern flounder, including effects on a number of immune-related genes, matching an observed oil-induced immunosuppression following bacterial challenge (Bayha et al. 2017).

29.4 Conclusions and Directions for Future Research

Although direct comparisons among species differences are scarce and further studies clearly are needed, the available literature demonstrates significant species-specific differences in sensitivity underscoring the importance of test organism choice. Similarly, life stage-dependent sensitivity differences are also obvious with early life stages generally being more susceptible. However, later life stages should not be ignored as some endpoints, including cardiac function, immune system function, and sensory and central nervous systems, may show effects even at relatively low exposure concentrations in later life stages.

In addition to choice of organisms and life stages to be tested, the endpoint examined will influence the impression of sensitivity. Cardiac development and cardiac function have long been viewed as central targets for oil exposure and are sensitive to oil exposure at low and environmentally realistic levels (<10 µg/L$\sum$PAH50). However, recent studies have revealed that eye development and immune system function can be impacted at lower concentrations (<1 µg/L$\sum$PAH50) illustrating the importance of considering multiple endpoints. Transcriptome wide responses have been employed successfully in recent years to shed light on additional relevant endpoints by examining sensitive biochemical pathways and have in some cases been matched with observations of phenotypic change.

To date, the laboratory studies discussed in this chapter examining the effects of exposure to DWH oil on the immune system of GoM fish clearly show evidence of immunosuppression at the gene level, in both the innate and adaptive immune responses. Conversely, field studies from DWH oil-contaminated areas showed an activation of immune response at the gene level, but immunosuppression at the cellular level. These field studies, however, did not provide oil concentrations for these contaminated collection sites which make direct comparison to laboratory-based studies difficult. The effects of oil on fish immunity can be contradictory and will depend on many variables including the concentration of PAHs, mode of exposure, life stage, and species (Reynaud and Deschaux 2006).

The field of ecological oxidative stress research is still relatively new and assessments of select oxidative markers appear to be valuable in determining the effects of oxidative processes on the fitness of individual organisms (Beaulieu 2013). Results from these different exposure route studies suggest decreased antioxidant defenses and increased oxidative damage (via LPO) in targeted fish species when exposed to PAHs, regardless of exposure route. A review of findings on pollution in aquatic environments by Birnie-Gauvin et al. (2017) shows similar oxidative stress responses have been reported in other fish species. More controlled mechanistic studies in tandem with field studies are needed to better connect oxidative stress with PAH contaminants and understand the possible implications of oxidative stress on wild populations.

In fish, there have been numerous studies associating PAHs with reproduction disruption resulting in either sex hormone modifications (endocrine disruption) or by physiological degradation of gonads and egg development (Vignet et al. 2016). In regard to the DWH oil spill, too few studies were done assessing at reproduction; however those conducted did observe a decrease in egg development genes and in fecundity. Therefore, more studies are warranted to better understand the long-term consequences of DWH oil exposure to GoM fish.

An inherent challenge associated with impact assessments of oil spills and other catastrophic events comes from working largely at the sub-organismal level to predict impacts at the population and ecosystem levels. The focus on understanding organismal level effects can be attributed to experimental feasibility and high-resolution outcomes one can obtain in carefully controlled experiments. In contrast, studies of wild populations and their interactions are logistically challenging and most often include variables other than the environmental insult under study which can lead to confounding results. Ecosystem level impact studies are obviously of great importance and may be guided, in an effort to optimize use of resources, by careful considerations of relevant endpoints in controlled laboratory studies. Ideally, such laboratory studies will lead to testable predictions of responses by wild populations which may or, may not, be subsequently validated. The most fruitful approach to impact assessment is not either laboratory studies or field studies but rather a coordinated integrative approach where biological levels of organization ranging from molecular to integrative organismal responses through ecosystem level impacts are examined.

The use of broad mortality indices such as LC50 is appealing in the sense that they provide a discrete quantification of the effects of exposure that can be compared across species, life stages, ecosystems, and stressors. They are also easily used to calculate economic effects of contaminant releases, a necessity given the NRDA legal framework that often accompanies such a release. However, a focus on mortality indices alone can obscure the fact that real harm, at both individual and population levels, can be a result of exposure, but not manifest as overt mortality within the available or selected observation window. Harm to organisms or ecosystems that does not result in mortality is still, nonetheless, harm. Using cellular, molecular, or sublethal endpoints as discussed in this chapter can provide useful information about the potential sublethal effects of a given exposure on selected species and ecosystems than can be provided with mortality alone.

Acknowledgments This research was made possible by a grant from The Gulf of Mexico Research Initiative through the C-IMAGE and RECOVER consortia.

References

Ali AO, Hohn C, Allen PJ, Ford L, Dail MB, Pruett S, Petrie-Hanson L (2014) The effects of oil exposure on peripheral blood leukocytes and splenic melano-macrophage centers of Gulf of Mexico fishes. Mar Pollut Bull 79(1–2):87–93

Arias CR, Koenders K, Larsen AM (2013) Predominant bacteria associated with red snapper from the northern Gulf of Mexico. J Aquat Anim Health 25:281–289

Bayha KM, Ortell N, Ryan CN, Griffitt KJ, Krasnec M, Sena J, Ramaraj T, Takeshita R, Mayer GD, Schilkey F, Griffitt RJ (2017) Crude oil impairs immune function and increases susceptibility to pathogenic bacteria in southern flounder. PLoS One 12(5):1–21

Bianchini A, Grosell M, Gregory SM, Wood CM (2002) Acute silver toxicity in aquatic animals is a function of sodium uptake rate. Environ Sci Technol 36:1763–1766

Birnie-Gauvin K, Costantini D, Cooke SJ, Willmore WG (2017) A comparative and evolutionary approach to oxidative stress in fish: A review. Fish and Fisheries 18:928–942

Bosker T, van Balen L, Walsh B, Sepulveda MS, DeGuise S, Perkins C, Griffitt RJ (2017) The combined effect of macondo oil and corexit on sheepshead minnow (*Cyprinodon Variegatus*) during early development. J Toxicol Environ Health Part A Current Issues 80(9):477–484

Brette F, Machado B, Cros C, Incardona JP, Scholz NL, Block BA (2014) Crude oil impairs cardiac excitation-contraction coupling in fish. Science 343:772–776

Brewton RA, Fulford R, Griffitt RJ (2013) Gene expression and growth as indicators of effects of the BP Deepwater Horizon oil spill on spotted seatrout (*Cynoscion Nebulosus*). J Toxicol Environ Health Part A 76(21):1198–1209

Brown-Peterson NJ, Krasnec M, Takeshita R, Ryan CN, Griffitt KJ, Lay C, Mayer GD, Bayha KM, Hawkins WE, Lipton I, Morris J, Griffitt RJ (2015) A multiple endpoint analysis of the effects of chronic exposure to sediment contaminated with Deepwater Horizon oil on juvenile southern flounder and their associated microbiomes. Aquat Toxicol 165:197–209

Cox GK, Crossley DA, Stieglitz JD, Heuer RM, Benetti DD, Grosell M (2017) Oil exposure impairs in situ cardiac function in response to beta-adrenergic stimulation in cobia (*Rachycentron canadum*). Environ Sci Technol 51:14390–14396

Dubansky B, Whitehead A, Rice CD, Galvez F (2013) Multi-tissue molecular, genomic, and developmental effects of the Deepwater Horizon oil spill on resident Gulf killifish (*Fundulus Grandis*). Environ Sci Technol 47(10):5074–5082

Duffy TA, Childress W, Portier R, Chesney EJ (2016) Responses of bay anchovy (*Anchoa Mitchilli*) larvae under lethal and sublethal scenarios of crude oil exposure. Ecotoxicol Environ Saf 134:264–272

Edmunds RC, Gill JL, Baldwin DH, Lindo TL, French BL, Brown TL, Esbaugh AJ, Mager EM, Stieglitz J, Hoeing R, Benetti D, Grosell M, Scholz NL, Incardona JP (2015) Corresponding morphological and molecular indicators of crude oil toxicity to the developing hearts of mahi mahi. Sci Rep 5:17326

Esbaugh AJ, Mager EM, Stieglitz JD, Hoenig R, Brown TL, , French BL, Lindo TL, Lay C, Forth H, Scholz NL, Incardona JP, Morris JM, Benetti DD, Grosell M (2016) The effects of weathering and chemical dispersion on Deepwater horizon crude oil toxicity to mahi-mahi (*Coryphaena hippurus*) early life stages. Sci Total Environ 543:644–651

Finch BE, Marzooghi S, Di Toro DM, Stubblefield WA (2017) Phototoxic potential of undispersed and dispersed fresh and weathered Macondo crude oils to Gulf of Mexico marine organisms. Environ Toxicol Chem 36:2640–2650

Garcia TI, Shen Y, Crawford D, Oleksiak MF, Whitehead A, Walter RB (2012) RNA-seq reveals complex genetic response to Deepwater Horizon oil release in *Fundulus grandis*. BMC Genomics 13(1):1–9

Grosell M, Nielsen C, Bianchini A (2002) Sodium turnover rate determines sensitivity to acute copper and silver exposure in freshwater animals. Comp Biochem Physiol C-Toxicol Pharmacol 133:287–303

Hazen TC, Dubinsky EA, DeSantis TZ, Andersen GL, Piceno YM, Singh N, Jansson JK, Probst A, Borglin SE, Fortney JL, Stringfellow WT, Bill M, Conrad ME, Tom LM, Chavarria KL, Alusi TR, Lamendella R, Joyner DC, Spier C, Baelum J, Auer M, Zemla ML, Chakraborty R, Sonnenthal EL, D'haeseleer P, Holman H-YN, Osman S, Lu Z, Van Nostrand JD, Deng Y, Zhou J, Mason OU (2010) Deep-Sea oil plume enriches indigenous oil-degrading bacteria. Science 330:204–208

Incardona JP, Carls MG, Teraoka H, Sloan CA, Collier TK, Scholz NL (2005) Aryl hydrocarbon receptor-independent toxicity of weathered crude oil during fish development. Environ Health Perspect 113(12):1755–1762

Incardona JP, Day HL, Collier TK, Scholz NL (2006) Developmental toxicity of 4-ring polycyclic aromatic hydrocarbons in zebrafish is differentially dependent on AH receptor isoforms and hepatic cytochrome P4501A metabolism. Toxicol Appl Pharmacol 217(3):308–321

Incardona JP, Linbo TL, Scholz NL (2011) Cardiac toxicity of 5-ring polycyclic aromatic hydrocarbons is differentially dependent on the aryl hydrocarbon receptor 2 isoform during zebrafish development. Toxicol Appl Pharmacol 257(2):242–249

Incardona JP, Garder LD, Linbo TL, Brown TL, Esbaugh AJ, Mager EM, Stieglitz JD, French BL, Labenia JS, Laetz CA, Tagal M, Sloan CA, Elizur A, Benetti DD, Grosell M, Block BA, Scholz NL (2014) Deepwater Horizon crude oil impacts the developing hearts of large predatory pelagic fish. Proc Natl Acad Sci 111(15):E1510–E1518. https://doi.org/10.1073/pnas.1320950111

Iwanowicz LR, Blazer VS (2009) An overview of estrogen-associated endocrine disruption in fishes: evidence of effects on reproductive and immune physiology. Conference proceedings of the third bilateral conference between the United States and Russia, pp 266–275

Johansen JL, Esbaugh AJ (2017) Sustained impairment of respiratory function and swim performance following acute oil exposure in a coastal marine fish. Aquat Toxicol 187:82–89

Johansen JL, Allan BJM, Rummer JL, Esbaugh AJ (2017) Oil exposure disrupts early life-history stages of coral reef fishes. Nat Ecol Evol 1:1146–1152

Jones ER, Martyniuk CJ, Morris JM, Krasnec MO, Griffitt RJ (2017) Exposure to Deepwater Horizon oil and Corexit 9500 at low concentrations induces transcriptional changes and alters immune transcriptional pathways in sheepshead minnows. Comp Biochem Physiol Part D Genomics Proteomics 23(March):8–16

Khursigara AJ, Perrichon P, Martinez Bautista N, Burggren WW, Esbaugh AJ (2017) Cardiac function and survival are affected by crude oil in larval red drum, *Sciaenops ocellatus*. Sci Total Environ 579:797–804

Kostka JE, Prakash O, Overholt WA, Green SJ, Freyer G, Canion A, Delgardio J, Norton N, Hazen TC, Huettel M (2011) Hydrocarbon-degrading bacteria and the bacterial community response in Gulf of Mexico beach sands impacted by the Deepwater Horizon oil spill. Appl Environ Microbiol 77:7962–7974

Larsen AM, Bullard SA, Womble M, Arias CR (2015) Community structure of skin microbiome of Gulf killifish, *Fundulus grandis*, is driven by seasonality and not exposure to oiled sediments in a Louisiana salt marsh. Microb Ecol 70:534–544

Lee S, Hong S, Liu X, Kim C, Jung D, Yim UH, Shim WJ, Khim JS, Giesy JP, Choi K (2017) Endocrine disrupting potential of PAHs and their alkylated analogues associated with oil spills. Environ Sci Processes Impacts 19(9):1117–1125

Mager EM, Esbaugh AJ, Stieglitz JD, Hoenig R, Bodinier C, Incardona JP, Scholz NL, Benetti DD, Grosell M (2014) Acute embryonic or juvenile exposure to Deepwater Horizon crude oil impairs the swimming performance of Mahi-Mahi (*Coryphaena hippurus*). Environ Sci Technol 48(12):7053–7061

Magnuson J, Khursigara AJ, Allmon EB, Esbaugh AJ, Roberts AP (2018) Effects of Deepwater Horizon crude oil on ocular development in two estuarine fish species, red drum (*Sciaenops ocellatus*) and sheepshead minnow (*Cyprinodon variegatus*). Ecotoxicol Environ Saf 166:186–191

Mason OU, Scott NM, Gonzalez A, Robbins-Pianka A, Bælum J, Kimbrel J, Bouskill NJ, Prestat E, Borglin S, Joyner DC, Fortney JL, Jurelevicius D, Stringfellow WT, Alvarez-Cohen L, Hazen TC, Knight R, Gilbert J, Jansson JK (2014) Metagenomics reveals sediment microbial community response to Deepwater Horizon oil spill. Int Soc Microbial Ecol J 8:464–475

Mitchelmore CL, Bejarano AC, Wetzel DL (2020) A synthesis of Deepwater Horizon oil, chemical dispersant and chemically dispersed oil aquatic standard laboratory acute and chronic toxicity studies (Chap. 28). In: Murawski SA, Ainsworth C, Gilbert S, Hollander D, Paris CB, Schlüter M, Wetzel D (eds) Deep oil spills – facts, fate and effects. Springer

Murawski SA, Hogarth WT, Peebles EB, Barbeiri L (2014) Prevalence of external skin lesions and polycyclic aromatic hydrocarbon concentrations in Gulf of Mexico fishes, post-Deepwater Horizon. Trans Am Fish Soc 143(4):1084–1097

McKim JM (1985) Early life stage toxicity tests. In: Rand GM (ed). (1995) Fundamentals of aquatic toxicology: effects, environmental fate, and risk assessment, 2nd edn. CRC press, Boca Raton, pp 974–1010

Nelson D, Heuer RM, Cox GK, Stieglitz JD, Hoenig R, Mager EM, Benetti DD, Grosell M, Crossley D (2016) Effects of crude oil on in situ cardiac function in young adult mahi-mahi (*Coryphaena hippurus*). Aquat Toxicol 180:274–281

Nelson D, Stieglitz JD, Cox GK, Heuer RM, Benetti DD, Grosell M, Crossley DA (2017) Cardio-respiratory function during exercise in the cobia, *Rachycentron canadum*: the impact of crude oil exposure. Comparative Biochemistry and Physiology C-Toxicology & Pharmacology 201:58–65

O'Shaughnessy KA, Forth H, Takeshita R, Chesney EJ (2018) Toxicity of weathered Deepwater Horizon oil to bay anchovy (*Anchoa mitchilli*) embryos. Ecotoxicol Environ Saf 148:473–479

Pasparakis C, Mager EM, Stieglitz JD, Benetti D, Grosell M (2016) Effects of Deepwater Horizon crude oil exposure, temperature and developmental stage on oxygen consumption of embryonic and larval Mahi-Mahi (*Coryphaena hippurus*). Aquat Toxicol 181:113–123

Pilcher W, Miles S, Tang S, Mayer G, Whitehead A (2014) Genomic and genotoxic responses to controlled weathered-oil exposures confirm and extend field studies on impacts of the Deepwater Horizon oil spill on native killifish. PLoS One 9(9):e106351

Portnoy DS, Fields AT, Greer JB, Schlenk D (2020) Genetics and oil: transcriptomics, epigenetics and population genomics as tools to understand animal responses to exposure across different time scales (Chap. 30). In: Murawski SA, Ainsworth C, Gilbert S, Hollander D, Paris CB, Schlüter M, Wetzel D (eds) Deep oil spills: facts, fate, effects. Springer, Cham

Pulster EL, Main K, Wetzel D, Murawski S (2017) Species-specific metabolism of naphthalene and phenanthrene in 3 species of marine teleosts exposed to Deepwater Horizon crude oil. Environ Toxicol Chem 36(11):3168–3176

Raimondo S, Hemmer BL, Lilavois CR, Krzykwa J, Almario A, Awkerman JA, Barron MG (2016) Effects of Louisiana crude oil on the sheepshead minnow (*Cyprinodon variegatus*) during a life-cycle exposure to laboratory oiled sediment. Environ Toxicol 31:1627–1639

Rauta PR, Nayak B, Das S (2012) Immune system and immune responses in fish and their role in comparative immunity study: a model for higher organisms. Immunol Lett 148(1):23–33

Reynaud S, Deschaux P (2006) The effects of polycyclic aromatic hydrocarbons on the immune system of fish: a review. Aquat Toxicol 77(2):229–238

Reynaud S, Raveton M, Ravanel P (2008) Interactions between immune and biotransformation systems in fish: a review. Aquat Toxicol 87(3):139–145

Rowsey LE, Johansen JL, Khursigara AJ, Esbaugh AJ (2019) Oil exposure impairs predator–prey dynamics in larval red drum (Sciaenops ocellatus). Mar Freshw Res. https://doi.org/10.1071/MF18263

Santana MS, Sandrini-Neto L, Filipak Neto F, Oliveira Ribeiro CA, Di Domenico M, Prodocimo MM (2018) Biomarker responses in fish exposed to polycyclic aromatic hydrocarbons (PAHs): systematic review and meta-analysis. Environ Pollut 242:449–461

Schirmer K, Fischer BB, Madureira DJ, Pillai S (2010) Transcriptomics in ecotoxicology. Anal Bioanal Chem 397:917–923

Starostenko LV, Rechkunova NI, Lebedeva NA, Lomzov AA, Koval VV, Lavrik OI (2017) Processing of the abasic sites clustered with the benzo[a]pyrene adducts by the base excision repair enzymes. DNA Repair 50:43–53

Stieglitz JD, Mager EM, Hoenig RH, Benetti DD, Grosell M (2016) Impacts of Deepwater Horizon crude oil exposure on adult mahi-mahi (*Coryphaena hippurus*) swim performance. Environ Toxicol Chem 35:2613–2622

van der Oost R, Beyer J, Vermeulen NP (2003) Fish bioaccumulation and biomarkers in environmental risk assessment: a review. Environ Toxicol Pharmacol 13(2):57–149

Vega-Retter C, Rojas-Hernandez N, Vila I, Espejo R, Loyola DE, Copaja S, Briones M, Nolte AW, Veliz D (2018) Differential gene expression revealed with RNA-Seq and parallel genotype selection of the ornithine decarboxylase gene in fish inhabiting polluted areas. Sci Rep 8(1):4820

Vignet C, Larcher T, Davail B, Joassard L, Le Menach K, Guionnet T, Lyphout L., Ledevin M, Goubeau M, Budzinski H, Bégout ML, Cousin X (2016) Fish reproduction is disrupted upon lifelong exposure to environmental PAHs fractions revealing different modes of action. Toxics 4(4):26

Whitehead A, Dubansky B, Bodinier C, Garcia TI, Miles S, Pilley C, Raghunathan V, Roach JL, Walker N, Walter RB, Rice CD, Galvez F (2012) Genomic and physiological footprint of the Deepwater Horizon oil spill on resident marsh fishes. Proc Natl Acad Sci 109(50):20298–20302

Xu EG, Mager EM, Grosell M, Pasparakis C, Schlenker LS, Stieglitz JD, Benetti D, Hazard ES, Courtney SM, Diamante G, Freitas J, Hardiman G, Schlenk D (2016) Time- and oil-dependent transcriptomic and physiological responses to Deepwater Horizon oil in Mahi-Mahi (*Coryphaena hippurus*) embryos and larvae. Environ Sci Technol 50(14):7842–7851

Xu EG, Mager EM, Grosell M, Hazard ES, Hardiman G, Schlenk D (2017a) Novel transcriptome assembly and comparative toxicity pathway analysis in mahi-mahi (*Coryphaena hippurus*) embryos and larvae exposed to Deepwater Horizon oil. Sci Rep 7:44546

Xu EG, Khursigara AJ, Magnuson J, Hazard ES, Hardiman G, Esbaugh AJ, Roberts AP, Schlenk D (2017b) Larval red drum (*Sciaenops ocellatus*) sublethal exposure to weathered Deepwater Horizon crude oil: developmental and transcriptomic consequences. Environ Sci Technol 51:10162–10172

Xu EG, Magnuson JT, Diamante G, Mager E, Pasparakis C, Grosell M, Roberts AP, Schlenk D (2018) Changes in the global micro-mRNA signatures agree with morphological, physiological and behavioral changes in larval mahi-mahi (*Coryphaena hippurus*) treated with Deepwater Horizon oil. Environ Sci Technol (Under review)

Xue W, Warshawsky D (2005) Metabolic activation of polycyclic and heterocyclic aromatic hydrocarbons and DNA damage: a review. Toxicol Appl Pharmacol 206(1):73–93

Zhang L, Nichols R, Patterson A (2018) The aryl hydrocarbon receptor as a moderator of host-microbiota communication. Current Opinion Toxicology 2:30–35

Zhang Y, Dong S, Wang H, Tao S, Kiyama R (2016) Biological impact of environmental polycyclic aromatic hydrocarbons (ePAHs) as endocrine disruptors. Environ Pollut 213:809–824

Chapter 30
Genetics and Oil: Transcriptomics, Epigenetics, and Population Genomics as Tools to Understand Animal Responses to Exposure Across Different Time Scales

David S. Portnoy, Andrew T. Fields, Justin B. Greer, and Daniel Schlenk

Abstract Exposure to oil causes basic organismal responses rooted in the genetic architecture of specific species. At the most basic level, exposure leads to immediate changes in gene expression related to xenobiotic metabolism but also changes in expression that may lead to impairment of important biological processes and abnormal development. While many responses amount to transient changes in function, some related to epigenetic gene regulation may be more permanent, persisting long after cessation of exposure. Further, some epigenetic change may be heritable, with oil-like responses persisting in later generations that were never directly exposed. Finally, exposure can create mass-mortality events that change contemporary levels of genomic variation and/or lead to selective regimes that favor individuals that carry allelic variants making them more resistant to the negative effects of oil exposure. The latter issue may be compounded if low levels of oil remain in the system leading to chronic exposure and continued selection. This chapter will review molecular techniques and "omics" tools that have allowed researchers a better understanding of the genetic underpinnings of organismal response to oil exposure, while highlighting future directions for this type of research.

Keywords Gene expression · Gene regulation · Selection · Drift · Sublethal effects

D. S. Portnoy (✉) · A. T. Fields
Texas A&M University-Corpus Christi, Department of Life Sciences,
Corpus Christi, TX, USA
e-mail: david.portnoy@tamucc.edu; andrew.fields@tamucc.edu

J. B. Greer · D. Schlenk
University of California, Department of Environmental Sciences, Riverside, CA, USA
e-mail: jgreer@ucr.edu; dschlenk@ucr.edu

© Springer Nature Switzerland AG 2020
S. A. Murawski et al. (eds.), *Deep Oil Spills*,
https://doi.org/10.1007/978-3-030-11605-7_30

30.1   Introduction

When organisms are exposed to oil, a number of subcellular responses occur to facilitate excretion of toxins and ensure maintenance of homeostasis. Metabolic pathways are induced to make xenobiotic compounds water soluble, facilitating excretion, but in the process often produce more bioactive toxic species (Schlenk et al. 2008a). Therefore, organisms may also mount responses meant to protect basic functioning and minimize direct DNA damage (Shimada 2006). These mechanisms are well suited to handle short-term response to low levels of contaminants; however, long-term (chronic) exposure and/or high concentrations of oil may over-burden pathways leading to the accumulation of bioactive species in tissues (Wills et al. 2009). Additionally, transcription factors involved in xenobiotic metabolism may disrupt processes important for normal cellular functioning (Androutsopoulos et al. 2009). Chronically exposed organisms, therefore, may differ in response to oil exposure due to heritable changes in gene regulation, population-level changes in gene frequencies, or a combination of these changes to genes that are involved in xenobiotic metabolism.

Because both immediate- and long-term responses to oil are a function of gene expression, gene regulation, and ultimately gene content, modern molecular techniques offer a set of tools for detecting exposure, understanding causes of lethal and sublethal responses, and assessing population-level effects. These techniques can be used to predict the long-term resilience of populations and the potential for residual oil in the system following a spill to cause permanent change in the genetic make-up of organisms, a phenomenon called anthropogenic evolution. In this chapter, we discuss three fields of molecular study, transcriptomics (gene expression), epigenetics (gene regulation), and population genomics (gene content within species), highlighting the latest advancements in the field, discussing what we have learned from the application of these techniques to exposure research, and suggesting future directions for research.

30.2   Transcriptomics

Transcriptomic studies use molecular technology to characterize gene expression in whole organisms and/or specific tissues and can be used to assess changes in gene expression among organisms in different experimental treatments or naturally occurring environments (Conesa et al. 2016). Early transcriptomic studies in the mid-1990s utilized microarrays to measure the abundance of several thousand targeted transcripts of interest at once via hybridization to complementary probes (Schena et al. 1995). In recent years, high-throughput transcriptome sequencing (RNASeq) has allowed researchers to identify gene expression changes without having to target specific genes. An advantage of non-targeted approaches, such as

RNASeq, is that they allow for non-biased assessments of all gene expression in an organism, which is important for fully understanding the range of subcellular responses an organism may have under a variety of conditions, including exposure to toxins. Consistent improvements in cost, versatility, and analysis methods have made RNASeq the current choice for transcriptomics studies.

For species with sequenced genomes, RNASeq reads can be aligned or "mapped" to an annotated genome to identify expressed transcripts. For non-model species without fully sequenced genomes, de novo assembly of reads can be used to create a transcriptome in a species of interest (methods reviewed in Conesa et al. 2016 and Chen et al. 2017), and/or a comparative genomics approach can be used to map reads to the genomes of several closely related taxa (O'Leary et al. 2018). After mapping, transcripts can be quantified and annotated using a variety of methods (e.g., Li and Dewey 2011; Anders et al. 2014; Patro et al. 2015) to identify log-fold changes and differentially expressed genes at a user-defined p-value (Robinson et al. 2010; Love et al. 2014). The characterization of differential gene expression and the processes/pathways they are involved in among experimental treatments can be used to identify and understand aspects of subcellular response to xenobiotic compounds such as oil.

Following the Deepwater Horizon (DWH) oil spill in the Gulf of Mexico, several whole-transcriptome studies assessed the biological response of native fish species to oil exposure. Impacts on early developmental stages have been studied in fish including mahi-mahi, *Coryphaena hippurus*; red drum, *Sciaenops ocellatus*; and haddock, *Melanogrammus aeglefinus* (Xu et al. 2016a, 2017; Sørhus et al. 2017). Most studies in adult fish used the Gulf killifish, *Fundulus grandis*, or mummichog, *Fundulus heteroclitus* (Whitehead et al. 2010, 2012a). The results of these studies confirmed previously identified response mechanisms to oil but in addition have identified novel changes in gene expression. First, this section will focus on the transcriptomic effects of oil on activation of aryl hydrocarbon receptor (AHR)-mediated pathways. The AHR pathway is a well-characterized response mechanism of oil exposure that contributes to cardiac malformations, including poor looping, arrhythmia, and reduced contractility (Incardona 2017). Secondly, this section will discuss emerging evidence that non-AHR-pathway gene expression changes also are a prevalent component of oil-induced organismal response.

One of the primary transcriptional responses to oil exposure is polycyclic aromatic hydrocarbon (PAH)-induced activation of aryl-hydrocarbon receptors (AHR) and their downstream response genes, known as the AHR pathway (Wirgin and Waldman 2004; Oleksiak et al. 2011). The AHR pathway regulates detoxification of planar aromatic hydrocarbons such as PAHs via activation of xenobiotic metabolism and stress-response pathways. Binding of PAHs to AHRs induces transcriptional changes in the AHR pathway. For example, activation of phase I enzymes such as cytochrome p450s (CYP) converts PAHs to intermediates, which are further processed by phase II enzymes into compounds that can be safely and easily eliminated from the body. Persistent overactivation of genes in the AHR pathway during prolonged oil exposure is related to developmental abnormalities, decreased hatching success, and decreased embryonic and larval survival (Whitehead et al. 2010, 2011; Lanham et al. 2014).

Activation of the AHR pathway has been consistently observed in developing embryos and larvae exposed to oil. When fertilized mahi-mahi embryos were exposed to slick oil from the DWH oil spill, *cyp1a1* was the most strongly upregulated gene at 24, 48, and 96 hours post fertilization (hpf), and downstream AHR pathway genes such as *cyp1b1*, udp-glucuronosyltransferase (*ugt*), and aryl-hydrocarbon receptor repressor (*ahrr*) were also upregulated (Xu et al. 2016a). In fact, a total of 127 AHR-responsive genes were differentially expressed at 96 hpf (Ibid); *cyp1a1*, *ahrr*, and *cyp3a4* were also among the most highly upregulated genes in slick-oil-exposed red drum embryos; and gene ontology analysis predicted significant activation of AHR pathways at both 24 and 72 hpf (Xu et al. 2017). Oil-exposed Atlantic haddock embryos also showed upregulation of many stress response genes mediated by the AHR pathway at all time points from 2 days post fertilization to 10 days post hatch (Sørhus et al. 2017).

Induction of AHR and stress response pathways also occurs in adult fish. In a field study of Gulf killifish, livers from fish at sites that had measurable PAH concentrations showed upregulation of *ahr1b* and *ahr2* receptors compared to non-oiled sites (Garcia et al. 2012), as well as upregulation of *cyp1a1* in the gills. Other AHR pathway genes such as cytochrome p450s and *ugt*s were also upregulated in the livers of killifish from contaminated sites. Upregulation of *ugt* has been observed in other *Fundulus* studies and in the gills of juvenile Japanese flounder, *Paralichthys olivaceus* (Whitehead et al. 2012a; Pilcher et al. 2014; Zhu et al. 2016). HEWAF exposed Japanese flounder also showed significant effects on stress genes related to AHR pathway activation (Zhu et al. 2016). Together, these studies have shown that upregulation of AHR-mediated genes, in particular *cyp1a1*, is consistent transcriptomic responses to oil in many species and life stages and supports previous research indicating the use of CYP as a biomarker for field studies assessing oil (Adams 1990; Schlenk et al. 2008b).

Transcriptomic studies performed in larval fish have also provided evidence that changes in gene expression not associated with the AHR pathway may in part explain abnormalities during early development. Altered cardiac gene expression, particularly for genes involved in intracellular Ca^{2+} homeostasis and excitation-contraction coupling, may directly contribute to cardiotoxicity in developing embryos and larvae. In mahi-mahi embryos at 48 and 96 hpf, transcripts of calsequestrins, which are involved in Ca^{2+} homeostasis, were upregulated after slick oil exposure (Xu et al. 2016a). Downregulation of T-type receptor 2 (*tnnt2*) transcripts was also observed. *Tnnt2* regulates intracellular Ca^{2+} concentrations (Metzger and Westfall 2004). Diminished expression of transcripts encoding L-type and P/Q-type voltage-dependent calcium channels was also observed following oil exposure, further suggesting alterations in Ca^{2+} signaling (Xu et al. 2016a). Similarly, in haddock, transcripts of the cardiac sarcoplasmic reticulum Ca^{2+}ATPase (*atp2a2*) and *ryr2*, a ryanodine receptor that mediates cellular calcium release, were downregulated, as was the gene encoding the delayed rectifying K^+ channel, *kcnh2*, which is important for repolarization of ventricular cells (Sørhus et al. 2017). Only diminished expression of *kcnh2* correlated with the cardiac phenotype in haddock. Subsequent studies by Diamante et al. (2017) indicated that the *kcnh2* regulating microRNA (miRNA) 133a was induced in mahi-mahi embryos exposed to oil, indicating dysregulation of miRNA may play a role in the downregulation of *kcnh2*. These data are consistent with evidence that disruption of Ca^{2+} and K^+ fluxes contribute to observed cardiotoxicity by disrupting excitation-contraction coupling in cardiomyocytes of bluefin tuna (Brette et al. 2014). Overexpression of other cardiac genes associated with severe heart defects (*bmp10*, *nkx25*, and *tbx3*) was also observed (Edmunds et al. 2015). Gene ontology categories for perturbations in cardiac function were altered at nearly all time points in mahi-mahi, red drum, and haddock (Xu et al. 2016a, 2017; Sørhus et al. 2017). Together, these studies have made it apparent that impaired cardiac function resulting from oil exposure is related to gene expression changes of both AHR- and non-AHR-mediated pathways.

Transcriptional changes not connected to AHR pathway genes may also underlie other oil-related morphological abnormalities, such as reduced eye area and craniofacial/jaw malformations. These abnormalities previously had been attributed to secondary effects of reduced circulation; however, some may be directly altered irrespective of cardiac changes (Incardona 2017). In haddock, expression of 12 genes with roles in craniofacial development were significantly altered, which corresponds to reduced eye area and underdevelopment of the jaw (Sørhus et al. 2017). In mahi-mahi, 35 of the 70 genes associated with visual perception were differentially expressed in 96 hpf exposed larvae (Xu et al. 2016a), and reductions in eye size and visual acuity were observed (Xu et al. 2018). Similar abnormalities in eye development following oil exposure were seen in red drum larvae (Magnuson et al. 2018; Xu et al. 2019). The resulting deficiencies in vision or the ability to feed would have direct impacts on survival in the wild. Other abnormalities that might affect survival include increased yolk size, corresponding to altered cholesterol biosynthesis (Xu et al. 2016a; Sørhus et al. 2017), which may reduce yolk absorption and result in underdeveloped embryos that are more susceptible to predation during the larval phase. Additionally, impaired cholesterol biosynthesis may also alter membrane structure and contribute to cardiac toxicity. Other predicted organ level effects of oil based on transcriptomic analyses were focused on nervous system degeneration (Xu et al. 2016a, 2017; Sørhus et al. 2017). Changes in expression of genes involved in nervous system development and functioning were identified, and brain size was significantly reduced in red drum larvae exposed to oil (Xu et al. 2017).

Overall, transcriptomic studies following exposure to oil have been instrumental in assessing the basic biological mechanisms of oil toxicity on species in the Gulf of Mexico. These studies have confirmed previously known pathways of toxicity but also shed light on novel mechanisms. Upregulation of AHR and its downstream stress response gene products is a common biological response across many fish species. In particular, *cyp1a1* and *ugt* were consistently upregulated, which supports the usage of them as biomarkers, though they are not specific to oil exposure (Nilsen et al. 1998; Hilscherova et al. 2000). Transcription of genes not related to the AHR pathway, such as those involved in cardiac and eye development, as well as nervous system function, has been found to change in oil-exposed individuals, leading to adverse outcomes. The summation of these changes in gene expression can lead to phenotypic defects and increased mortality of fishes exposed to oil.

30.3 Epigenetics

Epigenetics is the study of heritable modifications to the DNA molecule that affect the DNA's availability, but are not base changes, and as such are reversible. These changes, passed from mother cells to daughter cells, alter gene expression by inhibiting protein binding but also may reduce DNA damage by reducing availability of the DNA strand to things like reactive oxidative species (Ljungman and Hanawalt 1992; Takata et al. 2013) and ionizing radiation (Takata et al. 2013). Epigenetic changes include chromatin remodeling, such as histone modification, removal, replacement, or addition; DNA modifications, such as the addition of a methyl group to the carbon ring of cytosine; and non-coding RNA-mediated gene silencing. These modifications are often called epigenetic "tags" and serve important roles in cell line differentiation during development and phenotype-environment interactions, and are not only heritable down cell lines but can be passed from parent to offspring (Reik 2007; Labbé et al. 2017).

Many methods have been developed to assess and characterize epigenetic alterations. Patterns of DNA methylation are assessed by determining the methylation state of cytosines either in specific targeted areas of interest or on genome-wide scales. Instrumentation used for analyses ranges from high-performance liquid chromatography with an ultraviolet visible absorbance detector (HPLC-UV), to methylome sequencing, to enzyme-linked immunosorbent assay (ELISA; see Kurdyukov and Bullock 2016 for a review). The location of chromatin remodeling can be assessed through chromatin immunoprecipitation (ChIP) assays and can be complimented with DNA analysis methods such as quantitative PCR (qPCR), microarray, or bisulfite treatment of DNA followed by sequencing (see Das et al. 2004 and DeAngelis et al. 2008 for reviews).

When an organism is exposed to an environmental stimulus, cells may alter gene expression to account for the altered environment. During this process, epigenetic mechanisms can be used to modify gene expression (Kanherkar et al. 2014). Changes in epigenetic tags in one or a few cells can lead to rapid cascading effects when these changed cells export products that enter other cells and alter their epigenetic profile (Peschansky and Wahlestedt 2014). In this way, large-scale epigenetic changes can occur in hours (e.g., embryo reprogramming; see Jiang et al. 2013; Potok et al. 2013), but the results can last for a lifetime (e.g., Knecht et al. 2017).

While the majority of epigenetic marks are erased during gametogenesis and embryo development, some parental marks may persist across multiple generations (reviewed in Heard and Martienssen 2014; Skinner 2014; Hanson and Skinner 2016; Labbé et al. 2017). The genomic location of a mark, GC profile of the chromosomal region, mode of inheritance, and marker type may influence the likelihood of a mark persisting from germline cell through embryogenesis (Hackett et al. 2013; Tang et al. 2015; Szyf 2015; Blake and Watson 2016).

One stimulus that has been found to alter epigenetic state is toxin or toxicant exposure. While exposure to toxins creates a xenobiotic response, as described in the previous section, there are alterations to the epigenome that may also be important. For example, when zebrafish are exposed to 2,3,7,8-tetrachlorodibenzo-p-dioxin (TCDD), genes in the AHR pathway are upregulated (Handley-Goldstone et al. 2005; Carney et al. 2006). Aluru et al. (2015) found that TCDD exposure caused changes in methylation state for two out of the three AHR pathway promoter regions in concert with mRNA induction. There are many other examples of environmental toxins causing epigenetic alteration including changes in the methylation state of the *p53* and *p16* genes in humans correlated with arsenic exposure (Chanda et al. 2006), hypomethylation of the *Igf2r*, *Peg3*, and *H19* genes in mice exposed to the estrogen-like chemical diethylhexyl phthalate (Li et al. 2014), and altered global methylation patterns in insects and humans exposed to insecticides (Collotta et al. 2013; Oppold and Müller 2017).

Only one study to date has analyzed the epigenetic state of any wild organism with respect to the Deepwater Horizon (DWH) oil spill. Robertson et al. (2017) found that there were five differentially methylated loci between oil-exposed and non-oiled *Spartina alterniflora*, a salt marsh grass, but the genes associated with these loci were not identified. Thirty-four other differentially methylated loci between sampling sites (but not related to oil exposure) were also identified, so follow-up work is needed to determine if the differential methylation was caused by oil exposure. Other studies exist, however, that examine oil effects on fauna, environmental influence on epigenetic state, and transgenerational transmission of epigenetic marker. Oil-exposed fish embryos experience cardiac edema, yolk sac edema, skeletal abnormalities, fin fold abnormalities, and decreased growth (McEachern 2014; Xu et al. 2016b; Diamante et al. 2017). Transcriptional analysis indicates that many of these characteristics are associated with differential gene regulation during development (Xu et al. 2016b; Diamante et al. 2017). Exposure of adult fish to oil indicates decreased swimming performance, possibly due to alterations in the cardiovascular system (Stieglitz et al. 2016). There are currently no studies assessing whether phenotypic alterations are related to epigenetic modifications that could be passed on to future generations. There are also no known epigenetic marks that would function as biomarkers indicative of oil exposure though a wealth of candidate genes are suggested by transcriptional analysis. Knecht et al. (2017) did find global hypomethylation after benzo[a]pyrene (BaP) exposure, but no specific locations were analyzed, and F_2 generations were not tested for altered DNA methylation patterns.

When translating laboratory studies to the wild, it is imperative to consider that while some observed phenotypes in the laboratory would be fatal in the natural environment because of decreased performance, alterations to epigenomes are not likely to be lethal and therefore could serve as biomarkers of past oil exposure. For instance, Corrales et al. (2014) exposed zebrafish to varying levels of BaP, a known PAH contaminant, but did not observe altered phenotypes in the exposed generation (F_0); however, the F_1 and F_2 generations experienced increased mortality and higher incidence of deformities. Follow-up studies found that larvae exposed to BaP had neural degeneration (Gao et al. 2017; Knecht et al. 2017) and DNA methylation differences between the control and exposed zebrafish were found in several promoter regions of genes involved in neural development in adult brain tissue (Gao et al. 2017). Studies in human epigenetics indicate that some neural development genes are not reset during the demethylation of primordial germ cells (Tang et al. 2015), suggesting that there could be transgenerational BaP effects in wild fishes. Epigenetic biomarkers of exposure, discovered in controlled laboratory experiments, would be useful for monitoring wild populations for contaminant exposure and potential population effects. With chronic oil exposure in the Gulf of Mexico due to human and natural inputs (Murawski et al. 2014), it is very likely that some epigenetic marker diagnostic of exposure may already be present in Gulf of Mexico organisms. Further experimentation will be required to understand differences in biological response to acute exposures and long-term exposures of both naturally occurring and human-introduced hydrocarbons.

30.4 Population Genomics

Genomics in the context of this discussion refers to studies focused on characterizing the content of entire genomes (or reduced-representation genomes). In the past, such studies were cost prohibitive because of the inefficiency of Sanger sequencing, which relies on chain-terminating dideoxynucleotides. Massively parallel sequencing technologies (next-generation sequencing) have increased efficiency by allowing the simultaneous sequencing of millions of small fragments (Behjati and Tarpey 2013). The relative affordability and accessibility of these technologies has made genomics approaches practical for a variety of researchers investigating both model and non-model organisms alike.

At the most basic level, genomic studies focus on sequencing entire genomes from individual organisms and annotating functional portions of those genomes. Comparative genomic studies involve comparisons of genomic content across taxa to look for common aspects of genomic architecture evolution and function and to identify important differences (ICGSC et al. 2004; Lindblad-Toh et al. 2005). Both basic and comparative genomics studies are useful for understanding organismal responses to oil exposure at the subcellular level (e.g., Wirgin and Waldman

2004; Hahn et al. 2006; Oleksiak et al. 2011), contributing important background information for the types of transcriptomic and epigenetic studies discussed previously. Population genomic studies, the focus of this section, compare genomic content among individuals within the same species to understand how microevolutionary processes (drift, migration, selection, and mutation) interact to produce observed patterns of contemporary genomic variation and consider variants like single nucleotide polymorphisms (SNPs), insertions and deletions (indels), and larger chromosomal changes (e.g., inversions, recombination, duplications, etc.). Population genomics is an important field of study for understanding organismal response to oil spills, because exposure to toxins can lead to direct changes at the nucleotide level via mutation and changes in the frequency of alleles at the population level due to increased drift, changing levels and/or patterns of migration (population connectivity), and selection for resistant genotypes. These changes can affect individual organismal functioning, while having population-level effects (Belfiore and Anderson 2001; van Straalen and Timmermans 2002). Changes resulting from microevolutionary processes can be broadly divided into those that are manifest throughout the genome and those that influence only certain areas in the genome.

Chromosomal aberrations and point mutations associated with mutagenic properties of the components of crude oil (Pashin and Bakhitova 1979; Al-Sabti 1985) are changes at the sequence level that are specific to individuals and occur at specific sites or regions of chromosomes, which if replicated can become permanent. When affected sites/regions impact coding regions or regulatory elements, basic organismal functioning may be impaired, and negative health outcomes may result (Haensly et al. 1982; Wirgin and Waldman 1998). If mutations occur in germline cells or during early development, progeny may perform poorly or fail to develop properly (Frank 2010), and this can lead to population-level effects by decreasing the reproductive output of populations (Carls et al. 1999). While individual chromosomal aberrations are site specific, populations exposed to toxins, including oil, have been shown in some cases to present increased variation associated with accelerated mutation rates (e.g., oil, Klekowski et al. 1994; heavy metals and radiation, Eeva et al. 2006; radiation, Volkova et al. 2018). Alternatively, populations may lose genetic variation after exposure as deleterious and lethal mutations accumulate and interact with other microevolutionary processes (Bickham et al. 2000).

When populations of organisms experience mass-mortality events, the number of reproductive individuals decreases, and this can lead to reductions in effective size (N_E) that may persist across generations, a phenomenon often referred to as a bottleneck. Because the rate by which drift changes allele frequencies is inversely proportional to N_E, bottlenecks can lead to decreases in genetic diversity (Nei et al. 1975). Drift impacts the entire genome, so reductions in genetic diversity associated with bottlenecks are genome-wide phenomena, affecting neutral (non-coding) markers and coding regions as well. The latter is of concern because decreased variation at coding regions may equate to decreased adaptive variation (genetic erosion) that negatively impacts the resilience of organisms/populations (Bijlsma and Loeschcke 2011). The magnitude of diversity loss caused by a bottleneck is not only related to N_E but also to the length of time in generations that populations persist at small N_E and the amount of gene flow between populations (Nei et al. 1975). Exposure to contaminants can negatively affect connectivity (Puritz and Toonen 2011) causing a feedback loop between decreasing population size, decreasing connectivity (gene flow), and decreasing fitness that can lead to rapid extirpation (Frankham et al. 2014). This means that populations that are small and isolated to begin with are at greater risk of genetic erosion following an acute oil exposure. Direct assessment of horsefly populations following the DWH oil spill suggests that such a dynamic might be occurring; populations in heavily oiled habitat showed evidence of bottlenecks, reduced migrant rates, and reductions in the number of individuals (Husseneder et al. 2016). For red snapper and golden tilefish, evidence for decreased genome-wide variation is lacking, but shifts in allele frequencies at specific loci in areas impacted by oil have been observed (O'Leary pers. comm.). Because these species have long generation times and high potential connectivity, drift is expected to be weak, suggesting that observed changes in allele frequencies at specific loci may reflect losses of diversity caused by non-representative gamete sampling, resulting from high mortality and/or intense selection (Schiebelhut et al. 2018).

Prolonged selection pressure caused by human activities can lead to rapid phenotypic and underlying genetic change in wild populations, a phenomenon sometimes referred to as anthropogenic evolution (Palumbi 2001). Common examples of anthropogenic evolution often involve "resistance phenotypes" that evolve in response to toxin exposure; examples include plant resistance to heavy metals (Antonovics et al. 1971), microbial resistance to antibiotics (Andersson and Hughes 2010), and insect resistance to pesticides (Whalon et al. 2008). Because individual organisms have a range of tolerance to oil exposure, related to genetically determined phenotypes, both acute and chronic exposure can cause directional selection favoring resistance to oil-based contaminants. Phase I reactions of the AHR pathway, the first step in the xenobiotic metabolism of PAHs, may bioactivate or increase the toxicity of compounds (Denison and Nagy 2003; Harrigan et al. 2004; Shimada 2006; Shiizaki et al. 2017). Thus, activation of the AHR pathway itself presents a challenge to organisms, and prolonged activation can be a selective agent (Whitehead et al. 2017). Fishes living in environments heavily and/or chronically contaminated with industrial toxins show heritable resistance to levels of contaminants that are normally lethal and an associated repression of the AHR pathway (Wirgin et al. 2011; Oziolor et al. 2014; Di Giulio and Clark 2015). When analyzed at the nucleotide level, patterns consistent with strong, directional selection have been detected in specific regions of the genomes associated with the AHR pathway in killifish (Whitehead et al. 2012b; Nacci et al. 2016). Because of the polygenic nature of resistance, tolerant populations of killifish show distinctive variants at the same sets of genes involved in AHR response but also show evidence of selection at population-specific genomic locations associated with pollution tolerance as well (Reid et al. 2016). Taken as a whole, these studies suggest that tolerance to chronic exposure can evolve rapidly and repeatedly in these systems but requires adequate potential adaptive variation to begin with (Whitehead et al. 2017), a result that emphasizes the need to maintain/monitor genomic diversity to promote resilience in wild populations. Because of the AHR pathway's involvement in numerous fundamental cellular processes, resistance may come with a cost to absolute fitness and could potentially adversely affect populations by impacting fecundity, changing behavior, decreasing growth, and/or impairing the immune responses (Meyer and Di Giulio 2003; Frederick et al. 2007). Studies of copepods and Gulf killifish after the DWH oil spill have noted functional changes, similar to those seen in chronically impacted populations (Whitehead et al. 2012b; Lee et al. 2017), but long-term study will be needed to determine if these are permanent evolutionary changes.

30.5 Conclusions

Studies involving transcriptomics, epigenetics, and population genomics can provide information on organismal response to exposure across time scales from immediate to generational. These techniques compliment laboratory-based exposure experiments designed at evaluating lethality and impairment following exposure and provide a link for understanding how organismal response and population-level response are related. An important legacy of DWH oil spill research will be an increased understanding of the types of impacts acute oil exposure has on organisms in controlled laboratory settings, and molecular studies have been a vital component of that research. However, the results from laboratory experiments still need to be compared to observations made in wild populations across a variety of taxa. These field-based follow-up studies, which are ongoing, will not only provide valuable information on population response and resiliency but should aid in the development of novel biomarkers of oil exposure. Further, anthropogenic selection pressure, including exposure to hydrocarbons, has become omnipresent across ecosystems, and continued studies are needed to assess and understand patterns of evolutionary change, especially in living resources upon which humans depend.

Funding Information This research was made possible by grants from the Gulf of Mexico Research Initiative through its consortia: The Center for the Integrated Modeling and Analysis of the Gulf Ecosystem (C-IMAGE) and Relationships of Effects of Cardiac Outcomes in fish for Validation of Ecological Risk (RECOVER).

References

Adams SM (1990) Status and use of biological indicators for evaluating the effects of stress in fish. Am Fish Soc Symp 8:1–8

Al-Sabti K (1985) Frequency of chromosomal aberrations in the rainbow trout, *Salmo gairdneri* Rich., exposed to five pollutants. J Fish Biol 26(1):13–19. https://doi.org/10.1111/j.1095-8649.1985.tb04235.x

Aluru N, Kuo E, Helfrich LW, Karchner SI, Linney EA, Pais JE, Franks DG (2015) Developmental exposure to 2,3,7,8-tetrachlorodibenzo-p-dioxin alters DNA methyltransferase (dnmt) expression in zebrafish (*Danio rerio*). Toxicol Appl Pharmacol 284(2):142–151. https://doi.org/10.1016/j.taap.2015.02.016

Anders S, Pyl PT, Huber W (2014) HTSeq–a Python framework to work with high-throughput sequencing data. Bioinformatics 31:166–169

Andersson DI, Hughes D (2010) Antibiotic resistance and its cost: is it possible to reverse resistance? Nat Rev Microbiol 8:260. https://doi.org/10.1038/nrmicro2319

Androutsopoulos VP, Tsatsakis AM, Spandidos DA (2009) Cytochrome P450 CYP1A1: wider roles in cancer progression and prevention. BMC Cancer 9(1):187. https://doi.org/10.1186/1471-2407-9-187

Antonovics J, Bradshaw AD, Turner RG (1971) Heavy metal tolerance in plants. In: Cragg JB (ed) Advances in ecological research, vol 7. Academic Press, pp 1–85. https://doi.org/10.1016/S0065-2504(08)60202-0

Behjati S, Tarpey PS (2013) What is next generation sequencing? Arch Dis Child Educ Pract Ed 98(6):236

Belfiore NM, Anderson SL (2001) Effects of contaminants on genetic patterns in aquatic organisms: a review. Mutat Res 489(2):97–122. https://doi.org/10.1016/S1383-5742(01)00065-5

Bickham JW, Sandhu S, Hebert PDN, Chikhi L, Athwal R (2000) Effects of chemical contaminants on genetic diversity in natural populations: implications for biomonitoring and ecotoxicology. Mutat Res 463(1):33–51. https://doi.org/10.1016/S1383-5742(00)00004-1

Bijlsma R, Loeschcke V (2011) Genetic erosion impedes adaptive responses to stressful environments. Evol Appl 5(2):117–129. https://doi.org/10.1111/j.1752-4571.2011.00214.x

Blake GET, Watson ED (2016) Unravelling the complex mechanisms of transgenerational epigenetic inheritance. Curr Opin Chem Biol 33:101–107. https://doi.org/10.1016/j.cbpa.2016.06.008

Brette F, Machado B, Cros C, Incardona JP, Scholz NL, Block BA (2014) Crude oil impairs cardiac excitation-contraction coupling in fish. Science 343(6172):772–776

Carls MG, Rice SD, Hose JE (1999) Sensitivity of fish embryos to weathered crude oil: part I. Low-level exposure during incubation causes malformations, genetic damage, and mortality in larval pacific herring (*Clupea pallasi*). Environ Toxicol Chem 18(3):481–493. https://doi.org/10.1002/etc.5620180317

Carney SA, Chen J, Burns CG, Xiong KM, Peterson RE, Heideman W (2006) Aryl hydrocarbon receptor activation produces heart-specific transcriptional and toxic responses in developing zebrafish. Mol Pharmacol 70(2):549–561. https://doi.org/10.1124/mol.106.025304

Chanda S, Dasgupta UB, GuhaMazumder D, Gupta M, Chaudhuri U, Lahiri S, Das S, Ghosh N, Chatterjee D (2006) DNA hypermethylation of promoter of gene p53 and p16 in arsenic-exposed people with and without malignancy. Toxicol Sci 89(2):431–437. https://doi.org/10.1093/toxsci/kfj030

Chen G, Shi T, Shi L (2017) Characterizing and annotating the genome using RNA-seq data. Sci China Life Sci 60(2):116–125

Collotta M, Bertazzi PA, Bollati V (2013) Epigenetics and pesticides. Toxicology 307:35–41. https://doi.org/10.1016/j.tox.2013.01.017

Conesa A, Madrigal P, Tarazona S, Gomez-Cabrero D, Cervera A, McPherson A, Szcześniak MW, Gaffney DJ, Elo LL, Zhang X (2016) A survey of best practices for RNA-seq data analysis. Genome Biol 17(1):13

Corrales J, Thornton C, White M, Willett KL (2014) Multigenerational effects of benzo[a]pyrene exposure on survival and developmental deformities in zebrafish larvae. Aquat Toxicol 148:16–26. https://doi.org/10.1016/j.aquatox.2013.12.028

Das PM, Ramachandran K, vanWert J, Singal R (2004) Chromatin immunoprecipitation assay. BioTechniques 37(6):961–969. https://doi.org/10.2144/04376rv01

DeAngelis JT, Farrington WJ, Tollefsbol TO (2008) An overview of epigenetic assays. Mol Biotechnol 38(2):179–183. https://doi.org/10.1007/s12033-007-9010-y

Denison MS, Nagy SR (2003) Activation of the aryl hydrocarbon receptor by structurally diverse exogenous and endogenous chemicals. Annu Rev Pharmacol Toxicol 43(1):309–334. https://doi.org/10.1146/annurev.pharmtox.43.100901.135828

Di Giulio RT, Clark BW (2015) The Elizabeth River story: a case study in evolutionary toxicology. J Toxicol Environ Heal B 18(6):259–298. https://doi.org/10.1080/15320383.2015.1074841

Diamante G, Xu EG, Chen S, Mager E, Grosell M, Schlenk D (2017) Differential expression of microRNAs in embryos and larvae of Mahi-Mahi (*Coryphaena hippurus*) exposed to Deepwater Horizon oil. Environ Sci Technol Lett 4(12):523–529. https://doi.org/10.1021/acs.estlett.7b00484

Edmunds RC, Gill J, Baldwin DH, Linbo TL, French BL, Brown TL, Esbaugh AJ, Mager EM, Stieglitz J, Hoenig R (2015) Corresponding morphological and molecular indicators of crude oil toxicity to the developing hearts of mahi mahi. Sci Rep 5:17326

Eeva T, Belskii E, Kuranov B (2006) Environmental pollution affects genetic diversity in wild bird populations. Mutat Res 608(1):8–15. https://doi.org/10.1016/j.mrgentox.2006.04.021

Frank SA (2010) Somatic evolutionary genomics: mutations during development cause highly variable genetic mosaicism with risk of cancer and neurodegeneration. Proc Natl Acad Sci 107(suppl 1):1725–1730

Frankham R, Bradshaw CJA, Brook BW (2014) Genetics in conservation management: revised recommendations for the 50/500 rules, Red List criteria and population viability analyses. Biol Conserv 170:56–63. https://doi.org/10.1016/j.biocon.2013.12.036

Frederick LA, Van Veld PA, Rice CD (2007) Bioindicators of immune function in creosote-adapted estuarine killifish, *Fundulus heteroclitus*. J Toxic Environ Health A 70(17):1433–1442. https://doi.org/10.1080/15287390701382910

Gao D, Wang C, Xi Z, Zhou Y, Wang Y, Zuo Z (2017) Early-life benzo[a]pyrene exposure causes neurodegenerative syndromes in adult Zebrafish (*Danio rerio*) and the mechanism involved. Toxicol Sci 157(1):74–84. https://doi.org/10.1093/toxsci/kfx028

Garcia TI, Shen Y, Crawford D, Oleksiak MF, Whitehead A, Walter RB (2012) RNA-Seq reveals complex genetic response to Deepwater Horizon oil release in *Fundulus grandis*. BMC Genomics 13(1):474

Hackett JA, Sengupta R, Zylicz JJ, Murakami K, Lee C, Down TA, Surani MA (2013) Germline DNA demethylation dynamics and imprint erasure through 5-hydroxymethylcytosine. Science 339(6118):448

Hahn ME, Karchner SI, Evans BR, Franks DG, Merson RR, Lapseritis JM (2006) Unexpected diversity of aryl hydrocarbon receptors in non-mammalian vertebrates: insights from comparative genomics. J Exp Zool A Comp Exp Biol 305A(9):693–706. https://doi.org/10.1002/jez.a.323

Handley-Goldstone HM, Grow MW, Stegeman JJ (2005) Cardiovascular gene expression profiles of dioxin exposure in Zebrafish embryos. Toxicol Sci 85(1):683–693. https://doi.org/10.1093/toxsci/kfi116

Haensly WE, Neff JM, Sharp JR, Morris AC, Bedgood MF, Boem PD (1982) Histopathology of *Pleuronectes platessa* L. from Aber Wrac'h and Aber Benoit, Brittany, France: long-term effects of the Amoco Cadiz crude oil spill. J Fish Dis 5(5):365–391. https://doi.org/10.1111/j.1365-2761.1982.tb00494.x

Hanson MA, Skinner MK (2016) Developmental origins of epigenetic transgenerational inheritance. Environmental Epigenetics 2(1):dvw002–dvw002. https://doi.org/10.1093/eep/dvw002

Harrigan JA, Vezina CM, McGarrigle BP, Ersing N, Box HC, Maccubbin AE, Olson JR (2004) DNA adduct formation in precision-cut rat liver and lung slices exposed to benzo[a]pyrene. Toxicol Sci 77(2):307–314. https://doi.org/10.1093/toxsci/kfh030

Heard E, Martienssen RA (2014) Transgenerational epigenetic inheritance: myths and mechanisms. Cell 157(1):95–109. https://doi.org/10.1016/j.cell.2014.02.045

Hilscherova K, Machala M, Kannan K, Blankenship AL, Giesy JP (2000) Cell bioassays for detection of aryl hydrocarbon (AhR) and estrogen receptor (ER) mediated activity in environmental samples. Environ Sci Pollut Res 7(3):159–171. https://doi.org/10.1065/espr2000.02.017

Husseneder C, Donaldson JR, Foil LD (2016) Impact of the 2010 Deepwater Horizon oil spill on population size and genetic structure of horse flies in Louisiana marshes. Sci Rep 6:18968. https://doi.org/10.1038/srep18968

Incardona JP (2017) Molecular mechanisms of crude oil developmental toxicity in fish. Arch Environ Contam Toxicol 73(1):19–32

International Chicken Genome Sequencing Consortium (2004) Sequence and comparative analysis of the chicken genome provide unique perspectives on vertebrate evolution. Nature 432:695. https://doi.org/10.1038/nature03154

Jiang L, Zhang J, Wang J-J, Wang L, Zhang L, Li G, Yang X, Ma X, Sun X, Cai J, Zhang J, Huang X, Yu M, Wang X, Liu F, Wu C-I, He C, Zhang B, Ci W, Liu J (2013) Sperm, but not oocyte, DNA methylome is inherited by Zebrafish early embryos. Cell 153(4):773–784. https://doi.org/10.1016/j.cell.2013.04.041

Kanherkar RR, Bhatia-Dey N, Csoka AB (2014) Epigenetics across the human lifespan. Front Cell Dev Biol 2:49. https://doi.org/10.3389/fcell.2014.00049

Klekowski EJ, Corredor JE, Morell JM, Del Castillo CA (1994) Petroleum pollution and mutation in mangroves. Mar Pollut Bull 28(3):166–169. https://doi.org/10.1016/0025-326X(94)90393-X

Knecht AL, Truong L, Marvel SW, Reif DM, Garcia A, Lu C, Simonich MT, Teeguarden JG, Tanguay RL (2017) Transgenerational inheritance of neurobehavioral and physiological deficits from developmental exposure to benzo[a]pyrene in zebrafish. Toxicol Appl Pharmacol 329:148–157. https://doi.org/10.1016/j.taap.2017.05.033

Kurdyukov S, Bullock M (2016) DNA methylation analysis: choosing the right method. Biology (Basel) 5(1). https://doi.org/10.3390/biology5010003

Labbé C, Robles V, Herraez MP (2017) Epigenetics in fish gametes and early embryo. Aquaculture 472:93–106. https://doi.org/10.1016/j.aquaculture.2016.07.026

Lanham KA, Plavicki J, Peterson RE, Heideman W (2014) Cardiac myocyte-specific AHR activation phenocopies TCDD-induced toxicity in zebrafish. Toxicol Sci 141(1):141–154

Lee CE, Remfert JL, Opgenorth T, Lee KM, Stanford E, Connolly JW, Kim J, Tomke S (2017) Evolutionary responses to crude oil from the Deepwater Horizon oil spill by the copepod *Eurytemora affinis*. Evol Appl 10(8):813–828. https://doi.org/10.1111/eva.12502

Li B, Dewey CN (2011) RSEM: accurate transcript quantification from RNA-Seq data with or without a reference genome. BMC Bioinformatics 12(1):323. https://doi.org/10.1186/1471-2105-12-323

Li L, Zhang T, Qin X-S, Ge W, Ma H-G, Sun L-L, Hou Z-M, Chen H, Chen P, Qin G-Q, Shen W, Zhang X-F (2014) Exposure to diethylhexyl phthalate (DEHP) results in a heritable modification of imprint genes DNA methylation in mouse oocytes. Mol Biol Rep 41(3):1227–1235. https://doi.org/10.1007/s11033-013-2967-7

Lindblad-Toh K, Wade CM, Mikkelsen TS, Karlsson EK, Jaffe DB, Kamal M, Clamp M, Chang JL, Kulbokas Iii EJ, Zody MC, Mauceli E, Xie X, Breen M, Wayne RK, Ostrander EA, Ponting CP, Galibert F, Smith DR, deJong PJ, Kirkness E, Alvarez P, Biagi T, Brockman W, Butler J, Chin C-W, Cook A, Cuff J, Daly MJ, DeCaprio D, Gnerre S, Grabherr M, Kellis M, Kleber M, Bardeleben C, Goodstadt L, Heger A, Hitte C, Kim L, Koepfli K-P, Parker HG, Pollinger JP, Searle SMJ, Sutter NB, Thomas R, Webber C, Baldwin J, Abebe A, Abouelleil A, Aftuck L, Ait-zahra M, Aldredge T, Allen N, An P, Anderson S, Antoine C, Arachchi H, Aslam A, Ayotte L, Bachantsang P, Barry A, Bayul T, Benamara M, Berlin A, Bessette D, Blitshteyn B, Bloom T, Blye J, Boguslavskiy L, Bonnet C, Boukhgalter B, Brown A, Cahill P, Calixte

N, Camarata J, Cheshatsang Y, Chu J, Citroen M, Collymore A, Cooke P, Dawoe T, Daza R, Decktor K, DeGray S, Dhargay N, Dooley K, Dooley K, Dorje P, Dorjee K, Dorris L, Duffey N, Dupes A, Egbiremolen O, Elong R, Falk J, Farina A, Faro S, Ferguson D, Ferreira P, Fisher S, FitzGerald M, Foley K, Foley C, Franke A, Friedrich D, Gage D, Garber M, Gearin G, Giannoukos G, Goode T, Goyette A, Graham J, Grandbois E, Gyaltsen K, Hafez N, Hagopian D, Hagos B, Hall J, Healy C, Hegarty R, Honan T, Horn A, Houde N, Hughes L, Hunnicutt L, Husby M, Jester B, Jones C, Kamat A, Kanga B, Kells C, Khazanovich D, Kieu AC, Kisner P, Kumar M, Lance K, Landers T, Lara M, Lee W, Leger J-P, Lennon N, Leuper L, LeVine S, Liu J, Liu X, Lokyitsang Y, Lokyitsang T, Lui A, Macdonald J, Major J, Marabella R, Maru K, Matthews C, McDonough S, Mehta T, Meldrim J, Melnikov A, Meneus L, Mihalev A, Mihova T, Miller K, Mittelman R, Mlenga V, Mulrain L, Munson G, Navidi A, Naylor J, Nguyen T, Nguyen N, Nguyen C, Nguyen T, Nicol R, Norbu N, Norbu C, Novod N, Nyima T, Olandt P, O'Neill B, O'Neill K, Osman S, Oyono L, Patti C, Perrin D, Phunkhang P, Pierre F, Priest M, Rachupka A, Raghuraman S, Rameau R, Ray V, Raymond C, Rege F, Rise C, Rogers J, Rogov P, Sahalie J, Settipalli S, Sharpe T, Shea T, Sheehan M, Sherpa N, Shi J, Shih D, Sloan J, Smith C, Sparrow T, Stalker J, Stange-Thomann N, Stavropoulos S, Stone C, Stone S, Sykes S, Tchuinga P, Tenzing P, Tesfaye S, Thoulutsang D, Thoulutsang Y, Topham K, Topping I, Tsamla T, Vassiliev H, Venkataraman V, Vo A, Wangchuk T, Wangdi T, Weiand M, Wilkinson J, Wilson A, Yadav S, Yang S, Yang X, Young G, Yu Q, Zainoun J, Zembek L, Zimmer A, Lander ES (2005) Genome sequence, comparative analysis and haplotype structure of the domestic dog. Nature 438:803. https://doi.org/10.1038/nature04338

Ljungman M, Hanawalt PC (1992) Efficient protection against oxidative DNA damage in chromatin. Mol Carcinog 5(4):264–269. https://doi.org/10.1002/mc.2940050406

Love MI, Huber W, Anders S (2014) Moderated estimation of fold change and dispersion for RNA-seq data with DESeq2. Genome Biol 15(12):550

McEachern KL (2014) Toxicity of oil from BP Deepwater Horizon blowout on the early life stage of red drum, *Sciaenops ocellatus*. Faculty of the College of Arts and Sciences, Florida Gulf Coast University

Metzger JM, Westfall MV (2004) Covalent and noncovalent modification of thin filament action: the essential role of troponin in cardiac muscle regulation. Circ Res 94(2):146–158

Meyer JN, Di Giulio RT (2003) Heritable adaptation and fitness costs in killifish (*Fundulus heteroclitus*) inhabiting a polluted estuary. Ecol Appl 13(2):490–503. https://doi.org/10.1890/1051-0761(2003)013[0490:HAAFCI]2.0.CO;2

Murawski SA, Hogarth WT, Peebles EB, Barbeiri L (2014) Prevalence of external skin lesions and polycyclic aromatic hydrocarbon concentrations in Gulf of Mexico fishes, post-Deepwater Horizon. Trans Am Fish Soc 143(4):1084–1097. https://doi.org/10.1080/00028487.2014.911205

Magnuson JT, Khursigara AJ, Allmon EB, Esbaugh AJ, Roberts AP (2018). Effects of Deepwater Horizon crude oil on ocular development in two estuarine fish species, red drum (Sciaenops ocellatus) and sheepshead minnow (Cyprinodon variegatus). Ecotoxicol Environ Saf 166:186–191

Nacci D, Proestou D, Champlin D, Martinson J, Waits ER (2016) Genetic basis for rapidly evolved tolerance in the wild: adaptation to toxic pollutants by an estuarine fish species. Mol Ecol 25(21):5467–5482. https://doi.org/10.1111/mec.13848

Nei M, Maruyama T, Chakraborty R (1975) The bottleneck effect and genetic variability in populations. Evolution 29(1):1–10. https://doi.org/10.2307/2407137

Nilsen BM, Berg K, GoksøØr A (1998) Induction of cytochrome P4501A (*CYP1A*) in fish: a biomarker for environmental pollution. In: Phillips IR, Shephard EA (eds) Cytochrome P450 protocols. Humana Press, Totowa, NJ, pp 423–438. https://doi.org/10.1385/0-89603-519-0:423

O'Leary SJ, Hollenbeck CM, Vega RR, Gold JR, Portnoy DS (2018) Genetic mapping and comparative genomics to inform restoration enhancement and culture of southern flounder, *Paralichthys lethostigma*. BMC Genomics 19(1):163. https://doi.org/10.1186/s12864-018-4541-0

Oleksiak MF, Karchner SI, Jenny MJ, Franks DG, Welch DBM, Hahn ME (2011) Transcriptomic assessment of resistance to effects of an aryl hydrocarbon receptor (AHR) agonist in embryos of Atlantic killifish (*Fundulus heteroclitus*) from a marine Superfund site. BMC Genomics 12(1):263

Oppold A-M, Müller R (2017) Chapter nine - Epigenetics: a hidden target of insecticides. In: Verlinden H (ed) Advances in insect physiology, vol 53. Academic Press, pp 313–324. https://doi.org/10.1016/bs.aiip.2017.04.002

Oziolor EM, Bigorgne E, Aguilar L, Usenko S, Matson CW (2014) Evolved resistance to PCB- and PAH-induced cardiac teratogenesis, and reduced CYP1A activity in Gulf killifish (*Fundulus grandis*) populations from the Houston Ship Channel, Texas. Aquat Toxicol 150:210–219. https://doi.org/10.1016/j.aquatox.2014.03.012

Palumbi SR (2001) Humans as the World's greatest evolutionary force. Science 293(5536):1786–1790. https://doi.org/10.1126/science.293.5536.1786

Pashin YV, Bakhitova LM (1979) Mutagenic and carcinogenic properties of polycyclic aromatic hydrocarbons. Environ Health Perspect 30:185–189

Patro R, Duggal G, Kingsford C (2015) Salmon: accurate, versatile and ultrafast quantification from RNA-seq data using lightweight-alignment. Biorxiv:021592

Peschansky VJ, Wahlestedt C (2014) Non-coding RNAs as direct and indirect modulators of epigenetic regulation. Epigenetics 9(1):3–12. https://doi.org/10.4161/epi.27473

Pilcher W, Miles S, Tang S, Mayer G, Whitehead A (2014) Genomic and genotoxic responses to controlled weathered-oil exposures confirm and extend field studies on impacts of the jizon oil spill on native killifish. PLoS One 9(9):e106351

Potok ME, Nix DA, Parnell TJ, Cairns BR (2013) Reprogramming the maternal zebrafish genome after fertilization to match the paternal methylation pattern. Cell 153(4):759–772. https://doi.org/10.1016/j.cell.2013.04.030

Puritz JB, Toonen RJ (2011) Coastal pollution limits pelagic larval dispersal. Nat Commun 2:226. https://doi.org/10.1038/ncomms1238

Reid NM, Proestou DA, Clark BW, Warren WC, Colbourne JK, Shaw JR, Karchner SI, Hahn ME, Nacci D, Oleksiak MF, Crawford DL, Whitehead A (2016) The genomic landscape of rapid repeated evolutionary adaptation to toxic pollution in wild fish. Science 354(6317):1305–1308. https://doi.org/10.1126/science.aah4993

Reik W (2007) Stability and flexibility of epigenetic gene regulation in mammalian development. Nature 447:425. https://doi.org/10.1038/nature05918

Robertson M, Schrey A, Shayter A, Moss CJ, Richards C (2017) Genetic and epigenetic variation in *Spartina alterniflora* following the Deepwater Horizon oil spill. Evol Appl 10(8):792–801. https://doi.org/10.1111/eva.12482

Robinson MD, McCarthy DJ, Smyth GK (2010) edgeR: a Bioconductor package for differential expression analysis of digital gene expression data. Bioinformatics 26(1):139–140

Schena M, Shalon D, Davis RW, Brown PO (1995) Quantitative monitoring of gene expression patterns with a complementary DNA microarray. Science 270(5235):467–470

Schlenk D, Handy R, Steinert S, Depledge MH, Benson W (2008a) Biotransformation in fishes. In: Di Giulio R, Hinton D (eds) The toxicology of fishes, vol 1. CRC Press, Boca Raton, FL, pp 153–234

Schlenk D, Handy R, Steinert S, Depledge MH, Benson W (2008b) Biomarkers. In: Di Giulio R, Hinton D (eds) The toxicology of fishes, vol 1. CRC Press, Boca Raton, FL, pp 683–732

Schiebelhut LM, Puritz JB, Dawson MN (2018) Decimation by sea star wasting disease and rapid genetic change in a keystone species, *Pisaster ochraceus*. Proc Nat Acad Sci 115:7069. https://doi.org/10.1073/pnas.1800285115

Shimada T (2006) Xenobiotic-metabolizing enzymes involved in activation and detoxification of carcinogenic polycyclic aromatic hydrocarbons. Drug Metab Pharmacokinet 21(4):257–276. https://doi.org/10.2133/dmpk.21.257

Skinner MK (2014) Endocrine disruptor induction of epigenetic transgenerational inheritance of disease. Mol Cell Endocrinol 398(1):4–12. https://doi.org/10.1016/j.mce.2014.07.019

Sørhus E, Incardona JP, Furmanek T, Goetz GW, Scholz NL, Meier S, Edvardsen RB, Jentoft S (2017) Novel adverse outcome pathways revealed by chemical genetics in a developing marine fish. elife 6:e20707

Stieglitz JD, Mager EM, Hoenig RH, Benetti DD, Grosell M (2016) Impacts of Deepwater Horizon crude oil exposure on adult mahi-mahi (*Coryphaena hippurus*) swim performance. Environ Toxicol Chem 35(10):2613–2622. https://doi.org/10.1002/etc.3436

Szyf M (2015) Nongenetic inheritance and transgenerational epigenetics. Trends Mol Med 21(2):134–144. https://doi.org/10.1016/j.molmed.2014.12.004

Shiizaki K, Kawanishi M, Yagi T (2017) Modulation of benzo[a]pyrene-DNA adduct formation by CYP1 inducer and inhibitor. Genes Environ 39:14. https://doi.org/10.1186s41021-017-0076-x

Takata H, Hanafusa T, Mori T, Shimura M, Iida Y, Ishikawa K, Yoshikawa K, Yoshikawa Y, Maeshima K (2013) Chromatin compaction protects genomic DNA from radiation damage. PLoS One 8(10):e75622. https://doi.org/10.1371/journal.pone.0075622

Tang WW, Dietmann S, Irie N, Leitch Harry G, Floros Vasileios I, Bradshaw Charles R, Hackett JA, Chinnery PF, Surani MA (2015) A unique gene regulatory network resets the human germline epigenome for development. Cell 161(6):1453–1467. https://doi.org/10.1016/j.cell.2015.04.053

van Straalen NM, Timmermans MJTN (2002) Genetic variation in toxicant-stressed populations: an evaluation of the "genetic erosion" hypothesis. Hum Ecol Risk Assess Int J 8(5):983–1002. https://doi.org/10.1080/1080-700291905783

Volkova PY, Geras'kin SA, Horemans N, Makarenko ES, Saenen E, Duarte GT, Nauts R, Bondarenko VS, Jacobs G, Voorspoels S, Kudin M (2018) Chronic radiation exposure as an ecological factor: hypermethylation and genetic differentiation in irradiated Scots pine populations. Environ Pollut 232:105–112. https://doi.org/10.1016/j.envpol.2017.08.123

Whalon ME, Mota-Sanchez D, Hollingworth RM (2008) Global pesticide resistance in arthropods. CABI, Wallingford, Oxfordshire

Whitehead A, Clark BW, Reid NM, Hahn ME, Nacci D (2017) When evolution is the solution to pollution: key principles, and lessons from rapid repeated adaptation of killifish (*Fundulus heteroclitus*) populations. Evol Appl 10(8):762–783. https://doi.org/10.1111/eva.12470

Whitehead A, Dubansky B, Bodinier C, Garcia TI, Miles S, Pilley C, Raghunathan V, Roach JL, Walker N, Walter RB (2012a) Genomic and physiological footprint of the Deepwater Horizon oil spill on resident marsh fishes. Proc Natl Acad Sci 109(50):20298–20302

Whitehead A, Dubansky B, Bodinier C, Garcia TI, Miles S, Pilley C, Raghunathan V, Roach JL, Walker N, Walter RB, Rice CD, Galvez F (2012b) Genomic and physiological footprint of the Deepwater Horizon oil spill on resident marsh fishes. Proc Natl Acad Sci 109(50):20298–20302. https://doi.org/10.1073/pnas.1109545108

Whitehead A, Pilcher W, Champlin D, Nacci D (2011) Common mechanism underlies repeated evolution of extreme pollution tolerance. Proceedings of the Royal Society of London B: Biological Sciences:rspb20110847

Whitehead A, Triant D, Champlin D, Nacci D (2010) Comparative transcriptomics implicates mechanisms of evolved pollution tolerance in a killifish population. Mol Ecol 19(23):5186–5203

Wills LP, Zhu S, Willett KL, Di Giulio RT (2009) Effect of CYP1A inhibition on the biotransformation of benzo[a]pyrene in two populations of *Fundulus heteroclitus* with different exposure histories. Aquat Toxicol 92(3):195–201

Wirgin I, Roy NK, Loftus M, Chambers RC, Franks DG, Hahn ME (2011) Mechanistic basis of resistance to PCBs in Atlantic tomcod from the Hudson River. Science 331(6022):1322–1325. https://doi.org/10.1126/science.1197296

Wirgin I, Waldman JR (1998) Altered gene expression and genetic damage in North American fish populations. Mutat Res 399(2):193–219. https://doi.org/10.1016/S0027-5107(97)00256-X

Wirgin I, Waldman JR (2004) Resistance to contaminants in North American fish populations. Mutat Res 552(1):73–100

Xu EG, Khursigara AJ, Magnuson J, Hazard ES, Hardiman G, Esbaugh AJ, Roberts AP, Schlenk D (2017) Larval red drum (*Sciaenops ocellatus*) sublethal exposure to weathered Deepwater Horizon crude oil: developmental and transcriptomic consequences. Environ Sci Technol 51(17):10162–10172

Xu EG, Mager EM, Grosell M, Pasparakis C, Schlenker LS, Stieglitz JD, Benetti D, Hazard ES, Courtney SM, Diamante G (2016a) Time-and oil-dependent transcriptomic and physiological responses to Deepwater Horizon oil in mahi-mahi (*Coryphaena hippurus*) embryos and larvae. Environ Sci Technol 50(14):7842–7851

Xu X, Weber D, Martin A, Lone D (2016b) Trans-generational transmission of neurobehavioral impairments produced by developmental methylmercury exposure in zebrafish (*Danio rerio*). Neurotoxicol Teratol 53:19–23. https://doi.org/10.1016/j.ntt.2015.11.003

Xu EG, Magnuson JT, Diamante G, Mager E, Pasparakis C, Grosell M, Roberts AP, Schlenk D (2018) Changes in microRNA-mRNA signatures agree with morphological, physiological, and behavioral changes in larval mahi-mahi treated with Deepwater Horizon oil. Environ Sci Technol 52:13501–13510

Xu EG, Khursigara AJ, Li S, Esbaugh AJ, Dasgupta S, Volz DC, Schlenk D (2019) mRNA-miRNA-Seqreveals neuro-cardio mechanisms of crude oil toxicity in red drum (Sciaenops ocellatus). Environ Sci Technol. https://doi.org/10.1021/acs.est.9b00150

Zhu L, Qu K, Xia B, Sun X, Chen B (2016) Transcriptomic response to water accommodated fraction of crude oil exposure in the gill of Japanese flounder, *Paralichthys olivaceus*. Mar Pollut Bull 106(1–2):283–291

**Part VII
Ecosystem-Level Modeling of Deep Oil
Spill Impacts**

Tessa Wilson
Will You Be My Neighbor?
Acrylic paint on Canvas
18" × 26"

Chapter 31
A Synthesis of Top-Down and Bottom-Up Impacts of the Deepwater Horizon Oil Spill Using Ecosystem Modeling

Lindsey N. Dornberger, Cameron H. Ainsworth, Felicia Coleman, and Dana L. Wetzel

Abstract The *Deepwater Horizon* (DWH) oil spill in the Gulf of Mexico (GoM) triggered the largest response to a spill in US history (Levy and Gopalakrishnan, J Nat Resources Pol Res, 2(3):297–315, 2010; Barron, Toxicol Pathol 40(2):315–320, 2012). The cumulative research from this response has resulted in hundreds of publications describing the range of impacts from the DWH event on various components of the system. An ecosystem-based approach to assessing the consequences of the DWH oil spill can help to address non-linear and ecosystem-level interactions (reviewed by Curtin and Prellezo, Mar Policy 34(5):821–830, 2010) and would be a key step toward integrating the knowledge gained from research efforts. Whereas Ainsworth et al. (PLoS One 13(1):e0190840, 2018) tested top-down effects of the oil spill on fish abundance and mortality, this chapter represents a synthesis of bottom-up and top-down effects across a broader range of taxa. Bottom-up effects relate to the accumulation of detrital biomass and oil on the seafloor as a result of marine oil snow sedimentation and flocculent accumulation (MOSSFA).

Keywords Atlantis · Ecosystem modeling · Oil toxicity · Cumulative effects · Fishing mortality

L. N. Dornberger (✉) · C. H. Ainsworth
University of South Florida, College of Marine Science, St. Petersburg, FL, USA
e-mail: ainsworth@usf.edu

F. Coleman
Florida State University, Coastal and Marine Laboratory, St. Teresa, FL, USA
e-mail: fcoleman@fsu.edu

D. L. Wetzel
Mote Marine Laboratory, Sarasota, FL, USA
e-mail: dana@mote.org

31.1 Introduction

The *Deepwater Horizon* (DWH) oil spill in the Gulf of Mexico (GoM) triggered the largest response to a spill in US history (Levy and Gopalakrishnan 2010; Barron 2012; Lubchenco et al. 2012, Colwell et al. 2014). The cumulative research from this response has resulted in hundreds of publications describing the range of impacts from the DWH event on various components of the system. An ecosystem-based approach to assessing the consequences of the DWH oil spill can help to address non-linear and ecosystem-level interactions (reviewed by Curtin and Prellezo 2010) and would be a key step toward integrating the knowledge gained from research efforts. Whereas Ainsworth et al. (2018) tested top-down effects of the oil spill on fish abundance and mortality, this chapter represents a synthesis of bottom-up and top-down effects across a broader range of taxa. Bottom-up effects relate to the accumulation of detrital biomass and oil on the seafloor as a result of marine oil snow sedimentation and flocculent accumulation (MOSSFA).

Additionally, understanding of the cumulative effects of various human stressors on fish populations has been identified as a major gap in our knowledge (Pew Ocean Commission 2003; U.S. Commission on Ocean Policy 2004; Crain et al. 2008, 2009). In particular, Crain et al. (2008) note that the absence of controlled studies assessing the combined effects of fishing and other stressors on fish populations is a particular concern. Controlled field experiments of this type are inherently difficult to perform, requiring that they be addressed using a different means (Crain et al. 2009). For instance, by using established ecosystem-based modeling efforts, not only could the potential cumulative effects of fishing and an oil spill on fish populations be identified, but the synergistic effects of multiple stressors across a community of fished populations could be elucidated, providing critical insights for both scientists and managers. This chapter will address the combined effects of fishing mortality and an oil spill, using an Atlantis ecosystem-based fishery model.

31.2 Methods

31.2.1 Atlantis

Atlantis is a deterministic, end-to-end ecosystem model with a spatially explicit three-dimensional framework (Fulton 2001; Fulton et al. 2004a, 2004b, 2005, 2007). It operates at 12-hour time steps. Atlantis uses an irregular polygon structure reflecting bioregional features and political and regulatory boundaries. Each polygon has a sediment layer and up to six water column layers, depending on depth of the area described by the polygon. Depth strata are consistent across polygons (z-coordinate system).

Atlantis is also coupled to a hydrodynamic model that includes currents, temperature, and salinity per polygon and depth layer. For the GoM Atlantis model, the coupled hydrodynamic data are provided by the American Seas model, which is based on the National Research Laboratory Navy Coastal Ocean Model (Martin 2000). Nutrient cycling and subsequent primary productivity, as well as organism distribution, are all affected by the hydrodynamic inputs. Nitrogen is used as the currency of exchange between functional groups (Link et al. 2011), though several other nutrients are also tracked.

The GoM Atlantis model uses a total of 91 functional groups: 61 age-structured vertebrates and invertebrates, 19 non-age-structured invertebrates, 6 primary producers, 3 types of detritus, and 2 bacteria. All vertebrates and some exploited invertebrates that are age-structured are tracked in Atlantis by body weight in mg N and number of individuals. Non-exploited invertebrates are only tracked as biomass pools in mg N. Carrion detritus, labile detritus, and refractory detritus are the three detrital functional groups. Any functional group can add to detrital and nutrient pools through mortality, waste excretion, and remnants from feeding behavior. The detrital groups can also be consumed as a defined part of any functional groups' diet. A full review of the modeling system and equations is available in Link et al. (2011). The specific GoM implementation of Atlantis and model fitting is available in Ainsworth et al. (2015). For this research project, Atlantis simulations were run for 50 years, from 2010 to 2060.

31.2.2   Environmental Drivers of the Oil Spill

Oil concentration estimates were provided by a probabilistic framework for oil droplet tracking based on an open-source Lagrangian stochastic model, the Connectivity Modeling System (CMS) (Paris et al. 2012, 2013; Aman and Paris 2013). The most recent CMS simulations of the DWH blowout are described in an article in preparation (Perlin et al. 2020). Implementation of these oil data in Atlantis follows the methods of Ainsworth et al. (2018). Briefly, oil estimates were provided at 1/12th degree grid points and in four depth layers, averaged across 0–20 m, 20–50 m, 50–200 m, and 200–2000 m. Dose-response calculations were made at the grid point scale and averaged over all Atlantis polygons, adjusting for the proportion of grid points per polygon that do not contain oil. Transient biomass dynamics were allowed to settle in the first 110 days of the simulation before introducing oil forcing to the model, which then lasts for 167 days.

There is not enough available sediment oil data from the field to use directly in the model as the oil driver for the sediment layer in simulations. In addition, the precise distribution, concentration, and extent of DWH oil on the seafloor are unknown. Romero et al. (2015, 2017) provided sediment core concentrations, which were then compared to estimated time- and depth-integrated water column concentrations from CMS, to establish a ratio between sediment and water column concentrations (Ainsworth et al. 2018). Using an estimated ratio between sediment oil concentrations and modeled oil values allows for the generation of an estimated, continuous sediment concentration map used to drive sediment oil toxicity.

The sediment-to-water column ratio (K) is potentially up to 1000:1 (Ainsworth et al. 2018), so a $K = 1000$ (unitless) was used as an upper limit of sensitivity analyses, along with $K = 700$ and $K = 400$. K plays a critical role in the impact of the oil spill on benthic organisms.

Metabolic processing of oil for all fish and benthic invertebrates was accounted for by calculating uptake and depuration rates based on data from a fish exposure study. Miller et al. (2017) exposed sub-adult red drum to a DWH surrogate oil and Corexit 9500 CEWAF solution actual average 0.66 ± 0.092 ppm TPH for 4 days of uptake, followed by 6 days of contaminant free seawater depuration. During this experiment, two fish were sacrificed every day during the uptake period and then every other day during depuration to assess liver PAH concentrations. The $\sum$50PAH concentrations measured each day in the liver over the 4 days of uptake were used to calculate a linear regression uptake value based on the average 0.66 ± 0.092 ppm TPH exposure, resulting in an average uptake of 2.13 ppm $\sum$50PAH/day over the 4-day period. The use of nominal concentrations results in a conservative uptake calculation. The exposure to uptake ratio is 0.47. This can be used to estimate the uptake rate at any exposure level assuming a linear relationship. To reflect uptake in Atlantis, current oil exposure values were divided by 0.47 to generate the exposure-specific linear uptake rate μ. To calculate depuration rate, the peak average liver concentration and subsequent concentrations at 2, 4, and 6 days post exposure were used to calculate an exponential decay formula. When external exposure ends, the estimated decay formula returns the body load to 1% of peak levels through depuration in about 10 days (parameter p in the decay Eq. 31.1 = -0.46). In the Atlantis simulations, these rates are used for both water column and sediment exposure. These rates were combined with the oil concentration estimates to calculate an exposure concentration, $[\text{oil}]_{\text{exposed}}$.

$$[\text{oil}]_{\text{exposed}} = \left[\text{Oil}_{t-1} + \left(\frac{\text{Oil}_t}{\mu} \right) \right] * e^p \tag{31.1}$$

To drive simulated MOSSFA in the system, point source inputs for detrital loading were used. Detrital concentrations were targeted to increase by a factor of about 2–5 $\times$ in the seven polygons closest to the wellhead, based on observations of increased detritus after the DWH event (Brooks et al. 2015). The elevated detrital input values were held for the duration of the oil spill forcing (167 days), and then inputs were lowered to normal values.

31.2.3 Biological Drivers of the Oil Spill

To drive oil effects on organism health, dose-response models that were evaluated for mortality and growth in a previous publication were used (Dornberger et al. 2016). Four models were fit by maximum likelihood estimation and compared using an Akaike information criterion (AIC). For both mortality and growth, the "hockey-stick" equations best explained the relationship between oil exposure and fish health and growth. The applied version of this hockey-stick equation is shown in Eq. 31.2.

$$R_t = a * \log\left(K * \frac{[\text{oil}]_{\text{exposed}}}{[\text{oil}]_{\text{thresh}}}\right) * \omega^{-1} \quad \text{if } [\text{oil}]_{\text{exposed}} \geq [\text{oil}]_{\text{thresh}} \quad (31.2)$$

R_t is the response modifier for either growth or mortality at time t, and a is the fitted slope ($a_{\text{mortality}} = 0.2885$, $a_{\text{growth}} = 0.1051$). ω represents the average exposure period from the studies used to derive the dose-response relationships (15 days), so the inverse was used to determine an average daily effect. $[\text{oil}]_{\text{thresh}}$ is the threshold below which oil has no effect on the response. The mortality threshold has a large impact on the magnitude of the effect modifier, so sensitivity analyses included three variants of $[\text{oil}]_{\text{thresh}}$ for fishes, 362, 635, and 907 ppb based on validation and sensitivity done in Ainsworth et al. (2018), and two for benthic invertebrates: 1000 and 3000 ppb based on invertebrate analyses (Dornberger 2018). The oil threshold $[\text{oil}]_{\text{thresh}}$ affecting growth for both water and sediment exposure was estimated to be 2.94 ppb for fishes. Sensitivity analysis of this threshold would be of little value since it has a very low oil concentration. No growth modifier was applied to invertebrates.

Since some organisms are not constrained exclusively to the sediment or water column, a weighted response modifier is calculated for each group accounting for relative benthic association of the group. The total modifier R_{oil} for both growth and mortality at a given time step takes the relative modifiers for the water column R_{pelagic} and sediment R_{benthic}, scaling them by the proportion B of the affected group's diet that comes from benthic food sources. Note the symbol change, e.g., that $R_t = R_{\text{pelagic}}$ at time t when the exposure medium is the water column.

$$R_{\text{oil}} = R_{\text{pelagic}} * (1 - B) + R_{\text{benthic}} * B \quad (31.3)$$

The more heavily a group's diet depends on a pelagic or benthic food source, the more its response modifier will reflect the pelagic or benthic medium, respectively.

When R_{oil} represents the effect on mortality, the mortality modifier is ($R_{\text{oil}} = M_{\text{oil}}$). This is then applied in M1, which summarizes natural mortality from non-predation sources as.

$$M1 = m_l + m_q + m_{\text{oil}} \quad (31.4)$$

where m_l and m_q are linear and quadratic mortality, which are regular tuning parameters in Atlantis. When R_{oil} represents the effect on growth, the hockey-stick shape produces higher effect values at higher concentrations. Hence, in order to scale growth down from 100% to a reduced rate as oil exposure increases, the growth response modifier is subtracted from one ($1 - R_{\text{oil}} = G_{\text{oil}}$).

The growth modifiers G_{oil} and M1 are both applied to Eq. 31.4, the instantaneous rate of change of biomass (B, in mg N) for each polygon, depth layer, and age class of functional group i.

$$\frac{dB_i}{dt} = \left[\text{Rec}_i + T_{IMM,i} - T_{EM,i} - M1_i - \sum_j P_{i,j} - F_i \right] * G_{oil} \left[G_{i,s} + G_{i,r} \right] \quad (31.5)$$

Rec_i is recruitment of new individuals into the age class. $T_{IMM,i}$ is immigration of group i and $T_{EM,i}$ is emigration, F_i is fisheries mortality, and $P_{i,j}$ is predation term of predator j on group i. $G_{i,s}$ is growth in structural nitrogen per individual and $G_{i,r}$ is growth in reserve nitrogen per individual, which are functions of consumption, growth rate, and assimilation efficiency. No oil avoidance behavior modifier was applied.

Recruitment impacts were also included following the methods of Ainsworth et al. (2018). They used the results of a larval distribution map and surface oil comparison from DWH conducted by Chancellor (2015) to approximate the net recruitment impact from loss of larvae. They assumed that all larvae exposed to oil were killed.

The oil spill emergency fishery closures were implemented as spatiotemporally dynamic marine protected areas. The timing of the closures was reported by the National Centers for Environmental Information (2016). In Atlantis, the proportion of each polygon that overlapped a closure during the scheduled period was closed to the appropriate fishing fleet(s). This effectively simulates the local emergency closures; however, it does not include reallocation of fishing effort elsewhere. The closure dynamics were updated on a daily time step and were included for the full duration of the emergency fisheries closures.

To determine the effects of a toxic oil spill combined with theoretical varied fishing mortalities, the sensitivity of the system stressed by oil was also simulated by varying fishing mortality in three additional worst-case sensitivity scenarios. In these scenarios, the fleet-specific relative fishing mortality F of each functional group was scaled by 0.5, 2, and 10. These varied F values were in combination with the worst-case sensitivity scenario parameters of 1000 K, fish mortality threshold of 362 ppb, and invertebrate mortality threshold of 1000 ppb. Fishing mortality F by

Table 31.1 List of all parameter values used in Atlantis sensitivity and modified fishing simulations

Parameters	Sensitivity analyses (18)	Fishing mortality and oil (3)
K	400, 700, 1000	1000
Fish [oil]$_{thresh}$	362, 635, 907	362
Benthic invertebrate [oil]$_{thresh}$	1000, 3000	1000
Fishing mortality modifier	–	0.5, 2, 10

The number following the column title is the total number of simulations run using a combination of the below parameter values. K is the sediment to water column ratio. [oil]$_{thresh}$ is the oil concentration where mortality for that type of organism begins based on the hockey-stick dose-response model. The fishing morality modifier is the scalar used to adjust MFC matrices in theoretical fishing simulations functional group and fleet was originally fit to historical catch as described in Ainsworth et al. (2015). All scenario parameters are listed in Table 31.1.

31.2.4   Analysis

For consistency and cross comparison, analytical grouping by polygon and guild matched that of Ainsworth et al. (2018). All results presented here are averaged across 13 highly impacted polygons (Fig. 31.1) unless stated otherwise. These 13 polygons roughly correspond to the area targeted for injury assessment efforts (Deepwater Horizon Natural Resource Damage Assessment Trustees 2016). Results from Atlantis functional groups were averaged into nine guilds: snappers (family, Lutjanidae), groupers (family, Serranidae), Sciaenidae, elasmobranchs, large pelagic fish, small pelagic fish, small demersal and reef fish, large demersal fish, and invertebrates. The composition of these guilds is listed in Table 31.2. Guilds were used for clarity in assessing impacts across a broad range of species in the ecosystem than specific functional groups. In Atlantis, functional group is the term used to track predefined populations and can be either species specific or multiple related species. The Atlantis functional groups' species compositions are presented in Ainsworth et al. (2015). Outputs from Atlantis were analyzed across sensitivity analyses and compared to the results of Ainsworth et al. (2018) to determine the effects of including invertebrate mortality, MOSSFA detrital loading, and varying fishing mortality.

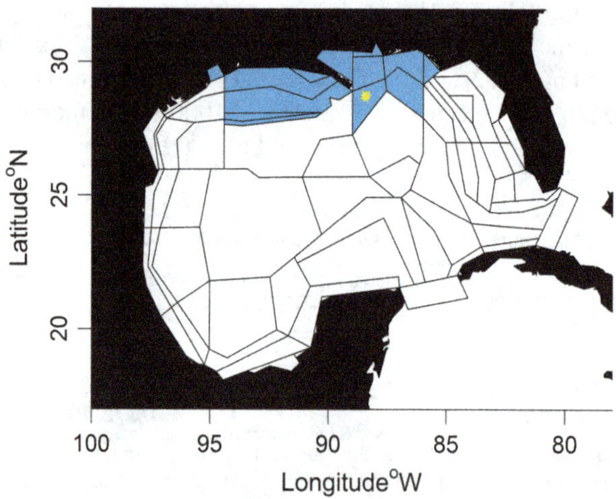

Fig. 31.1 Polygon geometry for Atlantis GOM. Shaded polygons represent the polygons used for oil spill analysis. The star is the relative location of the Deepwater Horizon site

Table 31.2 List of Atlantis functional groups contained within each guild

Guilds	Atlantis functional group names
Snappers	Red snapper, vermilion snapper, Lutjanidae
Groupers	Gag grouper, red grouper, scamp, shallow Serranidae, deep Serranidae
Large pelagics	Yellowfin tuna, bluefin tuna, little tunny, other tuna, swordfish, white marlin, blue marlin, other billfish, greater amberjack, jacks, king mackerel, Spanish mackerel
Small pelagics	Spanish sardine, menhaden, pinfish, small pelagic fish
Small demersal and reef	Small reef fish, small demersal fish
Sciaenidae	Black drum, red drum, Sciaenidae
Elasmobranchs	Blacktip shark, benthic feeding sharks, large sharks, filter feeding sharks, small sharks, skates and rays
Large demersal	Flatfish, other demersal fish
Invertebrates	Brown shrimp, white shrimp, pink shrimp, other shrimp, blue crab, stone crab, crabs and lobsters, stony corals, octocorals, sponges, carnivorous macrobenthos, infaunal meiobenthos, herbivorous echinoderms, oysters, bivalves, sessile filter feeders

31.3 Results

31.3.1 Population Impacts

In Fig. 31.2, guild biomass is presented relative to a control scenario, a no oil spill simulation. This figure shows the range of results observed for fishes and invertebrates from sensitivity analysis varying K and $[\text{oil}]_{\text{thresh}}$ as well as the average of those simulations. The results for snappers, groupers, Sciaenidae, and large demersal fishes agree with the trends from the oil spill simulations observed in Ainsworth et al. (2018), though the relative declines vary slightly. Relative changes qualitatively match remotely operated vehicle observations (W. Patterson, Pers. Comm.) in Ainsworth et al. (2018), which indicated as much as a 75% decrease in small reef fish and a 70% decrease in some large reef fish post spill. Small pelagics, small demersal and reef fish, and Sciaenidae exhibit declines during the spill period followed closely by recovery. However, the responses begin to differ with Ainsworth et al. (2018) around years 5–10. Small demersal and reef fishes and Sciaenidae average above the no-spill scenario from about year 2016 on, while the small pelagic guild declines again and maintains biomass at roughly half the no-spill scenario. This trend in small pelagics is driven by the "small pelagic" functional group within the guild. The small pelagic functional group goes virtually extinct after roughly 20 years due to fishing and increased predation pressure. In the small pelagic guild, species-specific functional groups menhaden and pinfish recover in 20 and 5 years, respectively.

When including oil effects on both fishes and benthic invertebrates, increased benthic invertebrate abundance is no longer evident at the invertebrate guild and polygon level in October 2010 (6 months after first spill), across all sensitivity

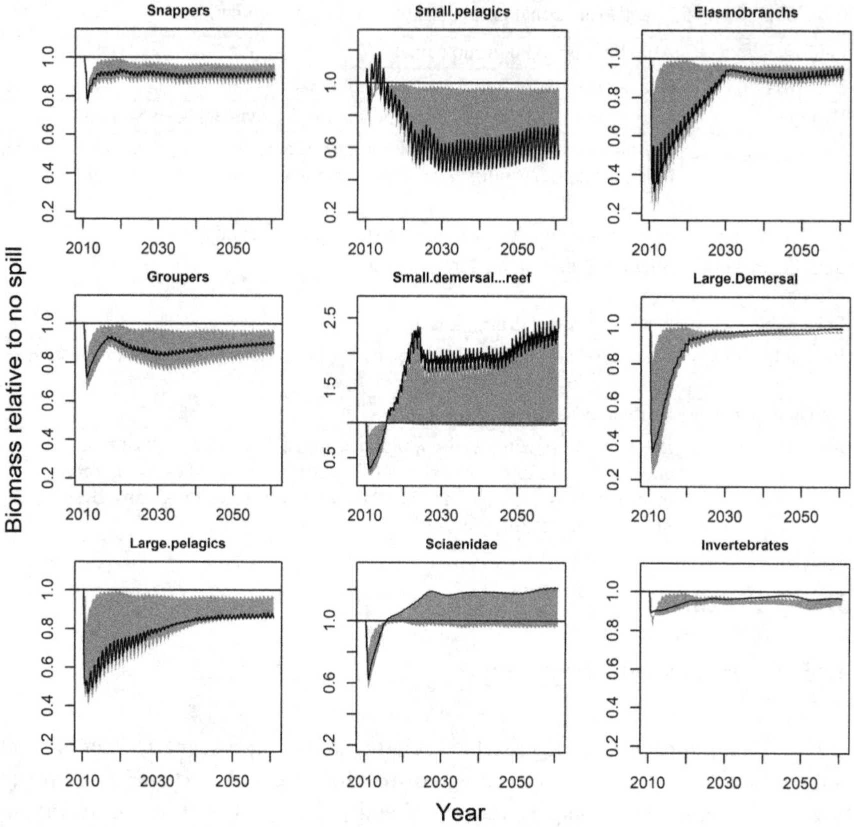

Fig. 31.2 Relative biomass trajectories of Atlantis guilds. Biomasses were averaged across 12 impacted polygons and divided by the biomass from the no-oil scenario. The shaded area represents the range of outcomes from sensitivity analysis on sediment concentration factor K and mortality [oil]$_{thresh}$ for fishes and benthic invertebrates. The solid black line shows the average biomass across all 18 sensitivity runs

analyses. However, a few functional groups within the benthic invertebrate guild do exhibit both increases and declines at the polygon level during the window of greatest impact (7–16 months): infaunal meiobenthos, sponges, sessile filter feeders, carnivorous macrobenthos, and blue crab.

Including benthic invertebrate mortality and MOSSFA forcing functions changes the recovery dynamics of functional groups compared to previous simulations seen in Ainsworth et al. (2018). However, there is still no distinguishable relationship between guild biomass minima and recovery time (Fig. 31.3). Of the functional groups included in the guild analysis within the highly impacted polygons, 27 did not recover to within 99% of original biomass levels in the worst-case scenario, roughly 51% of groups examined. In the best-case scenario, ten functional groups did not recover.

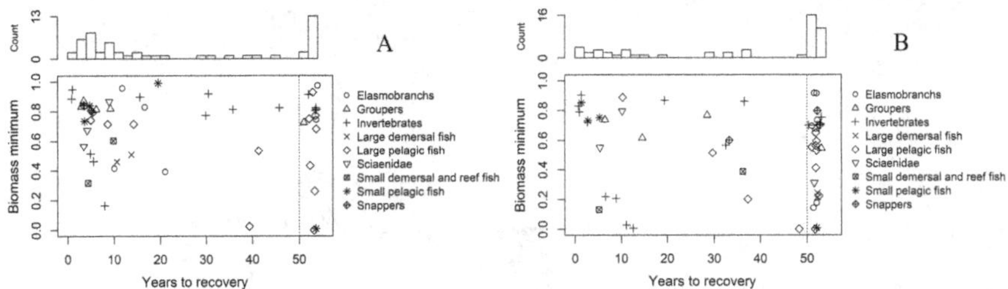

Fig. 31.3 Best-case and worst-case functional group recovery time and biomass minimum. (**a**) The "best-case" scenario, with sediment concentration factor K of 400, fish mortality threshold at 907 ppb, and invertebrate mortality threshold at 3000 ppb. (**b**) The "best-case" scenario, with sediment concentration factor K of 1000, fish mortality threshold at 362 ppb, and invertebrate mortality threshold at 1000 ppb. All functional groups that constitute a guild are shown. The count of functional groups by recovery time is shown in the count graph above

31.3.2   Combined Oil and Varied Fishing Mortality Scenarios

Relative biomass differences in the varied F scenarios from the base fishing scenario, for the nine guilds averaged across the 13 heavily impacted polygons, are shown in Fig. 31.4. The guilds most sensitive to fishing changes combined with "worst-case" scenario oil were snappers, groupers, and small demersal and reef fish. Snappers were most affected in the $F \times 10$ scenario. Groupers had the largest positive response to a release of fishing pressure. The guilds least affected by changes in F were small pelagics and benthic invertebrates.

In general, snappers, groupers, large pelagics, small pelagics, small demersal and reef, and Sciaenidae all exhibited lower biomass minima when F was increased during the oil spill simulation. Representative example guilds from this list are shown in Fig. 32.5. All guilds had different biomass trajectories with varied F relative to the no-oil scenario, and some reached reduced or elevated equilibrium states from the base F scenario. Snapper biomass oscillated around 0.35 relative to no oil for the $F \times 10$ scenario and 1.5 for the 0.5 F scenario (Fig. 31.5). Groupers reached a new equilibrium around 1.2 relative to no oil for the 0.5 F scenario (Fig. 31.5). Both guilds indicate that while the oil spill had an initial effect on biomass, the long-term biomass stabilization was dependent upon fishing effort, not magnitude of decline from the oil spill.

As a guild, benthic invertebrates indicated minor shifts in varied F scenarios. However, shrimps (brown shrimp, pink shrimp, white shrimp, other shrimps), blue crab, and lobsters were more heavily impacted by changes in F (Fig. 31.5). In the $F \times 10$ scenario, blue crab and lobsters exhibit similar trends and go extinct after 8–10 years. Shrimps eventually reach a new reduced equilibrium about 0.75 relative to no oil, though in all scenarios they initially indicate a severe decline in biomass. The long-term trends in shrimps indicate that while F plays a role in biomass stabilization, other food web dynamics that occur as a result of the oil spill allow the biomass to grow larger than in the control scenario for all scenarios except $F \times 10$.

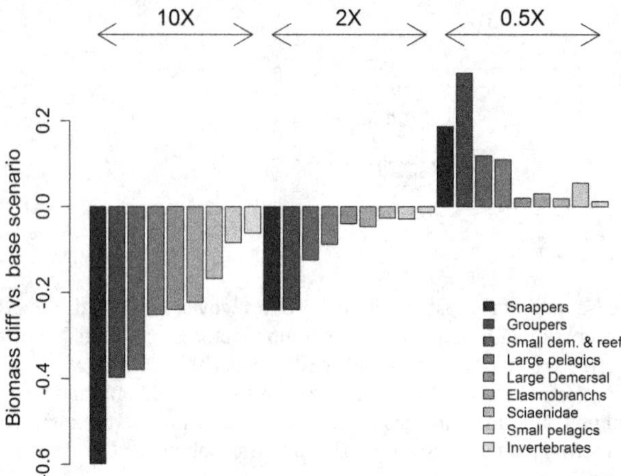

Fig. 31.4 *Average relative biomass changes from the base F scenario for Atlantis exploited guilds across three varied F scenarios.* F × 10, F × 2, and F × 0.5 mean increasing fishing mortality on all exploited groups by 10, 2, and 0.5 times, respectively. All scenarios have MOSSFA detritus forcing and the "worst-case" oil parameters of K = 1000, fish threshold = 363, and benthic invertebrate mortality threshold = 1000

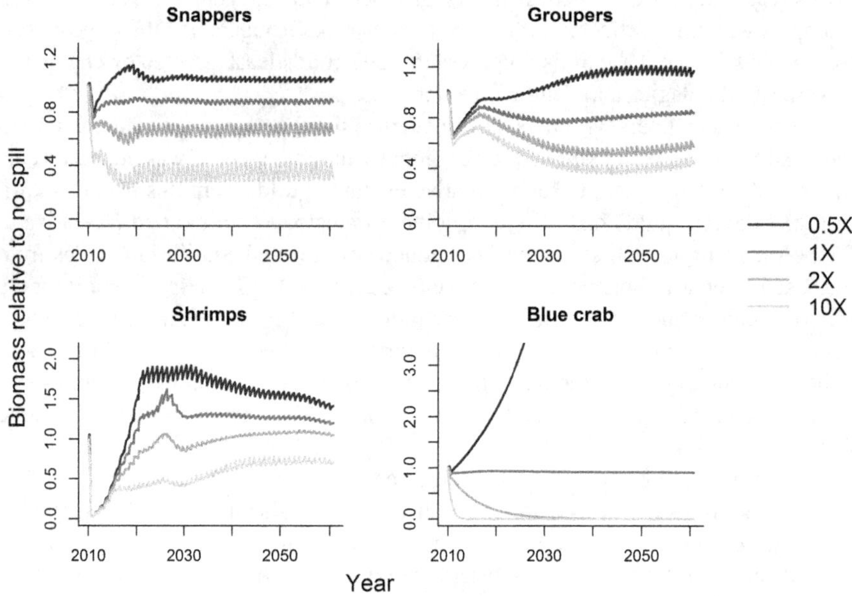

Fig. 31.5 Example groups of functional group relative biomass changes from no oil across three varied F and one base F scenarios. F × 10, F × 2, and F × 0.5 mean increasing fishing mortality F by a factor of 10, 2, and 0.5 across all fishing gears and species, respectively. Snappers and groupers are both guilds, shrimps are a sub-guild within the invertebrate guild, and blue crab is a functional group. The blue crab F × 0.5 relative biomass reached 70 times that of no oil, so the y-axis was limited to 2.5 for visibility of other scenarios' relative biomass

31.4   Discussion

The combined forcing of fish and benthic invertebrate health effects and MOSSFA detrital loading has proven to have a significant effect on Atlantis outputs. Since 1.8–14% of oil released is estimated to have reached the benthos (Valentine et al. 2014) and a key pathway of that oil transport has been identified in MOSSFA (Brooks et al. 2015; Daly et al. 2016; Passow 2016), accurate representation of benthic organisms and environmental forcing functions are important to capturing the dynamics of this habitat, as evidenced by Montagna et al. (2017). In the "worst-case" scenario, where fishes and invertebrates are most heavily impacted by the oil spill, over half of the examined functional groups never recover to pre-spill biomass in the targeted polygons. Although ten functional groups do not recover in the "best-case" scenario, the majority of these are less affected than in the "worst-case" scenario. This suggests that less severe oil spills cause fewer heavily reduced populations, and proportionally more functional groups reach a new stable state biomass that is slightly reduced from pre-spill conditions. These results highlight the importance of understanding oil field distribution, toxicity, uptake/depuration, and species-specific sensitivity.

Atlantis projections for large pelagic fish have not been realized in observations and sampling. Atlantis predicts large population declines in this guild due to the overlap of the sub-surface plume with migration corridors. However, oil avoidance is an important behavior in vertebrates (Rice et al. 1976; Gray 1990; Martin 2017; Ainsworth et al. 2018) that is currently not taken into account. Therefore, the results here likely overestimate the effect on migratory fishes like the large pelagic guild. Developing the ability to influence migration among Atlantis polygons temporarily during large biologically significant events could improve upon this and would provide better insights for this guild. Species-specific oil uptake rates (Eq. 31.1), limiting the maximum exposure a functional group would experience, could effectively differentiate species' abilities to avoid contaminated areas.

The other significant findings of this study are related to the synergistic effects of fishing and oil. A change in fishing pressure can make the difference between recovering post spill and not. Atlantis suggests that in targeted polygons, some of the affected guilds (e.g., snappers and groupers) will not recover without release from pre-spill fishing pressures. In fact, some guilds' long-term biomass is determined by the relative F value, not the magnitude of the decline due to the oil spill event. While the economic impacts of this type of policy decision are not examined here, it is a concept worth exploring in future research and considering as a mitigation response for future spills. The impact of fishing mortality F might be underestimated here or the impact of increased F in the sensitivity analyses. If fishermen moved outside of the closures to fish in areas that are not their normal target regions, this would not be captured in these simulations. If that were included, biomass impacts and catch estimates could differ.

This study provides additional important insights into the potential long-term implications of the DWH oil spill. Using ecosystem modeling to integrate results from coordinated research efforts highlights components critical to ecosystem resilience and helps direct future research efforts. Indirect food web and synergistic effects have significance both for injury assessment and recovery planning in large-scale ecosystem disturbances like DWH.

Funding Information This research was made possible by grants from the Gulf of Mexico Research Initiative through its consortia: the Center for the Integrated Modeling and Analysis of the Gulf Ecosystem (C-IMAGE) and the Deep Sea to Coast Connectivity in the Eastern Gulf of Mexico (Deep-C). Data are publicly available through the Gulf of Mexico Research Initiative Information and Data Cooperative (GRIIDC) at https://data.gulfresearchinitiative.org (doi: 10.7266/n7-dx3q-4y78).

References

Ainsworth CH, Schirripa MJ, Morzaria-Luna HN (2015) An Atlantis ecosystem model for the Gulf of Mexico supporting integrated ecosystem assessment. US Dept Comm NOAA Technical Memorandum NMFS-SEFSC-676, p 149

Ainsworth CH, Paris CB, Perlin N, Dornberger LN, Patterson WF III, Chancellor E, Murawski S, Hollander D, Daly K, Romero IC, Coleman F (2018) Impacts of the Deepwater Horizon oil spill evaluated using an end-to-end ecosystem model. PLoS One 13(1):e0190840

Aman ZM, Paris CB (2013) Response to comment on "Evolution of the Macondo well blowout: simulating the effects of the circulation and synthetic dispersants on the subsea oil transport". Environ Sci Technol 47(20):11906–11907

Barron MG (2012) Ecological impacts of the Deepwater Horizon oil spill: implications for immunotoxicity. Toxicol Pathol 40(2):315–320

Brooks GR, Larson RA, Schwing PT, Romero I, Moore C, Reichart GJ, Jilbert T, Chanton JP, Hastings DW, Overholt WA, Marks KP (2015) Sedimentation pulse in the NE Gulf of Mexico following the 2010 DWH blowout. PLoS One 10(7):e0132341

Chancellor E (2015) Vulnerability of larval fish populations to oil well blowouts in the Northern Gulf of Mexico. Master's Thesis, The University of South Florida, Tampa, FL

Colwell RR, Leinen M, Wilson C, Carron M (2014) The Gulf of Mexico Research Initiative: a new research paradigm. American Petroleum Institute. Int Oil Spill Conf Proc 2014(1):300123

Crain CM, Kroeker K, Halpern BS (2008) Interactive and cumulative effects of multiple human stressors in marine systems. Ecol Lett 11(12):1304–1315

Crain CM, Halpern BS, Beck MW, Kappel CV (2009) Understanding and managing human threats to the coastal marine environment. Ann N Y Acad Sci 1162(1):39–62

Curtin R, Prellezo R (2010) Understanding marine ecosystem-based management: a literature review. Mar Policy 34(5):821–830

Daly KL, Passow U, Chanton J, Hollander D (2016) Assessing the impacts of oil-associated marine snow formation and sedimentation during and after the Deepwater Horizon oil spill. Anthropocene 13:18–33

Deepwater Horizon Natural Resource Damage Assessment Trustees (2016) Deepwater Horizon oil spill: Final Programmatic Damage Assessment and Restoration Plan and Final Programmatic Environmental Impact Statement. Ocean Springs, MS. Accessed May 2018

Dornberger L (2018) Using ecosystem-based modeling to describe an oil spill and assess the long-term effects. PhD Dissertation. University of South Florida, Tampa, 78 pp

Dornberger L, Ainsworth C, Gosnell S, Coleman F (2016) Developing a polycyclic aromatic hydrocarbon exposure dose-response model for fish health and growth. Mar Pollut Bull 109(1):259–266

Fulton EA (2001) The effects of model structure and complexity on the behaviour and performance of marine ecosystem models. PhD Thesis, University of Tasmania, Hobart, Tasmania, Australia, p 427

Fulton EA, Smith A, Johnson C (2004a) Effects of spatial resolution on the performance and interpretation of marine ecosystem models. Ecol Model 176:27–42

Fulton EA, Fuller M, Smith A, Punt A (2004b) Ecological indicators of the ecosystem effects of fishing: Final Report. Australian Fisheries Management Authority, Canberra No. R99, p 1546

Fulton EA, Smith A, Punt A (2005) Which ecological indicators can robustly detect effects of fishing? ICES J Mar Sci 62:540–551

Fulton EA, Smith A, Smith DC (2007) Alternative management strategies for Southeast Australian Commonwealth fisheries: Stage 2: quantitative management strategy evaluation. Australian Fisheries Management Authority, Fisheries Research and Development Corporation, Canberra, ACT, Australia

Gray RH (1990) Fish behavior and environmental assessment. Environ Toxicol Chem 9(1):53–67

Levy JK, Gopalakrishnan C (2010) Promoting ecological sustainability and community resilience in the US Gulf Coast after the 2010 Deepwater Horizon oil spill. J Nat Resources Pol Res 2(3):297–315

Link JS, Gamble RJ, Fulton EA (2011) NEUS—Atlantis: construction, calibration, and application of an ecosystem model with ecological interactions, physiographic conditions, and fleet behavior. NOAA Tech Memo NMFS-NEFSC, p 218

Lubchenco J, McNutt MK, Dreyfus G, Murawski SA, Kennedy DM, Anastas PT, Chu S, Hunter T (2012) Science in support of the Deepwater Horizon response. Proc Nat Acad Sci 109(50):20212–20221

Martin PJ (2000) Description of the Navy Coastal Ocean Model Version 1.0, 749 NRL/FR/ 7322-00-9962. Naval Research Laboratory, p 42

Martin CW (2017) Avoidance of oil contaminated sediments by estuarine fishes. Mar Ecol Prog Ser 576:125–134

Miller C, Medvecky RL, Sherwood TA, Wetzel DL (2017) Uptake, depuration and residence time of polycyclic aromatic hydrocarbons in red drum (*Sciaenops ocellatus*) exposed to south Louisiana crude oil. Poster presented at the Gulf of Mexico Oil Spill & Ecosystem Science Conference, New Orleans, LA

Montagna PA, Baguley JG, Cooksey C, Hyland JL (2017) Persistent impacts to the deep soft-bottom benthos one year after the Deepwater Horizon event. Integr Environ Assess Manag 13(2):342–351

NCEI (2016) Fisheries Closures: Deepwater Horizon Support. National Centers for Environmental Information. National Oceanic and Atmospheric Administration. Accessed Sept 2016

Paris CB, Hénaff ML, Aman ZM, Subramaniam A, Helgers J, Wang DP, Kourafalou VH, Srinivasan A (2012) Evolution of the Macondo well blowout: simulating the effects of the circulation and synthetic dispersants on the subsea oil transport. Environ Sci Technol 46(24):13293–13302

Paris CB, Helgers J, Van Sebille E, Srinivasan A (2013) Connectivity modeling system: a probabilistic modeling tool for the multi-scale tracking of biotic and abiotic variability in the ocean. Environ Model Softw 42:47–54

Passow U (2016) Formation of rapidly-sinking, oil-associated marine snow. Deep Sea Res Part 2 Top Stud Oceanogr 129:232–240

Perlin N, Paris CB, Berenshtein I, Vaz AC, Faillettaz R, Aman ZM, Schwing PT, Romero IC, Schlüter M, Liese A, Noirungsee N, Hackbusch S (2020) Far-field modeling of a deep-sea blowout: sensitivity studies of initial conditions, biodegradation, sedimentation and subsurface dispersant injection on surface slicks and oil plume concentrations (Chap. 11). In: Murawski SA, Ainsworth C, Gilbert S, Hollander D, Paris CB, Schlüter M, Wetzel D (eds) Deep oil spills – facts, fate and effects. Springer, Cham

Pew Ocean Commission (2003) America's living oceans: charting a course for sea change. A report to the nation. Pew Oceans Commission, Arlington, VA

Rice SD, Short JW, Karinen JF (1976) Comparative oil toxicity and comparative animal sensitivity. In: Wolfe DA (ed) Fate and effects of petroleum hydrocarbons in marine ecosystems and organisms. Pergamon Press, New York, pp 78–94

Romero IC, Schwing PT, Brooks GR, Larson RA, Hastings DW, Ellis G, Goddard EA, Hollander DJ (2015) Hydrocarbons in deep-sea sediments following the 2010 Deepwater Horizon blowout in the Northeast Gulf of Mexico. PLoS One 10(5):e0128371

Romero IC, Toro-Farmer G, Diercks AR, Schwing P, Muller-Karger F, Murawski S, Hollander DJ (2017) Large-scale deposition of weathered oil in the Gulf of Mexico following a deep-water oil spill. Environ Pollut 228:179–189

U.S. Commission on Ocean Policy (2004) An ocean blueprint for the 21st century. Final report to the President and Congress, Washington, DC

Valentine DL, Fisher GB, Bagby SC, Nelson RK, Reddy CM, Sylva SP, Woo MA (2014) Fallout plume of submerged oil from Deepwater Horizon. Proc Natl Acad Sci 111(45):15906–15911

Chapter 32
Comparing Ecosystem Model Outcomes Between Ixtoc 1 and *Deepwater Horizon* Oil Spills

Joel G. Ortega-Ortiz, Cameron H. Ainsworth, and Adolfo Gracia

Abstract A spatially explicit biogeochemical Atlantis ecosystem model of the southern Gulf of Mexico was developed to simulate impacts from the Ixtoc 1 oil spill. Oil dispersion and concentration were estimated from satellite imagery and reported flow rates and a dose-response model applied to estimate impact on fish growth and mortality rates. Effects estimated for the Ixtoc 1 spill were compared with previously reported Atlantis ecological model simulations of the *Deepwater Horizon* (DWH) effects on the biomass of eight fish guilds. Biomass decreased sharply after the oil spill in both scenarios, compared to no-oil simulations, reaching minimum values within less than 2 years for most fish guilds. At the end of the 37-year simulation period, all fish guilds recovered to within 95% of the biomass of no-oil simulation for DWH, while all but one guild in the Ixtoc 1 simulation reached new equilibrium at less than 95% of the no-oil biomass. Comparison of the two oil spill simulations highlighted the important roles of starvation and juvenile mortality on fish stock biomass decrease.

Keywords Atlantis ecosystem model · Ixtoc 1 · Fish mortality · Fish biomass

32.1 Introduction

The blowout of the Macondo well and explosion of the *Deepwater Horizon* (DWH) drilling rig started the largest oil spill in the Gulf of Mexico on April 20, 2010. By the time the well was shut 87 days later, at least five million barrels of oil had been discharged into the Gulf (McNutt et al. 2012a). The DWH disaster generated a massive scientific and engineering response, including several studies to document

J. G. Ortega-Ortiz (✉) · C. H. Ainsworth
University of South Florida, College of Marine Science, St. Petersburg, FL, USA
e-mail: jortegaortiz@mail.usf.edu; ainsworth@usf.edu

A. Gracia
Universidad Nacional Autónoma de México, Instituto de Ciencias del Mar y Limnología, Ciudad de México, Mexico
e-mail: gracia@unam.mx

© Springer Nature Switzerland AG 2020
S. A. Murawski et al. (eds.), *Deep Oil Spills*,
https://doi.org/10.1007/978-3-030-11605-7_32

impacts to the Gulf's ecosystems (Lubchenco et al. 2012; McNutt et al. 2012b; Joye et al. 2016). Effects of the DWH oil spill have been described in plankton (Almeda et al. 2013, 2014), microbial communities (Bælum et al. 2012; Bik et al. 2012), foraminifera (Schwing et al. 2015), benthic invertebrates (Montagna et al. 2013), fish communities (Murawski et al. 2014; Tarnecki and Patterson 2015; Ainsworth et al. 2018), dolphins (Venn-Watson et al. 2015; Schwacke et al. 2014), salt marshes (Silliman et al. 2012), deep-water corals (White et al. 2012), and continental shelf ecosystems (Murawski et al. 2016).

While DWH is perhaps the most well-studied oil spill to date, it is not the first massive offshore spill to affect the Gulf of Mexico. On June 3, 1979, an explosion occurred on the exploratory oil well Ixtoc 1 in Campeche Sound. The drilling rig caught fire and sank, and the well spilled oil into the Gulf for nearly 10 months until it was capped on March 23, 1980. Oil and gas were discharged as a jet plume from the top of the well blowout preventer approximately 38 m below the surface and 14 m above the sea floor. Upwelling gas formed a large fire, 10 m in height and 25–30 m in diameter with occasional outburst up to 80 m, burning on the water surface directly above the stack (Haegh and Rossemyr 1980). Turbulent flow conditions caused by the jet plume thoroughly mixed spilled oil, gas, and seawater, creating a three-phase emulsion. The emulsion raised to the surface where most of the gas was burned (Jernelöv and Lindén 1981). During the first part of the blowout, the emulsified oil formed a surface layer 1–4 cm thick, 0.7–5 km wide, and about 60 km long (Jernelöv and Lindén 1981). Later reports describe an oil slick that varied in thickness from 0.2 to 0.5 cm and extended between 2 km and up to 10 km downstream depending on wind and waves (Haegh and Rossemyr 1980; Ross et al. 1980). At the end of the slick, windrows of very stable water-in-oil emulsion were formed and stretched out to 40 km downstream. The windrows gradually broke up downstream and formed oil patches, some of which washed ashore along the western and southern Gulf of Mexico, from Texas to Yucatan (Haegh and Rossemyr 1980). Estimates of the total volume of oil spilled from Ixtoc 1 vary from 0.44 to 1.4 million metric tons (National Research Council 1985; Jernelöv and Lindén 1981). Moreover, the rate of release varied during the 295 days of the spill (National Research Council 1985; PCEESC 1980; OSIR 1980; Ross et al. 1979; Jernelöv and Lindén 1981). Brooks et al. (1981) measured concentrations of volatile liquid hydrocarbons (VLH) as high as 400 µg/L in the surface water near the Ixtoc 1 wellhead. Concentration of both VLH and low-molecular-weight hydrocarbons (LMWH) in the water decreased by one, two, and three orders of magnitude at locations 6, 18, and 24 miles, respectively, downstream from the blowout site due to dilution and evaporation (Brooks et al. 1981). A subsurface plume of oil droplets suspended in the top 20 m of the water column extended 25 km away from the blowout. Subsurface concentrations of oil ranged from 10,600 µg/L within several hundred meters of the blowout to 5 µg/L at a distance of 40 km (Boehm and Flest 1982).

In comparison to DWH, assessments of Ixtoc 1 oil spill effects on the ecosystem were limited (Soto et al. 2014). While initial observations suggested a decline in zooplankton biomass (Guzmán del Próo et al. 1986) and changes to the composition of shrimp trawling bycatch (Soto et al. 1981), no conclusive evidence has been found of significant long-term environmental effects linked to the Ixtoc 1 oil spill (Soto et al. 2014; Jernelöv and Lindén 1981; Yáñez-Arancibia and Sánchez-Gil 1986; García-Cuéllar et al. 2004). Amezcua-Linares et al. (2015) estimated a decrease in diversity and changes in fish assemblages based on fisheries independent trawling surveys conducted before and after the Ixtoc 1 oil spill. They also identified changes in commercial fisheries landing composition pre- and post-spill and attributed these changes to impacts on larvae survival, although recognized that variations could also be related to ecological, biological, and economic factors (Amezcua-Linares et al. 2015). Natural variability in the ecosystem and lack of pre-spill baseline information are among the factors that limit the assessment of Ixtoc 1 effects (Soto et al. 2014; García-Cuéllar et al. 2004).

Applying a spatially explicit biogeochemical end-to-end ecosystem model, Ainsworth et al. (2018) simulated impacts from the DWH oil spill and subsequent recovery of fish guilds. Their results suggest that recovery of high-turnover populations may occur within 10 years, but some slower-growing populations may take 30+ years to fully recover (Ainsworth et al. 2018). Here we use a similar ecosystem model, an Atlantis model focused on the Southern Gulf of Mexico (SGoM), to simulate impacts and subsequent recovery of fish guilds in the area affected by the Ixtoc 1 oil spill. We compare our results to the simulations done for DWH in the northern Gulf to look at fisheries and ecosystem impacts.

32.2 Methods

32.2.1 *Ecological Model*

Atlantis is a deterministic modeling approach that simulates the structure and functioning of marine ecosystems by integrating physical conditions (salt, heat, and water circulation), biogeochemistry (nutrient input, detritus, and bacterial dynamics), biological processes (primary production, growth, reproduction, movement, mortality, and predator-prey interactions), and fisheries (Fulton et al. 2004, 2011). The SGoM Atlantis model is derived from the implementation of the Gulf of Mexico (GoM) model described in Ainsworth et al. (2015). The primary differences between the GoM and SGoM Atlantis models are in the polygon geometry, the partitioning of fisheries statistics between species groups and polygons, and the inclusion of a functional group for octopus due to the relevance of its fishery in the southern Gulf.

In the Atlantis modeling framework, ecosystems are resolved spatially in three dimensions using an irregular 3D polygon structure that saves computation time in homogeneous space. For the SGoM model, 42 polygons were designed to capture important biophysical and jurisdictional features (Fig. 32.1). Bathymetry, bottom type, sediments, and riverine discharge were taken into account. In general, the polygons reflect the continental shelf (from shore to the 200 m isobath), slope (200 m to 3000 m), and abyssal regions (> 3000 m). Protected areas and major

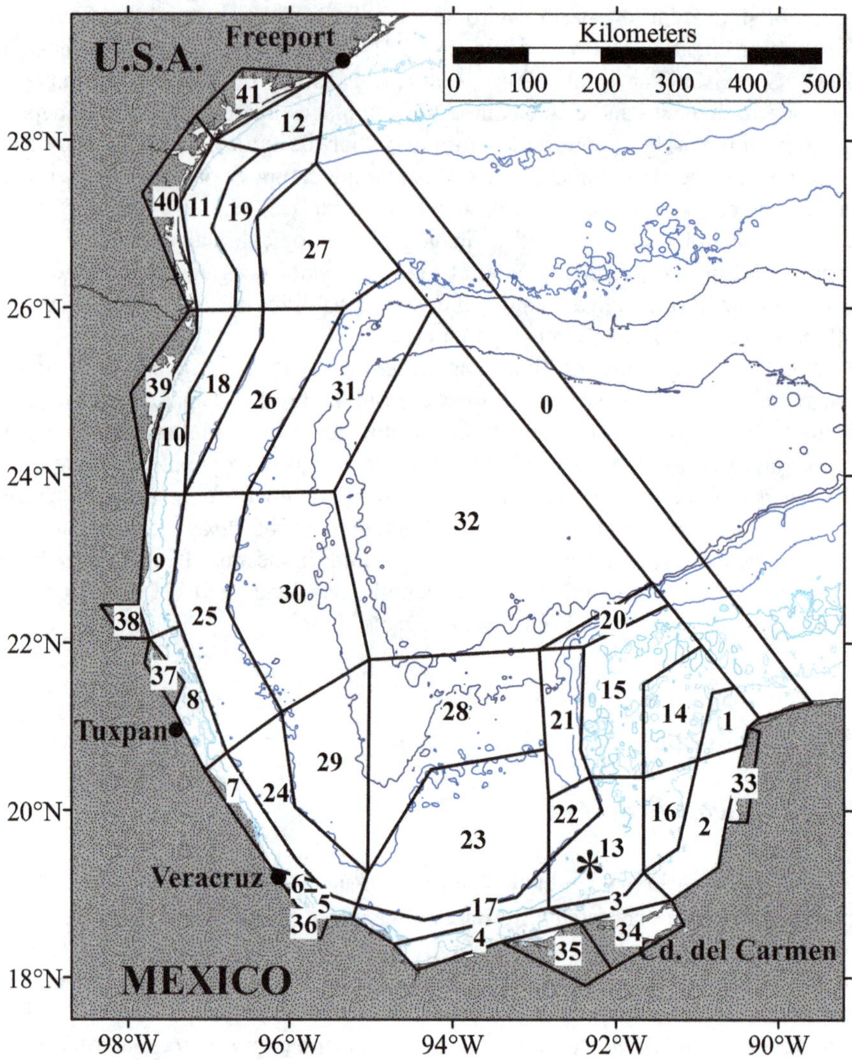

Fig. 32.1 Atlantis geometry for the southern Gulf of Mexico model showing polygon numbers. The asterisk indicates the location of the Ixtoc 1 oil well. Contour lines correspond to the 20 m, 50 m, 200 m, 2000 m, 3000 m, and 3500 m isobaths

estuaries are represented (polygon numbers in parenthesis): Sistema Arrecifal Veracruzano (6), Los Petenes and Celestún (33), Laguna de Términos (34), Pantanos de Centla (35), Laguna de Alvarado (36), Laguna de Tamiahua (37), Pánuco delta (38), and Laguna Madre (39, 40, and 41). Natural resource exploitation patterns, political boundaries, and economic exclusive zones (EEZ) of the USA and Mexico were also considered. Polygon 13 includes the location of the Ixtoc 1 well and the oil rig exclusion zone in Campeche Sound that was closed to fisheries from 1979 to

2016. Depending on their maximum depth, polygons were subdivided into up to six water column layers (0–10 m, 10–20 m, 20–50 m, 50–200 m, 200–2000 m, and 2000–4000 m), plus a sediment layer.

32.2.2 Hydrographic Data

Hydrodynamic data were obtained from the American Seas (AMSEAS) ocean prediction system operated by the USA Naval Oceanographic Office (NAVOCEANO) and based on the Navy Coastal Ocean Model (NCOM) developed by the Naval Research Laboratory (Martin et al. 2009). AMSEAS estimates ocean temperature, salinity, and eastward and northward currents on a regular latitude/longitude grid with a resolution of 1/36 degree. A 1-year AMSEAS data series with 12-hour steps was integrated over the Atlantis polygons using the Ocean Model Slicer tool created by the National Oceanographic and Atmospheric Administration (NOAA) National Coastal Data Development Center (Ainsworth et al. 2018). The resulting dataset included average temperature, salinity, and net water flux for each polygon and depth layer every 12 hours (Ainsworth et al. 2015). Initial conditions of nitrates, dissolved oxygen, and silicate concentrations correspond to climatological means obtained from NOAA's Gulf of Mexico Data Atlas (http://gulfatlas.noaa.gov/catalog/).

32.2.3 Functional Groups

Atlantis simulates essential ecosystem dynamics using functional groups of species aggregated by life history, habitat, and trophic similarities. Species of high fisheries importance or conservation interest are represented by single species groups. The SGoM model uses 91 functional groups, including primary producers (8 groups), nutrient cycle (5), structural species (4), filter feeders (3), pelagic invertebrates (4), macrobenthos (6), octopus (1), shrimp (4), elasmobranchs (6), forage fish (4), reef fish (11), demersal fish (12), pelagic fish (14), sea turtles (3), seabirds (2), and mammals (4). The model tracks flow of nitrogen as it passes between functional groups through the food web. Fish abundance data was obtained from the Southeast Area Monitoring and Assessment Program (Rester 2011; Ainsworth et al. 2015). Spatial distribution of biomass was based on estimates from generalized additive models (Drexler and Ainsworth 2013), and the diet matrix was determined using gut content and statistical analysis (Masi et al. 2014; Tarnecki et al. 2016). Vertebrates (fish, turtles, birds, and mammals), shrimp, and octopus groups are age-structured and tracked by number of individuals per polygon, structural nitrogen weight per individual (mg sN/individual), and reserve nitrogen weight per individual (mg rN/individual). Structural nitrogen represents hard body part such as bones and teeth, while reserve nitrogen represents somatic and gonadal soft body tissue. Invertebrates and primary producers are tracked as homogeneous biomass pools on a per-volume basis (mg N/m^3) for pelagic and infaunal groups and a per-area basis (mg N/m^2) for epibenthic groups.

32.2.4 Fisheries Catch

Historical fisheries catch data were included in the model through time series files describing the landings (in mg sec^{-1}) of each functional group for each polygon from 1979 to 2011. Catch time series files were constructed from species-specific fisheries landing data, collated by functional group and distributed across time and space (Ainsworth et al. 2015). Historical catch data was restricted to marine, wild-caught landings and, therefore, does not include landings associated with aquaculture or freshwater species. Catch for polygons within the US EEZ were obtained from Ainsworth et al. (2015) who compiled records from the Southeast Data, Assessment, and Review (SEDAR) stock assessments, from NOAA's Offices of Science and Technology Commercial Fisheries Statistics database and NOAA's Recreational Billfish Survey. For the SGoM model, monthly catch data for polygons within Mexico's EEZ were obtained from Arreguín-Sánchez and Arcos-Huitrón (2007) who compiled records from the annual reports (*Anuario Estadístico de Pesca*) published by the Mexican Federal Government (the database is currently managed by CONAPESCA – *Comisión Nacional de Acuacultura y Pesca; Secretaría de Agricultura, Ganadería Desarrollo Rural, Pesca y Alimentación*). Spatial allocation of species captured on artisanal fisheries was based on the local CONAPESCA office where landing was reported. Semi-industrial fisheries in Mexico use ships with larger range and can unload and report their catch in an office away from their fishing area. Therefore, landings for shelf trawl fleet were spatially allocated based on shrimp species habitat preferences (Gracia and Vázquez-Bader 1999; Diario Oficial de la Federación 2012), landings from the Yucatan-shelf longline fleet were allocated according to reported red grouper fishing areas (Hernandez and Seijo 2003), and landings of the pelagic longline fleet were allocated based on fishing locations reported to the International Commission for the Conservation of Atlantic Tunas (Ainsworth et al. 2015). Species in CONAPESCA fisheries landings reports were assigned to functional groups based on the list of taxa and common names provided by Arreguín-Sánchez and Arcos-Huitrón (2007). Spatial allocation accounted for catch restrictions in marine protected areas and the vessel exclusion zone around the oil platforms in Campeche Sound.

US fishing fleets and fleet distribution of landings were obtained from Ainsworth et al. (2015) who allocated the catch using the 2011 species-specific landings-by-gear data provided by NOAA. Mexican fishing fleets and fleet distribution of landings were determined from the landings and gear information in the *Carta Nacional Pesquera* (Diario Oficial de la Federación 2012) and CONAPESCA's *Anuario Estadístico* database.

32.2.5 *Growth and Mortality Effects*

Change in biomass (B, in units of mg N) for each polygon, depth layer, and age class of functional group i is determined following the Atlantis model application described by Ainsworth et al. (2018), as defined in Eq. 32.1:

$$\frac{dB_i}{dt} = \left[R_i + T_{IMM,i} - T_{EM,i} - M1_i - \sum_j P_{i,j} - F_i \right] \cdot g_{\text{oil}} \left[G_{i,s} + G_{i,r} \right] \tag{32.1}$$

R is entry of new individuals into the age class from recruitment or aging, $T_{IMM,i}$ and $T_{EM,i}$ are immigration and emigration, F_i is fisheries catch, and $P_{i,j}$ is predation by predator j on group i. $G_{i,s}$ and $G_{i,r}$ are growth in structural and reserve nitrogen, which are functions of assimilation efficiency, growth rate, and total consumption. $M1$ summarizes natural mortality from non-predation sources as in Eq. 32.2:

$$M1 = m_l + m_q + m_{\text{oil}} \tag{32.2}$$

where m_l and m_q are linear and quadratic mortality parameters set iteratively during model calibration and m_{oil} is mortality caused by oil exposure derived from the dose-response model in Eq. 32.3. g_{oil} is the modifier on growth rate caused by oil exposure defined in Eq. 32.4:

$$m_{\text{oil}} = m_{\text{pelagic}} \cdot (1 - B) + m_{\text{benthic}} \cdot B \tag{32.3}$$

$$g_{\text{oil}} = g_{\text{pelagic}} \cdot (1 - B) + g_{\text{benthic}} \cdot B \tag{32.4}$$

The m_{oil} term represents a weighted average of a pelagic mortality modifier (m_{pelagic}) and a benthic mortality modifier (m_{benthic}) that reflects the proportion of B in the affected group's diet that comes from either benthic or pelagic prey. Growth rate g_{oil} is similarly weighted based on pelagic and benthic diet. These modifiers were calculated from the dose-response models developed by Dornberger et al. (2016) to describe the effects of polycyclic aromatic hydrocarbons (PAH) on growth and mortality rates in fishes. The most parsimonious and best fit by Dornberger et al. (2016) were "hockey-stick" models that imply the existence of a PAH concentration threshold below which there are no adverse effects and above which there is a log-linear increase in mortality (Eq. 32.5) and decrease in growth (Eq. 32.6):

$$m_t = \alpha \cdot \log \left[\varphi_t \cdot \frac{1}{\beta} \right] \tag{32.5}$$

$$g_t = 1 - \left(\gamma \cdot \log \left[\varphi_t \cdot \frac{1}{\delta} \right] \right) \tag{32.6}$$

These dose-response models (Eqs. 32.5 and 32.6) estimate the mortality and growth as a function of the amount of bioavailable oil (φ_t) at time step t. The estimates, obtained separately for pelagic and benthic environments, were used for the modifiers in Eqs. 32.3 and 32.4, respectively. Parameters for slope (α = 0.2885 and γ = 0.0531) and PAH concentration threshold (β = 907.4306 and δ = 28.422 in ppb) were fit by maximum likelihood estimation (Dornberger et al. 2016; Ainsworth et al. 2018).

32.2.6 Estimation of Ixtoc 1 Spill Distribution and Oil Concentration

We estimated the amount of oil spilled during each calendar month, from June 1979 to March 1980, from the flow rates reported by the Coordinated Program of Ecological Studies in Campeche Sound (PCEESC 1980): 4770 tons per day (t·d^{-1}) during the first 41 days of the spill, 3180 t·d^{-1} for the next 32 days, then 1590 t·d^{-1} for 104 days, 318 t·d^{-1} for 35 days, and 229 t·d^{-1} for 83 days. The PCEESC (1980) report also indicates that approximately 58.1% of the spilled oil was burned or evaporated, 8.9% was recovered either at the surface or near the well head, and only 33% of the oil remained in the environment. Therefore, for estimation of oil effects, we only included 33% of the monthly estimates of spilled oil.

The oil was spatially allocated to the model area based on the results from Sun et al. (2015) who identified Ixtoc 1 oil slick in satellite images. A total of 82 satellite images processed by Sun et al. (2015) were grouped by calendar month, and the areas of Ixtoc 1 oil slick represented by polygon features were overlaid in ArcGIS (Fig. 32.2). The number of oil slick features overlaid in any particular area each month served as coefficient to determine relative oil concentration: an area with one oil slick feature had a coefficient of 1, while areas with two overlaying slick features in a month received a coefficient of 2. The estimated oil spilled each month was then divided over the entire area covered by the images and adjusted by the coefficients, resulting in estimates of oil volume by squared meter. These values were multiplied by a factor or 0.0001 to estimate the concentration of bioavailable oil (ppb) per feature. The oil concentration was used in Eqs. 32.5 and 32.6 to estimate mortality and growth effects in the feature layers. The total monthly mortality and growth effects in each Atlantis polygon were estimated by prorating the effect in any oil slick features present, as a proportion of the total area of the Atlantis polygon.

The highest oil concentration was estimated at 44183 ppb for sediments for a small area (2.8 hectares) of polygon 13, in the immediate vicinity of the Ixtoc 1 oil well, in December 1979. Oil concentrations greater than 8400 ppb were only estimated for small portions of polygon 13. Concentrations greater than the 907.4 ppb mortality effect threshold were estimated for portions of 36 Atlantis polygons during at least 1 month. Concentrations greater than the 28.4 in ppb growth effects threshold were estimated over a larger area in the same 36 polygons. The only polygons without estimated oil concentrations greater than the mortality or growth effect thresholds were six estuaries (polygons number 33, 35, 36, 37, 38 and 41) where oil

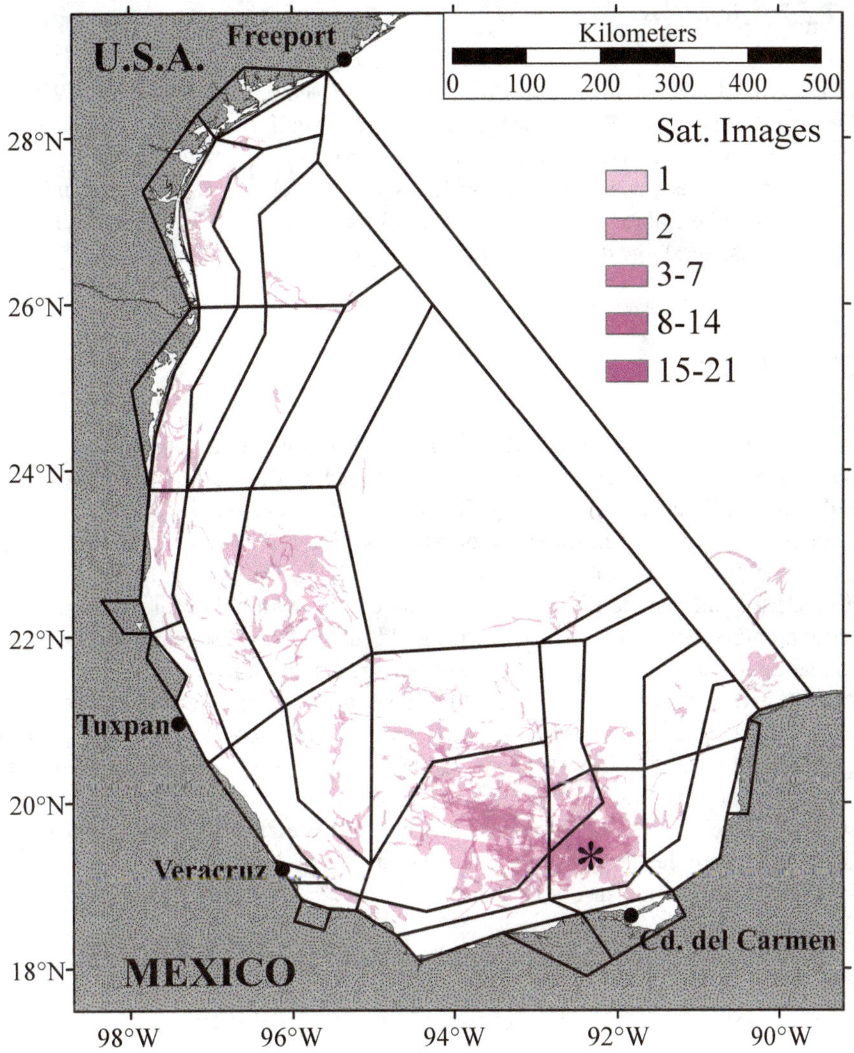

Fig. 32.2 Atlantis polygons and surface oil footprint of the Ixtoc 1 oil spill determined by satellite images. The asterisk indicates the location of the Ixtoc 1 oil well. Color contours indicate the number of overlapping satellite images. (From data described in Sun et al. 2015)

was not observed in the satellite images. However, after accounting for the extent of areas where oil was detected by satellite, the majority of the mortality effect occurred close to the spill site, in polygons 13, 22, and 23. During the summer of 1979, some areas with oil concentration above the mortality threshold were also estimated for polygons 9, 10, 11, 17, 28, and 30, located west and northwest.

32.2.7 Analysis

Simulations for the SGoM model were run for 37 years, from January 1, 1979, to December 31, 2013. For comparison purposes, we followed the approach by Ainsworth et al. (2018) and aggregated results from Atlantis fish functional groups into eight "guilds': snappers (family: Lutjanidae), groupers (family: Serranidae), Sciaenidae, elasmobranchs, large pelagic fish, small pelagic fish, small demersal and reef fish, and large demersal fish.

32.3 Results

A sharp decrease in biomass relative to no-oil spill simulations followed by a gradual increase (i.e., recovery) was observed on all fish guilds for both the Ixtoc 1 and DWH oil spill scenarios (Fig. 32.3). The biomass decrease was higher for the DWH scenario, with average minimum biomass of 40.2% relative to no-oil simulation, compared 70.2% for the Ixtoc 1 scenario (Table 32.1). Despite the larger decrease, all the fish guilds in the DWH scenario recovered to within 95% of the no-oil scenario biomass, while only half of the fish guilds in the Ixtoc 1 scenario recovered up to that threshold. Indeed, except for large pelagic fish, biomass for fish guilds in the Ixtoc 1 spill became asymptotic at values lower than the no-oil scenario toward the end of the 37-year simulation (Fig. 32.3). On the DWH scenario, only the grouper guild showed a similar biomass equilibrium at 94.8% of the no-oil simulation after 37 years. The magnitude of biomass decrease estimated in the Ixtoc 1 simulation was comparable to results from Ainsworth et al. 2018 in the southwest region of the Gulf. For example, the grouper guild had a decrease of about 5 t·km^{-2} in the Campeche Bay region in both simulations (Fig. 32.4). However, condition factor decreased in the DWH simulations.

In contrast to results from the DWH simulation described by Ainsworth et al. (2018), no meaningful changes in condition factor (Fig. 32.5) or shifts in age composition pre- and post-spill were observed in the Ixtoc 1 simulations (Fig. 32.6).

32.4 Conclusion

While it is practically impossible to determine the precise distribution and concentration of oil spilled from the Ixtoc 1 well, satellite imagery and flow rate estimates provide a reasonable scenario to use in the ecological model simulations. Differences in the depth of the wells and the use of oil dispersants at depth during the DWH spill may have resulted in more effects on the continental shelf during Ixtoc 1 and over the continental slope and abyssal region during DWH.

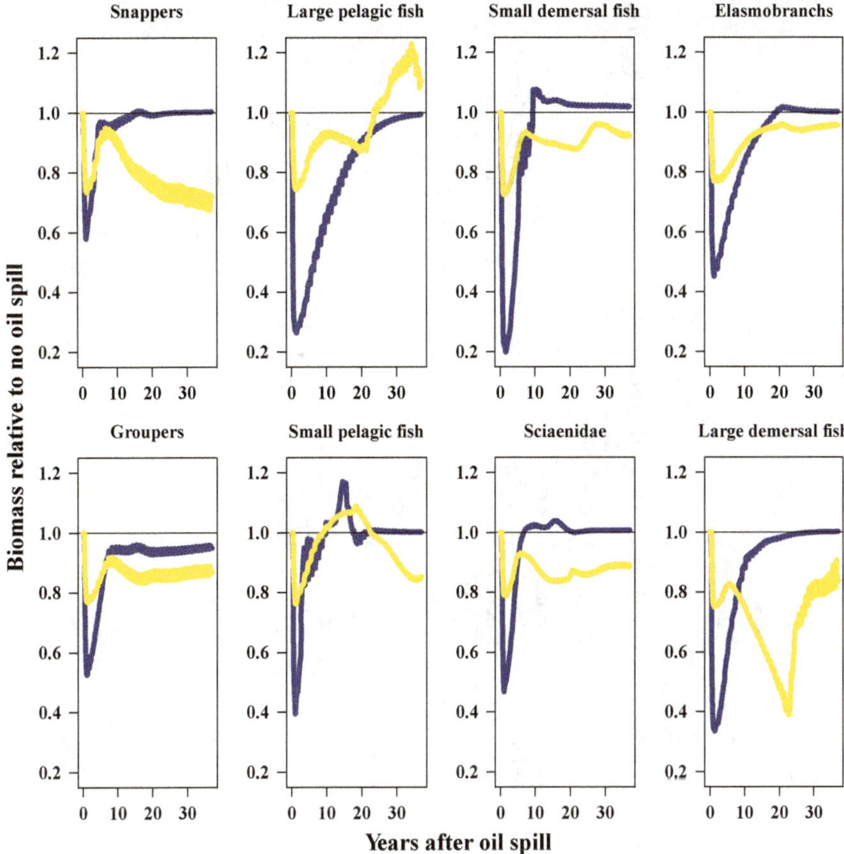

Fig. 32.3 Biomass trajectories for species guilds relative to no-spill scenario for Ixtoc 1 (yellow) and DWH (dark blue) 37-year simulations. Biomass is summed across all functional groups within these guilds

32.4.1 Difference in Impact on Juveniles and Adults

In contrast to results from the Ixtoc 1 simulations, Ainsworth et al. (2018) detected a large shift in the age composition of most reef and demersal guilds suggesting that young fish (juvenile and subadults) were impacted more heavily than older fish by the DWH spill. In their study, young fish were not represented as being any more vulnerable physiologically to oil than older fish. Therefore, any differences must be attributable to the degree of exposure, i.e., the relative overlap between the life stages' spatial distributions and the oil spills' footprint (also potentially influenced by the degree of association with sediment) (Ainsworth et al. 2018). The results of our simulations suggest that Ixtoc 1 oil did not linger in inshore juvenile habitats where it could have a cumulative effect with time but rather moved quickly throughout the system, exposing nearshore and offshore habitats to a similar degree. Less

Table 32.1 Biomass change, relative to no-oil simulation, in eight fish guilds for the 37-year Ixtoc 1 and DWH scenarios. DNR = fish guild did not recover to within 95% of the no-oil scenario

Fish guild	Ixtoc 1				DWH			
	Minimum relative biomass	Years to minimum biomass	Years to 95% recovery	Relative biomass after 37 years	Minimum relative biomass	Years to minimum biomass	Years to 95% recovery	Relative biomass after 37 years
Snappers	67.4%	36.25	DNR	71.9%	57.9%	0.82	4.52	100.3%
Large pelagic	74.3%	1.07	22.36	109.8%	26.4%	1.23	24.33	99.3%
Small demersal	72.4%	0.82	26.22	92.0%	19.9%	1.32	8.38	101.9%
Elasmobranchs	76.7%	1.81	17.51	95.6%	45.1%	0.99	15.37	100.2%
Groupers	76.8%	0.82	DNR	86.8%	52.6%	0.82	8.14	94.8%
Small pelagic	76.2%	0.66	6.25	85.2%	39.4%	0.74	4.11	100.2%
Sciaenidae	78.9%	1.15	DNR	88.7%	46.7%	0.82	5.42	100.7%
Large demersal	38.8%	22.27	DNR	83.5%	33.5%	1.15	13.89	100.2%
Average	70.2%	8.11	18.08	89.2%	40.2%	0.99	10.52	99.7%

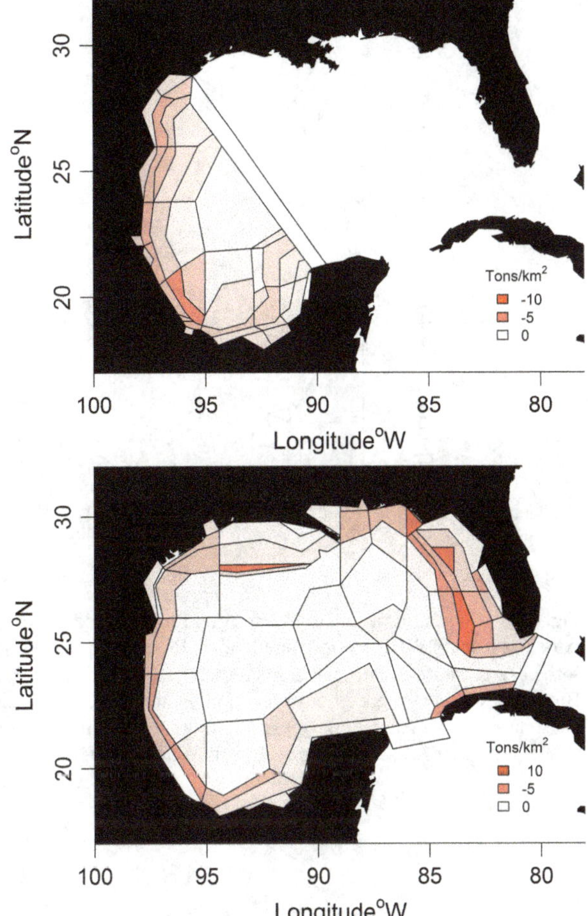

Fig. 32.4 Absolute biomass reduction for no-oil versus oil scenario for grouper guild for Ixtoc 1 (top) and DWH (bottom) simulations. Biomass minima is shown occurring at 10 months after the oil spill. (DWH data after Ainsworth et al. (2018, https://doi.org/10.1371/journal.pone.0190840.g005))

turbulent flow of oil is consistent with the overall lesser impact on sedentary reef-associated and demersal guilds that Ixtoc 1 had compared to DWH as predicted by Atlantis. In this case, the narrow shelf in the southwestern and western Gulf would not have allowed any significant gradient in oil concentration to establish perpendicular to shore. Our estimates of Ixtoc 1 oil distribution based on satellite images most likely underestimated the amount of oil that reached inshore juvenile habitat. Although up to 6.3% of Ixtoc 1 oil ended up on beaches (Jernelöv and Lindén 1981), it did not reach Laguna de Términos (PCEESC 1980), which is an important nursery ground in the southwestern Gulf. Thus, an alternative explanation as to why Ixtoc 1 had less of an effect on young fish compared to DWH may therefore be that Ixtoc 1 contaminated less the productive inshore habitats. This seems likely considering the well-documented oil intrusion on highly productive coastal wetlands of Louisiana and Mississippi: areas critical to the production of juvenile fish (Nixon et al. 2016).

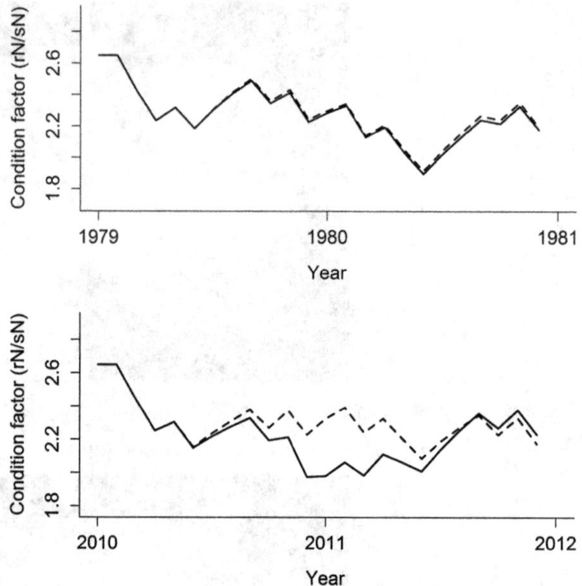

Fig. 32.5 Condition factor of grouper guild represented as reserve: structural nitrogen ratio for Ixtoc (top) and DWH (bottom) simulations. Reserve represents soft body tissue that can be reabsorbed (e.g., muscle, fat, gonads); structural represents hard tissues and structures (e.g., bone). High rN/sN indicates good body condition. Dotted line, no-oil scenario; solid line, oiled scenario. Seasonal saw-toothed pattern (present in both scenarios) reflects gonadal tissue loss in spawning. (DWH data after Ainsworth et al. (2018, https://doi.org/10.1371/journal.pone.0190840.s008))

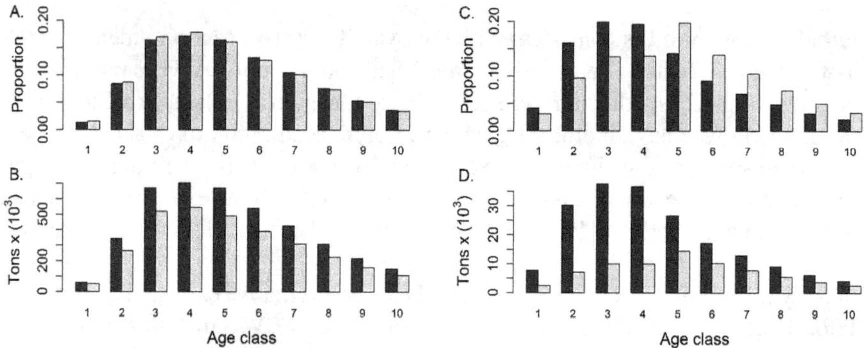

Fig. 32.6 Age composition of large demersal fish guild for Ixtoc (left side panels **a, b**) and DWH (right side panels **c, d**). No-oil (dark gray); oiled (light gray). Top panels (**a, c**) show relative proportion and bottom panels (**b, d**) absolute biomass. (DWH data after Ainsworth et al. (2018, https://doi.org/10.1371/journal.pone.0190840.g004))

32.4.2 *Recovery*

The trajectories of biomass recovery for fish guilds impacted by DWH are more monotonic than those impacted by Ixtoc 1. Monotonic recovery of DWH may be explained because oil mortality was much higher than predation mortality. The relatively quick release of oil contamination as a source of mortality enabled a quick and relatively linear recovery. The more complex biomass recovery trajectories after the Ixtoc 1 spill may be due to the relatively higher influence of predation mortality after the oil mortality effect was removed. Predation mortalities in Atlantis exhibit non-linearities typical of complex ecosystem dynamics, while oil mortality has a more uniform effect across all functional groups. The complex biomass recovery trajectories after the Ixtoc 1 oil spill could also explain the changes in fish assemblages composition reported by Amezcua-Linares et al. (2015).

32.4.3 *Trophic Impacts*

Ainsworth et al. (2018) concluded that a high degree of connectivity exists between populations in the Gulf. The loss of grouper biomass in Campeche Bay in DWH simulation, which was large and in the same order of magnitude as the Ixtoc 1 simulation, was communicated from over 1000 miles away. There is similar evidence from the current study of strong connectivity over large spatial scales. However, biomass decrease in the southern Gulf estimated by the DWH simulation was solely due to trophic interactions, including starvation, as evidenced by a decrease in condition factor estimates. Whereas, the biomass decrease estimated by Ixtoc 1 simulations was due to oil mortality. This highlights the important role of starvation on stock biomass decrease compared to direct oil mortality.

Acknowledgments Shaojie Sun provided GIS data of processed satellite images, and Francisco Arreguín-Sánchez provided fisheries data for the southern Gulf. This research was made possible by grants from the Gulf of Mexico Research Initiative through The Center for the Integrated Modeling and Analysis of the Gulf Ecosystem (C-IMAGE). Atlantis model data are publicly available through the Gulf of Mexico Research Initiative Information and Data Cooperative (GRIIDC) at https://data.gulfresearchinitiative.org (doi: 10.7266/N779435P; 10.7266/n7-24gt-6z95).

References

Ainsworth CH, Schirripa MJ, Morzaria Luna HN (eds) (2015) An Atlantis ecosystem model for the northern Gulf of Mexico supporting integrated ecosystem management. U.S. Dept. Commer., NOAA Tech. Memo. NMFS-SEFSC-TM-676. https://doi.org/10.7289/V5X63JVH

Ainsworth CH, Paris CB, Perlin N, Dornberger LN, Patterson WF III, Chancellor E, Murawski S, Hollander D, Daly K, Romero IC, Coleman F, Perryman H (2018) Impacts of the Deepwater Horizon oil spill evaluated using an end-to-end ecosystem model. PLoS One 13(1):e0190840. https://doi.org/10.1371/journal.pone.0190840

Almeda R, Wambaugh Z, Wang Z, Hyatt C, Liu Z, Buskey EJ (2013) Interactions between zooplankton and crude oil: Toxic effects and bioaccumulation of polycyclic aromatic hydrocarbons. PLoS One 8(6):e67212. https://doi.org/10.1371/journal.pone.0067212

Almeda R, Baca S, Hyatt C, Buskey EJ (2014) Ingestion and sublethal effects of physically and chemically dispersed crude oil on marine planktonic copepods. Ecotoxicology 23(6):988–1003. https://doi.org/10.1007/s10646-014-1242-6

Amezcua-Linares F, Amezcua F, Gil-Manrique B (2015) Effects of the Ixtoc I oil spill on fish assemblages in the Southern Gulf of Mexico. In: Alford JB, Peterson MS, Green CC (eds) Impacts of oil spill disasters on marine habitats and fisheries in North America. CRC Press, Boca Raton, FL, pp 209–236. https://doi.org/10.1201/b17633-15

Arreguín-Sánchez F, Arcos-Huitrón E (2007) Fisheries catch statistics for Mexico. In: Zeller D, Pauly D (eds) Reconstruction of marine fisheries catches for key countries and regions (1950-2005), Fisheries Centre Research Reports 15(2). Fisheries Centre, University of British Columbia, pp 81–103

Bælum J, Borglin S, Chakraborty R, Fortney JL, Lamendella R, Mason OU, Auer M, Zemla M, Bill M, Conrad ME, Malfatti SA, Tringe SG, Holman HY, Hazen TC, Jansson JK (2012) Deep-sea bacteria enriched by oil and dispersant from the Deepwater Horizon spill. Environ Microbiol 14(9):2405–2416. https://doi.org/10.1111/j.1462-2920.2012.02780.x

Bik HM, Halanych KM, Sharma J, Thomas WK (2012) Dramatic shifts in benthic microbial eukaryote communities following the Deepwater Horizon oil spill. PLoS One 7(6):e38550. https://doi.org/10.1371/journal.pone.0038550

Boehm PD, Flest DL (1982) Subsurface distributions of petroleum from an offshore well blowout. The Ixtoc I blowout, Bay of Campeche. Environ Sci Technol 16(2):67–74. https://doi.org/10.1021/es00096a003

Brooks JM, Wiesenburg DA, Burke RA, Kennicutt MC (1981) Gaseous and volatile hydrocarbon inputs from a subsurface oil spill in the Gulf of Mexico. Environ Sci Technol 15(8):951–959. https://doi.org/10.1021/es00090a011

Diario Oficial de la Federación (2012) Acuerdo por el que se da a conocer la Actualización de la Carta Nacional Pesquera. Secretaría de Agricultura, Ganadería, Desarrollo Rural, Pesca y Alimentación., vol DCCVII. Mexico

Dornberger L, Ainsworth C, Gosnell S, Coleman F (2016) Developing a polycyclic aromatic hydrocarbon exposure dose-response model for fish health and growth. Mar Pollut Bull 109(1):259–266. https://doi.org/10.1016/j.marpolbul.2016.05.072

Drexler M, Ainsworth CH (2013) Generalized additive models used to predict species abundance in the Gulf of Mexico: an ecosystem modeling tool. PLoS One 8(5):e64458. https://doi.org/10.1371/journal.pone.0064458

Fulton EA, Parslow JS, Smith ADM, Johnson CR (2004) Biogeochemical marine ecosystem models II: the effect of physiological detail on model performance. Ecol Model 173(4):371–406. https://doi.org/10.1016/j.ecolmodel.2003.09.024

Fulton EA, Link JS, Kaplan IC, Savina-Rolland M, Johnson P, Ainsworth C, Horne P, Gorton R, Gamble RJ, Smith ADM, Smith DC (2011) Lessons in modelling and management of marine ecosystems: the Atlantis experience. Fish Fish 12(2):171–188. https://doi.org/10.1111/j.1467-2979.2011.00412.x

García-Cuéllar JÁ, Arreguín-Sánchez F, Hernández Vázquez S, Lluch-Cota DB (2004) Impacto ecológico de la industria petrolera en la sonda de Campeche, México, tras tres décadas de actividad: Una revisión. Interciencia 29(6):311–319

Gracia A, Vázquez-Bader AR (1999) Shrimp fisheries in the south Gulf of Mexico: present and future management alternatives. In: Kumpf H, Steidinger K, Sherman K (eds) The Gulf of Mexico Large Marine Ecosystem. Blackwell Science, Malden, MA, pp 205–224

Guzmán del Próo SA, Chávez EA, Alatriste FM, de la Campa S, De la Cruz G, Gómez L, Guadarrama R, Guerra A, Mille S, Torruco D (1986) The impact of the Ixtoc 1 oil spill on zooplankton. J Plankton Res 8(3):557–581. https://doi.org/10.1093/plankt/8.3.557

Haegh T, Rossemyr LI (1980) A comparison of weathering processes of oil from the *Bravo* and the *Ixtoc* blowouts. OTC-3702-MS. In: Proceedings, 12th Annual Offshore Technology Conference, Houston, Texas, 5–8 May 1980. pp 237–244. doi:https://doi.org/10.4043/3702-MS

Hernandez A, Seijo JC (2003) Spatial distribution analysis of red grouper (*Epinephelus morio*) fishery in Yucatan, Mexico. Fish Res 63(1):135–141

Jernelöv A, Lindén O (1981) Ixtoc I: a case study of the world's largest oil spill. Ambio 10(6):299–306

Joye SB, Bracco A, Ozgokmen TM, Chanton JP, Grosell M, MacDonald IR, Cordes EE, Montoya JP, Passow U (2016) The Gulf of Mexico ecosystem, six years after the Macondo oil well blowout. Deep Sea Res Part II Top Stud Oceanogr 129:4–19. https://doi.org/10.1016/j.dsr2.2016.04.018

Lubchenco J, McNutt MK, Dreyfus G, Murawski SA, Kennedy DM, Anastas PT, Chu S, Hunter T (2012) Science in support of the *Deepwater Horizon* response. Proc Natl Acad Sci 109(50):20212–20221. https://doi.org/10.1073/pnas.1204729109

Martin PJ, Barron CN, Smedstad LF, Campbell TJ, Wallcraft AJ, Rhodes RC, Rowley C, Townsend TL (2009) User's Manual for the Navy Coastal Ocean Model (NCOM) Version 4.0. NRL/MR/7320-09-9151. Naval Research Laboratory, Stennis Space Center, MS 39529-5004

Masi MD, Ainsworth CH, Chagaris D (2014) A probabilistic representation of fish diet compositions from multiple data sources: a Gulf of Mexico case study. Ecol Model 284:60–74. https://doi.org/10.1016/j.ecolmodel.2014.04.005

McNutt MK, Camilli R, Crone TJ, Guthrie GD, Hsieh PA, Ryerson TB, Savas O, Shaffer F (2012a) Review of flow rate estimates of the *Deepwater Horizon* oil spill. Proc Natl Acad Sci 109(50):20260–20267. https://doi.org/10.1073/pnas.1112139108

McNutt MK, Chu S, Lubchenco J, Hunter T, Dreyfus G, Murawski SA, Kennedy DM (2012b) Applications of science and engineering to quantify and control the *Deepwater Horizon* oil spill. Proc Natl Acad Sci 109(50):20222–20228. https://doi.org/10.1073/pnas.1214389109

Montagna PA, Baguley JG, Cooksey C, Hartwell I, Hyde LJ, Hyland JL, Kalke RD, Kracker LM, Reuscher M, Rhodes ACE (2013) Deep-sea benthic footprint of the Deepwater Horizon blowout. PLoS One 8(8):e70540. https://doi.org/10.1371/journal.pone.0070540

Murawski SA, Hogarth WT, Peebles EB, Barbeiri L (2014) Prevalence of external skin lesions and polycyclic aromatic hydrocarbon concentrations in Gulf of Mexico fishes, post-Deepwater Horizon. Trans Am Fish Soc 143(4):1084–1097. https://doi.org/10.1080/00028487.2014.911205

Murawski SA, Fleeger JW, Patterson WF III, Hu C, Daly K, Romero I, Toro-Farmer GA (2016) How did the Deepwater Horizon oil spill affect coastal and continental shelf ecosystems of the Gulf of Mexico? Oceanography 29(3):160–173. https://doi.org/10.5670/oceanog.2016.80

National Research Council (1985) Oil in the sea: inputs, fates, and effects. The National Academies Press, Washington, DC. https://doi.org/10.17226/314

Nixon Z, Zengel S, Baker M, Steinhoff M, Fricano G, Rouhani S, Michel J (2016) Shoreline oiling from the Deepwater Horizon oil spill. Mar Pollut Bull 107(1):170–178. https://doi.org/10.1016/j.marpolbul.2016.04.003

OSIR (Oil Spill Intelligence Report) (1980) Special Report Ixtoc I. Newsletter; Center for Short-Lived Phenomena. Cambridge, MA

PCEESC (1980) Informe de los Trabajos Realizados para el Control del Pozo Ixtoc 1, el Combate del Derrame de Petróleo y Determinación de sus Efectos sobre el Ambiente Marino. Programa Coordinado de Estudios Ecológicos en la Sonda de Campeche (PC-EESC). Instituto Mexicano del Petróleo, México, DF

Rester JK (ed) (2011) SEAMAP Environmental and Biological Atlas of the Gulf of Mexico, 2009. Gulf States Marine Fisheries Commission, Ocean Springs, MS

Ross SL, Ross CW, Lepine F, Langtry EK (1979) Ixtoc I oil blowout. Spill Technol Newsl 4:245–256

Ross SL, Ross CW, Lepine F, Langtry RK (1980) Ixtoc-I oil blowout. In: Proceedings of a Symposium on Preliminary Results from the September, 1979 RESEARCHER/PIERCE IXTOC-I Cruise, Key Biscayne, Florida, June 9–10, 1980. Office of Marine Pollution Assessment, National Oceanic and Atmospheric Administration, U.S. Dept. of Commerce, Boulder, CO, pp 25–38

Schwacke LH, Smith CR, Townsend FI, Wells RS, Hart LB, Balmer BC, Collier TK, De Guise S, Fry MM, Guillette LJ, Lamb SV, Lane SM, McFee WE, Place NJ, Tumlin MC, Ylitalo GM, Zolman ES, Rowles TK (2014) Health of common bottlenose dolphins (*Tursiops truncatus*) in Barataria Bay, Louisiana, following the *Deepwater Horizon* oil spill. Environ Sci Technol 48(1):93–103. https://doi.org/10.1021/es403610f

Schwing PT, Romero IC, Brooks GR, Hastings DW, Larson RA, Hollander DJ (2015) A decline in benthic foraminifera following the Deepwater Horizon event in the Northeastern Gulf of Mexico. PLoS One 10(3):e0120565. https://doi.org/10.1371/journal.pone.0120565

Silliman BR, van de Koppel J, McCoy MW, Diller J, Kasozi GN, Earl K, Adams PN, Zimmerman AR (2012) Degradation and resilience in Louisiana salt marshes after the BP–*Deepwater Horizon* oil spill. Proc Natl Acad Sci 109(28):11234–11239. https://doi.org/10.1073/pnas.1204922109

Soto LA, Gracia A, Botello AV (1981) Study of the penaeid shrimp population in relation to petroleum hydrocarbons in Campeche Bank. Proc Gulf Caribb Fish Instit 33:81–100

Soto LA, Botello AV, Licea-Durán S, Lizárraga-Partida ML, Yáñez-Arancibia A (2014) The environmental legacy of the Ixtoc-I oil spill in Campeche Sound, southwestern Gulf of Mexico. Front Mar Sci 1:57. https://doi.org/10.3389/fmars.2014.00057

Sun S, Hu C, Tunnell JW Jr (2015) Surface oil footprint and trajectory of the Ixtoc-I oil spill determined from Landsat/MSS and CZCS observations. Mar Pollut Bull 101:632. https://doi.org/10.1016/j.marpolbul.2015.10.036

Tarnecki JH, Patterson WF (2015) Changes in red snapper diet and trophic ecology following the Deepwater Horizon oil spill. Mar Coast Fish 7(1):135–147. https://doi.org/10.1080/19425120.2015.1020402

Tarnecki JH, Wallace AA, Simons JD, Ainsworth CH (2016) Progression of a Gulf of Mexico food web supporting Atlantis ecosystem model development. Fish Res 179:237–250. https://doi.org/10.1016/j.fishres.2016.02.023

Venn-Watson S, Colegrove KM, Litz J, Kinsel M, Terio K, Saliki J, Fire S, Carmichael R, Chevis C, Hatchett W, Pitchford J, Tumlin M, Field C, Smith S, Ewing R, Fauquier D, Lovewell G, Whitehead H, Rotstein D, McFee W, Fougeres E, Rowles T (2015) Adrenal gland and lung lesions in Gulf of Mexico common bottlenose dolphins (*Tursiops truncatus*) found dead following the *Deepwater Horizon* oil spill. PLoS One 10(5):e0126538. https://doi.org/10.1371/journal.pone.0126538

White HK, Hsing P-Y, Cho W, Shank TM, Cordes EE, Quattrini AM, Nelson RK, Camilli R, Demopoulos AWJ, German CR, Brooks JM, Roberts HH, Shedd W, Reddy CM, Fisher CR (2012) Impact of the *Deepwater Horizon* oil spill on a deep-water coral community in the Gulf of Mexico. Proc Natl Acad Sci 109(50):20303–20308. https://doi.org/10.1073/pnas.1118029109

Yáñez-Arancibia A, Sánchez-Gil P (1986) Los peces demersales de la plataforma continental del sur del Golfo de México. 1. Caracterización ambiental, ecología y evaluación de las especies, poblaciones y comunidades. Publicaciones Especiales del Instituto de Ciencias del Mar y Limnología No. 9, vol 9. UNAM, Mexico

Chapter 33
Effects of the Deepwater Horizon Oil Spill on Human Communities: Catch and Economic Impacts

Christa Court, Alan Wade Hodges, Kara Coffey, Cameron H. Ainsworth, and David Yoskowitz

Abstract Economic activity associated with commercial and recreational fisheries in the northern Gulf of Mexico can be dampened when there are significant impacts to the biophysical environment and human communities due to natural or man-made disaster events. Due to its size and scope, the *Deepwater Horizon* incident caused extensive damage to marine and shoreline habitats, affecting activity in both the commercial and recreational fishing industries. The impacts on the Gulf of Mexico ecosystem during and after the *Deepwater Horizon* oil spill are well documented but it is also critical to evaluate impacts on economic activity in sectors that are reliant on this ecosystem. This chapter details estimates of the changes in economic activity supported by commercial and recreational fisheries in the U.S. regions of the northern Gulf of Mexico due to the *Deepwater Horizon* oil spill. Estimates are based upon biophysical model projections from the Atlantis model for scenarios that represent conditions with the oil spill and the counterfactual without the oil spill modeled over a 10-year time horizon, and regional economic impact analysis methods. Results indicate that the total economic impacts for the period 2010–2020 of foregone commercial fishing revenues and recreational fishing expenditures as a result of the *Deepwater Horizon* oil spill are a loss of over 25,000 jobs, \$2.3 billion in industry output, \$1.2 billion in total value added or gross regional product, \$700 million in labor income, \$160 million in state and local tax revenues, and \$160 million in federal tax revenues.

Keywords Ecosystem services · Atlantis · Fisheries modeling · Economic impact

C. Court (✉) · A. W. Hodges
University of Florida, Food and Resource Economics Department, Gainesville, FL, USA
e-mail: ccourt@ufl.edu; awhodges@ufl.edu

K. Coffey · D. Yoskowitz
Texas A&M University-Corpus Christi, Harte Research Institute for Gulf of Mexico Studies, Corpus Christi, TX, USA
e-mail: kara.coffey@tamucc.edu; david.yoskowitz@tamucc.edu

C. H. Ainsworth
University of South Florida, College of Marine Science, St. Petersburg, FL, USA
e-mail: ainsworth@usf.edu

© Springer Nature Switzerland AG 2020
S. A. Murawski et al. (eds.), *Deep Oil Spills*,
https://doi.org/10.1007/978-3-030-11605-7_33

33.1 Introduction

The impacts on the Gulf of Mexico ecosystem during and after the *Deepwater Horizon* (DWH) oil spill are well documented in this book and other research. While an understanding of those impacts on natural systems is important, it is as critical to evaluate impacts on human communities. Both commercial and recreational fisheries are important economic drivers in the northern Gulf of Mexico. In the latest Fisheries Economics of the United States (NMFS 2017), the Gulf of Mexico accounted for the greatest number of jobs (141,000) and sales ($24.7 billion) when compared to all other regions of the country. This economic activity can be dampened when there are significant impacts to the biophysical environment and human communities because of man-made or natural events. In this chapter, we will connect changes in the ecosystem, specifically modeled fisheries landings in the wake of the oil spill, to changes in the economic activity supported by Gulf Coast fisheries.

33.2 Methodology

Economic impact analysis measures changes in total economic activity for a specified region due to a particular event or activity. In the wake of an event such as the *Deepwater Horizon* oil spill, economic impact analysis can provide estimates of the foregone economic activity arising from changes in commercial fishing revenues or the recreational fishing expenditures in coastal areas in the Gulf of Mexico. These impacts are not only associated with changes in direct business revenues but also result in changes in consumer and industry spending, employment, and tax revenues. This section provides an overview of the methodology used for estimating both the direct and total economic impacts associated with changes in commercial and recreational fishing activity after the *Deepwater Horizon* oil spill.

33.3 Atlantis Model

The Atlantis marine ecosystem model was used to estimate changes in fisheries catch due to the DWH oil spill. Atlantis represents physical oceanography and nutrient and trophic dynamics from bacteria to apex predators in a three-dimensional spatially explicit domain and facilitates two-way coupling between environmental and anthropogenic effects, thus enabling examination of the cumulative or synergistic effects between oil spills and fishing (Fulton et al. 2004a, b). Technical specifications for the model may be found in Fulton et al. (2004a, b) and Link et al. (2011), and development of the Gulf of Mexico Atlantis model is documented in Ainsworth et al. (2015) and Tarnecki et al. (2016). The present study uses Atlantis simulation outputs of the *Deepwater Horizon* oil spill as documented in Ainsworth et al. (2018). Briefly, the hydrodynamic model, Connectivity Modeling System (CMS; Paris et al.

2013) was used to provide Atlantis with oil concentrations based on Lagrangian particle trajectories. Ainsworth et al. (2018) applied an uptake-depuration model to differentiate rates of uptake among benthic and pelagic fish and estimate bioavailability. Impacts on mortality and growth rates were estimated at 1° spatial resolution using dose-response relationships (Dornberger et al. 2016) and averaged to the scale of Atlantis polygons. Recruitment impacts were based on Chancellor (2015).

Fisheries in Atlantis are represented by 14 fishing fleet types with species and age class selectivity across ocean depths and habitats (Ainsworth et al. 2015). There are roughly 60 fishing closures in the model representing "status quo" seasonal management and conservation closures as present in the Gulf of Mexico in 2010. In addition, emergency closures related to the *Deepwater Horizon* oil spill were updated on a daily basis from April 20, 2010 to April 19, 2011 as reported by the National Centers for Environmental Information (NCEI 2016).

33.4 Catch

33.4.1 Commercial Fishing

Changes in fisheries yield in the oil spill simulations are assumed to represent changes in the abundance and availability of target species to fishing fleets. We did not consider a redistribution of fishing effort or any other types of adaptive behavioral response by fishermen such as changing target species/ages or investment in gear types. Estimated total commercial catch from two scenarios (a no oil spill scenario and an oil spill scenario) run through the Atlantis model was used to calculate the change in total commercial catch (metric tons) by year and by species between the two scenarios.

Data from the National Marine Fisheries Service (NMFS) on volume and value of commercial catch by species for the period 2009–2016 were used to estimate the average value ($/metric ton) (NOAA-NMFS 2009–2016). Forty-one species modeled within Atlantis were aggregated into 22 species groups to create a concordance between Atlantis species and NMFS species, as shown in Table 33.1. Species group average values were used to estimate the change in value associated with the change in commercial catch.

33.4.2 Recreational Fishing

The Atlantis model was used to calculate the change in total recreational catch (metric tons) by species, by year. As with commercial fishing, the change in total catch was calculated as the difference between the no oil scenario and the oil scenario. Data from the Marine Recreational Information Program (MRIP) on recreational catch and recreational fishing expenditures for the period 2011–2016 were used to estimate the average expenditure per catch ($1000/metric ton) by state and the change in total

Table 33.1 Commercial fishery species group concordance and average value for 2009–2016 period

Species group number	Species group name	Atlantis model	NMFS-FUS statistics	Average value, 2009–2016 ($/metric ton)
1	Jacks and pompano	Greater amberjack	Blue runner	$2,749
		Jacks	Crevalle jack	
		Pompano	Other marine finfishes (portion of)	
2	Crabs and lobsters	Blue crab	Crab, blue, hard	$4,418
		Stone crab	Crab, blue, soft/peeler	
		Crabs and lobsters	Crab, other	
			Lobster, spiny	
			Other shellfish (portion of)	
3	Bluefin tuna	Bluefin tuna	Tuna, bluefin	$10,524
4	Yellowfin tuna	Yellowfin tuna	Tuna, yellowfin	$8,032
5	Little Tunny	Little tunny	Tuna, little	$916
6	Menhaden	Menhaden	Menhaden	$188
7	Mullets	Mullets	Mullets	$1,587
8	Flatfish	Flatfish	Flounder, Atlantic,/Gulf	$5,447
			Flounder, summer	
9	Shrimp	Brown shrimp	Shrimp	$4,393
		Pink shrimp		
		White shrimp	Other shellfish (portion of)	
		Other shrimp		
10	Sharks	Blacktip shark	Shark, other	$1,640
		Benthic feeding sharks		
		Large sharks		
		Small sharks		
11	Red snapper	Red snapper	Snapper, Red	$8,020
12	Vermillion snapper	Vermillion snapper	Snapper, vermillion	$5,787
13	Other snapper	Lutjanidae	Snapper, other	$6,062
14	Swordfish	Swordfish	Swordfish	$5,857
15	King mackerel	King mackerel	Mackerel, king, and cero	$3,737
16	Spanish mackerel	Spanish mackerel	Mackerel, Spanish	$1,886
17	Spanish sardines	Spanish sardine	Sardine, Spanish	$370
18	Sheepshead	Sheepshead	Sheepshead	$1,140

(continued)

Table 33.1 (continued)

Species group number	Species group name	Atlantis model	NMFS-FUS statistics	Average value, 2009–2016 ($/metric ton)
19	Groupers and sea bass	Gag grouper	Groupers	$6,518
		Red grouper	Sea bass, black	
		Scamp		
		Deep Serranidae	Other marine finfishes (portion of)	
		Shallow Serranidae		
20	Drums	Black drum	Croaker, Atlantic	$2,140
		Red drum	Spot	
		Sciaenidae	Other marine finfishes (portion of)	
21	Sea trout	Sea trout	Sea trout, gray	$3,840
			Sea trout, sand	
			Sea trout, spotted	
22	Other	White marlin	N/A	$4,060[a]
		Blue Marlin		
		Other Billfish		
		Snook		

[a]No comparable species found in NMFS-FUS, average for all other species groups used

expenditures associated with the change in recreational catch from the Atlantis model (NOAA-NMFS 2011–2016). The MRIP data were also used to estimate the allocation of total recreational fishing expenditures to each of the Gulf of Mexico states for the period 2011–2016, as displayed in Table 33.2. Total fishing expenditures were allocated to trip expenditure categories (auto fuel, charter fees, boat rentals, etc.) using breakdowns specific to each Gulf state from Lovell et al. (2013). Purchases of durable goods by recreational fishermen were not considered in this analysis.

33.5 Economic Impact Methodology

The total economic impacts of the change in value of landed commercial catch and decreased recreational fishing expenditures due to the *Deepwater Horizon* oil spill were estimated using regional economic models constructed with the *IMPLAN©* software and associated 2013 state and county datasets (IMPLAN 2017). Input-output analysis is a standard technique for estimating the broad economic impacts resulting from changes in specific economic activities within a regional economy (Miller and Blair 2009), and *IMPLAN©* is an input-output/social accounting matrix modeling system that is widely used in both academic and private industry research. These models account for the transactions between industries, governments, employees, and households and provide detailed estimates of impacts on the

Table 33.2 Average recreational fishing expenditure per unit catch and allocation of total expenditure by state (2011–2016)

State	Average expenditure per catch ($1000/metric ton)	Allocation of total expenditures (%)
Alabama	$12.46	10.2
Florida (Gulf Coast)	$18.39	46.5
Louisiana	$8.92	18.2
Mississippi	$7.87	2.8
Texas	$11.91	22.4
Total	**$11.91**	**100.0**

Average expenditure per catch for Texas imputed as the average for the Gulf of Mexico region and missing values for Louisiana (2014–2016) were imputed as the average for the state in years 2011–2013

regional economy from changes in final demand, which is the value of goods and services produced and sold to final users (institutions) during a base year. Impacts are defined using a variety of economic indicators including:

- Employment: Number of full-time and part-time jobs.
- Industry output: Total value of production, comprised of both the value of intermediate inputs and value added.
- Value added: An industry's value of production over the cost of its purchased inputs of goods and services required to produce its goods or services. Total value added, also referred to as Gross Regional Product, is comprised of labor income, other property income, and indirect business taxes.
- Labor income: Wages, salaries, and benefits paid to workers.
- Other property income: Income received from investments, such as corporate dividends, royalties, property rentals, or interest on loans.
- Indirect business taxes: Payments to governments for excise taxes, sales taxes, customs duties, and various fees, except for payroll and income taxes.

Region-specific multipliers estimated within regional economic models represent the total impact on the economy resulting from a change in final demand in the affected industries. Output, value added, and labor income multipliers each represent the total dollar change in all industries due to a one-dollar change in final demand, in this case, the landed value of commercial catch or recreational fishing expenditures, while employment multipliers represent the change in number of jobs in all industries per million dollars change in final demand. Multipliers are comprised of direct, indirect, and induced components:

- Direct effects: The change in commercial fishing business revenues or recreational fishing trip spending. In determining the impact from the oil spill, direct effects estimate either the foregone value of commercial catch or the foregone expenditures associated with canceled recreational fishing trips.
- Indirect effects: The change in spending by supplying industries that result from the change in commercial catch or recreational fishing expenditures after the oil

spill (e.g., sales by baitfish suppliers within the region to commercial or recreational anglers).

- Induced effects: The change in expenditures by households of employees and business proprietors in directly and indirectly impacted businesses for personal consumption of housing, utilities, groceries, etc.

Results were estimated for two study areas: the five states bordering the Gulf of Mexico (Texas, Louisiana, Mississippi, Alabama, and Florida) and the 92 coastal counties with geographical centroids within 50 miles of the Gulf of Mexico. Models were constructed with the gravity model-based trade flows specification representing the share of commodities purchased from local sources and multipliers representing households as endogenous to the model.

Changes in commercial fishing revenues were entered into the models as changes in final demand (industry sales) in the commercial fishing sector. The associated values for direct changes in labor income, other property income, indirect business taxes, and employment were imputed by *IMPLAN©* using regional averages per unit industry output for 2013.

Changes in recreational fishing expenditures by category were entered into the models as changes in final demand. Retail margins were applied to purchases in the retail sectors using factors within the *IMPLAN©* software to adjust values for nonlocal manufacturing, transportation, and wholesaling and account for the difference between the price a consumer pays for products at a retail location (purchaser price) and the value of the same products at the factory door (producer price). Estimated expenditures across the 2011–2020 time period for each affected recreation and tourism sector were all entered in 2013 dollars (representing the midpoint of the baseline data period) allowing the *IMPLAN©* software to apply industry-specific deflators to express values in 2018 year dollars for reporting purposes.

33.6   Results

33.6.1   Commercial Fishing

The total volume of commercial catch for each scenario modeled is shown in Fig. 33.1. For the period 2010–2020, commercial catch under the *no oil* scenario is higher for all species modeled, except shrimp and crabs and lobsters. The catch of species from the lower trophic levels may increase after an oil spill since these species populations increase when predator species populations from upper trophic levels are adversely impacted by the oil spill. For all species modeled, total commercial catch is estimated to decrease by over 475,000 metric tons, largely driven by decreases in catch volume of menhaden and mullets. This change in volume is associated with a decrease of nearly $700 million in commercial catch value. Total commercial catch value in the *no oil* scenario was roughly 4.63% higher than the *oil* scenario.

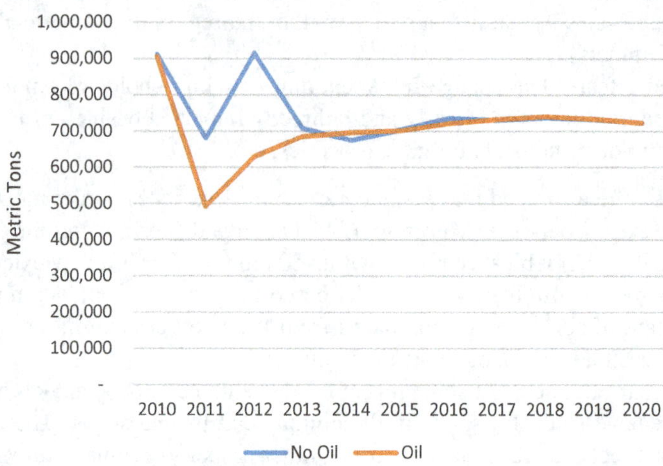

Fig. 33.1 Total commercial catch, no oil vs. oil scenario (2010–2020). Commercial catch under the *no oil* scenario is higher for all species modeled, except shrimp and crabs and lobsters

Table 33.3 Total economic impacts estimated for changes in commercial fishing because of the oil spill, by state (2010–2020)

		Alabama	Florida	Louisiana	Mississippi	Texas	Gulf states
Gulf states model	Employment	1289	3892	7470	925	5223	18,799
	Output	$103.4	$312.2	$599.1	$74.2	$418.9	$1,507.8
	Value added	$51.4	$155.3	$298.1	$36.9	$208.4	$750.1
	Labor income	$31.1	$93.9	$180.3	$22.3	$126.0	$453.7
	State and local taxes	$8.5	$25.6	$49.2	$6.1	$34.4	$123.7
	Federal taxes	$6.8	$20.6	$39.6	$4.9	$27.7	$99.6
Gulf counties model	Employment	1245	3760	7215	894	5045	18,158
	Output	$94.2	$284.5	$546.1	$67.6	$381.8	$1,374.3
	Value added	$46.9	$141.7	$272.0	$33.7	$190.2	$684.5
	Labor income	$29.1	$87.7	$168.4	$20.9	$117.7	$423.8
	State and local taxes	$7.8	$23.6	$45.3	$5.6	$31.7	$114.0
	Federal taxes	$6.3	$19.0	$36.6	$4.5	$25.6	$92.0

Employment is reported as full-time and part-time job-years and all other measures are reported in millions of 2018 dollars

The total economic impacts associated with the estimated change in value for commercial catch are displayed in Table 33.3 for both the Gulf states model and Gulf counties model. As expected, results for the Gulf counties model are slightly lower than that of the Gulf states model, reflecting the fact that counties are smaller, more open economies. Estimates across the two models suggest that the total economic impacts for the period 2010–2020 of foregone commercial fishing revenues as a result of the *Deepwater Horizon* oil spill are a loss of roughly 18,000 jobs,[1] $1.4 billion in industry output, $700 million in total value-added or Gross Regional Product, over $400 million in labor income, over $100 million in state and local tax revenues, and nearly $100 million in federal tax revenues. Table 33.4 displays the results on an average annual basis (dividing by the 11 years in the period).

Results from the Atlantis model suggest that disruptions in commercial fishing activity peak in 2011, and then largely return to the *no oil* activity levels by 2014, while actually reaching relatively small negative values indicating higher catch in the *oil* scenario. In the peak year, there was an estimated loss of over 6000 jobs. Annual impacts for both models and for all variables are expected to follow a similar temporal distribution pattern.

33.6.2 Recreational Fishing

For the period 2011–2020, recreational fishing catch and associated expenditures under the *no oil* scenario are higher for all years. Results indicate over $600 million in foregone recreational fishing expenditures over this period as a result of the oil spill and that recreational expenditures do not return to *no oil* levels by 2020.

Table 33.4 Average annual economic impacts estimated for changes in commercial fishing because of the oil spill, by state (2010–2020)

		Alabama	Florida	Louisiana	Mississippi	Texas	Gulf states
Gulf states model	Employment	117	354	679	84	475	1709
	Output	$9.4	$28.4	$54.5	$6.7	$38.1	$137.1
	Value added	$4.7	$14.1	$27.1	$3.4	$18.9	$68.2
	Labor income	$2.8	$8.5	$16.4	$2.0	$11.5	$41.2
	State and local taxes	$0.8	$2.3	$4.5	$0.6	$3.1	$11.2
	Federal taxes	$0.6	$1.9	$3.6	$0.4	$2.5	$9.1
Gulf counties model	Employment	113	342	656	81	459	1651
	Output	$8.6	$25.9	$49.6	$6.1	$34.7	$124.9
	Value added	$4.3	$12.9	$24.7	$3.1	$17.3	$62.2
	Labor income	$2.6	$8.0	$15.3	$1.9	$10.7	$38.5
	State and local taxes	$0.7	$2.1	$4.1	$0.5	$2.9	$10.4
	Federal taxes	$0.6	$1.7	$3.3	$0.4	$2.3	$8.4

Employment is reported as full-time and part-time jobs and all other measures are reported in millions of 2018 dollars

---

[1] Jobs include full-time, part-time, and seasonal positions and are not a measure of full-time equivalents (FTE). Jobs for the multi-year periods represent job-years, i.e., one job for 1 year.

Table 33.5 Total economic impacts associated with changes in recreational fishing expenditures, by state (2011–2020)

		Alabama	Florida	Louisiana	Mississippi	Texas	Gulf states
Gulf states model	Employment	701	3184	1245	192	1533	6855
	Output	$76.4	$347.4	$135.8	$20.9	$167.3	$747.9
	Value added	$41.9	$190.3	$74.4	$11.5	$91.6	$409.6
	Labor income	$27.1	$123.0	$48.1	$7.4	$59.2	$264.8
	State and local taxes	$4.6	$20.7	$8.1	$1.2	$10.0	$44.7
	Federal taxes	$6.1	$27.9	$10.9	$1.7	$13.4	$60.1
Gulf counties model	Employment	662	3006	1176	181	1448	6473
	Output	$70.2	$319.2	$124.8	$19.2	$153.7	$687.1
	Value added	$39.0	$177.4	$69.4	$10.7	$85.5	$382.0
	Labor income	$25.6	$116.2	$45.4	$7.0	$56.0	$250.2
	State and local taxes	$4.3	$19.4	$7.6	$1,	$9.4	$41.8
	Federal taxes	$5.8	$26.4	$10.3	$1.6	$12.7	$56.8

Employment is reported as full-time and part-time jobs and all other measures are reported in millions of 2018 dollars

The total economic impacts associated with the estimated change in recreational fishing expenditures are displayed in Table 33.5 for both the Gulf states model and Gulf counties model. Estimates across the two models suggest that the total economic impacts for 2011–2020 of foregone recreational fishing expenditures are a loss of over 6000 jobs, $700 million in industry output, $400 million in total value-added or Gross Regional Product, $250 million in labor income, $40 million in state and local tax revenues, and $60 million in federal tax revenues.

Table 33.6 displays the results on an average annual basis. The annual average decline in total output impacts represents roughly 5% of the average recreational fishing expenditures for the Gulf states between 2011 and 2016.

33.7 Discussion

Elucidating the connection between the biophysical and socioeconomic effects of the *Deepwater Horizon* oil spill is critical for generating a holistic picture of impacts and providing much needed information to resource managers and policy makers. Because of the economic and cultural importance of commercial and recreational fisheries in the Gulf of Mexico, overlooking the biophysical to economic connection would be limiting.

The results for both commercial and recreational fishing suggest significant economic harm due the oil spill – as generated by the biological output of the Atlantis

Table 33.6 Average annual economic impacts associated with changes in recreational fishing expenditures, by state (2011–2020)

		Alabama	Florida	Louisiana	Mississippi	Texas	Gulf states
Gulf states model	Employment	70	318	125	19	153	685
	Output	$7.6	$34.7	$13.6	$2.1	$16.7	$74.8
	Value added	$4.2	$19.0	$7.4	$1.1	$9.2	$41.0
	Labor income	$2.7	$12.3	$4.8	$0.7	$5.9	$26.5
	State and local taxes	$0.5	$2.1	$0.8	$0.1	$1.0	$4.5
	Federal taxes	$0.6	$2.8	$1.1	$0.2	$1.3	$6.0
Gulf counties model	Employment	66	301	118	18	145	647
	Output	$7.0	$31.9	$12.5	$1.9	$15.4	$68.7
	Value added	$3.9	$17.7	$6.9	$1.1	$8.5	$38.2
	Labor income	$2.6	$11.6	$4.5	$0.7	$5.6	$25.0
	State and local taxes	$0.4	$1.9	$0.8	$0.1	$0.9	$4.2
	Federal taxes	$0.6	$2.6	$1.0	$0.2	$1.3	$5.7

Employment is reported as full-time and part-time jobs and all other measures are reported in millions of 2018 dollars

modeling – for both the Gulf states model and the Gulf counties model. The aggregate economic impacts – commercial and recreational fishing – for the time period 2010–2020 are a loss of roughly 24,000 job-years, $2.1 billion in industry output, $1.1 billion in total value-added or Gross Regional Product, $650 million in labor income, $140 million in state and local tax revenues, and $160 million in federal tax revenues.

While these numbers are large, it must be noted that this is the estimated difference between a *spill* scenario and a *no spill scenario*. Commercial and recreational fishing continued in the Gulf of Mexico during and after the oil spill occurred. In 2015 the Gulf of Mexico recreational and commercial fisheries sectors employed over 141,000 individuals and accounted for $24.7 billion in sales (NMFS 2017).

The approach employed in this chapter of connecting the biological output of Atlantis modeling with the economic impact of recreational and commercial fisheries provides a reasonable first approximation. While a fully integrated model might be preferable in the long run, this initial step demonstrated the value of connecting biophysical modeling output with economic impact modeling.

Funding Information This publication was made possible, in part, by the *National Oceanic and Atmospheric Administration, Office of Education Educational Partnership Program award* (NA16SEC4810009). Its contents are solely the responsibility of the authors and do not necessarily represent the official views of *the US Department of Commerce, National Oceanic and Atmospheric Administration.*

References

Ainsworth CH, Schirripa MJ, Morzaria Luna HN (eds) (2015) An Atlantis ecosystem model for the northern Gulf of Mexico supporting integrated ecosystem management. U.S. Dept. Commerce, NOAA Technical Memorandum NMFS-SEFSC-TM-676. https://doi.org/10.7289/V5X63JVH

Ainsworth CH, Paris CB, Perlin N, Dornberger LN, Patterson WF III, Chancellor E, Murawski S, Hollander D, Daly K, Romero IC, Coleman F, Perryman H (2018) Impacts of the Deepwater Horizon oil spill evaluated using an end-to-end ecosystem model. PLoS One 13(1):e0190840. https://doi.org/10.1371/journal.pone.0190840

Chancellor E (2015) Vulnerability of larval fish populations to oil well blowouts in the Northern Gulf of Mexico. Master's thesis. Marine Science, University of South Florida, 108 pp

Dornberger L, Ainsworth C, Gosnell S, Coleman F (2016) Developing a polycyclic aromatic hydrocarbon exposure dose-response model for fish health and growth. Mar Pollut Bull 109(1):259–266. https://doi.org/10.1016/j.marpolbul.2016.05.072

Fulton EA, Parslow JS, Smith ADM, Johnson CR (2004a) Biogeochemical marine ecosystem models. 2. The effect of physiological data on model performance. Ecol Model 173:371–406. https://doi.org/10.1016/j.ecolmodel.2003.09.024

Fulton EA, Smith ADM, Johnson CR (2004b) Biogeochemical marine ecosystem models I: IGBEM – a model of marine bay ecosystems. Ecol Model 174:267–307. https://doi.org/10.1016/j.ecolmodel.2003.09.027

IMPLAN (2017) Software for Impact analysis and social accounting (Version 3.0) and Florida state and county data for 2013. IMPLAN Group, LLC, Huntersville, NC, http://www.implan.com

Link JS, Gamble RJ, Fulton EA (2011) NEUS-Atlantis: construction, calibration and application of an ecosystem model with ecological interactions, physiographic conditions and fleet behavior. NOAA Technical Memorandum, NMFS NE-218, 247 pp

Lovell SJ, Steinback S, Hilger J (2013) The economic contribution of marine angler expenditures in the United States, 2011. U.S. Department of Commerce, NOAA Technical Memorandum NMFS-F/SPO-134

Miller RE, Blair PD (2009) Input-output analysis: foundations and extensions, 2nd edn. Cambridge University Press, Cambridge, UK

National Center for Environmental Information (NCEI) (2016) Fisheries closures: Deepwater Horizon support. National Centers for environmental information. NOAA. Available: https://www.nodc.noaa.gov/deepwaterhorizon/fisheries_closures.html. Accessed Sept 2016

National Marine Fisheries Service (NMFS) (2017) Fisheries Economics of the United States, 2015. U.S. Dept. of Commerce, NOAA Technical Memorandum. NMFS-F/SPO-170, 247 p

National Oceanic and Atmospheric Administration – National Marine Fisheries Service (2009–2016) Commercial fisheries statistics. https://www.st.nmfs.noaa.gov/commercial-fisheries/commercial-landings/annual-landings/index

National Oceanic and Atmospheric Administration – National Marine Fisheries Service (2011–2016) Recreational fisheries statistics. https://www.st.nmfs.noaa.gov/recreational-fisheries/data-and-documentation/queries/index

Paris CB, Helgers J, Van Sebille E, Srinivasan A (2013) Connectivity modeling system: a probabilistic modeling tool for the multi-scale tracking of biotic and abiotic variability in the ocean. Environ Model Softw 42:47–54

Tarnecki J, Wallace A, Simons JD, Ainsworth CH (2016) Progression of a Gulf of Mexico food web supporting Atlantis ecosystem model development. Fish Res 179:237–250

Part VIII
Summary

Photo Credit: C-IMAGE Consortium

Chapter 34
Summary of Progress on Major Research Issues: Deep-Sea Oil Spills

Steven A. Murawski, Cameron H. Ainsworth, Sherryl Gilbert, David J. Hollander, Claire B. Paris, Michael Schlüter, and Dana L. Wetzel

Abstract How much has research on ultra-deep-sea (>1500 m) oil and gas blowout behavior progressed since the *Deepwater Horizon* (DWH) accident, and what additional key questions remain to be resolved? In this chapter, we summarize at a high level: (1) the state of knowledge of deep blowout behavior and mitigation prior to DWH, (2) essential information that has come to light in the 9+ years since DWH, and (3) important unresolved questions and additional key research bearing on the prevention of similar marine disasters, mitigation of their effects, and understanding their impacts.

Keywords Deepwater oil and gas · Ixtoc 1 · *Deepwater Horizon* · Oil spill research · Mitigation measures · Science priorities

34.1 Background

The preceding sections and chapters of this volume synthesize recent and historical information in a number of critical areas relevant to understanding and predicting the fate and effects of ultra-deep oil spills. Here we summarize the state of

S. A. Murawski (✉) · C. H. Ainsworth · S. Gilbert · D. J. Hollander
University of South Florida, College of Marine Science, St. Petersburg, FL, USA
e-mail: smurawski@usf.edu; ainsworth@marine.usf.edu; sherryl@usf.edu; davidh@usf.edu

C. B. Paris
University of Miami, Department of Ocean Sciences, Rosenstiel School of Marine & Atmospheric Science, Miami, FL, USA
e-mail: cparis@rsmas.miami.edu

M. Schlüter
Hamburg University of Technology, Institute of Multiphase Flows, Hamburg, Germany
e-mail: michael.schluter@tuhh.de

D. L. Wetzel
Mote Marine Laboratory, Sarasota, FL, USA
e-mail: dana@mote.org

S. A. Murawski et al. (eds.), *Deep Oil Spills*,
https://doi.org/10.1007/978-3-030-11605-7_34

information available previous to the *Deepwater Horizon* (DWH) accident, document critical knowledge acquired since the DWH spill, and briefly discuss priority science areas for the prevention of future deep blowouts and for minimizing, to the extent practicable, the impacts from them on the environment, to response workers, and to the public. This review is not intended to be comprehensive, but highlights some of the most significant and pressing issues in deep spill science.

34.2 What We Knew Then…

The schematic diagram presented in Chap. 1, (Fig. 1.1) provides a mechanistic summary of documented and potential relationships between nominal "blowout scenarios," ambient deep-sea conditions, the use of various mitigation techniques, and the ultimate impacts of the resulting contamination and corresponding response measures as they affect ecosystem components, processes, and people. While much was known previously about the top 1/3 of that diagram (e.g., related to surface oil spills), the bottom 2/3 (e.g., especially the deep sub-sea processes and outcomes) was largely unknown and unanticipated prior to DWH ("unknown-unknowns"). The Ocean Studies and Marine Boards of the National Research Council (Ocean Studies Board and Marine Board 2003) outlined a schematic of oil fate including sedimentation, but this diagram primarily described inter- and subtidal near coastal processes (their Fig. 4.1).

A number of experiments and integrated plume models (Johansen 2000; Chen and Yapa 2001) predicted the formation of subsurface plumes or layers of dissolved hydrocarbons and micro-droplets of liquid oil owing to the dynamics of blowouts prior to DWH. These results were summarized in a cautionary report by the Ocean Studies and Marine Boards of the National Research Council (Ocean Studies Board and Marine Board 2003) suggesting the formation of such plumes in the advent of a well control accident in the deep sea. In a quintessential understatement, however, the NRC noted: "The present understanding of deepwater releases suggests they are much more complicated than those in shallow water." Indeed.

Experiments conducted prior to DWH examined oil droplet formation, size composition, and impacts of dispersants from turbulent "jets" similar to a severed pipe scenario. While some theoretical work had been conducted on the behavior and fate of oil and gas under extreme pressure (Masutani and Adams 2000; Socolofsky and Adams 2005), research was largely disconnected from the rapid expansion of oil and gas exploration and production into the ultra-deep (>1500 m; Murawski et al. 2020a). It was not until the C-IMAGE consortium initiated experiments with methane-saturated ("live") oil, under relevant pressure and ambient temperature conditions (e.g., 15 MPa at 4.5 °C in the case of DWH), and simulating pressure drops as calculated from the actual well casualty (McNutt et al. 2011), that laboratory experiments became sufficiently sophisticated to mirror the likely important mechanisms leading to the formation of oil droplets small enough to remain suspended in the subsurface (Malone et al. 2018; Chap. 4; Pesch et al. 2018; Chap. 5) from ultra-deep spills. Much controversy remains, however, regarding the roles of physical processes vs. the use of SSDI (subsurface dispersant injection) in determining the proportions of small droplets forming from each, and this cannot be satisfactorily resolved from the imperfect field data arising from DWH and the small-scale laboratory experiments conducted to date (Murawski 2020; NASEM 2019).

The Ixtoc 1 spill, although much shallower (54 meters) than DWH (1500 meters), provides useful insights into deeper blowout phenomena both at the time of that spill and from subsequent reanalysis of historical data. The Ixtoc 1 subsurface blowout resulted in large quantities of methane and other gasses surfacing above the wellhead, where they were ignited, producing petrogenic polycyclic aromatic hydrocarbons (PAHs) and other materials (Soto et al. 2014), some of which were deposited on the sea floor (Chap. 13). The "DeepSpill" experiment conducted in 2000 (Johansen et al. 2003) assessed the behavior of an 800-meter deep release of diesel fuel and methane ("dead" oil and gas released simultaneously). That experiment determined that little if any methane gas, and by extension most of the BTEX compounds in crude oil, would surface in a blowout similar to DWH, which was subsequently validated with airborne sensors over the DWH well site (deGouw et al. 2011; Ryerson et al. 2012).

The formation of oiled marine snow (MOSSFA or Marine Oil Snow Sedimentation and Flocculent Accumulation), although apparent to researchers from previous tanker accidents, was never considered a significant sink or pathway for spilled oil sequestration at the ocean bottom. In particular, the location of Ixtoc 1 near a major deltaic river system off Ciudad del Carmen, Campeche, Mexico, provided a source of fine particulate clay material and, combined with the use of over one million gallons of dispersants, was a contributing factor to a large-scale, intense "dirty blizzard" of oiled marine snow emanating from Ixtoc 1 as documented in contemporary sediment cores (Chap. 13). It would be nearly 40 years later that N would be increased to two for oil blowouts leading to major MOSSFA events (Chap. 13).

Retrospective analyses of satellite imagery collected coincident with the Ixtoc 1 spill (e.g., using the Coastal Zone Color Scanner, CZCS, which was launched in 1978, the year previous to the Ixtoc 1 blowout, and Landsat imagery) document important features of the surface expression of that massive and sustained well blowout (Sun et al. 2015). Although not as precise as currently available satellite imagery products, the relationships between geographic extent and prevailing surface transport features, as well as the intermittent surface concentrations are apparent (Sun et al. 2015). Such data are important for calibrating three- and four-dimensional trajectory models (discussed later) required by oil spill responders.

Research conducted prior to DWH on the toxicity and use of dispersants (summarized in NRC 2005) was primarily undertaken in order to comply with government certification requirements for potential dispersant use in oil spill response. Significantly, this research evaluated the *relative* toxicity of various dispersant formulations, primarily in short-duration acute exposure trials using model laboratory species. The question of whether chemically dispersed oil is more toxic than dispersants or crude oil alone was evaluated, but issues with experimental protocols did not allow a definitive resolution of this essential question. These analyses almost exclusively used mortality-based toxic endpoints (e.g., LC_{50} or lethal concentrations resulting in 50% mortality to the species and life stage) being tested for short-duration (usually 96 h) exposures (NRC 2005).

The availability of baseline information on hydrocarbon and associated contaminants in water, sediments, and biota prior to DWH is important both for routine monitoring impacts of drilling operations and for assessing impacts of small and large accidental releases. Other than the NOAA Status and Trends ("Mussel Watch") program (Farrington et al. 2016), there were no organized, ongoing sentinel programs to collect contaminant data in a geographically synoptic fashion or over time, prior to DWH. While there had been several studies of PAHs collected in the Gulf of Mexico (GoM), most of these studies were site-specific (Santschi et al. 2001; Mitra and Bianchi 2003; Overton et al. 2004; Iqbal et al. 2007), used to evaluate issues such as near-field contaminant effects around oil and gas rigs and heavy metal contamination arising from crude oil pollution (Peterson et al. 1996; McDonald et al. 1996), evaluated impacts of produced water discharges in near-shore disposal (Boesch and Rabalais 1989), or assessed episodic large-scale contamination (Hom et al. 2008, following hurricane Katrina). More broadly, for some large-scale tanker accidents, periodic post-spill evaluations of contaminant effects and concentrations have been undertaken (e.g., *Exxon Valdez*, Rice et al. 1996), but post-spill monitoring is the exception not the rule, and thus understanding "return to baseline" when the pre-spill baselines were never established is problematic.

The GoM, as is the case with all oil-producing marine ecosystems, is subject to hydrocarbon inputs from a variety of natural and anthropogenic sources. Natural oil/gas seeps in the GoM account for the majority of oil inputs in most years, with smaller quantities from accidental releases, vessel traffic, atmospheric deposition, produced water discharges, riverine flows, and other sources. This oil "budget" is important for context when oil contaminant baselines are estimated. Several estimates of oil budgets for the GoM and other coastal systems off the USA have been developed (Ocean Studies Board and Marine Board 2003). Murawski et al. (2014) estimated that the DWH accident accounted for over seven times the annual sum of oil inputs from all other sources into the GoM, but that spill was confined to a portion of the northern Gulf, over about ¼ of the year. In coastal margins of the GoM, there are no areas completely free of oil derivatives (Pulster et al. 2020), and thus understanding spatial and temporal variation and trends in contaminant concentration baselines potentially allows attribution to the various sources contributing to observed contamination rates. This is particularly salient using chronological sediment cores, which document time histories of such episodes in relation to human and natural stressors.

Apart from the dynamics of ultra-deep spills, basic knowledge of the ecology of the pelagic, mesopelagic, and bentho-pelagic environments where oil exploration and production is now occurring was severely lacking prior to DWH and only slightly less so now (Cordes et al. 2008; Fisher et al. 2016; Sutton et al. 2020; Romero et al. 2018). These deep regions account for over half of the known species of fishes in the GoM (Sutton et al. 2020), and recent expeditions there continue to document range extensions and in some cases additional new species (Sutton et al. 2020). In order to adequately define environmental sensitivity to oil production operations and spills, a more complete accounting for the extant diversity of species and their productivity and sensitivity to oil contamination (Romero et al. 2018) is necessary, but such data were not routinely collected prior to DWH. This also extends to marine mammals, sea turtles, birds, and other ecosystem components extant in and dependent upon deep seas.

34.3   What We Know Now…

Subsequent to the DWH accident, about $1 billion has been spent at this writing to conduct impact assessments, to monitor species and human health impacts, and to develop fundamental science for many diverse aspects of deepwater blowouts and their effects. Additionally, the oil and gas industry and individual companies have underwritten technological innovations for various aspects of ultra-deep production, well control, and response protocols with the goals of minimizing the frequency of such accidents to as close to zero as is possible, containing uncontrolled blowouts in a more timely manner, and minimizing spill impacts to the ecosystem and people. The breath of new knowledge gained from this massive infusion of resources, partially synthesized herein and in a companion volume *"Scenarios and Responses to Future Deep Oil Spills – Fighting the Next War* (Murawski et al. 2020b) is impressive. This new research, conducted primarily since 2010, modernizes our view of the deep ocean environment and the interplay between the sub-bottom oil reservoir, the oil extraction infrastructure, and the pelagic realm. Complex physics and chemistry of a very hot oil (>100 °C at the base of the blowout preventer, BOP) under considerable pressure, discharging into a cold (~4.5 °C) environment create a chaotic transition from a single-phase hydrocarbon jet to a multiphase flow (oil, gas, water) that rapidly disperses both horizontally and vertically. The behavior of the multiphase flow and its components (e.g., dissolving oil and gas components and particlized oil droplets) interacts with mitigation techniques (e.g., the application of SSDI) in ways not yet fully understood. Nevertheless, laboratory studies at multiple scales confirm that both "natural" processes and the use of chemical dispersants each contribute to formation and sequestration of small oil droplets in deep-sea blowouts (Chaps. 2, 3, 4, and 5).

The DWH accident, occurring seaward of the Mississippi River delta, apparently had all the necessary antecedents for the formation of a large MOSSFA event (Chaps. 12, 13, 14, 15, 16, 17, and 18). Considerable experimentation and some modeling work have quantified the requirements for MOSSFA formation

(Chap. 12), and thus, these conditions can be assessed for future blowouts to predict if such events will occur and how significant a pathway they will represent under future blowout scenarios. Work on such prediction capabilities continues, including incorporation of MOSSFA modules into the government's oil spill response models.

An important consequence of the DWH spill has been the considerable investment documenting and evaluating biodiversity in the deepest parts of ocean basins including the GoM. Such research provides more context regarding impacts of deep spills and emphasizes the importance of using non-model species (e.g., those existing beyond coastal and estuarine or freshwater areas) in projecting likely toxicological outcomes and, importantly, in assessing trade-offs in oil spill response. Biota in the epi- and meso- and bentho-pelagic realms are not disconnected as research on diel vertical migrations of fishes, squids, marine mammals, and other species document (Fisher et al. 2016; Sutton et al. 2020; Chaps. 21, 22, 23, 24, 25, and 26). Also, deep coral communities are unique and sensitive and occur in deep regions of the GoM and elsewhere where oil and gas drilling is being directed (Cordes et al. 2008; Fisher et al. 2016). Thus, decision-making regarding the inherent "value" of various components of marine biodiversity in relation to oil spill response measures has become evermore complicated, reflecting the connected nature of ecosystems.

In addition to the collection of biodiversity baselines described above, as a consequence of the DWH and Ixtoc 1 spills, there are now contaminant baselines for sediments, waters, invertebrates, and fishes not only in the regions of these spills but more broadly for coastal areas, and deep waters of the GoM (*Deepwater Horizon* Natural Resource Damage Assessment Trustees 2016; Schwing et al. 2020; Chanton et al. 2020; Brooks et al. 2020; Pulster et al. 2020; Patterson III et al. 2020; Peebles and Hollander 2020; Romero et al. 2018). Chronological data from calibrated sediment cores document both invertebrate infaunal records pre- and post-spill and quantification of oil contaminant history, for areas where bioturbation or physical processes do not complicate successive layering. For non-benthic forms, few data were available prior to large spills (as noted above), but time series in fishes, in particular, now document declines in contamination for most species since DWH and the spatial patterns of PAH contamination (e.g., Murawski et al. 2014). Baselines for microbial community dynamics have also been complied (e.g., Kostka et al. 2020) and are useful for understanding and predicting biodegradation potential in the advent of future spills. With all of these data, they provide broad-scale context for the background contamination levels and are thus important in the event of future large-scale spills. Importantly, there are no commitments to long-term monitoring of oil and other contaminants anywhere in the Gulf, so as time progresses, the current baselines from DWH and Ixtoc 1-related research will become less useful for assessing future spill effects. In assessing the impacts of specific spills, the fact that facility-specific baselines are not required further complicates spill effects assessment.

Perhaps one of the most important research advances since DWH has been the obvious recognition that large, deepwater oil spills exist in four dimensions (horizontal, vertical, temporal; Paris et al. 2020). During the DWH spill, a variety of surface transport models were run, with varying degrees of success and sophistication, in an ad hoc ensemble ("mash up") of model predictions, much like hurricane "spaghetti" plots. While these transport models were appropriate for projecting surface expression of the spill necessary to inform oil spill response, very few operational 4-D models were available at the time (Liu et al. 2013). Since the DWH and continuing today, there have been considerable model-related efforts to (1) better describe in quantitative terms the "near-field" of deep marine blowouts at and above the broken wellhead (Chap. 2), (2) connect the results of near-field models with the far-field (i.e., 1000s of km^2; Chap. 9) in four dimensions to predict oil and gas fate and coincidence with biota, and (3) link near-far-field hydrodynamic and oil biodegradation models with spatially explicit ecosystem models to predict short- and long-term ecosystem outcomes (e.g., Peterson et al. 2003; Chaps. 31 and 32; Berenshtein et al. 2020). In the event of the next deepwater blowout, there will be significantly more capability to predict the fate and effects of oil in deep plumes, surfacing and sequestered on the bottom. As described in the schematic depicted in Chap. 1 (there Fig. 1.1), there are now linked quantitative models that describe what was only envisioned qualitatively in the early days DWH spill. What is not obvious is if there will be sufficient oceanographic observational data (current flow, temperature, salinity) with which to populate these models as the next crisis unfolds.

34.4 What Else We Need to Know…

In much the same way former US Secretary of Defense Donald Rumsfeld described known and unknown "unknowns," our knowledge of the deep ocean realm and how deep-sea blowouts may affect it has evolved from far too many "unknown-unknowns" to now mostly "known-unknowns." For a deep blowout of the magnitude of DWH, at that location, with that specific oil type, gas/oil ratio, etc., we are now able to provide more complete information with which to inform response efforts. This also includes the advisability of non-dispersant-related mitigation measures such as freshwater releases from impounded rivers and building of sand berms to contain drifting oil offshore. Nevertheless, there remain a number of critical "known-unknowns" that limit our present ability to confidently predict response, fate, and effects of particular accidents and mitigation measures used to fight the spill. Additional research targeted to close these remaining gaps is prioritized in Murawski (2020). Importantly, as the state of knowledge of environmental conditions has progressed and trends in the oil and gas industries have evolved, so too have our research priorities (e.g., Boesch et al. 1987).

While predicting impacts of a blowout analogous to DWH is difficult and idiosyncratic, there nevertheless is now an expanding body of empirical information and growing modeling sophistication upon which to build. The problem, however, lies in the fact that the marine oil and gas industries continue to press their frontiers both in terms of extreme depths and isolated locations of the world where oil and gas extraction has not heretofore occurred and where the baseline and capacity for relevant science may not exist. Predicting blowout behaviors in these situations requires us to understand better the fundamental physical, chemical, geological, and ecological processes at play and how they would impact decision-making. Volume two of this series (Murawski et al. 2020b) explores a variety of scenarios at frontier locations in the GoM and elsewhere (the Arctic, west Africa) and how decision-making might be different depending on the circumstances. We have learned a considerable amount in the intervening 9+ years since DWH, but clearly it is an evolving and unrelenting process to keep pace with an industry exploring at the margins of human knowledge.

References

Berenshtein I, Perlin N, Ainsworth C, Ortega-Ortiz J, Vaz AC, Paris CB (2020) Comparison of the spatial extent, impacts to shorelines and ecosystem, and 4-dimensional characteristics of simulated oil spills. In: Murawski SA, Ainsworth C, Gilbert S, Hollander D, Paris CB, Schlüter M, Wetzel D (eds) Scenarios and responses to future deep oil spills – fighting the next war. Springer, Cham

Boesch DF, Rabalais NN (1989) Produced waters in sensitive coastal habitats: an analysis of impacts, Central Coastal Gulf of Mexico. OCS Report/MMS 89-0031, U.S. Dept. of the Interior, Minerals Management Service, Gulf of Mexico OCS Regional Office, New Orleans, Louisiana, 157 pp

Boesch DF, Bulter JN, Cacchione DA, Geraci JR, Neff JM, Ray JP, Teal JM (1987) An assessment of the long-term effects of U.S. offshore oil and gas development activities: future research needs. In: Boesch DF, Rabalais NN (eds) Long-term environmental effects of offshore oil and gas development. Elsevier Applied Science, New York, pp 1–53

Brooks G, Larson RA, Schwing PT, Diercks AR, Armenteros-Almanza M, Diaz-Asencio M, Martinez-Suarez A, Sánchez-Cabeza JA, Ruiz-Fernandez AC, Herguera-García JC, Perez-Bernal LH, Hollander DJ (2020) Gulf of Mexico (GoM) bottom sediments and depositional processes: a baseline for future oil spills. In: Murawski SA, Ainsworth C, Gilbert S, Hollander D, Paris CB, Schlüter M, Wetzel D (eds) Scenarios and responses to future deep oil spills – fighting the next war. Springer, Cham

Chanton JP, Jaggi A, Radović J, Rosenheim BE, Walker BD, Larter SR, Rogers K, Bosman S, Oldenburg TBP (2020) Mapping isotopic and dissolved organic matter baselines in waters and sediments of Gulf of Mexico. In: Murawski SA, Ainsworth C, Gilbert S, Hollander D, Paris CB, Schlüter M, Wetzel D (eds) Scenarios and responses to future deep oil spills – fighting the next war. Springer, Cham

Chen F, Yapa PD (2001) Modeling gas separation from a bent deepwater oil and gas jet/plume. In: Proceedings of the 5th international marine environmental modeling seminar October 9–2001. New Orleans, LA

Cordes EE, McGinley MP, Podowski EL, Becker EL, Lessard-Pilon S, Viada ST, Fisher CR (2008) Coral communities of the deep Gulf of Mexico. Deep Sea Res I 55:777–787. https://doi.org/10.1016/j.dsr.2008.03.005

Deepwater Horizon Natural Resource Damage Assessment Trustees (2016) Deepwater Horizon oil spill: final programmatic damage assessment and restoration plan and final programmatic environmental impact statement. Retrieved from: http://www.gulfspillrestoration.noaa.gov/restoration-planning/gulf-plan

deGouw JA, Middlebrook AM, Warneke C, Ahmadov R, Atlas EL, Bahreini R, Blake DR, Brock CA, Brioude J, Fahey DW, Fehsenfeld FC, Holloway JS, Le Henaff M, Lueb RA, McKeen SA, Meagher JF, Murphy MD, Paris CB, Parrish DD, Perring AE, Pollack IB, Ravishankara AR, Robinson AL, Ryerson TB, Schwarz JP, Spackman JR, Srinivasan A, Watts LA (2011) Organic aerosol formation downwind from the *Deepwater Horizon* oil spill. Science 331:1295

Farrington JW, Tripp BW, Tanabe S, Subramanian A, Sericano JL, Wade TL, Knap AH (2016) Edward D. Goldberg's proposal of "the Mussel Watch": reflections after 40 years. Mar Pollut Bull 110:501–510

Fisher CR, Montagna PA, Sutton TT (2016) How did the *Deepwater Horizon* oil spill impact deep-sea ecosystems? Oceanography 29:182–195

Hom T, Collier TK, Krahn MM, Strom MS, Ylitalo GM, Nilsson WB, Paranjpye RN, Varanasi U (2008) Assessing seafood safety in the aftermath of Hurricane Katrina. Am Fish Soc Symp 64:73–93

Iqbal J, Portier RJ, Gisclair D (2007) Aspects of petrochemical pollution in coastal Louisiana. Mar Pollut Bull 54:792–819

Johansen Ø (2000) DeepBlow—a lagrangian plume model for deepwater blowouts. Spill Sci Technol Bull 6:103–111

Johansen Ø, Rye H, Cooper C (2003) DeepSpill—field study of a simulated oil and gas blowout in deep water. Spill Sci Technol Bull 8:433–443

Kostka JE, Overholt WA, Rodriguez-R LM, Huettel M, Konstantinidis K (2020) Toward a predictive understanding of the benthic microbial community response to oiling on the northern Gulf of Mexico coast. In: Murawski SA, Ainsworth C, Gilbert S, Hollander D, Paris CB, Schlüter M, Wetzel D (eds) Scenarios and responses to future deep oil spills – fighting the next war. Springer, Cham

Liu Y, MacFadden A, Ji, Z-G, Weisberg RH (2013) Monitoring and modeling the Deepwater Horizon oil spill: a record-breaking enterprise. Geophys Monogr Ser 195:271 pp

Malone K, Pesch S, Schlüter M, Krause D (2018) Oil droplet size distributions in deep-sea blowouts: influence of pressure and dissolved gases. Environ Sci Technol 52:6326–6333. https://doi.org/10.1021/acs.est.8b00587

Masutani SM, Adams EE (2000) Experimental study of multi- phase plumes with application to deep ocean oil spills., TA&R Project 377, Bureau of Ocean Energy Management, Regulation and Enforcement, Washington, DC. Available at: http://www.boemre.gov/tarprojects/377.htm

McDonald SJ, Willett KL, Thomsen J, Beatty KB, Connor K, Narasimhan TR, Erickson CM, Safe SH (1996) Sublethal detoxification responses to contaminant exposure associated with offshore oil production. Can J Fish Aquat Sci 53:2606–2617

McNutt MK, Camilli R, Crone TJ, Guthrie GD, Hsieh PA, Ryerson TB, Savas O, Shaffer F (2011) Review of flow rate estimates of the Deepwater Horizon oil spill. Proc Natl Acad Sci 109. www.pnas.org/cgi/doi/10.1073/pnas.1112139108

Mitra SS, Bianchi TS (2003) A preliminary assessment of polycyclic aromatic hydrocarbon distributions in the lower Mississippi River and Gulf of Mexico. Mar Chem 82:273–288

Murawski SA (2020) Perspectives on research, technology, policy and human resources for improved management of ultra-deep oil and gas resources and responses to oil spills. In: Murawski SA, Ainsworth C, Gilbert S, Hollander D, Paris CB, Schlüter M, Wetzel D (eds) Scenarios and responses to future deep oil spills – fighting the next war. Springer, Cham

Murawski SA, Hogarth WT, Peebles EB, Barbieri L (2014) Prevalence of external skin lesions and polycyclic aromatic hydrocarbon concentrations in Gulf of Mexico fishes, post-*Deepwater Horizon*. Trans Am Fish Soc 143:1084–1097

Murawski SA, Hollander D, Gilbert S, Gracia A (2020a) Deep-water oil and gas production in the Gulf of Mexico, and related global trends. In: Murawski SA, Ainsworth C, Gilbert S, Hollander D, Paris CB, Schlüter M, Wetzel D (eds) Scenarios and responses to future deep oil spills – fighting the next war. Springer, Cham

Murawski SA, Ainsworth C, Gilbert S, Hollander D, Paris CB, Schlüter M, Wetzel D (eds) (2020b) Scenarios and responses to future deep oil spills – fighting the next war. Springer, Cham

National Academies of Sciences, Engineering and Medicine (NASEM) (2019) Report of the committee on evaluation of the use of chemical dispersants in oil spill response. National Academies Press, Washington, DC

National Research Council (NRC) (2005) Oil spill dispersants: efficacy and effects. The National Academies Press, Washington, DC, 377 pp

Ocean Studies Board and Marine Board (2003) Oil in the sea III. Inputs, fates, and effects. National Academies Press, Washington, DC

Overton EB, Ashton BM, Miles MS (2004) Historical polycyclic and petroleum hydrocarbon loading in the Northern Central Gulf of Mexico shelf sediments. Mar Pollut Bull 49:557–563

Paris CB, Vaz AC, Berenshtein I, Perlin N, Faillettaz R, Aman ZM (2020) Simulating deep oil spills beyond the Gulf of Mexico. In: Murawski SA, Ainsworth C, Gilbert S, Hollander D, Paris CB, Schlüter M, Wetzel D (eds) Scenarios and responses to future deep oil spills – fighting the next war. Springer, Cham

Patterson WF III, Chanton JP, Barnett B, Tarnecki JH (2020) The utility of stable and radio isotopes in fish tissues as biogeochemical tracers of marine oil spill food web effects. In: Murawski SA, Ainsworth C, Gilbert S, Hollander D, Paris CB, Schlüter M, Wetzel D (eds) Scenarios and responses to future deep oil spills – fighting the next war. Springer, Cham

Peebles EB, Hollander DJ (2020) Combining isoscapes with tissue-specific isotope records to recreate the geographic histories of fish. In: Murawski SA, Ainsworth C, Gilbert S, Hollander D, Paris CB, Schlüter M, Wetzel D (eds) Scenarios and responses to future deep oil spills – fighting the next war. Springer, Cham

Pesch S, Jaeger P, Jaggi A, Malone K, Hoffmann M, Krause D, Oldenburg T, Schlüter M (2018) Rise velocity of live-oil droplets in deep-sea oil spills. Environ Eng Sci. https://doi.org/10.1089/ees.2017.0319, online ahead of print, 2018

Peterson CH, Kennicutt MC II, Green RH, Montagna P, Harper DE Jr, Powell EN, Roscigno PF (1996) Ecological consequences of environmental perturbations associated with offshore hydrocarbon production: a perspective on long-term exposure in the Gulf of Mexico. Can J Fish Aquat Sci 53:2637–2654

Peterson CH, Rice SD, Short JW, Esler D, Bodkin JL, Ballachey BE, Irons DB (2003) Long-term ecosystem response to the *Exxon Valdez* oil spill. Science 302:2082–2086

Pulster EL, Gracia A, Snyder SM, Romero IC, Carr B, Toro-Farmer G, Murawski SA (2020) Polycyclic aromatic hydrocarbon baselines in Gulf of Mexico fishes. In: Murawski SA, Ainsworth C, Gilbert S, Hollander D, Paris CB, Schlüter M, Wetzel D (eds) Scenarios and responses to future deep oil spills – fighting the next war. Springer, Cham

Rice SD, Spies RB, Wolfe DA, Wright BA (1996) Proceedings of the *Exxon Valdez* oil spill symposium. American Fisheries Society Symposium 19, Bethesda, MD, 931 pp

Romero IC, Sutton T, Carr B, Quintana-Rizzo E, Ross SW, Hollander DJ, Torres JJ (2018) Decadal assessment of polycyclic aromatic hydrocarbons in mesopelagic fishes from the Gulf of Mexico reveals exposure to oil- derived sources. Environ Sci Technol (online). https://doi.org/10.1021/acs.est.8b02243

Ryerson TB, Camilli R, Kessler JD, Kujawinski EB, Reddy CM, Valentine DL, Atlas E, Blake DR, de Gouw J, Meinardi S, Parrish DD, Peischla J, Seewald JS, Warneke C (2012) Chemical data quantify *Deepwater Horizon* hydrocarbon flow rate and environmental distribution. Proc Natl Acad Sci USA 109:20246–20253

Santschi PH, Presley BJ, Wade TL, Garcia-Romero B, Baskaran M (2001) Historical contamination of PAHs, PCBs, DDTs, and heavy metals in Mississippi River Delta, Galveston Bay and Tampa Bay sediment cores. Mar Environ Res 52:51–79

Schwing PT, Montagna PM, Machain-Castillo ML, Escobar-Briones E, Rohal M (2020) Benthic faunal baselines in the Gulf of Mexico: a precursor to evaluate future impacts. In: Murawski SA, Ainsworth C, Gilbert S, Hollander D, Paris CB, Schlüter M, Wetzel D (eds) Scenarios and responses to future deep oil spills – fighting the next war. Springer, Cham

Socolofsky SA, Adams EE (2005) Role of slip velocity in the behavior of stratified multiphase plumes. J Hydraul Eng 131:273–284

Soto LA, Botello AV, Licea-Durán S, Lizárraga-Partida M, Yáñez-Arancibia A (2014) The environmental legacy of the Ixtoc-I oil spill in Campeche Sound, Southwestern Gulf of Mexico. Front Mar Sci 1:1–9

Sun S, Hu C, Tunnell JW Jr (2015) Surface oil footprint and trajectory of the Ixtoc-I oil spill determined from Landsat/MSS and CZCS observations. Mar Pollut Bull 101:632–641. https://doi.org/10.1016/j.marpolbul.2015.10.036

Sutton T, Frank T, Judkins H, Romero IC (2020) As Gulf oil extraction goes deeper, who is at risk? Community structure, distribution, and connectivity of the deep-pelagic fauna. In: Murawski SA, Ainsworth C, Gilbert S, Hollander D, Paris CB, Schlüter M, Wetzel D (eds) Scenarios and responses to future deep oil spills – fighting the next war. Springer, Cham

Index

CPSIA information can be obtained
at www.ICGtesting.com
Printed in the USA
LVHW080028050819
626501LV00003B/33/P

9 783030 116040